HETCOR	Heteronuclear Correlation	MSD	Mean Square ~~Displacement~~
HF	Hydrofluoric Acid	MTBE	Methyl *tert*-Butyl
HFB	Hexafluorobenzene		
HILIC	Hydrophilic Interaction Liquid Chromatography	NAPL	Nonaqueous Phase Liquid
		NBA	Nitrobenzoic Acid
HMBC	Heteronuclear Multiple Bond Correlation	NMRD	Nuclear Magnetic Resonance Dispersion
HMDB	Human Metabolome Database	NOAEL	No-Observable Adverse-Effect Level
HMQC	Heteronuclear Multiple Quantum Coherence	NOESY	Nuclear Overhauser Effect Spectroscopy
		NOM	Natural Organic Matter
HOC	Hydrophobic Organic Contaminant	NP	Nitrophenol
HPLC	High Performance Liquid Chromatography	NT	Nitrotoluene
HPLC-MS	High Performance Liquid Chromatography-Mass Spectrometry	O/C	Oxygen-to-Carbon
		OM	Organic Matter
HR	High-Resolution	OPLS-DA	Orthogonal Partial Least Squares Discriminant Analysis
HR-FAB-MS	High-Resolution Fast Atom Bombardment Mass Spectroscopy		
HR-MAS	High-Resolution Magic-Angle Spinning	PA	Picric Acid
HSQC	Heteronuclear Single Quantum Coherence	PC	Principal Component
HSQCAD	Heteronuclear Single Quantum Coherence Adiabatic	PCA	Principal Component Analysis
		PCB	Polychlorinated Biphenyl
HSQMBC	Heteronuclear Single Quantum Multiple Bond Correlation	PCP	Pentachlorophenol
		PCR	Principal Components Regression
IA	Independent Action	PET	Positron Emission Tomography
ICP-MS	Inductively Coupled Plasma Mass Spectrometry	PFG	Pulsed-field Gradient
		PFOA	Perfluorooctanoic Acid
IEA	International Energy Agency	PFOS	Perfluorooctanesulfonic Acid
IR	Infrared	PFPE	Perfluoropolyether
IV	Iodine Value	PG	Peptidoglycan
		PIA	Polysaccharide Intracellular Adhesion
JRES	J-Resolved Spectroscopy	PLS	Partial Least Squares
		PLS-DA	Partial Least Squares Discriminant Analysis
K_d	Soil-Water Distribution Coefficient		
KEGG	Kyoto Encyclopedia of Genes and Genomes	PP	Pre-Polarized
		PSRE	Proton Spin Relaxation Editing
K_{oc}	Organic-Carbon Normalized Distribution Coefficient	PURGE	Presaturation Utilizing Relaxation Gradients and Echoes
K_{ow}	Octanol-Water Partition Coefficient		
		QCPMG	Quadrupolar Carr-Purcell-Meiboom-Gill
LB	Linebroadening		
LC_{50}	Lethal Concentration to 50% of the Population	RADE	Recovery of Relaxation Losses Arising from Diffusion Editing
LC	Liquid Chromatography	RAMP	Ramped Amplitude
LC-LC	Two-dimensional Liquid Chromatography	RARE	Rapid Acquisition with Refocused Echoes
LC-NMR	Liquid Chromatography Nuclear Magnetic Resonance	RESTORE	Restoration of Spectra via T_{CH} and T One Rho Editing
LC-SPE	Liquid Chromatography-Solid-Phase Extraction	RF	Radio Frequency
		RH-STD	Reverse Heteronuclear Saturation Transfer Difference
LMD	Laser Microdissection		
		RMSD	Root Mean Square Displacement
MAS	Magic-Angle Spinning	ROESY	Rotating-Frame Overhauser Effect Spectroscopy
MDLT	Material Derived From Linear Terpenoids		
MIC	Minimum Inhibitory Concentration		
MMM	Molecular Mixing Model	SCV	Small Colony Variant
MOA	Mode of Action	SDBS	Spectral Database for Organic Compounds
MOUSE	Mobile Universal Surface Explorer	SHY	Statistical Heterospectroscopy
MPNBG	Methyl-4,6-*O*-(*p*-Nitrobenzylidene)-α-D-Glucopyranoside Gel	S/N	Signal-to-Noise
		SNIF	Site-specific Nuclear Isotope Fractionation
MRI	Magnetic Resonance Imaging	SOA	Secondary Organic Aerosols
MRS	Magnetic Resonance Spectroscopy	SOM	Soil Organic Matter
MS	Mass Spectrometry	SON	Soil Organic Nitrogen

SOP	Standard Operation Procedure		TRAPDOR	Transfer of Populations in Double Resonance
SPE	Solid Phase Extraction		TSP	3-Trimethylsilyltetradeuteropropionate
SSB	Spinning Side Band			
SST	Sample Shuttle Technique			
STD	Saturation Transfer Difference		UPEN	Uniform PENalty regularization
STDD	Saturation Transfer Double Difference		UPLC	Ultra Performance Liquid Chromatography
STMAS	Satellite Transition Magic-Angle Spinning		UTE	Ultrashort Echo Time
STOCSY	Statistical Total Correlation Spectroscopy		UV	Ultraviolet
STWL	Simulated Tank Waste Leachate		UV-VIS	Ultraviolet/Visible
SUS	Shared and Unique Structures			
SUVA	Specific Ultraviolet Absorbance		VCT	Variable Contact Time
SWIFT	Sweep Imaging with Fourier Transform		VIP	Variable's Influence on Projection
SWT	Switching Time		VSL	Variable Spin Lock
			VT	Variable Temperature
T_1	Spin-Lattice Relaxation Time			
T_2	Spin-Spin Relaxation Time		WAF	Water-Accommodated Fraction
TAME	*tert*-Amyl Methyl		WaMB	Water Molecule Bridges
TCA	Tricarboxylic Acid		WEFT	Water-Eliminated Fourier Transform
TCI	Triple-resonance Inverse		WET	Water Suppression Enhanced through T_1 Effects
TEMPO	2,2,6,6-Tetramethylpiperidine-1-Oxyl			
TFS	(3,3,3-Trifluoropropyl)dimethylchlorosilane		WISE	Wideline Separation
TMA	Thermomechanical Analysis		WS	Wrinkly Spreader
TMDP	2-Chloro-4,4,5,5-Tetramethyl-1,3,2- Dioxaphospholane		WS-FSFA	Water-Soluble Forest Soil Fulvic Acid
			WSOM	Water-Soluble Organic Matter
TNB	Trinitrobenzene		WURST	Wide, Uniform Rate, and Smooth Truncation
TNBA	Trinitrobenzoic Acid			
TNT	2,4,6-Trinitrotoluene		WWTP	Wastewater Treatment Plant
TOCSY	Total Correlation Spectroscopy			
TOSS	Total Sideband Suppression		XOC	Xenobiotic Organic Contaminant
TPPM	Two-Pulse Phase-Modulated		XRD	X-Ray Diffraction

NMR Spectroscopy: A Versatile Tool for Environmental Research

eMagRes Handbooks

Based on *eMagRes* (formerly the Encyclopedia of Magnetic Resonance), this monograph series focuses on hot topics and major developments in modern magnetic resonance and its many applications. Each volume in the series will have a specific focus in either general NMR or MRI, with coverage of applications in the key scientific disciplines of physics, chemistry, biology or medicine. All the material published in this series, plus additional content, will be available in the online version of *eMagRes*, although in a slightly different format.

Previous eMagRes Handbooks

NMR Crystallography
Edited by Robin K. Harris, Roderick E. Wasylishen, Melinda J. Duer
ISBN 978-0-470-69961-4

Multidimensional NMR Methods for the Solution State
Edited by Gareth A. Morris, James W. Emsley
ISBN 978-0-470-77075-7

Solid-State NMR Studies of Biopolymers
Edited by Ann E. McDermott, Tatyana Polenova
ISBN 978-0-470-72122-3

NMR of Quadrupolar Nuclei in Solid Materials
Edited by Roderick E. Wasylishen, Sharon E. Ashbrook, Stephen Wimperis
ISBN 978-0-470-97398-1

RF Coils for MRI
Edited by John T. Vaughan, John R. Griffiths
ISBN 978-0-470-77076-4

*MRI of Tissues with Short T_2s or T_2*s*
Edited by Graeme M. Bydder, Gary D. Fullerton, Ian R. Young
ISBN 978-0-470-68835-9

eMagRes

Edited by Robin K. Harris, Roderick E. Wasylishen, Edwin D. Becker, John R. Griffiths, Vivian S. Lee, Ian R. Young, Ann E. McDermott, Tatyana Polenova, James W. Emsley, George A. Gray, Gareth A. Morris, Melinda J. Duer and Bernard C. Gerstein.

The *eMagRes* (formerly EMR) is based on the original printed *Encyclopedia of Nuclear Magnetic Resonance*, which was first published in 1996 with an update volume added in 2000. *eMagRes* was launched online in 2007 with all the material that had previously appeared in print. New updates have since been and will be added on a regular basis throughout the year to keep the content up to date with current developments. Nuclear was dropped from the title to reflect the increasing prominence of MRI and other medical applications. This allow the editors to expand beyond the traditional borders of NMR to MRI and MRS, as well as to EPR and other modalities. *eMagRes* covers all aspects of magnetic resonance, with articles on the fundamental principles, the techniques and their applications in all areas of physics, chemistry, biology and medicine for both general NMR and MRI. Additionally, articles on the history of the subject are included.

For more information see: www.wileyonlinelibrary.com/ref/eMagRes

NMR Spectroscopy: A Versatile Tool for Environmental Research

Editors

Myrna J. Simpson
University of Toronto, Toronto, Ontario, Canada

André J. Simpson
University of Toronto, Toronto, Ontario, Canada

WILEY

This edition first published 2014
© 2014 John Wiley & Sons Ltd

Registered office
John Wiley & Sons Ltd, The Atrium, Southern Gate, Chichester, West Sussex,
PO19 8SQ, United Kingdom

For details of our global editorial offices, for customer services and for information about
how to apply for permission to reuse the copyright material in this book please see our
website at www.wiley.com.

Library of Congress Cataloging-in-Publication Data

NMR spectroscopy : a versatile tool for environmental research /
editors Myrna J. Simpson, Andre J. Simpson.
        pages cm
    Includes bibliographical references and index.
    ISBN 978-1-118-61647-5 (cloth)
  1. Nuclear magnetic resonance spectroscopy. 2. Environmental chemistry.
I. Simpson, Myrna J., editor. II. Simpson, Andre J., editor.
    TD193N57 2014
    577'.140154366--dc23

                                                                        2014012695

A catalogue record for this book is available from the British Library.

ISBN-13: 978-1-118-61647-5

Cover image: The central image on the front cover is reprinted with permission from
A. J. Simpson, M. J. Simpson, and R. Soong, Nuclear Magnetic Resonance Spectroscopy
and Its Key Role in Environmental Research, Environ. Sci. technol. 46(21), 2012.
Copyright 2012 American Chemical Society.

Set in 9.5/11.5 pt Times by Laserwords (Private) Limited, Chennai, India
Printed and bound in Singapore by Markono Print Media Pte Ltd

International Advisory Board

John S. Waugh
Massachusetts Institute
  of Technology (MIT)
Cambridge, MA
USA

Bernd Wrackmeyer
Universität Bayreuth
Bayreuth
Germany

Kurt Wüthrich
The Scripps Research
  Institute
La Jolla, CA
USA
and
ETH Zürich
Zürich
Switzerland

Contents

Part C: NMR and Environmental Metabolomics   **345**

Contributors

Giuseppe Alonzo
Dipartimento di Scienze Agrarie e Forestali, Università degli Studi di Palermo, 90128 Palermo, Italy
Chapter 8: Environmental NMR: Fast-field-cycling Relaxometry

Brian Andrew
Bruker BioSpin Corporation, Billerica, MA 01821–3991, USA
Chapter 6: Environmental Comprehensive Multiphase NMR

Gregory A. Barding Jr
Department of Chemistry, University of California, Riverside, Riverside, CA 92521, USA
Chapter 27: Plant Metabolomics

Thomas Baumann
Institute of Hydrochemistry, Technische Universität München, D-81377 Munich, Germany
Chapter 7: Environmental NMR: Magnetic Resonance Imaging

Daniel W. Bearden
Analytical Chemistry Division, National Institute of Standards and Technology, Hollings Marine Laboratory, Charleston, SC 29412, USA
Chapter 22: Environmental Metabolomics
Chapter 23: Environmental Metabolomics: NMR Techniques

Anne E. Berns
Forschungszentrum Jülich GmbH, Jülich 52425, Germany
Chapter 13: Soil Organic Matter

Marko Bertmer
Institute of Experimental Physics II, Faculty of Physics and Earth Sciences, Leipzig University, Leipzig, D-04103, Germany
Chapter 18: Soil–Water Interactions

Bernhard Blümich
Institute for Technical Chemistry and Macromolecular Chemistry, RWTH Aachen University, Aachen D-52074, Germany
Chapter 9: Mobile NMR

Sean Booth
Department of Biological Sciences, University of Calgary, Calgary, Alberta, T2N 1N4, Canada
Chapter 26: Metabolomics in Environmental Microbiology

Paul T. Callaghan
MacDiarmid Institute, Victoria University of Wellington, Wellington 6012, New Zealand
Chapter 10: Terrestrial Magnetic Field NMR: Recent Advances

Federico Casanova

Institute for Technical Chemistry and Macromolecular Chemistry, RWTH Aachen University, Aachen D-52074, Germany
Chapter 9: Mobile NMR

Gabriela Chilom

Department of Chemistry and Biochemistry, South Dakota State University, Brookings, SD 57007-0896, USA
Chapter 20: Organic Pollutants in the Environment

Pellegrino Conte

Dipartimento di Scienze Agrarie e Forestali, Università degli Studi di Palermo, 90128 Palermo, Italy
Chapter 8: Environmental NMR: Fast-field-cycling Relaxometry

Denis Courtier-Murias

Department of Chemistry, University of Toronto Scarborough, Toronto, Ontario, M1C 1A4, Canada
Chapter 6: Environmental Comprehensive Multiphase NMR

Ernesto Danieli

Institute for Technical Chemistry and Macromolecular Chemistry, RWTH Aachen University, Aachen D-52074, Germany
Chapter 9: Mobile NMR

Daniel A. Dias

Metabolomics Australia, School of Botany, The University of Melbourne, Parkville, Victoria 3010, Australia
Chapter 24: Environmental Metabolomics of Soil Organisms

Armando C. Duarte

CESAM & Department of Chemistry, University of Aveiro, 3810–193 Aveiro, Portugal
Chapter 12: Atmospheric Organic Matter

Regina M.B.O. Duarte

CESAM & Department of Chemistry, University of Aveiro, 3810–193 Aveiro, Portugal
Chapter 12: Atmospheric Organic Matter

Hashim Farooq

Department of Chemistry, University of Toronto Scarborough, Toronto, Ontario, M1C 1A4, Canada
Chapter 6: Environmental Comprehensive Multiphase NMR

Antonio G. Ferreira

Department of Chemistry, Federal University of São Carlos, São Carlos, 13565-905, Brazil
Chapter 16: Biofuels

Michael Fey

Bruker BioSpin Corporation, Billerica, MA 01821–3991, USA
Chapter 6: Environmental Comprehensive Multiphase NMR

Markus Godejohann

Bruker BioSpin GmbH, 76274 Rheinstetten, Germany
Chapter 3: Environmental NMR: Hyphenated Methods

Meghan E. Halse

MacDiarmid Institute, Victoria University of Wellington, Wellington 6012, New Zealand
Chapter 10: Terrestrial Magnetic Field NMR: Recent Advances

Karen M. Hammer *SINTEF Materials and Chemistry/Environmental Technology, Trondheim NO-7465, Norway*
Chapter 25: Environmental Metabolomics of Aquatic Organisms

Norbert Hertkorn *Department of Environmental Sciences (DES), Helmholtz Zentrum Muenchen (HMGU), Neuherberg 85758, Germany*
Chapter 1: Environmental NMR: Solution-State Methods

Natalia Homan *Laboratory of Biophysics and Wageningen NMR Centre, Wageningen University, Dreijenlaan 3, 6703 HA Wageningen, The Netherlands*
Chapter 21: Soil–Plant–Atmosphere Continuum Studied by MRI

Howard Hutchins *Bruker BioSpin Corporation, Billerica, MA 01821–3991, USA*
Chapter 6: Environmental Comprehensive Multiphase NMR

Oliver A.H. Jones *School of Applied Sciences, RMIT University, Melbourne, Victoria 3001, Australia*
Chapter 24: Environmental Metabolomics of Soil Organisms

Heike Knicker *Department of Geoecology and Biogeochemistry – Organic Biogeochemistry, Instituto de Recursos Naturales y Agrobiología de Sevilla (IRNAS-CSIC), Sevilla 41012, Spain*
Chapter 13: Soil Organic Matter

Sridevi Krishnamurthy *Bruker BioSpin Corporation, Billerica, MA 01821–3991, USA*
Chapter 6: Environmental Comprehensive Multiphase NMR

Rajeev Kumar *Bruker BioSpin Corporation, Billerica, MA 01821–3991, USA*
Chapter 6: Environmental Comprehensive Multiphase NMR

Leayen Lam *Department of Chemistry, University of Toronto Scarborough, Toronto, Ontario, M1C 1A4, Canada*
Chapter 6: Environmental Comprehensive Multiphase NMR

Cynthia K. Larive *Department of Chemistry, University of California, Riverside, Riverside, CA 92521, USA*
Chapter 27: Plant Metabolomics

Luciano M. Lião *Federal University of Goiás, Goiânia, 74001-970, Brazil*
Chapter 16: Biofuels

James G. Longstaffe *Department of Chemistry, University of Toronto Scarborough, Toronto, Ontario, M1C 1A4, Canada*
Chapter 6: Environmental Comprehensive Multiphase NMR

Werner E. Maas *Bruker BioSpin Corporation, Billerica, MA 01821–3991, USA*
Chapter 6: Environmental Comprehensive Multiphase NMR

Hussain Masoom *Department of Chemistry, University of Toronto Scarborough, Toronto, Ontario, M1C 1A4, Canada*
Chapter 6: Environmental Comprehensive Multiphase NMR

Perry J. Mitchell	*Department of Chemistry and Environmental NMR Centre, University of Toronto, Toronto, Ontario, M1C 1A4, Canada* Chapter 11: Dissolved Organic Matter
Martine Monette	*Bruker Ltd, Milton, Ontario, L9T 1Y6, Canada* Chapter 6: Environmental Comprehensive Multiphase NMR
Marcos R. Monteiro	*Department of Chemistry, Federal University of São Carlos, São Carlos, 13565-905, Brazil* Chapter 16: Biofuels
Robert Morris	*College of Arts and Science, School of Science & Technology, Nottingham Trent University, Nottingham, NG1 4BU, UK* Chapter 7: Environmental NMR: Magnetic Resonance Imaging
Karl T. Mueller	*Department of Chemistry, Pennsylvania State University, University Park, PA 16802, USA* *Pacific Northwest National Laboratory, Environmental Molecular Sciences Laboratory, Richland, WA 99352, USA* Chapter 17: Clay Minerals
Nikolaus Nestle	*BASF SE, D-67056 Ludwigshafen, Germany* Chapter 7: Environmental NMR: Magnetic Resonance Imaging
Charlotte E. Norris	*Department of Renewable Resources, University of Alberta, Edmonton, Alberta T6G 2E3, Canada* Chapter 15: Forest Ecology and Soils
Daniel J. Orr	*Department of Chemistry, University of California, Riverside, Riverside, CA 92521, USA* Chapter 27: Plant Metabolomics
Caroline M. Preston	*Natural Resources Canada, Pacific Forestry Centre, Victoria, British Columbia, V8Z 1M5, Canada* Chapter 4: Environmental NMR: Solid-State Methods
William S. Price	*Nanoscale Organisation and Dynamics Group, University of Western Sydney, Penrith, New South Wales 2751, Australia* Chapter 2: Environmental NMR: Diffusion Ordered Spectroscopy Methods
Sylvie A. Quideau	*Department of Renewable Resources, University of Alberta, Edmonton, Alberta T6G 2E3, Canada* Chapter 15: Forest Ecology and Soils
James A. Rice	*Department of Chemistry and Biochemistry, South Dakota State University, Brookings, SD 57007-0896, USA* Chapter 20: Organic Pollutants in the Environment
Rebecca L. Sanders	*Department of Geosciences, Princeton University, Princeton, NJ 08544, USA* Chapter 17: Clay Minerals

Gabriele E. Schaumann

Department of Environmental and Soil Chemistry, University Koblenz-Landau, Landau D-76829, Germany
Chapter 18: Soil–Water Interactions

Bernd Schneider

Max Planck Institute for Chemical Ecology, Jena 07745, Germany
Chapter 14: Chemical Ecology

André J. Simpson

Department of Chemistry and Environmental NMR Centre, University of Toronto, Toronto, Ontario, M1C 1A4, Canada
Chapter 6: Environmental Comprehensive Multiphase NMR
Chapter 11: Dissolved Organic Matter
Chapter 19: Metals in the Environment

Myrna J. Simpson

Department of Chemistry and Environmental NMR Centre, University of Toronto, Toronto, Ontario, M1C 1A4, Canada
Chapter 6: Environmental Comprehensive Multiphase NMR
Chapter 11: Dissolved Organic Matter
Chapter 23: Environmental Metabolomics: NMR Techniques

Ronald Soong

Department of Chemistry and Environmental NMR Centre, University of Toronto Scarborough, Toronto, Ontario, M1C 1A4, Canada
Chapter 6: Environmental Comprehensive Multiphase NMR

Laure N. Soucémarianadin

Department of Renewable Resources, University of Alberta, Edmonton, Alberta T6G 2E3, Canada
Chapter 15: Forest Ecology and Soils

Trond R. Størseth

SINTEF Materials and Chemistry/Environmental Technology, Trondheim NO-7465, Norway
Chapter 25: Environmental Metabolomics of Aquatic Organisms

Ruth E. Stark

Department of Chemistry, City University of New York, New York, NY 10031, USA
Chapter 5: Environmental NMR: High-resolution Magic-angle Spinning

Henry J. Stronks

Bruker Ltd, Milton, Ontario, L9T 1Y6, Canada
Chapter 6: Environmental Comprehensive Multiphase NMR

Jochem Struppe

Bruker BioSpin Corporation, Billerica, MA 01821–3991, USA
Chapter 6: Environmental Comprehensive Multiphase NMR

Andre Sutrisno

Department of Chemistry and Environmental NMR Centre, University of Toronto, Toronto, Ontario, M1C 1A4, Canada
Chapter 6: Environmental Comprehensive Multiphase NMR
Chapter 19: Metals in the Environment

Shiying Tian

Department of Chemistry, City University of New York, New York, NY 10031, USA
Chapter 5: Environmental NMR: High-resolution Magic-angle Spinning

Raymond J. Turner

Department of Biological Sciences, University of Calgary, Calgary, Alberta, T2N 1N4, Canada
Chapter 26: Metabolomics in Environmental Microbiology

Henk Van As

Laboratory of Biophysics and Wageningen NMR Centre, Wageningen University, Dreijenlaan 3, 6703 HA Wageningen, The Netherlands
Chapter 21: Soil–Plant–Atmosphere Continuum Studied by MRI

Frank J. Vergeldt

Laboratory of Biophysics and Wageningen NMR Centre, Wageningen University, Dreijenlaan 3, 6703 HA Wageningen, The Netherlands
Chapter 21: Soil–Plant–Atmosphere Continuum Studied by MRI

Nancy M. Washton

Pacific Northwest National Laboratory, Environmental Molecular Sciences Laboratory, Richland, WA 99352, USA
Chapter 17: Clay Minerals

Roderick E. Wasylishen

Gunning-Lemieux Chemistry Centre, Department of Chemistry, University of Alberta, Edmonton, Alberta, T6G 2G2 Canada
Chapter 15: Forest Ecology and Soils

Aalim Weljie

Department of Pharmacology, Institute for Translational Medicine and Therapeutics, University of Pennsylvania, Philadelphia, PA 19104, USA
Chapter 26: Metabolomics in Environmental Microbiology

Carel W. Windt

Forschungszentrum Jülich GmbH, Jülich 52425, Germany
Chapter 21: Soil–Plant–Atmosphere Continuum Studied by MRI

Guohua Wu

Department of Chemistry, City University of New York, New York, NY 10031, USA
Chapter 5: Environmental NMR: High-resolution Magic-angle Spinning

Bin Yan

Department of Chemistry, City University of New York, New York, NY 10031, USA
Chapter 5: Environmental NMR: High-resolution Magic-angle Spinning

Bingwu Yu

Department of Chemistry, City University of New York, New York, NY 10031, USA
Chapter 5: Environmental NMR: High-resolution Magic-angle Spinning

Gang Zheng

Nanoscale Organisation and Dynamics Group, University of Western Sydney, Penrith, New South Wales 2751, Australia
Chapter 2: Environmental NMR: Diffusion Ordered Spectroscopy Methods

Junyan Zhong

Department of Chemistry, City University of New York, New York, NY 10031, USA
Chapter 5: Environmental NMR: High-resolution Magic-angle Spinning

Series Preface

The *Encyclopedia of Nuclear Magnetic Resonance* was published, in eight volumes, in 1996, in part to celebrate the fiftieth anniversary of the first publications in NMR in January 1946. Volume 1 contained an historical overview and 200 articles by prominent NMR practitioners, whilst the remaining seven volumes were constituted by 500 articles on a wide variety of topics in NMR (including MRI). A ninth volume was brought out in 2000 and two "spin off" volumes incorporating the articles on MRI and MRS (together with some new ones) were published in 2002. In 2006 the decision was taken to publish all the articles electronically (i.e. on the world-wide web) and this was carried out in 2007. Since then, new articles have been placed on the web every three months and some of the original articles have been updated. This process is continuing and to recognize the fact the Encyclopedia of Magnetic Resonance is a true online resource, the web site has been redesigned and new functionalities added, with a relaunch in January 2013 in a new Volume and Issue format, under the new name eMagRes. In December, 2012, a print edition of the Encyclopedia of NMR was published in ten volumes (6200 pages). This, much needed update of the 1996 edition of the Encyclopedia, encompasses the entire field of NMR with the exception of medical imaging (MRI).

The existence of this large number of articles, written by experts in various fields, is enabling a new concept to be implemented, namely the publication of a series of printed handbooks on specific areas of NMR and MRI. The chapters of each of these handbooks will be constituted by a carefully chosen selection of Encyclopaedia articles relevant to the area in question. In consultation with the Editorial Board, the handbooks are coherently planned in advance by specially-selected editors, and new articles written (together with updating of some already existing articles) to give appropriate complete coverage of the total area. The handbooks are intended to be of value and interest to research students, postdoctoral fellows and other researchers learning about the topic in question and undertaking relevant experiments, whether in academia or industry. Consult the eMagRes web site (http://onlinelibrary.wiley.com/book/10.1002/9780470034590) for the latest news on magnetic resonance Handbooks.

Robin K. Harris
University of Durham, Durham, UK

Roderick E. Wasylishen
University of Alberta, Edmonton, Alberta, Canada

June 2014

Preface

The challenges faced by environmental scientists today are vast, complex, and multi-faceted. For instance, predicting the fate of an environmental pollutant or understanding ecosystem responses to climate change, necessitate a firm understanding of molecular structure and dynamics of environmental media as well as the components that exist and interact within this media. Furthermore, linking information obtained at the molecular-scale to ecosystem-level processes is a major pursuit of modern environmental research. As such, NMR spectroscopy and its scalability from the molecular-scale to the macroscopic-scale, is facilitating rapid growth in environmental science. In addition, the versatility of NMR spectroscopy has resulted in the development and implementation of different types of NMR techniques to examine the structure of various types of environmental samples, living and non-living, as well as the study of critical environmental processes. This handbook is a collection of chapters that span from methods to how NMR is used in environmental research to gain insight into various ecosystem properties. These chapters also highlight the immense potential of NMR spectroscopy which has expanded our fundamental understanding of environmental processes and will likely continue to do so well into the future.

NMR spectroscopy has been used to study environmental samples since the early 1960s. One of the first applications, by Barton and Schnitzer in 1963 (*Nature* 198:217–218), used solution-state ^{1}H NMR to study the composition of isolated soil humic substances. Since this time, the advancements in NMR have resulted in a wide range of methods to be applied to environmental samples and required the development of environment-specific methods, such as comprehensive multi-phase NMR. This handbook is organized into three parts—Part A focuses on methods used in environmental NMR which span from solution-state to magnetic resonance imaging. Part B emphasizes how NMR spectroscopy has played an essential role in understanding various types of environmental components and related processes. These include different forms of organic matter found in soil, water, and air as well as how NMR is used to probe the fate of water, organic pollutants, and metals in the environment. Part C focuses on the growing field of environmental metabolomics which uses NMR as its main discovery platform. NMR-based environmental metabolomics is reshaping the understanding of ecotoxicity of problematic environmental pollutants in different environments.

We sincerely hope that the reader will benefit from the overviews written by experts in the growing and diverse field of environmental NMR spectroscopy. We thank the authors for their important and excellent contributions to this handbook. We also thank Professor Robin K. Harris and Professor Roderick Wasylishen (Editors in Chief for *eMagRes*) for their guidance and support. This handbook would not have been possible without the support and assistance from the *eMagRes* team at Wiley which consists of Elke Morice-Atkinson, Stacey Woods, and Martin Rothlisberger. It is also important to note that many influential people nurtured our early interest in environmental NMR. We thank our PhD mentors (Professor Marvin Dudas, Professor William McGill and Professor Michael H. B. Hayes) and our Postdoctoral research supervisors (Professor William Kingery and Professor Pat Hatcher). Our collaborators at Bruker BioSpin, especially Dr. Manfred Spraul, Dr. Werner Mas, and Dr. Henry Stronks, are also acknowledged because of their keen interest in the co-development of environmental NMR methods and their long-term support of environmental NMR development and application. We must also thank the generous support for environmental NMR research from the Canada Foundation for Innovation, Krembil Foundation, Natural Sciences and Engineering Research Council of Canada, Ontario Research Foundation, and the University of Toronto. Lastly, we dedicate this book to our four year old twins, Sam and Sophie, who are great admirers of both the environment and our NMR "rockets" (a.k.a. spectrometers).

Myrna J. Simpson
University of Toronto, Canada

André J. Simpson
University of Toronto, Canada

June 2014

PART A
Fundamentals of Environmental NMR

Chapter 1

Environmental NMR: Solution-State Methods

Norbert Hertkorn

Department of Environmental Sciences (DES), Helmholtz Zentrum Muenchen (HMGU), Neuherberg 85758, Germany

1.1 INTRODUCTION

Magnetic resonance spectroscopy is concerned with the splitting of magnetic spins of electrons (ESR) and that of atomic nuclei (NMR) in an external magnetic field B_0.[1–7] The splitting of the NMR transition is not solely an intrinsic atomic and molecular property but also depends on the magnitude of an external magnetic field B_0. Here, an increase in B_0 results in NMR sensitivity enhancement and improved spectral resolution.[3–5] The special role of NMR spectroscopy in the molecular-level characterization of complex mixtures and amorphous materials resides in its ability to provide unsurpassed in-depth, isotope-specific information about short-range molecular order.[7]

NMR offers the capability for quantitative and nondestructive determination of chemical environments across the periodic table with a very few exceptions: only the elements Ar and Ce lack any stable, magnetically active isotope with nuclear spin >0. Quantitative relationships between number of spins and area of NMR resonances operate in the absence of differential NMR relaxation.[1,2,5,6] This key feature of NMR in the de novo analysis of complex systems implies the use of NMR spectroscopy as a quantitative reference for other, complementary analytical methods, in particular, when complex unknowns are to be characterized with molecular precision.[7,8] However, low intrinsic overall NMR sensitivity compared with other analytical methods restricts the accessible signal-to-noise (S/N) ratio in NMR spectra and accuracy of signal definition.

When performed properly on any environmental sample, NMR spectroscopy will provide isotope-specific information in unsurpassed detail on the arrangements of chemical bonds, including connectivities, stereochemistry, and spatial proximity as well as meaningful clues about their dynamics and reactivity.[1–3,9–11] However, elaborate sample

NMR Spectroscopy: A Versatile Tool for Environmental Research
Edited by Myrna J. Simpson and André J. Simpson
© 2014 John Wiley & Sons, Ltd. ISBN: 978-1-118-61647-5

preparation with attentive consideration of the physical processes initiated by the NMR pulse sequence might become essential to obtain meaningful data from polydisperse mixtures. Common environmental samples are mostly complex mixtures of small and large molecules, related by a continuous range of weak to strong interactions. Typically, formation history is poorly constrained, whereas polydispersity and molecular heterogeneity across various size scales is the norm rather than the exception. NMR spectroscopy offers uniquely versatile options to study liquids, gels (see Chapter 5), solids (see Chapter 4), gases, and any combination thereof, a very beneficial prerequisite to study environmental samples in their native state[10-13] (see Chapter 6).

The impressive contributions of, e.g., solid-state (see Chapter 4) and comprehensive multiphase NMR spectroscopy (see Chapter 6) have been addressed in excellent reviews[10,11] and are not considered in this account, which is concerned with solution-state NMR spectroscopy in environmental sciences. This focus implies a stronger emphasis on studies of extracts and, hence, environmental sample preparation. The often unavoidable extraction selectivity will become an asset when purposeful decrease in heterogeneity and impurities will improve sensitivity and S/N ratio in NMR spectra, which is of perpetual concern in NMR spectroscopy.

Current scientific exploration of biochemical organic molecular complexity in which clearly resolved patterns (and their alterations) are readily observed appears more attractive to many than investigation of the vastly more complex biogeochemical mixtures. Here, analytical data are subject to far more extensive intrinsic averaging and necessarily produce less resolved signatures.[7,14] Here, NMR spectroscopy shows the most unambiguous relationship between NMR observable (NMR with chemical shift, line shape, and couplings) and atomic process (reorientation of nuclei spinning with individual precession frequencies in an external magnetic field B_0). It cannot be overemphasized that NMR spectroscopy alone will provide the most direct evidence on molecular structure of any unknown (amorphous) organic substance and mixture; this degree of immediacy of NMR–structure relationships is not available by any other analytical technique.[8] This allows one to define the relative quantities and remarkable structural detail of fundamental building blocks.[15,16] Here, multinuclear quantitative one-dimensional (1D) NMR spectroscopy provides the key margin for any structural model of a complex unknown environmental sample. The unique

capability to generate and analyze data from multiple higher dimensional and multinuclear NMR experiments obtained from a single sample[1,2,8,15-19], serves to enhance the reliability of NMR assignments and allows definition of rather extended substructures in environmental organic mixtures.[8,18]

The most prominent obstacles to implement the potential intrinsic to NMR in environmental sciences are an often novercal attitude of chemistry toward environmental complexity and a factual and technical inaccessibility of competitive NMR resources to virtually all environmental sciences. This has led to a widespread perception that NMR spectroscopy is not sensitive enough to cope with the low concentrations and diversity of matrices encountered in many environmental reactions.[20] This chapter highlights some of the encouraging conceptual and hardware developments in modern NMR spectroscopy, which may provide about one order of magnitude improved resolution and sensitivity compared with equipment commonly used in environmental sciences. This game-changing evolution should place this most powerful analytical method at the heart of most environmental studies that are aimed at a molecular understanding of archetypical complex environmental unknowns that are poorly amenable to any target analyses.

1.2 GENERAL NMR CHARACTERISTICS OF NUCLEI ACROSS THE PERIODIC TABLE

NMR spectroscopy measures the precession frequencies of individual nuclear magnetic moments in an external magnetic field B_0 and the rate of nuclear spin reorientation after excitation (NMR relaxation).[1,2,5,6,21,29] NMR characteristics depend on the isotope-specific nuclear properties γ_N: gyromagnetic ratio, B_0 applied, and the local chemical environments (Figure 1.1). Therefore, any atom within a molecule will show individual NMR-relevant properties, allowing the assembly of molecular structures from NMR spectra in the case of resolvable mixtures and the reconstruction of key structural principles in the case of nonresolvable biogeochemical mixtures.[7] An increase in B_0 results in NMR sensitivity enhancement and improved spectral resolution (Figure 1.1, panel B). The Larmor equation of NMR [$\omega_i = \gamma_N \cdot B_0 \cdot (1 - \delta_i)$; γ_N: gyromagnetic ratio; δ_i: chemical

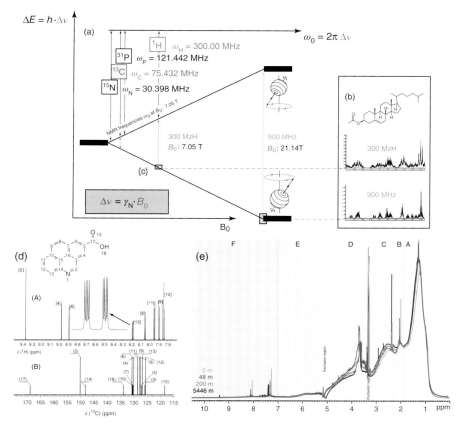

Figure 1.1. Dependence of some magnetic resonance spectral characteristics on the magnetic field B_0.[7] (a) At constant magnetic field B_0, the resonance frequency depends on the relative gyromagnetic ratios of the nuclei γ_N (NMR). (b) Sections of proton NMR spectra of cholesterol acetate at $B_0 = 7.05\,\text{T}$ (300 MHz ^{1}H frequency) and at $B_0 = 21.14\,\text{T}$ (900 MHz ^{1}H frequency); note the *qualitative* difference of high-resolution proton NMR spectra acquired at various B_0. This variation remains the most significant obstacle for automated NMR assignment in organic molecules. Pattern recognition in 2D NMR spectra can be more successfully automated. (c) The relative energy, represented by the chemical shift range, covers a miniscule (ppm) range of the (already tiny) NMR energy transition energy; the ratio of total chemical shift range to total NMR transition energy ranges from ~20 ppm (^{1}H, diamagnetic molecules) up to ~20 000 ppm (^{59}Co and ^{195}Pt NMR; Figure 1.2). Owing to the near equality of the Boltzmann factors for the NMR energy levels, out of 1 000 000 proton nuclei, only 81 participate in the ^{1}H NMR experiment at $B_0 = 11.7\,\text{T}$ and 283 in the NMR experiment at $B_0 = 21.14\,\text{T}$ (at room temperature: 300 K). All other proton nuclei remain silent throughout the NMR experiment. This ratio is even worse for other nuclei, explaining the relative insensitivity of NMR spectroscopy when compared to higher energy spectroscopic methods. (d) Mid-sized molecules produce individual NMR signatures for any atom as shown for ^{1}H (A) and ^{13}C NMR (B) NMR spectra of carboxyazapyrene, which allow reconstruction of unambiguous structures; the coupling of H13 is shown in expansion: the vicinal coupling $^3J(\text{H12}-\text{H13}) = 7.9\,\text{Hz}$ generates the large dublett splitting; further splitting is effected by $^4J(\text{H12}-\text{H14}) = -1.3\,\text{Hz}$ with favorable W-shaped geometry, which reflects a succession of two near 180° dihedral angles (cf. Karplus equation) and even $^6J(\text{H13}-\text{H9}) = 0.6\,\text{Hz}$ is observable, in which two separate long-range coupling pathways coadd to result in a detectable splitting of the H13 NMR. ((a-c) Reproduced from Ref. 32.) (e) Complex environmental mixtures similar to this marine organic matter depth profile (800 MHz ^{1}H NMR spectra), normalized to identical integral area, exhibit low-resolution NMR signatures resulting from extensive superposition of NMR resonances.[8] Integration provides quantities of coarse substructure regimes; NMR spectra of environmental samples acquired at increased B_0 show enhanced sensitivity and resolution as well as improved assignment options from combinations of multinuclear and higher dimensional NMR spectra. (Reproduced from Ref. 8. Distributed under the Creative Commons Attribution 3.0 License)

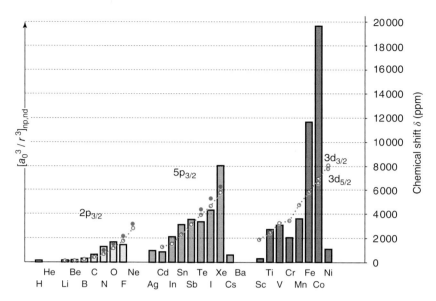

Figure 1.2. Range of chemical shift of selected main group and transition elements;[21] the atomic radii $r^{-3}_{np,nd}$ (with relativistic correction) values are shown as dimensionless ratios a_0/r^3, where a_0 is the atomic unit of length ($a_0 = 52.9$ pm). Owing to large quadrupolar moment, [135/137]Ba NMR studies are rare, and [61]Ni NMR is restricted to rather symmetrical complexes of the type NiL_4, $XNiL_3$, and X_2NiL_2, explaining the restricted chemical shift range observed so far. The most commonly measured NMR parameters, chemical shift δ and indirect coupling constant (J), largely relate to the chemical properties of molecules and are comparable between the nuclei of different metals, provided simple adjustments for nuclear parameters (mainly γ_N) are made. For instance, advantage can be taken from the extensive studies of [59]Co and [195]Pt NMR to aid in the interpretation of NMR data from other metal nuclei[21].

shift] implies that actual NMR frequencies ω_i are proportional to the external magnetic field B_0. Similarly, NMR total chemical shift ranges as well as frequency differences $\Delta\omega_{ij}$ between dissimilar chemical environments ω_i and ω_j scale with B_0. Nevertheless, NMR spectra are visualized as plots of line intensity versus frequency ω_i, expressed as dimensionless, magnetic field independent units of chemical shift δ_i.

The chemical shift δ_i denotes the fractional change in frequency induced by the variance in chemical environments normalized to B_0; the accessible chemical shift range varies from about 15 ppm for ^{1}H NMR up to 22 000 ppm in the case of [59]Co NMR spectroscopy (Figure 1.2).[21,23,24]

Mid-sized molecules provide resolved individual NMR signatures (corresponding to unique NMR frequencies) for any NMR-active atom as shown here for a ^{1}H NMR and a ^{13}C NMR spectrum of azacarboxypyrene (computed 800 MHz ^{1}H, ^{13}C, and ^{15}N NMR spectra); the solution-state ^{17}O NMR spectrum

consists of a single (broadened) resonance because of fast chemical exchange between the two different binding states of oxygen (Figure 1.3), whereas the ^{15}N nucleus resonates at δ (^{15}N): −71 ppm with a singlet. NMR spectra of complex environmental materials, such as the 800 MHz ^{1}H NMR spectrum of a dissolved marine organic matter shown, exhibit low-resolution signatures because of extreme overlap of individual NMR resonances.[8] However, extensive and far-reaching structural detail not available from any other analytical methods at present regularly results from even these ill-resolved NMR spectra in the case of 'small' nuclei (e.g., ^{1}H, ^{13}C, ^{15}N, and ^{31}P) because of the plausible correspondence between chemical shift and alignment of extended substructures and the quantitative relationships between given chemical environments and NMR integral (Figure 1.3). Nevertheless, near identity of chemical shift not necessarily implies similarity of chemical structures.[10,11,15]

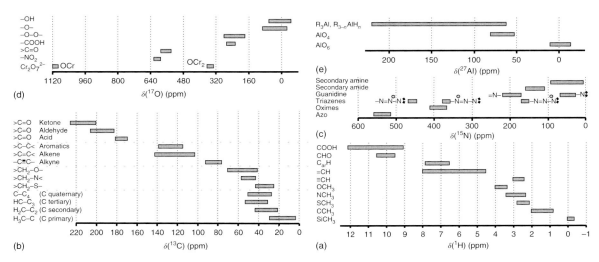

Figure 1.3. Chemical shift ranges of some major substructures in molecules for the nuclei (a) ^{1}H, (b) ^{13}C, (c) ^{15}N, (d) ^{17}O, and (e) ^{27}Al.[25] For the 'small-sized' spin-$1/2$ nuclei ^{1}H, ^{13}C, ^{15}N, and ^{31}P, which constitute the backbone of organic molecules, the relationship between δ, J, and the corresponding structural features is rather well understood. $\delta(^{1}$H) is chiefly governed by diamagnetic shielding contributions and, therefore, electron withdrawal and CSA contributions are substantial (cf. alkynes, carbonyl derivatives, and aromatic ring currents). Paramagnetic shielding dominates δ of the other nuclei, providing large overall chemical shift ranges. Orbital contraction and the ΔE^{-1} term play substantial roles in defining $\delta(X)$. $\delta(^{27}$Al) is strongly affected by coordination geometry (tetrahedral against octahedral coordination).

NMR spectroscopy has always suffered from poor sensitivity, largely because of the low Boltzmann factors involved (Figure 1.1). Nuclear properties of magnetic isotopes with nuclear spin >0 govern the NMR sensitivity S and receptivity R as follows:

$$R_{N} = n_{N} \cdot S_{N} = n_{N} \cdot \gamma_{N}^{3} \cdot I(I + 1)$$

where n_{N} is the (natural) isotopic abundance, γ_{N} the gyromagnetic ratio, and I the nuclear spin (Figure 1.1). Thus, the NMR receptivity at natural abundance ranges across nearly seven orders of magnitude (from ^{1}H to ^{187}Os; receptivity ratio $1 : 2.43 \times 10^{-7}$).[21,25] Provided that the signal to noise (S/N) ratio in NMR spectra scales with the square root of the number of acquired scans (S/N $\sim$ NS$^{1/2}$), it is obvious that every attainable measure has to be taken to improve the accessible S/N ratio in NMR spectra. In recent years, increase in sensitivity and resolution was realized by high-field magnets, cryogenic, micro (nanoliter) probes, and spinning microcoil probes.[10,11] In the case of better sample availability, improved throughput was realized by fast higher dimensional spectroscopy with superior sensitivity and parallel acquisition of NMR spectra.[7] Nevertheless, NMR acquisition conditions for every

single environmental NMR sample should be carefully optimized; key parameters are optimal tuning and matching of the probe and accurately determined 90° pulses and fairly decent sample concentrations in rather small volumina (the quality factor Q in cryogenic probeheads increases for smaller sample size). Method-oriented NMR signal enhancements such as dynamic nuclear polarization, signal amplification by reversible exchange, and parahydrogen[10] offer potentially huge amplification factors with however distinct structural selectivity (which may be advantageous in dedicated target analysis).

1.3 CLASSICAL NMR OBSERVABLES

1.3.1 The Basics of NMR Chemical Shift

The external magnetic field B_0 induces circulating currents in the electron cloud, which, in turn, generate an induced magnetic field B_i, which is in the order of $10^{-4} B_0$. The individual spins sense the sum of the external and locally induced magnetic fields $(B_0 + B_i)$, resulting in distinguishable magnetic fields (and NMR frequencies) for any atomic environment in molecules.[1,3,5,26] In solution-state NMR, a motionally averaged isotropic chemical shift δ_i is observed. The contrasting (positive) charge of the nuclei and the (negative) charge of the electrons imply that chemical shift and electronic shielding, which is related to the electron density, are of opposite sign $\delta \sim -\sigma$: A lesser shielding of the nucleus by electron-withdrawing substituents leads to a higher local magnetic field B_i and to a higher NMR frequency v_i and vice versa. Several mechanisms of chemical shift operate in parallel with relative contributions depending on the positioning of the nucleus within the periodic table and specific molecular characteristics.[3,6,21,24,27]

The two main contributors to B_i are the field-induced circulation of electrons in the electronic ground state (diamagnetic term) and electron circulation from mixing of excited electronic states into the electronic ground state (paramagnetic term).[3,6,21,27] Both terms are of opposite sign, and the paramagnetic term is usually much larger than the diamagnetic term. The diamagnetic shielding is determined by the core electrons and nearly constant for any nucleus, except for atoms where the outer orbitals are s-orbitals such as hydrogen and the alkali metals ($^{1/2/3}$H, $^{6/7}$Li, ^{23}Na, $^{39/40/41}$K, $^{85/87}$Rb, and ^{133}Cs). Here, the chemical shift range is small, in the order of tens of parts per million (also a consequence of restricted overall chemical diversity in alkali metal coordination motifs). σ^{dia} dominates the hydrogen chemical shift ($^{1/2/3}$H) and is rationalized in terms of electron density and hybridization.[26]

$$\sigma_A^d(\text{loc}) \propto \sum_{\psi}^{A} P_{\psi\psi} \left\langle \psi \left| \frac{1}{r_{A\psi}} \right| \psi \right\rangle$$

where $P_{\psi\psi}$ is the charge density of wave function ψ and $\langle \psi | r_{A\psi}^{-1} | \psi \rangle$ the expectation value of ψ at $r_{A\psi}$. The neighboring group chemical shift anisotropy (CSA) σ^{CSA} is concerned with uneven electron distributions around certain double and triple bonds as well as aromatic rings (aromatic ring current), causing a few parts per million displacement of the chemical shift of any atom. Hence, σ^{CSA} contributions similar to those of C=C and C=O double and C≡C and C≡N triple bonds as well as aromatic ring currents are important in hydrogen (and sometimes lithium) NMR but otherwise commonly rather insignificant for larger nuclei in which other shielding factors dominate the positioning of the experimental chemical shift δ.[6,26]

The paramagnetic shielding σ^{para} refers to electron excitation between ground and excited states and is affected by the energy gap (ΔE) of commonly singlet–triplet transitions, the average radius (r) of the concerned p, d, and f orbitals and the bond order (P_ψ, D_ψ, F_ψ), which defines the charge density of the p, d, and f valence electrons [$Z_{np(nd,nf)}$, effective nuclear charge of np (or nd, nf) orbital; n, orbital number; and a_0, Bohr radius: 53 pm]:

$$\sigma_A^p(\text{loc}) \propto -\frac{1}{\Delta E} \left[\left\langle \frac{1}{r^3} \right\rangle_{np} P_\psi + \left\langle \frac{1}{r^3} \right\rangle_{nd} D_\psi \right]; \ \langle r^{-3} \rangle_{np}$$
$$= \frac{1}{3} \left(\frac{Z_{np}}{na_0} \right)^3$$

Qualitatively, low lying electronic excited states contribute most to the chemical shift; in heavy atoms, low lying electronic states are more closely spaced and more easily accessible, leading to a larger chemical shift range.[3,21] Small energy gaps (ΔE) and orbital contraction caused by increasing effective nuclear charge will provoke large shielding contributions to the observed chemical shift.[21] This refers, in particular, to a series of nuclei when moving from left to right within a period (row) of the periodic table, where successive addition of protons and electrons draws the electrons closer to the nucleus, decreasing atomic radii and increasing, e.g., the ionization energy (Figure 1.2).[21] Electron-withdrawing substituents enlarge local fields of neighboring atoms, thereby increasing the chemical shift. Furthermore, nuclear magnetic and orbital characteristics, oxidation state, coordination geometry, and ligand properties are decisive parameters of chemical shift.[21,23,24,28]

Magnetic properties and factual chemical diversity govern the attainable NMR chemical shift range of any given nucleus. Hence, isotope mass and available NMR chemical shift dispersion correlate only very loosely. An instructive example is the second period from Li to Ne with mainly s- and p-electrons engaged in chemical bonds, which nevertheless show a clearly noncontinual dispersion of chemical shifts (Figure 1.2).

The unique capacity of NMR for structural elucidation of unknowns independent of prior or auxiliary information critically depends on understandable relationships between observed NMR chemical shift δ and respective chemical environments.[2,6,21,27,29,30] As many NMR-relevant interactions are transmitted via bond electrons, molecules connected solely by covalent bonds will provide the most straightforward relationship between NMR parameters and chemical structure. Here, the routinely familiar relationships between NMR parameters and chemical environments

Table 1.1. 'Indirect' effects on the chemical shift in environmental materials (cf. text)

Type of effect	NMR nuclei most likely to become affected
Differential longitudinal relaxation causes integral distortions from incomplete relaxation	Low-frequency X-nuclei and hydrogen-deficient environments (alkylated fused rings, condensed aromatics, largely unsaturated molecules, and possibly with high content of heteroatoms: O, N, P, and S)
Differential transverse relaxation causes disappearance of NMR resonances because of excessive line broadening	^{1}H, high-frequency nuclei in anisotropic chemical environments
Charge effects, hydrogen bonding, coordination chemical shift, association, and aggregation	^{1}H, ^{15}N, ^{17}O, ^{19}F, ^{33}S, and all nuclei with free electron pairs
Chemical shift anisotropy imposes differential transverse NMR relaxation, especially at high magnetic fields B_0	^{199}Hg (linear coordination), X-nuclei with a large range of chemical shift in low-symmetry environments
Paramagnetics (radicals and metal ions) and chemical exchange	Metals, ^{1}H, ^{15}N, ^{17}O, ^{33}S, and any nucleus
Severe background resonance from probe	^{11}B, ^{17}O, ^{23}Na, ^{27}Al, and ^{29}Si

relate to electron density, bond order, and basic geometry (hybridization, CSA). This eminently applies to ^{13}C NMR spectra of almost any organic compound.[6] These fairly unambiguous criteria and their broadband applicability for expert structural elucidation have decisively contributed to the popularity of NMR. However, already disposable electron lone pairs common in small nuclei ($^{14/15}$N, ^{17}O) compounds offer more latitude in the relationships of bond properties and NMR parameters (Figure 1.3, Table 1.1).[27,29] Transition metals, which offer latitude for facile alteration of charge (oxidation number) as well as coordination number and geometry, produce in part vastly larger chemical shift ranges than accessible to main group elements.[21,23,24,28]

In the NMR-based structural analysis of complex environmental mixtures, the most widely used approach is to relate chemical shifts and NMR integrals as deduced from NMR spectra to structural features in these materials (Figure 1.1e).[7,8,16,17] However, intricate physics and chemistry of intra- and intermolecular interactions in complex mixtures may interfere with direct relationship between chemical shift and molecular structure and (because of relaxation-induced variable linewidth) quantification, too.[8,31]

The so-called direct effects directly and quantitatively relate chemical shifts and resonance integrals in experimental NMR spectra with respective chemical environments. The so-called indirect effects are caused by additional interactions, which may affect chemical shifts and resonance amplitudes in experimental NMR spectra and therefore obscure the facile conclusions drawn by the sole analysis of 'direct' effects. A list of 'indirect' effects is provided in Table 1.1 together with the NMR nuclei most likely to become affected. Many environmental samples show appreciable molecular heterogeneity and polydispersity (Figure 1.4), which implies that a dependable NMR analysis has to properly address both 'direct' and 'indirect' effects on the chemical shift.

1.3.2 NMR Scalar Couplings

B_0-independent interaction of nuclear spins occurs via a direct through-space dipole–dipole coupling without involving the electrons (D) and indirect interactions (J) that are mediated by electrons.[1–3,21] Dipole–dipole couplings may be intra- and intermolecular, whereas indirect J-couplings are intramolecular, a key asset to identify connected nuclei in molecules. In an oriented system, both D and J contribute to the observed NMR spectrum. In solution, D is commonly averaged to zero; however, purposeful minor alignment in specifically designed liquids leaves small residual dipolar couplings observable, which are valuable for elucidation of stereochemistry and other molecular properties.[33]

If J is larger than the linewidth Δv, then a splitting will be observed in NMR spectra, provided that J (in Hertz) is larger than the relaxation rate of the connected nuclei. In environmental mixtures, which often are composed of many compounds with a continual distribution of concentrations, superposition of many tiny NMR resonances of about similar amplitude will make splitting by J-couplings difficult to observe in 1D

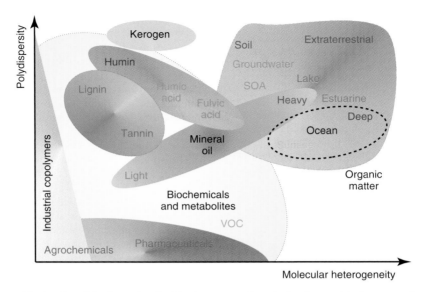

Figure 1.4. Hierarchical order of intricacy invoked in the structural analysis of materials in terms of polydispersity and molecular heterogeneity (cf. text).[7,32] The structures (connectivities and stereochemistry) of monodisperse molecules are readily accessible (provided that sufficient amounts of materials are available) by organic structural spectroscopy. Supramolecular structures require an adequate definition of covalently bonded molecules and of their noncovalent interactions. The structural analysis of nonrepetitive complex unknowns, which feature substantial extents of both polydispersity and molecular heterogeneity, is most demanding with respect to methodology and concepts. Any highly resolved three-dimensional structure of a monodisperse molecule is based on a precise description of the unique chemical environment of any single atom. Currently, the molecular-level structural analysis of complex systems is primarily focused on the definition of the covalent bonds. Future high-quality structural analyses of these materials will have to assess both (classes of) individual molecules and the mechanisms of their interactions.

NMR spectra. Here, two-dimensional (2D) J-resolved NMR spectroscopy[34] will often retrieve (especially) ^{1}H,^{1}H couplings in biological (see Chapter 23) and biogeochemical mixtures. Spin–spin couplings differ for chemically equivalent nuclei in case of magnetic nonequivalence. In alkenes $H_2C{=}CX_2$, $^3J_{XH}$(trans) is commonly larger than $^3J_{XH}$(cis) because of difference in coupling pathways.

The three main mechanisms for J couplings are all mediated by the valence electrons. Fermi-contact interaction operates via interactions of s-electrons at the nucleus (p, d, and f orbitals have no probability density at the nucleus).[3,6,21] The orbital interaction J_{OD} arises from the perturbation of the nuclear magnetic moment by the magnetic field caused by the electron orbital motion, whereas the spin–dipolar interaction J_{SD} originates from the direct interaction of the magnetic dipole of the nuclei with that of the orbital electron. The observed coupling constant J is the sum of $J_{FC} + J_{OD} + J_{DD}$, with J_{FC} usually dominating (in the case of ^{1}H,^{1}H couplings, only J_{FC} operates, whereas for couplings of larger nuclei, the other mechanisms have to be considered).[21]

For comparisons involving different nuclei, a reduced coupling constant $^nK_{AB}$ is useful: the effects of variable gyromagnetic ratios are removed and the influence of electrons on the couplings can be directly compared.

$$^nJ(A,B)\frac{4\pi^2}{h\gamma_A\gamma_B} = {}^nK(A,B)$$

J-couplings range from <1 Hz to >100 kHz $[^1J(^{205}\text{Tl}{-}^{195}\text{Pt})]$ and provide a wealth of empirical information about structural features. Typically, $^1J_{XY} \gg {}^3J_{XY} > {}^2J_{XY} > {}^4J_{XY}$ is valid, but mutually nonexclusive ranges and exceptions prevail (especially in ^{19}F NMR and metal NMR spectroscopy).[21,23,24] Multiple coupling paths, especially in cyclic geometries, may contribute to the observed J-coupling (Figure 1.1d). One bond couplings $^1J_{XY}$ depend on

n	$I = 1/2$	$I = 1$	$I = 3/2$
0	1	1	1
1	1 1	1 1 1	1 1 1 1
2	1 2 1	1 2 3 2 1	1 2 3 4 3 2 1
3	1 3 [3 1]	1 3 6 7 [6 3 1]	1 3 6 10 12 12 [10 6 3 1]
4	1 4 6 [4] 1	1 4 10 16 19 16 [10] 4 1	1 4 10 20 31 40 44 40 31 [20] 10 4 1

Figure 1.5. In general, the coupling of A nuclei to n equivalent I spins gives rise to a multiplett with $(2 \cdot n \cdot I + 1)$ lines in the A NMR spectrum. For $I = 1/2$, the relative intensities follow the Pascal triangle, and corresponding relationships operate for higher spin numbers.[21] Each number in the triangles (for $n > 0$) can be obtained as the sum of the $(2 \cdot n \cdot I + 1)$ nearest neighbors in the line above, with zero for empty spaces. These simple rules break down when the difference in chemical shift in frequency units becomes comparable to J and then complex higher order NMR spectra may result.

the nominal hybridization of the respective nuclei and general coordination geometry, oxidation state, inductive substituent effects, and the gyromagnetic ratio of the concerned elements.[6,21] For larger nuclei, variance in effective nuclear charge, s-orbital and relativistic contributions are important as well.[21,23] Geminal couplings $^2J_{XY}$ are commonly rather small, often from cumulative contributions of opposite signs and more difficult to interpret in terms of structures; here, orbital orientations and bond angles play a critical role. In addition to influences active in $^2J_{XY}$, vicinal $^3J_{XY}$ couplings exhibit a strong dependence of the dihedral angle, which is very useful for structural elucidation, including stereochemistry (Karplus equation: $^3K = C \cdot \cos 2\phi + B \cdot \cos \phi + A$, with A, B, C being nuclei-dependent empirical constants and ϕ: dihedral angle).[6,21,26]

1.3.3 Relevant Characteristics of Multinuclear NMR Relaxation

The minute spacing of energy levels in NMR ($\Delta E = h \cdot v_0 = h \cdot g_N \cdot B_0/2\pi \leq 5$ µeV; v_0, Larmor frequency; γ_N, gyromagnetic ratio; B_0, magnetic field) entails at first a rather low NMR sensitivity S (Figure 1.1). In addition, the NMR relaxation operates during and in-between NMR pulse sequences and is governed by nuclear magnetic moment and spin as well as local symmetry and mobility rather than by spontaneous emission (which is virtually absent in NMR spectroscopy). Longitudinal relaxation, with spin–lattice relaxation time T_1, reestablishes the Boltzmann distribution of spin states in NMR at thermal equilibrium (z-magnetization parallel to B_0).

Very fast longitudinal NMR relaxation such as commonly found in quadrupolar nuclei and paramagnetic compounds, in which the very efficient paramagnetic relaxation is mediated by the large electron magnetic moment, $\mu_e \sim 658.2\mu_H$, not only enables fast NMR repetition rate (and hence increased S/N ratio per unit of acquisition time) but also implies a rather large linewidth ($T_1 \approx T_2$).[21] Transverse relaxation, with spin–spin relaxation time T_2, effects loss of coherence of xy-magnetization (plane of observation, perpendicular to B_0) and determines the NMR linewidth $\Delta v_{1/2}$ ($\Delta v_{1/2} = 1/\pi \cdot T_2$). Return of magnetization into z-direction inherently causes loss of xy-magnetization, and therefore, $T_1 \geq T_2$ is commonly valid.[1,2,5,21]

An interesting comparison is that between ^{13}C and ^{17}O NMR spectroscopy of natural organic matter (NOM).[32] Common elemental compositions of isolated solid, near ash-free NOM range near 52% carbon and 42% oxygen by weight, which transposes into an elemental O/C (oxygen-to-carbon) ratio 0.61. The nominal ratio of ^{13}C/^{17}O NMR receptivity is 16.4:1 (Figure 1.9), suggesting a low S/N ratio for ^{17}O NMR spectra of NOM. In practice, fast quadrupolar NMR relaxation enables a ^{17}O NMR repetition rate of (at least) 100 ms between scans, whereas full longitudinal ^{13}C NMR relaxation in NOM ranges near 15 s; accordingly, the relative S/N ratio based on the number of scans acquired for given NMR acquisition times is $(150)^{1/2} = 12.2$. Basically, ^{17}O NMR spectra will produce about half of the nominal NMR integrals as ^{13}C NMR spectra of NOM for a given NMR acquisition time. However, the key difference is the faithful depiction of virtually all carbon chemical environments by ^{13}C NMR spectroscopy,[6] whereas ^{17}O NMR spectroscopy of NOM suffers from functional group-selective chemical exchange as well as from differential quadrupolar transverse NMR relaxation that accelerates for ever larger molecules.[21,29,32]

Both T_1 and T_2 decisively constrain quantification of NMR resonances.[2,3,5,6,21] A very fast spin–spin relaxation in molecular segments with large CSA and/or low internal mobility will eventually broaden NMR resonances beyond recognition but even in less precarious situations, large linewidths will correspond to low S/N ratio at given NMR integrals. NMR resonances in noisy NMR spectra are ill defined in terms of amplitude, linewidth, line shape, and position, limiting the accuracy of integrating these often weak NMR resonances. Notably, common Lorentzian NMR resonances carry ~3% of their integral outside the $\pm 3 \Delta v_{1/2}$ window. Slow longitudinal NMR relaxation defines a lower limit on the rate at which an NMR experiment can be repeated: equilibrium magnetization has to be reestablished in-between successive scans; otherwise, respective NMR resonances will be reduced in area.[3,5,6]

Differential and slow longitudinal relaxation, which is a characteristic property of nuclei of restricted (region of spin diffusion) and very high flexibility (extreme narrowing limit), might impair quantification of NMR spectra. In a crude approximation, the NMR longitudinal relaxation of heteronuclei is inherently related to their NMR frequencies and to their distance to adjacent protons and free electrons.[6,21,32] NMR-active nuclei behaving according to the extreme narrowing limit and to spin diffusion (or in between) may be present in the same environmental mixture.

1.4   PRACTICAL ASPECTS RELATED TO SENSITIVITY AND RESOLUTION IN COMPLEX MIXTURES

Solution-state NMR spectroscopy at high field (e.g., $B_0 = 18.8$ T, 800 MHz proton NMR frequency) with cryogenic detection offers ultimate sensitivity and dispersion of NMR resonances. Both items are very useful for analysis of common environmental organic mixtures, which often show low-resolution signatures in 1D proton and carbon NMR spectra because of prevalent superposition of many NMR resonances at any data point.[7] Notably, individual NMR experiments carry specific intrinsic nominal resolution (Table 1.2).[8,10] Here, 2D NMR spectra not only exceed the nominal resolution of 1D NMR spectra considerably (Table 1.2) but 2D NMR pulse sequences also act as filters to selectively emphasize individual transfers

of magnetization such as certain homonuclear and heteronuclear spin–spin couplings.[1,2] Hence, 2D NMR spectra of NOM commonly exhibit a vastly superior effective resolution even if (in most cases: precisely because) they might represent a lesser overall number of NMR resonances than those which define a 1D NMR spectrum.[8,15,16,35] Analogously, individual NMR experiments will carry intrinsic characteristics with respect to spectral resolution and/or S/N ratio, which can be selectively pronounced and attenuated by judicious choice of acquisition and apodization parameters (see Section 8.7).[36]

Increased NMR detection sensitivity at high magnetic field B_0 might elevate characteristics intrinsic to NMR to the level of observation, which otherwise would be lost in noise. NMR spectra represent operating quantum mechanics with macroscopic observables, which carry information about chemical environments of atoms as well as physical phenomena, such as atomic and molecular mobility.[9] The different NMR experiments will react unequal to these challenges. For example, NMR experiments with long intrinsic duration and variable magnetization transfer/relaxation delays (such as higher dimensional NMR experiments) will be more susceptible to effects of fast and differential relaxation than rather short duration NMR pulse sequences (such as 1D NMR experiments).[8] Effects of differential NMR relaxation, which itself depend on atomic and molecular mobility on frequency timescales, are magnetic field dependent and proportional to $B_0{}^m$ ($m \geq 1$, depending on the actual relaxation mechanism).[21] Fast transverse relaxation will attenuate cross-peak intensities at higher F1 increments, resulting in lesser overall S/N ratio at long acquisition times and often noticeable diminished F1 resolution. Larger molecules exhibit higher proportions of low-frequency atomic and molecular motions that impose efficient transverse relaxation. Differential transverse NMR relaxation in the case of complex environmental materials might lead to significant line broadening for molecular units of low mobility and low local symmetry because of a significant contribution of CSA to transverse relaxation.[2,6,21] Therefore, NMR resonances derived from sp²-hybridized carbon (commonly $\delta_C > 110$ ppm) are expected to suffer more from [HSQC (heteronuclear single quantum coherence) and HMBC (heteronuclear multiple bond correlation)] cross-peak attenuation at higher F1 increments in 2D NMR spectra than other NMR resonances derived from sp³-hybridized carbon (commonly $\delta_C < 110$ ppm).[8]

Table 1.2. Fundamental characteristics of key solution-state NMR experiments in the analysis of organic environmental materials (nominal bandwidth for ^{1}H NMR spectroscopy at $B_0 = 18.8$ T: 10 ppm $\times$ 800 Hz ppm^{-1} = 8000 Hz; ^{13}C NMR: 235 ppm $\times$ 200 Hz ppm^{-1} = 47 kHz nominal spectral width)

NMR experiment	Nominal NMR matrix size (Hz^n) (apparent pixel/voxel resolution at $B_0 = 18.8$ T)	Intrinsic nominal sensitivity* ($\gamma_{ex}\gamma_{det}^{3/2}$)	Transfer mechanism and nominal time delay for magnetization transfer	General characteristics for the analysis of environmental samples
1D ^{1}H	$10 \times 800 = 8 \times 10^3$	1	10 μs (spin echo delay)	Sensitive, quantitative, and strong signal overlap
1D ^{13}C	$235 \times 200 = 4.7 \times 10^4$	1.7×10^{-4}	10 μs (spin echo delay)	Insensitive, informative, strong signal overlap, and coverage of quaternary carbons
2D ^{1}H,^{1}H JRES (J-resolved spectroscopy)	$8000 \times 25 = 2 \times 10^5$	1	$^nJ_{HH}$ 100 ms	Good to reveal abundant molecular signatures; strong attenuation in the case of fast transverse relaxation because of long duration of F1 increments
2D ^{1}H,^{1}H Correlation Spectroscopy (COSY)	$8000^2 = 6.4 \times 10^7$	1	$1/^{2-3}J_{HH}$ 133 ms	Transfer amplitude $\sim 1/J_{HH}$; improved resolution of smaller couplings J_{HH} at higher number of F1 increments with attenuation from differential relaxation; cancelation of antiphase COSY cross peaks at higher linewidth is possible
2D ^{1}H,^{1}H Total Correlation Spectroscopy (TOCSY)	$8000^2 = 6.4 \times 10^7$	1	$1/2 \times ^{2-5}J_{HH}$ 100 ms	Transfer amplitude $\sim 1/2 \times J_{HH}$; sensitive for detection of minor signatures with absorptive line shape; series of TOCSY NMR spectra with variable mixing time allow alignment of extended spin systems
2D ^{1}H,^{1}H NOESY/ROESY (rotating frame Overhauser effect spectroscopy)	$8000^2 = 6.4 \times 10^7$	1 (in reality weaker cross peaks than observed in COSY and TOCSY NMR spectra)	$1/2 \times ^{2-5}J_{HH}$ 100 ms	Interactions through space and via chemical exchange appear as phase-sensitive cross peaks in NOESY and ROESY NMR spectra; NOE for small molecules is positive, whereas for large (slow tumbling) molecules, and chemical exchange negative NOESY cross peaks result. ROESY cross peaks are positive for all interactions through space, whereas ROESY cross peaks resulting from chemical exchange are negative

(continued overleaf)

Table 1.2.   *(Continued)*

NMR experiment	Nominal NMR matrix size (Hz^n) (apparent pixel/voxel resolution at $B_0 = 18.8$ T)	Intrinsic nominal sensitivity* ($\gamma_{ex}\gamma_{det}^{3/2}$)	Transfer mechanism and nominal time delay for magnetization transfer	General characteristics for the analysis of environmental samples
2D ^{1}H,^{13}C HSQC family (heteronuclear single quantum coherence; HMQC: heteronuclear multiple quantum coherence)	$8000 \times 200^2 = 3.2 \times 10^8$	2.5×10^{-3}	$1/^1J_{CH}$ 3.5 ms	Absorptive line shape; good combination of sensitivity and large information content; very informative NMR spectral editing according to (carbon) multiplicity is feasible
2D ^{1}H,^{13}C HMBC (heteronuclear multiple bond correlation)	$8000 \times 200 \times 235 = 3.8 \times 10^8$	2.5×10^{-3}	$1/^{2-4}J_{CH}$ 150 ms	Discriminates in favor of abundant molecular signatures because of low sensitivity; excellent peak dispersion; allows assembly of extended spin systems across heteroatoms and quaternary carbon
2D ^{13}C,^{13}C (INADEQUATE)	$47\,000^2 = 2.3 \times 10^9$	3.1×10^{-6}	$1/^1J_{CC}$ 25 ms	Ultimate information about ^{13}C,^{13}C connectivities; not practical for many environmental samples at present because of limited sensitivity
3D ^{1}H,^{13}C,^{1}H HMQC-TOCSY	$8000^2 \times 200^2 = 2.6 \times 10^{12}$	2.5×10^{-3}	$1/2 \times {}^{2-5}J_{HH} + 1/^1J_{CH}$ 100 ms	Excellent resolution with appreciable S/N ratio is now within realistic reach for high-field NMR; in practice lesser utilizable resolution for ^{13}C (e.g., 1.7×10^7 volumetric pixels)[18]

ᵃ Intrinsic nominal sensitivity refers to natural isotopic abundance (^{1}H: 99.985%; ^{13}C: 1.1%) and full relaxation. The unique capability to generate and analyze data from multiple NMR experiments obtained from a single sample enables an assessment of the significance and authenticity of individual spectra. Higher dimensional NMR spectra (>4D) seem impractical at present for environmental mixtures at natural isotopic abundance because coherence transfer via low abundance ^{13}C ($n = 1.07\%$), and ^{15}N ($n = 0.37\%$) nuclei will severely impair NMR cross-peak integrals.

1.5 GENERAL CHARACTERISTICS OF HIGHER DIMENSIONAL NMR SPECTRA OF ENVIRONMENTAL MIXTURES

The invaluable contribution of quantitative 1D NMR spectra for environmental sample structural analysis resides in the ability to define the relative quantities of fundamental building blocks with already appreciable structural resolution. The main significance of the uniquely versatile 2D NMR spectroscopy is to enhance the reliability of these preliminary NMR assignments, as cross peaks in 2D NMR spectra indicate molecule fragments rather than individual atoms (Tables 1.2, Figure 1.6). A combined analysis of several 2D NMR spectra ultimately leads to the assemblage of extended substructures within complex environmental mixtures (Figure 1.6).[8,15–17,19,35]

An increased signal dispersion of NMR cross peaks into two frequency dimensions greatly reduces resonance overlap, especially in the case of heteronuclear correlated NMR spectra, where the resonance frequencies of heteronuclei cover a substantial range. In addition, 2D NMR experiments act as filters, which emphasize specific forms of binding (cf. Table 1.2) and strongly discriminate against other resonances, thereby allowing to selectively probe bonding interactions, spatial relationships, and chemical exchange.[1,2] This simplification of spectra enables a detailed analysis of heavily crowded regions of chemical shift. The sensitivity of proton-detected homo- and heteronuclear 2D NMR spectra is commonly much higher than that of heteronuclear 1D NMR spectra (Table 1.2).[1,2]

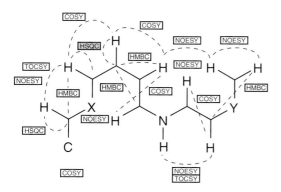

Figure 1.6. The contributions of different 2D NMR experiments in the assembly of extended substructures of organic compounds (cf. text)[32].

1.6 THE NATURE OF ENVIRONMENTAL SAMPLES

All molecules found in the environment may be termed 'environmental' to begin with; a viable operational classification defines three groups of environmental materials according to their molecular characteristics rather than with respect to their relationship to anthropogenic pollution.

1.6.1 Common Organic Mixtures

Natural organic molecules are commonly evolutionary tested, functional biomolecules eventually deriving from a genetic code, whereas many anthropogenic chemicals inadvertently or by purposeful design will interfere with biological processes in the environment. Therefore, environmental NMR of common organic mixtures is concerned with the elucidation of evolution and degradation of anthropogenic chemicals as well as the interactions of those with biota. This implies the NMR characterization of natural products and metabolites for which the entire suite of multinuclear and higher dimensional solution-state NMR spectroscopy is available.

1.6.2 Biogeochemical Materials

Complex biogeochemical nonrepetitive materials such as NOM are formed according to the general constraints of thermodynamics and kinetics from molecules of geochemical or ultimately biogenic origin.[7] The molecular signatures of these supermixtures often approach the limitations imposed by the laws of chemical binding.[7,37] Nevertheless, modern organic structural spectroscopy is capable of revealing meaningful molecular detail out of most complex natural mixtures. Here, the key role of NMR spectroscopy is quantification and in-depth elucidation of key structural principles.

1.6.3 Metal Compounds

Adverse effects of metal exposure to human health are known from the early days of civilizations. In recent times, large-scale mining and processing has increased environmental exposure to metals through point source and diffuse input. The environmental

Table 1.3. Key characteristics of common environmental samples

Class of compounds	Key environmental characteristics	NMR properties
Sample group A (biological and industrial mixtures). Organic mixtures of commonly small and defined molecules, which are amenable to successful separation into unambiguously defined molecular fractions or individual compounds		
Agrochemicals: fertilizers pesticides: herbicides, fungicides, and insecticides	Designed to show biological activity, use in environment loosely regulated; point source and diffuse input	The entire suite of high-resolution multinuclear and higher dimensional NMR experiments is available to eventually provide complete (de novo) structural elucidation, provided that sufficient material is available; otherwise, elucidation of environmental degradation mechanisms and products is feasible after preconcentration
Pharmaceuticals and drugs, hormones, and microplastics	Human and veterinary use, designed for strong biological activity; no concern with respect to environmental behavior, continual diffuse input in aquatic ecosystems from sewage; strong and incidental adverse (e.g., endocrine disruption), biological interaction at low environmental concentration. Analogously, microplastics with long residence time in oceans accumulate pollutants through sorption and are ingested by marine organisms	
Industrial and domestic chemicals	Loosely regulated with respect to environmental impact and huge diversity of degradation products	
Metabolites	Organism-specific and ecosystem-specific metabolite distribution might help to define 'healthy natural' status and anthropogenic disturbance in organisms and ecosystems	NMR-based metabolomics and flux analysis serves to determine metabolic integrity and disturbance in organisms and ecosystems
Nanoparticles	Engineered nanomaterials (<100 nm) offer large surface area per mass, facilitating organism uptake, retention, unusual bioactivity, and in vivo surface modifications (nanotoxicology)	
Sample group B (biogeochemical complex materials). Complex biogeochemical nonrepetitive materials and supermixtures that cannot be purified in the conventional sense because of an extreme intricacy (size-reactivity continuum), which often includes pronounced organo-mineral interactions. The abundance of group B compounds in the ecosphere exceeds that of group A compounds by several orders of magnitude		
Natural organic matter, global carbon, and other element cycles	Natural organic matter is the most abundant fraction of the global carbon and several other element cycles and continually interacts with a broad range of terrestrial, limnic, and marine ecosystems. Common to all these environments are the fundamental molecular aspects of life, and the dynamic equilibrium of biotic and abiotic NOM formation and decomposition, which ranges across timescales of many orders of magnitude. Molecular-level NOM composition and structure at first depends on ecosystem characteristics and exhibit far more variance than anticipated from often rather uniform bulk parameters. However, all NOM on earth is connected in a boundless carbon cycle	Successive acquisition of multiple complementary multinuclear and multidimensional NMR spectra will define quantity and considerable structural detail of fundamental building blocks of any biogeochemical mixture

Crude oil and petroleum	Exceedingly complex organic mixture ranging from liquid hydrocarbons (light oil) to heteroatom-rich (N, O, S, and trace metals) larger molecules, e.g., in oil sands; extensive mining activities in the Gigaton range leave hazardous tailings and waste; recent increase in fracking activities might damage groundwater resources	
Combustibles: grass, wood, and coal	Biomass and geopolymer extraction and burning causes erosion and widespread contamination of soil and waters; long-range transport of secondary organic aerosols (SOA) imposes global warming and climate change	

Sample group C (metal compounds). Sensitive detection methods such as inductively coupled plasma mass spectrometry (ICP-MS) and fluoresecence spectroscopy have facilitated detection of trace quantities of virtually any metal in the environment. However, species-conserving, molecule-specific metal analysis is highly demanding and comparatively rare for environmental samples. Persistence of metals alleviates bioaccumulation and biomagnification in microorganisms and food chains. Recent estimates propose that up to one-third of all biologically active molecules might coordinate to metals (metallome and metallomics)

Mining and metal processing residues; electronic waste (multielement mixture with rare elements)	Point source pollution from tailings and inappropriate practices (cyanide and mercury use in artisanal operations) and leachates as well as diffuse input from long-range transport	Coordination chemical shift at functional groups of ligands helps to identify mode and site of coordination; direct metal NMR is available for many nuclei (Table 1.4) On the NMR timescale, pH- and concentration-dependent superposition of slow, intermediate, and fast chemical exchange may require mapping of organic-metal systems through NMR acquisition at different B_0 and systematic variation of metal/ligand ratio and pH values
Mobilization of metals through acid rain	Indirect effects on biodiversity through large-scale diffuse input into watersheds	
Environmental bioinorganic and geochemical metal chemistry	Widely unexplored: long-range atmospheric metal transport provokes transcontinental sustenance and adversity in terrestrial and oceanic ecosystems	
Radioactive waste	Metal-rich: toxic and often highly radioactive. Many radioactive isotopes are available, some with spin-$1/2$ (^{209}Po, ^{211}Rn, ^{239}Pu, and ^{251}Cf)	^{99}Tc NMR spectra of inorganic and coordination compounds are available.

Table 1.4.   Nuclei sorted with respect to NMR receptivity at natural isotopic abundance in descending order

Nuclear spin	High NMR receptivity	Appreciable NMR receptivity	Low NMR receptivity
$I = {}^1/_2$	^{1}H, ^{19}F, ^{205}Tl, ^{31}P, ^{203}Tl, ^{129}Xe, ^{119}Sn, ^{117}Sn, ^{195}Pt, and ^{125}Te	^{207}Pb, ^{113}Cd, ^{111}Cd, ^{199}Hg, ^{171}Yb, ^{169}Tm, ^{77}Se, ^{29}Si, and ^{13}C	^{123}Te, ^{115}Sn, ^{89}Y, ^{109}Ag, ^{107}Ag, ^{103}Rh, ^{183}W, ^{15}N, ^{57}Fe, ^{3}He, and ^{187}Os
$I > {}^1/_2$	^{93}Nb, ^{51}V, ^{115}In, ^{45}Sc, ^{59}Co, ^{7}Li, ^{27}Al, ^{165}Ho, ^{55}Mn, ^{209}Bi, ^{11}B, ^{127}I, ^{23}Na, ^{121}Sb, ^{187}Re, ^{151}Eu, ^{63}Cu, ^{139}La, ^{159}Tb, ^{71}Ga, ^{185}Re, ^{87}Rb, ^{81}Br, ^{133}Cs, ^{69}Ga, ^{79}Br, ^{181}Ta, ^{65}Cu, ^{175}Lu, and ^{75}As	^{123}Sb, ^{113}In, ^{9}Be, ^{153}Eu, ^{85}Rb, ^{10}B, ^{35}Cl, ^{91}Zr, ^{14}N, ^{176}Lu, ^{137}Ba, ^{37}Cl, ^{6}Li, ^{131}Xe, ^{95}Mo, ^{39}K, ^{143}Nd, ^{189}Os, ^{97}Mo, ^{135}Ba, ^{163}Dy, ^{25}Mg, ^{105}Pd, ^{101}Ru, ^{147}Sm, ^{83}Kr, ^{173}Yb, ^{49}Ti, ^{201}Hg, and ^{87}Sr	^{47}Ti, ^{50}V, ^{177}Hf, ^{67}Zn, ^{167}Er, ^{73}Ge, ^{149}Sm, ^{53}Cr, ^{157}Gd, ^{138}La, ^{161}Dy, ^{145}Nd, ^{61}Ni, ^{155}Gd, ^{179}Hf, ^{197}Au, ^{99}Ru, ^{193}Ir, ^{33}S, ^{17}O, ^{191}Ir, ^{43}Ca, ^{21}Ne, ^{41}K, ^{2}H, and ^{235}U

Spin-$^1/_2$ nuclei with 'appreciable' receptivity range from 2‰ (^{207}Pb) down to 0.16‰ (^{13}C) of ^{1}H NMR receptivity. Quadrupolar spin $>^1/_2$ nuclei with mid receptivity range from 2% (^{123}Sb) down to 0.2% (^{87}Sr) of ^{1}H NMR receptivity.

speciation of metals governs their bioavailability, and many organisms have developed sophisticated mechanisms to extract trace metals. Apart from minerals and rocks, environmental metal binding occurs as a quasi-continuum of metal binding sites to natural organic matter (NOM). Metallomics studies imply that up to a third of all biologically active molecules may coordinate to metals at some stage of activity. Metals as elements cannot be degraded but their environmental speciation can be changed into less bioavailable and less toxic forms (remediation and detoxification). Here, NMR appears as an ideal metal speciation tool. However, low NMR sensitivity, complex chemical exchange phenomena, and potential interferences from fast quadrupolar and paramagnetic NMR relaxation make metal NMR studies of environmental metal binding a complex and extended exercise.

In practice, attribution of environmental compounds to one of these three groups might remain ambiguous at times because all common biogeochemical cycles on earth are by definition continually interacting and 'boundless'.[38] For instance, a marine oil spill with heavy crude oil will disperse crude oil in salt water, create large amounts of collateral microbial and abiotic degradation products, each with individual chemoselective partitioning with respect to marine organisms, mineral phases, and sea-spray formation at the sea air interface as well as species-specific toxicity profiles with regard to marine organisms. Intentionally added chemicals for oil 'removal' will exacerbate the complexity of this anthropogenic ecosystem interactions. Here, environmental NMR has its place in the elucidation of crude oil (structure-selective) partitioning and transformation as well as in contaminant enrichment (bioaccumulation, biomagnification,

and especially biotransformation through trophic levels) in organisms. Sublethal exposure can be assessed by NMR-based metabolomics (see Chapters 23, 24, 25, and 26) to elucidate biochemical fluxes and molecular organism-specific toxic modes of action. Furthermore, toxicity-induced species displacements might shift substrate utilization and production within the aquatic marine food chain, eventually altering the composition and structure of marine organic matter on regional scales.

Owing to the complexity of environmental mixtures, any NMR parameter will be characterized by a weighed average and a distribution of values. NMR shieldings and resonance integrals of individual chemical environments will be superimposed to produce a broad envelope of overlapping resonances in the NMR spectra, resulting in a rather low S/N ratio with respect to weight unit (when compared, e.g., to natural products). The overall resolution will become insufficient to clearly resolve *J-couplings* under routine conditions in 1D NMR spectra. NMR relaxation parameters and the linewidths will also be described by rather elaborate distribution functions.

1.7   NMR ANALYSIS OF FUNCTIONAL GROUPS IN ENVIRONMENTAL ORGANIC MIXTURES

The composition of environmental organic mixtures refers to the elemental composition and to the quantity and chemical environment of functional groups and the molecular backbone. Functional groups comprise from 25% to 50% (*w/w*) of common biogeochemical materials and decisively affect their physicochemical properties (e.g., polarity, nonvolatile behavior, metal binding, and the surface activity) and ecological effects. They are primarily derived from heteroatoms in the order $O \gg N \approx S > P >$ others. Oxygen-containing functional groups include the dominant aliphatic and aromatic carboxy and hydroxyl groups, but methoxy, quinoid, keto, ether, acetal, and ketal functions are also present. Nitrogen frequently occurs in the form of peptides and amines, and, less common, as five- and six-membered heterocyclic nitrogen. Sulfur-derived functional groups found, e.g., in NOM include sulfhydryl, thiophenes (in crude oil), sulfonates, and sulfates.

The combination of chemical derivatization of acidic protons in functional groups (e.g., $-OH$, $-NH$, $-SH$, $-COOH$, $-CONH$, $-OSO_3H$) with NMR-active labels and consecutive 1D and 2D NMR spectroscopy is a powerful method to determine the composition of functional groups in these complex and polydisperse materials.[39] Acidic protons of functional groups are reacted with reagents composed of a suitable NMR-active label and a leaving group (cf. Figure 1.7); ideally, the tag would substitute any labile proton of a given (class of) functional groups without side reaction – this condition is not necessarily met in the case of heavily functionalized materials owing to differential reactivity and extensive steric hindrance.

A very wide range of reagents can be reacted with environmental mixtures for an NMR-based analysis of functional groups. This method complements existing techniques of functional group analysis such as infrared, ultraviolet/visible (UV−vis), fluorescence spectroscopy, acid−base titration, and

microcalorimetry but provides unprecedented resolution with respect to the structural details of their environment. Owing to the very high sensitivity of proton-detected 2D NMR experiments (Table 1.2), these NMR analyses of functional groups can be realized with submilligram amounts of environmental materials.

1.7.1 The Silylation of Environmental Mixtures

(Trimethyl)silylation substitutes the exchangeable protons with rather nonpolar trimethylsilyl groups; carboxylic acids are transformed into silyl esters, alcohols into silyl ethers, and amides into silyl amides.

$$-C - X - H + (H_3C)_3Si - Y$$
$$= -C - X - Si(CH_3)_3 + HY \quad (X : O, N, S)$$

In ^1H, ^{29}Si heteronuclear single quantum coherence (HSQC) NMR spectra of silylated organic mixtures, the aromatic silyl esters exhibit downfield proton chemical shift compared with aliphatic silyl esters. Furthermore, $\delta(^{29}Si)$ of the silyl esters correlates with the pK_a of the parent carboxylic acid. The aliphatic and aromatic silyl esters, the primary, secondary, and tertiary silyl ethers as well as phenol-derived silyl ethers, and silyl amides cover distinct areas in the $^2J(^1H,^{29}Si)$- and $^1J(^1H,^{13}C)$-HSQC NMR spectra and show within their range of chemical shift in the 2D matrix appreciable detail and resolution.[39] Accordant detailed structural information about parent functional groups in complex unknowns is currently not accessible by any other analytical method. With a mathematical procedure of 2D NMR subtraction, the minimum and difference NMR spectra can be computed from experimental NMR spectra (cf. Figure 1.8). Minimum spectra [humic acid (HA) = fulvic acid (FA)] show the common structural elements within two and more humic materials; difference NMR spectra show their specific characteristics, respectively.

1.7.2 Methylation and Other Derivatization Methods

Methylation in conjunction with 1D and 2D NMR spectroscopy is another independent method for the characterization of functional groups according to structural details. Replacing acidic protons with nonpolar residues might reduce association tendency,

Figure 1.7. Substitution of an acidic proton in complex environmental materials, such as natural organic matter (NOM), with an NMR-active functional group (tag); auspicious tag nuclei are ^{29}Si, ^{77}Se, and ^{119}Sn.[32] Permethylated $-SiMe_3$, $SeMe_2$, and $SnMe_3$ produce singlet resonances in ^1H, ^{13}C, and X NMR spectra (X: ^{29}Si, ^{77}Se, and ^{119}Sn) as well as absorptive line shape HSQC cross peaks, which are commonly placed outside of typical ^1H and ^{13}C NMR chemical shift windows (Figure 1.8). (Reproduced from Ref. 32, with kind permission from Springer Science and Business Media)

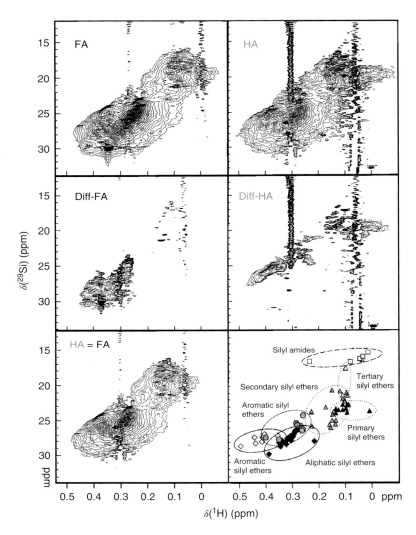

Figure 1.8. ^2J (^{29}Si,^{1}H) HSQC NMR spectra of a silylated fulvic (FA, left) and a humic acid (HA, right) from a bog lake (Holohsee, Black forest, Germany); *upper trace*: full spectra, *middle trace*: difference NMR spectra of HA and FA showing preferential substructures in FA (left) and HA (right); *lower left*: conforming structures in FA and HA; *lower right*: ^{1}H,^{29}Si HSQC–NMR chemical shift data of model compounds[39].

mediated via hydrogen bonds and metal coordination, thereby prolonging T_2 relaxation, leading to improved resolution in 1D and 2D NMR spectra. Methylation of NOM effects a functionalization of not only acidic protons (OH → OCH$_3$) but also side reactions, such as selective cleavages, C-methylation, and the 1,3-dipolar cycloaddition, occur that allow a better comprehension of the substitution patterns of the aliphatic backbone structure of NOM.[32] Methyl

is sterically less demanding than any silyl group and may therefore be more appropriate to derivatize less accessible acidic protons in environmental mixtures. ^{13}C labeling of methylating agents offers a very high sensitivity in proton-detected (and) carbon nD NMR spectra (n = 1, 2, 3) and additional NMR assignment options via ^{13}C, ^{13}C coupling constants. In the methoxyl region of the HSQC NMR spectrum, aromatic and aliphatic methyl esters and methyl ethers are represented in four distinctly separated regions of chemical shift.

Analogously, tagging with ^{13}C-enriched acetic anhydride produced carboxylic acids from amines.[40] Alternatively, reaction of carboxylic acids with ^{15}N-ethanolamin ($H_2{}^{15}NCH_2CH_2OH$) produced amides,[41] whereas formylation with ^{13}C-enriched formic acid ($H^{13}COOH$) resulted in formamides.[42] The detection of these labeled products with ^{1}H, ^{13}C, or ^{1}H, ^{15}N HSQC NMR spectroscopy enabled the quantification of several dozens of educts in the submicromolar range in biological samples directly out of mixtures.

1.8 EXAMPLES

1.8.1 Analysis of Natural Organic Matter (NOM)

While all NOM on earth is connected in a boundless carbon cycle[38] and other element cycles as well, the ecosystem-imposed molecular-level structural characteristics of NOM are vastly more diverse than anticipated from the often rather uniform bulk parameters (see Chapters 11 and 13). In fact, the narrow distribution of ecosystem-specific NOM bulk parameters is an inevitable outcome of any low-resolution characterization in which many different chemical environments project on any single data point, often with ill-defined relationships between measured variable and atomic or molecular process.

In all ecosystems, the amount of NOM exceeds that of biomolecules and living biomass up to several orders of magnitude.[7] Aliphatic-rich freshwater and groundwater NOM is frequently subject to initial large-scale processing from plant-derived into microbial-derived NOM[35] and then progressively deoxygenated on longer timescales when detached from fresh input, reflecting an ever-decreasing availability of reactive functional groups. Soil-derived NOM is aromatic rich from vascular plant input of lignins,[43,44] whereas oxygen-deficient polyaromatic NOM arises from natural and anthropogenic biomass and fossil fuel burning; the latter is transported such as atmospheric NOM[45] and distributed within various ecosystems across regional and continental distances. Internal lake, riverine, and estuarine NOM turnover, which has been largely overlooked until recently, now ranges among the most active biogeochemical processes on earth with respect to

turnover rate per surface area. From the surface to deep ocean, aliphatic-rich marine dissolved organic matter (DOM) gets progressively more complex and less characterized;[8] credible evidence from latest studies indicates alignment of marine NOM composition and structure with the previously established Longhurst biogeochemical provinces of the ocean. The ultimate molecular diversity is observed in NOM of extreme environments such as hot springs, which harbor widely variable pH range, temperature, and occurrence of trace elements and even more so, in extraterrestrial NOM, similar to that common to carbonaceous chondrites.[46] This organic matter originates from the earliest days of the solar system and probably, even before the sun formed 4.65 billion years ago. In contrast to terrestrial conditions of temperate climate and oxic atmosphere, carbonaceous chondrites have sampled across a huge variety of spatial, compositional, temperature, and high-energy irradiation regimes, which initiated nearly unmatched spatial physical and chemical heterogeneity on nano-, micro-, and macroscales. Here, primitive meteorites deviate from almost all terrestrial materials, which display more uniform histories of formation.

Ever since NMR spectroscopy has been available to NOM research, it has critically shaped our perception of these intricate mixtures and without any doubt, it will do so in any foreseeable future. Owing to the huge molecular diversity of NOM on earth (and beyond), the multifaceted NOM structural elucidation appears as one of the most attractive future playgrounds for modern discovery-oriented environmental NMR spectroscopy. NMR spectroscopy with its unique versatility and unmatched capacity to retrieve in-depth structural elucidation of complex unknowns will be indispensable in these endeavors. High-performance Fourier transform infrared cyclotron resonance (FT-ICR) mass spectrometry (MS) is capable of retrieving several thousands of elemental formulas directly out of mixtures[8,17,35,37] with very high sensitivity but sample and method-dependent ionization selectivity and will aid in these studies. Similarly, an educated high-performance separation and NOM fractionation with an eye on purposeful decreased mixture complexity[47,48] will contribute to produce subsets of the NOM chemical space available for dedicated NMR and MS analysis. Eventually, a better molecular understanding of NOM ecosystem services and functions will contribute to a conceptual convergence of biodiversity- and biogeochemistry-oriented perception of our environment.

1.8.2 Hydrogen NMR Spectra (^{1}H, ^{2}H, ^{3}H NMR)

Hydrogen NMR appears in three complementary variants: ^{1}H NMR spectroscopy combines utmost sensitivity with clear relationships between NMR parameters δ and J and chemical environments; it is the most widespread NMR method in use, in particular, if proton-detected higher dimensional NMR experiments are counted as well.[26] ^{2}H (deuterium) is a quadrupolar nucleus with a small quadrupolar moment, enabling ^{2}H NMR spectra with narrow lines.[49] The proportion of the ^{1}H/^{2}H gyromagnetic ratios is 6.51 : 1, reducing the available frequency bandwidth of ^{2}H NMR spectra related to ^{1}H NMR spectra accordingly. ^{2}H NMR spectroscopy is used to elucidate the often remarkable position-specific, nonstatistical deuterium distribution of organic molecules [^{2}H site-specific nuclear isotope fractionation (SNIF) NMR] to detect position-selective isotope effects of (bio)chemical reactions. SNIF NMR is often used to detect fraudulent food adulteration. In addition, the motional dependence of the quadrupolar ^{2}H NMR relaxation can be used for interaction studies of pollutants with organics and minerals.[49] ^{2}H NMR spectroscopy is employed in studies of deuterated paramagnetic complexes that exhibit attenuated (1/6.51) line broadening from reduced electron–nuclear coupling. The radioactive (half-life 12.5 years) ^{3}H (tritium) is a spin-$^1/_2$ nucleus and shows the largest gyromagnetic ratio of all nuclei (^{1}H/^{3}H receptivity ratio at full abundance is 1 : 1.21; Figure 1.9). ^{3}H NMR spectra of tritiated organic compounds, produced by microwave-enhanced procedures to limit radioactive waste and purified by radiochromatographic methods, were acquired in pharmaceutical and life sciences; in an ^{3}H-cryoprobe, an ^{3}H activity of 11 μCi produced an S/N ratio of 21 : 1 overnight.

Hydrogen in organic molecules is bonded to carbon atoms, forming kinetically stable covalent bonds (the molecular backbone), and to heteroatoms, commonly in the abundance O > N > S > P >> others, forming labile bonds (functional groups), which are susceptible to chemical exchange on the NMR timescale. Proton NMR spectra of environmental samples acquired in acidic or alkaline (e.g., NaOD) solution, acquired under routine conditions, combine all of the labile proton species into a single NMR, which represents the weighted average chemical shift of all chemically exchanging protons. Efficient selective suppression of this rather strong NMR is available and diverse experimental protocols have been applied to obtain the relative quantities of nonexchangeable hydrogen in various substructures of environmental samples, in particular, distinct fractions of NOM (see Chapter 13). Among the miscellaneous main methods (selective presaturation, jump-return sequences, gradient-based selective nonexcitation, and diffusion editing of faster moving solvent molecules), classical simple and robust presaturation works surprisingly well, also in combination with a short mixing time (e.g., 1 ms) to accommodate for differential transverse NMR relaxation in a 1D variant of NOESY or the recently introduced gradient-assisted presaturation using relaxation gradients and echoes (PURGE) sequence. Sample-dependent performance of NMR solvent suppression is likely influenced by spectrometer and probehead as well as differential chemical exchange characteristics and sample microheterogeneity in complex environmental mixtures.

1.8.3 Spectral Editing of Carbon and Nitrogen NMR Spectra

Spectral editing, in particular, sorting carbon atoms according to their number of directly bound protons, is a very powerful assignment tool in 1D and 2D NMR spectra: the subspectra of methyl, methylene, and methine carbon atoms are computed from linear combinations of three individual Distortionless Enhancement by Polarization Transfer (DEPT) spectra in each case.[8,16] The two main applications of spectral editing are the evaluation of branching in aliphatic units and the determination of the average degree of carbon substitution in aromatic rings. The depiction of aliphatic branching in ^{1}H and ^{13}C NMR spectra differs considerably: a relative insensitivity of ^{1}H NMR chemical shift with respect to aliphatic branching (negligible aliphatic substituent effects on ^{1}H NMR chemical shift beyond directly bond carbon) contrasts with sizable aliphatic increments in ^{13}C NMR chemical shifts (substituent ^{13}C NMR chemical shift increments for carbon substitution Cα: +9.1 ppm; Cβ: +9.4 ppm; Cγ: −2.5 ppm; Cδ: 0.3 ppm).[6] This results in strong ^{13}C NMR chemical shift effects of aliphatic substitution at least up to two bonds away in

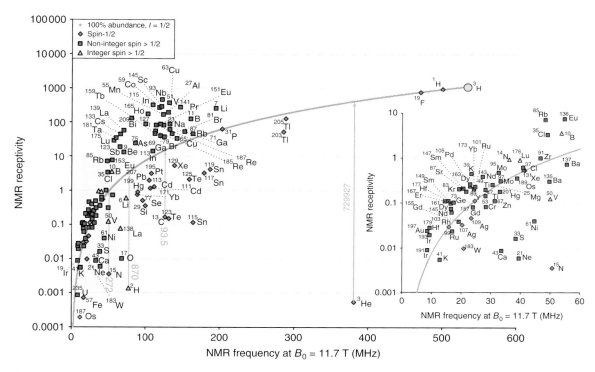

Figure 1.9. Display of NMR receptivity R against NMR frequency (proportional to γ_N) at $B_0 = 11.7$ T (500 MHz proton NMR frequency), sorted according to spin-$\frac{1}{2}$ nuclei (blue diamonds) and integer (red squares) and noninteger (yellow triangles) quadrupolar nuclei ($I > \frac{1}{2}$). The continuous green line indicates R for spin-$\frac{1}{2}$ nuclei at 100% abundance (all fully enriched spin-$\frac{1}{2}$ nuclei show this NMR receptivity; see enrichment factors for ^{2}H (870), ^{3}He (729927), ^{13}C (93.49), and ^{15}N (272) from natural to 100% isotopic abundance. Large nuclear spins I and appreciable natural abundance elevate many quadrupolar nuclei to even higher receptivity; purple shaded area: tuning range of common multinuclear NMR probeheads (^{109}Ag$-^{31}$P). Yellow shade: expanded insert of some low-frequency nuclei. (Reproduced from Ref. 32, with kind permission from Springer Science and Business Media)

any direction. Carbon multiplicity information in conjunction with chemical shift data will enable detailed conclusions about the remote aliphatic substitution in complex environmental materials to be further refined by the analysis of HSQC NMR spectra.[8,16]

In the absence of very fast transverse relaxation, the average number of protons per aromatic ring is obtained by comparing the ratios of the (normalized) areas of a ^{13}C NMR DEPT spectrum (showing protonated carbon only) and the single-pulse ^{13}C NMR spectrum (showing all carbon atoms). For the analysis of complex environmental mixtures, it is advisable to use short duration DEPT pulse sequences, as they are the least susceptible to differential relaxation.

However, the subspectral editing according to quaternary carbon atoms via routine quaternary-only carbon spectra (QUAT)-sequences visibly suffers from the variance of the ^1J(CH) values and cross talk from protonated carbon atoms.

1.8.4 Metal NMR

All metals except Ce feature at least one NMR-active isotope and thereby provide the opportunity to directly observe metal binding to environmental matrices (see Chapter 19). However, the understanding of the NMR chemical shift–structure relationship of coordination and organometallic compounds, where alterations in coordination number, symmetry, and oxidation state can readily occur, remains rather restricted for all

NMR spectra of metal nuclei until today.[21,23,24,28,31,50] Any of these effects and, in addition, pronounced concentration, pH, and temperature dependencies very strongly influence the metal NMR chemical shift – and all of these effects are highly specific for any metal NMR nucleus. Fast quadrupolar and paramagnetic NMR relaxation (if applicable) might broaden metal NMR resonances beyond recognition. Furthermore, paramagnetic metal NMR resonances show large shift displacements from those of diamagnetic counterparts and are commonly broadened beyond detection because of very efficient electron–nuclear dipole–dipole interactions. Accordingly, the utility of the (admittedly low receptivity; Figure 1.9) spin-$1/2$ nucleus ^{57}Fe to elucidate the immensely relevant environmental $Fe_xO_yS_zL_n$ (L: organic ligands) chemistry by means of NMR spectroscopy will remain elusive. Here, Mössbauer spectroscopy has become a viable alternative to determine electronic environments of iron atoms.

The lack of understanding how all these processes will translate into metal NMR chemical shifts and linewidths impedes current research: so far, many of the metal nuclei with high NMR receptivity (Figure 1.9) have never been used in the investigation of metal binding in environmental compounds.

Physical and chemical processes, such as internal rotations in molecules, valence isomerization, chemical exchange, and chemical reactions can lead to exchange of nuclei between nonequivalent electronic surroundings. Kinetic processes on timescales ranging from microseconds to seconds are effective in causing chemical exchange effects in NMR spectra. The near continuous distribution of binding constants and rates of chemical exchange of environmental ligands to metals will produce strong effects of chemical exchange on the metal NMR spectra.[31] Hence, chemical shifts and resonance integrals in NMR spectra of metal nuclei will not directly indicate the quantity and the nature of the chemical environment of the respective nuclei but most likely will reflect a weighed average of superimposed slow, intermediate, and fast chemical exchange characteristics, produced by groups of coordination environments, related by chemical shift differences and exchange rates. Metal NMR spectra will invariably show broadened resonances and a limited S/N ratio when metal coordination to (mixtures of) environmental ligands comes into effect, and it is not feasible to define an unambiguously defined status of slow chemical exchange, which is an initial requirement for significant mathematical description

of chemically exchanging systems. As metal ions are partners in the chemical exchange, the regime of slow chemical exchange is reached only at extremely low metal ion concentrations, when the S/N ratio of the metal NMR spectra may become prohibitively low. Even conditions of pure fast chemical exchange in environmental metal–ligand systems are difficult to prove experimentally, as conditions of a dynamic equilibrium will prevail, defined by the superposition of the temperature-dependent binding constants and exchange rates of each individual metal coordination environment. A very low metal ion concentration, only the strongest binding sites will be occupied, and selectivity of metal binding will vastly differ from that observed at elevated metal-to-ligand ratios; a familiar example is that of mercury binding to humic substances.

However, with these particulars under consideration, metal NMR spectroscopy offers a direct and detailed access to structure and dynamics of coordination compounds in environmental materials. Any reliable analysis of metal binding to environmental mixtures by metal NMR spectroscopy requires a systematic mapping of the exchanging system by variation of the metal-to-ligand (sample) ratio, pH, temperature, and – possibly – concentration in conjunction with an independent determination of the total metal content and the fraction of the (free) metal ions.[31] The acquisition of the metal NMR spectra at different field strengths B_0 is advantageous to further characterize the NMR exchange characteristics. Several of the environmentally significant and toxic metal isotopes are spin-$1/2$ nuclei (^{113}Cd, ^{195}Pt, ^{199}Hg, $^{203/205}$Tl, and ^{207}Pb) with a reasonable NMR receptivity (cf. Figure 1.9) and, therefore, are suitable targets for direct observation of metal coordination to environmental matrices and determination of metal coordination in (affected) biological systems.

1.8.5 NMR Spectra of Quadrupolar Nuclei with Spin Quantum Numbers $I > 1/2$

All nuclei with a spin quantum number $I > 1/2$ exhibit an ellipsoidal charge distribution, an electric quadrupolar moment eQ, and a quadrupolar coupling constant ($e^2q_{zz}Q/h$). Q is positive, if the nucleus is prolate (lengthened) in the direction of its spin angular momentum, negative if it is oblate (flattened) (Figure 1.10). The energy of a nuclear quadrupole is quantized according to its orientation in the *efg*

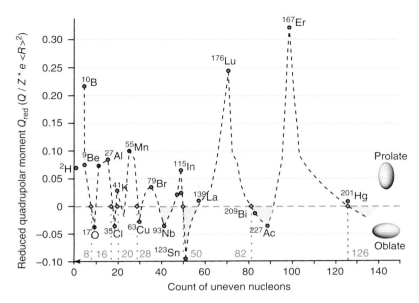

Figure 1.10. Reduced quadrupolar moments Q_{red} according to atomic number, with isotope mass sorted according to count of uneven nucleons (protons and neutrons), scaled to the atomic number.[22] Here, a continual progression of Q in between nuclei results; crossover nuclei are that with magic numbers, e.g., 2, 8, 20, 28, 50, 82, and 126 (double magic nuclei are spherical and not NMR active, e.g., ^{4}He, ^{16}O, ^{40}Ca, ^{48}Ni, and ^{208}Pb). (Reproduced from Ref. 32, with kind permission from Springer Science and Business Media)

even in the absence of an external magnetic field B_0. Similar to the chemical shift δ, the scalar coupling J, and the dipolar coupling D, *efg* is a tensor, and as D, it is traceless: the isotropic average of energy terms involving the *efg* is zero. Therefore, in the liquid phase, the resonance positions in the NMR spectrum are not affected by the nuclear quadrupolar coupling and only the averaged value of the shielding tensor δ and the (indirect) spin–spin coupling constants J contribute to the positions of the NMR spectral lines. Reorientation of the nucleus itself in a nonsymmetrical chemical environment (characterized by the asymmetry parameter η and the electric field gradient q_z) will cause the very efficient quadrupolar relaxation with large nuclear spins showing a rather tendency for slower quadrupolar relaxation (compare, e.g., ^{47}Ti/^{49}Ti NMR).

$$\pi\Delta\nu_{1/2} = T_{1Q}^{-1} = T_{2Q}^{-1}$$
$$= \frac{3\pi^2}{10}\frac{2I+3}{I^2(2I-1)}\left(\frac{e^2Qq}{h}\right)^2\left(1+\frac{\eta^2}{3}\right)\tau_c$$

where I is the nuclear spin quantum number, Q the quadrupole moment (square femtometers), η the asymmetry parameter [$0 < \eta < 1$; $\eta = (q_{xx} - q_{yy})/q_{zz}$], and q_{zz} the maximum component of the *efg* tensor with principal components $|q_{zz}| > |q_{yy}| > |q_{xx}|$.

With the exception of low Q nuclei such as ^{2}H, ^{6}Li, and ^{133}Cs, $T_1 = T_2 = T_{1Q} = T_{2Q}$ operates in the case of active quadrupolar relaxation. Quadrupolar nuclei such as ^{2}H, ^{11}B, ^{23}Na, ^{27}Al, ^{51}V, and ^{133}Cs have been used in environmental studies of pollutant interaction, metal speciation, and ion binding to NOM and its precursors.

1.8.6 Pollutant Interactions (Remediation)

Accurate molecular-level information about mechanisms and dynamics of the interactions of environmental organic and inorganic pollutants with any other biogeomolecules and minerals (see Chapters 17 and 20) is largely available from the NMR spectra of the main constituent nuclei by methods of chemical shift mapping, nuclear-spin relaxation, nuclear Overhauser effect, and the study of exchange and diffusion phenomena.[10-12] When isotopic enrichment of ^2H, ^{13}C, and ^{15}N in small molecules is used, the low natural abundance of ^2H (0.00015%), ^{13}C (1.1%), and ^{15}N (0.37%) in environmental materials is of considerable advantage, as the behavior of specific chemical environments within these small molecules can be probed by an array of 1D, 2D, and three-dimensional (3D) NMR spectra at rather low concentrations. The NMR relaxation in ^2H, ^{13}C, and ^{19}F NMR spectra of small aromatic molecules has been used to analyze the dynamics of their interaction with NOM.[32]

The interaction of environmental matrices with fluorinated organic compounds can be probed with ^{19}F NMR spectroscopy. Owing to a very strong back bonding contribution in the carbon–fluorine bond, the effects of fluorine substitution of hydrogen on representative bulk properties of the organic molecule and its reactivity are surprisingly small. The exceptional sensitivity of ^{19}F NMR (83% receptivity of ^1H; Figure 1.9) is even more useful as organically bound fluorine is absent in about any environmental sample: fluorinated organic molecules can be readily investigated by ^{19}F NMR spectroscopy down to the parts per billion range.[32]

Interaction of environmental samples with nonmetal (e.g., ^{11}B and ^{77}Se) and metal ions (most appropriately: ^{27}Al, ^{51}V, ^{113}Cd, ^{195}Pt, ^{199}Hg, $^{203/205}Tl$, and ^{207}Pb) can be followed by direct observation of these NMR-active nuclei. In NMR spectra of quadrupolar nuclei, both chemical exchange and quadrupolar relaxation obscure direct relationships between the appearance of the metal NMR spectra, the coordination geometry, and the nature of the NOM ligands (see section titled 'NMR Spectra of Quadrupolar Nuclei with Spin Quantum Numbers $I > {}^1/_2$').

1.8.7 NMR Data Processing

The common NMR signal acquired is a discrete digital time domain sampling sequence (interferogram; free induction decay (FID)), and the most widely used approach to convert this time domain signal into a frequency domain NMR spectrum is by applying a Fourier transformation. Other processing techniques, such as (forward and backward) linear prediction, maximum entropy reconstruction, and Bayesian analysis, are available to expand truncated FIDs in one-and higher dimensional NMR spectra.[1,36] Alternative NMR acquisition schemes such as nonuniform sampling (for higher dimensional

NMR spectra with sufficient sample availability) and band-selective Hadamard encoding enable dedicated NMR acquisition from samples with prior knowledge of composition. Shortcomings common to experimental NMR spectra, such as truncation artifacts, low S/N ratios, limited resolution, or undesirable line shapes can be partially addressed by educated apodization aimed at improving the S/N ratio, resolution enhancement, line shape conversion, or a combination of those. Typically, resolution enhancement causes a decrease in the S/N ratio and vice versa.

Individual NMR experiments will carry intrinsic characteristics with respect to spectral resolution and/or S/N ratio (Table 1.2), which can be selectively enhanced and attenuated by judicious choice of acquisition and apodization parameters.[36] One-dimensional NMR spectra are commonly apodized to enhance S/N ratio by easily adjustable exponential multiplication (EM), which leaves spectral line area ratios invariant but not amplitude ratios (Figure 1.11). Hence, the matched filter concept is not unambiguously defined in environmental mixtures because of variable linewidth. Often, several different apodization functions serve to selectively enhance slow and fast relaxing components in environmental mixtures. Proton NMR spectra, acquired in dilute solution at optimum B_0 homogeneity, may benefit from a modest resolution enhancement by a Lorentz-to-Gaussian transformation. Reliability and robustness in similarity assessment of NMR spectra by mathematical methods, e.g., bucket analysis, is commonly achieved by S/N enhancement with concomitant decrease in resolution.

1.8.8 Biogenic Small Molecules (BSMs) in Environmental Metabolomics and Chemical Ecology

Biogenic small molecules (BSMs)[51] with molecular weights <1.5 kDa mediate interactions within and between organisms and other critical life functions in the ecosphere such as metabolism, signal transduction, mating attraction, and chemical defence (Chapters 14, 22–26, and 27).[51-53] Hence, many of the drugs currently in use derive from BSMs. BSMs occur as complex mixtures and identification of structures and function of BSMs is the goal of natural products chemistry, metabolomics, and chemical ecology, among others. NMR is indispensable for structural elucidation of these highly diverse compounds, and the full array of multinuclear and multidimensional

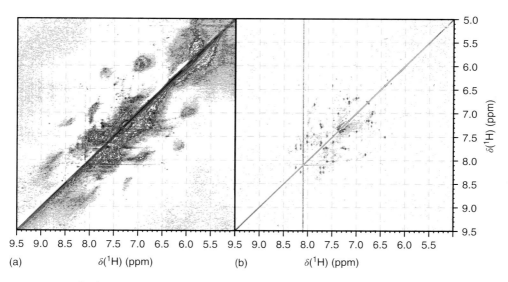

Figure 1.11. 800 MHz ^{1}H,^{1}H TOCSY NMR spectra of Atlantic Ocean surface marine solid-phase extracted dissolved organic matter (48 m depth, chlorophyll fluorescence maximum) with (a) sensitivity-enhanced apodization (exponential multiplication: EM = 7.5 Hz in F2; π/2.5 shifted sine bell in F1) to emphasize faint and extended NMR signatures of unsaturated chemical environments in marine-dissolved organic carbon; (b) resolution-enhanced apodization (Gaussian multiplication: EM: −0.4 Hz; GB: 0.6 Hz in F2; π/6 shifted sine bell in F1) to emphasize depiction of (bio)molecular complexity in marine DOM.[8] (Reproduced from Ref. 8. Distributed under the Creative Commons Attribution 3.0 License)

solution NMR spectroscopy as well as Diffusion Ordered Spectroscopy (DOSY) NMR (see Chapter 2) are available for the study of BSMs. The intrinsic gain in sensitivity, dynamic range, and resolution of high-field NMR as well as elaborate chemometric and statistical data analyses have greatly facilitated large-scale measurements of BSMs and even advanced the de novo characterization of unstable compounds in situ, which are difficult to isolate from mixtures.[51–53] Enhanced information recovery from statistical total correlation spectroscopy (STOCSY) is available from single NMR spectra (covariance NMR), for series of alike NMR spectra (such as 1D ^{1}H NMR spectra from several samples), from sets of different NMR spectra [e.g., combined ^{1}H and ^{19}F NMR for environmental studies of fluorinated organic compounds,[54] from concerted analysis of correlation spectroscopy (COSY) (provide connectivity) and heteronuclear multiple bond correlation (HMBC) NMR spectra (provide carbon chemical shifts) for general connectivity assessment], and even across methods (joint analysis of NMR, MS, and/or separation data).[53] These correlation techniques provide higher resolution and sensitivity compared with direct experimental observation and an extended range of positive and negative correlations. Structural

correlations across entire molecules irrespective of the distance of relevant nuclei and weaker pathway correlations of compounds associated by functional networks, such as metabolism, biological, or environmental stress, are available. Extended NMR spectral features and MS^n fragments as deduced from statistical heterospectroscopy (SHY)[7,53] are highly specific and correspond to a merely few isomers in existence. Information-rich detection enhanced by statistical spectroscopy and pattern recognition with (web-based) databases will further increase the dependability of this innovative complementary approach to complex organic mixture analysis. Bioactivity may be subject to synergistic and antagonistic interactions with environmental toxicants, and high-resolution analytical characterization of environmental pollution will decisively advance our understanding of molecular environmental mechanisms.

1.8.9 Hyphenation

In general, the degree of significant detail generated by a certain analytical technique will depend on the relationship between the resolving power of the method and intrinsic analyte properties. Investigations of near-featureless materials with methods of supreme resolution may be wasteful, but insufficient resolution of any analytical method with respect to the properties of the analyte will inevitably result in intrinsic averaging and low-resolution signatures, with limited variance in bulk and spectral properties.[7] Entropy gain from ever increased molecular diversity in mixtures will cause progressive disappearance of analytical signature patterns, which might invite wrong conclusions; for instance, some invariance in analytical data of certain organic matter fractions may be mistaken as an apparent recalcitrance. Eventually, discrimination capacity to enable meaningful differentiation of extremely complex supermixtures has to be restored by hyphenation of analytical methods such as combining separation and organic structural spectroscopy.[7,48] The significant resolution of each NMR spectroscopy, FT-ICR-MS and separation technologies, such as ultraperformance liquid chromatography (UPLC) and capillary electrophoresis (CE), jointly defines a discrete analytical volumetric pixel space in the range 10^{8-14} voxels, which corresponds to our current capacity to depict molecular dissimilarity in complex organic mixtures.[7] This nominal resolution is clearly sufficient to reveal significant molecular level detail from the most complex environmental mixtures available. Analysis of these correlated data is feasible at the level of the direct hyphenation of separation and spectroscopy (e.g., LC-NMR and LC-MS or CE-MS)[55,56] and by means of SHY.[53] Any joint mathematical analysis of these correlated data will enhance the effective resolution and significance of the molecular-level information of complex environmental unknowns. This nominal voxel space can be readily expanded to higher dimensions by including complementary data, as those derived from genomic and proteomic analyses, or by recognizing selective chemical reaction products.

1.9 CONCLUSION AND FUTURE TRENDS

The uniquely versatile NMR spectroscopy will assist in the interdisciplinary quest for molecular level understanding of environmental matrices and processes on several levels. Miniaturization of environmental sampling and NMR detection will enable identification of minute samples and environmental reactivity

gradients. Increased signal dispersion and sensitivity from high-field NMR combined with enhanced information recovery from mathematical analysis of NMR spectra will facilitate complex environmental mixture analysis. Critical bottlenecks are scarcity of competitive NMR resources in environmental sciences and a funding policy of science that often favors short-term optimization of known over long-term discovery-oriented research. NMR 'mapping' studies of complex environmental unknowns are often time consuming and require quite some method development but, in the end, will provide unmatched detail and significance of otherwise unattainable molecular level information. The introduction of cheap low-field NMR benchtop spectrometers with permanent magnets and modern electronics is an opportunity for outreach to students and research fields unfamiliar with NMR at present, which might eventually change an observant position into active participation. Both low-field and high-end NMR instrumentation will have a place in future interdisciplinary environmental sciences.

REFERENCES

1. G. A. Morris and J. W. Emsley, Multidimensional NMR Methods for the Solution State, Wiley: Chichester, 2010.

2. J. Cavanagh, W. J. Fairbrother, A. G.Palmer III, and N. J. Skelton, Protein NMR-Spectroscopy, Principles and Practice, Academic Press: San Diego, 1995.

3. M. H. Levitt, Spin Dynamics, Basics of Nuclear Magnetic Resonance, Wiley: Chichester, 2001.

4. F. J. M.van de Ven, Multidimensional NMR in Liquids, Basic Principles and Experimental Methods, VCH-Verlagsgesellschaft: Weinheim, 1995.

5. J. Keeler, Understanding NMR Spectroscopy, Wiley: Chichester, 2005.

6. H.-O. Kalinowski, S. Berger, and S. Braun, ^{13}C-NMR-Spektroskopie, Georg Thieme Verlag Stuttgart: New York, 1984.

7. N. Hertkorn, C. Ruecker, M. Meringer, R. Gugisch, M. Frommberger, E. M. Perdue, M. Witt, and P. Schmitt-Kopplin, *Anal. Bioanal. Chem.*, 2007, **389**, 1311.

8. N. Hertkorn, M. Harir, B. P. Koch, B. Michalke, and P. Schmitt-Kopplin, *Biogeosciences*, 2013, **10**, 1583.

9. O. F. Lange, N. A. Lakomek, C. Fares, G. F. Schroder, K. F. A. Walter, S. Becker, J. Meiler, H. Grubmuller,

C. Griesinger, and B. L.de Groot, *Science*, 2008, **320**, 1471.

10. A. J. Simpson, D. J. McNally, and M. J. Simpson, *Prog. Nucl. Mag. Res. Spectrosc.*, 2011, **58**, 97.

11. A. J. Simpson, M. J. Simpson, and R. Soong, *Environ. Sci. Technol.*, 2012, **46**, 11488.

12. L. A. Cardoza, A. K. Korir, W. H. Otto, C. J. Wurrey, and C. K. Larive, *Prog. Nucl. Mag. Res. Spectrosc.*, 2004, **45**, 209.

13. C. M. Preston, *Soil Sci.*, 1996, **161**, 144.

14. N. Mahieu, D. S. Powlson, and E. W. Randall, *Soil Sci. Soc. Am. J.*, 1999, **63**, 307.

15. N. Hertkorn, A. Permin, I. Perminova, D. Kovalevskii, M. Yudov, V. Petrosyan, and A. Kettrup, *J. Environ. Qual.*, 2002, **31**, 375.

16. B. Lam, A. Baer, M. Alaee, B. Lefebvre, A. Moser, A. Williams, and A. J. Simpson, *Environ. Sci. Technol.*, 2007, **41**, 8240.

17. N. Hertkorn, R. Benner, M. Frommberger, P. Schmitt-Kopplin, M. Witt, K. Kaiser, A. Kettrup, and J. I. Hedges, *Geochim. Cosmochim. Acta*, 2006, **70**, 2990.

18. A. J. Simpson, W. L. Kingery, and P. G. Hatcher, *Environ. Sci. Technol.*, 2003, **37**, 337.

19. A. J. Simpson, W. L. Kingery, M. H. B. Hayes, M. Spraul, E. Humpfer, P. Dvortsak, R. Kerssebaum, M. Godejohann, and M. Hofmann, *Naturwissenschaften*, 2002, **89**, 84.

20. M. A. Nanny, R. A. Minear, and J. A. Leenheer, Nuclear Magnetic Resonance Spectroscopy in Environmental Chemistry, Oxford University Press: New York, 1997.

21. J. Mason, Multinuclear NMR, Plenum Press: New York, 1987.

22. W. Demtröder, Experimentalphysik 4, Kern- Teilchen- und Astrophysik, 3rd edn, Springer: Dordrecht, 2010, pp. 25 and 145.

23. M. Gielen, R. Willem, and B. Wrackmeyer, Advanced Applications of NMR to Organometallic Chemistry, Wiley: Chichester, 1996.

24. P. S. Pregosin, Studies in Inorganic Chemistry 13, Transition Metal Nuclear Magnetic Resonance, Elsevier: Amsterdam, 1991.

25. Bruker Almanac 2010, Bruker Biospin, Rheinstetten, Germany (www.bruker.com).

26. H. Günther, NMR-Spektroskopie, 3rd edn, Thieme: Stuttgart, 1992.

27. S. Berger, S. Braun, and H.-O. Kalinowski, NMR-Spektroskopie von Nichtmetallen, Grundlagen, ^{17}O-, ^{33}S- und ^{129}Xe-NMR-Spektroskopie, Georg Thieme Verlag Stuttgart: New York, 1992.

28. J. M. Ernsting, S. Gaemers, and C. J. Elsevier, *Magn. Reson. Chem.*, 2004, **42**, 721.

29. S. Berger, S. Braun, and H.-O. Kalinowski, NMR-Spektroskopie von Nichtmetallen, Band 2, ^{15}N-NMR-Spektroskopie, Georg Thieme Verlag Stuttgart: New York, 1992.

30. S. Berger, S. Braun, and H.-O. Kalinowski, NMR-Spektroskopie von Nichtmetallen, Band 3, ^{31}P-NMR-Spektroskopie, Georg Thieme Verlag Stuttgart: New York, 1993.

31. N. Hertkorn, E. M. Perdue, and A. Kettrup, *Anal. Chem.*, 2004, **76**, 6327.

32. N. Hertkorn and A. Kettrup, in Use of Humic Substances to remediate Polluted Environments: From Theory to Practice, eds I. V. Perminova, K. Hatfield and N. Hertkorn, Springer: Dordrecht, 2005, Chap. Molecular level structural analysis of natural organic matter and of humic substances by multinuclear and higher dimensional NMR spectroscopy.

33. R. R. Gil, *Angew. Chem. Int. Edit.*, 2011, **50**, 7222.

34. C. Ludwig and M. R. Viant, *Phytochem. Anal.*, 2010, **21**, 22.

35. F. Einsiedl, N. Hertkorn, M. Wolf, M. Frommberger, P. Schmitt-Kopplin, and B. P. Koch, *Geochim. Cosmochim. Acta*, 2007, **71**, 5474.

36. J. C. Hoch and A. S. Stern, NMR Data Processing, Wiley-Liss: Chichester, 1996.

37. N. Hertkorn, M. Frommberger, M. Witt, B. P. Koch, P. Schmitt-Kopplin, and E. M. Perdue, *Anal. Chem.*, 2008, **80**, 8908.

38. T. J. Battin, S. Luyssaert, L. A. Kaplan, A. K. Aufdenkampe, A. Richter, and L. J. Tranvik, *Nat. Geosci.*, 2009, **2**, 598.

39. N. Hertkorn, A. Günzl, D. Freitag, and A. Kettrup, in Refractory Organic Substances in the Environment, eds F. H. Frimmel, G. Abbt-Braun, K. G. Heumann, B. Hock, H.-D. Lüdemann and M. Spiteller, Wiley-VCH: Weinheim, 2002, Chap. Nuclear Magnetic Resonance Spectroscopy Investigations of Silylated Refractory Organic Substances.

40. N. Shanaiah, M. A. Desilva, G. A. N. Gowda, M. A. Raftery, B. E. Hainline, and D. Raftery, *Proc. Natl. Acad. Sci. U. S. A.*, 2007, **104**, 11540.

41. T. Ye, H. P. Mo, N. Shanaiah, G. A. N. Gowda, S. C. Zhang, and D. Raftery, *Anal. Chem.*, 2009, **81**, 4882.

42. T. Ye, S. C. Zhang, H. P. Mo, F. Tayyari, G. A. N. Gowda, and D. Raftery, *Anal. Chem.*, 2010, **82**, 2303.

43. M. J. Simpson and A. J. Simpson, *J. Chem. Ecol.*, 2012, **38**, 768.

44. H. Kim and J. Ralph, *Org. Biomol. Chem.*, 2010, **8**, 576.

45. P. Schmitt-Kopplin, A. Gelencser, E. Dabek-Zlotorzynska, G. Kiss, N. Hertkorn, M. Harir, Y. Hong, and I. Gebefugi, *Anal. Chem.*, 2010, **82**, 8017.

46. P. Schmitt-Kopplin, Z. Gabelica, R. D. Gougeon, A. Fekete, B. Kanawati, M. Harir, I. Gebefuegi, G. Eckel, and N. Hertkorn, *Proc. Natl. Acad. Sci. U. S. A.*, 2010, **107**, 2763.

47. G. C. Woods, M. I. Simpson, B. P. Kelleher, M. McCaul, W. L. Kingery, and A. J. Simpson, *Environ. Sci. Technol.*, 2010, **44**, 624.

48. G. C. Woods, M. J. Simpson, P. J. Koerner, A. Napoli, and A. J. Simpson, *Environ. Sci. Technol.*, 2011, **45**, 3880.

49. M. A. Nanny and J. P. Maza, *Environ. Sci. Technol.*, 2001, **35**, 379.

50. X. Q. Lu, W. D. Johnson, and J. Hook, *Environ. Sci. Technol.*, 1998, **32**, 2257.

51. S. L. Robinette, R. Bruschweiler, F. C. Schroeder, and A. S. Edison, *Acc. Chem. Res.*, 2012, **45**, 288.

52. R. R. Forseth and F. C. Schroeder, *Curr. Opin. Chem. Biol.*, 2011, **15**, 38.

53. S. L. Robinette, J. C. Lindon, and J. K. Nicholson, *Anal. Chem.*, 2013, **85**, 5297.

54. H. C. Keun, T. J. Athersuch, O. Beckonert, Y. Wang, J. Saric, J. P. Shockcor, J. C. Lindon, I. D. Wilson, E. Holmes, and J. K. Nicholson, *Anal. Chem.*, 2008, **80**, 1073.

55. K. Albert, On-line LC-NMR and Related Techniques, Wiley: Chichester, 2002.

56. M. V. Silva Elipe, LC-NMR and other Hyphenated NMR Techniques, Overview and Applications, Wiley: Chichester, 2011.

Chapter 2

Environmental NMR: Diffusion Ordered Spectroscopy Methods

Gang Zheng and William S. Price

Nanoscale Organisation and Dynamics Group, University of Western Sydney, Penrith, New South Wales 2751, Australia

2.1 INTRODUCTION

Understanding environmental processes requires a great deal of information including what molecules are present, their environment and context, and their dynamics. Thus, for example, it is desirable to know whether a species exists in isolation or it is aggregated (self-association or to another species?). Or perhaps is the aggregated state stable or is part or all of the species undergoing exchange? Indeed, the situation could be even more complicated: there could be many types of aggregates—even of just the one species.

NMR provides a great many tools for attacking chemical problems; however, in general, these are

NMR Spectroscopy: A Versatile Tool for Environmental Research
Edited by Myrna J. Simpson and André J. Simpson
© 2014 John Wiley & Sons, Ltd. ISBN: 978-1-118-61647-5

based on NMR observables that reflect the properties of a single nucleus (e.g., chemical shift, relaxation time, and Nuclear Overhauser Effect (NOE)). Consequently, if the sample contains more than one solute—as environmental samples almost inevitably do—the applicability and, perhaps more importantly, the interpretation is greatly hampered. Thus, what is most beneficial in complex samples is to be able to project the NMR information onto another dimension that reflects molecular size. Fortunately, such a dimension exists and is based on measuring self-diffusion. Providentially for environmental applications, NMR diffusion measuring hardware,[1-4] techniques and data interpretation,[5,6] and ancillary techniques such as solvent suppression[7-9] have seen enormous improvements over the past five decades.

In addition to merely being able to separate otherwise overlapping NMR data in a diffusion dimension, being able to probe diffusive motion is a rich source of physical chemical information in a system. Neglecting exchangeable groups (e.g., −OH and −NH protons) and extremely short measurement timescales that would result in root mean square displacements (RMSDs) approaching molecular size such that reorientational motions become confused with bulk motion of the molecule,[10] a single diffusion coefficient describes the translational motion of the entire molecule. No other single NMR observable can claim to represent an entire molecule. As a technique, NMR also has the huge advantage of being noninvasive, information rich, and at least the possibility of being able to make some measurements in situ. However, environmental samples present particular challenges for NMR diffusion measurements with the species of interest often being present at very low concentrations.

This chapter provides an overview of using NMR-based diffusion studies for environmental research. Some basic concepts and relations for translational diffusion are presented in Section 2. How NMR can be used to measure diffusion is considered in Section 3. Applications of NMR diffusion measurements to environmental problems are considered in Section 4. The future of NMR diffusion measurements in this field is considered in Section 5.

2.2 TRANSLATIONAL DIFFUSION

Without further definition, the term diffusion is ambiguous. In the chemical literature, two types of diffusion are commonly encountered: mutual diffusion and self- or translational diffusion. Mutual diffusion is the process in which species move to remove concentration gradients and is thus driven by chemical potentials. Self-diffusion, on the other hand, is a consequence of random thermal motion (i.e., Brownian motion). Despite having the same units (i.e., $m^2 s^{-1}$), they are two entirely different physical phenomenon and only agree at infinite dilution.[11] As we are mainly concerned with NMR diffusion measurements, which measure self-diffusion, the term diffusion will hereinafter refer to self-diffusion. Although we note that by obtaining standard magnetic resonance images as a function of time, it is possible to study mutual diffusion (see Chapter 7).

It is convenient to define diffusion in terms of the mean square displacement (MSD) of a species undergoing free isotropic diffusion with diffusion coefficient D in a time Δ, viz.

$$\langle R^2 \rangle = \langle (r_1 - r_0)^2 \rangle = nD\Delta \qquad (2.1)$$

where the angled brackets denote an average over all of the same species, r_0 and r_1 denote the initial and final positions of the species, and $n = 2$, 4, or 6, depending on whether diffusion is occurring in one, two, or three dimensions, respectively. It is instructive to calculate the RMSD of some representative molecules. The diffusion coefficients of xenon (gas),[12] water,[13] and lysozyme[14] are 5.7×10^{-6}, 2.3×10^{-9}, and 1.1×10^{-10} $m^2 s^{-1}$ at $25\,°C$, respectively. Thus, using equation (2.1) and a measurement timescale of $\Delta = 50$ ms and then taking the square root gives RMSDs

of 755, 15, and 3.3 μm for xenon, water, and lysozyme, respectively.

For diffusion to be a useful probe of environmental processes, it is important to develop a framework connecting diffusing species and their environment to what is measured in the NMR diffusion experiment. The Stokes–Einstein–Sutherland equation[15,16] relates the diffusion coefficient to the solution environment through the solvent viscosity (η) and the effective hydrodynamic radius (r_S), viz.

$$D = \frac{kT}{c\pi\eta r_S} \qquad (2.2)$$

where k is the Boltzmann constant and c a constant ranging from 4 to 6, which corresponds to the slip and stick boundary conditions, respectively. Although in theory only holding when the solute molecule sees the solvent as a continuum, equation (2.2) provides an intuitive framework for interpreting NMR diffusion data to molecular level processes. Of particular interest is the constant c, as this relates to how the diffusing species interacts with the solvent, but the Stokes radius can also be extremely informative. Simplistically, the volume of a molecule can be taken as $V = 4\pi r_S^3/3$, which in combination with equation (2.2) and assuming that $V \propto$ MW reveals that

$$D \propto MW^{-\frac{1}{3}} \qquad (2.3)$$

It is important to reflect on some additional points: what can affect r_S and finally, why does the Stokes–Einstein–Sutherland equation only hold in infinitely dilute solution? Of course, apart from some ions and to a good approximation Buckminsterfullerenes, no molecule is truly spherical. Thus, the concept of a Stokes radius is somewhat idealistic, and it is important to keep in mind that r_S is the effective hydrodynamic radius and not the van der Waals radius and can reflect any process that changes the conformation[17,18] of a molecule or the effective charge on a molecule.[14] At finite solute concentrations, the effects of obstruction must also be considered. For example, a small molecule diffusing in a solution of larger molecules will have its diffusive path altered by the presence of larger molecules. Thus, even though at short times the diffusive behavior of a species is the same, at longer times, the RMSD will be reduced.

Equations (2.1) and (2.2) provide insight into the sorts of information that a diffusion measurement can provide. Equation (2.1) indicates that provided diffusion can be measured over a suitable timescale, diffusion can provide information on surrounding structures that might constrain the RMSD–for example, a species constrained by cell membranes will at long times have its RMSD limited by the cell dimensions, whereas equation (2.2) indicates that diffusion can provide cogent information on molecular size and the surrounding environment (e.g., binding and hydration).

2.3 PULSED-GRADIENT SPIN-ECHO (PGSE) AND DIFFUSION ORDERED SPECTROSCOPY (DOSY)

2.3.1 Measuring the Diffusion of a Single Component System

NMR diffusion measurements are referred to by various names in the literature including NMR diffusometry, pulsed-field gradient (PFG) NMR, PGSE NMR, and diffusion ordered spectroscopy (DOSY). Of these, NMR diffusometry and PGSE NMR are to be preferred, as they emphasize that diffusion is being measured (noting that in NMR, an echo is sensitive to diffusion). DOSY refers to a two-dimensional (2D) display mode of PGSE data with the diffusion coefficient on one axis and the chemical shift on the other.

Magnetic resonance imaging (MRI) (also known as k-space imaging; see Chapter 7) and NMR diffusion measurements (also known as q-space imaging) share great commonality with both using purposely applied magnetic field gradient pulses to impart spatial localization and provide information on the spin density at a particular position [i.e., $\rho(\mathbf{r})$] in the case of MRI and to measure displacement or the diffusion propagator in the case of PGSE NMR. Indeed, MRI and PGSE NMR are often combined together to determine localized diffusion information. This review is primarily involved with PGSE NMR measurements of self-diffusion.

The PGSE technique is conceptually straightforward and originates from the Larmor equation and has been covered in detail elsewhere.[16,19] Ordinarily, in NMR, when a homogeneous static magnetic field, $\mathbf{B}_0$, is used, a resonance of the same species will be at the same frequency irrespective of position. However, if the static field varies in a spatially well-defined manner, the resonance frequency will become correlated with position. For example, assuming a magnetic gradient along the z-direction, g_z, the Larmor frequency will be given by

$$\omega(z) = \gamma g_z z \qquad (2.4)$$

where γ is the gyromagnetic ratio and the contribution from the static field has been ignored, as it is common to the entire sample. Thus, following from equation (2.4), an free induction decay (FID) acquired in the presence of a gradient will, upon Fourier transformation, produce a one-dimensional (1D) image [i.e., $\rho(z)$].

The ability of a gradient pulse of duration δ to change the phase of a spin is related to the area of the gradient pulse, which scaled according to the gyromagnetic ratio becomes

$$q = \frac{1}{2\pi}\gamma g\delta \ (\text{m}^{-1}) \qquad (2.5)$$

Note, when a gradient is used to locate a position because of the correlation with frequency, it is normal to use k instead of q in equation (2.5)–but the definition is entirely the same.

To see how diffusion is measured, consider an NMR sample in a typical cylindrical NMR tube containing a single species (e.g., H_2O) and the spin-echo pulse sequence shown in Figure 2.1.

Immediately after a $\pi/2$ RF pulse, all of the spins will be coherent and pointing in one direction (imagine a ribbon of transverse arrows equally spaced along the z-direction). According to equation (2.4), a gradient pulse along the direction of the ribbon will result in this ribbon of magnetization winding into a helix, with the pitch of the helix being given by q^{-1}, as each spin will acquire a phase angle depending on its position. If a gradient pulse of the opposite sign was applied, which would have the effect of unwinding the ribbon to its original condition, the 'echo' signal then acquired would be a maximum although (albeit smaller than the initial transverse magnetization because of the effects of spin relaxation). However, if a delay, usually denoted by $\varDelta$, is allowed between the winding and unwinding gradient pulses for diffusion to significantly randomly move the spins along the direction of the ribbon, the effect will be to attenuate the helix so that after the second gradient pulse, the magnitude of the signal acquired is diminished. It is possible to formulate the relationship between echo attenuation and diffusion in numerous ways, but the most conceptually straightforward approach is the short-gradient-pulse approximation, viz.[20]

$$E = \iint \rho(\mathbf{r}_0)P(\mathbf{r}_0,\mathbf{r}_1,\varDelta)\mathrm{e}^{-i2\pi q \cdot (\mathbf{r}_1-\mathbf{r}_0)}\mathrm{d}\mathbf{r}_0\mathrm{d}\mathbf{r}_1 \quad (2.6)$$

where $P(\mathbf{r}_0,\mathbf{r}_1,\varDelta)$ is the diffusion propagator, which gives the probability of a spin moving from $\mathbf{r}_0$ to $\mathbf{r}_1$ in time $\varDelta$, and is determined by solving the diffusion

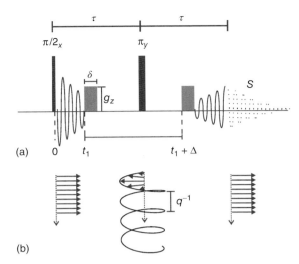

Figure 2.1. (a) The Stejskal and Tanner PGSE sequence. This is a modified Hahn spin-echo pulse sequence containing two equal gradient pulses. If the gradient pulses were infinitely short, the timescale of the diffusion would be given by Δ. (b) A conceptual idea of the effect of the pulse sequence on spin magnetization (imagine a cylindrical sample). Initially, after the $\pi/2$ RF pulse, the magnetization is coherent as denoted by the arrows all pointing in the same direction. The first gradient pulse winds this magnetization into a helix of pitch q^{-1} m along the direction of the gradient—in this case, z. The π RF pulse then reverses the direction of the helix. The second gradient pulse then unwinds the helix resulting in the reappearance of coherent transverse magnetization (i.e., the echo). The key to understanding this measuring technique is to note that diffusion during Δ will move spins, which have different phase angles depending on their initial position, resulting in attenuation of this helix, which in turn results in attenuated transverse magnetization at the end of the sequence (i.e., a smaller echo). It is the correspondence between echo intensity and diffusive motion, which is at the heart of technique. The delay from the beginning of acquisition (i.e., $t = 2\tau$) until the application of the next $\pi/2$ pulse is termed the *recycle delay*. If a spoiler gradient pulse is included in the recycle delay, it is possible to perform the sequence with a delay shorter than $5T_1$. (Reprinted with permission from Ref. 16. © 2009 Cambridge University Press)

equation with the appropriate initial and boundary conditions. Following equations (2.4) and (2.5), the exponential term reflects the phase change undergone by a spin, as it moves from r_0 to r_1. Thus, the signal attenuation is obtained by integrating over all initial and final positions. If the propagator for free diffusion is

inserted into equation (2.6), the signal attenuation for a freely diffusing species is a simple exponential decay given by,

$$E = \exp(-(2\pi q)^2 D\Delta) \qquad (2.7)$$

However, when effects of the finite duration of the gradient pulses (NB infinitely short gradient pulses are an experimental impossibility) on the signal attenuation are included, equation (2.7) becomes[21]

$$E = \exp\left(-\gamma^2 g^2 D\delta^2 \left(\Delta - \frac{\delta}{3}\right)\right) \qquad (2.8)$$

Experimentally, a diffusion measurement proceeds by collecting a number of different echo intensities measured as a function of Δ, δ, or (most commonly) g. Normally, the total length of the sequence is fixed to allow the relaxation contribution to the echo signal intensity to be normalized out. Equation (2.8) is then regressed on to the signal attenuation data to give D.

In a normal diffusion measurement, the diffusion behavior measured is an average over the whole sample. However, in diffusion MRI, the measurement is averaged over each volume element (i.e., voxel). In the case of diffusion-weighted imaging (DWI), the diffusion attenuation is just used to discriminate voxels on the basis of diffusion. However, by collecting two or more images acquired with different q values, it is possible to be quantitative and assign a diffusion coefficient to each voxel and use this as the source of contrast.[22]

As NMR diffusion measurements measure diffusion along the direction of q, it is possible to conduct diffusion measurements in a number of directions in order to study the anisotropy of the diffusion—as would, for example, be expected in an ordered environment such as a liquid crystal or biological tissue sample. Under such circumstances, a diffusion tensor D is obtained. Again, this can be combined with imaging, in which case, a diffusion tensor for each voxel is obtained and the method is known as diffusion tensor imaging (*DTI*).[22]

2.3.2   Analysis of Multicomponent Systems

As noted in Section 3.1, the ever present spin relaxation attenuation of a single diffusing species (or equivalently, spectrally separated species in a mixture) presents no problem and can be normalized out from the PGSE NMR signal to leave only the attenuation because of diffusion. However, complexity arises when the diffusing species is polydisperse (e.g., appearing as monomers and dimers) because, as is usually the case, the signals of each of the oligomeric species overlap. For a polydisperse sample containing N_D types of freely diffusing species, the PGSE NMR signal is given by,

$$S = \sum_i^{N_D} M_{0,i} \exp\left(\frac{-2\tau}{T_{2i}}\right) \exp\left(-\gamma^2 g^2 D_i \delta^2 \left(\Delta - \frac{\delta}{3}\right)\right)$$

(2.9)

where $M_{0,i}$ ($\propto Mw_i n_i$) and T_{2i} denote the equilibrium magnetization and spin–spin relaxation time for the ith species, respectively. Clearly, $M_{0,i} \propto Mw_i n_i$, where n_i is the number of molecules of the ith species present. Equation (2.9) can be written in integral form for a truly polydisperse system.

If at least some of the resonances from each individual species are distinct, then the analysis of a multicomponent species is not much more difficult than for a single diffusing species with the exception that normalization to remove the signal attenuation because of relaxation may become problematic, as the relaxation time of the same resonance on different oligomeric states may differ. Further, some modeling will need to be conducted to account for the distribution of the oligomeric species.

The theory can be extended to encompass the appearance of the spectrum. For example, if each species contributes a 1D NMR spectrum, $S_D(\nu)_i$, where ν is the frequency, the final spectrum, $S(\nu, q)$ would be given by

$$S(\nu, q) = \sum_i^{N_D} \exp\left(-\gamma^2 g^2 D_i \delta^2 \left(\Delta - \frac{\delta}{3}\right)\right) S_D(\nu)_i$$

(2.10)

or upon considering a continuum of diffusion coefficients, this becomes

$$S(\nu, q) = \int_0^\infty \exp\left(-\gamma^2 g^2 D_i \delta^2 \left(\Delta - \frac{\delta}{3}\right)\right) S_D(\nu, D) \mathrm{d}D$$

(2.11)

While traditional NMR experiments typically involve measuring a single species or a single species dissolved in a solvent. The ability to measure diffusion coupled with the realization that a single diffusion coefficient is linked to one entire molecule opens up the possibility of spectrally separating mixtures of species on the basis of the respective diffusion coefficients[23] with some terming it 'NMR chromatography'.

Equation (2.11) shows that $S(\nu, q)$ is the Laplace transform of the Laplace spectrum of diffusion coefficients $S_D(\nu, D)$. Thus, in theory, it is possible to perform an inverse Laplace transform of equation (2.11)

to obtain a spectrum of diffusion coefficients. This approach has now become known as DOSY with the data displayed in a 2D format with chemical shift on one axis and diffusion on the other.[24] Despite early promise and widespread availability of the software for generating DOSY plots, the difficulties in performing the inverse Laplace inversion (a mathematically ill-posed problem) mean that the application of the DOSY approach to samples that contain neither distinct resonances nor widely differing diffusion coefficients is not straightforward and will almost certainly require some degree of prior knowledge to reach meaningful results.[25]

2.3.3 Practical Issues

Compared to many techniques, NMR diffusion methods have only modest requirements in terms of sample preparation. It is instructive to consider the practical issues in conducting an NMR diffusion measurement from what is implicit in the pulse sequence.

2.3.3.1 *Concentration and Measuring Time*

Sensitivity is always an issue in NMR diffusion measurements and, especially, so in some environmental samples and even the determination of initial parameters may require special techniques if the optimization is to be completed within a reasonable time.[26] Ideally, a data set containing echo intensities obtained with something like 12 or more q values are required with reasonable signal-to-noise (S/N) ratio. Despite being a quantitative measurement, by including a 'spoiler' gradient pulse, it is generally possible to run NMR diffusion measurements with far less than a relaxation delay sufficient for full thermal relaxation (i.e., $5T_1$ of the resonance with the longest T_1)[27,28] and, therefore, the required S/N ratio can be achieved in a shorter time. For some samples, signal enhancement techniques such as hyperpolarization might be applicable, but these require substantial modification to the sample. In theory, gas phase measurements are possible; however, in practice, the low spin density in such samples makes them all but impossible to conduct unless the sample contains hyperpolarized species.[12]

Although spin–spin (T_2) and spin–lattice (T_1) relaxation times do not directly enter into the diffusion measurement, they nevertheless limit the possible length of the PGSE sequence. For example, for the spin-echo sequence depicted in Figure 2.1, 2τ (and thus, Δ) must be less than several T_2. If a stimulated echo-based PGSE (i.e., PGSTE) sequence is used, then the limitations depend on both T_1 and T_2. A PGSTE sequence is preferable to a PGSE sequence only when T_1 is considerably greater than T_2. Sometimes, it can be preferable to use a PGSE sequence if one wishes to conduct diffusion measurements in which a broad background (i.e., with a short T_2) is edited out leaving only the peaks of interest. Environmental samples containing paramagnetic ions will be a especially challenging because of their very short T_2 values.

2.3.3.2 *Solvent Suppression*

A particular problem with environmental samples is that they almost invariably contain water and, in general, it is preferable to perform the NMR diffusion measurement on the sample in its native state (i.e., avoiding procedures such as freeze drying and solvent exchange). In addition to overlapping peaks of interest and complicating signal acquisition by saturating the spectrometer receiver, large solvent peaks can cause insidious side effects through radiation damping and the distant dipolar field.[29,30] Consequently, the PGSE sequence must be adapted to give good solvent suppression.[31] An example of a highly optimized PGSE sequence with solvent suppression[32] is shown in Figure 2.2.

2.3.3.3 *Background Gradients*

All of the PGSE theory presented in Section 3 assumes that the only magnetic gradients are those purposely applied. However, in situ magnetic ('background') gradients arise because of spatial heterogeneities in magnetic susceptibility in a sample when it is placed in a magnetic field. Consequently, most porous samples (e.g., rocks) will have internal gradients when placed in a magnetic field.[33] Such internal gradients can be substantial and even, as is commonly the case, they are smaller than the applied gradients; the cross terms between the applied and the internal gradients necessitate the application of appropriately modified pulse sequences to remove their effects.[34,35]

2.4 APPLICATIONS

2.4.1 Toxicity

Toxicological studies on many long-existing pollutants are far from mature. DWI studies on methylmercury chloride-affected rats showed the disturbance in the integrity of microtubules and neurofilaments in the optical nerves.[36] DWI was also applied to studies on basal ganglia abnormalities caused by the exposure to manganese.[37] DTI was applied to studies on white matter integrity[38] and altered myelination and axonal integrity[39] of individuals with lead exposure.

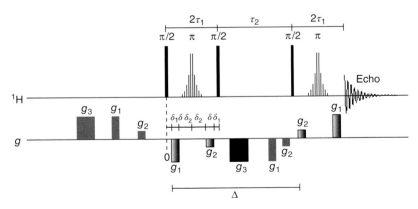

Figure 2.2. An example of a highly optimized PGSE sequence with solvent suppression: the modified PGSTE-WATERGATE method. $\pi/2$ Pulses and selective π pulses are represented by bars and bar groupings, respectively; g_1 (with a duration of δ), g_2, and g_3 are pulsed magnetic field gradients; the gradient filled, black, and gray rectangles represent diffusion weighting, spoiler, and compensating gradients; τ_1, τ_2, δ_1, and δ_2 are delays. (Reprinted with permission from Ref. 32. © 2012 American Chemical Society)

2.4.2 Biofilms

Biofilms are ubiquitous in natural and artificial environments. They not only mediate geochemical processes but also find applications in numerous fields such as wastewater treatment, remediation of polluted sites, and microbial fuel cells. Most studies have been focused on water diffusivity, which can be used for predicting diffusivities of other low-molecular-weight molecules. ^{1}H PGSE NMR was applied to the measurement of the self-diffusion of water in anaerobic methanogenic and sulfidogenic aggregates in bioreactors[40] and anaerobic aggregates of an expanded granular sludge bed reactor.[41] Spatially and temporally resolved diffusion in biofilms was investigated by the use of a novel NMR microimaging probe, which afforded simultaneous electrochemical and PGSE NMR measurements.[42] Structure and water diffusivity in a phototrophic biofilm[43] and contaminant-induced structural and hydrodynamic property changes in biofilms during bioremediation[44] were studied with DWI.

For large molecules (e.g., molecular weight > 1000) and colloids, water diffusivity can no longer serve as a reliable proxy of their diffusivity.[45] Owing to their ability to mimic the diffusion of real pollutants, paramagnetically tagged molecules [e.g., a complex of gadolinium and diethylenetriamine pentaacetic acid (Gd-DTPA)] have found wide applications in mass-transport-related environmental processes. Mass transport through a phototrophic biofilm composed of the cyanobacterium *Phormidium* sp. was studied by using Gd-DTPA-based T_1-weighted MRI and fitting the solution of 1D model of Fickian diffusion to temporal Gd-DTPA concentration data.[45]

2.4.3 Natural Organic Matter

Natural organic matter (NOM) [e.g., soil organic matter (SOM), see Chapter 13, and dissolved organic matter (DOM), see Chapter 11] exists ubiquitously in the environment and can be found in soils, sediments, ponds, lakes, rivers, and oceans. NOM is a complex mixture of inhomogeneous chemical components, which take part in many environmental processes. As a potentially large labile carbon reservoir, NOM plays an important role in the global carbon cycle;[46] interacting with a variety of organic (see Chapter 20) and inorganic anthropogenic (see Chapter 19) pollutants, it determines the fate, transport, bioavailability, toxicity,

and degradation of these compounds;[47–52] further, the nature of these interactions determine the design of remediation processes in aquatic and terrestrial environments.

Diffusion NMR has been widely applied to the studies on NOM because of its noninvasiveness and chemical information richness. These studies include pollutant/sorbate–NOM interactions, NOM structure and aggregation, and magic-angle spinning (MAS) PGSE-based studies.

2.4.3.1 NOM Structure and Aggregation

Diffusion-weighted/ordered NMR provides structural information of NOM components and the oligomeric state of NOM. The structural information furthers our understanding of pollutant/sorbate–NOM interactions, and the oligomeric state of NOM directly affects many NOM-related environmental processes such as the transport of many anthropogenic pollutants[53] and light absorption in natural waters.[54] Studies have focused on both DOM and SOM.

The polydispersity of humic acid (HA) and fulvic acid (FA) was studied using CONTIN (a general purpose constrained regularization program for inverting noisy linear algebraic and integral equations)[55]-based DOSY analysis.[56] Diffusion of the structural components of Suwannee River FA was studied using ^{1}H PSGE NMR, fluorescence correlation spectroscopy, and flow field-flow fractionation, and the obtained data indicated the existence of size and chemical heterogeneity (Figure 2.3) and most Suwannee River FA components having hydrodynamic radii between 1.5 and 2.5 nm.[57] ^{1}H DOSY NMR studies on the aggregation of the Suwannee River DOM revealed that the DOM was composed of a group of smaller components (behaving like maltodextrins of approximately 180–950 Da or proteins of 100–1000 Da) and a group of larger components/aggregates (behaving like maltodextrins of approximately 1000–21 000 Da or proteins of 1050–70 000 Da), and the findings also suggested the involvement of weak dispersive associations of low-molecular-weight materials and carboxyl-rich alicyclic molecules in the formation of aggregates (Figure 2.4).[58] DOMs from pond, river, and ocean were directly studied using PGSE NMR (Figure 2.5), and the obtained average hydrodynamic radii of the DOM components were between 0.42 and 0.54 nm for unconcentrated pond and river waters, which were significantly smaller than those for concentrated pond water.[32]

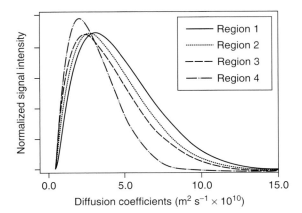

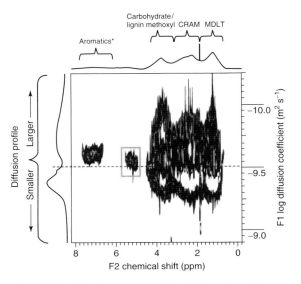

Figure 2.3. Normalized distribution of diffusion coefficients determined by ^{1}H PGSE NMR at an ionic strength of 27 mM and pD of 8.5 for region 1 (0.8–1.9 ppm, corresponding to protons on terminal methyl groups of methylene chains, aliphatic carbons bonded to other carbons, and protons on methyl groups of branched aliphatic structures), region 2 (1.9–3.5 ppm, corresponding to protons on aliphatic carbons, which are two or more carbons from an aromatic ring or polar functional groups), region 3 (3.5–4.3 ppm, corresponding to protons on carbons adjacent to aromatic rings or electronegative functional groups), and region 4 (6.3–8.1 ppm, corresponding to aromatic protons). (Reprinted with permission from Ref. 57. © 2000 American Chemical Society)

Figure 2.4. ^{1}H DOSY map of a 20 g L^{-1} sample of Suwannee River dissolved organic matter. The signals in the boxed region are largely baseline distortions caused by water-signal suppression. Most aromatic resonances are mainly lignin derived. CRAM and MDLT stand for carboxyl-rich alicyclic molecules and material-derived from linear terpenoids, respectively. (Reproduced with permission from Ref. 58. © 2009 SETAC)

Sugar, aliphatic, aromatic, and amino acid components were identified with sugar having the largest molecular weight ~1500 Da in two extracts of SOM from the surface horizon of an oak forest soil by using ^{1}H DOSY NMR and the findings also supported the aggregate and colloidal nature of FA and HA (Figure 2.6).[59] Solubilized humin fractions from grassland and cultivated land were identified as macromolecules or stable aggregates using multidimensional and ^{1}H diffusion-edited (DE) NMR.[60] Arctic permafrost zone SOM degradation stimulated by active layer detachments was indicated by the enrichment of bacterial-derived peptidoglycan (PG) observed by ^{1}H–^{13}C heteronuclear multiple quantum correlation (HMQC), ^{1}H DE NMR (Figure 2.7), and ^{13}C solid-state NMR.[61] The organic matter composition of the density and particle fractions of agricultural, grassland, and grassland–forest transition soil was studied. High concentrations of microbial-derived PG and peptide

side chains were found by ^{1}H DE NMR in clay-sized fractions.[62]

2.4.3.2 Pollutant/Sorbate–NOM Interactions

Most pollutant–NOM interaction studies were focused on the interactions between organic pollutants and NOM. The association of 4′-fluoro-1′-acetonaphthone with the Suwannee River FA has been investigated by performing ^{19}F PGSE diffusion experiments, and the p$_\alpha$H* dependence of F-acetonaphthone diffusion indicates the involvement of two competing binding mechanisms (i.e., hydrophobic interactions and hydrogen bonding) in the association.[63] The association of 2,4-dichlorophenol with a soil FA and HA and a lignite HA was investigated using ^{1}H DOSY and relaxation NMR, and the obtained association constants were 3.1 ± 0.3 M^{-1} (fulvic acid-2,4-dichlorophenol complex), 15 ± 3 M^{-1} (humic acid-2,4-dichlorophenol complex), and 11 ± 1 M^{-1} (lignite humic acid-2,4-dichlorophenol complex).[64]

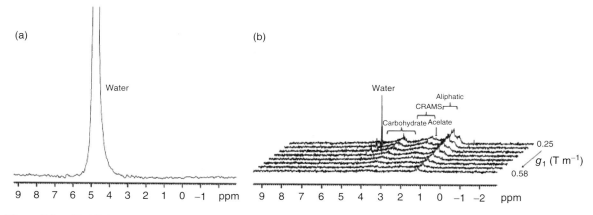

Figure 2.5. ^{1}H NMR spectra of the unconcentrated pond water (a) without water-signal suppression and (b) with PGSTE-WATERGATE-based water-signal suppression. (Reprinted with permission from Ref. 32. © 2012 American Chemical Society)

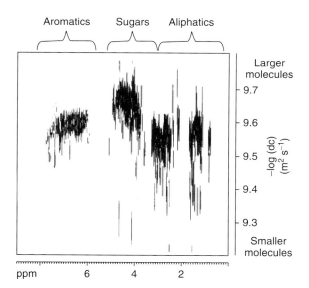

Figure 2.6. [1]H DOSY map of the whole-soil extract in D_2O. (Reprinted with permission from Ref. 59. © 2001 American Chemical Society)

The associations of phenol, 2,4-dichlorophenol, 2,4,6-trichlorophenol, and 2,4,6-trifluorophenol with a dissolved HA were investigated using ^{1}H and ^{19}F DOSY and relaxation NMR, and it was found that the associations of 2,4-dichlorophenol and 2,4,6-trichlorophenol to HA were much more significant than that of phenol and 2,4,6-trifluorophenol.[65]

Recently, ^{1}H and ^{31}P DOSY and relaxation NMR experiments were performed to probe the associations of glyphosate (*N*-phosphonomethylglycine) herbicide with soluble soil FA and HA at pH 5.2 and 7, and the obtained association constants and Gibbs free energies of transfer for glyphosate–humic complex formation indicated that noncovalent interactions were greatest for FA at pH 5.2 (Figure 2.8).[66]

Metal–DOM interactions have also attracted the attention of many researchers. For example, complexes formed between a HA and either Al^{3+} or Ca^{2+} were investigated using ^{13}C cross polarization magic-angle spinning (CPMAS) and ^{1}H DOSY NMR and it was found that the addition of Al^{3+} and Ca^{2+} decreased the diffusion of humic alkyl components and increased the diffusion of aromatic and hydroxyalkyl components (Figures 2.9 and 2.10).[67]

2.4.3.3 MAS PGSE NMR Studies

Magic-angle spinning (MAS) PGSE NMR combines both NMR spectroscopy and diffusometry to selectively monitor the diffusion of compounds sorbed in porous media or polymer matrices. The mobility of organic contaminants (e.g., toluene) in SOM (i.e., HA) was probed in the solid state using MAS PGSE NMR.[3] Mobile aliphatic domains in humic substances and their impact on contaminant mobility within the matrix was also studied by the use of MAS PGSE NMR and it was confirmed that the aliphatic domains in humic

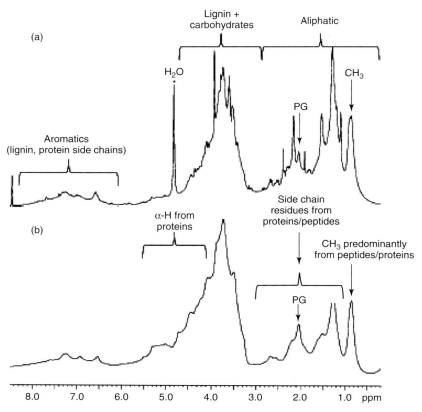

Figure 2.7. ^{1}H (a) PURGE and (b) diffusion-edited NMR spectra of the soil organic matter humic extract from an undisturbed site. The application of diffusion-edited NMR highlighted the existence of microbial peptide/protein such as peptidoglycan (PG). (Reprinted with permission from Ref. 61. © 2010 American Chemical Society)

substances play an important role in the mobility of sorbed contaminants within this matrix.[4]

2.4.4   Natural Porous Media

The flow and diffusion of liquids in natural porous media (e.g., soil) play a crucial role in many environmental processes such as groundwater flow, contaminant or colloidal transport, and bioconcentration and degradation of contaminants.

Flow and self-diffusion of water in a column filled with quartz sand and calcareous gravel were measured by using DWI and the measured self-diffusion coefficient was $1.2–2.1 \times 10^{-9}$ m^2 s^{-1} at measured flow velocity between 0.15 and 6.67 mm s^{-1}; however, the

effects of background gradients were not discussed (Figure 2.11).[68,69]

To mimic the traveling of surface bacteria through natural porous media, T_2-weighted MRI was used to study the mutual diffusion of MnCl$_2$ and diffusive migration of immunomagnetically labeled *Pseudomonas putida* through a column packed with glass-coated polystyrene beads and the tortuosity of the column was found to be 1.7 ± 0.3 (for MnCl$_2$) or 87 ± 16 (for *P. putida*).[70] T_2-relaxation experiments were also performed on water-saturated sand column to detect the mutual diffusion of 2,2,6,6-tetramethylpiperidine-1-oxyl (TEMPO), which had a variety of derivatives covering a range of physical properties and the obtained mutual diffusion coefficient for TEMPO was $1.4 \pm 0.5 \times 10^{-10}$ m^2 s^{-1}, which was in line with the literature values ranging

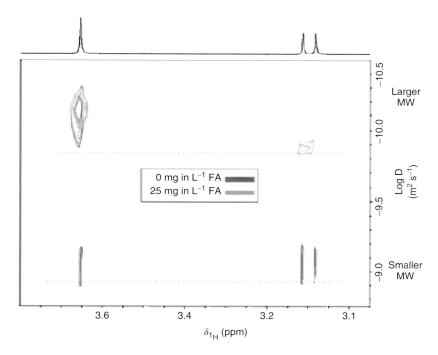

Figure 2.8. ¹H DOSY spectra of GLY with pH = 5.2 at two different FA concentrations (0 and 0.025 mg L⁻¹). (Reprinted with permission from Ref. 66. © 2012 American Chemical Society)

from 1.3 to 2.3×10^{-10} m² s⁻¹;[71] the mutual diffusion coefficient was obtained by fitting equation (2.12) to the temporal T_2-relaxation data (Figure 2.12).[71]

$$
\begin{aligned}
\frac{M(t_{CPMG})}{M_0} &= \int_0^h \sum_j p_j \exp\left[-t_{CPMG}\left(\frac{1}{T_{2,c_0}^{bulk,j}} + \left[\frac{c_S}{2}\left[\mathrm{erf}\left(\frac{h'+r}{2\sqrt{D_{sand}t}}\right) + \mathrm{erf}\left(\frac{h'-r}{2\sqrt{D_{sand}t}}\right)\right]\right]R_{TEMPO}\right)\right]\mathrm{d}h' \\
&+ \int_h^r \sum_k p_k \exp\left[-t_{CPMG}\left(\frac{1}{T_{2,c_0}^{sand,k}} + \left[\frac{c_S}{2}\left[\mathrm{erf}\left(\frac{h+r'}{2\sqrt{D_{sand}t}}\right) + \mathrm{erf}\left(\frac{h-r'}{2\sqrt{D_{sand}t}}\right)\right]\right]R_{TEMPO}\right)\right]\mathrm{d}r'
\end{aligned}
$$
(2.12)

2.5 FUTURE OF NMR DIFFUSION METHODS IN THIS FIELD

2.5.1 Better Solvent/Water-Signal Suppression

As discussed earlier, solvent (i.e., water)-signal suppression is crucial for PGSE NMR diffusion studies on dissolved organic pollutants and DOM. WATERGATE-based solvent-signal suppression methods (e.g., Refs 7, 9, 72–78 have attracted lots of applications because of their effectiveness and easiness of setup). All WATERGATE-based suppression methods rely on the dephasing of transverse magnetizations by the application of pulsed magnetic field gradients, and, therefore, the efficiency of solvent-signal suppression depends on the duration and strength of the applied pulsed gradients. For a typical WATERGATE-based pulse sequence (e.g., PGSTE-WATERGATE[76]), a typical gradient duration of 3 ms is needed to achieve desirable water-signal suppression for a natural water sample

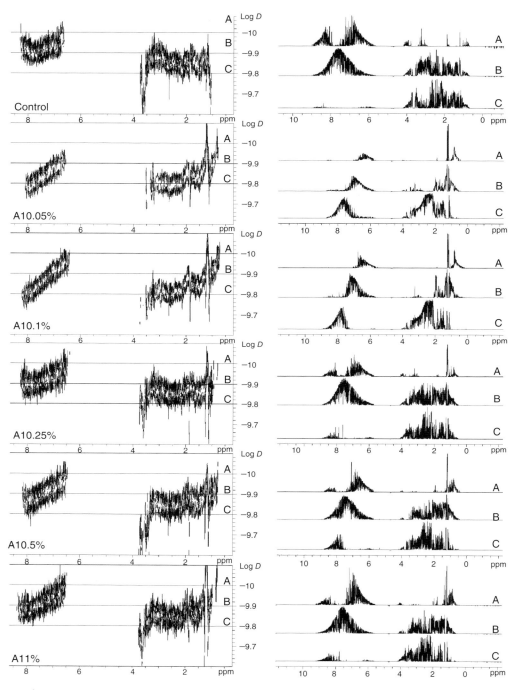

Figure 2.9. ^{1}H DOSY spectra of control HA and Al-HA at different Al concentrations. The ^{1}H chemical shift profiles shown on the left were obtained by cutting through the DOSY spectra at three different diffusion levels (i.e., A, B, and C). (Reprinted with permission from Ref. 67. © 2009 American Chemical Society)

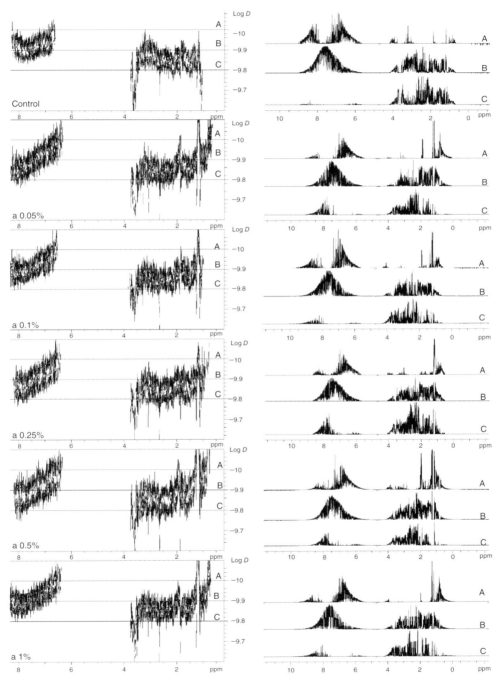

Figure 2.10. ^{1}H DOSY spectra of control HA and Ca-HA at different Ca concentrations. The ^{1}H chemical shift profiles shown on the left were obtained by cutting through the DOSY spectra at three different diffusion levels (i.e., A, B, and C). (Reprinted with permission from Ref. 67. © 2009 American Chemical Society)

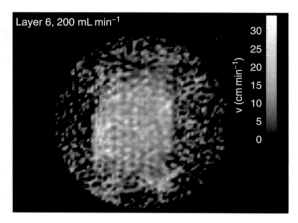

Figure 2.11. Spatial distribution of the flow velocities in the medium/coarse sand layer at 200 mL min^{-1}. (Reprinted with permission from Ref. 69. © 2000 American Chemical Society)

with a gradient strength of $0.3\,T\,m^{-1}$, which results in a T_2-relaxation exposure of >6 ms and thus, the undesirable attenuation of fast-relaxing chemical components (e.g., aromatic protons of DOMs). To minimize the T_2-based attenuation, a high-resolution NMR probe with relatively high gradient strength (e.g., $3\,T\,m^{-1}$) can be utilized, so that the required gradient duration and thus the T_2 exposure time can be significantly shortened (e.g., $\sim$2 ms).

In spite of many new developments in NMR solvent-signal suppression, radiation damping is still hampering the suppression process especially in the case of extreme suppression for detecting trace compounds in environmental samples (e.g., ocean water).

2.5.2 Increasing S/N Ratio

As solute concentrations are often low in environmental samples, the obtaining of adequate S/N ratios in a reasonable experimental time generally limits the samples that can be studied. Of the methods available for increasing the S/N ratio, conducting experiments at higher magnetic field strengths is very expensive and the increase in S/N ratio is only $\propto$ $\sim B_0^{3/2}$. A cheaper 'hardware' approach is the use of cooled probes and cryoprobes. More advanced signal processing (i.e., alternatives to Fourier transformation) is also a possibility–but this generally

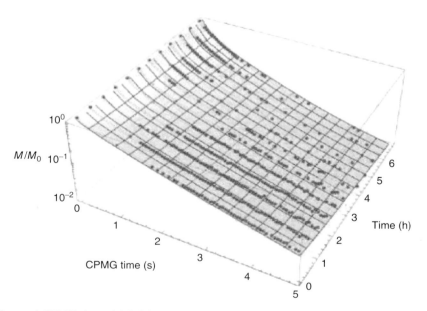

Figure 2.12. Measured CPMG data of 2,2,6,6-tetramethylpiperidine-1-oxyl (TEMPO) mutual diffusion experiment (red dots) and fitted data according to equation 14 (semitransparent surface, blue cross-hatching). (Reprinted with permission from Ref. 71. © 2011 American Chemical Society)

requires prior knowledge on the nature of the sample (i.e., the number of components/resonances). Another alternative, as mentioned earlier, although it involves some alteration of the sample, is the use of hyperpolarization.

2.5.3 Separation of Mixtures

The toxicity of many important organic pollutants (e.g., 2,4-dichlorophenol and 2,4,6-trichlorophenol) has isomer and congener dependence (i.e., different isomers or congeners show different toxicological effects) and these compounds usually exist as complex mixtures in the environment. Therefore, it is important to quantify the highly toxic isomers/congeners in the mixtures. Matrix-assisted DOSY/PGSE NMR[79–82] enables the diffusion-based separation of different isomers and/or congeners on a DOSY map and thus, the quantitation of these compounds. The intertwinement with a 2D NMR technique (e.g., HMQC) can significantly enhance the separation efficiency of matrix-assisted DOSY/PGSE NMR.

RELATED ARTICLES IN EMAGRES

Cryogenic NMR Probes: Applications

Diffusion-Ordered Spectroscopy

Dipolar Field and Radiation Damping: Collective Effects in Liquid-State NMR

Diffusion Measurements by Magnetic Field Gradient Methods

Diffusion-Ordered Spectroscopy

Diffusion-Based Studies of Aggregation, Binding and Conformation of Biomolecules: Theory and Practice

Diffusion Weighted Imaging in Cancer

Diffusion in Porous Media

Hyperpolarized Gas Imaging

Methods and Applications of Diffusion MRI

Microsample Cryogenic Probes: Technology and Applications

Water Signal Suppression in NMR of Biomolecules

REFERENCES

1. P. Mansfield and B. Chapman, *J. Phys. E Sci. Instrum.*, 1986, **19**, 540.

2. P. Mansfield and B. Chapman, *J. Magn. Reson.*, 1986, **66**, 573.

3. K. W. Fomba, P. Galvosas, U. Roland, J. Kärger, and F. D. Kopinke, *Environ. Sci. Technol.*, 2009, **43**, 8264.

4. K. W. Fomba, P. Galvosas, U. Roland, J. Kärger, and F. D. Kopinke, *Environ. Sci. Technol.*, 2011, **45**, 5164.

5. P. Stilbs, *Anal. Chem.*, 1981, **53**, 2135.

6. K. F. Morris and C. S. Johnson Jr, *J. Am. Chem. Soc.*, 1992, **114**, 3139.

7. M. Piotto, V. Saudek, and V. Sklenár, *J. Biomol. NMR*, 1992, **2**, 661.

8. A. J. Simpson and S. A. Brown, *J. Magn. Reson.*, 2005, **175**, 340.

9. B. Lam and A. J. Simpson, *Analyst*, 2008, **133**, 263.

10. A. J. Baldwin, J. Christodolou, P. D. Barker, and C. M. Dobson, *J. Chem. Phys.*, 2007, **127**, 114505.

11. H. J. V. Tyrrell and K. R. Harris, Diffusion in Liquids: A Theoretical and Experimental Study, 1st edn, Butterworths: London, 1984.

12. R. W. Mair, D. G. Cory, S. Peled, C.-H. Tseng, S. Patz, and R. L. Walsworth, *J. Magn. Reson.*, 1998, **135**, 478.

13. H. Weingärtner, *Z. Phys. Chem.*, 1982, **132**, 129.

14. W. S. Price, F. Tsuchiya, and Y. Arata, *J. Am. Chem. Soc.*, 1999, **121**, 11503.

15. G. A. Truskey, F. Yuan, and D. F. Katz, Transport Phenomena in Biological Systems, 1st edn, Prentice Hall: New York, 2003.

16. W. S. Price, NMR Studies of Translational Motion: Principles and Applications, 1st edn, Cambridge University Press: Cambridge, 2009.

17. R. E. Moll, *J. Am. Chem. Soc.*, 1968, **90**, 4739.

18. W. S. Price, F. Tsuchiya, C. Suzuki, and Y. Arata, *J. Biomol. NMR*, 1999, **13**, 113.

19. P. T. Callaghan, Translational Dynamics & Magnetic Resonance, 1st edn, Oxford University Press: Oxford, 2011.

20. E. O. Stejskal, *J. Chem. Phys.*, 1965, **43**, 3597.

21. E. O. Stejskal and J. E. Tanner, *J. Chem. Phys.*, 1965, **42**, 288.

22. D. K. Jones (ed), Diffusion MRI, Oxford: Oxford University Press, 2011.

23. P. Stilbs, *Prog. NMR Spectrosc.*, 1987, **19**, 1.

24. C. S. Johnson Jr, *Prog. NMR Spectrosc.*, 1999, **34**, 203.

25. P. Stilbs, *J. Magn. Reson.*, 1998, **135**, 236.

26. H. Farooq, D. Courtier-Murias, R. Soong, H. Masoom, W. Maas, M. Fey, R. Kumar, M. Monette, H. Stronks, M. J. Simpson, and A. J. Simpson, *Magn. Reson. Chem.*, 2013, **51**, 129.

27. T. Stait-Gardner, P. G. Anil Kumar, and W. S. Price, *Chem. Phys. Lett.*, 2008, **462**, 331.

28. G. H. Sørland, H. W. Anthonsen, K. Zick, J. Sjöblom, and S. Simon, *Diffus. Fundam.*, 2011, **15**, 1.

29. J. Jeener, *Concepts Magn. Reson.*, 2002, **14**, 79.

30. V. V. Krishnan and N. Murali, *Prog. NMR Spectrosc.*, 2013, **68**, 41.

31. G. Zheng and W. S. Price, *Prog. NMR Spectrosc.*, 2010, **56**, 267.

32. G. Zheng and W. S. Price, *Environ. Sci. Technol.*, 2012, **46**, 1675.

33. Y.-Q. Song, *Concepts Magn. Reson. A*, 2003, **18**, 97.

34. F. Stallmach and P. Galvosas, in Annual Reports on NMR Spectroscopy, ed G. A. Webb, Elsevier: London, 2007 Vol. 61.

35. G. Zheng and W. S. Price, *Concepts Magn. Reson. A*, 2007, **30**, 261.

36. Y. Kinoshita, A. Ohnishi, K. Kohshi, and A. Yokota, *Environ. Res. A*, 1999, **80**, 348.

37. S. Criswell, J. S. Perlmutter, J. L. Huang, N. Golchin, H. P. Flores, A. Hobson, M. Aschner, K. M. Erikson, H. Checkoway, and B. A. Racette, *Occup. Environ. Med.*, 2012, **69**, 437.

38. T.-J. Hsieh, H.-Y. Chuang, Y.-C. Chen, C.-L. Wang, S.-H. Lan, G.-C. Liu, C.-K. Ho, and W.-C. Lin, *Radiology*, 2009, **252**, 509.

39. C. J. Brubaker, V. J. Schmithorst, E. N. Haynes, K. N. Dietrich, J. C. Egelhoff, D. M. Lindquist, B. P. Lanphear, and K. M. Cecil, *NeuroToxicology*, 2009, **30**, 867.

40. P. N. L. Lens, R. Gastesi, F. Vergeldt, A. C.van Aelst, A. G. Pisabarro, H. Van As, and A. C. V. Aelst, *Appl. Environ. Microbiol.*, 2003, **69**, 6644.

41. G. Gonzalez-Gil, P. N. L. Lens, A.van Aelst, H. Van As, A. I. Versprille, and G. Lettinga, *Appl. Environ. Microbiol.*, 2001, **67**, 3683.

42. R. S. Renslow, J. T. Babauta, P. D. Majors, and H. Beyenal, *Energy Environ. Sci.*, 2013, **6**, 595.

43. V. R. Phoenix and W. M. Holmes, *Appl. Environ. Microbiol.*, 2008, **74**, 4934.

44. B. Cao, P. D. Majors, B. Ahmed, R. S. Renslow, C. P. Silvia, L. Shi, S. Kjelleberg, J. K. Fredrickson, and H. Beyenal, *Environ. Microbiol.*, 2012, **14**, 2901.

45. B. Ramanan, W. M. Holmes, W. T. Sloan, and V. R. Phoenix, *Appl. Environ. Microbiol.*, 2010, **76**, 4027.

46. N. Jiao, G. J. Herndl, D. A. Hansell, R. Benner, G. Kattner, S. W. Wilhelm, D. L. Kirchman, M. G. Weinbauer, T. Luo, F. Chen, and F. Azam, *Nat. Rev. Microbiol.*, 2010, **8**, 593.

47. Q. S. Fu, A. L. Barkovskii, and P. Adriaens, *Environ. Sci. Technol.*, 1999, **33**, 3837.

48. M. R. Servos, D. C. G. Muir, and G. R. B. Webster, *Aquat. Toxicol.*, 1989, **14**, 169.

49. M. R. Servos and D. C. G. Muir, *Environ. Toxicol. Chem.*, 1989, **8**, 141.

50. S. Frankki, Y. Persson, M. Tysklind, and U. Skyllberg, *Environ. Sci. Technol.*, 2006, **40**, 6668.

51. G. R. B. Webster, D. H. Muldrew, J. J. Graham, L. P. Sarna, and D. C. G. Muir, *Chemosphere*, 1986, **15**, 1379.

52. L. A. Cardoza, C. W. Knapp, C. K. Larive, J. B. Belden, M. Lydy, and D. W. Graham, *Water Air Soil Pollut.*, 2005, **161**, 383.

53. A. Shirzadi, M. J. Simpson, Y. Xu, and A. J. Simpson, *Environ. Sci. Technol.*, 2008, **42**, 1084.

54. M. L. Pace, I. Reche, J. J. Cole, A. Fernández-Barbero, I. P. Mazuecos, and Y. T. Prairie, *Biogeochemistry*, 2012, **108**, 109.

55. S. W. Provencher, *Comput. Phys. Commun.*, 1982, **27**, 229.

56. K. F. Morris, B. J. Cutak, A. M. Dixon, and C. K. Larive, *Anal. Chem.*, 1999, **71**, 5315.

57. J. R. Lead, K. J. Wilkinson, E. Balnois, B. J. Cutak, C. K. Larive, S. Assemi, and R. Beckett, *Environ. Sci. Technol.*, 2000, **34**, 3508.

58. B. Lam and A. J. Simpson, *Environ. Toxicol. Chem.*, 2009, **28**, 931.

59. A. J. Simpson, W. L. Kingery, M. Spraul, E. Humpfer, P. Dvortsak, and R. Kerssebaum, *Environ. Sci. Technol.*, 2001, **35**, 4421.

60. A. J. Simpson, G. Song, E. Smith, B. Lam, E. H. Novotny, and M. H. B. Hayes, *Environ. Sci. Technol.*, 2007, **41**, 876.

61. B. G. Pautler, A. J. Simpson, D. J. McNally, S. F. Lamoureux, and M. J. Simpson, *Environ. Sci. Technol.*, 2010, **44**, 4076.

62. J. S. Clemente, E. G. Gregorich, A. J. Simpson, R. Kumar, D. Courtier-Murias, and M. J. Simpson, *Environ. Chem.*, 2012, **9**, 97.

63. A. M. Dixon, M. A. Mai, and C. K. Larive, *Environ. Sci. Technol.*, 1999, **33**, 958.

64. D. Šmejkalová and A. Piccolo, *Environ. Sci. Technol.*, 2008, **42**, 8440.

65. D. Šmejkalová, R. Spaccini, B. Fontaine, and A. Piccolo, *Environ. Sci. Technol.*, 2009, **43**, 5377.

66. P. Mazzei and A. Piccolo, *Environ. Sci. Technol.*, 2012, **46**, 5939.

67. A. Nebbioso and A. Piccolo, *Environ. Sci. Technol.*, 2009, **43**, 2417.

68. T. Baumann, R. Petsch, G. Fesl, and R. Niessner, *J. Environ. Qual.*, 2002, **31**, 470.

69. T. Baumann, R. Petsch, and R. Niessner, *Environ. Sci. Technol.*, 2000, **34**, 4242.

70. M. S. Olson, R. M. Ford, J. Smith, and E. J. Fernandez, *Environ. Sci. Technol.*, 2005, **39**, 149.

71. F. Furtado, P. Galvosas, F. Stallmach, U. Roland, J. Kärger, and F. D. Kopinke, *Environ. Sci. Technol.*, 2011, **45**, 8866.

72. V. Sklenár, M. Piotto, R. Leppik, and V. Saudek, *J. Magn. Reson. A*, 1993, **102**, 241.

73. M. Liu, X.-A. Mao, C. Ye, H. Huang, J. K. Nicholson, and J. C. Lindon, *J. Magn. Reson.*, 1998, **132**, 125.

74. T.-L. Hwang and A. J. Shaka, *J. Magn. Reson. A*, 1995, **112**, 275.

75. W. S. Price, F. Elwinger, C. Vigouroux, and P. Stilbs, *Magn. Reson. Chem.*, 2002, **40**, 391.

76. G. Zheng, T. Stait-Gardner, P. G. Anil Kumar, A. M. Torres, and W. S. Price, *J. Magn. Reson.*, 2008, **191**, 159.

77. J. Wang, X. Zhang, P. Sun, X. Jiang, B. Jiang, C. Cao, and M. Liu, *J. Magn. Reson.*, 2010, **206**, 205.

78. R. W. Adams, C. M. Holroyd, J. A. Aguilar, M. Nilsson, and G. A. Morris, *Chem. Commun.*, 2013, **49**, 358.

79. R. Evans, S. Haiber, M. Nilsson, and G. A. Morris, *Anal. Chem.*, 2009, **81**, 4548.

80. C. F. Tormena, R. Evans, S. Haiber, M. Nilsson, and G. A. Morris, *Magn. Reson. Chem.*, 2010, **48**, 550.

81. J. Cassani, M. Nilsson, and G. A. Morris, *J. Nat. Prod.*, 2012, **75**, 131.

82. D. J. Codling, G. Zheng, T. Stait-Gardner, S. Yang, M. Nilsson, and W. S. Price, *J. Phys. Chem. B*, 2013, **117**, 2734.

Chapter 3
Environmental NMR: Hyphenated Methods

Markus Godejohann

Bruker BioSpin GmbH, 76274 Rheinstetten, Germany

3.1 INTRODUCTION

Emerging organic contaminants (EOCs) and other xenobiotic organic contaminants (XOCs), e.g., pesticides, pharmaceuticals, explosives, or industrial chemicals, are released to the environment at high concentrations from point sources, such as landfills without water protection, wastewater treatment plants (WWTPs), leakages of tanks, or releases from

accidents,[1] or from nonpoint sources where contaminants are released into the environment in a diffused way resulting in a much lower concentration. XOCs are further chemically modified in the environment. These transformation products are new chemicals for which reference standards are typically not available. XOCs and their transformation products released into the environment pose a threat especially to surface- and groundwater from which, typically, drinking water for public use is produced. It is, therefore, extremely important to monitor this environmental compartment frequently. While known and expected chemicals, e.g., pesticides in an agricultural area, can be monitored using standard instrumental analytical techniques in a targeted approach,[2] unexpected XOCs and completely unknown transformation products will be completely overlooked, even though concentrations can be in the microgram per liter range.

Chromatography coupled to high-resolution mass spectrometry (MS) can be used in a nontargeted approach for the detection of ionizable organic micropollutants.[3] This approach results in a tentative structure proposal of an unknown, which needs to be verified by comparison with reference standards. Alternatively, NMR spectroscopy is the method of choice for structural elucidation of unknown organic molecules. Further, NMR techniques provide quantitative results even when reference standards are not available. However, NMR techniques are inherently less sensitive in comparison to MS methods. Accordingly, samples often need concentration before analysis, especially in the case of environmental samples. Despite this, there are other advantages of using NMR.

Written by Markus Godejohann, in his capacity as employee of Bruker BipSpin GmbH.

The signal-to-noise ratio of the NMR signal of an analyte is inversely proportional to its molecular mass and directly proportional to its mass concentration and the number of nuclei giving rise to the NMR signal. It also strongly depends on the type of nucleus under investigation. For ^{1}H and ^{19}F nuclei, the lowest detection limits can be achieved because of the high natural abundance of their magnetically active nuclei and their high gyromagnetic ratios. The sensitivity of the NMR technique also depends on the magnetic field strength and the type of probe head used. Higher field strengths provide higher signal intensity and the use of cryogenically cooled probe heads decreases the noise. Both lead to an increase in the signal-to-noise ratio of the NMR experiment. Absolute detection limits for small molecules are in the mid-to-upper nanogram range of analyte present in the active volume of the NMR coil. This means that for environmental samples where contaminants in the lower microgram per liter range are expected, an enrichment factor between 1000 and 10 000 is needed depending on the injection volume and/or end volume of the final extract. This extract can be used with conventional NMR analysis of samples in NMR tubes and subsequently, to high-performance liquid chromatography (HPLC)-NMR/MS analysis.

In hyphenated NMR experiments, the outlet of the chromatography is fed to NMR and (high-resolution) MS as shown in Figure 3.1. This allows the correlation of the mass response to the NMR spectrum obtained from an isolated contaminant eluting from the column.

If enough material is present (20–50 μg), all necessary NMR spectra for complete structure elucidation can be acquired with modern NMR equipments including sensitive electronics and cryoprobe technology.[4] This has led to this technique being extensively used in the analysis of pharmaceutical and natural products.[5–7] However, the amounts present in environmental extracts are typically in a low concentration range and may even be close to the detection limit. In this case, the molecular formula obtained from the accurate mass measurement can be used to search for possible structures in public libraries and the ^{1}H-NMR spectrum can be used to exclude structures that are not consistent with the chemical shift values and/or coupling patterns. This is, of course, possible only if the contaminant is not completely unknown. Should that be the case, the only possibility to obtain enough material is by the enrichment of a sufficient amount of sample during the preparation step or by identifying point sources where the concentration is high enough for NMR analysis. Routine analysis techniques can take over the task of identifying point sources showing highest concentration of the unknowns.

3.1.1   Requirements

Owing to the low sensitivity of the NMR detection, a rather high quantity of analyte needs to be injected on column. For some chromatographic systems, these amounts lead to an overload of the column. This is especially the case for microbore/capillary columns and/or columns filled with sub-2 μm particles, monolithic or nonporous material and capillary gas chromatographic columns. This restricts the chromatography mainly to analytical scale, i.e., HPLC columns with column diameters of 2–4.6 mm and particle sizes between 3 and 5 μm. Both reversed phase and hydrophilic interaction liquid chromatography (HILIC) separations are possible in hyphenation with NMR.[8] For direct LC-NMR (liquid chromatography nuclear magnetic resonance) analysis, the solvent system should be kept as simple as possible using either acetonitrile or methanol with deuterium oxide and volatile buffers for MS detection. As the sample preparation delivers extracts in an organic solvent, the maximum injection volume is limited to about 20 μL without compromising the chromatography. If this volume is injected several times at high aqueous mobile phase conditions, the contaminants are trapped on the head of the chromatographic column without broadening the peaks. This multiple injection technique can be accomplished with injection programs and allows the administration of the complete extract ($200\,\mu L = 10 \times 20\,\mu L$) on column. A step gradient after the sample loading period initiates the separation by gradient elution.

The type of mass spectrometer used should be able to determine the molecular formula of an unknown contaminant to aid in structural elucidation.

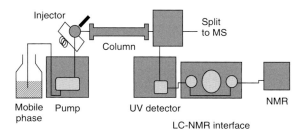

Figure 3.1.   Schematic representation of an LC-NMR system.

High-resolution mass spectrometers available for this purpose are time-of-flight, Orbitrap, or FT-ion cyclotron resonance (ICR)-MS systems. Electrospray or atmospheric pressure chemical ionization (APCI) ion sources are employed for the ionization of polar contaminants. The mass spectrometer is coupled to the hyphenated system using a splitter. This splitter only transfers 2–5% of the chromatographic flow to the MS ion source. The main portion is guided to the interface for direct analysis in the NMR or intermediate storage/trapping.

An NMR spectrometer of 5007 MHz proton frequency or higher should be used because of the sensitivity required. For the same reason, cryogenically cooled probe heads with inverse geometry (optimized for proton sensitivity) are also advisable. These probe heads can be used either in tube mode or converted to a flow probe using a special adapter. The interface between the chromatography and the NMR spectrometer is needed to store or trap the peaks before measurement. The direct on-flow technique, however, does not need any interface; the outlet of the primary detector [typically, an ultraviolet (UV) detector or a diode array detector (DAD)] can be directly connected to the flow probe or flow insert.

3.1.2 Targeted Analysis

Targeted analysis of XOCs is within the classical domain of instrumental analytical techniques employing MS detection. In particular, a triple quadrupole mass spectrometer coupled to HPLC or ultra performance liquid chromatography (UPLC) systems provides the highest sensitivity and specificity. Modern systems do not require sample enrichment before analysis to monitor the legal drinking water limits of 0.1 µg L^{-1} for common pesticides. NMR-based techniques cannot compete against these rugged and proven analytical solutions. However, with sample enrichment, detection limits of 0.1 µg L^{-1} are possible even with NMR detection.[9] A very promising approach to investigate an unknown environmental extract is the acquisition of a ^{1}H-NMR spectrum of the extract in a 3 mm NMR tube, as here, the volume corresponds to the injection volume capable for HPLC. This gives a fast first overview of the proton-carrying constituents present in the extract. Already at this stage, a direct comparison with NMR spectra of pure reference compounds measured in the same NMR solvent allows identification of known contaminants. Their quantification

can be accomplished by single one-point calibration using an external standard of known composition and concentration. The volume needed to fill a 3 mm tube is around 200 µL. After recovery of this extract into an HPLC vial, the sample can be injected onto an HPLC column in case the sample is too complex for mixture analysis or if contaminants cannot be readily identified.

3.1.3 Nontargeted Analysis

For nontargeted analysis, an ideal detector should comply with the following requirements:

1. detection of all chemicals present in a sample showing identical response factors independent of chemical properties;
2. provide information for structure elucidation of unknowns;
3. no interference with background matrix;
4. good resolution to detect a large number of chemicals without superimposition;
5. inherent robustness, accuracy, precision, repeatability, and reproducibility of results;
6. sufficient sensitivity to detect chemicals at environmentally relevant concentration levels.

Neither MS nor NMR can fulfill all these requirements. While point 6 is the weakest of NMR, the MS detection fails in particular at points 1, 3, and 5. A combination of both techniques, however, gives complementary information and can be realized by coupling chromatography to NMR and MS in a hyphenated setup.

To examine an environmental sample in the hyphenated setup using a nontargeted approach requires measuring the complete chromatographic run using NMR, MS, and UV detections. This can only be achieved by running the sample in an on-flow or time-slice mode, as only in these modes is the complete chromatogram measured by NMR. This is extremely time-consuming because every chromatographic run is subdivided into 64–128 individual NMR measurements and cannot be applied to each sample. It is, therefore, extremely beneficial to run a simple ^{1}H-NMR measurement of the environmental extract before the hyphenated run in order to select samples with sufficiently high concentration of unknown contaminants. In nontargeted analysis, each NMR spectrum of an environmental extract acts as a 'fingerprint'. This 'fingerprint' contains not only qualitative but also quantitative information about components present in the extract.

Using well-defined standard operation procedures (SOPs) for sample enrichment and measurement, statistical models can be built to automatically find outlying samples with unexpected concentrations of contaminants, as has already been shown for the analysis of fruit juices.[10] A combined covariance analysis can be employed along with the LC-MS data of the same samples to correlate the MS response of a contaminant to their NMR resonances in a statistical manner. This is only possible if the analyte of interest shows a response in both NMR and MS.[11]

3.2 PREPARATION OF ENVIRONMENTAL SAMPLES

As mentioned earlier, the absolute detection limits for modern NMR techniques, including cryogenically cooled probe heads and field strengths above 500 MHz proton frequency, is in the upper nanogram range. For structural elucidation, an amount in the lower microgram range needs to be present in the active region of the NMR magnet. As injection volumes are in the range 20–100 μL and the active volume in a 3 mm tube is around 100 μL, the contaminants present in an aqueous environmental sample need to be concentrated from the microgram per liter range to the microgram per milliliter range, which correlates to an enrichment factor between 1000 and 10 000. The techniques to achieve these enrichment factors are based on standard enrichment techniques used in environmental analytical chemistry. The NMR technique, however, has special requirements when extracts are to be measured directly using tube-NMR. This concerns the use of deuterated solvents and is discussed in detail in the following sections.

3.2.1 Solid-Phase Extraction (SPE)

Solid-phase extraction (SPE) is the method of choice for the enrichment and clean-up of aqueous environmental samples, such as surface, ground, and drinking water. Cartridges are available with different types of phases. In addition to classical silica-based reversed phase material, pH-stable copolymer materials, in particular, are used for the extraction of polar contaminants.[12] Cartridges are typically prepacked

with sorbent amounts of 200, 500, or 1000 mg and provide a very good batch-to-batch reproducibility. Many manufacturers offer ion-exchange SPE cartridges, which enable preseparation into neutral analytes, strong acids, strong bases, weak acids, and weak bases. This reduces the complexity of the extracts to some extent and also, therefore, the risk of coelution or superimposition of signals originating from different contaminants.

Owing to the enhanced sensitivity of the analytical techniques employed (especially, MS techniques), the sample volume taken for SPE is decreased dramatically. For NMR detection, however, the typical volume needed is 1 L of sample enriched on 500 mg of sorbent material after conditioning and equilibration of the SPE cartridge. The elution is typically performed using methanol or a methanol/acetonitrile mixture. Before elution, the SPE bed must be dried in order to remove water from the cartridges and guarantee reproducible recovery values for the analytes of interest. After elution, the extract needs to be evaporated until dry to remove the protonated solvent. The reconstitution is done with deuterated methanol or acetonitrile. The final volume should be exactly 200 μL, which is the volume needed to fill a 3 mm NMR tube for the initial investigation of the sample. If the concentration is still too low for NMR detection, the extracts of several SPE cartridges can be combined before evaporation. This is the case especially if a contaminant is close to the detection limit and the structure needs to be elucidated.

3.2.2 Liquid–Liquid Extraction

Another method for sample enrichment and clean-up is liquid–liquid extraction. The drawback of this technique over SPE is the high volumes of organic solvents used for the extraction step. This in turn needs to be removed by evaporation before NMR analysis. This technique is, therefore, typically not used for sample pretreatment in an environmental context.

3.3 LIQUID CHROMATOGRAPHY NUCLEAR MAGNETIC RESONANCE

The early examples have focused on the use of LC and NMR separately, but recent advances have directly coupled LC to the NMR. LC-NMR is mainly restricted to reverse phase, HILIC, and ion-exchange separations where the solvent composition should be kept as simple as possible. As a default, the aqueous part of the mobile phase is substituted with deuterium oxide or a deuterated buffer. To reach the required sensitivity for environmental extracts, the NMR spectrometer should be equipped with a cryogenically cooled probe head converted to a flow probe using an exchangeable flow cell. Section 3.3.1 provides an overview of LC-NMR operation and hyphenation with SPE and MS.

3.3.1 Operation Modes

The direct coupling of LC and NMR can be realized in an on-flow or stop-flow operation mode using special interfaces.[13] The on-flow technique is known to have the lowest sensitivity because of the short residence time of the peaks in the flow cell at standard LC flow rates. However, the great advantage of this mode is that all peaks eluting from the column are detected in a completely nontargeted way. If the chromatography is developed at slow flow rates (10–100 µL min⁻¹), the residence time in the flow cell and, therefore, the signal-to-noise ratio of the NMR measurements can be increased drastically. It is important to note that on-flow measurements should entail an extremely shallow solvent gradient or even run under isocratic conditions to avoid line broadening in the pseudo-two-dimensional (2D) NMR spectrum where individual FIDs are combined in a serial file and Fourier transformed only in the spectroscopic dimension. This results in a contour plot where one axis shows the NMR spectrum and the other dimension shows the retention time.

The maximum gradient slope depends on the type of probe head used. Figure 3.2 shows the pseudo-2D NMR spectrum after injection of 5 µg for each compound of a test mixture on column under isocratic conditions (32 scans per NMR spectrum present in the serial file were accumulated). On the top, the projection shows the spectrum of the last eluting

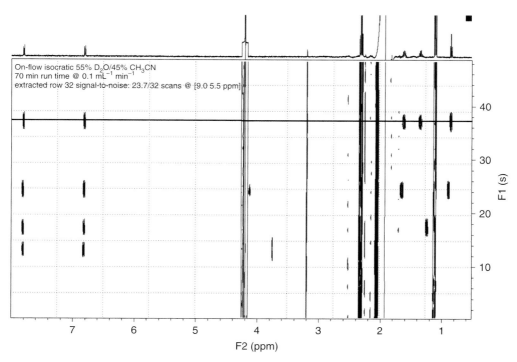

On-flow isocratic 55% D_2O/45% CH_3CN
70 min run time @ 0.1 mL⁻¹ min⁻¹
extracted row 32 signal-to-noise: 23.7/32 scans @ [9.0 5.5 ppm]

Figure 3.2. On-flow chromatogram after injecting 5 µg (each compound) of a mixture of *para*-hydroxybezoic acid esters on an analytical HPLC column under isocratic conditions. The flow rate was 0.1 mL min⁻¹; mobile phase: 45% acetonitrile and 55% deuterium oxide. The projection on top was extracted from row 32 of the on-flow chromatogram. The signal-to-noise ratio after 32 scans was 23.7 : 1.

compound. Along with the resonances originating from the analyte, solvent signals can be observed. Acetonitrile at $\delta_H = 2$ ppm and the residual water signal at $\delta_H = 4.2$ ppm are suppressed using the WET (water suppression enhanced through T_1 effects) sequence. Two strong signals around $\delta_H = 1.2$ and 2.2 ppm originate from propionitrile, a known impurity present in acetonitrile. The small signal at $\delta_H = 3.2$ ppm is from methanol. The total run time of the on-flow run is 70 min. The signal-to-noise ratio of the last eluting ester is 23.7 : 1.

Stop-flow measurements can be used either as a direct approach or after intermediate storage of chromatographic peaks in loops followed by a transfer into the NMR flow cell. Both types of stop-flow operation modes are designed to be used in a targeted way and with the use of another predetector (either UV or MS) for the selection of chromatographic peaks that will also be analyzed using NMR. For example, the time-slice technique in the LC-NMR can be used to obtain a complete overview of the protonated constituents in an environmental extract. In this mode,

the LC pump stops every 20–30 s at a flow rate of 0.5 mL min^{-1}. This corresponds to fractions of 167–250 µL for each stop of the pump, which roughly matches the active volume of the flow cell (120 µL). The advantage of this mode over the on-flow mode is that for each stop, the parameters for NMR detection can be perfectly adjusted. This adjustment includes tuning and matching, shimming and pulse calibration, and optimization. In addition, the number of scans can be increased and also, therefore, the overall sensitivity of the NMR measurement. After acquisition, the individual one-dimensional (1D) spectra can be combined in a serial file and processed like an on-flow experiment as shown in Figure 3.3. The comparison to Figure 3.2 reveals an increase in signal-to-noise ratio of the last eluting compound from 23.7 : 1 to 46.9 : 1. This can be explained by the increase in the number of scans from 32 to 128 per increment, as this factor of 4 is equivalent to a factor of 2 in increase of signal-to-noise ratio, which is in good agreement with the observed results. From Figure 3.3, it can clearly be seen that a steep gradient elution was applied to

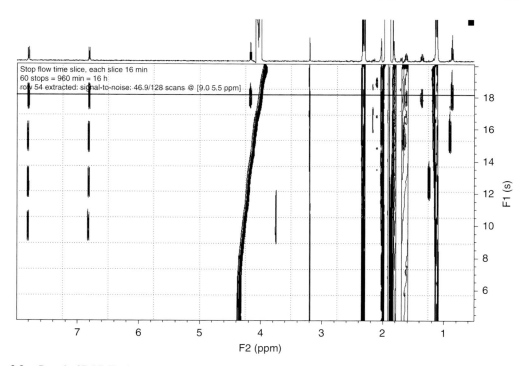

Figure 3.3.   Pseudo-2D NMR chromatogram after a stop-flow time-slice experiment combining the spectra of 60 stops. Each spectrum was obtained after 128 scans. The flow was stopped every 15 s during the gradient elution run at 0.5 mL min^{-1}. The projection on top was obtained from the last peak eluting at the horizontal line.

achieve a fast separation. The shift of the water signal from $\delta_H = 4.4$ to 4 ppm also highlights this change. The shift is caused by the change of solvent composition in the flowcell because of the gradient elution. With high water content, the water signal is close to 4.7 ppm, whereas the shift changes toward acetonitrile when the water content decreases. The resonances of the analytes are not affected by this shift because at each stop, the sample is individually shimmed to achieve optimum line shape. The total time required for the complete experiment was 960 min, which is an increase of a factor of 14 when compared to the on-flow experiment.

3.3.2 Introduction of Mass Spectrometry

For LC-NMR applications, MS can be coupled using a splitter and doing so provides additional information about analytes that can be used for structural elucidation such as accurate mass and MS fragmentation patterns. The splitter allows only 2–5% of the total flow to the MS ion source, such that the majority of the flow still goes into the NMR flow cell. In order to get UV and MS traces aligned in the final chromatogram, the split needs to be installed before the flow enters the UV detector. The length of the capillary from the splitter to the UV detector needs to be adjusted to result in coincident chromatographic traces for direct comparison. The direct coupling of MS to the LC-NMR set up is only possible in on-flow applications, as the MS acquisition cannot be halted.

It is important to note that if deuterium oxide is used instead of water, acidic or basic protons, e.g., from phenols, amines, and amides, will be exchanged for deuterium, which leads to a higher mass than expected. This information can be used to gain additional structural information about exchangeable proton sites within a molecule. However, when unknown compounds are being investigated, the determination of the molecular formula is complicated, as deuterium needs to be considered along with hydrogen and other potential nuclei. To circumvent this complication, the mass spectrometric information can be obtained by running a separate LC-MS analysis of the extract studied in conjunction with the LC-NMR hyphenated approach. In this case, the chromatography runs under fully protonated conditions, which allow the determination of the correct accurate mass. This approach, however, can make the correlation between the MS response and the NMR spectrum more challenging, as the retention behavior of compounds might differ slightly

from the separation obtained with deuterium oxide. As such, the hyphenation of LC-NMR with MS ensures consistency between all analyses because they are obtained under the same chromatographic conditions.

3.4 LIQUID CHROMATOGRAPHY SOLID-PHASE EXTRACTION NUCLEAR MAGNETIC RESONANCE

Recent developments in LC-NMR hyphenation have led to the availability of an automated postcolumn SPE interface, where chromatographic peaks are diluted with an aqueous phase and trapped on SPE cartridges. After analyte trapping and enrichment, the cartridges are dried with nitrogen gas and eluted to the NMR with deuterated organic solvents. The peaks can be eluted either into a flow cell or into NMR tubes. The advantage of this technique over direct LC-NMR is the fact that peaks of several chromatographic runs can be collected on the same cartridge for multiple peak trapping. This is particularly advantageous when analytes are present in ultralow concentrations. The maximum number of multiple trappings is limited by the breakthrough of the analytes.[14]

If the number of cartridges available for trapping is high enough, fixed time intervals of the chromatogram can be stored on individual cartridges in a completely nontargeted time-slice mode. Typically, 35 s to 1 min portions of the chromatogram are trapped on SPE cartridges starting after the sample loading period. The MS acquires spectra in parallel to ensure a correlation of MS and NMR data. Figure 3.4 shows the results of such a time-slice LC-SPE-NMR run after injection of 5 µg for each component of the test mixture, which also had been used for the time-slice LC-NMR run shown in Figure 3.3. Because the elution of each cartridge takes place with acetonitrile-d_3, no elution profile from the solvent gradient is visible. The residual solvent signals from acetonitrile and water at $\delta_H = 2$ ppm are suppressed and not visible anymore. In addition, propionitrile is removed from the cartridge by the washing and drying steps after the trapping process. The signal-to-noise ratio for the latest eluting peak is 79.3 : 1 after 128 scans, which is an increase of 69% when compared to the LC-NMR time-slice experiment.

This increase can be explained by the fact that the elution band from the cartridge has a volume of only 20–30 µL. This band is transferred into an empty flow cell with an active volume of 30 µL. The cell is

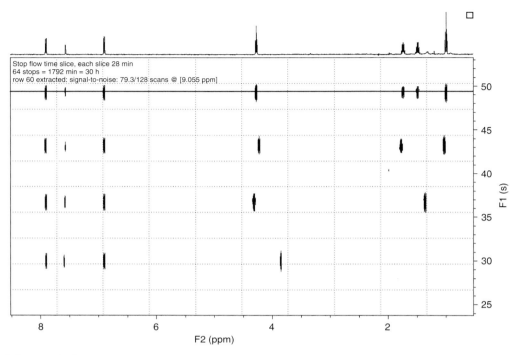

Figure 3.4.   Pseudo-2D NMR chromatogram of an SPE-time-slice experiment using a 5 µL injection of the test mixture. For each transfer, 128 scans were accumulated. The maximum signal-to-noise ratio for the latest eluting peak was 79.3 : 1.

empty because between each elution of consecutive cartridges, the flow cell is emptied with nitrogen gas. The time needed for the LC-SPE-NMR time-slice experiment is about twice that needed for the LC-NMR time-slice experiment (Figure 3.3). It is worth mentioning that the signals at $\delta_H = 4.4$ ppm are not obscured by the water signal and are visible for all test compounds. In addition, the phenolic –OH signal at $\delta_H = 7.6$ ppm is not exchanged with deuterium and is also visible.

In LC-SPE-NMR mode, the NMR spectra are acquired in well-defined deuterated solvents. Typically, methanol-d_4 or acetonitrile-d_3 is used for the elution process from SPE cartridges. If the same solvent is used, the chemical shifts are identical to those of structures published in the literature. On the other hand, if a tentative identification based on the molecular formula has already been obtained, the acquisition of a ^{1}H-NMR spectrum of the reference compound dissolved in the same deuterated solvent helps determine that the structure is present in the

environmental extract. This allows a fast identification of already known compounds.[15]

3.4.1   Chemistry of SPE Phases

For postcolumn trapping applications, a huge number of different SPE phases are available. In reverse phase separations, the elution strength is reduced by postcolumn addition of an aqueous buffer. Classical silica-based C8 or C18 chemistries are the phase of choice for the predominantly nonpolar analytes eluting with high organic content in the mobile phase from the separation column. For medium to polar analytes, crosslinked polystyrene–divinylbenzene copolymers are used, very similar to the off-line SPE material employed for sample concentration and enrichment. This material is the best choice for nontargeted applications, as the same SPE chemistry is involved for both steps resulting in comparable recovery values. Besides these classical SPE materials, ion-exchange cartridges can be used for postcolumn enrichment of analytes carrying a charge.[16] This approach, however, needs a special sample preparation technique, which divides the environmental sample into subfractions as described in Section 3.2.1. For very polar nonionizable analytes, no postcolumn SPE method has been described for environmental applications.

3.4.2 Introduction of Mass Spectrometry

The technical realization of the hyphenation of an MS to an LC-SPE-NMR system is the same as described in Section 3.3.2. In this case, however, if a high-resolution mass spectrometer is used, the mass response obtained leads directly to the generation of the molecular formula, as deuterium exchange is no longer an issue because deuterated solvents are only used to elute samples from the SPE to the NMR. The mass spectrometer can guide the trapping process based on the mass response of an unknown contaminant. The response factor of an organic xenobiotic is not universal; however, it is not clear from the MS detection alone that if there is enough material present to obtain an NMR spectrum. In addition, compounds with a low MS response factor can easily be overlooked by NMR. As such, the MS can be invaluable for assisting with LC-SPE-NMR method development.

3.5 APPLICATIONS OF LC-NMR TO ENVIRONMENTAL SAMPLES

Applications reported in the literature so far are obtained solely from point sources where the concentration is high enough for NMR detection.[9,17–25] In all cases, sample enrichment before NMR detection was mandatory. Both direct LC-NMR coupling and LC-SPE-NMR were employed. If available, mass spectrometric information was either obtained from a separate LC-MS run or acquired during parallel detection using a splitter.

3.5.1 Former Ammunition Sites

The first application of chromatography directly coupled to NMR detection for the investigation of an environmental extract was the on-flow LC-NMR measurements of groundwater extracts taken from a former ammunition production site in Elsnig, Sachsen, Germany. In this kind of point source contamination, waste from the trinitrotoluene (TNT) production during World War II was directly deposited into the environment. Over decades, these contaminants were further modified by photolytic or microbiological transformation to polar, water-soluble transformation products present in different layers of the groundwater. After SPE of the contaminants, the sample was injected on a short HPLC column at very low flow rates (<0.017 mL min^{-1}) using isocratic conditions. This allowed the acquisition of 128 scans for each row of the on-flow chromatogram with detection limits below 5 µg of compound injected on column. Along with previously identified contaminants, new transformation products could be detected as shown in Figure 3.5. Owing to the lack of mass spectrometric information, however, it was not possible to elucidate the structure of these unknowns.

Before the direct coupling of a mass spectrometer to an existing LC-NMR system was realized, a separate LC-MS measurement was acquired to correlate the mass response of an unknown contaminant to NMR spectra obtained from an on-flow chromatogram. This approach was used in investigations of photolytic degradation products of TNT, a hexogen-contaminated waste site, a TNT-contaminated waste site, and a mononitrotoluene-contaminated waste site. For transformation products with very low concentrations and for completely unknown transformation products, HPLC peaks were fractionated several times, evaporated, and reconstituted in deuterated solvent for 1D and 2D NMR measurements. This approach revealed the presence of several previously unidentified transformation products in the extracts for which no reference compounds were available.

3.5.2 Groundwater Downgradient of Municipal Solid Waste (MSW) Landfill

The same approach as described for the TNT-contaminated waste site in Section 3.5.1 was applied to the nontargeted analysis of groundwater extracts downgradient of an old landfill. Owing to the complexity of the sample, two extraction steps were applied: liquid–liquid extraction with methylene chloride followed by SPE on LiChrolut EN cartridges. The samples were subjected to gas chromatography (GC)-MS, LC-MS, and LC-NMR in the on-flow

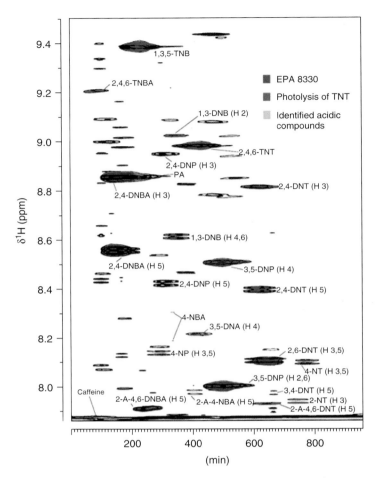

Figure 3.5. On-flow chromatogram of the extract of the upper groundwater layer of the former ammunition site at Elsnig. The color coding identifies compounds, which are part of the official EPA (Environmental Protection Agency) method 8330 (green), formed during the photolysis of trinitrotoluene (TNT, red) and other acidic contaminants. TNBA, trinitrobenzoic acid; DNBA, dinitrobenzoic acid; NBA, nitrobenzoic acid; DNP, dinitrophenol; NP, nitrophenol; TNB, trinitrobenzene; DNB, dinitrobenzene; DNT, dinitrotoluene; NT, nitrotoluene; DNA, dinitroaniline; A-DNT, amino-dinitrotoluene; and PA, picric acid.

mode for identification of potential targets for further off-line analysis by cutting the contaminants from an analytical HPLC run followed by characterization using tube-NMR and LC-MS. In the extracts investigated, a variety of polar XOCs could be identified and their concentration was estimated in a semi-quantitative approach. It needs to be mentioned that this approach is very time-consuming because of the multiple fractionation of individual HPLC peaks, subsequent evaporation, and reconstitution before analysis.

3.5.3 Effluent of a Textile Company

Another example for the analytical approach described earlier is the analysis of the effluent from a textile company. Again, the extract was measured by LC-MS and LC-NMR separately. Compounds identified were anthraquinone-type dyes and their byproducts, a fluorescent brightener, a byproduct from polyester production, and anionic and nonionic surfactants. This study clearly demonstrated how important the combination of MS and NMR data is, as it provides complementary structural information.

3.5.4 Tar Oil-Contaminated Waste Sites

Tar oil-contaminated sites are known to release *N*-heterocyclic compounds, especially, quinolines and methylquinolines, into the groundwater.[2] They undergo biotransformation to yield quinolones as well as their methylated, hydroxylated, and hydrated isomers. The study reported by Reineke *et al.*[24] used LC-NMR in the on-flow mode to identify different isomers in groundwater extracts of tar oil-contaminated waste sites. Before identification in environmental samples, degradation studies using anoxic redox conditions in microcosms were also conducted. LC-MS and ^{1}H-NMR were used to elucidate the structures of the unknown isomers during the biodegradation studies.

3.5.5 Ammunition Destruction Site

In contrast to TNT production sites, where waste was released into the environment, ammunition storage and destruction sites contain additional contaminants. The investigation of a groundwater extract from a point source at a Swiss destruction site located in the vicinity of the Sustenpass revealed the presence of toxic phenylurea derivatives originating from additives in the propellant charges of the ammunition and also residues of explosives, their transformation products, and other chemicals. The LC-SPE-NMR/MS coupling was used here for the first time to identify previously unknown contaminants present in an extract obtained from an environmental sample. The peak trapping was directed by MS and UV responses after the entire extract had first been measured by tube-NMR. This tube-NMR measurement had already identified the presence of explosives and related contaminants within the extract. In addition, resonances originating from highly concentrated unknown contaminants were detected as shown in Figure 3.6.

After chromatographic separation, two late eluting peaks were present showing UV response but no MS response in negative ionization mode. In positive ionization mode, however, the molecular formulae were calculated indicating nitrogen-containing contaminants (Figure 3.7). The low number of nitrogen and oxygen nuclei present in the unknown molecules, however, suggested that these compounds were nonexplosive.

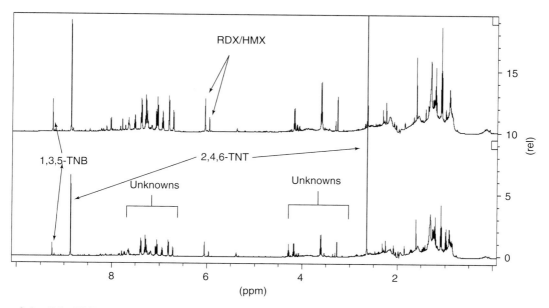

Figure 3.6. Tube-NMR measurement of groundwater extracts obtained from a former ammunition destruction site. The upper spectrum was taken in July 2008 and the lower spectrum is from October 2008.

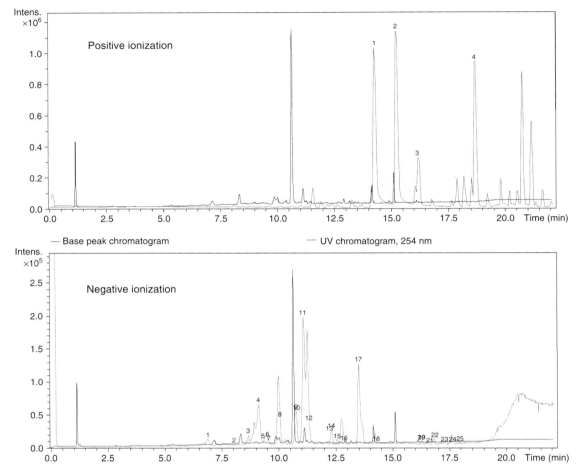

Figure 3.7. HPLC-UV-MS chromatograms of the groundwater sample taken in July 2008.

A PubChem library search for the molecular formulae lists 3584 for the first and 6801 possible isomers for the later eluting peak (http://pubchem.ncbi.nlm.nih.gov/search/search.cgi). After trapping the late eluting peaks postcolumn on SPE cartridges, drying, and elution with CD_3OD into a 600 MHz cryoprobe equipped with a flow insert, 1D-^{1}H-NMR spectra of the isolated contaminants could be obtained as displayed in Figure 3.8.

Compared to the results proposed by the PubChem search, the lower spectrum is in agreement to the hit number 17 (*N*,*N*-diphenylurethane) and the upper spectrum corresponds to hit number 1 (*N*,*N'*-diethylcarbanilide or centralite I). Comparison

to the ^{1}H-NMR spectra of the pure reference compounds allowed the unequivocal identification and quantification of the two contaminants, which are structurally related to phenylurea herbicides exhibiting the same toxicological potential. The nontargeted approach used in this study is based on the response obtained from the UV and MS detectors. Compounds that do not contain a chromophor and that have low ionization efficiency will not be covered and submitted to the NMR acquisition. The risk of overlooking these types of contaminants was limited in this study because the entire extract was already measured by tube-NMR beforehand. Furthermore, direct comparison of the NMR spectra obtained after separation and

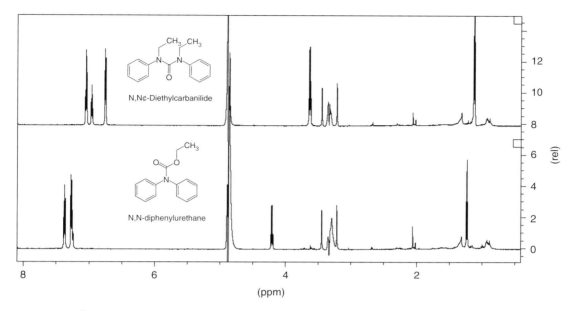

Figure 3.8. 1D-^{1}H-NMR spectra of the two late eluting peaks labeled with 1 (lower spectrum) and 2 (upper spectrum) in the upper chromatogram of Figure 3.7.

postcolumn trapping with the spectrum of the extract measured in the same deuterated solvent allows the estimation of the coverage of all contaminants present in the sample found by the LC-SPE-NMR approach.

3.5.6 Effluents of Wastewater Treatment Plants (WWTPs)

A nontargeted LC-SPE-NMR/MS approach based only on NMR detection is possible in the time-slice mode as discussed in Section 3.4. This method has been applied to the investigation of effluents from WWTPs. Effluent samples from WWTPs at Ins and Lyss, Switzerland, located in an area rich in intensive agriculture, were obtained during the main application period of pesticides to farms. After SPE, the samples were reconstituted in 200 μL methanol-d_3 and measured in a 3 mm NMR tube to obtain the overview spectrum shown in Figure 3.9. Apart from the well-resolved resonances of benzoic acid, no individual signals were detected (Figure 3.9).

After injection of the sample onto a classical 250 × 4.6 mm analytical HPLC column with parallel time of flight (TOF)-MS detection, a huge number of individual MS traces can be extracted. In Figure 3.10, only the most abundant mass peaks above 20 000 ion counts are displayed. As the response factors for the unknown compounds are not identical, the peak intensities do not reflect the real concentration distribution.

The exact mass information for the contaminants eluting, however, is indispensible for later identification of the compound. The parallel detection allows the correlation of MS to the NMR response of unknowns. For each fraction trapped on the SPE cartridge postcolumn, a set of mass-detected peaks corresponds to the NMR resonance lines present in the SPE extract. Owing to the coelution of several peaks during the minute of the chromatography trapped on one cartridge, this correlation sometimes becomes cumbersome.

Figure 3.11 shows the pseudo-2D NMR chromatogram of the WWTP sample taken in Lyss. Although the concentration of the analytes is quite low, many signals can be detected illustrating the complexity of the sample. It is worth mentioning

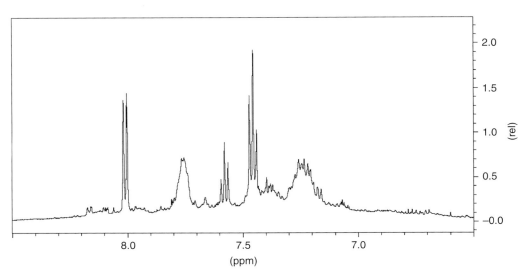

Figure 3.9. 1D-^{1}H-NMR spectrum (downfield part) of the extract obtained from the WWTP in Lyss after accumulation of 32 scans.

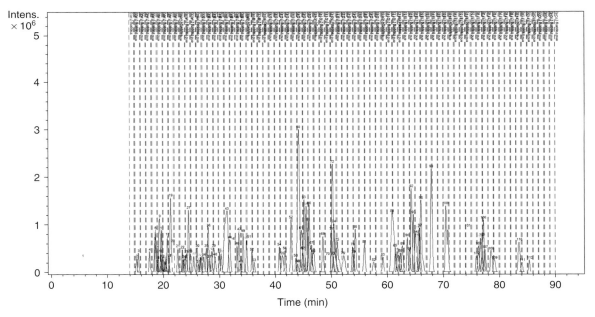

Figure 3.10. HPLC-SPE-NMR/MS time-slice chromatogram of the WWTP extract from Lyss after injection of $4 \times 12.5\ \mu$L of the sample. The dotted lines indicate 1 min portions trapped on SPE cartridges. Individual MS traces are extracted using the DISSECT algorithm.

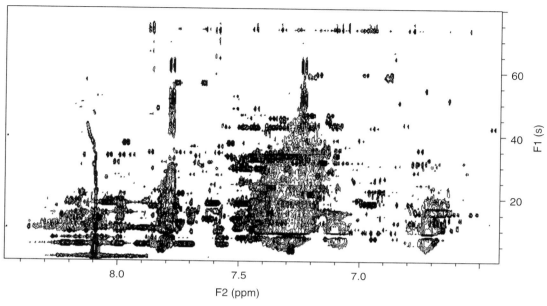

Figure 3.11. LC-SPE-NMR/MS time-slice pseudo-2D-NMR chromatogram of the aromatic region of the WWTP extract from Lyss.

that most compounds elute at the beginning of the chromatogram. This is in agreement with the fact that biological, chemical, or photochemical transformations in the environment usually yield more polar compounds.[26]

As this approach provides a huge number of individual NMR spectra, especially when several samples are investigated, manual inspection of each time-slice spectrum would mean a risk of overlooking contaminants. Therefore, the data can be processed using the nonsupervised multivariate statistical principal components analysis (PCA). Here, the loading plot for a given pair of principal components (PCs) directly points to NMR resonances present in the corresponding SPE slice as shown in Figure 3.12. In case of large intensities, i.e., resonances, originating from highly concentrated contaminants, the representing loading will be far away from the origin of the plot of a given PC pair. Starting with the PCs explaining the highest degree of variance, e.g., PC1 vs PC2, the corresponding loadings belong to the highest concentrated contaminants. By selecting PCs explaining less variance, contaminants are supposed to be less concentrated.

The correlation of the NMR resonances obtained during these processing steps to the corresponding MS data is exemplified in Figure 3.13. Figure 3.13(a) and (b) displays extracted ion traces of the extract from the WWTP in Ins in negative (a) and positive (b) ionization modes in the time slice between 45 and 46 min. The first peak eluting in this time window ionizes in positive (m/z 272, mass spectrum shown in Figure 3.13(d)) and negative modes (m/z 270, mass spectrum shown in Figure 3.13(c)) with a neutral loss of two CO_2. The second peak at 45.45 min only shows a response in positive mode with m/z 207.15 (mass spectrum shown in Figure 3.13(e)).

A PubChem search for the molecular formulas associated with these pseudo-molecular ions lists 6146 compounds for m/z 207 ($C_{12}H_{19}N_2O$) and 2038 compounds for m/z 272 ($C_{15}H_{14}NO_4$). Comparison of the proton spectrum shown in Figure 3.13(g) to the possible compounds results in a tentative identification of hit number 9 for m/z 207 (Isoproturon, NMR spectrum of the reference compound is shown in Figure 3.13(f)) and hit number 61 for m/z 272 (3-carboxymefenamic acid). While Isoproturon was readily identified by comparison with a reference standard as shown in

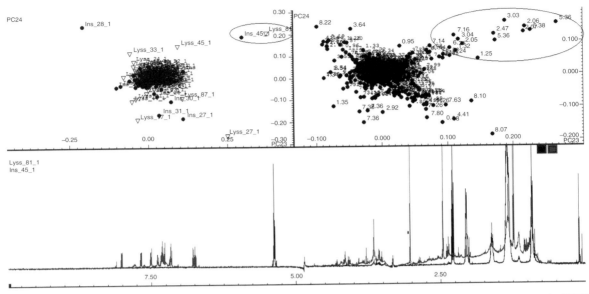

Figure 3.12. Loadings and scores of PC23 vs PC24 of the PCA of the time-slice spectra obtained for the samples of WWTP at Ins and Lyss. The spectra below belong to the NMR slice 45 of the Ins sample and 81 of the Lyss sample.

Figure 3.13(f), 3-carboxymefenamic acid was identified by detailed spin system analysis displayed in Figure 3.13(h), as no reference compound was commercially available.

3.6 CONCLUSIONS

Even though NMR is an inherently insensitive method for ultralow levels of XOCs, modern instrumentation and appropriate sample preparation enable the detection of environmentally significant concentrations present in aqueous samples. For very complex extracts, or in the case of completely unknown contaminants, the coupling of chromatography to NMR and MS spectrometers allows isolation and structural elucidation using high accurate mass and 2D-NMR spectrum acquisition of the unknowns.

For the hyphenation of chromatography to NMR, two techniques are used: direct coupling and post-column SPE. For purely NMR-driven nontargeted analysis of environmental extracts, the hyphenation needs to be run in the on-flow or time-slice mode where everything eluting from the column is subjected to NMR detection. While direct LC-NMR coupling only shows some advantages for very polar analytes,

the method of choice is LC-time-slice-SPE-NMR/MS. This method is superior, as it provides a higher sensitivity when compared to direct coupling, and the analytes are measured in well-defined deuterated organic solvents. Besides this, the mass spectra obtained from high-resolution MS allow the generation of the molecular formula of unknown contaminants. As the hyphenated NMR techniques show low sensitivity and are quite time-consuming, their main field of application is in the examination of point sources where the concentrations are high enough for NMR detection. However, future advancements in probe technology and related NMR hardware will likely enable the expansion of LC-NMR technology to other environmental applications. In addition, in cases where analytes of interest can be concentrated, NMR analysis enables the structural elucidation of novel compounds and their transformation products. As such, LC-NMR will likely be a key tool for future studies involving the fate of XOCs in the environment.

ACKNOWLEDGMENT

David Kilgour is kindly acknowledged for proofreading the manuscript.

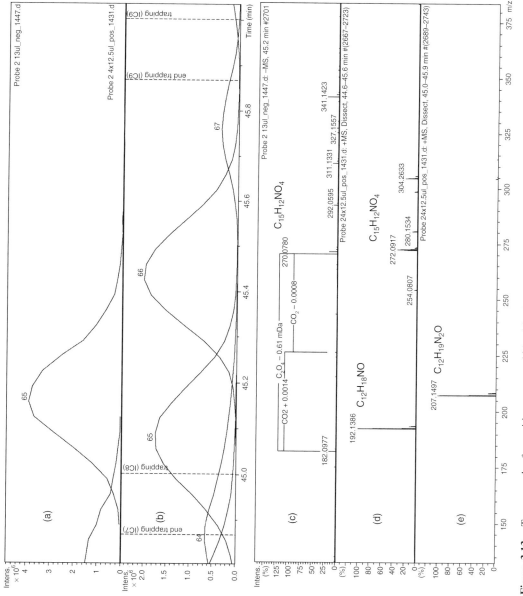

Figure 3.13. Two examples for positive compound identification: (a) isoproturon and (b) 3-carboxymefenamic acid identified based on MS, MS/MS (c–e), and NMR responses (f–h). (Reprinted with permission from Ref. 25. © 2011 Elsevier).

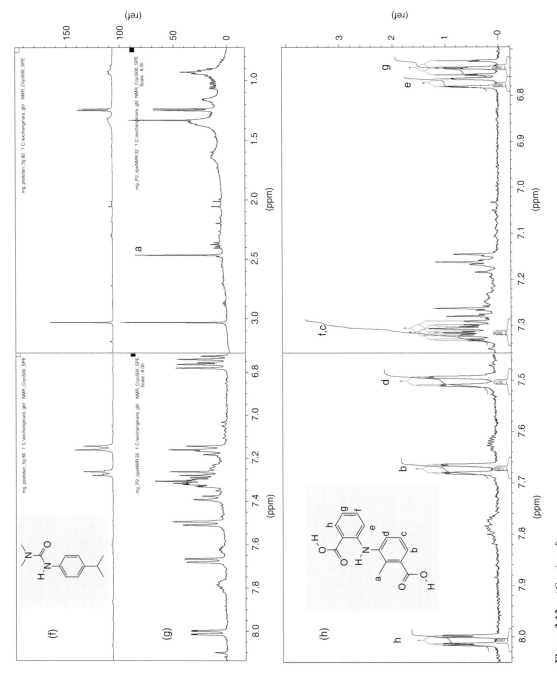

Figure 3.13. *(Continued)*

RELATED ARTICLES IN EMAGRES

Chemometric Exploration of Quantitative NMR Data

Quantitative NMR for Solutions

REFERENCES

1. D. J. Lapworth, N. Baran, M. E. Stuart, and R. S. Ward, *Environ. Pollut.*, 2012, **163**, 287.

2. L. G. Freitas, C. W. Götz, M. Ruff, H. P. Singer, and S. J. Müller, *J. Chromatogr. A*, 2004, **1028**, 277.

3. M. Krauss, H. Singer, and J. Hollender, *Anal. Bioanal. Chem.*, 2010, **397**, 943.

4. S. Sturm and C. Seger, *J. Chromatogr. A*, 2012, **1259**, 50.

5. Z. Yang and Z. J. Pharm, *Biomed. Anal.*, 2006, **40**, 516.

6. O. Corcoran and M. Spraul, *Drug Discov. Today*, 2003, **8**, 624.

7. J. W. Jaroszewski, *Planta Med.*, 2005, **71**, 691.

8. M. Godejohann, *J. Chromatogr. A*, 2007, **1156**, 87.

9. M. Godejohann, L. Heintz, C. Daolio, J.-D. Bersert, and D. Muff, *Environ. Sci. Technol.*, 2009, **43**, 7055.

10. M. Spraul, E. Humpfer, H. Schäfer, B. Schütz, M. Mörtter, and P. Rinke, in NMR Spectroscopy in Pharmaceutical Analysis, eds U. Holzgrabe, I. Wawer and B. Diehl, Elsevier: Amsterdam, 2008, Chapter 1, pp. 317, 319–339.

11. D. J. Crockford, E. Holmes, J. C. Lindon, R. S. Plumb, S. Zirah, S. J. Bruce, P. Rainville, C. L. Stumpf, and J. K. Nicholson, *Anal. Chem.*, 2006, **78**, 363.

12. D. Bratkowska, N. Fontanals, F. Borrull, P. A. G. Cormack, D. C. Sherrington, and R. M. Marcé, *J. Chromatogr. A*, 2010, **1217**, 3238.

13. K. Albert, On-Line LC-NMR and Related Techniques, John Wiley & Sons Ltd: Chichester, 2002.

14. J. W. Jaroszewski, *Planta Med.*, 2005, **71**, 795.

15. C. Seger, M. Godejohann, L.-H. Tseng, M. Spraul, A. Girtler, S. Sturm, and H. Stuppner, *Anal. Chem.*, 2005, **77**, 878.

16. K. T. Johansen, S. J. Ebild, S. B. Christensen, M. Godejohann, and J. W. Jaroszewski, *J. Chromatogr. A*, 2012, **1270**, 171.

17. M. Godejohann, A. Preiss, C. Mügge, and G. Wünsch, *Anal. Chem.*, 1997, **69**, 3832.

18. M. Godejohann, M. Astratov, A. Preiss, K. Levsen, and C. Mügge, *Anal. Chem.*, 1998, **70**, 4104.

19. A. Preiss, M. Elend, S. Gerling, and S. Träckner, *Magn. Reson. Chem.*, 2005, **43**, 736.

20. A. Preiss, M. Elend, S. Gerling, E. Berger-Preiss, and K. Steinbach, *Anal. Bioanal. Chem.*, 2007, **389**, 1979.

21. A. Preiss, E. Berger-Preiss, M. Elend, J. Hollender, and A.-K. Reineke, *Chemosphere*, 2011, **84**, 1650.

22. A. Preiss, E. Berger-Preiss, M. Elend, S. Gerling, S. Kühn, and S. Schuchardt, *Anal. Bioanal. Chem.*, 2012, **403**, 2553.

23. A. Preiss, U. Sänger, N. Karfich, and K. Levsen, *Anal. Chem.*, 2000, **72**, 992.

24. A.-K. Reineke, A. Preiss, M. Elend, and J. Hollender, *Chemosphere*, 2008, **70**, 2118.

25. M. Godejohann, J.-D. Berset, and D. Muff, *J. Chromatogr. A*, 2011, **1218**, 9202.

26. A. B. A. Boxall, Transformation Products of Synthetic Chemicals in the Environment, Springer-Verlag: Berlin Heidelberg, 2009.

Chapter 4

Environmental NMR: Solid-State Methods

Caroline M. Preston

Natural Resources Canada, Pacific Forestry Centre, Victoria, British Columbia, V8Z 1M5, Canada

4.1 INTRODUCTION

4.1.1 Principles of NMR

The basis of NMR spectroscopy is that only certain atomic nuclei have a nuclear spin I, and thus a magnetic moment μ, which interacts with a static magnetic field. These nuclei include ^1H, ^{13}C, ^{15}N, and ^{31}P (Table 4.1[1]). In an external magnetic field B_0, which is represented as a vector in the z direction, such nuclei split into $2I + 1$ energy levels, with the splitting proportional to the field strength B_0 and γ, a constant

NMR Spectroscopy: A Versatile Tool for Environmental Research
Edited by Myrna J. Simpson and André J. Simpson

for each nucleus. Nuclei with $I = {}^1/_2$, such as ^1H and ^{13}C, therefore, split into two populations. The very slight surplus of nuclei in the lower energy level, as determined by the Boltzmann equation, results at equilibrium in a macroscopic sample magnetization M_0 (Figure 4.1a). In fact, each nuclear spin is not statically aligned, but precessing around B_0 at the Larmor frequency ω, which corresponds to the energy gap ΔE:

$$\omega_0 = \gamma B_0 \qquad (4.1)$$

Rather than the laboratory (static) frame of reference, Figure 4.1 uses a reference frame rotating at the Larmor frequency, which simplifies the picture. In NMR, transitions between the energy levels are induced by a very brief pulse of a second oscillating magnetic field B_1, oriented perpendicular to B_0, which rotates the magnetization M_0 into the direction of the y-axis (Figure 4.1b). The angle of rotation is determined by the product of the strength and the duration of B_1; a 90° pulse rotates M_0 into the y-axis, a 180° to the z direction, and so on. The magnetization continues to precess around the y-axis, and the resulting periodic oscillation of the y-component of M_0 is picked up as an electrical signal induced in the coil surrounding the sample.

When a collection of nuclei is placed into the static magnetic field, the z-magnetization increases to M_0 exponentially according to the spin–lattice relaxation time (T_1). (T_1 and other relaxation times that we shall meet is the inverse of a rate constant.) Conversely, after a disturbance, such as a 90° pulse, the spins return to thermal equilibrium, as the excess spin energy is returned to the lattice at a rate similarly governed by

Table 4.1. Properties of NMR nuclei used in solid-state environmental studies[1]

Nucleus	Spin	Abundance %	Frequency MHz @ 9.4 T	Sensitivity @ Nat. abun.	Chemical shift Reference
^{1}H	1/2	99.99	400.0	1.000	Me_4Si
^{13}C	1/2	1.07	100.6	1.7×10^{-4}	Me_4Si
^{15}N	−1/2	0.368	40.5	3.8×10^{-6}	$MeNO_2$
^{31}P	1/2	100	161.9	6.6×10^{-2}	85% H_3PO_4

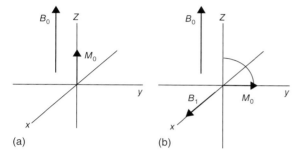

Figure 4.1. (a) Equilibrium magnetization M_0 in the rotating reference frame and (b) effect of a 90° pulse.

T_1. At the same time, energy is redistributed among the spins, and the signal in the $x–y$ plane is lost exponentially at the rate determined by T_2, the spin–spin relaxation time.

NMR is different from many other forms of spectroscopy. First, only nuclei with a magnetic moment are able to participate; for example, we may look at ^{13}C, with 1.1% natural abundance, but not the abundant isotope ^{12}C. Table 4.1 lists some nuclei of interest for organic matter (OM) studies; this chapter will focus on ^{13}C, with a brief introduction to ^{15}N and ^{31}P NMR. Second, the energies involved in NMR are very low compared to infrared, or visible and ultra-visible light. This means that the equilibrium population differences are extremely small, so that sensitivity is inherently limited. Third, the processes of spin energy transfer are much slower, which limits the rate of signal buildup and thus the repetition rate of NMR experiments. In higher energy spectroscopies, after a perturbation to put energy into the system, relaxation mainly occurs by re-emission of photons corresponding to the relevant energy gaps. The probability of this process basically increases as the third power of the energy (frequency)[2] and is completely inconsequential for NMR. To reach M_0, energy is transferred to (or acquired from) the lattice by several processes based largely on motional fluctuations corresponding to the

frequency of the spin energy gaps. On the other hand, these slow processes allow much creative manipulation of magnetization in different types of experiments. Descriptions of basic NMR theory are widely available and basic summaries are found in several of the references cited herein.[3–5]

4.1.2 NMR Chemical Shifts

While the primary Larmor frequency of a nucleus is determined by its nuclear γ and the magnetic field strength, the local fields of nearby electrons act in opposition to shield the nuclei from B_0. This means that the signals are spread out across a range of chemical shifts (δ values), which generally increase with increasing electronegativity of carbon substituents and bond unsaturation, and are expressed in ppm from a reference frequency. For ^{1}H and ^{13}C, this is usually tetramethylsilane with δ of zero. The ppm scale is used as it is independent of the magnetic field strength permitting direct comparison of data collected on instruments at different fields. For example, with a 9.4 T magnet (400 MHz for ^{1}H), the ^{13}C Larmor frequency is 100.6 MHz, and 1 ppm would be 100.6 Hz.

Characteristic chemical shift ranges for ^{13}C with assignments typical for OM are listed in Table 4.2.[6,7] These are a general guide, and detailed studies should be consulted for specific compounds, including condensed tannins.[8,9] Alkyl C occurs at 0–50 ppm, with often distinct signals for CH_3 at 15 ppm and for long-chain CH_2 at 30 and 33 ppm. The sources of this alkyl C include waxes and cutin, which form the outer coating of leaves and fruit; the more complex biopolymer suberin, found in roots and bark; C4 of condensed tannins; and the sidechains of proteins. The region for alkyl C with O and N substitutions stretches from 50 to 112 ppm and is often subdivided into three regions: methoxyl C with a sharp peak at 57 ppm, O-alkyl C from 60 to 90 ppm, mainly dominated by

Table 4.2. Main assignments for solid-state ^{13}C spectra referenced to tetramethylsilane at 0 ppm[6,7]

Chemical shift range (ppm)	Assignment
0–25	Methyl groups bound to C and CH_3 of acetate
25–45	Methylene groups in aliphatic rings and chains
45–60	Methoxyl groups, C6 of carbohydrates, and $C\alpha$ of most amino acids
60–90	C-2–C-6 of carbohydrates, $C\alpha$ of some amino acids, and higher alcohols
90–110	Anomeric C of carbohydrates, syringyl lignin C-2, C-6
110–140	Aromatic C–H, guaiacyl C-2, C-6, olefinic C
140–160	Aromatic COR or CNR groups
160–220	Carboxyl/carbonyl/amide groups

carbohydrate C with single oxygen substitution, and 90–112 ppm for di-*O*-alkyl C, typical of the anomeric C-1 of carbohydrates. However, the *O*-alkyl region also includes the 3-carbon sidechain of lignin, C2 and C3 of condensed tannins, and primary and secondary alcohol esters of cutin. Aromatic C with C and H substitutions falls mainly within 112–140 ppm and phenolic C (aromatic with O-substitution) from 140 to 160 ppm. The aromatic and phenolic regions may be reported as a combined aryl region. For soil OM, aryl carbons are mainly associated with lignins, condensed and hydrolyzable tannins, and black carbon produced by fire. Carboxyl/carbonyl C, including free carboxyls, amides, and esters, is at 160–185 ppm, with aldehyde and ketone signals found as far as 220 ppm. Proteins contribute to intensity in several regions: β, δ, and γ-CH_2 and CH_3 are found at 15–40 ppm; α-CH at 52–60 ppm, except glycine at 42 ppm; aromatic CH of phenylalanine and tyrosine at 130–155 ppm; O–C of tyrosine at 155 ppm; and C=O of amide linkages at 175 ppm.[9,7,10] With decomposition, signals from plant components decline and are increasingly replaced by those of the corresponding structures (lipids, carbohydrates, proteins, etc.) of microbial biomass and the 'molecularly uncharacterized components'.[11]

4.1.3 Development of Environmental Solid-State NMR

Early NMR spectrometers were based on electromagnets for field strengths up to 100 MHz for ^1H, and spectra were acquired through 'continuous-wave' processes as recounted in an interesting historical reflection.[12] The frequency of the oscillating magnetic field was varied slowly enough to avoid signal saturation, and the signals were detected when the magnetic field corresponded to an energy gap of the spin system; the whole process usually converted to movement of an ink-filled pen. Sometimes, cumbersome integrators summed the signals from multiple scans. Modern NMR is based on application of brief broadband pulses of radiofrequency energy covering the whole chemical shift range of interest. The decaying signal along the *y*-axis is called the free induction decay (FID), the sum of the time-domain signals from all frequencies. Fast Fourier transformation (FFT) from the time-to-frequency domain results in a spectrum in which, ideally, the intensity at each chemical shift is proportional to the number of corresponding nuclei.

For many decades, applications of NMR in organic chemistry were limited to soluble samples with regular structures, whereas attempts to obtain information from solids faced seemingly insurmountable obstacles.[13] First, in many molecules, the electron distribution around a nucleus is anisotropic, so that the chemical shift depends on the orientation of each molecule to B_0. In liquids, this chemical shift anisotropy (CSA) averages to zero because of molecular tumbling and a sharp line is observed (extreme-narrowing limit). In solids, however, CSA results in very broad asymmetric signals. Second, the interactions among nuclei result in splitting of energy levels and multiple lines. In molecules rotating freely in solution, these interactions similarly average to zero, and the resulting intramolecular (e.g., ^1H–^1H and ^{13}C–^1H) coupling patterns can be directly related to the molecular structure. However, in solids, where spins are held rigidly with respect to each other, dipolar interactions result in significant broadening. That is, one spin 'sees' each neighbor's spin as a perturbation of B_0, resulting in a different chemical shift. Finally, ^{13}C typically has long relaxation times for soil and geochemical samples, of the order of seconds to minutes, as well as low natural sensitivity, because of both low γ and low natural abundance (Table 4.1).

Some three decades later, it is hard to imagine the excitement generated by the initial demonstration of a new combination of techniques to generate chemically meaningful solid-state ^{13}C NMR spectra of complex organic samples,[14] and quickly applied to fossil fuels,[5] humic fractions,[15] peat,[16] and mineral soils.[17] Broadening due to dipolar coupling and CSA was overcome by high-power decoupling and magic-angle spinning (MAS), respectively, and sensitivity was enhanced by cross polarization (CP), which counteracts both the long relaxation times of ^{13}C and its inherently low sensitivity. Applications to OM developed rapidly,[6,18,19] and recent years have seen continuing increases in magnetic field strength, spinning speed, and the complexity of pulse sequences.

4.2 BASIC TECHNIQUES FOR ENVIRONMENTAL SOLID-STATE NMR

4.2.1 High-power Decoupling

Because of the low natural abundance of both ^{13}C and ^{15}N, homonuclear coupling is inconsequential compared to heteronuclear interactions with ^{1}H. The basis of heteronuclear decoupling is the application of another continuous B_2 field over the whole range of ^{1}H frequencies. This perturbs the ^{1}H spins by inducing fast transitions, so that the ^{13}C or ^{15}N nuclei 'see' only an average interaction. Broadband ^{1}H decoupling has long been used for solution ^{13}C NMR, but solids require much higher power because of the greater spread of ^{1}H chemical shifts, and care must be taken to avoid overheating or equipment damage by excessive acquisition times, or too-short relaxation delays. The pulse sequences shown later depict an invariant ^{1}H power level during decoupling, as was the case for many older instruments, including some still functioning. However, more complex decoupling schemes use reduced power while improving sensitivity and resolution,[20] the most widely used being two-pulse phase-modulated (TPPM) decoupling.[21]

4.2.2 Magic-angle Spinning (MAS)

MAS was independently reported in 1959 for removal of ^{23}Na homonuclear dipolar broadening in a crystal of NaCl[22] and of ^{19}F interactions in CaF_2 and

teflon.[23] The dipolar interactions are proportional to $3\cos^2\beta - 1$, where β is the angle between the static magnetic field B_0 and the internuclear vector. At $\beta = 54.7°$, the result is zero, so that spinning at the magic angle effectively cancels dipolar interactions, substituting for the motional averaging, which occurs in liquids. To be effective, the spinning rate has to be at least as high as the width of the broadening caused by the interactions, so that MAS does not in fact compensate for the much larger $^{1}H-^{13}C$ or $^{1}H-^{15}N$ dipolar interactions, which are mainly dealt with by high-power decoupling; however, in the same way, MAS effectively counteracts the broadening due to CSA. If the spinning rate is not high enough, however, the peak breaks up into a series of spinning side bands (SSBs) on each side of the central peak at multiples of the spinning speed. Thus, if the MAS rate is 5000 Hz, and the Larmor frequency for ^{13}C is 25 MHz, the SSB appear at intervals of 200 ppm and at low intensity compared to the central peak. If the Larmor frequency is 100 MHz, the SSBs are now both larger and only 50 ppm from the central peak, enough to cause serious interference.[24] The intensity of SSB is proportional to the CSA, so that they generally increase with increasing chemical shift; i.e., with increasing O and N substitutions and bond unsaturation, and thus are usually inconsequential for alkyl C with its more symmetric distribution of electron bond density.

4.2.3 Single-pulse Acquisition

Figure 4.2(a) shows the pulse program for a simple acquisition, usually referred to as direct polarization (DP) and also described as single-pulse excitation (SPE) or Bloch decay (BD). A 90° pulse (typically, <5 μs) rotates the ^{13}C magnetization into the y-axis, and immediately following, ^{1}H decoupling is applied along the y-axis during FID acquisition, after which the decoupler is turned off to allow the spins to return to the equilibrium. This duration is determined by the ^{13}C T_1, with full relaxation requiring five times the T_1 value. For most OM samples, the FID decays rapidly, within 10–20 ms, and there is no point in recording for much longer than its length. Once the FID has decayed to essentially noise level, longer acquisition produces little additional information for the broad signals of OM samples, while unnecessarily increasing the duration of most high-power decoupling. For all acquisitions, the FID should be examined to verify that the acquisition time is appropriate, and to check for

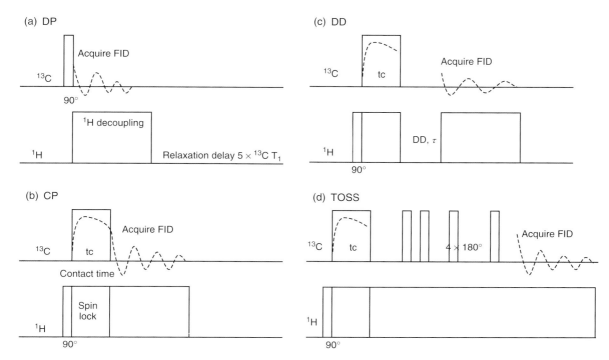

Figure 4.2. Pulse sequences for acquisition by (a) direct polarization (DP), (b) cross polarization (CP), (c) dipolar dephasing (DD), and (d) total sideband suppression (TOSS).

low signal that indicates a malfunction, or for artifacts such as spikes due to excessive sample conductivity, as can happen with soot, for example. Delay times required for complete relaxation are often 100 s and/or longer.[25] A background-corrected DP spectrum of a mineral soil with 29 mg C g^{-1} (Figure 4.3b[26]) shows the poor signal-to-noise ratio (S/N) usually obtained with a limited number of scans. DP spectra of OM usually have broader peaks and more rolling baselines as a result of better detection of carbon close to paramagnetics such as iron. This can include some very broad lines with short ^{13}C T_2; the resulting FID is characterized by a short initial spike of intensity, which causes baseline distortion (see Section 4.3.5).

4.2.4 CP and CP with TOSS

In environmental samples, protons are often much more abundant than ^{13}C nuclei. For ^{1}H, the γ is four times higher and their T_1 values in OM samples are typically <2 s. By transferring magnetization from ^{1}H to ^{13}C before acquisition, CP can enhance ^{13}C signal

intensity to a theoretical maximum of fourfold[27] and reduce the relaxation delay to that determined by the T_1 of protons. For magnetization to be transferred between spins of different γ, their ΔE must be equal; i.e., the Hartmann–Hahn condition must be fulfilled:[28]

$$\Delta E_H = \Delta E_C = \gamma_H B_{1H} = \gamma_C B_{1C} \qquad (4.2)$$

This is accomplished by rotating both spins into the y-axis, and applying simultaneous B_1 fields adjusted to satisfy equation 4.2. During this spin-locked condition, both spins precess around z with the same frequency, and magnetization is transferred from ^{1}H to ^{13}C spins during the contact time (tc). At the end of tc, which is typically 0.5–1.5 ms for OM, the ^{13}C B_1 field is turned off and the ^{13}C signal acquired with ^{1}H decoupling, as for DP (Figure 4.2b).

More detailed theory and the equations describing the transfer of magnetization are available elsewhere,[25,29,30] but basically, the changes in the ^{13}C magnetization are controlled by two competing mechanisms. As shown schematically in Figure 4.2(b), this generally results in a fairly rapid increase in ^{13}C intensity, followed by a slower decline. The former is

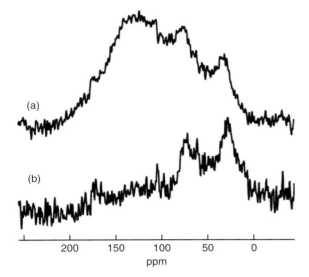

Figure 4.3. Solid-state ^{13}C direct polarization MAS spectra of (a) A-horizon mineral soil with 29 mg C g^{-1} and (b) after background correction. (Reproduced with permission from Ref. 26. © 2002 Natural Resources Canada)

controlled by the CP transfer time constant T_{CH}, but at the same time, ^{1}H magnetization is lost exponentially during the spin-locked state. The time constant for this is $T_{1\rho H}$, the spin–lattice relaxation time in the rotating frame. Considering only the effect of distance from protons, T_{CH} is the shortest for CH$_2$ and CH$_3$, slower for CH, and the slowest for nonprotonated ^{13}C, especially for those remote from protons, as in highly condensed aromatic structures or carbonate.[19,29,31] However, CP is also weakened by molecular motions, which can occur even in the solid state and also have the effect of decreasing dipolar interactions.[32] Rotation of CH$_3$ groups and movement of CH$_2$ in long chains result in less-efficient CP for carbons in methyl and methoxyl groups, and long-chain hydrocarbons. Because of the variations in T_{CH}, and internal motion, the resulting CP enhancements are variable, resulting in a lack of quantitative reliability in the relative areas of the spectrum. For carbons remote from protons, CP can be very weak, as the pool of ^{1}H magnetization available for transfer is largely lost through $T_{1\rho H}$ before it can be transferred. As discussed in Section 4.3.2, the CP efficiency is also affected by paramagnetics such as Fe^{3+}, which can drastically shorten $T_{1\rho H}$.

For most OM samples, the intensity enhancement due to CP is closer to 2 than the theoretical maximum of 4.[4] For a set of fresh and decomposed foliar litter samples, CP/DP ratios were determined by the comparison of CP and DP intensities for spectra acquired with the same number of scans.[33] Mean CP/DP ratios were alkyl = 1.9, methoxyl = 2.0, O-alkyl = 2.3, di-O-alkyl = 2.0, aromatic = 1.5, phenolic = 1.3, carboxyl = 1.4, and 1.8 for overall intensity. Enhancements were lowest for carboxyl, phenolic, and aromatic C, and highest for O-alkyl C, as expected from their relative ease of CP.[34] The intermediate value for alkyl C may reflect variable CP efficiency as a result of the relative proportions of rigid and mobile C. The CP advantage thus results from the combination of intensity enhancement and greatly reduced relaxation delays. Figure 4.4(a)[19] shows the good S/N easily obtained for a sample with 56% C.

With increasing magnetic field strengths, higher spinning speeds are required to overcome SSBs, so that the broad H–H matching profile breaks up into a series of narrow bands separated by the spinning speed. To compensate for this, the fixed ^{1}H power level during CP is replaced with a ramped amplitude (RAMP) CP ^{1}H power profile.[4,35,36] RAMP-CP is now in wide use, although like TPPM decoupling, may not be available on older instruments, and also requires additional steps in the experimental setup. Especially at higher spinning rates (ca >6000 Hz), RAMP-CP should be set up on the −1 sideband rather than the center band, and after initial setup on a standard such as glycine, it should be optimized, as far as possible, on a representative OM sample. A recent study presents a very detailed analysis of the effects of ramp size and spinning speed on CP intensity for two humic acids and recommendations for OM.[37]

If it is not possible to remove or minimize SSBs by fast spinning, they can be suppressed using the TOSS variant of CP, as shown for the same forest floor sample in Figure 4.4(b). TOSS incorporates a series of four ^{13}C 180° pulses with variable spacing between the contact time and signal acquisition, during which time the decoupler remains on (Figure 4.2d).[38] Calculation of these delay times is based on the spinning speed, as reviewed in more detail elsewhere,[39] but usually this is done automatically once the MAS rate is entered. During this time, some intensity is lost because of T_2 relaxation of ^{13}C, so that T_2 variation could result in selective intensity loss.[40]

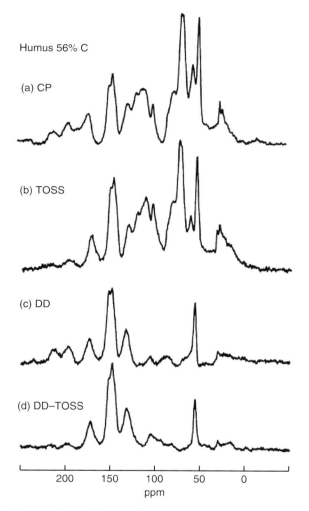

Humus 56% C

(a) CP

(b) TOSS

(c) DD

(d) DD–TOSS

200 150 100 50 0
ppm

Figure 4.4. Solid-state ^{13}C MAS spectra of forest floor from Vancouver Island (560 mg C g^{-1}) for acquisition with (a) CP, (b) TOSS, (c) DD, and (d) DD–TOSS. (Reproduced with permission from Ref. 19)

For a wide range of fossil fuel samples, aromaticity was only slightly lower for TOSS than for normal CP, but one peat sample showed greater loss of alkyl C.[41]

Intensity loss has been reported with TOSS,[20,39,42] but published quantitative evaluations of TOSS-induced intensity losses for OM have not. In additional unpublished work on the foliar litters used in a previous study,[33] it was observed that the

overall TOSS intensity was 63% of the corresponding CP spectra for ^{13}C at 75.47 MHz with spinning at 4300 Hz. The differentials [(TOSS-CP)/CP] were positive for alkyl (4.2%) and O- and di-O-alkyl C (6.8%) and negative for methoxyl (−12%), aryl (−11%), and carboxyl (−16%) C. However, TOSS was less successful for highly aromatic char samples, as the resulting spectra are prone to artifacts, and difficult to phase, and comparison of intensity distributions is further complicated by the difficulty of making the SSB corrections for the CP spectra.

4.2.5 Dipolar Dephasing (DD) and DD–TOSS

With dipolar dephasing (DD), which is a variant of CP,[43] a slight delay without decoupling is inserted between T_{CH} and signal acquisition (Figure 4.2c). During this delay, signal is rapidly lost from carbons that have the strongest dipolar interactions with protons; i.e., the same groups with the shortest T_{CH}. Signal is lost more slowly for carbons with weak dipolar coupling, thus with no attached protons, or weakened because of internal motions. Like CP, DD is also slower at higher spinning speeds. A dephasing time around 45–50 μs typically results in complete loss of carbon in rigid structures with attached protons, so that the resulting DD spectrum retains carbon without attached protons, and those with some motional freedom. The dephasing time should be adjusted for complete loss of the O-alkyl C signal. A series of similar samples should be analyzed using similar dephasing times. For the DD–TOSS combination, the delays are similarly calculated based on the MAS speed. Because DD spectra are dominated by carbons with greater CSA, the SSBs can be very prominent as shown in Figure 4.4(c). Especially, if the goal is qualitative structural information, the combination of DD with TOSS provides an unobstructed view of peak shapes and positions (Figure 4.4d).

The DD–TOSS forest floor spectrum (Figure 4.4d) is dominated by carboxyl and aryl C without attached hydrogen and also has peaks for methoxyl C at 57 ppm and long-chain alkyl C at 30 ppm. A small peak or shoulder may also be observed at 15 ppm for terminal methyl C. The CH$_2$ peak is often observed to have two maxima, at 30 and 33 ppm, with the latter lost rapidly under DD, whereas the 30 ppm peak represents long-chain CH$_2$ with greater mobility.[44–46]

The 50–60 ppm region, often attributed to methoxyl C in the standard division of spectra into chemical shift regions, is also the region where C–N from amino acids resonates. During DD, the broad C–N intensity is rapidly lost, leaving the sharper peak of methoxyl at 57 ppm, a characteristic feature of lignin.[9,10,47] The DD spectra also help clarify the relative importance of lignin and condensed tannins. In the di-*O*-alkyl region, condensed tannins have a broad peak at 106 ppm (C10 and C8 in C4 → C8 linkages), in a region usually otherwise lacking intensity in the DD spectrum. In the phenolic region, condensed tannins also have characteristic sharp peaks of 144 and 154 ppm, compared to the broader signal of lignin at 147–153 ppm.[9] The DD–TOSS spectrum is also useful to ascertain whether intensity beyond around 190 ppm in the CP spectrum is due to carbonyl C, or to SSBs, thus providing information to guide SSB correction.

While much qualitative information can be obtained using the basic DD pulse sequence, in addition to dipolar interactions, the intensities are also strongly modulated by effects due to CSA, MAS, and static field inhomogeneity, resulting in faster intensity loss, nonexponential decays, and even oscillations between negative and positive signals.[32,48,49] These artifacts are partially eliminated by a pulse sequence with a 180° refocusing pulse on both ^{1}H and ^{13}C channels.[32,50] Further isolation of dipolar interactions is obtained by using only the ^{13}C refocusing pulse, but with the interval between T_{CH} and FID acquisitions corresponding to two full MAS rotations, or multiples thereof,[51,52] although more intensity is lost because of ^{13}C T_2 relaxation during the longer delay times required by rotor synchronization.

Because carbons lose intensity at different rates during dephasing, a series of spectra run with different dephasing times can be used to determine the contribution of each peak to the normal CP at zero dephasing time.[32,47] However, time sequences using the rotor-synchronized ^{13}C refocusing pulse yield DD rate constants ($1/T_{2DD}$), which can provide valid information on distance to nearest protons, proportion of nonprotonated aromatics, and molecular mobility.[52] Using intervals up to eight rotor periods (1600 μs) and dephasing times up to 750 μs, the same authors[52] used linear combinations of CP and DD spectra to generate subspectra of three classes of OM carbon: nonprotonated, methyl, and nonmethyl protonated.

4.3 ACQUIRING AND PROCESSING SPECTRA

4.3.1 Sample Preparation and Packing

For many samples, the only preparation required is drying and grinding into a fine powder. Samples should be stored to minimize moisture sorption, or redried before use, with extra care for hygroscopic samples such as isolated OM extracts.[25,53,54] Changes in relaxation times and increased mobility due to sorption of water result in reduced sensitivity and distorted intensity distribution. Coarse or uneven particle sizes make it difficult to achieve a high spinning speed, especially with larger rotors, and the sample must be carefully packed down in layers. With the larger 7 mm rotors, a little glass wool on the top may prevent particles from attaching themselves to the cap and causing imbalance. Especially, in larger rotors with manual MAS controls, speed is increased slowly to allow the particles to arrange themselves, which typically occurs around 1000 Hz. The centrifugal forces leave an empty cone in the center with particles packed toward the bottom and sides.[55] The consequences of this for signal strength are discussed later. Dilution with Al_2O_3 may work for samples unwilling to spin, and is also essential for highly conductive samples, such as soot and some charcoals.

4.3.2 Sample Pretreatment with Hydrofluoric Acid (HF)

Even with CP, it can be difficult or impossible to obtain spectra with high S/N for mineral soil samples because of their low total carbon and the impacts of paramagnetic species such as Fe^{3+} and Cu^{2+}. Interactions with unpaired electrons drastically reduce $T_{1\rho H}$, thus causing rapid loss of ^{1}H magnetization and hindering its transfer to ^{13}C.[34,56–59] Even for DP spectra, broadening of ^{13}C signals due to proximity of paramagnetic ions may also render them invisible. Ferromagnetic domains also degrade signal quality by broadening the ^{13}C signals due to distortions of the local field.[60] Size or density fractionation of mineral soils often yields silt or clay fractions with higher C content, which still may require pretreatment with hydrofluoric acid (HF). HF reacts with silica to form soluble fluoride complexes, dissolving the mineral matrix and magnetic impurities, and usually, increasing the carbon concentration of the remaining OM. Samples rich in carbonates should be pretreated with hydrochloric acid (HCl) to increase the relative concentration of organic C.

Mineral soil samples are generally treated with 5–10% v/v HF by shaking or stirring several times at room temperature,[61] followed by washing with water. HF/HCl mixtures are also used, especially if carbonates are present,[57,60,62–64] and removal of ferromagnetic particles may be aided by shaking with a magnetic stir bar.[65] HF is extremely hazardous and safety procedures must be followed;[63] HF-treated samples may also damage analytical instruments, as similarly found by the author.[61,66]

Pretreatment of samples with acids may result in OM losses, and for some soils, such as acidic forest subsoils, carbon losses can be quite high sometimes with little enhancement of organic C concentration.[67,68] However, quality of spectra generally still improves because of the removal of paramagnetic minerals. This can be especially important for *O*-alkyl C of carbohydrates because of their close interactions with paramagnetic ions such as Fe^{3+}, Mn^{2+}, Cu^{2+}, and VO^{2+}.[56,69–71] Of course, a primary concern is whether the remaining carbon is representative of the original total carbon, but these changes are difficult to quantify for low-C samples with poor or nonexistent pretreatment spectra. Most studies indicate only small effects of HF treatment, with some indications that carbohydrates are more likely to be lost.[60,61,64,66,67,72,73] Recently, HF treatment has been utilized as another approach to probe the nature of mineral-associated OM and mechanisms of C stabilization.[58,74]

4.3.3 Instrument Setup

Choice of acquisition conditions includes consideration of magnetic field, rotor size, and available MAS rate. Small rotors hold less amount of sample, and due to the smaller coil, exhibit higher mass sensitivity, larger excitation bandwidth, and spinning may be easier to achieve, albeit with a reduced sample volume. They are also often the only choice at high spinning speeds (i.e., ca >5000 Hz). The combination of higher magnetic field and low spin rate means more interference from SSBs, and may necessitate more use of TOSS and DD–TOSS. Each instrument and probe has its own characteristics, which need to be optimized. For most probes, the magic-angle position of the rotor is quite stable and need only be checked occasionally, or after some modification such as coil replacement.

This is conveniently done by maximizing the number and intensity of sidebands on the ^{79}Br signal of KBr.[20,39] The Larmor frequency of this quadrupolar nucleus is very close to that of ^{13}C (50.12 MHz compared to ^{13}C at 50.306 MHz), so only a quick probe tuning is necessary.

As for ^{1}H, chemical shifts of ^{13}C are reported with respect to tetramethylsilane at 0 ppm. The COOH signal of glycine at 176.49 ppm[73] is often used as an external chemical shift reference, to establish the CP match, optimize shimming and measure the ^{1}H 90° pulse length. This is done indirectly by looking for the null of the CP ^{13}C signal at the ^{1}H 180° pulse. The ^{13}C 90° pulse length is also checked using DP acquisition, and both should be <5 μs. Measuring the actual CP enhancement of this signal is also a good practice. Using previously reported optimum acquisition parameters,[34] the author obtained similar CP/DP ratios of 3.3 for C1 and 3.5 for C2 (32 ppm) of glycine, and 3.5 (overall) for cellulose. Especially at higher spinning speeds, it is important to establish CP at the spinning rate to be used, and also on an actual sample, rather than glycine, but this is not always practical. For ^{13}C spectra of charcoal at 75 MHz and MAS >7000 Hz on a system without RAMP, good results were obtained by setting up CP on the −1 sideband,[4] using ^{13}C1-enriched glycine to improve the S/N. A mixture of ^{13}C1- and ^{13}C2-enriched glycine provides an additional peak for testing spectrometer performance. The chemical shift reference setting should also be checked occasionally using a sample of adamantane (42 ppm). The narrow linewidth of this signal also makes it useful for optimizing shimming, even if many OM samples have much broader peaks.

4.3.4 Background Signal

Especially with older instruments, a broad background signal may be produced mainly from probe ringdown, resulting from very broad lines with short-lived FIDs, general instrument noise, and C-containing components in the probe. The latter are mainly produced from components outside the coil volume, and in particular, the signal from caps made of Kel-F (poly(chlorotrifluoroethylene)) is negligible for CP spectra.[33,75] Background signal is not usually a problem for CP spectra of samples high in C, but of more consequence for low carbon and DP spectra (Figure 4.3). Background was equivalent to 1 mg of observable C (C_{obs}) for CP acquisition,[75] but for DP, amounted to 69[75] and 100 mg,[33] comparable to the C_{obs} in the sample. To correct for this, the FID is acquired of an empty rotor under the same conditions as the sample. The rotor may be filled with Al_2O_3 if required for spinning stability. It is useful to record the background FID at a series of increasing number of scans, and such a series may be applied to many subsequent acquisitions. The background FID is subtracted from the sample FID, with scaling if necessary to match the number of scans. If FID subtraction does not work (e.g., FIDs were acquired with different sweep widths and computer glitches), the final phased, baseline-corrected spectra can be subtracted, although this is subject to more operator influence.

Another approach is to use a background generated by the actual sample.[76] Background signals that appear in the spectrum mostly arise from probe components outside the coil (usually, Kel-F or Teflon), which thus effectively experience a much weaker pulse than the sample. The sample signal increases with pulse length as a cosine function, whereas the background only increases linearly. For CP spectra, at a ^1H pulse length corresponding to 180° for the sample (a null signal), the background signal should be twice that of its value at a 90° pulse. Therefore, subtracting one-half the 180° signal from the 90° signal should leave a background-free signal. The 180° spectrum also provides a sensitive check of the ^1H 90° pulse for the actual sample, as any strong peak produces a residual signal in the background, and often, the pulse length can be adjusted after observing a few acquisitions. Similar to the empty-rotor background, this background may be used for similar samples and acquisition conditions. This method can also be used for DP spectra with 180° and 90° carbon pulses. An alternative approach suggested for DP spectra is a pulse sequence incorporating $^{19}F-^{13}C$ dipolar dephasing.[75]

4.3.5 Processing the FID

For most OM spectra, before Fourier transformation (FT), the FID is modified to enhance S/N. The simplest method is to apply a decaying exponential, which emphasizes the early part of the signal, an operation corresponding to 30–100 Hz linebroadening (LB). The amount of LB should be varied to get the best result, as excessive LB results in loss of detail and even baseline distortion. Spectra of OM, especially those with low C, or DP spectra, may be distorted by low-frequency artifacts, including probe ringdown, which appear as

a spike at the beginning of the FID. Sometimes, the resulting spectrum can be improved by removing the first two or four points by setting their value to zero, or left-shifting the FID. Soil OM spectra usually need more aggressive phasing than usual, as well as baseline correction; examples are shown elsewhere.[19,63]

4.3.6 Some Approaches to Spectral Analysis

Aside from questions of quantitative reliability, OM spectra may need more care in integration. Integrals may be distorted by noise and curved or rising baselines, and if values seem unreliable, integral areas should be measured by hand. Copying, cutting, and weighing of spectra may occasionally be useful to obtain relative areas.[9,42] Measurement of relative areas is also complicated by peak overlap. For example, the broad wings of aryl C may extend into carboxyl and O-alkyl regions for charcoal samples, so that some sketching in of shoulders ('manual deconvolution') is more useful than applying the usual chemical-shift boundaries. It may be necessary to correct for the SSB intensity of the aromatic, phenolic, and carboxyl regions, and this is done by assuming equal intensity of the upfield and downfield SSB peaks. For many spectra, SSB can be weak and difficult to measure; for a large series of litter spectra, mean percent SSBs from the CP spectra were then used to correct both CP and DP intensities.[33] Deconvolution could probably be applied much more widely to ^{13}C OM spectra, especially for poorly decomposed samples, such as litter and organic horizons, which have a good peak definition.[19,72,77,78]

A molecular mixing model (MMM) has been developed to analyze NMR spectra as a mixture of five biochemical components: carbohydrate, protein, lignin, lipid, and black carbon and an extra pure carbonyl component to account for uronic acids and oxidation of other components.[79] The MMM uses the distribution of signal intensity in seven chemical-shift regions and the elemental composition of the representative biopolymers, and applications include soil[80] and peat.[81] The model is an extremely useful tool for understanding OM formation and decomposition over a wide range of situations. However, it is important to remember that the results are subject to the usual caveats regarding quantitative reliability of the spectra. Furthermore, as noted previously, with decomposition, the proportion of C that can be identified from chemical analysis decreases,[11] so that intensity identified as, e.g., carbohydrate or protein, might yield only a small proportion of that C as monosaccharides or amino acids upon hydrolysis.

Some suggestions could be made for further development. Plant litters can be very high in condensed tannins,[9,33] of which at least the more soluble portion is rapidly lost. Studies with a focus on litter decomposition could benefit by adding a CT component to the model. Most NMR data analyses, including the MMM or widely used principal component analysis (PCA), are also still based on the distribution of intensity into chemical-shift slabs. It would be interesting to see the results of an MMM using linear combinations of actual spectra of the representative biopolymers, or PCA with the NMR spectra as inputs, so that the loadings for each component come out as subspectra, as used for solution spectra in metabolomics.[82] As most solid-state OM ^{13}C NMR spectra can be well described with 1024 or even 516 data points, these analyses are quite feasible, as already demonstrated for peat in 1989.[83]

4.4 TOWARD QUANTITATIVE RELIABILITY

The benefits of CP acquisition remain essential for most routine environmental applications of solid-state NMR, but its quantitative reliability has always been an issue.[3,25,34,57,73,78,84] In an ideal world, a DP spectrum would detect 100% of the C (see 'spin counting' discussion later in this section) and quantitatively represent the distribution of carbon structures. As long as the ^{13}C relaxation delay is long enough, this is the case for simple organic molecules and even for more complex substrates such as wood, which are low in paramagnetic species.[34,85] In practice, however, the long acquisition times, problematic baseline distortions, and poor S/N of DP spectra generally make them impractical for routine application, even with carbon-rich samples. For OM samples, C_{obs} may still be quite low because of excessive broadening of C close to paramagnetic centers. For pyrogenic C (charcoal), however, with low CP efficiency and high total C, DP may sometimes be the better option.

Moreover, in an ideal world, CP enhancement would be the same for all carbon, even if not the theoretical maximum. However, as discussed previously in Section 4.3, CP intensities are determined by T_{CH} and

$T_{1\rho H}$ both of which vary with C structure and proximity to paramagnetic centers, and their opposing effects during the contact time tc. Ideally, if $T_{CH} \ll tc \ll T_{1\rho H}$, CP is complete before much ^{1}H polarization is lost because of $T_{1\rho H}$ relaxation. This situation is only met for a small proportion of sample, i.e., rigid CH_2 and CH_3 free from paramagnetic influence. However, contact times around 1 ms generally have been found to give the best overall CP intensities for OM.[25,34,69] Longer contact times may be used to enhance the aryl intensity of pyrogenic C, but there is usually little benefit beyond 2–3 ms, as the losses due to $T_{1\rho H}$ increasingly predominate.

Spin counting has long been used to determine C_{obs} for both CP and DP acquisitions, the assumption being that 100% observability corresponds to quantitative intensity distribution. For spin counting, the total signal due to the sample is compared with that of a standard such as glycine or cellulose often run separately.[34,57,65,85,86] Spin counting is also possible with internal standards mixed in with the sample, but these introduce other complexities, including overlap with sample peaks or alterations in the relaxation properties of the standard because of interactions with the sample.[87] Sensitivity also varies with position within the rotor,[34,55,88] and even well-mixed materials may migrate to different parts of the rotor with spinning. Inserts have been used to center a capillary of standard along the axis of a large rotor[72] or to place samples at different points along the rotor axis;[34] presumably, the latter could be used to position a layer of standard separated from the sample. If sample recovery is not a requirement, a simpler approach might be to first weigh in a small amount of standard, such as ^{13}C-enriched glycine or silicone rubber, and then pack the sample on top. This avoids the complexities of inserts, and a calibration could be developed to compensate for the lower response at the end of the rotor. The effect of sample position was determined rather simply by measuring signal intensity after successive additions of adamantane and measuring the depth occupied by the sample.[55]

Several approaches have been devised to overcome or compensate for the variability in CP intensities. If CP/DP enhancements are measured on a representative sample, correction factors can be applied to a series of similar samples.[89] For higher field instruments, the combination of RAMP and high-speed spinning may yield CP intensity distributions very close to those obtained by BD acquisition.[4,90] However, this may not overcome the very limited potential of CP for highly condensed aromatic samples; also such tests are usually carried out on relatively clean and high-C samples. As such, the extension of such comparisons; also to low-C samples requires further examination.

The most reliable (and time-consuming) approach is to run a series of spectra with variable contact times (VCT) and fit the resultant intensities to the equation describing the CP dynamics.[3,4,25,77,87] This yields the theoretical maximum CP intensity at zero tc if T_{CH} were infinitely short and $T_{1\rho H}$ infinitely long. A simplified version can be used to extrapolate intensities back to tc = 0 or to determine $T_{1\rho H}$ using only the decaying part of the curve mainly controlled by $T_{1\rho H}$ relaxation.[34,69,91] If not practical for routine application, such VCT data provide insights into the factors affecting CP for specific C structures or environments, guidance for choosing the best tc for routine CP acquisition, or correction factors applicable to a series of related samples.[77]

4.5 USING MORE COMPLEX PULSE SEQUENCES

4.5.1 ^{1}H Relaxation- and T_{CH}-based Spectral Editing

To this point, relaxation processes have been considered as simple exponentials characterized by a single rate constant, as would be the case for most pure chemicals. However, many OM samples exhibit spatial heterogeneity at the nanometer scale with separation into domains that present barriers to spin diffusion, resulting in two- or even three-component relaxation behavior.[84] Differences in proton relaxation rates can be used to separate the subspectra of C in domains on the order of 30–100 nm for T_{1H} and around 2–30 nm for $T_{1\rho H}$. Proton spin relaxation editing (PSRE) separates C in domains with different proton T_1 values.[92,93] If the ^{1}H magnetization is initially inverted by a 180° pulse and a CP spectrum taken after different delay intervals, the ^{13}C signal recovers from negative to positive intensity. In a structurally homogeneous solid with strong dipolar ^{1}H–^{1}H coupling, spin diffusion leads to an average value for T_{1H} and signal recovery follows a simple exponential. If ^{1}H relaxation is a sum of two components, at a certain delay interval, the signal from the ^{13}C associated with slowly relaxing protons is passing through the null point, whereas that of the fast-relaxing component has already almost recovered to full intensity, so that the slow component is

missing from the resulting ^{13}C spectrum. In practice, spectra need to be acquired at only two delay intervals, roughly corresponding to the null of the fast-relaxing component and to the full recovery of the slow component, and the subspectra are then generated by linear combination of the two. Generally, the slowly relaxing component includes less-decomposed plant residues or long-chain alkyl C, whereas fast-relaxing components may include more highly decomposed material, microbial residues, or C structures more closely associated with paramagnetics. PSRE has been applied to humin,[65] soil,[86,90] and decomposing wood,[85] and even extended to three components for dairy pond sludge.[93]

Restoration of spectra via T_{CH} and T one rho editing (RESTORE[94]) is a process that can largely correct for inefficient CP and also generates subspectra of three components: C_{SS} with short T_{CH} and $T_{1\rho H}$, C_{SL} with short T_{CH} and long $T_{1\rho H}$, and C_{LL} with long T_{CH} and $T_{1\rho H}$. C_{SL} cross-polarizes efficiently, whereas C_{SS} is underrepresented because of its short $T_{1\rho H}$ and slow CP. In addition to a series of VCT spectra to determine T_{CH} values, three series are run with fixed T_{CH} and variable spin lock (VSL), which isolates the $T_{1\rho H}$ decay from the CP dynamics.[84] In this process, the proton magnetization is spin-locked for a variable delay before CP. The VCT curves are then fitted to a model containing components with both short and long T_{CH}, with $T_{1\rho H}$ fixed as determined from the VSL results. Corrections from these calculations resulted in a large increase in C_{obs}, especially for samples with high charcoal content (C_{LL}). Subspectra representing the three components are then generated from linear combinations of spectra taken under specified conditions, with C_{LL} giving 'virtual fractionation' of soil charcoal. The technique is very demanding with 39 acquisitions and complex data analysis, but again, for limited numbers of samples, it provides valuable insight into distinct C pools, restoration of relative intensities comparable to those obtained by DP, and reliable T_{CH} and $T_{1\rho H}$ values.

4.5.2 Advanced Methods for Extracting Structural Information

Subspectra of CH, CH$_2$, and CH$_3$ + C (quaternary) carbons of a humic acid were generated by linear combination of four types of CP spectra.[95] Application of DP ^{13}C NMR has always been limited by the long recycle delays required, essentially $5 \times T_{1C}$. However, a

Table 4.3. Main assignments for solid-state ^{15}N spectra referenced to nitromethane at 0 ppm[6,7,101]

Chemical shift range (ppm)	Assignment
25 to −25	Nitrate, nitrite, and nitro groups
−25 to −90	Imine, phenazines, pyridines, and Schiff bases
−90 to −145	Purine (N-7) and nitrile groups
−145 to −220	Heterocyclic N (chlorophyll, purines/pyrimidines, imidazoles, N-substituted pyrroles, and Maillard products)
−220 to −285	Amides/peptides, N-acetyl derivatives of amino sugars, tryptophanes, prolines, unsubstituted pyrroles, indoles, and carbazoles
−285 to −325	NH in guanidines, NH$_2$- and NR$_2$-groups (N_δ, N_ε-arginine, urea, nucleic acids, and aniline derivatives)
−325 to −350	Free amino groups in amino sugars and sugars
−350 to −375	NH$_4^+$

CP/T$_1$-TOSS method has been used to reduce recycle delays to around $1.3 \times T_{1C}$, by measuring a correction factor for each part of the spectrum.[78] This was combined with a CSA filter to isolate aromatic C, which is particularly important to separate alkyl O–C–O around 90–120 ppm, and DD was additionally used to separate protonated and nonprotonated aromatic C.[24] The CP/T$_1$-TOSS correction was also used to combine DP with DD for OM samples.[96] The two-dimensional (2D) NMR spectroscopy has been very successful for solid-state NMR, and again while not routine, further structural information on ^{1}H–^{13}C through-space interactions can be obtained from 2D heteronuclear correlation (HETCOR) NMR.[97] The systematic approach to the characterization of humic acid by these techniques, especially HETCOR,[98] has also applied to OM fractions from an agricultural study.[99]

4.6 NITROGEN-15 NMR

Applications of ^{15}N NMR to OM in the solid state have been limited by its combination of low natural abundance and low sensitivity (Table 4.1). Chemical shifts are usually referenced to nitromethane at 0 ppm, which results in negative δ values for most of the spectrum, but some studies reference anhydrous liquid ammonia at −380.23 ppm.[100] The chemistry of organic N is complex, and the potential chemical shift range very large (Table 4.3[6,7,101]). In general, however, studies of plant decomposition and OM,[7,10,101–106] often with ^{15}N enrichment, result in rather similar-looking spectra dominated by the broad peak of amide N from peptides, small signals of amino groups of terminal amino acids at −347 ppm, and occasional sharper signals because of nitrate and ammonium ions. Heterocyclic N may appear as a broad shoulder of the amide peak (approximately, −145 to −230 ppm). However, some plants such as tobacco have naturally high levels of heterocyclic and pyridinic N (in nicotine), resulting in a more diverse spectrum[107] (Figure 4.5). With increased biodegradation, pyrolysis, or geological processing, there is an increase in the proportions of heterocyclic and eventually aromatic (pyridinic) N.[25,105,108–110] Much effort

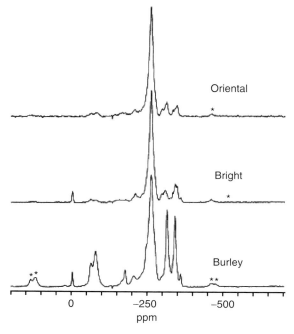

Figure 4.5. Solid-state ^{15}N CPMAS spectra of three varieties of tobacco leaves, referenced to nitromethane at 0 ppm (contact time = 5 ms, * = spinning side band). The lower spectrum is a variety high in nicotine (pyridine-type N at −62 and −77 ppm and pyrrolidine N at −338 ppm). (Reprinted with permission from Ref. 107. © 2004 American Chemical Society)

Table 4.4. General ranges of chemical shifts for ^{31}P referenced to 85% H_3PO_4 at 0 ppm.[119] Values are for solutions at high pH; chemical shifts in solids are highly variable and original papers should be consulted

Chemical shift range (ppm)	Assignment
20	Phosphonate (C–P bond)
5–7	Orthophosphate
6–3	Orthophosphate monoesters, one C moiety per P
2.5 to −1	Orthophosphate diesters, two C moiety per P
−4 to −5	Pyrophosphate, terminal P group of polyphosphates
−20	Polyphosphate

has been expended to optimize acquisition parameters, especially in the search for possibly undetected heterocyclic and aromatic N in soil.[7,106,107,111,112] In particular, CP for nonprotonated aromatic N is limited by very long T_{CH}, and acid treatment of a kerogen to protonate pyridinic N improved its detectability.[109]

However, some innovative approaches have been demonstrated. In double CP NMR,[113] intensity is first transferred from ^{1}H to the ^{15}N spin system, and then to ^{13}C where it is detected, so that the spectrum only shows ^{13}C in very close proximity to ^{15}N. It was applied to study the decomposition of doubly enriched plant residues, with the results supporting the assignment of the main ^{15}N intensity to peptides.[114] Increasing aniline N (N bonded to aromatic C) with continuous lowland rice cropping may contribute to declining N availability but is difficult to detect in OM; however, anilides were detected in a humic fraction using an indirect method based on their dipolar coupling to the naturally abundant ^{14}N nucleus.[115] Labeled substrates can be used to investigate the fate of contaminants such as 2,4,6-trinitrotoluene (TNT).[116]

Regarding detectability of pyridinic N, there do not appear to be any studies of such compounds mixed with OM, or even of behavior of sorbed ^{15}N-pyridine as done with coal[117] and clay.[118] As caffeine, nicotine, and other alkaloids may occur at high levels in some plants, interesting studies could be done by mixing such compounds with OM, rather than studying their spin dynamics in isolation. Incubations of high-alkaloid ^{15}N-enriched plants would start with less dominance of amide N, and one might ask if the effects

could be detected in soils from long-term tobacco cultivation. Some ^{15}N NMR studies would benefit from additional chemical analyses, even of extractable ammonium and nitrate.

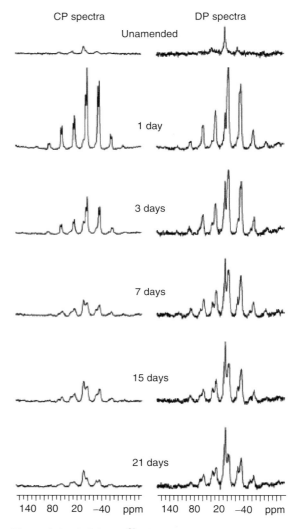

Figure 4.6. Solid-state ^{31}P CP and direct polarization (DP) spectra of soil before and after amendment with pyrophosphate (2000 mg P kg^{-1}). The vertical scales allow direct comparison of CP and DP intensities. Pyrophosphate occurs at −8.5 ppm, and the time series shows increasing conversion to orthophosphate at 2.1 ppm. (Reproduced with permission from Ref. 128. © 2006 Soil Science Society of America)

4.7 PHOSPHORUS-31 NMR

With spin-$1/2$, 100% natural abundance, and high sensitivity (Table 4.1), ^{31}P NMR has been widely used to characterize and quantify P structures in extracts of soils and sediments,[119] but applications to solids have been much more limited. Large CSA and interactions with paramagnetics can result in substantial peak broadening, large higher order sidebands, and complete loss of signal, even with DP acquisition,[120–126] and HF pretreatment did not produce large gains, partly because of high losses of P.[127] The challenges are increased by the small chemical-shift range for P structures typically found in OM (Table 4.4[119]), sensitivity of chemical shifts to structural and environmental factors, its low concentration in most samples, limited potential for CP, occasional very long [31]P T_1 values, and lack of opportunity for isotopic enrichment. High-speed spinning is required for optimum results (i.e., at least 10 000 Hz for a 400 MHz system with ^{31}P at 161.9 MHz), and deconvolution is widely used to interpret broad, overlapping peaks.[119,121,123,124]

Chemical shifts for ^{31}P are reported with respect to external 85% H_3PO_4 at 0 ppm. The values in Table 4.4 are for solution spectra of OM extracts at high pH, and reports for specific phosphorus compounds and minerals, as solids should be consulted.[119,120,124,128] However, ^{31}P NMR can still provide valuable insight into nature and availability of phosphate minerals and organic P in soils,[122,124,128] sewage sludge,[129,130] manure,[93] and composts.[103] Figure 4.6 shows the transformation of pyrophosphate to orthophosphate after addition to soil.[128] With highly organic samples low in paramagnetic metals, SSB are not always a large problem, and the increased availability of high-speed spinning should facilitate applications of ^{31}P NMR. It has also been combined with ^{27}Al NMR to investigate soil Al–P interactions[123] and used in model systems to investigate the fate of environmental contaminants.[131,132]

RELATED ARTICLES IN EMAGRES

Environmental Comprehensive Multiphase NMR

Soil Organic Matter

Forest Ecology and Soils

Organic Pollutants in the Environment

REFERENCES

1. R. K. Harris, E. D. Becker, S. M. C.de Menezes, R. Goodfellow, and P. Granger, *Solid State NMR*, 2002, **22**, 458.

2. D. I. Hoult, *Concepts Magn. Reson.*, 2009, **34A**, 193.

3. P. Kinchesh, D. S. Powlson, and E. W. Randall, *Eur. J. Soil Sci.*, 1995, **46**, 125.

4. K. J. Dria, J. R. Sachleben, and P. G. Hatcher, *J. Environ. Qual.*, 2002, **31**, 393.

5. V. J. Bartuska, G. E. Maciel, J. Schaefer, and E. O. Stejskal, *Fuel*, 1977, **56**, 354.

6. I. Kögel-Knabner, *Geoderma*, 1997, **80**, 243.

7. H. Knicker and H.-D. Lüdemann, *Org. Geochem.*, 1995, **23**, 329.

8. I. Kögel-Knabner, *Soil Biol. Biochem.*, 2002, **34**, 139.

9. C. M. Preston, J. A. Trofymow, and CIDET Working Group, *Can. J. Bot.*, 2000, **78**, 1269.

10. H. Knicker, *J. Environ. Qual.*, 2000, **29**, 715.

11. J. I. Hedges, G. Eglinton, P. G. Hatcher, D. L. Kirchman, C. Arnosti, S. Derenne, R. P. Evershed, I. Kögel-Knabner, J. W.de Leeuw, R. Littke, W. Michaelis, and J. Rullkötter, *Org. Geochem.*, 2000, **31**, 945.

12. J. S. Waugh, *Anal. Chem.*, 1993, **65**, 725A.

13. G. E. Maciel, *Science*, 1984, **226**, 282.

14. J. Schaefer and E. O. Stejskal, *J. Am. Chem. Soc.*, 1976, **98**, 1031.

15. P. G. Hatcher, D. L. VanderHart, and W. L. Earl, *Org. Geochem.*, 1980, **2**, 87.

16. C. M. Preston and J. A. Ripmeester, *Can. J. Spectrosc.*, 1982, **27**, 99.

17. M. A. Wilson, R. J. Pugmire, K. W. Zilm, K. M. Goh, S. Heng, and D. M. Grant, *Nature*, 1981, **294**, 648.

18. C. M. Preston, *Soil Sci.*, 1996, **161**, 144.

19. C. M. Preston, *Can. J. Soil Sci.*, 2001, **81**, 255.

20. D. L. Bryce, G. M. Bernard, M. Gee, M. D. Lumsden, K. Eichele, and R. E. Wasylishen, *Can. J. Anal. Sci. Spectrosc.*, 2001, **46**, 46.

21. A. E. Bennett, C. M. Rienstra, M. Auger, K. V. Lakshmi, and R. G. Griffin, *J. Chem. Phys.*, 1995, **103**, 6951.

22. E. R. Andrew, A. Bradbury, and R. G. Eades, *Nature*, 1959, **183**, 1802.

23. I. J. Lowe, *Phys. Rev. Lett.*, 1959, **2**, 285.

24. J.-D. Mao and K. Schmidt-Rohr, *Environ. Sci. Technol.*, 2004, **38**, 2680.

25. H. Knicker, *Org. Geochem.*, 2011, **42**, 867.

26. C. M. Preston, C. H. Shaw, J. S. Bhatti, and R. M. Siltanen, in Soil C and N Pools in Forested Upland and Non-forested Lowland Sites Along the Boreal Forest Transect Case Study in Central Canada, eds C. H. Shaw and M. J. Apps, Natural Resources Canada: Edmonton, AB, 2002, 155.

27. A. Pines, M. G. Gibby, and J. S. Waugh, *J. Chem. Phys.*, 1973, **59**, 569.

28. S. R. Hartmann and E. L. Hahn, *Phys. Rev.*, 1962, **128**, 2042.

29. L. B. Alemany, D. M. Grant, R. J. Pugmire, T. D. Alger, and K. W. Zilm, *J. Am. Chem. Soc.*, 1983, **105**, 2133.

30. W. Kolodziejski and J. Klinowski, *Chem. Rev.*, 2002, **102**, 613.

31. L. B. Alemany, D. M. Grant, R. J. Pugmire, T. D. Alger, and K. W. Zilm, *J. Am. Chem. Soc.*, 1983, **105**, 2142.

32. L. B. Alemany, D. M. Grant, T. D. Alger, and R. J. Pugmire, *J. Am. Chem. Soc.*, 1983, **105**, 6697.

33. C. M. Preston, J. R. Nault, and J. A. Trofymow, *Ecosystems*, 2009, **12**, 1078.

34. R. J. Smernik and J. M. Oades, *Geoderma*, 2000, **96**, 101.

35. G. Metz, X. Wu, and S. O. Smith, *J. Magn. Reson. Ser. A*, 1994, **110**, 219.

36. R. L. Cook, C. H. Landford, R. Yamdagni, and C. M. Preston, *Anal. Chem.*, 1996, **68**, 3979.

37. A. E. Berns and P. Conte, *Org. Geochem.*, 2011, **42**, 926.

38. W. T. Dixon, J. Schaefer, M. D. Sefcik, E. O. Skejskal, and R. A. McKay, *J. Magn. Reson.*, 1982, **80**, 341.

39. D. E. Axelson, Solid State Nuclear Magnetic Resonance of Fossil Fuels: An Experimental Approach, Ministry of Supply and Services Canada: Ottawa, 1985, 226.

40. J. Hirschinger, *Solid State NMR*, 2008, **34**, 210.

41. D. E. Axelson, *Fuel*, 1987, **66**, 195.

42. S. A. Quideau, M. A. Anderson, R. C. Graham, O. A. Chadwick, and S. E. Trumbore, *For. Ecol. Manage.*, 2000, **138**, 19.

43. S. J. Opella and M. H. Frey, *J. Am. Chem. Soc.*, 1979, **101**, 5854.

44. W. L. Earl and D. L. Vanderhart, *Macromolecules*, 1979, **12**, 762.

45. W.-G. Hu, J. Mao, B. Xing, and K. Schmidt-Rohr, *Environ. Sci. Technol.*, 2000, **34**, 530.

46. I. Kögel-Knabner and P. G. Hatcher, *Sci. Total Environ.*, 1989, **81/82**, 169.

47. P. G. Hatcher, *Org. Geochem.*, 1987, **11**, 31.

48. R. H. Newman, *J. Magn. Reson.*, 1990, **86**, 176.

49. C. M. Preston, P. Sollins, and B. G. Sayer, *Can. J. For. Res.*, 1990, **20**, 1382.

50. M. A. Wilson, R. J. Pugmire, and D. M. Grant, *Org. Geochem.*, 1983, **5**, 121.

51. R. H. Newman, *J. Magn. Reson.*, 1992, **96**, 370.

52. R. J. Smernik and J. M. Oades, *Eur. J. Soil. Sci.*, 2001, **52**, 103.

53. P. G. Hatcher and M. A. Wilson, *Org. Geochem.*, 1991, **17**, 293.

54. R. J. Smernik, *Eur. J. Soil. Sci.*, 2006, **57**, 665.

55. F. Ziarelli and S. Caldarelli, *Solid State NMR*, 2006, **29**, 214.

56. R. J. Smernik and J. M. Oades, *Geoderma*, 1999, **89**, 219.

57. R. J. Smernik and J. M. Oades, *Geoderma*, 2000, **96**, 151.

58. I. Schöning, H. Knicker, and I. Kögel-Knabner, *Org. Geochem.*, 2005, **36**, 1378.

59. J.-M. Séquaris, H. Philipp, H.-D. Narres, and H. Vereecken, *Eur. J. Soil Sci.*, 2008, **59**, 592.

60. C. M. Preston, M. Schnitzer, and J. A. Ripmeester, *Soil Sci. Soc. Am. J.*, 1989, **53**, 1442.

61. M. W. I. Schmidt, H. Knicker, P. G. Hatcher, and I. Kögel-Knabner, *Eur. J. Soil Sci.*, 1997, **48**, 319.

62. M. J. Simpson and P. G. Hatcher, *Org. Geochem.*, 2004, **35**, 923.

63. M. J. Simpson and C. Preston, in Soil Sampling and Methods of Analysis, 2nd edn, eds M. R. Carter and E. G. Gregorich, CRC Press: Boca Raton, FL, 2008, Chapter 53.

64. C. Rumpel, N. Rabia, S. Derenne, K. Quenea, K. Esterhues, I. Kögel-Knabner, and A. Mariotti, *Org. Geochem.*, 2006, **37**, 1437.

65. C. M. Preston and R. H. Newman, *Geoderma*, 1995, **68**, 229.

66. M. W. I. Schmidt and G. Gleixner, *Eur. J. Soil Sci.*, 2005, **56**, 407.

67. K. H. Dai and C. E. Johnson, *Geoderma*, 1999, **93**, 289.

68. C. N. Gonçalves, R. S. D. Dalmolin, D. P. Dick, H. Knicker, E. Klamt, and I. Kögel-Knabner, *Geoderma*, 2003, **116**, 373.

69. C. M. Preston, R. L. Dudley, C. A. Fyfe, and S. P. Mathur, *Geoderma*, 1984, **33**, 245.

70. M. Schilling and W. T. Cooper, *Environ. Sci. Technol.*, 2004, **38**, 5059.

71. E. H. Novotny, H. Knicker, L. A. Colnago, and L. Martin-Neto, *Org. Geochem.*, 2006, **37**, 1562.

72. C. Keeler and G. E. Maciel, *Anal. Chem.*, 2003, **75**, 2421.

73. X. Fang, T. Chua, K. Schmidt-Rohr, and M. L. Thompson, *Geochim. Cosmochim. Acta*, 2010, **74**, 584.

74. K. Esterhues, C. Rumpel, and I. Kögel-Knabner, *Org. Geochem.*, 2007, **38**, 1356.

75. R. J. Smernik and J. M. Oades, *Solid State NMR*, 2001, **20**, 74.

76. Q. Chen, S. S. Hou, and K. Schmidt-Rohr, *Solid State NMR*, 2004, **26**, 11.

77. M. F. Davis, H. R. Schroeder, and G. E. Maciel, *Holzforschung*, 1994, **48**, 99.

78. J.-D. Mao, W.-G. Hu, K. Schmidt-Rohr, G. Davies, E. A. Ghabbour, and B. Xing, *Soil Sci. Soc. Am. J.*, 2000, **64**, 873.

79. J. A. Baldock, C. A. Masiello, Y. Gélinas, and J. I. Hedges, *Mar. Chem.*, 2004, **92**, 39.

80. L. Cécillon, G. Certini, H. Lange, C. Forte, and L. T. Strand, *Org. Geochem.*, 2012, **46**, 127.

81. J. Kaal, J. A. Baldock, P. Buurman, K. G. J. Nierop, X. Pontevedra-Pombal, and A. Martínez-Cortizas, *Org. Geochem.*, 2007, **38**, 1097.

82. M. J. Simpson and J. R. McKelvie, *Anal. Bioanal. Chem.*, 2009, **394**, 137.

83. B. Nordén and C. Albano, *Fuel*, 1989, **68**, 771.

84. K. Abelmann, K. U. Totsche, H. Knicker, and I. Kögel-Knabner, *Solid State NMR*, 2004, **25**, 252.

85. C. M. Preston, R. J. Smernik, R. F. Powers, J. G. McColl, and T. M. McBeath, *Org. Geochem.*, 2011, **42**, 936.

86. C. M. Preston, R. H. Newman, and P. Rother, *Soil Sci.*, 1994, **157**, 26.

87. M. A. Wilson, A. M. Vassallo, E. M. Perdue, and J. H. Reuter, *Anal. Chem.*, 1987, **59**, 551.

88. A. E. Berns and P. Conte, *Open Magn. Reson. J.*, 2010, **3**, 75.

89. I. Bergman, P. Lundberg, C. M. Preston, and M. Nilsson, *Soil Sci. Soc. Am. J.*, 2000, **64**, 1368.

90. R. J. Smernik, *Geoderma*, 2005, **125**, 249.

91. J. A. Baldock, J. M. Oades, A. M. Vassallo, and M. A. Wilson, *Aust. J. Soil Res.*, 1990, **28**, 213.

92. R. H. Newman and K. R. Tate, *J. Soil Sci.*, 1991, **42**, 39.

93. R. H. Newman and L. M. Condron, *Solid State NMR*, 1995, **4**, 259.

94. R. J. Smernik and J. M. Oades, *Eur. J. Soil Sci.*, 2003, **54**, 103.

95. C. Keeler and G. E. Maciel, *J. Mol. Struct.*, 2000, **550–551**, 297.

96. G. Ding and J. A. Rice, *Geoderma*, 2012, **189–190**, 381.

97. J.-D. Mao, B. Xing, and K. Schmidt-Rohr, *Environ. Sci. Technol.*, 2001, **35**, 1928.

98. J.-D. Mao, N. Chen, and X. Cao, *Org. Geochem.*, 2011, **42**, 891.

99. X. Cao, D. C. Olk, M. Chappell, C. A. Cambardella, L. F. Miller, and J. Mao, *Soil Sci. Soc. Am. J.*, 2011, **75**, 1374.

100. P.-H. Hsu and P. G. Hatcher, *Org. Geochem.*, 2006, **37**, 1694.

101. N. Maie, H. Knicker, A. Watanabe, and M. Kimura, *Org. Geochem.*, 2006, **37**, 12.

102. L. Benzing-Purdie, J. A. Ripmeester, and C. M. Preston, *J. Agric. Food Chem.*, 1983, **31**, 913.

103. C. M. Preston, J. A. Ripmeester, S. P. Mathur, and M. Lévesque, *Can. J. Spectrosc.*, 1986, **31**, 63.

104. P. W. Clinton, R. H. Newman, and R. B. Allen, *Eur. J. Soil Sci.*, 1995, **46**, 551.

105. H. Knicker, P. G. Hatcher, and F. J. González-Vila, *J. Environ. Qual.*, 2002, **31**, 444.

106. R. J. DiCosty, D. P. Weliky, S. J. Anderson, and E. A. Paul, *Org. Geochem.*, 2003, **34**, 1635.

107. Z. Ma, D. H. Barich, M. S. Solum, and R. J. Pugmire, *J. Agric. Food Chem.*, 2004, **52**, 215.

108. H. Knicker, G. Almendros, F. J. González-Vila, F. Martin, and H.-D. Lüdemann, *Soil Biol. Biochem.*, 1996, **28**, 1053.

109. S. R. Kelemen, M. Afeworki, M. L. Gorbaty, P. J. Kwiatek, M. S. Solum, J. Z. Hu, and R. J. Pugmire, *Energy Fuels*, 2002, **16**, 1507.

110. S. R. Kelemen, M. Afeworki, M. L. Gorbaty, P. J. Kwiatek, M. Sansone, C. C. Walters, and A. D. Cohen, *Energy Fuels*, 2006, **20**, 635.

111. R. J. Smernik and J. A. Baldock, *Biogeochemistry*, 2005, **75**, 507.

112. R. J. Smernik and J. A. Baldock, *Plant Soil*, 2005, **275**, 271.

113. J. Schaefer, E. O. Stejskal, J. R. Garbow, and R. A. McKay, *J. Magn. Reson.*, 1984, **59**, 150.

114. H. Knicker, *Org. Geochem.*, 2002, **33**, 237.

115. K. Schmidt-Rohr, J.-D. Mao, and D. C. Olk, *Proc. Natl. Acad. Sci. U. S. A.*, 2004, **101**, 6351.

116. H. Knicker, C. Achtnich, and H. Lenke, *J. Environ. Qual.*, 2001, **30**, 403.

117. J. A. Ripmeester, R. E. Hawkins, J. A. MacPhee, and B. N. Nandi, *Fuel*, 1986, **65**, 740.

118. L. Ukrainczyk and K. A. Smith, *Environ. Sci. Technol.*, 1996, **30**, 3167.

119. B. J. Cade-Menun, *Talanta*, 2005, **66**, 359.

120. Z. He, C. W. Honeycutt, T. Zhang, P. J. Pellechia, and W. A. Caliebe, *Soil Sci. Soc. Am. J.*, 2007, **71**, 940.

121. P. Conte, D. Šmejkalová, A. Piccolo, and R. Spaccini, *Eur. J. Soil Sci.*, 2008, **59**, 584.

122. W. J. Dougherty, R. J. Smernik, and D. J. Chittleborough, *Soil Sci. Soc. Am. J.*, 2005, **69**, 2058.

123. R. Lookman, P. Grobet, R. Merckx, and W. H. Van Riemsdijk, *Geoderma*, 1997, **80**, 369.

124. R. W. McDowell, L. M. Condron, N. Mahieu, P. C. Brookes, P. R. Poulton, and A. N. Sharpley, *J. Environ. Qual.*, 2002, **31**, 450.

125. C. A. Shand, M. V. Cheshire, C. N. Bedrock, P. J. Chapman, A. R. Fraser, and J. A. Chudek, *Plant Soil*, 1999, **214**, 153.

126. B. Sutter, R. E. Taylor, L. R. Hossner, and D. W. Ming, *Soil Sci. Soc. Am. J.*, 2002, **66**, 455.

127. W. J. Dougherty, R. J. Smernik, E. K. Bünemann, and D. J. Chittleborough, *Soil Sci. Soc. Am. J.*, 2007, **71**, 1111.

128. T. M. McBeath, R. J. Smernik, E. Lombi, and M. J. McLaughlin, *Soil Sci. Soc. Am. J.*, 2006, **70**, 856.

129. Z. R. Hinedi, A. C. Chang, and J. P. Yesinowski, *Soil Sci. Soc. Am. J.*, 1989, **53**, 1053.

130. Z. R. Hinedi and A. C. Chang, *Soil Sci. Soc. Am. J.*, 1989, **53**, 1057.

131. D. M. Mizrahi and I. Columbus, *Environ. Sci. Technol.*, 2005, **39**, 8931.

132. J. A. Shaw, R. K. Harris, and P. R. Norman, *Langmuir*, 1998, **14**, 6716.

FURTHER READING

A. J. Simpson, D. J. McNally, and M. J. Simpson, *Prog. NMR Spectrosc.*, 2011, **58**, 97.

J. Schaefer, E. O. Stejskal, and R. Buchdahl, *Macromolecules*, 1977, **10**, 384.

D. L. VanderHart, W. L. Earl, and A. N. Garroway, *J. Magn. Reson.*, 1981, **44**, 361.

D. L. VanderHart and G. C. Campbell, *J. Magn. Reson.*, 1998, **134**, 88.

M. A. Wilson, NMR Techniques and Applications in Geochemistry and Soil Chemistry, Pergamon Press: Oxford, 1987.

Chapter 5

Environmental NMR: High-resolution Magic-angle Spinning

Ruth E. Stark, Bingwu Yu, Junyan Zhong, Bin Yan, Guohua Wu and Shiying Tian

Department of Chemistry, College of Staten Island, City University of New York, New York, NY 10031, USA

5.1 INTRODUCTION

Traditionally, most NMR investigations of molecular structure and dynamics have been conducted on samples in solution or the solid state. However, these well-developed and information-rich spectroscopic approaches are frequently ill suited to intermediate states such as liquid crystals, bulk polymers, or colloidal dispersions. Materials of interest in environmental science, such as soils, plant tissues, melanin pigments, and urban surface films, have major swollen or swellable constituents that are not studied ideally using either solution- or solid-state NMR alone. On the one hand, partial orientational order or overall tumbling that is restricted in terms of rate and/or angular excursion may render motional narrowing incomplete, compromising both spectral resolution and sensitivity in solution-state NMR experiments. These same characteristics may compromise the efficiency of cross polarization (CP) in solid-state experiments; moreover, the resolution and consequent structural details of the latter spectra may prove disappointing, even with magic-angle spinning (MAS), if the samples are amorphous rather than microcrystalline. Thus, to examine the molecular structure and motion of such materials in their intact 'native' states, strategies that select or combine the advantages of particular solution- and solid-state methods to obtain high-quality NMR data are desirable.

^{1}H MAS NMR approaches were reported as early as the 1970s for solvent-swollen cross-linked polystyrene gels[1,2] and extended a decade later to obtain ^{13}C spectra using direct polarization (DP) in combination with MAS methods for solvated, mobile macromolecular moieties that exhibit partially averaged dipolar interactions and chemical-shift anisotropy.[3] In contrast to the weak CPMAS ^{13}C

NMR Spectroscopy: A Versatile Tool for Environmental Research
Edited by Myrna J. Simpson and André J. Simpson

NMR signals observed for such materials,[4] DPMAS experiments conducted with very modest decoupling strengths (~3 kHz) and spin rates (~2 kHz) yielded well-resolved resonances for both backbone chains and cross-link junctions. The latter approach opened up new ways to monitor chemical functionalization, in situ gelation kinetics, and cross-link morphology.[5]

The late 1980s also witnessed the introduction of MAS NMR methods for spectral observation of lyotropic liquid crystals such as model biological membranes,[6,7] in which rapid lateral diffusion and axial rotation reduce inter- and intramolecular dipole–dipole interactions, respectively. This breakthrough made it possible to obtain high-resolution ^1H, ^{13}C, and ^{31}P spectra of multibilayers without sonication, laying the groundwork for subsequent extensions of this strategy. For instance, ^{13}C MAS NMR spectra were resolved and assigned for cell-wall glucans in whole yeast cells,[8] and two-dimensional (2D) MAS-assisted Nuclear Overhauser effect spectroscopy (NOESY) measurements were used to investigate the bilayer organization of phospholipid multibilayers[9] and the partitioning of ethanol within such model biological membranes.[10]

Spurred in part by the high-throughput demands of combinatorial chemistry, this period also saw a significant advance that was dubbed high-resolution (HR)-MAS, achieved by the engineering of NMR probes to average discontinuities in bulk magnetic susceptibility. The HR-MAS technology was applied first in a Nano-NMR® probe to characterize peptides and drug candidates that were still covalently attached to the solvent-swelled polymeric resins used in solid-phase organic synthesis[11] and subsequently extended with pulsed field gradients (PFGs) and diverse probe designs to obtain 2D homo- and heteronuclear correlated NMR spectra that permitted structural elucidation.[12–14] Beginning later in the 1990s, HR-MAS capabilities were also exploited extensively to assist in characterization of diseased human tissue samples and other biomedical targets.[15]

In the environmental science and agricultural fields, early applications of one-dimensional (1D) and 2D HR-MAS NMR included studies of stress-resistant plant tissues[16] dispersed in aqueous and organic solvents, respectively.[16] The usefulness of HR-MAS NMR methods has also been recognized for structural investigations of soil samples[17] and melanin pigments.[18] Recent reviews including HR-MAS NMR applications have focused on applications

to live bacterial cells,[19] food science,[20] biomedical metabonomics,[21] and environmental science.[22] Unprecedented molecular information, often with diagnostic or biochemical significance, has thus been forthcoming without the need for, e.g., isolation of cellular extracts or chemical degradation of the constituent biopolymers. The NMR strategies required to achieve such results are the principal subject of this overview, in which both fundamentals of the HR-MAS technique and the resulting atomic-level findings are illustrated for environmentally important targets including soil organic matter, protective plant polymers, and ultraviolet (UV)-resistant melanin pigments.

5.2 EXPERIMENTAL METHODS

The following section outlines the experimental procedures required to obtain well-resolved, high-sensitivity HR-MAS NMR spectra that fairly represent the chemical moieties in swelled solids, using plant biopolymers and biomass, fungal melanin, and whole soil samples to illustrate these guidelines.

5.2.1 Validation of ^1H HR-MAS NMR Spectra for Intractable Environmental Solids

The impressive resolution enhancement achieved by solvent swelling combined with MAS-assisted NMR acquisition is illustrated in Figure 5.1, which compares ^1H MAS NMR spectra of an insoluble plant polymer

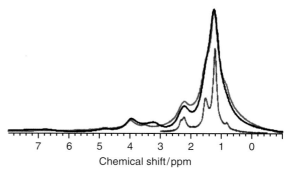

Chemical shift/ppm

Figure 5.1. 300 MHz ^1H HR-MAS NMR spectra of an intact lime fruit cutin polymer, as either a dry powder spinning at 12 kHz (red) or a dispersion of 11 mg cutin swelled in ~60 mg DMSO-d_6 and spinning at 8 kHz (green, no artificial line broadening; blue, 70 Hz exponential line broadening). The data were obtained using a Doty XC-5 probe and presented, in part, previously.[24]

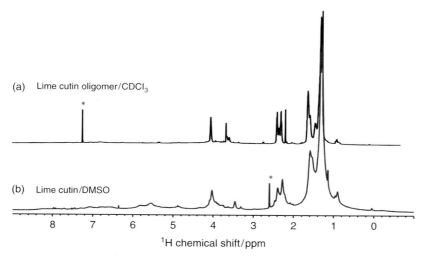

Figure 5.2. An oligomer obtained by chemical degradation of the polymer (a) and 600 MHz ^{1}H HR-MAS NMR spectra of an intact lime fruit cutin polymer swelled in DMSO-d_6 at 50 °C (b)[24]. The spectra were obtained using a Varian (Agilent) nanoprobe with MAS at 2.500 ± 0.005 kHz. The * designations indicate sharp solvent resonances.

from intact lime fruit (*Citrus aurantifolia*) cuticle in dry and swelled states.[23,24] The relatively featureless 300 MHz spectrum of dry cutin (>100 Hz full width at half height for $(CH_2)_n$ groups) obtained with what was, at the time, considered moderately rapid 12 kHz MAS (red trace) is narrowed by at least 70% if the sample is swelled in DMSO-d_6, but observed using slower 8 kHz HR-MAS (green inset). Moreover, the quantitative reliability of spectra obtained for this 'soggy' solid may be validated by establishing that all of the resonances observed using MAS on dry samples are represented in their correct amounts and proportions using the alternative of MAS applied to solvent-swelled materials. Thus, if 70 Hz of artificial broadening is applied to the spectrum of the swelled lime cutin sample, it becomes nearly coincident with dry cutin (blue trace). This result shows that solvent swelling has reduced the linewidths but preserved both the absolute integrated signal intensity and the relative ratios of the components derived from the spectrum. Stated differently, solvent swelling increases the effective ^{1}H spin–spin relaxation times (T_2) but yields a similar spectroscopic fingerprint corresponding to the macromolecular structure of the cutin polymer.

Using a probe with optimized ^{1}H lineshape characteristics and collecting the NMR data at 600 MHz, it is possible to observe aliphatic resonances for the swelled polymer with resolution that approaches that achieved for a soluble constituent trimeric fragment

(Figure 5.2).[24] Both spectra provide straightforward estimates of the types and numbers of functional groups that comprise this intractable plant material. For instance, in the polymer, a variety of aliphatic components [$-CH_3$ at 0.9 ppm, 3H; $-(CH_2)_n-$ at 1.2 and 1.5 ppm, 42H; $-CH_2COO-$ and $-CH_2C(O)CH_2-$ at 2.2–2.4 ppm, 6H; $-CH_2OR$ or $-CHOH$ at 3.4 ppm (set to 1H); and $CH_2OC(O)$ at 4.0 ppm, 3H] are fairly well resolved. It is also possible to discern modest spectral contributions from multiply bonded groups (5.5–6.0 ppm). Other than the latter moieties, these results fit the compositional profile derived from an extensive series of ^{13}C solid-state NMR measurements on the dry material,[25–27] thus confirming that most functional groups in lime fruit cutin are rendered flexible enough to display ^{1}H NMR signals under conditions of solvent swelling and slow MAS. Similar approaches have also been adopted for structural investigations of cutins from tomato fruits and *Agave americana* leaves.[28,29]

The potential of HR-MAS experiments conducted on swelled solids is also demonstrated by the PFG-assisted 2D heterocorrelated multiple-quantum coherence (HR-MAS-gHMQC) spectra[30] in Figure 5.3, which are acquired as straightforwardly as their solution-state analogs and again show excellent resolution.[16,24] The somewhat broader contours of the polymer spectrum, which parallel the broader

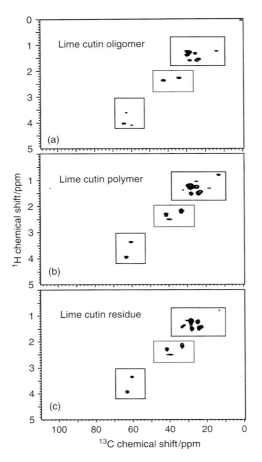

Figure 5.3. 600 MHz ^{1}H HR-MAS-gHMQC NMR spectra of intact lime fruit cutin and an associated trimer derived from low-temperature HF degradation,[24,46] displayed as contour plots to obtain a fingerprint for each material. The spectra were obtained with 2.500 kHz MAS and *z*-axis PFGs of 11 G cm^{-1}. (a) ~1 mg of oligomer dissolved in CDCl$_3$ at room temperature. (b) 1.5 mg of polymer swelled in 40 μl of DMSO-d_6 at 50 °C. (c) 1.5 mg of insoluble residue from HF treatment swelled in 40 μl of DMSO-d_6 at 50 °C. Boxes highlight the oxygenated, carboxylated, bulk methylene, and methyl groups, respectively, as labeled. (Reproduced with permission from Ref. 24. © 2001 Elsevier)

linewidths noted in the 1D spectra of Figure 5.2, are attributed to chemical-shift heterogeneity rather than incomplete motional averaging, because the intact cutin likely contains a greater variety of chemically and magnetically similar moieties as compared with a trimeric fragment obtained by low-temperature

hydrofluoric acid treatment. A discussion of how HR-MAS can complement solution-state NMR to monitor the associated chemical transformations is presented in Section 5.3.

5.2.2   Sample Preparation of Swelled Solids

The encouraging results illustrated previously for fruit cuticular polymers nevertheless raise a variety of questions regarding the choice of experimental conditions in HR-MAS NMR experiments. First and foremost, care must be observed regarding the physical state of the solid under study. As noted earlier, near-solution-state resolution in HR-MAS NMR spectra requires partial motional averaging by rapid liquid-like molecular tumbling ($\geq 10^6$ s^{-1}); vigorous motions are also expected to lengthen ^{1}H and ^{13}C spin–spin relaxation times (T_2) under these circumstances. In practice, attaining values of T_2^{H} that exceed ~50 ms is especially crucial to the success of gradient-assisted 2D heterocorrelated multiple-bond correlation (MAS-gHMBC) experiments: their three-bond and four-bond correlations are richly informative regarding molecular structure, but the $1/2J_{\mathrm{CH}}$ mixing period is vulnerable to signal losses if transverse relaxation occurs too efficiently. This same challenge is well established for HMBC experiments in large protein complexes in solution,[31] for which selective or perdeuteration is used to lengthen the transverse spin relaxation times.

5.2.2.1   *Surface Area and Solvent Accessibility*

To enhance the overall molecular motion and HR-MAS signal intensity, for instance, the maximum possible surface area of the sample should be exposed to the solvent. Figure 5.4 compares MAS-assisted HMBC results for the defensive biopolymer suberin formed in potato (*Solanum tuberosum* L.) tuber wound periderms,[16,23] showing differences depending on whether the sample is in flaked or powder form and whether one examines the light- or heavy-powder fractions produced by benchtop centrifugation of a slurried sample. In each case, the resulting HR-MAS NMR spectra are expected to reflect only solvent-accessible molecular moieties; the soluble portion of the swelled sample should also be examined to rule out contributions from extractable small molecules or degradation products.

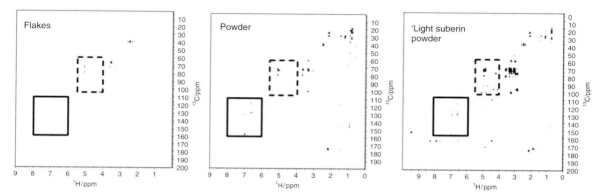

Figure 5.4. 750 MHz ^{1}H HR-MAS-gHMBC NMR spectra of suberized potato wound periderm in flaked, powder, and 'light' forms, each swelled in DMSO-d_6 and run at 50 °C. Suberized tissue, either as flakes or ground into a powder with a liquid N$_2$-cooled freezer mill, was treated with cellulase and pectinase as described earlier;[54] the 'light' designation refers to the floating solid layer obtained after centrifugation of powdered wound periderm. The data were obtained with 10.000 ± 0.001 kHz MAS and 40 G cm^{-1} magic-angle PFGs using a Bruker HR-MAS probe; the 65 ms defocusing period in HMBC was optimized for a multiple-bond ^{1}H–^{13}C scalar coupling of 8 Hz.

The trend for wound-healing suberized tissues is clear-cut: the smaller the particles that are dispersed in solution are, the more ^{1}H–^{13}C HMBC NMR cross-peaks are observed with significant intensity. The beneficial effect of increasing surface area-to-volume ratio is especially pronounced for aromatic moieties (solid box: $\delta_H = 6$–8 ppm, $\delta_C = 110$–160 ppm), which are expected to occur in regions of the polymer-cell-wall matrix that are heavily cross-linked and marginally accessible to solvent. Major enhancements are also observed for the polysaccharide resonances (dashed box: $\delta_H = 4$–5.5 ppm, $\delta_C = 60$–105 ppm): although their hydrophilic hydroxyl groups could have an affinity for DMSO-d_6, their ability to cross-link with the biopolyester could limit solvent accessibility. As detailed in Section 5.3, such optimization measures make it possible, although still challenging, to obtain otherwise inaccessible information regarding macromolecular structure from MAS-assisted 2D NMR experiments.

5.2.2.2 Sample Concentration

A different potato biopolymer material illustrates a seemingly counterintuitive guideline for the preparation of samples to enhance the signal intensity in these HR-MAS NMR experiments. Although it is tempting to use the highest possible concentration except in sample-limited circumstances, the series of gHMBC NMR spectra in Figure 5.5 belie this strategy. While raising the concentration at first gives the expected rise in cross-peak intensity, increases beyond 5 mg/50 μl have the opposite effect for this sample, an aliphatic–aromatic polymer associated with textural defects in potatoes (*S. tuberosum* L.).[32] The deleterious impact of high concentration on signal strength is especially severe in the aromatic region; as noted for potato suberin, a falloff in intensity is expected for functional groups that exhibit limited motional freedom.

To put the particle size and concentration trends on a more quantitative footing, spin relaxation measurements were conducted on an insoluble swelled biological solid before and after mechanical grinding; in this case, a ^{13}C-enriched sample of melanized cell walls from the pathogenic fungus *Cryptococcus neoformans*.[18,33] Like other melanin pigments with roles in soil bioremediation and animal camouflage, this intractable amorphous material is an energy transducer that can either protect living organisms or damage their DNA under different circumstances.[34] Preparation of *C. neoformans* melanin 'ghosts' as fine powders enhances the signal-to-noise ratio of MAS-gHMBC NMR spectra by an order of magnitude (data not shown), suggesting explanations that invoke faster tumbling of smaller particles and/or more complete solvent infusion.

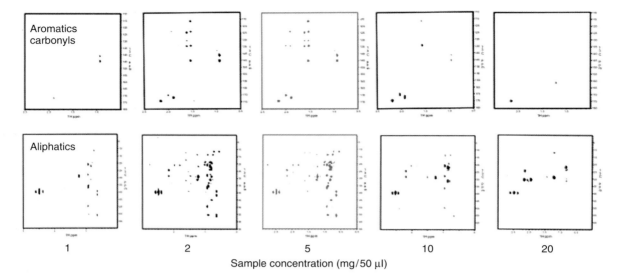

Figure 5.5. 750 MHz ^{1}H HR-MAS-gHMBC NMR spectra of intracellular adhesion-strengthened ('hardened') potato tissue swelled in DMSO-d_6 at different concentrations and run at 50 °C. The data were obtained with 6.00 ± 0.001 kHz MAS and 40 G cm^{-1} magic-angle PFGs; the 65 ms defocusing period was optimized for a multiple-bond ^{1}H–^{13}C scalar coupling of 8 Hz.

Table 5.1. ^{1}H spin relaxation times for melanin samples[a]

Chemical shift (ppm)	T_1 (s)		T_2 (ms)	
	Unground	Ground	Unground	Ground
0.84	1.7	1.0	162	171
1.23	0.9	0.7	79	89
2.16	1.0	0.6	158	119
5.30	1.2	0.7	40	35
5.31	1.2	0.8	42	45

[a] *Cryptococcus neoformans* melanin ghosts were prepared from [2,3-^{13}C]-L-dopa and [U-^{13}C]-glucose as described earlier.[33] 'Ground' denotes 20 min of manual grinding with a mortar and pestle. Errors are estimated as 10% based on repeated trials.

Table 5.2. ^{13}C spin relaxation times for melanin samples[a]

Chemical shift (ppm)	T_1 (s)		T_2 (ms)	
	Unground	Ground	Unground	Ground
28.8	1.1	1.0	5.0	62
28.6	1.2	1.0	1.3	23
129.5	1.6	1.3	2.4	30

[a] *Cryptococcus neoformans* melanin ghosts were prepared from [2,3-^{13}C]-L-dopa and [U-^{13}C]-glucose as described earlier.[33] 'Ground' denotes 20 min of manual grinding with a mortar and pestle. Errors are estimated as 15% based on repeated trials.

The T_2^{H} values are surprisingly insensitive to the grinding procedure (Table 5.1), however, suggesting that faster molecular reorientation of smaller particles is balanced by detection of an additional population of semi-mobile macromolecular species or intermolecular association effects in these 'soggy' samples. Nonetheless, the T_2^{H} values for aliphatic groups associated with fungal melanins are long enough with respect to typical HMBC defocusing periods of 65 ms to avoid devastating losses in signal intensity,

although the aromatic moieties of the pigment (and of plant suberins, data not shown) are more vulnerable to signal losses in such experiments. By contrast, grinding of the melanin sample lengthens T_2^{C} values by factors of 12–18 to approach the defocusing time (Table 5.2), offering possible additional contributions to improved HMBC performance.

5.2.2.3 Solvent Choice

A third consideration in sample preparation concerns the choice of solvent. If the hydrophobic–hydrophilic balance of an intractable solid is uncertain, then DMSO is viewed as a safe compromise choice that is compatible with, e.g., hydrocarbon chains, peptides, polysaccharides, and phenolic ring structures—presuming that such molecular groupings are actually accessible to the solvent. To maximize molecular mobility, a solvent with low viscosity is preferred; this rationale has been exploited, for instance, in high-resolution NMR studies of proteins encapsulated in reverse micelles dissolved in liquid alkanes.[35,36] If a low-viscosity solvent is not suitable for swelling the solid of interest, then it may be practical to enhance molecular mobility by selecting a nonvolatile solvent and running the experiments at the highest temperatures consistent with the probe specifications and sample integrity.

In our experience with the primarily aliphatic cross-linked polyester in lime fruit cutin, both DMSO-d_6 and CDCl$_3$ were better swelling solvents than D$_2$O, DMF, methanol, benzene, or toluene. For the aliphatic–aromatic potato biopolyester suberin in polysaccharide cell walls, only DMSO-d_6 yielded well-resolved HR-MAS NMR spectra spanning several spectral regions. As detailed later, whole soil HR-MAS NMR spectra acquired in contrasting solvents can be used to favor polar compounds at the soil–water interface or nonpolar compounds at more buried locations[17]; similar strategies could be useful to swell particular polymeric or networked regions within a plant or soil assembly. Mixed solvents are expected to do a complete job of swelling a sample that contains chemical moieties with a wide range of polarities, although we observed no improvements for, e.g., DMSO-d_6 with CDCl$_3$ or CD$_3$OD with benzene.

5.2.2.4 *Swelling Protocol*

A rough measure of swelling efficacy can be made by immersing a given mass of solid sample in each of several solvents and comparing the volumes of the precipitates at equilibrium in closed containers,[37] although this procedure cannot reveal which constituents of a complex soil or plant material are swollen or inaccessible to solvent. NMR samples may be prepared by swelling the solid of interest in an excess volume of solvent, with vigorous stirring or bath sonication, for >24 h at ambient or elevated temperature. Then, excess solvent is removed by filtration, and the swollen material is packed immediately into the rotor (or rotor insert). Alternatively, when the amount of sample is limited, it is possible to mix the solid with solvent

inside the 40 to 50 μl rotor, allow swelling in situ, and then mix further during MAS-assisted acquisition of the NMR spectra. Liquid-tight sample cells are essential to maintain the solid-to-solvent ratio for the duration of one or more HR-MAS NMR measurements.

5.2.2.5 *Choice of Temperature*

Certainly molecular mobility is enhanced as the temperature rises, so provided that the solid of interest is chemically stable and the solvent remains in the liquid phase, more complete motional averaging should make it easier to achieve liquid-like spectral resolution with MAS methods. In addition, elevated temperatures will benefit experiments such as HMBC by lengthening T_2 of the liquid-like molecular moieties, a trend confirmed for our potato wound suberin samples at 25 vs 50 °C (data not shown). However, raising the temperature of these samples is also likely to compromise the overall spectral resolution by increasing resonance contributions from the solid-like chemical moieties (as proposed earlier in connection with particle size and concentration considerations). In practice, it is often necessary to simply evaluate a series of ^{1}H spectra at different temperatures empirically and then proceed with more comprehensive 2D experiments.

5.2.3 Spectral Acquisition Strategies for Swelled Solids

In addition to the precautions recommended earlier for sample preparation, care is required with the choice of acquisition conditions for HR-MAS NMR experiments on swelled solids. By monitoring the impact of MAS and specific spin rate on ^{1}H NMR spectra for plant and soil samples, it is possible to deduce the factors determining the observed linewidth and to devise strategies for optimization. In addition, it is possible to improve spectral quality for many solvent-swelled biomaterials through the use of solvent suppression, diffusion editing,[38] and spin-echo techniques that are standard for HR-MAS NMR of aqueous-based tissue samples.[21]

5.2.3.1 *Choice of Magic-angle Spinning Rate*

Although ^{1}H NMR spectra of solvent-swelled biopolymer samples are typically broad ($\Delta\nu_{1/2} \geq 100$ Hz) in the absence of MAS, trials conducted with MAS at 1–10 kHz evidence dramatic resolution improvements for lime cutin[16,24] (Figures 5.1 and 5.2), potato tissues,[32] and fungal melanins[18,33] that are essentially complete at modest spin rates of ~2.5 kHz. These observations suggest that dipole–dipole interactions are reduced (or abolished) for any molecular entities that are exposed to the solvent and thus able to tumble more freely. Any remaining dipole–dipole couplings could then be averaged by MAS; however, the rate requirements should depend on the size of the residual coupling for a given pair of proton spins. As modest MAS rates often serve to improve spectral resolution for all ^{1}H resonances in these plant samples, it is likely instead that swelling allows complete motional averaging and that MAS serves to remove magnetic susceptibility anisotropy.[13] The ^{1}H resonance linewidths nonetheless remain somewhat broader, for instance, for solvent-swelled lime fruit cutin than a soluble trimer constituent trimer (Figure 5.3); we attribute this trend to chemical-shift heterogeneity—partially overlapped resonances from a greater number of chemically similar groups in the intact biopolymer as compared with a particular aliphatic trimer.

Can we benefit by using higher MAS rates? As indicated earlier, if residual dipolar couplings are removed by solvent swelling and magnetic susceptibility anisotropy is abolished by MAS, then increasing the spin rate will produce no further line narrowing. Nonetheless, it may be advantageous to spin at, say, 10 kHz in order to place spinning sidebands outside the spectral window of interest,[17] provided that sample integrity is not compromised. Although current commercial HR-MAS probe specifications include ^{1}H linewidths of 1.5/15/25 Hz at 50/1/0.05% height for solution-state standards such as chloroform in acetone, half-height linewidths of 10–1500 Hz are observed in practice for amorphous swelled-solid environmental materials in which chemically similar constituents exhibit overlapping resonances.

Do FastMAS®, UltraFastMAS®, or Very High Speed DVT MAS® probe technologies offer an alternative to doing HR-MAS NMR experiments on solvent-swelled environmental materials? If the goal is characterization of soil–aqueous interfaces[17] or urban surface films,[39] then swelling followed by HR-MAS NMR offers a measure of spatial selectivity that would be lost by FastMAS experiments on dry samples. If the target is a region of suberized plant cell walls that remains rigid because of dense crosslinking or obstructed solvent accessibility, then FastMAS methods may be the only

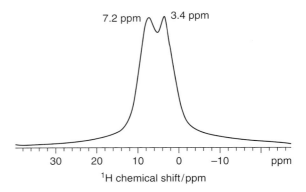

Figure 5.6. 600 MHz ^{1}H NMR spectrum obtained at a spinning speed of 35.00 ± 0.01 kHz using a Varian (Agilent) FastMAS probe for pyrogenic organic matter from ponderosa pine wood, showing spectral features from oxygenated aliphatic and aromatic moieties at 3.4 and 7.2 ppm, respectively.[40] (Reproduced with permission from Ref. 40. © 2012 Elsevier)

NMR-based investigative options that can augment the organizational and molecular information forthcoming from FT-infrared (IR) and X-ray diffraction measurements. Even so, [1]H FastMAS experiments are likely to exhibit limited spectral resolution, as illustrated by aryl and alkyl functional groups in a pyrogenic carbon residue (Figure 5.6).[40]

5.2.3.2 Water Suppression

In the presence of an abundance of residual ^{1}H nuclei from the swelling solvent, the spectrum of the sample of interest will be dominated by solvent signals; nuclear spins that resonate at similar chemical shifts will also be lost in the shoulders of the solvent signal. The techniques for removal of interfering solvent resonances from solution-state NMR spectra, which are often suitable for environmental or biomedical HR-MAS NMR applications, run the gamut from simple presaturation to WATERGATE, excitation sculpting, and presaturation utilizing relaxation gradients and echoes (PURGE).[21,22]

5.2.3.3 Spectral Editing for Mobile or Immobilized Environmental Constituents

The use of Carr-Purcell-Meiboom-Gill (CPMG) spin-echo sequences (CPMG T_2 filters)[31] is well

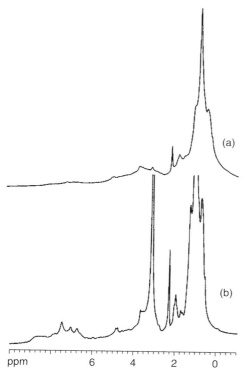

Figure 5.7. 600 MHz ^{1}H HR-MAS spectra of a whole forest soil swelled in DMSO-d_6, obtained with a Bruker HR-MAS probe.[17] (a) 12 kHz spinning with presaturation of water present in the solvent; (b) 12 kHz spinning (with partial presaturation of the water signal at 3.3 ppm) and T_2 filtering achieved by insertion of a CPMG sequence with 1.2 ms delays between 180° pulses before NMR signal acquisition. (Reprinted with permission from Ref. 17. © 2001 American Chemical Society)

established for the suppression of broad NMR signals that arise from rapidly relaxing immobilized species such as human bone; the offending species have also included polymer resins in HR-MAS NMR spectra of covalently bound combinatorial peptide libraries,[13] and intact organs in HR-MAS NMR experiments designed to observe small-molecule metabolites.[21] Figure 5.7 shows a striking example of the resulting spectral improvement for a whole soil environmental sample,[17] in which the broad background signals were attributed either to macromolecules or to components associated with mineral surfaces or paramagnetic substances. Thus, such T_2 filter methods are efficacious when the investigator wishes to observe relatively

mobile species present in an immobilized matrix. The chemical information forthcoming from such investigations is summarized in Section 5.3.

Conversely, an editing strategy that favors rigid constituents has been implemented for wheat grass residues from simulated degradation studies using ^{1}H diffusion-gated NMR experiments.[38,41] Using bipolar pulse pairs refined in connection with diffusion-ordered 2D NMR spectroscopy, it was possible to remove mobile components from the spectra of grass blades (Figure 5.8), demonstrating that the

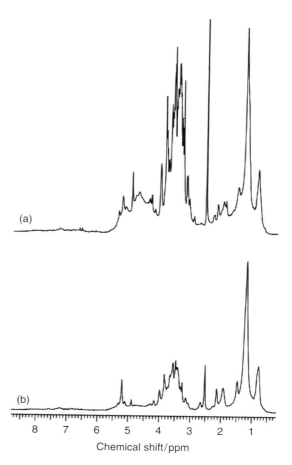

Figure 5.8. 500 MHz ^{1}H HR-MAS spectra of wheat grass blades swelled in DMSO-d_6 after 1 month of modeled environmental degradation.[41] (a) Standard HR-MAS with a Bruker probe at 10 kHz; (b) HR-MAS at 10 kHz with diffusion editing. (Reproduced with permission from Ref. 41. © 2006 Elsevier)

sharp carbohydrate resonances appearing in the ^{1}H HR-MAS NMR spectrum after a month's microbial decomposition come from chemical entities displaying rapid translational diffusion. A more complete discussion of the chemical transformations appears in Section 5.3.

5.2.4 Hardware for HR-MAS NMR Experiments

Each of the major commercial NMR vendors offers specialized probes for HR-MAS; no additional high-power amplifiers are required, and probes are currently in operation at ^{1}H frequencies up to 900 MHz. As noted earlier, many MAS probes have similarly excellent lineshape specifications to solution-state probes, although attainable resolution is often limited by the heterogeneity or motional freedom of the environmental sample. Currently available HR-MAS probes are typically outfitted with a ^{2}H field-frequency lock and PFG capabilities to allow for tailored solvent suppression and coherence selection in multidimensional experiments. The range of typical probe temperatures is typically 20–50 °C, which is sufficient for examination of most environmental and biomedical samples of interest. Inverse detection and three-channel capabilities are available; these features make it possible, for instance, to collect 2D ^{1}H–^{13}C and ^{1}H–^{15}N data with ease, particularly if the sample is present at high concentration in natural abundance or is enriched with NMR-active isotopes. Typical rotor diameters are 4–5 mm, with spinning capabilities of 2–15 kHz depending on the rotor material and other design parameters. Although rotors can be costly to purchase, many vendors offer inexpensive inserts that also ensure sealing of wet samples. It should also be noted that HR-MAS NMR experiments can be performed in a recently introduced probe for comprehensive multiphase NMR spectroscopy.[42] The reader is referred to the vendor websites and literature for additional information.

5.3 REPRESENTATIVE APPLICATIONS TO ENVIRONMENTAL SCIENCE

Although HR-MAS NMR applications are presented elsewhere in this book, in this section, we highlight both the methodology and caveats related to such environmental science investigations. While 1D HR-MAS NMR procedures have been described earlier, we now emphasize the use of HR-MAS in conjunction with various 2D NMR methods (COSY, TOCSY, HMQC, HMBC, NOESY, HMQC-TOCSY, etc.) to delineate structurally informative through-bond and through-space interactions for plant and soil systems. Comparisons are made with solution- and solid-state NMR investigations of representative environmental materials, including critical assessments of the capabilities of HR-MAS NMR to provide reliable and otherwise unavailable structural or dynamic information.

5.3.1 Soil Science: HR-MAS NMR at Solid–Solvent Interfaces

Whole soil is comprised of soluble, solid, and swellable constituents, each requiring a tailored spectroscopic approach for molecular characterization. The first ^{1}H HR-MAS NMR experiments of these complex systems[17] were shown to provide otherwise inaccessible information regarding molecular structure at the solid–aqueous interface. HR-MAS NMR experiments can avoid concerns about artifacts associated with the spectra of soluble extracts, and they also have the potential to refine generic compositional information derived from CPMAS ^{13}C NMR of powdered samples.[43] The spectra exhibit relatively sharp resonances and good sensitivity, particularly if presaturation is used to attenuate the solvent signal and spin-echo filtering is used to remove broad contributions from rapidly relaxing high-molecular-weight and paramagnetic species. The ^{1}H HR-MAS NMR data reflect the solvent-accessible components of the solid soil; independent examination of the supernatant verifies that the spectra contain no contributions from soluble constituents.

In D_2O, the observed chemical shifts implicate predominantly (oxy)aliphatic moieties, which were confirmed by 2D MAS-total correlation spectroscopy (TOCSY) NMR (Figure 5.9) to arise from acids, esters, ethers, and sugars found previously in soil organic extracts. Additional resonances from aromatic structures are evident when the sample is swelled in DMSO-d_6, a 'penetrating' solvent for both hydrophilic and hydrophobic domains that may also break hydrogen bonds. The inference drawn from such studies is that aliphatic moieties are most involved in the binding of the aqueous-phase contaminants and heavy

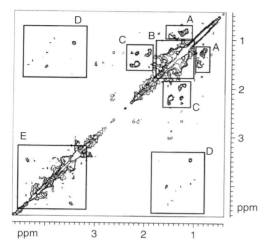

Figure 5.9. Aliphatic region of a 600 MHz 2D HR-MAS-TOCSY spectrum recorded for a whole soil sample swollen in D_2O, showing scalar couplings between pairs of CH_3, CH_2, and sugar groups designated by the letters A–E.[17] (Reprinted with permission from Ref. 17. © 2001 American Chemical Society)

metals that raise environmental concerns. By contrast, the aromatic moieties may exist in hydrophobic environments such as phyllosilicate clays.

5.3.2 Environmental Carbon and Nitrogen Cycling: HR-MAS NMR for Chemical Conversions

HR-MAS NMR approaches also have excellent potential for studies of terrestrial plant litter decomposition, a process in which a variety of carbohydrates, tannins, proteins, lignins, waxes, and cutins have differing susceptibilities to environmental degradation and capabilities for carbon or nitrogen sequestration.[44] An example that illustrates several useful techniques for such work is provided by Kelleher *et al.*,[41] who devised a model system to monitor the degradation of lodgepole pine (*Pinus contorta*) and Western wheat grass (*Agropyron smithii*) plant materials during a year's time.

^{1}H HR-MAS NMR provided sufficient spectral resolution to identify regions for aromatic, carbohydrate, oxymethine, oxymethylene, methylene, and methyl groups, and as described earlier, the rigid moieties could be selected by a diffusion-edited experiment.[38]

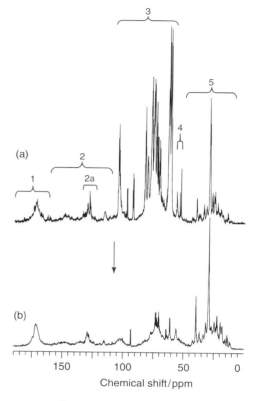

Figure 5.10. [13]C HR-MAS spectra of wheat grass leaves obtained with a spinning rate of 10 kHz at an NMR frequency of 125 MHz, showing chemical differences between materials that are fresh (a) and decomposed for 12 months (b). The numbered spectral regions were assigned structurally by analogy with published literature to carbonyls (1), aromatic and doubly bonded moieties (2), carbohydrates (3), methoxyls (4), and aliphatics (5).[41] (Reproduced with permission from Ref. 41. © 2006 Elsevier)

By enriching these materials in the ^{13}C isotope, the investigators were able to observe ^{13}C spectra directly, obtaining the superior chemical-shift dispersion and resolution illustrated in Figure 5.10. Enrichment with both ^{13}C and ^{15}N also facilitated acquisition of 2D heteronuclear single-quantum coherence (HSQC) NMR spectra, further improving the resolution of the ^{1}H spectra and allowing identification of covalently bound [1]H–[13]C and [1]H–[15]N structural fragments.

In the wheat grass, for instance, fresh blades were demonstrated to consist of predominantly rigid (macromolecular) components, whereas decomposed materials had a greater proportion of mobile (small) carbohydrate units that also displayed sharper resonances. Conversely, the relative intensity of aliphatic resonances increased with time, likely reflecting the more modest decomposition rates of cutin and wax components. Notably, it was possible to demonstrate disappearance and buildup of particular compound classes by analysis of the HR-MAS-assisted ^{1}H–^{13}C HSQC NMR fingerprints of the pine needle samples: cross-peaks from tannins had nearly disappeared after 12 months, whereas features attributable to triterpenoids became more prominent in the spectra because of their evident recalcitrance. Finally, HR-MAS ^{1}H–^{15}N HSQC NMR of the pine needles and stem/root materials revealed a decrease in protein that was accompanied by buildup of as-yet-unidentified nitrogen-containing molecules that could originate either from the fresh plants or new microbe-assisted synthesis. Although it remains difficult to discriminate quantitatively between swelled and solid materials and to identify specific chemical compounds as well as compound classes involved in particular transformations, the superior resolution and covalent bonding information provided by HR-MAS NMR makes it a powerful tool on its own and as a complement to CPMAS approaches for associated solids.

5.3.3 Plant Protective Biopolymers: nD HR-MAS NMR for Macromolecular Architecture

5.3.3.1 HR-MAS-HMQC NMR of Citrus Fruit Cutins: Capabilities and Caveats

The 1D ^{1}H HR-MAS spectra shown in Figures 5.1 and 5.2 display dramatic NMR resolution enhancement of the lime fruit cutin biopolymer with solvent swelling. Moreover, 2D data such as the well-resolved ^{1}H–^{13}C Heteronuclear Multiple Quantum Coherence (HMQC) NMR spectra in Figure 5.3 offer the potential for monitoring the mild chemical or enzymatic transformations used to produce oligomeric fragments that preserve elements of the biopolymer architecture.[45] The HMQC NMR spectrum of the swelled-solid residue in Figure 5.3(c) suggests loss of some aliphatic chain groups but resembles the original polyester as expected, as only ~20% of the lime cutin is broken down by this particular treatment. Moreover, the highlighted regions of the ^{1}H–^{13}C HMQC NMR fingerprints shown in Figure 5.3(a) and (b) offer insight into the chemical

transformations that accompany depolymerization: the CH$_2$OR group of the cutin (61, 3.4 ppm) is cleaved to CH$_2$OH (63, 3.6 ppm) in a major soluble product. The oligomer also shows new CH$_3$CH$_2$OC=O (60, 4.1 ppm) and CH$_3$CH$_2$OC=O (14, 1.2 ppm) groups consistent with termination by an ethyl ester[24,46] and was subsequently elucidated as a linear trimer ester of 10-oxo-16-hydroxyhexadecanoic acid using traditional solution-state NMR and mass spectrometry methods.[45]

However, neither midchain alcohol nor ester groups are observed in 2D HR-MAS NMR spectra of the cutin polymer or the unreacted residue (Figure 5.3b and c),[24] despite evidence for such structural features in both soluble monomeric breakdown products[47,48] and intact dry solids examined by CPMAS ^{13}C NMR.[25] In retrospect, then, the 1D HR-MAS ^{1}H spectral comparisons in Figures 5.1 and 5.2 do not provide sufficiently stringent support for the completeness of solvent swelling in this material; the same factors that pose challenges for polymer degradation[49] are likely to defeat efforts to swell the samples with DMSO-d_6.

5.3.3.2 HR-MAS-HMQC with HR-MAS-COSY NMR of Potato Tuber Stress Response: Tracing Polysaccharide Architecture

A better prospectus for macromolecular investigation is offered by potato biopolymers formed on cell-wall surfaces in response to environmental stresses such as wounding in order to prevent dehydration and microbial invasion. Figure 5.11 illustrates the coordinated use of 2D HR-MAS-assisted HMQC with correlation spectroscopy (COSY) NMR experiments to make structural assignments for the cell-wall polysaccharides protected by the aliphatic–aromatic polyester suberin. For these DMSO-d_6-swelled samples, resolution is excellent and all of the expected functional groups are swelled successfully following the sample preparation guidelines outlined earlier. To assign the polysaccharide resonances, the ^{1}H peaks at 4.4, 5.4, and 5.5 ppm were first identified tentatively as hydroxyl substituents on sugar rings of the polysaccharides, based on their chemical-shift values and susceptibility to exchange with D$_2$O. Using a strategy similarly to that reported in a prior investigation of intracellular adhesion-strengthened potato tissues,[32] these sugar hydroxyl assignments were confirmed and linked to their respective sugar ring positions by comparing the through-bond connectivities in double-quantum-filtered (DQF) COSY and ^{1}H–^{13}C

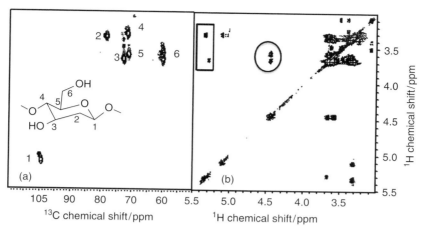

Figure 5.11. Upfield portion of the HR-MAS-assisted $^1H-^{13}C$ gHMQC (a) and $^1H-^1H$ COSY (b) contour plots from NMR experiments conducted on DMSO-d_6-swelled suberized potato wound periderm at 50 °C, obtained with a proton frequency of 600 MHz and a spinning frequency of 3.000 ± 0.001 kHz. The highlighted COSY cross-peaks designate through-bond correlations between protons on the sugar rings and their attached OH groups; the HMQC correlations confirm the ring assignments and locate the hydroxyl substituents (Table 5.3).

Table 5.3. NMR resonance assignments for carbohydrate signals in suberized potato wound periderm tissues

Glucose structural position	1	2	3	4	5	6
^1H shift (ppm)	5.01, 5.04	3.35	3.66	3.31	3.60	3.55, 3.67
^{13}C shift (ppm)	99.9	78.7	72.7	71.7	71.4	60.1

HMQC experiments (Table 5.3). As reported for the texture-spoiling potato polymer described earlier,[32] the lack of a hydroxyl group at ring position 2 suggests that other parts of the biopolymer are attached at that site.

5.3.3.3 HR-MAS-HMBC NMR of Potato Tuber Stress Response: Mapping Inter-domain Connections

How are the aliphatic and aromatic domains of the suberin polyester linked together, and what is the nature of the covalent architecture or noncovalent associations that occur with the underlying cell-wall polysaccharides? Till date, proposals of covalent bonding within this assembly have relied on indirect evidence from ^{13}C spin–lattice relaxation and $^{13}C-^{13}C$ spin diffusion experiments.[50,51] Current solid-state NMR protocols for measuring carbon–carbon spatial proximities or through-bond

connections require isotopic enrichment and may display limited spectroscopic resolution for amorphous materials. By contrast, a measure of detail regarding inter-domain connections can be derived from HR-MAS-assisted $^1H-^{13}C$ HMBC NMR experiments if the plant tissue samples are swelled in DMSO-d_6 to enhance molecular motions (Figure 5.12).

Firstly, the spectrum shows multiple-bond connections between carboxyl carbons and aliphatic chain methylene groups (dashed rectangle), as expected from longstanding observations of C_{18} hydroxy fatty acid monomer building blocks within the aliphatic domain of the potato suberin polyester.[47] Secondly, HMBC demonstrates long-range bonding between aliphatic and aromatic groups (solid rectangle), consistent with suberin's proposed aliphatic–aromatic polyester structure.[47,52,53] Finally, the contour plot shows connections between aliphatic and carbohydrate resonances (solid circle), augmenting NOESY and TOCSY evidence of their proximities (data not

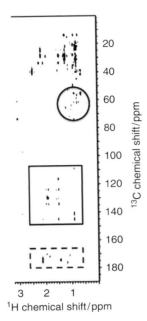

Figure 5.12. Upfield portion of the contour plot from 750 MHz HR-MAS-gHMBC NMR of suberized potato wound periderm swelled in DMSO-d_6 at 50 °C, showing multiple-bond connectivities within the suberin aliphatic domain (dashed rectangle), between suberin aliphatic and aromatic domains (solid rectangle), and between the suberin aliphatics and cell-wall polysaccharides (solid circle). The data were obtained with 10.000 ± 0.001 kHz MAS and 40 G cm^{-1} magic-angle PFGs; the 65 ms defocusing period was optimized for a multiple-bond scalar coupling of 8 Hz.

shown) and offering the first direct evidence of covalent linkages between the protective suberin polymer and the cell-wall periderm (phellem) tissues. Efforts to enhance the sensitivity and completeness of these encouraging experiments are ongoing.

5.4   CONCLUSIONS

HR-MAS NMR techniques are well suited for atomic- and molecular-level investigations of heterogeneous environmental samples such as soils, plant biomass, protective plant biopolymers, and urban surface films. With suitable optimization of sample preparation and data acquisition, this technology can be coupled usefully with a wealth of powerful 1D spectral editing methods and multidimensional structure elucidation experiments and is generally applicable to

NMR-active nuclei present at natural abundance. Neither soil or tissue extraction is required nor is it necessary to subject the macromolecular entities of interest to chemical degradation. HR-MAS NMR can often provide a unique route to inaccessible but functionally important information about intractable amorphous solids, solid–solvent interfaces, and chemical transformations of importance to environmental science and metabolic engineering.

ACKNOWLEDGMENTS

The work presented herein was supported by grants from the National Science Foundation (MCB-0843627, MCB-0741914, and DEB-1127253), the National Institutes of Health (R01-AI052733), and the US–Israel Binational Agricultural Research and Development Fund (USIS-3368-02). NMR resources were supported by the College of Staten Island, the City College of New York, and the CUNY Institute for Macromolecular Assemblies. Additional infrastructural support was provided by NIH 2G12RR03060 from the National Center for Research Resources and 8G12MD007603 from the National Institute on Minority Health and Health Disparities. Prof. R.E.S. is a member of the New York Structural Biology Center (NYSBC); data collected at the NYSBC was made possible by a grant from NYSTAR. We gratefully acknowledge Drs Hsin Wang and Boris Itin for technical assistance with the NMR experiments; Dr Subhasish Chatterjee and Ms Lauren Gohara also provided patient assistance with the preparation of this chapter.

RELATED ARTICLES IN EMAGRES

NMR Probes for Small Sample Volumes

High-resolution MAS for Liquids and Semisolids

Agriculture and Soils

REFERENCES

1.   D. Doskocilova, D. D. Tao, and B. Schneider, *Czech. J. Phys.*, 1975, **B25**, 202.

2.   D. Doskocilova, B. Schneider, and J. Jakes, *J. Magn. Reson.*, 1978, **29**, 79.

3. H. D. H. Stover and J. M. J. Frechet, *Macromolecules*, 1989, **22**, 1574.

4. W. T. Ford, S. Mohanraj, and H. Hall, *J. Magn. Reson.*, 1985, **65**, 156.

5. H. D. H. Stover and J. M. J. Frechet, *Macromolecules*, 1991, **24**, 883.

6. E. Oldfield, J. L. Bowers, and J. Forbes, *Biochemistry*, 1987, **26**, 6919.

7. J. Forbes, C. Husted, and E. Oldfield, *J. Am. Chem. Soc.*, 1988, **110**, 1059.

8. E. Krainer, R. E. Stark, F. Naider, K. Alagramam, and J. M. Becker, *Biopolymers*, 1994, **34**, 1627.

9. H. N. Halladay, R. E. Stark, S. Ali, and R. Bittman, *Biophys. J.*, 1990, **58**, 1449.

10. L. L. Holte and K. Gawrisch, *Biochemistry*, 1997, **36**, 4669.

11. W. L. Fitch, G. Detre, and C. P. Holmes, *J. Org. Chem.*, 1994, **59**, 7955.

12. R. C. Anderson, M. A. Jarema, M. J. Shapiro, J. P. Stokes, and M. Ziliox, *J. Org. Chem.*, 1995, **60**, 2650.

13. P. A. Keifer, L. Balthusis, D. M. Rice, A. A. Typiak, and J. N. Shoolery, *J. Magn. Reson. A*, 1996, **119**, 65.

14. W. E. Maas, F. H. Laukien, and D. G. Cory, *J. Am. Chem. Soc.*, 1996, **118**, 13085.

15. D. Moka, R. Vorreuther, H. Schicha, M. Spraul, E. Humpfer, M. Lipinski, P. J. Foxall, J. K. Nicholson, and J. C. Lindon, *J. Pharm. Biomed. Anal.*, 1998, **17**, 125.

16. R. E. Stark, B. Yan, A. K. Ray, Z. Chen, X. Fang, and J. R. Garbow, *Solid State Nucl. Magn. Res.*, 2000, **16**, 37.

17. A. J. Simpson, W. L. Kingery, D. Shaw, M. Spraul, E. Humpfer, and P. Dvortsak, *Environ. Sci. Technol.*, 2001, **35**, 3321.

18. S. Tian, J. Garcia-Rivera, B. Yan, A. Casadevall, and R. E. Stark, *Biochemistry*, 2003, **42**, 8105.

19. W. Li, *Analyst*, 2006, **131**, 777.

20. F. Bertocchi and M. Paci, *J. Agric. Food Chem.*, 2008, **56**, 9317.

21. J. C. Lindon, O. P. Beckonert, E. Holmes, and J. K. Nicholson, *Prog. Nucl. Magn. Reson. Spectrosc.*, 2009, **55**, 79.

22. A. J. Simpson, D. J. McNally, and M. J. Simpson, *Prog. Nucl. Magn. Reson. Spectrosc.*, 2011, **58**, 97.

23. R. E. Stark, B. Yan, and G. Wu, 40th Experimental NMR Conference, Orlando, FL, 1999, p 20.

24. X. Fang, F. Qiu, B. Yan, H. Wang, A. J. Mort, and R. E. Stark, *Phytochemistry*, 2001, **57**, 1035.

25. T. Zlotnik-Mazori and R. E. Stark, *Macromolecules*, 1988, **21**, 2412.

26. R. E. Stark, T. Zlotnik-Mazori, L. M. Ferrantello, and J. R. Garbow, *ACS Symp. Ser.*, 1989, **399**, 214.

27. J. R. Garbow and R. E. Stark, *Macromolecules*, 1990, **23**, 2814.

28. A. P. Deshmukh, A. J. Simpson, and P. G. Hatcher, *Phytochemistry*, 2003, **64**, 1163.

29. A. P. Deshmukh, A. J. Simpson, C. M. Hadad, and P. G. Hatcher, *Org. Geochem.*, 2005, **36**, 1072.

30. R. E. Hurd and B. K. John, *J. Magn. Res.*, 1991, **91**, 648.

31. J. Cavanagh, W. J. Fairbrother, A. G. Palmer III, M. Rance, and N. J. Skelton, Protein NMR Spectroscopy: Principles and Practice, 2nd edn, Academic Press: New York, 2007.

32. B. Yu, G. Vengadesan, H. Wang, L. Jashi, T. Yefremov, S. Tian, V. Gaba, I. Shomer, and R. E. Stark, *Biomacromolecules*, 2006, **7**, 937.

33. J. Zhong, S. Frases, H. Wang, A. Casadevall, and R. E. Stark, *Biochemistry*, 2008, **47**, 4701.

34. H. Z. Hill, *BioEssays*, 1992, **14**, 49.

35. K. G. Valentine, R. W. Peterson, J. S. Saad, M. F. Summers, X. Xu, J. B. Ames, and A. J. Wand, *Structure*, 2010, **18**, 9.

36. J. M. Kielec, K. G. Valentine, C. R. Babu, and A. J. Wand, *Structure*, 2009, **17**, 345.

37. I. Shomer, R. Vasiliver, and P. Lindner, *Carbohydr. Polym.*, 1995, **26**, 55.

38. D. Wu, A. Chen, J. Johnson, and S. Charles, *J. Magn. Reson. A*, 1995, **115**, 260.

39. A. J. Simpson, B. Lam, M. L. Diamond, D. J. Donaldson, B. A. Lefebvre, A. Q. Moser, A. J. Williams, N. I. Larin, and M. P. Kvasha, *Chemosphere*, 2006, **63**, 142.

40. S. Chatterjee, F. Santos, S. Abiven, B. Itin, R. E. Stark, and J. A. Bird, *Org. Geochem.*, 2012, **51**, 35.

41. B. P. Kelleher, M. J. Simpson, and A. J. Simpson, *Geochim. Cosmochim. Acta*, 2006, **70**, 4080.

42. D. Courtier-Murias, H. Farooq, H. Masoom, A. Botana, R. Soong, J. G. Longstaffe, M. J. Simpson, W. E. Maas, M. Fey, B. Andrew, J. Struppe, H. Hutchins, S. Krishnamurthy, R. Kumar, M. Monette, H. J. Stronks, A. Hume, and A. J. Simpson, *J. Magn. Reson.*, 2012, **217**, 61.

43. M. A. Wilson, NMR Techniques and Applications in Geochemistry and Soil Chemistry, Pergamon Press: Oxford, UK, 1987.

44. P. Rovira and V. Ramo, *Geoderma*, 2002, **107**, 109.

45. S. Tian, X. Fang, W. Wang, B. Yu, X. Cheng, F. Qiu, A. J. Mort, and R. E. Stark, *J. Agric. Food Chem.*, 2008, **56**, 10318.

46. R. E. Stark and S. Tian, in Biology of the Plant Cuticle, eds M. Riderer and C. Muller, Blackwell Publishing Co: Oxford, UK, 2006, 126.

47. P. E. Kolattukudy, *Adv. Biochem. Eng. Biotechnol.*, 2001, **71**, 1.

48. T. J. Walton, *Meth. Plant Biochem.*, 1990, **4**, 105.

49. A. K. Ray and R. E. Stark, *Phytochemistry*, 1998, **48**, 1313.

50. R. E. Stark and J. R. Garbow, *Macromolecules*, 1992, **25**, 149.

51. B. Yan and R. E. Stark, *J. Agric. Food Chem.*, 2000, **48**, 3298.

52. M. A. Bernards, *Can. J. Bot.*, 2002, **80**, 227.

53. J. Graca and S. Santos, *Macromol. Biosci.*, 2007, **7**, 128.

54. R. A. Pacchiano, W. Sohn, V. L. Chlanda, J. R. Garbow, and R. E. Stark, *J. Agric. Food Chem.*, 1993, **41**, 78.

FURTHER READING

A. J. Simpson, M. J. Simpson, and R. Soong, *Environ. Sci. Technol.*, 2012, **46**, 11488.

Chapter 6
Environmental Comprehensive Multiphase NMR

André J. Simpson[1], Denis Courtier-Murias[2], James G. Longstaffe[2], Hussain Masoom[2], Ronald Soong[2], Leayen Lam[2], Andre Sutrisno[1], Hashim Farooq[2], Myrna J. Simpson[1], Werner E. Maas[3], Michael Fey[3], Brian Andrew[3], Jochem Struppe[3], Howard Hutchins[3], Sridevi Krishnamurthy[3], Rajeev Kumar[3], Martine Monette[4] and Henry J. Stronks[4]

[1]*Department of Chemistry and Environmental NMR Centre, University of Toronto, Toronto, Ontario, M1C 1A4, Canada*
[2]*Department of Chemistry and Environmental NMR Centre, University of Toronto Scarborough, Toronto, Ontario, M1C 1A4, Canada*
[3]*Bruker BioSpin Corporation, Billerica, MA 01821-3991, USA*
[4]*Bruker Ltd, Milton, Ontario, L9T 1Y6, Canada*

NMR Spectroscopy: A Versatile Tool for Environmental Research
Edited by Myrna J. Simpson and André J. Simpson
© 2014 John Wiley & Sons, Ltd. ISBN: 978-1-118-61647-5

6.1 INTRODUCTION

Samples in the environment in their natural state commonly contain an array of different chemical components in different phases. Often it is synergy between these phases that give rise to overall structure and reactivity in natural samples. Consider for example a plant. The epidermis provides mechanical strength and protection to the plant that itself is covered by a waxy cuticular layer that acts as a water barrier. The vascular tissue provides structural support and contains xylem and phloem tissues that transport fluids and nutrients. Ground tissue often constitutes the pith of stems or pulp of fruits. This tissue has thin cellulose walls but contains large water-filled vacuoles, which store and regulate water and waste. As such, a range of different structural components are present that in turn exhibit a range of different rigidities from true solids (e.g., cross-linked ligno-celluloses in the cell walls), 'gels' (various swollen biopolymers), to true solutions (liquids in vascular conduits). Indeed, it is a complex relationship between the chemical components as well the assimilations of components from one phase to the other that ultimately determines plant growth. To truly understand such processes, researchers need high-resolution (HR) molecular information from all the various materials in their natural state.

A similar argument can be made for soil. The material in plant debris or organic material tightly associated with mineral surfaces may be relatively solid, swollen organic matter more 'gel-like', while components in soil pores may be fully dissolved. As such, it is the complex interactions between various chemical constitutes in various phases that are fundamental to the formation, stability, and fertility of soils. The situation gets even more interesting if you consider the fate of a molecule interacting within a soil. In the case of a spill, a hydrophobic contaminant may enter the soil as a solution. As it interacts with the soil organic matter (SOM), it may take on properties of a gel before being sequestered into hydrophobic pockets with more solid-like character. To fully elucidate the process of chemical sequestration, it is important to be able to investigate and differentiate the molecular interactions from various phases over time. In turn, it is critical to understand how and why molecules become sequestered in soil in order to design the most effective ways to remediate them. This was summarized by Bertsch and Seaman who stated 'a complete understanding of the chemistry of complex soil assemblages is prerequisite to accurately assessing environmental and human health risks of contaminants or in designing environmentally sound, cost-effective chemical and biological remediation strategies'.[1] Consequently, there is a great need in environmental research to study samples in their natural state and be able to extract and differentiate structure/interaction information from liquids, gels, and solids. Processes such as flocculation, drying, swelling, and melting involve the conversion of one phase to another and are key to larger environmental processes such as erosion, chemical transport, and cycling of carbon. Technology that can provide molecular-level information, as to the conversion of a liquid to a gel to a solid (or any combination thereof), has considerable widespread potential for not just environmental research but also other fields including biology, materials research, medicine, and industrial catalysis.

6.2 NMR SPECTROSCOPY OF NATURAL SAMPLES

6.2.1 Traditional NMR Approaches

Traditionally, NMR spectroscopy has developed into two distinct fields, in large part because of the different hardware requirements of each. The first deals with soluble materials (solution-state NMR spectroscopy) and the other with solids (solid-state NMR spectroscopy). Solution-state NMR spectroscopy generally employs probes using low-power electronics, a lock channel, and a gradient, and provides excellent lineshape on dissolved samples. ^{1}H detection is most commonly used because of its high sensitivity, excellent solution lineshape, and abundance in common organic structures. On the other hand, most solid-state NMR experiments need sufficient B_1 strength for high-power decoupling and cross polarization (CP) (common elements of most solid-state experiments) and are commonly performed while spinning at the magic angle. ^{1}H–^{1}H dipolar interactions in the solid state are considerable and reduce the information that can be obtained by ^{1}H NMR in solids. As such, ^{13}C is often the most commonly studied nucleus for solids. Solid-state NMR spectroscopy has been predominantly reserved for the study of true solids and as such, a spectrometer lock and gradients are not required. The line separating the two fields was blurred somewhat in 1996 with the introduction of HR-magic-angle spinning (MAS) NMR.[2] HR-MAS NMR targets gel or swellable materials. The probes are fitted with a lock and a gradient, and the sample is spun at the magic angle. Dipolar interactions dominant in the solid phase are reduced by either adding a swelling solvent or simply studying a matrix that is naturally 'gel-like'. Common applications involve the study of plant materials,[3] soils,[4] biological tissues,[5] and synthetic chemistry using solid supports.[6,7] Excellent ^{1}H lineshape can be obtained in HR-MAS, and for mobile components, it can be close to that observed in solution-state NMR. HR-MAS NMR is able to utilize the full range of experiments available for solution-state NMR spectroscopy and represents a very powerful technique for the study of swollen materials. HR-MAS probes are, however, designed using low-power circuitry and cannot handle the B_1 fields required for CP or high-power decoupling, and as a result, the vast majority of solid-state experiments cannot be performed on an HR-MAS probe.

In addition to the differences in the hardware and specific NMR experiments related to each technology, these technologies also deviate on the pretreatment procedures required for samples. Analysis using solid-state NMR spectroscopy requires the least pretreatment. For many samples with high organic content, they are simply dried. Soils and sediments, however, are often pretreated with hydrofluoric acid[8] to remove the inorganic minerals; this both concentrates the organic matter and helps remove paramagnetic metals. Samples are commonly ground to be <100 μm (if possible) after drying to avoid problems with spinning.

Solution-state NMR can be more invasive depending on the type of sample being studied. If the material is completely and naturally soluble, such as dissolved organic matter, it can be studied in its entirety and simply needs to be dissolved.[9] In the case of dissolved organic matter, as it is isolated from water in the first place, an aqueous NMR solvent is the obvious choice for dissolution. Solution-state NMR can also be very useful and provide complimentary information to solid-state NMR for SOM.[10] However, in this case, the SOM must be extracted, dried, and reconstituted for NMR analysis. Before dissolution, the use of a cation-exchange resin can help remove paramagnetic metals and sodium (sodium hydroxide is most commonly used for the isolation of SOM) and replace them with protons. This is especially important if a solvent such as DMSO-d_6 is used, which is an excellent solvent for SOM but only when the SOM is in the protonated form.[10] With such treatments, there is always the concern that constituents may be altered from their natural state at each step. Furthermore, as only the soluble fraction is targeted, it is possible that different structures remain in the solid phase and are nonextractable. One approach is to use very aggressive solvent systems to dissolve the material that cannot be extracted in aqueous solvents. In the case of soil, the nonextractable material is termed humin and studies using a mix of DMSO/sulfuric acid can be used to dissolve 70% of the humin fraction. While multidimensional NMR studies have shown this material to be relatively similar to more readily extractable components in soil with high contributions from cellulose and microbial cells walls,[11] the use of chemical extractions likely change conformations and aggregation-state of molecules. The result is that while structural information can be elucidated, studies involving molecular interactions may be less reliable, as the material will be somewhat altered with respect to its natural state.

Sample preparation for HR-MAS NMR depends on not only the type of sample but also the purpose of the study.[12] In some cases, samples are studied as is. Bunescu *et al.*[13] reported on the in vivo HR-MAS of *Daphnia magna* (freshwater crustaceans), while Miglietta and Lamanna[14] studied raw vegetables. Both studies employed the addition of D_2O for the spectrometer lock. In these cases, the sample is very close to its natural state and has not even been dried. This is important, as biological function is retained permitting a range of studies such as metabolism, biotransformation of contaminants, and drug binding. In other HR-MAS NMR studies, it may be of interest to deliberately change the sample. For example, in 2001, Simpson *et al.*[4] employed HR-MAS NMR to investigate different soil domains. When using an aqueous solvent, they observed the chemical components at the soil–water interface. However, by titrating DMSO-d_6 (an aprotic solvent that can break hydrogen bonds), they were able to investigate the hydrophobic domains that exist underneath the soil surface, which are known to be important for contaminant sorption.[15]

As such, it is clear that solid-state, gel-phase (HR-MAS), and solution-state NMRs are very powerful approaches in their own right. If applied to their full potential, they can unravel a huge plethora for molecular information from environmental samples arguably more than any other single analytical approach.[16] The main problem arises if information is required from the sample in its natural state and information from all bonds in their natural state need to be studied. In the case of a soil in its natural state, it will be swollen with water and contain water-filled pores. For solid-state NMR, samples must first be dried, thus it is not possible to study the hydrophobic soil domains, the reactive water interface, soluble compounds in pore spaces, or interactions and dynamics between these phases. Not only this but the components that were naturally dissolved could have given HR proton NMR signals in a well-shimmed, locked spectrometer. Indeed, two-dimensional (2D) solution-state NMR could have been acquired on these soluble components and a large amount of information is potentially lost by drying the sample and treating it as a pure solid. Similar arguments can be made for the gel-phase components, which could be selectively studied with improved resolution over solid-state NMR using HR-MAS NMR. Conversely, using HR-MAS NMR alone is also not ideal. While it will give excellent information on the soluble and gel components, its inability to handle CPMAS experiments means that little to no information can be obtained on the components in soil that are true solids in nature. One option is, therefore, to attempt such a study but use different NMR probes on the same sample to try and extract information from all the components present. Simpson *et al.*[17] came close to this in 2006 in a study of atmospheric deposits through combining solution, HR-MAS, and solid-state NMR on the same sample. The problem was that the study was extremely laborious, took many months, and requires access to solution-, HR-MAS, and solid-state NMR hardware. Furthermore, if molecular interactions are the focus, separate probes do not allow the transfer, kinetics, and interactions between phases to be accessed. Using comprehensive multi-phase (CMP) technology novel experiments can be designed that build upon advantages of both solution- and solid-state NMRs such as gradient-enhanced solid experiments or diffusion-weighted CPMAS. Theoretically, it is possible by integrating concepts from different fields of NMR spectroscopy to produce a novel comprehensive approach that can be used to study and differentiate between the various phases (solids, liquids, and gels) in a heterogeneous sample as well as identify the key structures and their molecular interactions within each phase. Indeed, there has been a collaboration between the Environmental NMR Centre (University of Toronto) and Bruker Biospin over recent 6 years. The resulting hardware has been termed CMP-NMR spectroscopy and it permits the full range of solid-state, HR-MAS, and solution-state experiments to be performed on sample without pretreatment or extraction.

6.2.2 CMP-NMR Spectroscopy

CMP-NMR technology incorporates all of the aforementioned methods, including MAS, a magic-angle gradient, a lock, full susceptibility matching, and solid-state circuitry to permit high-power handling. Therefore, it is built to observe all bonds, in all phases in natural and unaltered samples, resulting in a universal approach. For example, HR-MAS NMR has been applied previously and the experiments showed that it can be used to monitor sorption kinetics of contaminants across the soil–water interface in an intact whole soil.[18] However, once the molecules penetrate into pure solid domains, they are no longer observable and information as to the sequestered state could not

be obtained. CMP-NMR changes this, permitting the full-range NMR experiments available today to be applied to the solution, gel, and solids components. In turn, this permits studies following the molecular interactions and kinetics of a contaminant molecule, as it transgresses between different phases including the final sequestered state. In many ways, CMP-NMR can be thought of as changing NMR technology to match natural samples, rather than changing the sample to match a specific NMR technique. Alterations to NMR hardware also change the pretreatments required for samples. With CMP-NMR, samples are deposited directly in the sample rotor and a lock solvent is added. The lock can be added directly to the sample or separately from the sample in an external capillary so as not to perturb the natural state. The goal of CMP-NMR is to permit the uncompromised study of structure and interactions in situ.

The first CMP-NMR probe was described by Courtier-Murias *et al.*[19] and was based on a four-channel ^{1}H/^{2}H/^{19}F/^{13}C design. An HR-MAS style stator and gradient were used for improved lineshape along with transmission line circuitry from solid-state probe design to permit high-power handling. The ^{2}H channel was purely for locking purposes and the lock tuning was placed on the main circuit. In subsequent designs, a separate lock circuit has been introduced that removes the need for an additional tuning on the main coil, which in turn improves performance and sensitivity of the main circuit. The most challenging aspect of CMP-NMR is the considerable difficulty in fitting all the circuitry into the rather tight confines of a narrow bore probe. Complications can arise from interactions of the gradient coil with high-power RF. A simple solution is to reduce the high-power decoupling bandwidth by ~20% over a dedicated solids probe, which reduces performance slightly but does not prevent the application of solid-state experiments, even very demanding ones such as combined rotation and multiple-pulse sequence (CRAMPS).[19] However, it is important to stress that the additional gradient and lock tuning will affect overall sensitivity over a dedicated solid-state probe or HR-MAS probe (between 10 and 40% loss depending on the probe design). As such, CMP-NMR probes are ideal for the study of intact multiphase samples but should not be thought of as a 'wonder' probe that replaces the need for separate dedicated probes for more traditional 'single-phase' work.

The spectral editing capabilities of CMP-NMR were first demonstrated on a model multiphase sample and a contaminated soil,[19] which also served as evidence that binding preferences between contaminants and soil can also be studied in situ. This study showed that different solution-, gel-, and solid-like NMR spectra for ^{1}H, ^{13}C, and ^{19}F nuclei can be obtained, and clearly demonstrated that the approach is capable of studying all components in intact heterogeneous mixtures such as those commonly studied in environmental research. Editing approaches were also introduced that permit the signals from solution, gel, and solid components to be differentiated. These filters exploit various properties inherent in different phases (nuclear tumbling, dipolar interactions, and molecular diffusion) to obtain physical-phase edited NMR spectra mainly targeting basic structural information.

Examples of these editing experiments were presented on a model sample (composed of soluble nicotinamide, 1-aminohexylamide gel, and insoluble cholesterol acetate in aqueous solvent[19]). The simplest filter based on [1]H T_2-filtering highlights signals from liquid components (Figure 6.1). An alternative approach can be used to get a similar result. This is based on post-processing difference spectroscopy of diffusion-weighted spectra (see Ref. 19 for further details). The result is only components that have unrestricted self-diffusion (i.e., truly soluble

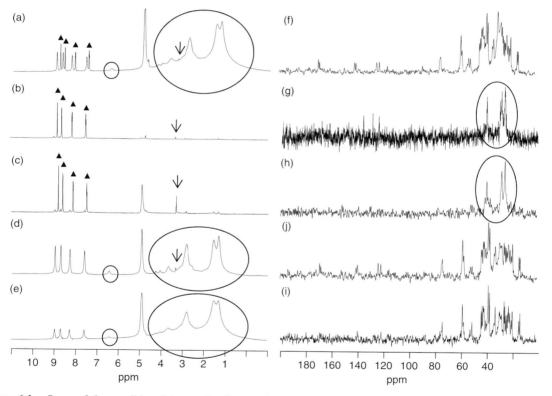

Figure 6.1. Some of the possible editing applications performed on the model sample, which produce liquid-, gel-, and solid-like spectra for both ^{1}H and ^{13}C nuclei. All these experiments were run on the model sample including the reference (a) single-pulse excitation ^{1}H NMR and (f) ^{13}C CPMAS experiments, which are shown on the top of the figure. Both (b) [1]H T_2-filtered and (c) 'inverse' ^{1}H diffusion-edited NMR experiments were used to show liquid-like molecules only. Spectra of gel-like components can be acquired using different approaches such as (d) ^{1}H diffusion-edited and (e) 'inverse' ^{1}H T_2-filtered NMR experiments. In the case of ^{13}C NMR experiments, spectra of gel-like components were acquired with (g) 'inverse' [13]C T_2-filtered and (h) [1]H T_2-filtered ^{13}C CPMAS experiments. Finally, (i) 'inverse' [1]H T_2-filtered ^{13}C CPMAS and (j) [1]H $T_{1\rho}$-filtered ^{13}C CPMAS spectra are presented, showing solid-like molecules only. Signals from 1-aminohexylamide gel are highlighted with ovals, nicotinamide free in solution with ▲, when in a gel-like environment with *, and the degradation product methanol is marked with an arrow. (Reproduced with permission from Ref. 19. © 2012 Elsevier)

components) that are retained in the final edited spectrum (Figure 6.1c).

When gel-like components are of interest, one option is to exploit T_2 relaxation as shown in Figure 6.1(e), which consists of an 'inverse' ^{1}H T_2-filtered NMR experiment created by difference. An alternative approach to obtain a gel-only spectrum is diffusion editing using a strong gradient (Figure 6.1d). However, when diffusion editing-based approaches are applied to natural samples, components with rapid relaxation will relax during the relatively long delays potentially leading to some components not be observed. This can be avoided if a recovery of relaxation losses arising from diffusion editing (RADE) experiment is performed. RADE is discussed in detail by Courtier-Murias *et al.*,[19] but essentially accounts for any components that preferentially relax during diffusion-based experiments and fully recovers the signals. The spectrum contains ^{1}H signals from semisolid components and when inverse diffusion editing (soluble components), diffusion editing (restricted components), and RADE (fast relaxing semisolids) filters are combined, signals in a sample can be fully differentiated without signal loss.

Figure 6.1 also shows some of the possible approaches for spectral editing for ^{13}C nuclei, which in this case are mainly based on a ^{13}C CPMAS experiment (Figure 6.1f). Gel-like spectra were acquired with 'inverse' ^{13}C T_2-filtered (Figure 6.1g) and ^{1}H T_2-filtered (Figure 6.1h) ^{13}C CPMAS experiments. The solid fraction of the sample is selectively observed either using 'inverse' ^{1}H T_2-filtered ^{13}C CPMAS (Figure 6.1i) or ^{1}H $T_{1\rho}$-filtered ^{13}C CPMAS (Figure 6.1j) experiments. The examples can be further extended to the ^{1}H nucleus in solids using CRAMPS for detection.[19] While the ^{1}H signal even after narrowing via CRAMPS is relatively broad, ^{1}H detection is more sensitive than the ^{13}C detection and, therefore, has potential for kinetic experiments where a solid may form over time (for example, freezing or a chemical reaction) and the goal is simply to follow the formation of the solid components.

One important characteristic of CMP-NMR spectroscopy is the ability to use magic-angle gradients, as employed in some of the one-dimensional editing approaches already discussed (Figure 6.1). However, if numerous components are present or fine interactions between the components are of importance, 2D NMR may be more effective. For example, 2D diffusion ordered spectroscopy is able to separate components

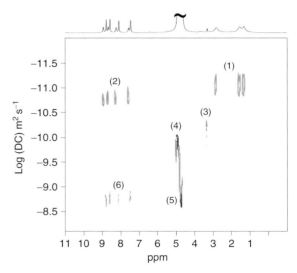

Figure 6.2. An example of DOSY NMR spectroscopy used to differentiate the ^{1}H NMR signals for liquid and gels. The 2D DOSY ^{1}H experiment was carried out in the model sample, where signals on the top of the spectrum are from molecules in a gel phase and those on the bottom correspond to molecules in solution phase. (1) 1-aminohexylamide gel, (2) nicotinamide in gel, (3) methanol, (4) water interacting with the gel, (5) free water, and (6) free nicotinamide. (Reproduced with permission from Ref. 19. © 2012 Elsevier)

with various arrays of diffusivity in a single experiment. Figure 6.2 shows that, when applied to the model sample, the ^{1}H DOSY experiment provides a continuous diffusion-based separation and is extremely useful for characterizing and separating the components of the mixture (see Figure 6.2 and Ref. 19 for details).

6.3   MOLECULAR STRUCTURE IN NATURAL MATERIALS USING CMP-NMR

6.3.1   Soil Structure

CMP-NMR has great potential for the study of soil structure in its natural state. Previously, a number of SOM composition studies have been performed on extracts and solution-state 2D NMR has been used to identify the most abundant major components in SOM as carbohydrates, lignin, protein, and aliphatic components.[20] However, there is a concern that extractions may alter the structural components or lead to a biased fractionation where insoluble components are unaccounted for. Preliminary CMP-NMR data (not shown) indicate that when applied to whole soil, 2D NMR experiments demonstrate that the majority of components swollen by DMSO-d_6 are consistent with a mixture of carbohydrates, lignin, protein, and aliphatics. However, the organization of these components in a whole soil is much more challenging to determine especially using traditional NMR approaches. Using comprehensive multi-phase (CMP)-NMR solid-state, solution-state, and HR-MAS, experiments can be applied simultaneously to investigate which components are solvent accessible under various conditions. Figure 6.3(a) shows the ^{1}H NMR spectrum of a soil swollen in water. In this case, a range of carbohydrates and aliphatic components (mainly lipids) are at the soil–water

interface. Interestingly, when swollen in acidic conditions (Figure 6.3b), the S/N ratio drops indicating that less organic components are available to aqueous solvent, in particular the carbohydrates that are preferentially suppressed. This is consistent with highly negatively charged nature of SOM that expands at high pH because of charge–charge repulsion, but can become more condense at lower pH making it more difficult for solvent to penetrate.[21] Figure 6.3(c) shows the same soil at alkaline conditions; here, the signal-to-noise (SN) ratio is much higher indicating that the SOM is expanded and highly swollen. Note at high pH values, some aromatic signals are also present, which are not apparent at neutral and acidic pH. This is important, as aromatics are known to be key binding sites for hydrophobic contaminants.[22] The data suggest that at neutral pH, the aromatics are buried and not accessible at the soil–water interface, which is consistent with previous HR-MAS studies.[4] Figure 6.3(d) shows the same soil swollen in DMSO-d_6, an aprotic solvent capable of breaking hydrogen bonds and swelling hydrophobic domains. When used as a solvent to swell the soil, a wide range of components become accessible, which include strong signals from aromatics including lignin and protein (Figure 6.3d). Interestingly, these components are not easily accessible to aqueous solvent systems and their lack of accessibility to water is likely critical in the stabilization and overall properties of soil.

In addition to ^{1}H HR-MAS experiments, CMP-NMR spectroscopy permits the full range of solid-state NMR experiments. Figure 6.4 compares the ^{13}C CPMAS spectra of the same soil in its dry and swollen states. CPMAS is sensitive to molecular mobility and rigid components will efficiently cross polarize, while very mobile species show no signal in the experiment.[19] As such, the loss of signal with the addition of water is a simple approach to determine which components swell and become mobile. The aromatic region is not affected by the addition of water indicating that these species are not water accessible in soil (Figure 6.4). Conversely, some of carbohydrates and aliphatic are swollen, which is consistent with the ^{1}H results (Figure 6.3). However, it is important to stress that from the ^{1}H data alone while it was possible to conclude that aliphatic components and carbohydrates are indeed present at the soil–water interface, it was not possible to determine if any nonswellable aliphatic and carbohydrates domains also exist. From the ^{13}C CPMAS data, it is clear that there are considerable aliphatic and carbohydrate

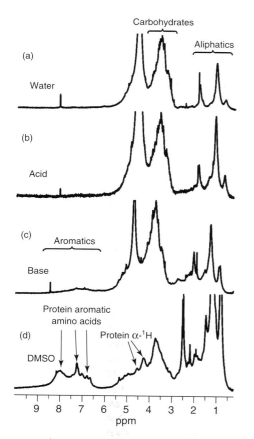

Figure 6.3. ^{1}H CMP-NMR spectra of a grassland soil swollen at different pH (a–c) and in DMSO-d_6 (d).

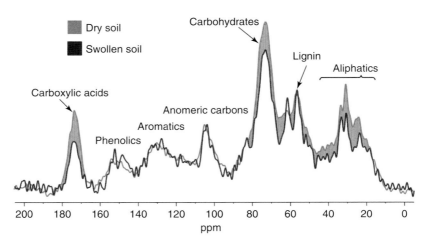

Figure 6.4.   Comparison of ^{13}C CPMAS spectra in a dry soil and soil swollen with water using a CMP-NMR probe.

signals that remain after swelling indicating that there are indeed large fractions of carbohydrates and aliphatics that are not water accessible in soil. This is logical and at least some of these domains likely result from cellulose in plant materials and structural carbohydrates in cell walls as well as lipid membranes in soil microbes and plant cuticles (waxy covering of leaves).[23]

6.3.2   Seed Structure and Growth

The application of CMP-NMR can include other facets of the environment and go beyond soil structure. Most biological processes, for example, involve interactions of various phases, be it the interaction of tissue with bone, the absorption of nutrients in the gut, or plant growth. As an example, the application of CMP-NMR to seeds that have both environmental and biological connotations is briefly considered.

Seeds are complex entities comprised of the embryo, accumulated food storage reserves, and an outer protective layer or 'seed coat', which allows them to survive harsh external conditions such as extreme cold or drought for prolonged periods in the absence of external nutrients. CMP-NMR can be used to probe these various components in their native state. It is also important to mention if seeds are spun in a well-sealed rotor filled with aqueous solvent they actually survive the spinning process; as such, CMP-NMR also has potential for in vivo applications. However, as the examples

here focus solely on seed structure, D_2O is used as a solvent that does not permit germination.[24]

Figure 6.5 compares the conventional ^{13}C NMR spectrum of a corn seed (*Zea mays*) to that of the inverse diffusion-edited and the diffusion-edited spectra. The conventional ^{13}C NMR spectrum was purposely acquired with low-power waltz-16 decoupling such that swollen/gel and soluble components are emphasized. Carbon nuclei in true solids will be attached to proton with a very broad lineshape and will not be efficiently decoupled and be strongly attenuated (see Ref. 19 for example on standards). The seed contains a range of structural components including (i) carbonyl, (ii) double bonds, (iii) aromatics, (iv) anomeric from carbohydrate, (v) signals arising from carbon beside oxygen (mainly from carbohydrates signals, and alcohol/ether/ester in lipids, and (vi) various aliphatics. In the inverse diffusion-edited spectrum, molecules with unrestricted diffusion are emphasized. A range of (vii) lipids and free sugars are detected and is consistent with molecules used for energy production in seeds.[25] The diffusion-edited spectrum is dominated by lipid signals consistent with seed membranes and higher-molecular-weight fat/oil storage.[25] To further probe the gel/soluble structures, $^1H-^{13}C$ heteronuclear single-quantum coherence (HSQC) correlations can also be performed (Figure 6.6). The additional dispersion and connectivity information permits the clear identification of triglyceride, which is the main component of most seed oils.[26] For comparison, a spectral prediction of triglyceride using Advanced

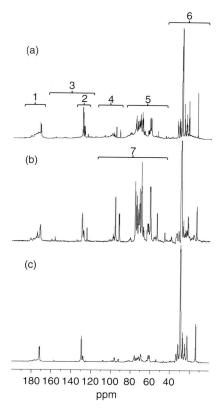

Figure 6.5. CMP-NMR spectra of a ^{13}C-labeled corn (*Zea mays*) seed. (a) conventional carbon experiments with low-power decoupling containing all components except true solids. (b) Inverse diffusion editing highlighting only molecules that have unrestricted diffusion. (c) Diffusion editing emphasizing molecules with restricted diffusion.

Chemistry Development (ACD) version 12.5 is shown in Figure 6.6(b).

The semisolid and solid components of the seeds can be probed by solid-state approaches and, in part, are based on difference spectroscopy to recover fast-relaxing components. The later technique termed RADE (and discussed earlier in this section) recovers signal that relax during diffusion editing and represents semisolid components. Figure 6.7(a) shows the RADE spectrum for the seed. Signals from L are from lipids and likely represent fats in lipid stores or membranes. Signals labeled as C on figure 6.7(a) arise from carbohydrates and may come from structural components in cell walls and the husk of the seed. These carbohydrate signals are prominent in the CPMAS

(Figure 6.7b), which emphasizes the solid in the seed husk.[27] When combined, CMP-NMR spectroscopy permits the comprehensive and detailed overview of all components within the seed. The preliminary data presented here is simply to demonstrate the concept and more advanced studies could provide information on layering,[28] molecular distances,[29] dynamics,[30] swelling,[31] metabolite profiling,[32] and theoretically be used to monitor germination and growth in real time.[33] In summary, CMP-NMR permits the study of all components in natural sample in their unaltered state. As questions in modern science become even more complex, techniques such as CMP-NMR that can provide a unique molecular insight into in situ and in vivo structures are likely to become more common place and will likely become an invaluable tool in many fields.

6.4   MOLECULAR INTERACTIONS

6.4.1   Contaminant Interactions in Soils and Sediments

Only a few NMR studies aimed at probing contaminant interactions in complete soils or sediments have been reported.[4,15,18,34] In most studies, conventional solid-state NMR hardware is employed to reduce line-broadening because of anisotropy, and provide high-powered pulses to extract information such as the number of distinct binding sites and the rigidity of those sites or domains.[15,34] HR-MAS hardware has also been applied to study contaminant interactions in whole soils; however, without the ability to employ high-powered pulses, only relatively mobile species can be observed.[4,18] Real-time analysis of changes in the signals from mobile compounds provides a direct mechanism to observe the transfer of contaminants between mobile and immobile phases and domains.[15,18]

In CMP-NMR, the combination of solid-state and solution-state NMR hardware overcomes the need for separate studies to probe the interactions that occur in solid and fluid phases of a whole soil. Till date, there have only been two initial CMP-NMR investigations of contaminant–soil interactions, all focusing on organofluorine compounds sorbed into D_2O-saturated peat soil.[19,35] The overall goal of this work has been to characterize the types of sites where interactions occur and the strengths of the associations at those sites

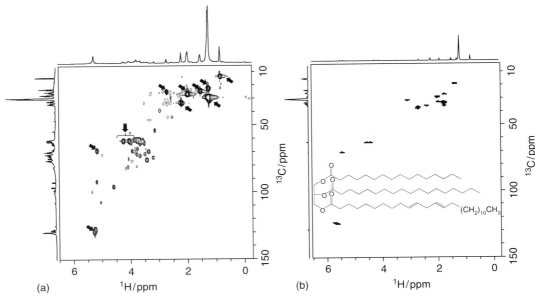

Figure 6.6.   CMP-NMR spectra of a ^{13}C-labeled corn (*Zea mays*) seed. (a) ^{1}H–^{13}C HSQC spectrum with assignments from triglycerides highlighted. (b) NMR prediction of the HSQC spectrum of the triglyceride shown.

in a whole, unmodified soil sample. Figure 6.8 displays an overview of some basic experiments that can be used to study molecular interactions. Figure 6.8(a) and (b) compares the lineshape of perfluorooctanoic acid (PFOA) before and after addition to soil. Figure 6.8(c) demonstrates various T_2 filters that can be used to differentiate relatively mobile vs bound species. The more dynamic interactions can be investigated using ^{19}F NOESY experiments (Figure 6.8d), whereas the strong interactions and binding sites further probed via ^{1}H–^{19}F CP (Figure 6.8e and f), which demonstrates different contaminants interact with different components in a whole soil.

Longstaffe *et al.*[35] further investigated CP in a fully swollen soil via CMP-NMR. In this study, CP is used to transfer ^{19}F magnetization to the ^{1}H nuclei of components of an intact peat soil where organofluorine compounds have formed associations after sorption. The natural complexity of SOM combined with the strong homonuclear ^{1}H dipolar couplings present in the less-mobile components of the peat makes it difficult to resolve signals from discrete binding sites, as shown in Figure 6.9 for PFOA and heptafluoronaphthol. Nevertheless, the general appearance of the spectrum produced by interactions with PFOA is different than that of the spectrum

produced by interactions with heptafluoronaphthol. In an earlier study using the same organofluorine compounds and the dissolved humic acid fraction of this same peat soil, it was shown that PFOA interacts only with protein-like structures of SOM; whereas heptafluoronaphthol interacts with all components, but preferentially with lignin-like structures, as is shown in Figure 6.10. The CMP-NMR study, looking at a whole peat sample, produces similar results; the heptafluoronaphthol-peat spectra closely resemble that of heptafluoronaphthol-lignin, while the PFOA-peat spectra closely resemble that of PFOA-protein. A clearer demonstration of different binding sites within the peat is the relationship between CP signal intensity and contact time,[35] which are shown in Figure 6.11 for different organofluorine compounds. These curves differ in both their overall magnitudes, which relate to the strength of the observed interactions, and in their profiles, which relate to the dynamics of magnetization transfer, which is strongly connected to nature of the binding sites. Figure 6.12 shows that the profiles of curves produced by interactions in peat match almost exactly those produced by lignin for heptafluoronaphthol, and protein for PFOA, in agreement with the results found in the solution-state investigation of humic acid.

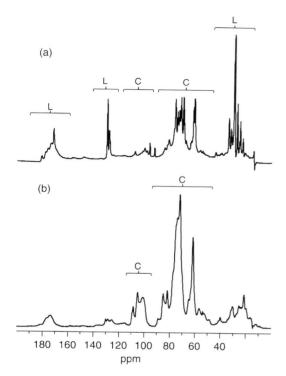

Figure 6.7.   CMP-NMR spectra of a ^{13}C-labeled corn (*Zea mays*) seed. (a) RADE spectrum and (b) CPMAS spectrum. L = lipids and C = carbohydrates.

Another alternative to study organofluorine–soil interactions is the use of double CPMAS (^{19}F–^{1}H–^{13}C) NMR. This experiment is discussed in more detail in the section titled 'Contaminant Interactions with Plant Materials', but is mentioned here in the context of soil interactions. Figure 6.13 shows a preliminary example of the double CPMAS result for the interaction of perfluorooctanesulfonic acid (PFOS) in a whole peat soil. The result shows that double CPMAS is possible in soil at natural abundance and the result is a ^{13}C spectral profile of just the components in soil responsible for the interaction. However, even after 21 days of acquisition, the S/N is still rather low because of the relatively insensitive nature of the experiment, and while this may prohibit its routine application at natural abundance in the near future, the experiments hold considerable potential for the study of isotopically enriched samples or more widespread application with future improvements in NMR sensitivity.

Using organofluorine compounds, it has been demonstrated that CMP-NMR spectroscopy can provide a powerful analysis of the binding interactions between small compounds and geological materials in their natural state. These types of measurements have a significant potential to impact the methods through which risk assessments are made for contaminated soils and sediments. While shown only initially for fluorinated compounds, nearly any compound possessing an NMR-active nucleus has potential as a target for CMP-NMR analysis of binding in soils and sediments.

6.4.2   Contaminant Interactions with Plant Materials

As previously discussed, the use of CMP-NMR technology has considerable implications for understanding structure and interactions in plants. Interactions of heavy metals, nutrients, and pesticides are of considerable importance in environmental research and are central process behind both phytoremediation and food production. Another considerable advantage for CMP-NMR is that plant material can be isotopically labeled and provides an excellent matrix to develop and optimize CMP-NMR experiments that may be challenging to optimize in natural abundance samples. One key experiment is ^{19}F→^{1}H→^{13}C double CPMAS NMR, which is mentioned earlier and described in more detail here. In ^{19}F→^{1}H→^{13}C double CPMAS NMR, the source of ^{19}F magnetization comes from a perfluorinated molecule. The resulting spectrum contains only carbon signals from the plant matrix that interact with the fluorinated molecule. As ^{13}C in the solid state exhibits considerably better spectral resolution compared to ^{1}H, the experiment is extremely important for identifying 'what' components in a matrix bind to a xenobiotic.

The contaminant chosen as an example here is PFOS, which is a man-made surfactant and is extremely persistent in the environment that can be transported long distances in air. As a result, it is widely distributed and found in soil, air, and groundwater.[36] Figure 6.14 shows an overlay of the ^{1}H→^{13}C CPMAS NMR spectrum (solid line), which highlights only components in the broccoli interacting with PFOS. The corresponding spectrum is mainly dominated by signals from protein-like molecules (see thin dashed spectrum for comparison). Moreover, signals from carbohydrates (likely

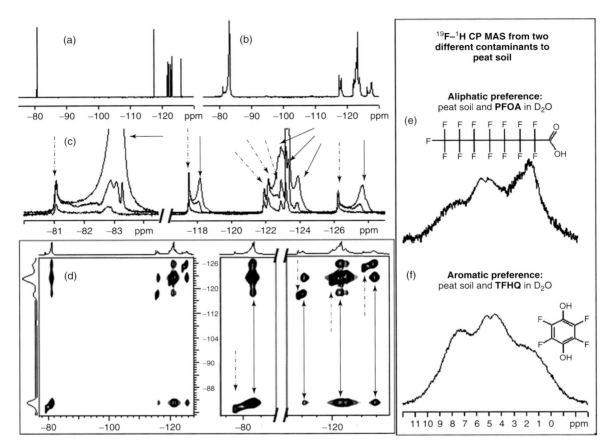

Figure 6.8. This shows some of the ^{19}F NMR experiments that can be performed in CMP-NMR to differentiate contaminant behavior in soil. This includes ^{19}F spectra using a short T_2-filter (1.2 ms) to remove the insert background of (a) PFOA in D_2O and (b) PFOA and soil in D_2O. Using longer T_2-filters, it is possible to differentiate molecules that are strongly interacting with soil. (c) Various T_2-filtered (1.2, 24, and 120 ms) spectra for the sample shown in (b), where it can be observed that some peaks are highly attenuated, suggesting that they correspond to PFOA in a strong contact with the soil. This can be corroborated in (d), a ^{19}F 2D NOESY experiment, where resonances from PFOA interacting with the soil show negative off-diagonal peaks; however, those from PFOA in a less-restricted forms only show the cross-peak signals. Different binding preferences between contaminants and soil can be studied using ^{19}F→^{1}H CPMAS as shown in (e) and (f), where PFOA and TFHQ show different binding preference when interacting with soil. Note that solid arrows denote PFOA interacting with soil and dashed arrows indicate PFOA in less-restricted conditions. (Reproduced with permission from Ref. 19. © 2012 Elsevier)

cellulose) are also observed but less intense compared to the reference spectrum. Interestingly, on additional washing (not shown), the signals from carbohydrates in the ^{19}F→^{1}H→^{13}C double CPMAS NMR experiment disappear indicating that the interactions from carbohydrates are nonpermanent and may arise from physical entrapment in the fibers of the plant. The protein interactions, however, remain consistent with the permanent binding of

perfluorinated chemicals to protein, which is consistent with previous reports.[37] This experiment opens the potential of other similar approaches in the form of X→^{1}H→^{13}C double CPMAS NMR experiments, where X could be ^{2}H in the case of a per-deuterated contaminant, silicon in the case of understanding clay–organic interactions, or mercury, lead, platinum, or cadmium to understand heavy metal binding or sequestration. Deuterium is especially

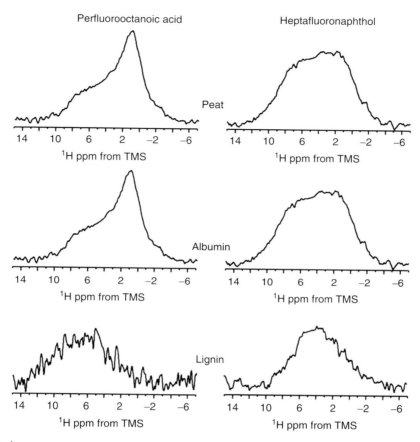

Figure 6.9. $^{19}F \rightarrow {}^1H$ CPMAS NMR spectra of mixtures of peat, albumin, or lignin with perfluorooctanoic acid (left) and heptafluoronaphthol (right). (Reproduced with permission from Ref. 35. © 2012 American Chemical Society)

interesting, as it can be used as a replacement for ^1H with a little alteration on the chemistry of the molecule. As such, ^2H represents a universal label that can be potentially employed in the vast majority of structures. If ^2H is used for observation or CP, the simplest solution for a lock would be to use an alternative nucleus such as ^{19}F and contain the lock solvent in an external lock capillary within the rotor.

6.5 CONCLUSIONS AND FUTURE DIRECTIONS

CMP-NMR spectroscopy provides the ability to study *all bonds* in *all phases* in a natural unaltered sample. As such, it represents a key step forward

in environmental research, where, for example, the overall properties of a soil are of environmental relevance and understanding complex associations, aggregations, interactions, and reactions in situ are key to unravel its environmental properties. CMP-NMR will likely find widespread application in environmental research because of its considerable versatility. Materials with complex multiphase structure such as plants, air particles, sediments, and soils are ideally suited to CMP-NMR analysis. Processes that convert one phase into another, for example, the conversion of biomass to bioethanol, can be probed in detail by CMP-NMR. In this example, solid-state experiments could be used to understand which components are not accessible to the liquid reactants, and HR-MAS NMR approaches to follow swelling and conversion with molecular resolution. At present, only ~30% of

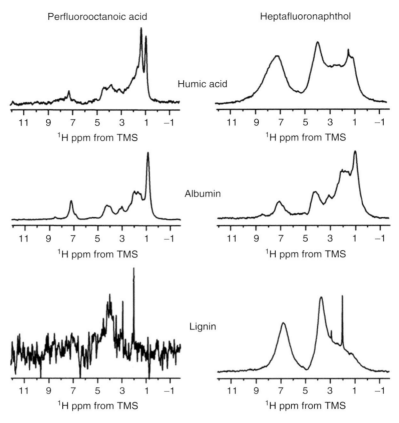

Figure 6.10. $^1H-^{19}F$ reverse heteronuclear saturation transfer difference solution-state spectra of humic acid, albumin, or lignin interacting with perfluorooctanoic acid (left) and heptafluoronaphthol (right). (Reproduced with permission from Ref. 22. © 2010 American Chemical Society)

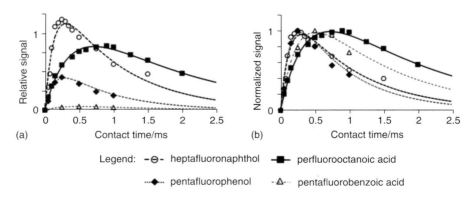

Figure 6.11. $^{19}F\rightarrow{}^1H$ CPMAS buildup curves of mixtures for four organofluorine compounds in Pahokee peat soil using the relative signal intensity for each mixture (a) and signals normalized with respect to the strongest signal for each mixture (b). (Reproduced with permission from Ref. 35. © 2012 American Chemical Society)

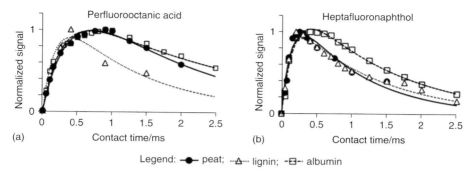

Figure 6.12. Overlaid $^{19}F{\rightarrow}^{1}H$ CPMAS buildup curves of mixtures of Pahokee peat, albumin, and lignin for perfluorooctanoic acid (a) and heptafluoronaphthol (b). (Reproduced with permission from Ref. 35. © 2012 American Chemical Society)

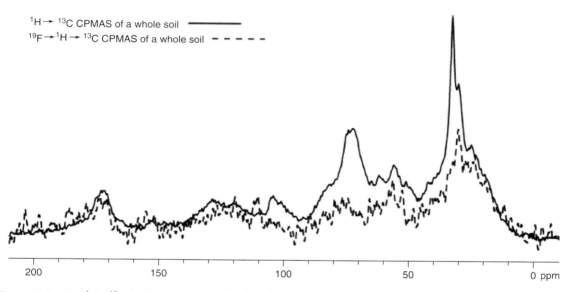

Figure 6.13. The $^{1}H{\rightarrow}^{13}C$ (solid line) and the $^{19}F{\rightarrow}^{1}H{\rightarrow}^{13}C$ (dashed line) CPMAS NMR experiments of the reference and 'contaminated' moieties in a whole peat soil, respectively. The double CPMAS spectrum highlights only components in the soil interacting with the perfluorinated chemical, whereas the solid represents all components in the soil. The general profiles indicate that protein has a key role in the binding of PFOS consistent with previous studies.[35]

cellulose is converted to bioethanol and understanding how and why cellulose is physically protected and not accessible for conversion is a key question in the field.[38]

CMP-NMR also has power capabilities for the study of molecular binding. A chemical can be tracked, as it moves from one phase to another, and the molecular orientation and interactions from each phase can be determined. Studies are not restricted to organic chemical interactions, and various heavy metal interactions

on mineral surfaces can also be potentially evaluated. It has been recently demonstrated that with proper preparation, even whole organism *D. magna* can be studied in vivo under MAS.[13] These preliminary studies were reported using an HR-MAS NMR probe; however, using CMP-NMR spectroscopy, all components including outer shell, cell walls, membranes, and crystalline proteins could also be studied. Considered *D. magna* is a key global species for aquatic toxicity tests,[39] the ability to study all components with molecular

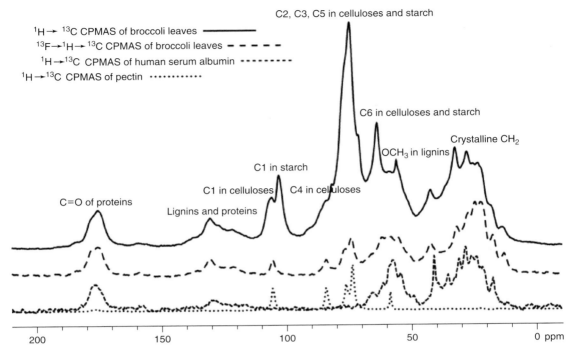

Figure 6.14. The $^1H \rightarrow ^{13}C$ (solid line) represents a conventional CPMAS spectrum for reference. The long dashed-line spectrum is the result from the $^{19}F \rightarrow ^1H \rightarrow ^{13}C$ double CPMAS experiment and shows only components in the broccoli plant that interact with PFOS. Through comparison with HSA (short dashed line) and pectin (dotted line), it can be seen that carbohydrate and protein components in the whole plant interact with the PFOS molecule.

resolution in vivo holds particular promise. Studies of the contaminant/metal can provide insight into the processes behind biotransformation, bioaccumulation, binding sites, and excretion, whereas studies of the metabolic profile of the organisms itself can help determine the mode of action of the chemical, bio-indicators of environmental stress, recovery profiles, and potentially the first molecular indicators that are precursors to subsequent disease.

The drawbacks of the technology itself are relatively minor in comparison with the scientific gains. The probes are somewhat challenging to build because of the considerable array of components required in such a small space, and there is a loss (roughly 10–40%) in performance over a dedicated solids, HR-MAS, or solution-state NMR probes. However, the main drawback arises not from the technology but from the samples themselves. Consider for example that a researcher needs to obtain NMR information on broccoli. Normally, for solid-state NMR, the

sample would be freeze dried, ground, and then packed (~100 mg in a 4 mm rotor). However, if the sample is now studied by CMP-NMR (intact sample is used), then 92% of the sample is water. Even if a 100 mg section can be placed in the rotor that is only 8 mg of that contains carbon. Now consider that in an intact sample, only 50% maybe present as a true solid, it is clear that a spectrum is going to take a lot longer to acquire than using the conventional approach. Of course, it is important to remember that the information is not the same, and that in CMP-NMR, the components are studied in their natural state and the higher resolution information from the gel and solution phases is often very useful in making assignments in the less resolved solid-state data. However, it is equally important to stress that if the goal of a project is simply to obtain the overall carbon composition of broccoli and information on the different phases is not required, then CMP-NMR would not be the sensible choice. Future 7 mm CMP-NMR designs will permit approximately five times the sample to be introduced into the rotor, which should considerably help offset the sample limitations of a 4 mm rotor for intact samples.

In conclusion, CMP-NMR should be reserved for studies that ask complex questions that can only be answered by studying intact samples. In this case, CMP-NMR represents a powerful and complementary tool to conventional NMR approaches and represents a key step forward by providing a unique range of information. While environmental research is the focus of this chapter, CMP-NMR technology is likely to have widespread applicability in field such as medicine, biochemistry, biology, heterogeneous catalysis, or any other area that deal with samples containing more than one phase.

ACKNOWLEDGMENTS

We would like to thank the Natural Sciences and Engineering Research Council of Canada (NSERC, Strategic and Discovery Grants Programs), the Canada Foundation for Innovation (CFI), the Ontario Ministry of Research and Innovation (MRI), and the Krembil Foundation for providing funding. Andre Simpson would like to thank the Government of Ontario for an Early Researcher Award. We would like to thank Ries de Visser and his colleagues at IsoLife for donation of the ^{13}C-labeled seeds. We would like to thank everyone at Bruker BioSpin (Germany, USA, and Canada) for providing continued support and collaboration that made this work possible.

RELATED ARTICLES IN EMAGRES

Microcoils

Radiofrequency Power Amplifiers for NMR and MRI

Radiofrequency Heating Models and Measurements

Agriculture and Soils

Coal Structure from Solid State NMR

Geological Applications

REFERENCES

1. P. M. Bertsch and J. C. Seaman, *Proc. Natl. Acad. Sci. USA*, 1999, **96**, 3350.

2. W. E. Maas, F. H. Laukien, and D. G. Cory, *J. Am. Chem. Soc.*, 1996, **118**, 13085.

3. R. E. B. Lee, W. Li, D. Chatterjee, and R. E. Lee, *Glycobiology*, 2005, **15**, 139.

4. A. J. Simpson, W. L. Kingery, D. R. Shaw, M. Spraul, E. Humpfer, and P. Dvortsak, *Environ. Sci. Technol.*, 2001, **35**, 3321.

5. V. Righi, A. Mucci, L. Schenetti, A. Bacci, R. Agati, M. Leonardi, R. Schiavina, G. Martorana, G. Liguori, C. Calabrese, E. Boschetti, S. Bonora, and V. Tugnoli, *Oncol. Rep.*, 2009, **22**, 1493.

6. R. Warrass, J. M. Wieruszeski, and G. Lippens, *J. Am. Chem. Soc.*, 1999, **121**, 3787.

7. A. D. Roy, K. Jayalakshmi, S. Dasgupta, R. Roy, and B. Mukhopadhyay, *Magn. Reson. Chem.*, 2008, **46**, 1119.

8. M. W. I. Schmidt, H. Knicker, P. G. Hatcher, and I. Kogel-Knabner, *Eur. J. Soil Sci.*, 1997, **48**, 319.

9. N. Hertkorn, R. Benner, M. Frommberger, P. Schmitt-Kopplin, M. Witt, K. Kaiser, A. Kettrup, and J. I. Hedges, *Geochim. Cosmochim. Acta*, 2006, **70**, 2990.

10. A. Simpson, *Soil Sci.*, 2001, **166**, 795.

11. A. J. Simpson, G. X. Song, E. Smith, B. Lam, E. H. Novotny, and M. H. B. Hayes, *Environ. Sci. Technol.*, 2007, **41**, 876.

12. A. J. Simpson, M. J. Simpson, W. L. Kingery, B. A. Lefebvre, A. Moser, A. J. Williams, M. Kvasha, and B. P. Kelleher, *Langmuir*, 2006, **22**, 4498.

13. A. Bunescu, J. Garric, B. Vollat, E. Canet-Soulas, D. Graveron-Demilly, and F. Fauvelle, *Mol. Biosys.*, 2010, **6**, 121.

14. M. L. Miglietta and R. Lamanna, *Magn. Reson. Chem.*, 2006, **44**, 675.

15. S. D. Kohl, P. J. Toscano, W. H. Hou, and J. A. Rice, *Environ. Sci. Technol.*, 2000, **34**, 204.

16. A. J. Simpson, D. J. McNally, and M. J. Simpson, *Prog. Nucl. Magn. Reson. Spectrosc.*, 2011, **58**, 97.

17. A. J. Simpson, B. Lam, M. L. Diamond, D. J. Donaldson, B. A. Lefebvre, A. Q. Moser, A. J. Williams, N. I. Larin, and M. P. Kvasha, *Chemosphere*, 2006, **63**, 142.

18. A. Shirzadi, M. J. Simpson, R. Kumar, A. J. Baer, Y. P. Xu, and A. J. Simpson, *Environ. Sci. Technol.*, 2008, **42**, 5514.

19. D. Courtier-Murias, H. Farooq, H. Masoom, A. Botana, R. Soong, J. G. Longstaffe, M. J. Simpson, W. E. Maas, M. Fey, B. Andrew, J. Struppe, H. Hutchins, S. Krishnamurthy, R. Kumar, M. Monette, H. J. Stronks, A. Hume, and A. J. Simpson, *J. Magn. Reson.*, 2012, **217**, 61.

20. B. P. Kelleher and A. J. Simpson, *Environ. Sci. Technol.*, 2006, **40**, 4605.

21. C. E. Clapp, M. H. B. Hayes, A. J. Simpson, and W. L. Kingery, in Chemical Processes in Soils, eds M. A. Tabatabai and D. L. Sparks, Soil Science Society of America: Madison, WI, 2005, 1.

22. J. G. Longstaffe, M. J. Simpson, W. Maas, and A. J. Simpson, *Environ. Sci. Technol.*, 2010, **44**, 5476.

23. I. Kogel-Knabner, *Soil Biol. Biochem.*, 2002, **34**, 139.

24. C. R. Bhatia and H. H. Smith, *Planta*, 1968, **80**, 176.

25. I. A. Graham, *Annu. Rev. Plant Biol.*, 2008, **59**, 115.

26. T. A. El-Adawy, E. H. Rahma, A. A. El-Bedawy, and A. M. Gafar, *Nahrung*, 1999, **43**, 385.

27. J. F. Kennedy, J. S. Sandhu, and D. A. T. Southgate, *Carbohydr. Res.*, 1979, **75**, 265.

28. S. E. C. Whitney, J. E. Brigham, A. H. Darke, J. S. G. Reid, and M. J. Gidley, *Plant J.*, 1995, **8**, 491.

29. L. Cegelski, R. D. O'Connor, D. Stueber, M. Singh, B. Poliks, and J. Schaefer, *J. Am. Chem. Soc.*, 2010, **132**, 16052.

30. R. Lutterbach and J. Stockigt, *Phytochemistry*, 1995, **40**, 801.

31. P. J. OConnor, S. S. Cutie, P. B. Smith, S. J. Martin, R. L. Sammler, W. I. Harris, M. J. Marks, and L. Wilson, *Macromolecules*, 1996, **29**, 7872.

32. J. Kikuchi, *Plant Cell Physiol.*, 2005, **46**, S16.

33. C. Hinse, C. Richter, A. Provenzani, and J. Stockigt, *Bioorgan. Med. Chem.*, 2003, **11**, 3913.

34. R. J. Smernik, *J. Environ. Qual.*, 2005, **34**, 1194.

35. J. G. Longstaffe, D. Courtier-Murias, R. Soong, M. J. Simpson, W. E. Maas, M. Fey, H. Hutchins, S. Krishnamurthy, J. Struppe, M. Alaee, R. Kumar, M. Monette, H. J. Stronks, and A. J. Simpson, *Environ. Sci. Technol.*, 2012, **46**, 10508.

36. C. Chaemfa, J. L. Barber, S. Huber, K. Breivik, and K. C. Jones, *J. Environ. Monitor.*, 2010, **12**, 1100.

37. J. C. D'eon, A. J. Simpson, R. Kumar, A. J. Baer, and S. A. Mabury, *Environ. Toxicol. Chem.*, 2010, **29**, 1678.

38. L. Laureano-Perez, F. Teymouri, H. Alizadeh, and B. E. Dale, *Appl. Biochem. Biotech.*, 2005, **121**, 1081.

39. H. C. Poynton, J. R. Varshavsky, B. Chang, G. Cavigiolio, S. Chan, P. S. Holman, A. V. Loguinov, D. J. Bauer, K. Komachi, E. C. Theil, E. J. Perkins, O. Hughes, and C. D. Vulpe, *Environ. Sci. Technol.*, 2007, **41**, 1044.

Chapter 7

Environmental NMR: Magnetic Resonance Imaging

Nikolaus Nestle[1], Robert Morris[2] and Thomas Baumann[3]

[1]*BASF SE, Ludwigshafen, D-67056, Germany*
[2]*College of Arts and Science, School of Science & Technology, Nottingham Trent University, Nottingham, NG1 4BU, UK*
[3]*Institute of Hydrochemistry, Technische Universität München, Munich, D-81377, Germany*

7.1 INTRODUCTION

Soil and groundwater aquifers are of paramount importance for ensuring sustainable supply of water and maintaining environmentally sound conditions. While the top soil is considered highly reactive and serves as a reactive filter to all kinds of anthropogenic impacts, groundwater aquifers are the main source of drinking water. Aquifers do act as filters but only in a limited capacity. Processes in soil and groundwater are highly dynamic, both spatially and temporally. As reactions take place at biogeochemical interfaces, which are localized hotspots in soil, reaction rates are dependent on the position of the reactants. Together with the reaction kinetics, this adds an additional level of complexity to the description of the function of a soil–groundwater system.[1] The quantitative description of transport, retardation, and degradation in subsurface environments is typically examined using laboratory experiments, mainly batch and column tests.[2] Batch tests provide equilibrium conditions and almost complete accessibility of interfaces, which may not be relevant to in situ soil or aquifer conditions. Both conditions are not met in natural systems; therefore, the results of batch tests tend to overestimate the efficiency of the filter. Column tests, on the other hand, although seemingly very intuitive, often give no access to the spatial heterogeneity, which is then lumped into the transport parameters and reaction rates.[2]

Without access to the pore space, it seems very difficult, if not impossible, to provide a resilient link between theory and processes on the continuum scale. This situation has changed for better with the availability of highly sophisticated imaging techniques. Many of these developments have

NMR Spectroscopy: A Versatile Tool for Environmental Research
Edited by Myrna J. Simpson and André J. Simpson
© 2014 John Wiley & Sons, Ltd. ISBN: 978-1-118-61647-5

been triggered by medical applications. Noninvasive scanning techniques such as micro X-ray tomography (μCT) are used to visualize the pore topology.[3] Here, the resolution with modern equipment is down to 0.5 μm. As imaging contrast is given by density differences, there is a little possibility to visually discriminate different environments, for example, water and microorganisms in porous media. The spatial resolution and the dynamic range can significantly be increased using synchrotron radiation.[3] With these techniques, however, the accessible sample sizes are really small and the acquisition times are too long to perform dynamic experiments. Microfluidic structures provide access to processes on a single interface level and are open to a number of spectroscopic methods.[4,5] Although the materials and flow patterns are close to natural systems, the complete heterogeneity cannot be mapped. Thus, these micromodels provide a link from batch systems to small-scale columns.

Magnetic resonance imaging (MRI) in environmental sciences, which is detailed in this book, picks up the baton from μCT and micromodels and takes it to the column scale. One of the most important features of MRI for environmental applications is its sensitivity to protons, which are abundant in water, the primary target, and other proton-containing liquids. The distribution, diffusion, and velocity of water in aquifer materials are directly accessible using MRI techniques. It is also possible to differentiate between free water within pores and adhesion of water at the pore throats using relaxation time measurements (see Chapter 18). Transport processes can be studied directly for any contaminant, which has a strong influence on local NMR properties such as relaxation times.[6] For those contaminants that do not alter the MRI signal, magnetically labeled probes can be used to elucidate their position and transport dynamics.[7] As with any imaging technique, the operation mode is limited by the ternary system given by image resolution versus field of view versus acquisition time. Nevertheless, MRI is a rather fast imaging technique, which is compatible with the timescale of processes in aquifers.[6]

7.2 NMR PROPERTIES OF ENVIRONMENTAL MATERIALS

In both medical and nonmedical MRI applications, MRI is almost always performed on hydrogen nuclei (protons). Hydrogen may be present in environmental matrices in the following forms: (i) water and other pore liquids, (ii) biomass (living or dead organisms and degradation products such as natural organic matter; see Chapter 11), and (iii) chemically bound in inorganic matter such as hydrate phases or hydroxides (see Chapter 17). In the following sections, the NMR properties of these different forms of hydrogen are reviewed. Furthermore, possibilities for heteronuclear MRI of environmental samples are also discussed.

7.2.1 Water and Other Liquids in Pores

NMR properties of liquids in porous materials have been studied for more than 50 years.[8-10] Some key phenomena determining their behavior include surface relaxation, interaction with paramagnetic centers, and motion relative to internal magnetic field variations inside the porous medium.

7.2.1.1 Surface Relaxation

As paramagnetic centers are usually located at the pore surfaces, surface relaxation due to paramagnetic interactions and relaxation due to the different molecular dynamics of adsorbed species on the surface. Owing to the exchange between adsorbed water undergoing surface relaxation and free liquid in the bulk of the pore, surface relaxation influences the NMR behavior of all the liquid inside the pore. In the case of 'fast exchange' (i.e., if residence times of adsorbed molecules on the pore surface are short and the pores are sufficiently small to allow diffusive mixing of all the pore liquid on the timescale of the NMR relaxation), the relaxation rate (i.e., the reciprocal value of the relaxation time) of the liquid in the pore is directly related to the surface/volume ratio of the pore system as shown in equation (1) with T_i denoting the relaxation time of the liquid in the pore system, $T_{i,bulk}$ the relaxation time of the bulk liquid, S/V the surface-to-volume ratio of the pore system, and the factor ρ_i the so-called relaxivity.

$$\frac{1}{T_i} = \frac{1}{T_{i,bulk}} + \rho_i \frac{S}{V} \qquad (7.1)$$

This ρ_i factor is determined by the magnetic field strength, the temperature, and the chemical properties of the liquid and the surface. Usually, water is more sensitive to relaxation on mineral surfaces is stronger than nonaqueous liquids.[11] For the sensitivity of surface relaxivities on the magnetic field, see Ref. 8. Surface relaxivities at lower magnetic field strengths

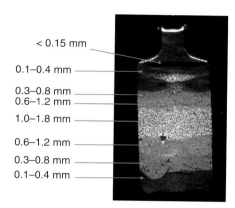

Figure 7.1. T_2-weighted MRI scan of a bottle filled with water-saturated quartz sand of different grain sizes. The higher the signal intensity in the image is, the longer is the local transverse relaxation time.

are typically stronger (i.e., leading to shorter relaxation times) than at higher magnetic fields. For larger pores, more complex models considering incomplete diffusive exchange of the liquid inside the pore have been developed.[9,10] An intuitive visualization of the sensitivity of relaxation time contrast to different pore sizes is given in Figure 7.1.

7.2.1.2 Paramagnetic Effects

While paramagnetic species such as transition or rare earth metal ions or dissolved oxygen in solution usually exhibit a simple, linear influence on the relaxation rate of the surrounding liquid,[12] the presence of paramagnetic ions on pore walls may lead to more complex relaxation behavior of the liquid phase. While in some cases a similar linear behavior is observed as in solutions,[6,12] in other cases the nonlinear relationships between concentration of paramagnetic material and relaxation rates were observed.[12] Furthermore, relaxation rates of adsorbed paramagnetic ions may strongly differ from those of the respective free ions.[12]

As the concentration of paramagnetic species is quite high in many natural porous materials, relaxation times for water in those materials are rather short. Additionally, also the presence of high quantities of fine fraction materials may lead to short relaxation times. The sensitivity of other liquids (i.e., nonaqueous phases such as oil or other organic solvents) to surface relaxation and the presence of paramagnetic materials are often considerably different to that of water. This is the

case for both solutions of paramagnetic species[13] and liquids inside pores. In magnetically clean materials, the sensitivity of relaxation rates of water to adsorbed paramagnetic species can be high enough to detection concentrations in the range of a few ppm.[14]

7.2.1.3 Motion Relative to Internal Magnetic Field Variations

While spin–lattice or longitudinal relaxation is typically not influenced by magnetic field gradients, the sensitivity of spin–spin or transverse relaxation times to inhomogeneous magnetic fields has been documented.[15] In heterogeneous materials, even comparably small susceptibility variations on a micrometer length scale may lead to considerable magnetic field gradients, which result in a significant shortening of apparent spin–spin relaxation times in materials of interest.[16] Owing to larger susceptibility variations, mineral porous media and unsaturated systems are especially sensitive to magnetic field inhomogeneity. In unsaturated media, this may lead to unwanted signal losses for MRI experiments especially when working with long echo times (see also comparison between biomedical and environmental MRIs in Section 7.3.1). The inner magnetic field gradient is directly proportional to the magnetic field strength; however, the influence on the apparent relaxation rate is dependent on the square of the gradient. For this reason, MRI becomes increasingly sensitive to gradient effects with increasing field strength.

7.2.2 Biomass and Its Degradation Products

Both living and dead biomass are present in most environmental materials. Living biomass such as roots and biofilms has been studied previously (see Chapter 21). Nonliving biomass undergoes various decomposition steps, finally leading to a complex mixture of natural organic matter (see Chapter 13). These materials are of great importance for the water-retention capabilities of soils and have been studied to great detail by NMR relaxation methods (see Chapter 18). Dry, nonliving plant-derived biomass, such as wood fragments, usually has very short spin–spin relaxation times and is difficult to detect using MRI. With moist biomass, the water present is detectable using NMR. In some cases, the polymeric components exhibit sufficiently long relaxation times and contribute to the signal in MRI.

7.2.3 Chemically Bound Water

As already mentioned in the context of solid components in biomass and its degradation products, protons in solid chemical environments usually exhibit very short spin–spin relaxation times. This often means that a number of water populations are difficult to detect using standard MRI techniques. Conditions are slightly more favorable for materials such as water-intecalated swelling clay minerals such as bentonite. Here, diffusive exchange of light versus heavy water can be studied by means of continuous wave NMR imaging, which is less sensitive to short transverse relaxation times than more standard MRI methods.[17]

7.2.4 Heteronuclear MRI Options in Environmental Science

Heteronuclear MRI (i.e., MRI for nuclei other than protons) is usually more affected by sensitivity issues than proton MRI. The reasons for this are not only lower gyromagnetic ratios and lower isotopic abundances of most heteronuclei but also their much lower volume concentration in practically relevant systems. The only notable exception in these considerations is ^{19}F MRI of perfluorinated liquid phases[18] where sensitivities almost equal to those in proton MRI are reached and where the absence of a natural background signal offers an excellent selectivity to tracer fluids into the porous medium. Recent advances in heteronuclear MRI for biomedical applications may also open up novel techniques for MRI in environmental matrices. The most likely candidate nucleus for such experiments is [23]Na. From the hydrogeological viewpoint, this nucleus is highly relevant when it comes to salt propagation in aquifers where considerable Na concentrations may be present in the liquid phase. In medical MRI, quantitative sodium imaging at concentrations down to as little as $50 \, \text{mmol} \, \text{l}^{-1}$ (i.e., roughly $1 \, \text{g} \, \text{l}^{-1}$ sodium) has been recently reported.[19]

Heteronuclear imaging may become even more relevant in the context of hyperpolarized imaging techniques many of which rely on heteronuclear systems. While the introduction of hyperpolarized molecular species or [129]Xe into environmental samples would suffer even more than in medical applications from the inherently short relaxation times of the hyperpolarized species inside the sample, this problem will not be there for particulate hyperpolarized species such as [29]Si, which has recently been demonstrated to exhibit a relaxation time of more than 30 min even inside a living organism.[20]

7.3 CONSIDERATIONS OF EQUIPMENT AND METHODOLOGY

The development of MRI over the last 30 years has essentially been driven by its fascinating potential in biomedical applications, especially in the clinical field. Like that, most of the available MRI equipment is optimized to the NMR properties of human tissue and whole-body dimensions. In the following section, these parameters and the needs in environmental applications shall be compared. After that, requirements and methods for dedicated equipment shall be discussed. As a discussion of magnetic resonance methodology as such would be beyond the scope of this chapter, the reader is referred to the general literature in this field.[21–23]

7.3.1 Environmental NMR Samples Versus Clinical Samples

With the notable exception of the lungs, all tissues inside vertebrate organisms are essentially water saturated and exhibit only minor variations of the magnetic susceptibility and thus of the local magnetic field on the micrometer scale. In contrast, environmental samples usually exhibit rather short apparent spin–spin relaxation times because of much stronger local magnetic field variations leading to rapid signal loss by diffusive echo attenuation; this is especially problematic in unsaturated samples where considerable amounts of the NMR signal may be lost even at moderate echo times in the range of a few microseconds. Furthermore, the volume fraction of typical materials (i.e., water, fat, and mobile polymers) containing MRI-detectable protons is usually on the order of 85% or more. In environmental samples, the volume fraction of MRI-detectable protons is typically in the range of 30% or less even for saturated media.

Another difference between medical and environmental samples is the structure of the features of interest. While the anatomic structure usually consists of well-delineated, known features, the structures of interest in environmental samples may be irregularly shaped and not necessarily well delineated. Hence, a qualitative interpretation of the features in MRI scans on environmental samples is not always straightforward, and quantitative evaluations of MRI data will often need much more sophisticated approaches than volume rendering (i.e., the determination of volume regions with a given range of NMR signal intensity). Last but not least, the volume studied by MRI may not represent the overall sample in heterogeneous sediments. Typical flow velocities are different in medical and environmental applications: in the medical case, they may be around several millimeters to centimeters, whereas in aquifers, they may be in the range of micrometers.

It is clear that the mostly used clinical MRI systems, which are quite easily accessible to environmental research groups through cooperations with university hospitals, are optimized for samples with NMR properties differing from those of environmental samples. This leads to a range of limitations in the application of clinical MRI systems to environmental samples. For example, in order to obtain meaningful results, samples have to be pretreated to minimize the susceptibility variations.[24] Even in such pretreated samples, some water may remain inaccessible to the MRI scanner, for example, water in partially saturated sediments or water in clay gels. Despite these limitations, water-saturated neat sand beds can be studied in a clinical MRI scanner quite well and surprisingly high sensitivities for paramagnetic and superparamagnetic materials have been reported.[14] Nevertheless, many other samples will lead to unsatisfying results in clinical MRI scanners. Some of these problems are intrinsic to MRI and many others can be avoided or at least reduced when working with MRI systems developed for general applications or porous media instead of clinical MRI systems.

7.3.2 Dedicated MRI Equipment for Environmental Science

There are various aspects by which MRI equipment can be optimized for environmental science applications. Owing to technical reasons, not all desirable features can be combined with each other: for example, the detection of signals from fast-relaxing components such as clay gels requires short RF excitation pulses with duration less than about 10 μs, which can be only realized in sample coils with dimensions of about 2 cm or less.[25] Similarly, fast-switching, strong

gradient systems (up to about $40\,T\,m^{-1}$ within less than 100 μs) are needed for encoding slow translational movements and work best for small sample diameters in the range of a few centimeters or less.[26] Achieving good sensitivity to such materials, therefore, suggests using instrumentation, which can be easily accommodated in a standard wide-bore NMR spectroscopy magnet. Often, such spectrometers are not really used for imaging but rather for studies of material properties such as relaxation times or diffusion coefficients.[25,26] Nevertheless, imaging of fast-relaxing samples can be done in such instrumentation at a very high spatial resolution.[27]

Despite their advantages with respect to achievable pulse lengths and gradient strength, such small-diameter MRI systems are only of limited use for many environmental science applications because of the small available sample sizes that lead to serious limitations with respect to how representative a sample is. From this viewpoint, sample on the length scale of lysimeters (i.e., meters) or even larger would be desirable. However, such length scales come with severe problems with respect to pulse length and magnetic field and gradient strength. Priority in terms of the development of dedicated MRI equipment was, therefore, in systems with intermediate sample diameters up to around 10 cm.[28] Magnets for such sample sizes are in most cases dedicated special systems or medical imaging magnets. A further consideration is the orientation of the bore of the magnet. While clinical MRI systems and small animal scanners usually are built with horizontally oriented magnets, a vertical sample orientation is preferable for environmental science applications. With horizontal magnets, gravity effects are parallel to the sample axis, and it even is possible to study phenomena over a length scale larger than the length of the coil by moving the sample relative to the NMR coil. Similarly, favorable geometrical conditions for MRI systems can also be achieved in setups with a permanent magnet or an electromagnet where the magnetic field is oriented horizontally but the sample volume is located in a vertically oriented gap.[29] Generally speaking, design choices for such instruments are similar to those made for chemical engineering applications. Typically, gradient strengths around $1\,T\,m^{-1}$ are available in such instruments (compared to just $40\,mT\,m^{-1}$ or less in clinical systems); the magnetic field strengths in such systems are typically between 0.25 (10 MHz) and 4.7 T (200 MHz).

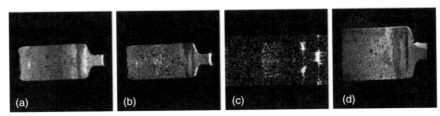

Figure 7.2. Gradient echo (FLASH) images obtained on a bottle with water-saturated sands of different grain sizes in a 1.5 T MRI scanner at: (a) 1.8, (b) 5, and (c) 10 ms echo time and in a 200 mT MRI scanner at 12 ms echo time (d, only upper part of the bottle). While the signal intensity in image c has deteriorated almost completely for the echo time of 10 ms, the low-field image d is still of good quality for an echo time of 12 ms.

7.3.3 MRI Sequences for Environmental Science

7.3.3.1 Weighted Versus Quantitative Images

For reasons of speed and simplicity, most clinical MRIs are performed with weighted contrast. In this case, each k-space data point results from the acquisition of an NMR signal with a single set of pulse sequence parameters (such as echo time, relaxation delay, gradient strength, gradient spacing, or gradient duration). While such images are very helpful in separating well-delineated compartments with different NMR properties from each other, the information content in each voxel (volume element) is quite limited and usually quantification of the NMR parameters of the voxel is not possible.

In order to achieve quantification, a systematic variation of one or several of the sequence parameters is needed for each point within a k-space. As such, a systematic variation results in considerably longer measurement times (both because of the parameter variation and because of the need of a better signal/noise ratio to allow quantitative evaluations). Quantitative MRI maps are not frequently used in clinical applications (or if they are, they are performed with a considerable reduction in the number of experimental points). Another issue to consider in quantitative MRI is the fact that changing parameters in simple, relaxation-weighted imaging sequences may result in simultaneous unwanted contrast variations. For example, changing the echo time in a sample with internal magnetic field gradients will not only lead to a change in T_2 weighting (which might be intended) but as well in the diffusive attenuation of the signal (which usually is unwanted, as it is hard to quantify and may even lead to a loss of most of the signal, see Figure 7.2).

As this demonstrates, quantitative MRI needs contrast preparation and encoding of spatial information to be separated.[30] This is usually achieved by replacing the initial excitation pulse with a robust NMR sequence to achieve relaxation time, diffusion, or flow weighting, which is followed by spatial encoding module, which is chosen to be fast and result in minimal intrinsic contrast weighting. After that, for each voxel, an evaluation of the changes in signal intensity as a function of the parameters in the preparation sequence is performed.

7.3.3.2 Contrasts Available Using Clinical MRI Sequences

Most clinical MRI sequences are optimized with respect to scanning speed on samples with rather long relaxation times and moderate internal magnetic field gradients. In contrast, many of these sequences, especially those relying on EPI-type[31] modules for detection, fail when applied to environmental samples. Much better results can be obtained using rapid acquisition with refocused echoes (RARE)[32] sequences or other spin echo or gradient echo sequences with sufficiently short echo times.

While standard spin echo sequences usually result in good signal/noise ratios, the time taken to acquire each image is considerable, as a relatively long relaxation delay has to be observed for each k-space line of the image. Even when working with relaxation delays of about 1 s (usually far from full relaxation in water-saturated sediments) and using standard tricks to reduce the number of phase-encoding steps, this will usually result in acquisition times over 1 min for a 128×128 pixel image (with multislicing, around 20 images from different slices can be acquired simultaneously within this time). Longer relaxation delays or better image resolution can easily lead to imaging times around 10 min for a single image.

Faster imaging is possible using RARE where a Carr-Purcell-Meiboom-Gill (CPMG)[33]-like train of echoes is read out with each echo corresponding to a different line in k-space. Usually, such sequences read out about 5–10 echoes within one excitation cycle. As the acquisition of an echo train takes more time than a single spin echo, there will be usually less potential for multislicing in a RARE sequence than in an ordinary spin echo unless one works with long relaxation delays.

Among the gradient echo sequences, fast low-angle shot (FLASH)[34] is especially interesting for applications in environmental science, as short echo times make it relatively immune to internal field gradients and short recycle delays offer the potential for relatively fast imaging. If relaxation times are sufficiently short, large excitation angles can be used and good signal/noise ratios can be achieved. In such conditions, 3D datasets with a volume resolution in the 10 µl range can be acquired about every 12 s.[13]

7.3.3.3 Additional Options Using NonClinical Imaging Sequences

If high temporal resolution is not required, quantitative NMR parameter mapping is relatively easy to achieve when using a contrast-preparation approach.[30] Spatial encoding in such experiments is most straightforwardly done with a spin echo or a gradient echo with a short echo time. An advantageous aspect of environmental samples over medical diagnostics is the absence of macroscopic incoherent motion in the sample. This enables the use of sequences with very long echo times (e.g., stimulated echoes), which usually fail with in vivo applications because of macroscopic motion problems.

7.3.3.4 Using Contrast Agents

Toxicity is not a consideration in MRI studies of contaminant transport in environmental samples. Therefore, the range of contrast agents available for environmental science applications is much greater than in the medical field. Like in medical diagnostics, different contrast agents can be used to obtain different types of information on the sample. For example, if the contrast agent behaves as a conservative tracer, following the spatial distribution of contrast agents injected into the sample can help to visualize flow patterns inside the sample (see Section 7.4.1). Gd-DTPA that is widely used in medical applications behaves very close to a conservative tracer, which also can be seen from the way it has distributed in surface waters since its introduction in medical MRI.[35] In addition, D_2O is well suited as a contrast agent for such purposes. Imaging the distribution of conservative tracers with time is complementary to direct NMR encoding of velocity distributions, as it allows also the observation of slow flow processed and in addition gives insights into dispersive processes. Furthermore, imaging of tracers is much less susceptible to unfavorable NMR properties such as short relaxation times or strong internal magnetic field gradients than direct NMR velocimetry.

Hydrophilic contrast agents may be applied to enhance the natural contrast between water and nonaqueous phase liquids (NAPLs), as they only affect relaxation behavior of the water phase in which they are dissolved. Depending on the relaxivity of the contrast agent, a range of different contrasting options are available. Imaging at long echo times in the presence of a contrast agent in the water will highlight NAPL phases. Like this, even relatively small quantities of NAPL (e.g., present in small droplets) can still be visualized against a large background of water. By contrast, the use of sequences with short repetition times is very well fitted to visualize a NAPL phase present in larger quantities against a heavily contrast-agent-doped water background (see Section 7.4.3.1).

Contrast agents that bind to specific target molecules may open up further options. Such contrast agents might help to visualize other adsorbed species that are elusive to direct MRI detection because of small quantities or unfavorable NMR relaxation properties. The development of such contrasting options may benefit from the ongoing trend in medical imaging toward 'molecular imaging.' Possible candidate materials might be superparamagnetic nanoparticles with appropriate immunochemical functionalizations or cross-relaxation agents for electron-nuclear double resonance imaging.[36]

7.4 SOME SELECTED APPLICATIONS OF MRI IN ENVIRONMENTAL SCIENCE

7.4.1 MRI Visualization of Water Flow in Porous Media

As described in Section 7.3.3.4, visualization of tracer flow is an excellent method to examine flow velocity distributions inside a sample. This approach was chosen to visualize the flow distribution inside a quartz sand column after the actual adsorption/desorption experiment.[14] Figure 7.3 shows a few snapshots from such an experiment. A major drawback of tracer imaging is the fact that the experiment has to be conducted as the last of the experiments in a test regiment of a given sample to avoid contamination of the sample from possible residues of the tracer (no tracer is really conservative, no sample exhibits perfect Darcy flow).

While the visualization of externally applied flow using tracers represents a workaround against poor direct NMR velocimetry imaging, the tracer itself may also lead to flow processes under other conditions. This can be seen in Figure 7.4 where the propagation of two Mn-containing tracer boluses with different density under the action of gravity is shown. Note also the very small density differences that are sufficient to trigger such a density-driven flow phenomenon. The white upper rim of the sinking boluses is mainly related to dispersion: in these tails, the density difference driving the flow is smaller than at the center of the bolus, so the solution sinks slower and the tail widens. The inversion of the contrast is due to the use of sequence with both T_2 and T_1 weighting (T_2 weighting dominates in the center of the bolus, whereas T_1 at the smaller concentrations on the upper rim).

7.4.2 NonConservative Paramagnetic Materials in Sediments

Many paramagnetic materials do not behave like conservative tracers inside environmental matrices and actually it is those materials that are much more interesting to study in MRI, as their behavior inside a sample can be much more complex and much harder to follow via conventional, nonimaging studies. In the following sections, two very different types of nonconservative behavior are discussed.

7.4.2.1 Distribution of Oxygen

Molecular oxygen (O_2) dissolved in water is a well-known troublesome molecule in NMR relaxation experiments because of its paramagnetic behavior.[9] At the same time, it is also possible to make use of these properties and use NMR relaxation times to analyze the dissolved oxygen content in water.[37] In NMR studies of sufficiently clean sediments, it is also possible to image the distribution of dissolved oxygen injected into the matrix as well as the oxygen depletion due to outgassing.[38] Despite this successful proof of principle, one has to be aware that the combination of very clean sediment model materials and oxygen supersaturation is quite far away from natural conditions. A way to increase the sensitivity of MRI to oxygen saturation even in less ideal model materials might be through the use of contrast agents

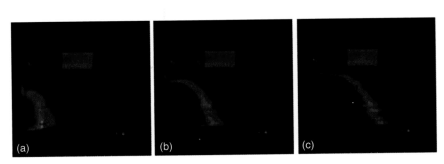

Figure 7.3. (a–c) Distribution of a 'spike' with pure water in a quartz sand column fed with a flow of paramagnetic tracer ions with negligible sorption (Mn^{2+} ions at pH 3). The front shape reveals the inhomogeneous flow velocity distribution inside the column.

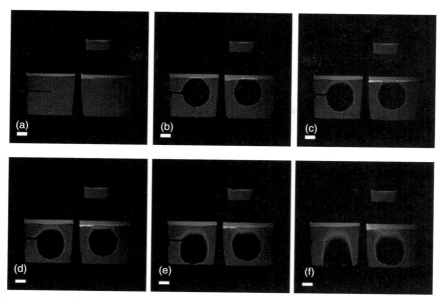

Figure 7.4. 'Rat race' of Mn^{2+}-containing salt solutions of different density (right side: 0.9991 g ml^{-1} and left side: 1.0001 g ml^{-1}) descending under the action of gravity in water-saturated sand-filled columns (grain size: 0.3–0.8 mm). (a) Columns before injection of the solution, (b) directly after injection, (c) 30 min after injection, (d) 60 min after injection, (e) 1 h 30 min after injection, and (f) 2 h 30 min after injection. Length of the white bar: 2 cm. The density of neat water at 22 °C is 0.9978 g cm^{-3}.

undergoing a strong change in relaxivity depending on the dissolved oxygen content.

7.4.2.2 Adsorption of Paramagnetic Ions and Filtration of Superparamagnetic Colloids

In sediments free of magnetic impurities, such as acid-washed quartz sand, the adsorption of even small amounts of paramagnetic ions such as Cu^{2+}, Cr^{3+}, or Gd^{3+} leads to strong relaxation time contrasts[14,24] that allow a good localization of adsorbed ions inside a macroscopic sediment column. The sensitivity of the experiments could be estimated to 4 µg of adsorbed ions, which can be localized with a resolution in the range of 2 mm over a sediment volume of about 1 l. In an energy-dispersive X-ray spectroscopy (EDX) experiment on water–wet sand in an environmental scanning electron microscope (ESEM), the same amount of Gd on the sediment surface could not be clearly detected.

The physics behind the relaxation time contrast due to adsorbed ions on the sediment surface is complicated and has not been systematically studied. On the basis of current research, it is clear that solution-phase relaxivities of the ions may grossly differ from those of the adsorbed ions. For example, adsorbed Dy^{3+} ions produce no notable contrast in the pore water, whereas their solution-state relaxivity for water is similar to that of Cu^{2+}, which leads to very clear contrast in the adsorbed state.[39] The reason behind this is the much shorter electron spin relaxation time of Dy compared to the ions providing good contrast in the adsorbed state. A more quantitative understanding of the relaxivity of adsorbed ions on sediments requires field-cycling NMR studies (see Chapter 8) and additional theoretical work.

Similar to paramagnetic ions, superparamagnetic colloidal particles can be followed inside a magnetically clean model sediment by means of MRI. Figure 7.5 provides an example for such an experiment in which the propagation of $CoFe_2O_4$ nanoparticles (size: 9.8 nm, obtained from Sustech, Darmstadt) in a column with quartz sand or 0.3–0.8 mm grain size is shown. The fading of the contrast is due to filtration of the colloidal particles in the sand bed.

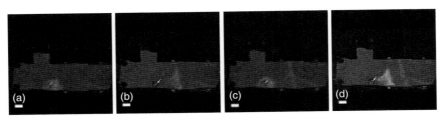

Figure 7.5. Snapshots (T_1-weighted FLASH images) from an experiment of colloidal transport with two subsequent injections of 40 µg of uncoated $CoFe_2O_4$ colloid suspended in 10 ml/water. (a) Shortly after first injection, (b) 6 min after first injection, (c) 10 min after first injection and immediately after second injection, and (d) 5 min after second injection. White bar: 2 cm.

7.4.2.3 Spin-labeled Antibodies to Visualize Organic Contaminants

While the quantitative analysis of organic contaminants in soil is the state of the art, macroscopic processes such as accumulation or degradation are still not readily accessible to the state-of-the-art analytical tools. It is hypothesized that biogeochemical interfaces in soil,[40] which are transient in space and time, are controlling the overall fate of contaminants in soil. MRI is a good choice to visualize and quantify dynamic processes on the scale of centimeters and in natural systems with a spatial resolution down to 200 µm and a temporal resolution in the seconds range.[41] While only few contaminants such as heavy metals are directly accessible with MRI, others can be visualized with MRI-labeled tracers. For instance, bacterial chemotaxis in porous media has been studied using immunomagnetic labeled monoclonal antibodies.[42]

The concept of magnetically labeled antibodies can be extended to organic contaminants.[7] Here, superparamagnetic nanoparticles have been coated with an antibody against benzo[a]pyrene. The antibody is able to detect and visualize benzo[a]pyrene adsorbed to silica gel using a change of the relaxation times. Visualization is limited to hotspots, which are accessible to the labeled antibodies. Furthermore, to remove antibodies that have been attached unspecifically to the matrix, a washing step has to be introduced. This limits the use of this technique to detailed process studies, monitoring of benzo[a]pyrene transport in situ and in real time is not yet possible. However, the limited accessibility of narrow pores comes in handy, as it provides a tool to visualize the accessibility of contaminants for microbial decay and thus a selective visualization of biogeochemical interfaces. The method for detection and visualization of benzo[a]pyrene in soil at

the biogeochemical interfaces is applicable for a larger group of organic pollutants using tailored nanoparticles and monoclonal antibodies.

7.4.3 Nonaqueous Phase Liquids in Saturated and Unsaturated Sediments

In contrast to paramagnetic trace contaminants in an aqueous environment that can only be detected by their natural relaxation time contrast, there is a range of different possibilities to discern nonaqueous phases from water in sediments. Natural relaxation time contrast[13,43] can be used for long-chained hydrocarbons. The T_1 times tend to be shorter than those of water, whereas the T_2 times may be longer because of lower surface relaxivities on most mineral surfaces. In addition, diffusion contrast can be applied for many hydrocarbon phases where self-diffusion coefficients tend to be considerably smaller than in water. Another option is spin density contrast:[18] the ultimate form of spin-density contrast is the use of perfluorinated organic phases, which can be unambiguously discerned from everything else in ^{19}F MRI; contrast can also be achieved between water and solvent phases with lower proton densities such as aromatics or partially chlorinated solvents. The most versatile contrast options are available when using paramagnetic contrast agents.[13,44] Both simple salts of paramagnetic ions such as Mn^{2+} or Gd^{3+} and most medical contrast agents are hydrophilic, so that they affect only the relaxation time in the water phase but not in the hydrocarbon phase. Owing to different relaxivities in the liquid phase, contrast agents are even useful to localize water-soluble hydrocarbons in solution.[13,44]

7.4.3.1 Imaging NAPL Using Relaxation and Spin-Density Contrasts

Relaxation of water and nonaqueous phases in sediments is affected by many different phenomena. Relaxation of oil and water has been extensively studied in natural reservoir rock in the context of well-logging and core characterization.[45]

There are several factors affecting relaxation time contrasts between different liquid phases in a sediment. One possible source of contrast is due to the different intrinsic relaxation times of the liquids (related to the viscosity and typical proton–proton distances of the respective phases); this difference can be used for T_1 to provide a straightforward contrast between water and higher-viscosity liquids such as *n*-octanol.[13] An alternative option is the exploitation of the different relaxivities for water and nonaqueous phases at the pore walls: usually nonpolar liquids are much less sensitive to surface relaxation on polar sediment surfaces than water. In reservoir rocks, this often leads to water populations with very short T_2 that can be used to differentiate between water and oil. In order to undergo surface relaxation, the liquid phase of interest must of course be in contact with the surface. Depending on the saturation history of the sediment, either the water or the nonaqueous phase may be in contact with the wall (terminology in the oil industry 'water-wet' or 'oil-wet'). This may have a great impact on the relaxation times observable for the same amounts of water and NAPL in the same type of sediment. Furthermore, diffusion and flow contrasts can be applied. In contrast to relaxation times, which are strongly sensitive to the surface properties in a material, the self-diffusion coefficient of continuous liquid phases is mainly sensitive to the pore geometry but not significantly sensitive to specific surface interactions between the pore material and the liquid phase. Like this, diffusion contrasts can be interpreted in a more straightforward way than relaxation time contrast. In order to avoid problems with internal magnetic field gradients, diffusion weighting sequences with a compensation for static field gradients should be used.[46,47] As the example in Figure 7.6(a) and (b) shows, such a contrast allows the differentiation between water ($D = 2.3\ 10^{-9}\ \mathrm{m^2\,s^{-1}}$) and hexadecane ($D = 3.61\ 10^{-10}\ \mathrm{m^2\,s^{-1}}$) in a quartz sand model sediment where almost no contrast due to relaxation times is visible. If one of the phases is stagnant, the same type of imaging sequence can also be used for suppression of the signal from the flowing phase leading to an even better differentiation between the signals from the two phases than in diffusion weighting (Figure 7.6c and d).

While water and NAPLs can be separated by their relaxation time contrast, in porous media, this has a high degree of ambiguity, as the relaxation times are influenced by a number of factors. However, with the use of fluorinated NAPLs, water and NAPLs can be visualized concurrently.[48] Using MRI, it is possible to visualize and quantify the shape and the surface area of NAPLs, which control the dissolution rate, directly. Evaluation is usually a two-step procedure, where the flow field around the NAPL blobs is simulated whenever the size or the shape of the blobs changes by a predefined value. With this flow field, the dissolution rates are then calculated.[49]

7.4.3.2 Experiments with Contrast Agents

Compared to natural relaxation time contrast that is sensitive to a range of different effects such as surface relaxation or wetting history, contrast agents allow

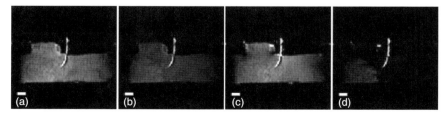

Figure 7.6. Diffusion/flow contrast in a water-saturated quartz sand column (grain size 0.1–0.5 mm) with 35 ml hexadecane injected into the left half. Imaging was performed with a multislice-stimulated echo sequence (five slices of thickness 8 mm) with additional refocusing pulses for internal magnetic field gradients. The white bar in the images corresponds to 2 cm, the in-plane pixel size is 1.95 mm². (a) Static column without gradient pulses, (b) static column with two bipolar pairs of gradients (20 mT m^{-1}, 3 ms per gradient lobe), (c) column under water flow 0.3 mm s^{-1} without gradient pulses, and (d) under flow with gradient pulses like in (b).

a relatively straightforward discrimination between different phases regardless of surface interactions and connectivity. Several experimental options using such contrasts are available.[13,44] Processes accessible in such kinds of studies are dissolution of NAPLs (either directly in the water phase or in an alcohol flooding-type situation) or dislocation of NAPLs (e.g., by changing the level of the water table). Depending on the contrast agent used, the pH and ionic strength of the water phase is modified more or less strongly. In this respect, the use of medical contrast agents such as Gd-DTPA is less invasive than the use of simple paramagnetic ions that usually behave as near-conservative tracers only at low pH. Sorptive effects should be avoided in the use of a contrast agent, as changes in the pore-wall relaxivity may also affect the relaxation of dislocated NAPL and also the pore-wetting properties may change because of adsorption of the contrast agent.

7.4.4   Filtration and Clogging

Filtration processes are of great relevance in environmental technology. Both technical filter systems and natural sediments are applied in this context. The general principle of a filtration system is to remove one or more components of an inflowing fluid from others that are present. Regardless of the influents' nature, it is driven through a physical matrix which by chemical or physical processes separates the influent mixture. Over time, this matrix becomes less efficient through either chemical binding or physical clogging of the pore space, which will eventually render the filter unusable. MRI provides a high contrast between protonated liquids and solids allowing determination of the particulate content and mobility of the fluid, which has a number of environmental applications. In contrast to filters for applications in chemical and water technologies, which have been also extensively studied using magnetic resonance (e.g., Ref. 50 who used spin echo imaging to investigate the dewatering of sewage sludge cakes dynamically, using one-dimensional porosity profiles), filtration in environmental media such as wetlands has only recently been studied by MRI.[51]

In most filter systems, clogging occurs mostly at the inlet and toward the lowest part of the system owing to particulate hold up and gravity-induced settling, respectively. In systems that utilize loose filtration

media, the type of media determines the hydraulic properties of the system and hence the extent of clogging. MRI is a powerful tool to locate the clogged matter within such a system and visualize it in more than one dimension to allow the media to be chosen to optimize various parameters of the system. An example of such a measurement is shown in Figure 7.7 where the clogging of a model-constructed wetland system was tested using a single bead size as a function of time.[52] The effective transverse relaxation time T_2^{eff} of the system is evaluated at 2.35 T using a multislice spin echo sequence. The T_2 shortens over time because of the presence of increasing numbers of field perturbers and is well correlated to clogging.[53] In this study, it was found that there is an optimum grain size for such a filtration system between 8 and 14 mm diameter that will maximize the lifetime of the system while maintaining filtration efficiency. This information is highly beneficial and demonstrates the power of MRI to enhance environmental filtration systems. The growth of biofilms in porous media may also lead to clogging (in this case often called biofouling). A range of MRI studies have been conducted to map both the spatial distribution and the growth of biofilms in model sediments and their impact on the flow patterns inside the model sediment.[54,55]

7.4.5   Interactions of Roots and Soil

Another important aspect of environmental MRI is the effect of contaminations or other environmental stresses on living organisms such as plants. Using dedicated MRI equipment,[29] the root system of plants and its reactions to environmental effects can be studied with increasingly high degree of details (see Chapter 21).

7.5   FUTURE TRENDS AND CHALLENGES

There has been a great deal of investigation into environmental systems using magnetic resonance, and a number of techniques have been demonstrated as very powerful tools for such experiments. Two significant developments have been made in recent years, which will most likely influence future advances. The first is the improvement in console electronics, which has seen a reduction in the minimum echo time, which can be set allowing imaging to be performed on samples

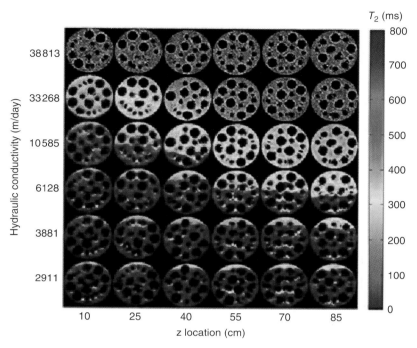

Figure 7.7.   Visualization of clogging in a column filled with a bead model for a wetland filtration system. (Reprinted with permission from Ref. 52. © 2013 Elsevier)

with significantly shorter T_2 values. Sequences such as ultrashort echo time (UTE)[56] or sweep imaging with Fourier transform (SWIFT)[57] are demanding on hardware and require very stable digital consoles, which are now commonplace and allow for high-quality images in the presence of significant field perturbation. Furthermore, all digital receiver chains greatly improve the signal-to-noise ratio and long-term stability making heteronuclear MRI much easier even for less-abundant nuclei. The second significant development is the availability of high-quality rare earth permanent magnets at lower costs and higher fields. This has facilitated the development of a number of probes, which can be deployed into real systems rather than laboratory models. Similarly, this has made it practical to optimize all elements of a small MRI system to a single application. The coming years will see more applications of MRI systems designed for specific applications and measurements of more challenging systems in terms of field inhomogeneity, and greater exploitation of heteronuclear MRI.

Apart from technical issues, one of the major challenges for the use of MRI in environmental sciences seems to be to resolve the ambiguity of the MRI signal. Here, hyphenated techniques could provide the required information. In a combination of μ CT and MRI, the pore topology is visualized by X-ray tomography and the flow and transport processes inside the pore space are visualized by MRI or nonimaging NMR techniques.[58] With this combination, the high resolution of μCT allows for a better tessellation of the pore space and provides data to resolve partial volume effects in the larger voxels of the MRI. Another feature of X-ray CT is the possibility to discern solid sediment materials from gas-filled voids, which cannot be reliably achieved from MRI measurements.[59] Likewise, other imaging techniques could be coupled, such as γ-ray radiography and MRI or positron emission tomography (PET) and MRI.

Imaging the activity of biofilms inside environmental samples is a further field that is likely to receive additional attention in the future. In this field, spectrally resolved MRI would be one of the most attractive options to identify metabolic activity of the biofilms. However, this is again complicated by the much stronger internal gradients present in environmental samples compared to medical specimens. Separating resonance offsets due to chemical shift and due to magnetic field inhomogeneity is possible for samples with pores considerably larger than the individual voxel size. For samples with intravoxel heterogeneity, however, separating chemical and susceptibility information cannot be achieved with the presently available techniques. However, target-specific labeling of superparamagnetic agents or hyperpolarized particles may provide a way to obtain information on specific metabolites even in the presence of strong internal field heterogeneities.

Imaging the fate of colloidal particles inside sediments will greatly profit from the availability of hyperpolarized particles, as this allows a positive identification of particles by signal enhancement, whereas superparamagnetic particles manifest themselves by signal reduction, which is a much less reliable option especially at low concentrations. Furthermore, hyperpolarized particles introduce a 'time stamp' into the system, which offers additional potential for time-resolved studies at timescales that might be even shorter than the actual image acquisition time.

ACKNOWLEDGMENTS

Financial support for the work performed by N. Nestle and M. Rieger from the DFG under grants Ba 1592/1-1 and Ba 1592/5-1 is gratefully acknowledged.

RELATED ARTICLES IN EMAGRES

Environmental NMR: Fast Field Cycling Relaxometry

Soil-Plant Atmosphere Continuum Studied by MRI

Soil-Water Interactions

REFERENCES

1. K. U. Totsche, T. Rennert, M. H. Gerzabek, I. Kögel-Knabner, K. Smalla, M. Spiteller, and H. J. Vogel, *J. Plant Nutr. Soil Sci.*, 2010, **173**, 88.

2. C. A. J. Appelo and D. Postma, Geochemistry, Groundwater and Pollution, Balkema: Leiden, 2006.

3. C. J. Werth, C. Zhang, M. Brusseau, M. Oostrom, and T. Baumann, *J. Contam. Hydrol.*, 2010, **113**, 1.

4. T. Baumann and C. J. Werth, *VZJ*, 2004, **4**, 434.

5. T. Baumann and R. Niessner, *Water Resour. Res.*, 2006, **42**. DOI: 10.1029/2006WR004893.

6. N. Nestle, T. Baumann, and R. Niessner, *Environ. Sci. Technol.*, 2002, **36**, 154A.

7. M. Rieger, G. E. Schaumann, Y. Kunhi Mouvenchery, R. Niessner, M. Seidel, and T. Baumann, *Anal. Bioanal. Chem.*, 2012, **403**, 2529.

8. R. Kimmich, NMR-Tomography, Diffusometry, Relaxometry, Springer: Heidelberg, 1997.

9. D. Michel and H. Pfeifer, *Z. Naturforsch.*, 1965, **20a**, 200.

10. K. R. Brownstein and C. E. Tarr, *Phys. Rev. A: At. Mol. Opt. Phys.*, 1979, **19**, 2446.

11. D. Allen, M. Andreani, R. Badry, C. Flaum, P. Gossenberg, J. Horkowitz, J. Singer, and J. White, *Oilfield Rev.*, 1997, **9**, 34.

12. T. R. Bryar, C. J. Doughney, and R. J. Knight, *J. Magn. Reson.*, 2000, **142**, 74.

13. N. Nestle, A. Wunderlich, and T. Baumann, *Eur. J. Soil Sci.*, 2008, **59**, 559.

14. N. Nestle, A. Wunderlich, R. Niessner, and T. Baumann, *Environ. Sci. Technol.*, 2003, **37**, 3972.

15. H. Y. Carr and E. M. Purcell, *Phys. Rev.*, 1954, **94**, 630.

16. N. Nestle, A. Qadan, P. Galvosas, W. Süss, and J. Kärger, *Magn. Reson. Imag.*, 2002, **20**, 567.

17. A. J. Fagan, N. Nestle, and D. J. Lurie, *Magn. Reson. Imag.*, 2005, **23**, 317.

18. C. Zhang, C. J. Werth, and A. G. Webb, *Environ. Sci. Technol.*, 2002, **36**, 3310.

19. S. Haneder, S. Konstandin, J. Morelli, A. Nagel, F. Zöllner, L. Schad, S. Schoenberg, and H. Michaely, *Radiology*, 2011, **260**, 857.

20. M. C. Cassidy, H. R. Chan, B. D. Ross, P. K. Bhattacharya, and C. M. Marcus, *Nat. Nanotechnol.*, 2013, **8**, 363.

21. P. T. Callaghan, Principles of Magnetic Resonance Microscopy, Clarendon Press: Oxford, 1991.

22. B. Blümich, NMR Imaging of Materials, Clarendon Press: Oxford, 2000.

23. E. H. Hardy, *Chem. Eng. Technol.*, 2006, **29**, 785.

24. N. Nestle, T. Baumann, A. Wunderlich, and R. Niessner, *Magn. Reson. Imag.*, 2003, **21**, 345.

25. Y. Nakashima, *J. Nucl Sci. Technol.*, 2004, **41**, 981.

26. P. Galvosas, F. Stallmach, G. Seiffert, J. Kärger, U. Kaess, and G. Majer, *J. Magn. Reson.*, 2001, **151**, 260.

27. S. D. Beyea, B. J. Balcom, I. V. Mastikhin, T. W. Bremner, R. L. Armstrong, and P. E. Grattan-Bellew, *J. Magn. Reson.*, 2000, **144**, 255.

28. S. Haber-Pohlmeier, M. Bechtold, S. Stapf, and A. Pohlmeier, *VZJ*, 2010, **9**, 834.

29. N. M. Homan, C. W. Windt, F. J. Vergeldt, E. Gerkema, and H. Van As, *Appl. Magn. Reson.*, 2007, **32**, 157.

30. A. Haase, M. Brandl, E. Kuchenbrod, and A. Link, *J. Magn. Reson., Ser. A*, 1993, **105**, 230.

31. R. Turner, D. Le Bihan, and A. S. Chesnick, *Magn. Reson. Med.*, 1991, **19**, 247.

32. J. Hennig, A. Nauerth, and H. Friedburg, *Magn. Reson. Med.*, 1986, **3**, 823.

33. S. Meiboom and D. Gill, *Rev. Sci. Instrum.*, 1958, **29**, 688.

34. J. Frahm, A. Haase, and D. Matthaei, *Magn. Reson. Med.*, 1986, **3**, 321.

35. P. Möller, P. Dulski, M. Bau, A. Knappe, A. Pekdeger, and C. Sommer-von Jarmerstedt, *J. Geochem. Explor.*, 2000, **69–70**, 409.

36. N. Nestle, K. Shet, and D. J. Lurie, *Magn. Reson. Imag.*, 2005, **23**, 183.

37. N. Nestle, T. Baumann, and R. Niessner, *Water Res.*, 2003, **37**, 3361.

38. N. Nestle, T. Baumann, A. Wunderlich, and R. Niessner, *Magn. Reson. Imag*, 2003, **21**, 411.

39. N. Nestle, T. Baumann, A. Wunderlich, and R. Niessner, *Magn. Reson. Imag.*, 2003, **21**, 345.

40. K. U. Totsche, T. Rennert, M. H. Gerzabek, I. Kögel-Knabner, K. Smalla, M. Spiteller, and H. J. Vogel, *J. Plant Nutr. Soil Sci.*, 2010, **173**, 88.

41. C. J. Werth, C. Zhang, M. Brusseau, M. Oostrom, and T. Baumann, *J. Contam. Hydrol.*, 2010, **113**, 1.

42. M. S. Olson, R. M. Ford, J. A. Smith, and E. J. Fernandez, *Environ. Sci. Technol.*, 2004, **38**, 3864.

43. J. A. Chudek and A. D. Reeves, *Biodegradation*, 1998, **9**, 443.

44. N. Nestle, in Soil and Sediment Remediation, eds P. Lens, T. Grotenhuis, G. Malina and H. Tabak, IWAP Publishing: London, 2005, 16.

45. S. Godefroy, J.-P. Korb, M. Fleury, and R. G. Bryant, *Phys. Rev. E*, 2001. DOI: 10.1103/PhysRevE.64.021605.

46. R. M. Cotts, M. J. R. Hoch, T. Sun, and J. T. Markert, *J. Magn. Reson.*, 1989, **83**, 252.

47. A. J. Lucas, S. J. Gibbs, E. W. G. Jones, M. Peyron, J. A. Derbyshire, and L. D. Hall, *J. Magn. Reson.*, 1993, **104**, 273.

48. C. Y. Zhang, C. J. Werth, and A. G. Webb, *Environ. Sci. Technol.*, 2007, **41**, 3672.

49. C. Y. Zhang, C. J. Werth, and A. G. Webb, *J. Contam. Hydrol.*, 2008, **100**, 116.

50. E. J. La Heij, P. J. A. M. Kerkhof, K. Kopinga, and L. Pel, *AIChE J.*, 1996, **42**, 953.

51. R. H. Morris, M. I. Newton, P. R. Knowles, M. Bencsik, P. A. Davies, P. Griffin, and G. McHale, *Analyst*, 2011, **136**, 2283.

52. M. F. Shamim, M. Bencsik, R. H. Morris, and M. I. Newton, *Microporus Mesoporus Mater.*, 2013, **178**, 48.

53. M. Bencsik, M. F. Shamim, R. H. Morris, and M. I. Newton, *Int. J. Environ. Sci. Technol.*, 2013. DOI: 10.1007/s13762-013-0336-7.

54. J. D. Seymour, S. L. Codd, E. L. Gjersing, and P. S. Stewart, *J. Magn. Reson.*, 2004, **167**, 322.

55. E. O. Fridjonsson, J. D. Seymour, L. N. Schultz, R. Gerlach, A. B. Cunningham, and S. L. Codd, *J. Contam. Hydrol.*, 2011, **120–121**, 79.

56. D. J. Tyler, M. D. Robson, R. M. Henkelman, I. R. Young, and G. M. Bydder, *J. Magn. Reson. Imag.*, 2007, **25**, 279.

57. D. Idiyatullin, C. Corum, J. Y. Park, and M. Garwood, *J. Magn. Reson.*, 2006, **181**, 342.

58. C. A. Arns, *Phys. Stat. Mech. Appl.*, 2004, **339**, 159.

59. V. Jelinkova, M. Snehota, A. Pohlmeier, D. Van Dusschoten, and M. Cislerova, *Org. Geochem.*, 2011, **42**, 991.

Chapter 8

Environmental NMR: Fast-field-cycling Relaxometry

Pellegrino Conte and Giuseppe Alonzo

Dipartimento di Scienze Agrarie e Forestali, Università degli Studi di Palermo, 90128 Palermo, Italy

8.1 INTRODUCTION

NMR relaxometry refers to the ensemble of techniques applied to monitor how fast nuclear spin magnetization switches from a nonequilibrium state to the

NMR Spectroscopy: A Versatile Tool for Environmental Research
Edited by Myrna J. Simpson and André J. Simpson
© 2014 John Wiley & Sons, Ltd. ISBN: 978-1-118-61647-5

equilibrium distribution. This phenomenon, referred to as relaxation, arises from fluctuating local magnetic or electrical fields that are generated by nuclear dipoles, unpaired electrons, electric charges interacting with nuclear quadrupole moments for $> 1/2$ spin nuclei, anisotropy of the chemical shielding tensor, fluctuating scalar coupling interactions, and molecular rotations. Mostly, fluctuations are the result of molecular motions. For this reason, measurement of the relaxation times (or relaxation rates that are the inverse of the relaxation times) can be related to molecular dynamics.[1–4] The frequency range of the microscopic molecular fluctuations is comprised within the interval of $10–10^9$ Hz.[5] Because of this, relaxation time values range from few picoseconds to seconds.

Two different relaxation times are associated to nuclear relaxation. The first one, indicated as longitudinal or spin–lattice relaxation time (T_1), measures the time needed for the recovery of the longitudinal component of the magnetization along the z-axis. The second relaxation time, known as *transversal or spin–spin relaxation time* (T_2), measures the decaying for the transversal component of the magnetization to zero. In particular, frequency motions from <10 to $\approx10^4$ Hz are monitored by diffusimetry, transverse relaxometry, and residual spin coupling NMR techniques; motions with frequency ranging between $\approx10^4$ and $\approx10^5$ Hz are investigated by rotating frame NMR relaxometry experiments; motion frequencies from $\approx10^5$ to $\approx10^8$ Hz are observed by field-cycling NMR relaxometry; and frequencies of $>10^8$ Hz are monitored using traditional high field NMR spectrometers.[5,6]

Field-cycling NMR investigates molecular motions through the measurement of the variations of the longitudinal relaxation times, as the external magnetic field intensity (B_0) is changed. Field cycling can be achieved by the 'sample shuttle technique' (SST). This consists in moving pneumatically the sample through different positions in a magnetic field, where different flux densities can be recognized.[6] Typical SST field switching times (SWTs) are of the order of hundreds of milliseconds. This limits the application of the technique to samples with longitudinal relaxation time values >100 ms. A better solution for measuring samples with T_1 values <100 ms is the application of fast-field cycling (FFC). In this case, field switching is achieved electronically by modulating the current passing through the coil generating the magnetic field. Owing to the development of modern electronics, field switching can be automatically achieved in 2–3 ms (that is around 30–50 times lower than in SST method), thereby allowing to monitor longitudinal relaxation times of the order of 2 ms.[6,7] The high-frequency limits for FFC NMR relaxometry depend on the available high field magnets. Conversely, the low-frequency limits are governed by small fluctuations of magnetic field because of, for example, the Earth's field or any other magnetic stray field, and T_1 values that are of the order of the SWT.[6,8] Relaxation

times shorter than the SWTs can also be measured if special electronic precautions are applied.[9]

This chapter highlights FFC NMR relaxometry. The nature of the FFC NMR relaxometry experiment will be described and recent advances in environmental applications will be provided. It must be stated that the technique can be applied, in principle, on all the NMR visible nuclei.[10] However, until now, to the best of our knowledge, all published studies focus only on ^{1}H when FFC NMR relaxometry is applied in environmental investigations. For this reason, from now on, we refer only to FFC NMR behavior of protons. Finally, the readers are referred to wider and more specific review papers to deep knowledge on the theory and the instrumental details behind the technique.[5,6,8,11]

8.2   THE FFC NMR RELAXOMETRY EXPERIMENT IN PRACTICE

Figure 8.1 shows the typical FFC NMR experimental design based on a pre-polarized (PP) and a nonpolarized (NP) sequences.[6] Namely, three steps can be recognized: polarization, relaxation, and acquisition. During the first step of the PP sequence (Figure 8.1a), the longitudinal magnetization is generated through the application of a polarization field (B_{POL}) for a

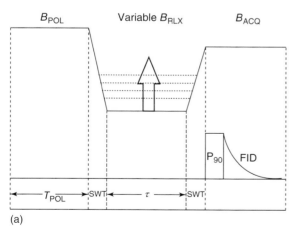
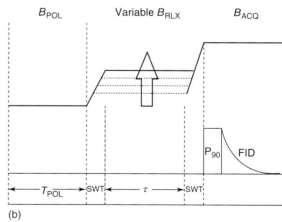

Figure 8.1.   Pre-polarized (a) and nonpolarized (b) sequences used in fast field cycling NMR relaxometry experiments. B_{POL} is the polarization field applied for the period T_{POL}, B_{RLX} is the relaxation field applied for the variable τ time, and B_{ACQ} is the acquisition field. FID is the free induction decay, P_{90} is the 90° pulse, and SWT is the switching time needed to change the intensity of the magnetic field. B_{POL} intensity is null in the NP sequence.

limited and fixed period of time (indicated as polarization time, T_{POL}). Afterwards, the magnetic field is switched to a new one (the relaxation field, B_{RLX}), applied for a period τ during which the magnetization intensity relaxes to reach a new equilibrium condition. Finally, the application of a ^{1}H 90° pulse into an acquisition magnetic field (B_{ACQ}) held for a fixed time, makes the magnetization observable and the free induction decay (FID) acquirable. In the NP sequence, B_{POL} is null (Figure 8.1b). The PP sequence is applied when the relaxation field becomes very low in intensity and enhancement of sensitivity is needed for FID achievement.[11] The crossover field between NP and PP sequences is approximately retrieved when the relaxation field intensity is half of that of the polarization field.[11]

The FIDs obtained by arraying the τ values applied during the relaxation step provide the block scheme shown in Figure 8.2(a) and (b) for the NP and the PP sequences, respectively. Each block represents the FID at any of the arrayed τ values. After measurement of FID areas, the block schemes are transformed in a recovery (Figure 8.2c) or decay (Figure 8.2d) curve according to the nature of the sequence applied (NP and PP, respectively).

8.3 ELABORATION OF THE DECAY/RECOVERY CURVES FOR NATURAL SYSTEMS

Both the recovery and decay curves (Figure 8.2) describe the evolution of the ^{1}H longitudinal magnetization [$M(\tau)$] in the relaxation field when this is held for an arrayed time τ. The magnetization in the relaxation field is aligned along the B_{RLX} direction and its value equals that in the polarization field when $\tau = 0$[5]:

$$M(0) = M_z(0) = M_0(B_{POL}) \tag{8.1}$$

Equation (8.1) is valid only when the evolution of the magnetization is negligible during the SWT. When $\tau \neq 0$, the magnetization evolves according to equation (8.2)[5]:

$$M_z(\tau) = M_0(B_{RLX}) + [M_0(B_{POL}) - M_0(B_{RLX})]$$
$$\times \exp\left[\frac{-t}{T_1(B_{RLX})}\right] \tag{8.2}$$

At the end of the relaxation period, the magnetization is detected by applying a third magnetic field and a 90° pulse, as outlined earlier. According to equation (8.2), the evolution of the magnetization is a recovery curve in the NP condition (Figure 8.2c), whereas it is a decay curve in the PP condition (Figure 8.2d).

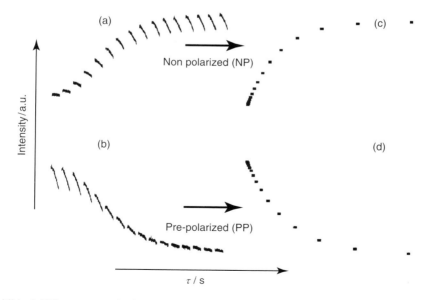

Figure 8.2. Multiblock FID sequences obtained by applying (a) the NP and (b) the PP schemes. The integration of the FID in each block provides (c) the recovery and (d) the decay curves, whose fitting allows achievement of T_1 values.

In equation (8.2), T_1 is the spin–lattice relaxation time, which depends on B_{RLX} intensity. T_1 represents the lifetime of the first-order rate process that returns the spin magnetization to the Boltzmann equilibrium along the z-axis.[12] Its value depends on both the molecular dynamics of the investigated system[12] and the frequency of the applied magnetic field.[6] In fact, the equilibrium recovery of the longitudinal magnetization (and, hence, the time, T_1, needed for that) is determined by the presence in the lattice of magnetic fields oscillating at a frequency corresponding to that of the transition between the nuclear energetic levels. All the nuclear Larmor frequencies are affected by the applied magnetic field. For this reason, only the molecular motions oscillating at the frequency of the specific magnetic field value are effective in promoting nuclear relaxation. The same motions could not be effective for a different value of the applied magnetic field. Molecular motions strongly affect dipolar interactions. In particular, the faster the motions are, the lower is the dipolar interaction efficiency, thereby favoring longer T_1 values. Conversely, slower molecular dynamics can be associated with shorter spin–lattice relaxation times because of stronger nuclear dipolar interactions.[12]

According to Canet,[13] the relaxation decay/recovery curves at each relaxation field intensity are described by the sum of exponential functions as indicated in equation (8.3):

$$M(\tau) = a + \sum_{i=1}^{N} \left[b_i \times \exp\left(\frac{-\tau}{T_{1i}} \right) \right] \quad (8.3)$$

Here, a is the offset and b_i is the magnetization intensity at the Boltzmann equilibrium of the ith component of the molecular motion at each fixed B_{RLX} intensity.

Equation (8.3) is monoexponential only when homogeneous molecular fluctuations are present in simple systems such as pure solvents.[14] Conversely, a multiexponential behavior must be accounted for when complex mixtures are investigated.[13,15–25] As an example, Figure 8.3 reports a decay curve obtained at 0.25 mT for a water-saturated poplar char whose dynamics has been described by De Pasquale *et al.*[22] and Conte *et al.*[23] The monoexponential form of equation (8.3) (dashed line in Figure 8.3) failed in fitting the data for τ values >0.15 ms, whereas the biexponential form (continuous line in Figure 8.3) was successful within the whole range of τ values applied during the analysis. The presence of two different T_1 components indicates that the motion of the water molecules on

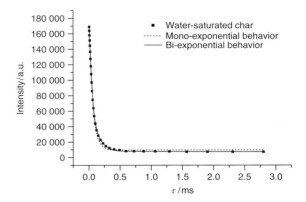

Figure 8.3. Example of a decay curve obtained at 0.25 mT for a water-saturated biochar. The dashed and continuous lines are the fitting retrieved by applying the mono- and biexponential forms of equation (8.3), respectively.

the surface of the porous solid phase is more restricted than that of the bulk water (see also the section titled 'From τ Domain to T_1 Domain: the Inverse Laplace Transformation and the Relaxograms' for further details).

The limiting factor regarding equation (8.3) is the necessity to make assumptions on the number of exponential components to be used for the fitting of the experimental data. In fact, the larger the number of components is, the better is the quality of the fitting (e.g., higher R^2 and lower χ^2 values). However, relaxometry measurements are also affected by an experimental noise that, in turn, influences the fitting error on the estimated T_1 values. In particular, the higher is the number of exponential components in equation (8.3), the larger is the fitting T_1 error, thereby making difficult the evaluation of the molecular dynamics of complex mixtures.

The stretched function given in equation (8.4) has been used instead of equation (8.3) in some papers dealing with environmental samples.[20,22–26] Here, a and b have the same meaning as in equation (8.3), and k is a heterogeneity parameter related to the stretching of the decay/recovery process:

$$M(\tau) = a + b \times \exp\left[-\left(\frac{\tau}{T_1} \right)^k \right] \quad (8.4)$$

This function, which accounts for the large sample heterogeneity resulting in a multiexponential behavior of the decay/recovery curves,[27] can be considered as a superposition of exponential contributions, thereby describing the likely physical picture of some distribution in T_1. Equation (8.4) has the advantage that it is able to handle a wide range of behaviors within a single model. For this reason, assumptions about the number of exponentials to be used in modeling relaxometry data are not necessary.

8.4 FROM τ DOMAIN TO T_1 DOMAIN: THE INVERSE LAPLACE TRANSFORMATION AND THE RELAXOGRAMS

The recovery/decay curves (Figure 8.2c and d) and the related distributions of exponential components [equations (8.3) and (8.4)] vary greatly in character and complexity in multiphase systems such as soils,[17,19,24] sediments,[25] plant tissues,[15,18,20] natural organic matter,[16,28] char,[22,23] and new pesticide formulations.[21] When different components of the molecular dynamics in these multiphase frames are described by longitudinal relaxation times with values very close to each other, a better representation of their distribution is attained by applying an inverse Laplace transformation in the form of equation (8.5)[29-33]:

$$M(\tau) = \int_{T_{1\min}}^{T_{1\max}} D(T_1) \times \exp\left[-\left(\frac{\tau}{T_1}\right)\right] d(T_1) + \sigma$$
(8.5)

Here, $T_{1\min}$ and $T_{1\max}$ are the slowest and the longest spin–lattice relaxation times within which all the T_1 values fall; $D(T_1)$ is the distribution function, which must be determined by solving equation (8.5); and σ is an unknown noise component. The latter term renders impossible to find the exact distribution of relaxation times, thereby allowing infinite possible $D(T_1)$ solutions for equation (8.5). In order to find the most probable $D(T_1)$ distribution, some constraints must be used such as smoothness of the solution and variance of the experimental data. The algorithms developed to account for equation (8.5) are CONTIN[29,30] (CONTINuous distribution) and UPEN[31-33] (Uniform PENalty regularization). They differ between each other for the smoothing procedure used during the application. A detailed description of the two algorithms is out of the aims of the present chapter. The readers are addressed to specific reports[29-33] that provide more details for the different approaches.

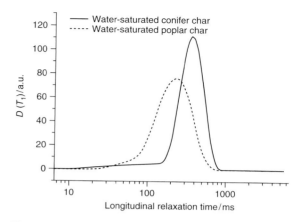

Figure 8.4. Relaxograms of water-saturated conifer and poplar chars obtained by applying the UPEN algorithm. (Reprinted with permission from Ref. 22. © 2012 Springer-Verlag)

Regardless of the procedure used to reveal the most probable distribution of relaxation times, either CONTIN or UPEN algorithms provide similar T_1 distributions also referred to as relaxograms. Figure 8.4 shows, as an example, the relaxograms retrieved by applying the UPEN algorithm on the recovery curves acquired at a relaxation field of 200 mT on conifer and poplar water-saturated chars.[22] Both relaxograms are broad and complex, spanning several T_1 decades with the width of the signal (A) decreasing in the order $A_{\text{(poplar char)}} > A_{\text{(conifer char)}}$. Spin–lattice relaxation time values of water on porous media surfaces are related to porosity. In fact, the motion of water confined in small-sized porous is more restricted than that moving in larger porous. For this reason, inter- and intramolecular interactions are stronger in smaller sized pores than in larger ones, thereby leading to the conclusion that pore-size distribution in the poplar char is wider than in conifer char.[22]

Relationships between pore sizes and T_1 distributions have been used to distinguish among different natural soils,[19,24,34] to evaluate properties of natural and synthetic hydrated sands,[17] to reconstruct the environmental evolution of saltmarsh sediments,[25] to identify differences in plant tissues involved in soil development,[20] to investigate the interactions between natural organic matter and clay minerals,[28] and to monitor interactions between soil organic matter and cation-treated soils.[35] Moreover, ^1H T_1 relaxograms were also successfully applied to follow cellulose degradation in phosphoric acid with the aim to suggest new possible solvents for biomass transformations to biofuels.[36]

8.5 THE NUCLEAR MAGNETIC RESONANCE DISPERSION PROFILES: QUANTITATIVE ASPECTS IN ENVIRONMENTAL APPLICATIONS

Nuclear magnetic resonance dispersion (NMRD) curves are obtained by reporting longitudinal relaxation rates ($R_1 = 1/T_1$) versus the B_{RLX} intensities applied during the FFC NMR experiment. NMRD profiles reflect the spectrum of the reorientational and the diffusional molecular dynamics. The latter are described by Lorentzian functions with the form reported in equation (8.6) through which information about correlation functions of the microscopic fluctuations can be achieved[6,37]:

$$J(\omega_L) = \frac{\tau_C}{1 + (\omega_L \tau_C)^2} \quad (8.6)$$

In equation (8.6), $J(\omega_L)$ is the spectral density function describing the distribution of the motion frequencies in a molecular system, ω_L is the proton Larmor frequency, and τ_C is the correlation time. Correlation time describes the random molecular motions of molecular systems either in solution or in porous media.[6] Namely, τ_C is the time taken for a molecule to rotate one radian or to move a distance of the order of its own dimension.[12] The longer the τ_C value is, the slower are the molecular motions, thereby revealing restrictions in the motional freedom degrees of spatially restrained molecular systems. Conversely, as a molecule encompasses faster motions because of higher degrees of freedom in larger spaces, shorter correlation time values are expected.

Equation (8.6) was redefined by Halle *et al.*,[38] as in equation (8.7), to account for the stretching of the NMR dispersion profiles as a consequence of the complexity of the reorientational dynamics within the molecular system, the heterogeneous distributions of proton exchange rates, and the heterogeneous distribution of intermolecular dipole couplings. Equation (8.7) is also referred to as model-free-analysis[38]:

$$J(\omega_L) = \sum c_i \frac{\tau_{Ci}}{1 + (\omega_L \tau_{Ci})^2} \quad (8.7)$$

In equation (8.7), the subscript 'i' refers to the different components of the motion and c_i is a fitting parameter. The summation of c_i values represents the mean

square fluctuation that contains information about the equilibrium structure of the system, thereby being independent of the system dynamics.[38,39]

The parameters of the multi-Lorentzian equation (8.7) can be determined by any nonlinear parameter estimation method, while the number N of Lorentzians to be included can be objectively determined by means of any statistical procedure. As an example, the *F*-test has been applied in some papers.[22,23,25,38] Briefly, a fit of N Lorentzian terms to M data points $J(\omega_i)$ with errors σ_i provides a $\chi^2(N)$ value that must be compared to the $\chi^2(N+1)$ value obtained by applying $N+1$ Lorentzians. If the fit improves, the ratio $F_{(N,N+1)} = \chi^2(N)/\chi^2(N+1)$ increases [i.e., $\chi^2(N+1) < \chi^2(N)$]. The procedure must be iterated until $F_{(N+m-1,N+m)} > F_{(N+m,N+m+1)}$, where m is integer and ≥ 0. When the latter condition is satisfied, the acceptable number of Lorentzians to be applied in equation (8.7) is $N + m$.

It must be emphasized that the model proposed by Halle *et al.*[38] is only a useful and convenient method to fit the experimental data. The set of parameters {c_i, τ_i} obtained by equation (8.7) does not have any physical significance, unless independent information suggests that the system can actually be modeled by a fixed number of Lorentzians. In the latter case, a direct physical interpretation of the parameters can be attempted. As an example, Průšová *et al.*[40] reported the dynamics properties of three water hydration layers (bound, weakly bound, and bulk) surrounding hyaluronan. Presence of these layers around hyaluronan has been previously hypothesized by rheology and thermal analyses.[41,42]

Halle *et al.*[38] also showed that each fitting parameter, c_i, can be considered as a weighing factor for the ith correlation time value, thereby leading to the calculation of a weight-average correlation time ($\langle \tau \rangle$) according to expression (8.8):

$$\langle \tau \rangle = \frac{\sum_i c_i \tau_{Ci}}{\sum_i c_i} \quad (8.8)$$

The weight-average correlation time retrieved by applying equation (8.8) describes the average dynamics of the whole complex system under investigation.

Following application of the biexponential form of equations (8.3), equations (8.7) and (8.8) have been used to monitor the dissolution mechanism of crystalline cellulose in phosphoric acid (H_3PO_4).[14] In particular, two forms of H_3PO_4 have been recognized. The first one showed the fastest longitudinal relaxation rates in the whole range of B_{RLX} intensities (i.e., the shortest spin–lattice relaxation times). This H_3PO_4 form was (i) considered strongly bound to cellulose because of the formation of phosphor–ester linkages and (ii) responsible for cellulose dissolution. Formation of C–O–P bridges allowed cellulose swelling and solvation. The second H_3PO_4 form showed the slowest R_1 values in the NMRD profile, thereby being considered as bulk phosphoric acid. The $\langle\tau\rangle$ values reported by Conte *et al.*[14] supported the aforementioned findings. In fact, the correlation time for the bulk phosphoric acid resulted shorter than that retrieved for the strongly bound form of H_3PO_4.[14]

It has been proved that the Bloembergen–Purcell–Pound (BPP) model reported in equation (8.9) can be successfully applied to fit NMRD profiles in order to reveal the dynamics properties of liquids interacting to the surface of crystalline organic solids.[6,43] This model applies only to pure compounds when isotropic rotational diffusion of molecules and intramolecular interaction of two-spin-$\frac{1}{2}$ system with fixed internuclear distances can be hypothesized.[6] Equation (8.7) can be included in equation (8.9) in order to account for dynamics inhomogeneity in more heterogeneous systems[44]:

$$R_1 = \frac{1}{T_1} = \alpha + \beta[0.2J(\omega_L) + 0.8J(2\omega_L)] \quad (8.9)$$

In equation (8.9), all the terms have been already defined with the exception of α and β. The former represents the high-field relaxation rate, whereas the latter is a constant related to the dipolar interactions. Increasing dipolar strengths, due to reduced molecular mobility, produce longer α and larger β values. Conversely, weak dipolar couplings are generated by unbound (or freely moving) molecules, thereby providing shorter α and smaller β values.[38,44] As correlation times are a measure of the rate of molecular reorientation in the time unit (see earlier), low α and β values due to unconstrained molecular motions are also associated to short τ_C values. On the other hand, the larger the α and β values are, the longer result the correlation times, because of restrictions in molecular mobility.

The model-free BPP analysis was used to differentiate among water-saturated chars obtained from a gasification process applied to marc, poplar, and

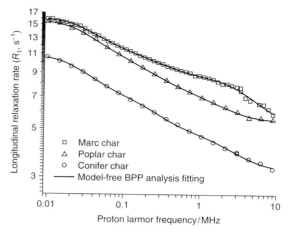

Figure 8.5. NMRD profiles for water-saturated marc, poplar, and conifer chars fitted by applying the model-free BPP analysis. (Reprinted with permission from Ref. 22. © 2012 Springer-Verlag)

conifer wastes.[22] In particular, results revealed that: (i) the water relaxation rates in the NMRD profiles varied in the order $R_{1(marc\ char)} > R_{1(poplar\ char)} > R_{1(conifer\ char)}$ (Figure 8.5); (ii) the α values were in the order $\alpha_{(marc\ char)} \approx \alpha_{(poplar\ char)} > \alpha_{(conifer\ char)}$; (iii) the β values changed as $\beta_{(marc\ char)} \gg \beta_{(conifer\ char)} \approx \beta_{(poplar\ char)}$; and (iv) the weight-average correlation times varied in the order $\langle\tau\rangle_{(marc\ char)} < \langle\tau\rangle_{(conifer\ char)} < s\langle\tau\rangle_{(poplar\ char)}$.[22] The fastest $R_{1(marc\ char)}$ and the longest $\beta_{(marc\ char)}$ values were due to the presence of paramagnetic impurities. The latter fasten spin–lattice relaxation times and raise up β values, thereby altering the results achievable by combining equations (8.7) and (8.9).[12] For this reason, dynamics of water on the surface of marc char could not be compared to that of water moving on poplar and conifer chars. On the other hand, differences between α, β, and $\langle\tau\rangle$ values for the latter two water-saturated chars were attributed to the different porosities that were, in turn, affected by the nature of the biomass feedstocks.[22]

Porosity was also responsible for the differences among the parameters retrieved by applying the model-free BPP analysis on water-saturated soils[24] and sediments.[25] In the first case, NMRD profiles were able to reveal that texture of a gypsic haploxerept soil was affected by the nature of three different plant species used in the afforestation of a Mediterranean area. In the second case, porosity variations associated to different α, β, and $\langle\tau\rangle$ values in the water-saturated layers of a sedimentary core allowed to reconstruct the environmental evolution of a Sicilian (Italy) saltmarsh.

Equations (8.6)–(8.9) apply either to liquid mixtures or to liquids moving on the surface of porous media. Conversely, NMRD profiles of solid-state systems are better described by the power-law model reported in equation (8.10)[45]:

$$R_1 = \frac{1}{T_1} = A\omega_0^{-b} + C \qquad (8.10)$$

Here, A is the size of the low-frequency relaxation rate that includes differences in spin density, structural organization, and local internal dynamics,[46–48] ω_0 is the proton Larmor frequency, b is a power factor, and C is the high-field relaxation rate. This model was applied to study the transformation of leaves to litters in a reafforestated soil located in Sicily (Italy).[20] Results showed that while A changed in the order $A_{leaves} > A_{litters}$, C varied in the order $C_{leaves} < C_{litters}$. The trend inversion was attributed to the changes in the motion regimes (see Section 8.6), which flip from fast to slow, as the proton Larmor frequency is raised up.[12] In the slow-motion regime, the faster the rotational motions are (as in litters), the slower is the relaxation efficiency (i.e., $A_{leaves} > A_{litters}$). Conversely, in the fast-motion regime, the relaxation efficiency increases with the swiftness of the rotational motions (i.e., $C_{leaves} < C_{litters}$). In addition, as A includes also differences in spin density, the results revealed that spin density in litters is lower than in leaves as a consequence of the degradation mechanisms. The values of the power factor, b, changed in the order $b_{leaves} > b_{litters}$. In general, b ranges from 0.45 to 0.75 in solid proteins.[6] As a matter of fact, hydrated solid proteins reveal b values close to 0.45; whereas the lower the hydration degree is, the higher is the power factor value.[6,45,49,50] In polymer science, it is also reported that changes in b values can be associated to intra- or intersegment relaxation contributions according to the range of values within which b falls. Namely, if $0.20 < b < 0.33$, intrasegment relaxations prevail, whereas when $0.40 < b < 0.50$, intersegment relaxations become predominant.[6] However, Berns et al.[20] pointed out that, at the moment, a complete theory on the power dependence of NMRD profiles of heterogeneous complex solid systems (such as leaves, litters or, more generally, biomasses and environmental matrices) is still missing. For this reason, the physical meaning of b values from equation (8.10) applied to environmental systems cannot be provided yet.

8.6   THE MOTION REGIMES AND THE QUALITATIVE ASPECTS OF NMRD PROFILES IN ENVIRONMENTAL APPLICATIONS AND GREEN CHEMISTRY

Figure 8.6 shows the evolution of T_1 as affected by correlation time and temperature variations.[12,51] This behavior comes by the Lorentzian shape of the spectral density as reported in equation (8.6).

The fast-motion regime is typical for liquid systems whose τ_C values lay generally in the interval 10^{-8} to 10^{-12} s.[51] In this condition, $(\omega_0\tau_C)^2 \ll 1$ and $J(\omega_L) \propto \tau_C$. For this reason, the longer the τ_C is, the shorter is the spin–lattice relaxation time. T_1 reduction follows a monotonic behavior (Figure 8.6). Then, it turns around a minimum, where $(\omega_0\tau_C)^2 \approx 1$ (intermediate-motion regime) to attain the slow-motion regime, where very large T_1 values for very large correlation times are retrieved (Figure 8.6). The condition $(\omega_0\tau_C)^2 \gg 1$, valid for rigid solids, is satisfied in the slow-motion regime. As correlation times are directly related to temperature, switch among the three motion regimes can be also described in terms of T variations (Figure 8.6).

Variable temperature (VT) FFC NMR relaxometry experiments have been conducted to evaluate the motion regime of water on the surface of two polymorphs of a commercial photo-catalyst (rutile and anatase

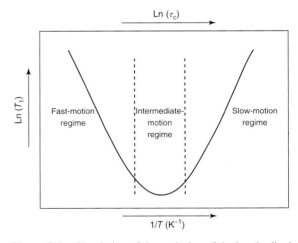

Figure 8.6.   Simulation of the evolution of the longitudinal relaxation time as a function of the temperature ($1/T$) and the correlation time (τ_C) for a fixed value of the measurement magnetic field frequency.

TiO$_2$) generally used in green chemistry reactions.[26] Results revealed that the motion regime changed according to the nature of the two titanium oxides. In particular, water moving on the surface of rutile was subjected to fast-motion regime, whereas that on anatase surface was moving according to the slow-motion regime. The dynamics properties of water on the surface of a nonparamagnetic porous system can be described by equation (8.11)[3,6,52–54]:

$$R_1 = \frac{1}{T_1} = \frac{f_M}{T_{1M} + \tau_M} \qquad (8.11)$$

Here, f_M represents the molar fraction of water chemically bound to porous material, T_{1M} is the proton spin–lattice relaxation time of the chemically bound water, and τ_M is the exchange correlation time, which measures the mean residence time of the bound water. When fast-motion regime prevails (such as on the surface of rutile), the proton spin–lattice relaxation rate is proportional to $1/T_{1M}$, thereby allowing the consideration that water is not chemically retained on the rutile surface. Conversely, as predominance of slow-motion regime is achieved (such as in the case of water moving on anatase surface), the proton spin–lattice relaxation rate depends on $1/\tau_M$ and water lays longer on the anatase surface because of the formation of H-bonds[26] (Figure 8.7). The FFC NMR relaxometry findings explained why the two TiO$_2$ polymorphs differ in their photo-reactivity yield. In fact, the mechanism of the photocatalytic oxidation reactions foresees the formation of ·OH radicals that are believed to be responsible for the primary oxidant attack to

the substrate. The interaction between adsorbed water and TiO$_2$ surface plays the major role to produce the oxidant species. Consequently, the absence of chemical interactions between H$_2$O and the surface of rutile, as highlighted by NMRD results, suggests that the formation of ·OH radicals from H$_2$O is highly improbable, thereby making such a catalyst poorly active.[26]

Formation of hydrogen bonds was suggested also to explain the slow-motion regime observed by VT FFC NMR relaxometry experiments for water moving on the surface of a char obtained from an industrial thermochemical process.[23] In this case, hydrogen bonding involved the electron-deficient water hydrogen atoms and the electron-rich graphite-like systems, thereby leading to the realization of nonconventional H-bond interactions. The presence of weak H-bonds between char and water can explain the large water capacity and structural stability of char amended soils.[55]

Promising information that is relevant for mineral oil industry has also been obtained from VT FFC NMR relaxometry experiments performed to monitor the dynamics of oil in bulk and reservoir rocks.[56] Results revealed that oil in bulk rocks was subjected to fast-motion regime. Conversely, the presence of hydrophobic sites in the reservoir rocks was hypothesized to explain the more restricted motion of oil, thereby resulting in a slow-motion regime. The hydrophobic oil molecules can be trapped on the surface of reservoir rocks because of Van der Waals interactions.

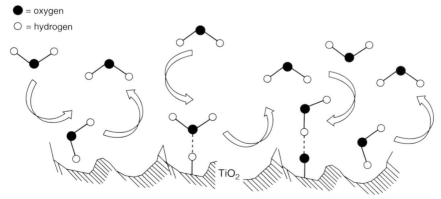

Figure 8.7. Graphical representation of the water dynamics on the surface of titanium oxide. Water is either retained on the porous surface by hydrogen bonds or not chemically constrained. H-bonds are responsible for the slow-motion regime, whereas freely moving water undergoes fast-motion regime. (Reprinted with permission from Ref. 26. © 2013 American Chemical Society)

Temperature and magnetic field variations in relaxometry experiments allowed to reveal solvent–gelator interactions in a mixture made by chlorobenzene and methyl-4,6-*O*-(*p*-nitrobenzylidene)-α-D-glucopyranoside gel (MPNBG).[21] In particular, the slow-motion regime observed between 273 and 340 K was explained by the confinement of chlorobenzene in the MPNBG tri-dimensional network. This confinement was driven by the interactions between the electric dipolar moment of chlorobenzene and the polar MPNBG surface. Importance of such study resides in the possibility to formulate new systems for pesticide synthesis and applications.

8.7 CONCLUSIONS AND FUTURE PERSPECTIVES

While doing literature search for the present chapter, we realized that the number of FFC NMR relaxometry applications in environmental field is, at the moment, very few and mostly those applications have been cited here. The lack of studies involving FFC NMR relaxometry to evaluate environmental problems and possibly suggest solutions is probably due to the better worldwide distribution of the traditional high-field NMR spectroscopy instruments with emphasis on answering questions that are related to the structure of organic matter in soils, water, and atmosphere.[57,58] Information about molecular dynamics and molecular interactions between contaminants and constituents of environmental compartments can also be achieved in high-field NMR spectroscopy by application of VT experiments. These may induce phase transitions and sample decomposition, thereby altering the experimental results. In addition, the frequency of the molecular motions investigated by high-field NMR spectroscopy downs only to 10^8 Hz. As evidenced throughout the present chapter, FFC NMR relaxometry overcomes all the aforementioned problems, although it lacks in resolution and cannot provide structural information concerning the chemical nature of the investigated environmental system.[6] In addition, model theories for FFC NMR relaxometry have been developed mainly for polymer dynamics such as lipids, proteins, and carbohydrates,[5,6,44,59,60] material science,[53,61–63] and contrast agents for magnetic resonance imaging applications.[64,65] For this reason, FFC NMR relaxometry should be supported by results from other analytical techniques when it is applied in environmental field.

Additional advantages of FFC NMR relaxometry as compared to traditional high-field NMR spectroscopy lie in the nonnecessity of deuterated solvents (locking of magnetic fields is unnecessary); in the cryogen-free technology (the necessity to switch among three different magnetic fields makes cryogen gasses superfluous); in the possibility to monitor molecular dynamics at magnetic field intensities close to that of the Earth's magnetic field (this makes possible investigation of very slow dynamics and, hence, very strong molecular interactions); and in the ability to monitor dynamics of quadrupolar nuclei through indirect detection of NMR sensitive spin-$^1/_2$ nuclei in intact systems.[6,8,11] The latter point is very important in environmental investigations because of the possibility to monitor fate of chlorinated contaminants by observing the quadrupolar effects of chlorine in polluted soils. In particular, ^{35}Cl quadrupoles are observed in NMRD profiles as quadrupolar dips in the proton Larmor frequency interval 0.3–4.5 MHz.[61] Shape and position of the quadrupolar dips depend on temperature of the system and nature of the interactions between organo-chlorine compounds and their chemical environment.[61]

Quadrupolar effects on relaxometry behavior of hydrogen nuclei in environmental systems can be generated also by ^{14}N nuclei.[6,66] The possibility to monitor the environmental behavior of nitrogen-containing molecules through the influence of ^{14}N on the dynamics of ^{1}H overcomes the classical limit for the high-field NMR spectroscopy. In fact, although relative abundance of ^{14}N is larger than that of ^{15}N, ^{14}N NMR spectroscopy is resolutionless as compared to ^{15}N NMR spectroscopy because of ^{14}N quadrupole properties. For this reason, high-field NMR spectroscopy on environmental samples is traditionally performed on ^{15}N-enriched materials.[67] The ^{15}N enrichment is achieved either by purification[67] or by growing biomasses with ^{15}N-containing nutrients.[68] Both procedures are not needed when molecular dynamics investigation is performed by FFC NMR relaxometry. Future advances in FFC NMR relaxometry can be achieved by the high-temperature superconducting technology.[69] In fact, the use of high-temperature superconductors may allow to span a very wide interval of magnetic fields with the possibility to combine together multinuclear relaxometry and spectroscopy in only one instrument.

ACKNOWLEDGMENT

The authors are very grateful to Dr Salvatore Bubici (Stelar srl) for the fruitful discussions on the relaxometry technique.

REFERENCES

1. A. Abragam, 'Principles of Nuclear Magnetism', 1st edn, Oxford Science Publications, 1962.

2. D. Neuhaus and M. Williamson, 'The Nuclear Overhauser Effect in Structural and Conformational Analysis', VCH Publishers, Inc, 1989.

3. J.-P. Korb, *Magn. Reson. Imag.*, 2001, **19**, 363.

4. J. V. Bayer, F. Jaeger, and G. E. Schaumann, *Open Magn. Reson. J.*, 2010, **3**, 15.

5. R. Kimmich and N. Fatkullin, *Adv. Polymer Sci.*, 2004, **170**, 1.

6. R. Kimmich and E. Anoardo, *Progr. Nucl. Magn. Reson. Spectr.*, 2004, **44**, 257.

7. R. Kimmich, *Bull. Magn. Reson.*, 1980, **1**, 195.

8. E. Anoardo, G. Galli, and G. Ferrante, *Applied Magn. Reson.*, 2001, **20**, 365.

9. S. V. Dvinskikh, *Instrum. Exp. Tech.*, 1996, **39**, 709.

10. S. Bubici, G. Ferrante, and R. Steele Application note at http://www.stelar.it/2010/download/free/files/AN_120901-NMRD-HeteroNUc.pdf

11. G. Ferrante and S. Sykora, *Adv. Inorg. Chem.*, 2005, **57**, 405.

12. V. I. Bakhmutov, 'Practical NMR Relaxation for Chemists', Wiley, 2004.

13. D. Canet, *Progr. Nucl. Magn. Reson. Spectr.*, 1989, **21**, 237.

14. P. Conte, A. Maccotta, C. De Pasquale, S. Bubici, and G. Alonzo, *J. Agric. Food Sci.*, 2009, **57**, 8748.

15. F. Vaca Chàvez and B. Halle, *Magn. Reson. Med.*, 2006, **56**, 73.

16. J. R. Melton, A. Kantzas, and C. H. Langford, *Anal. Chim. Acta*, 2007, **605**, 46.

17. C. L. Bray, R. G. Bryant, M. J. Cox, G. Ferrante, Y. A. Goddard, S. K. Sur, and J. P. Hornak, *J. Environ. Eng. Geoph.*, 2009, **14**, 49.

18. P. Conte, S. Bubici, E. Palazzolo, and G. Alonzo, *Spectr. Lett.*, 2009, **42**, 235.

19. A. Pohlmeier, S. Haber-Pohlmeier, and S. Stapf, *Vadose Zone J.*, 2009, **8**, 735.

20. A. E. Berns, S. Bubici, C. De Pasquale, G. Alonzo, and P. Conte, *Org. Geochem.*, 2011, **42**, 978.

21. J. Tritt-Goc, M. Bielejewski, and R. Luboradzki, *Tetrahedron*, 2011, **67**, 8170.

22. C. De Pasquale, V. Marsala, A. E. Berns, M. Valagussa, A. Pozzi, G. Alonzo, and P. Conte, *J. Soils Sediments*, 2012, **12**, 1211.

23. P. Conte, V. Marsala, C. De Pasquale, S. Bubici, M. Valagussa, A. Pozzi, and G. Alonzo, *GCB Bioenergy*, 2013, **5**, 116.

24. V. A. Laudicina, C. De Pasquale, P. Conte, L. Badalucco, G. Alonzo, and E. Palazzolo, *J. Soils Sediments*, 2012, **12**, 1222.

25. A. Maccotta, C. De Pasquale, A. Caruso, C. Cosentino, G. Alonzo, and P. Conte, *Environ. Sci. Poll. Res.*, 2013. On-line-first. DOI: 10.1007/s11356-012-1445-4.

26. P. Conte, V. Loddo, C. De Pasquale, V. Marsala, G. Alonzo, and L. Palmisano, *J. Phys. Chem. C*, 2013, **117**, 5269.

27. L. A. Morozova-Roche, J. A. Jones, W. Noppe, and C. M. Dobson, *J. Mol. Biol.*, 1999, **289**, 1055.

28. P. Conte, C. Abbate, A. Baglieri, M. Nègre, C. De Pasquale, G. Alonzo, and M. Gennari, *Org. Geoch.*, 2011, **42**, 972.

29. W. Provencher, *Computer Phys. Commun.*, 1982, **27**, 213.

30. W. Provencher, *Comp. Phys. Commun.*, 1982, **27**, 229.

31. G. C. Borgia, R. S. J. Brown, and P. Fantazzini, *J. Magn. Reson.*, 1998, **132**, 65.

32. G. C. Borgia, R. S. J. Brown, and P. Fantazzini, *J. Magn. Reson.*, 2000, **147**, 273.

33. G. C. Borgia, R. S. J. Brown, and P. Fantazzini, *Magn. Reson. Imag.*, 2001, **19**, 473.

34. S. Haber-Pohlmeier, S. Stapf, D. van Dusschoten, and A. Pohlmeier, *Open Magn. Reson. J.*, 2010, **3**, 57.

35. G. E. Schaumann, D. Diehl, M. Bertmer, A. Jaeger, P. Conte, G. Alonzo, and J. Bachmann, *J. Hydrol. Hydromech.*, 2013, **1**, 50.

36. G. Butera, C. De Pasquale, A. Maccotta, G. Alonzo, and P. Conte, *Cellulose*, 2011, **18**, 1499.

37. A. M. Albano, P. A. Beckmann, M. E. Carrington, F. A. Fusco, A. E. O'Neil, and M. E. Scott, *J. Phys. C: Solid State Phys.*, 1983, **16**, L979.

38. B. Halle, H. Johannesson, and K. Venu, *J. Magn. Reson.*, 1998, **135**, 1.

39. M. Dobies, M. Kozak, and S. Jurga Solid, *State Nucl. Magn. Reson.*, 2004, **25**, 188.

40. A. Průšová, P. Conte, J. Kučerik, and G. Alonzo, *Anal. Bioanal. Chem.*, 2010, **397**, 3023.

41. E. Fouissac, M. Milas, and M. Rinaudo, *Macromolecules*, 1993, **26**, 6945.

42. M. K. Cowman and S. Matsuoka, *Carbohydr. Res.*, 2005, **340**, 791.

43. Y. L. Wang and P. S. Belton, *Chem. Phys. Let.*, 2000, **325**, 33.

44. C. Luchinat and G. Parigi, *Applied Magn. Reson.*, 2008, **34**, 379.

45. W. Nusser, R. Kimmich, and F. Winter, *J. Phys. Chem.*, 1988, **92**, 6808.

46. E. E. Burnell, D. Capitani, C. Casieri, and A. L. Segre, *J. Phys. Chem. B*, 2000, **104**, 8782.

47. E. Murray, D. Carty, P. C. Innis, G. G. Wallace, and D. F. Brougham, *J. Phys. Chem.*, 2008, **112**, 17688.

48. N. Fatkullin, A. Gubaidullin, and S. Stapf, *J. Chem. Phys.*, 2010, **132**, 094903.

49. R. G. Bryant, D. A. Mendelson, and C. C. Lester, *Magn. Reson. Med.*, 1991, **21**, 117.

50. J.-P. Korb and R. G. Bryant, *Magn. Reson. Med.*, 2002, **48**, 21.

51. D. Canet, *Adv. Inorg. Chem.*, 2005, **57**, 3.

52. R. B. Lauffer, *Chem. Rev.*, 1987, **87**, 901.

53. J.-P. Korb and R. G. Bryant, *Adv. Inorg. Chem.*, 2005, **57**, 293.

54. S. Laurent, D. Forge, M. Port, A. Roch, C. Robic, L. V. Elst, and R. N. Muller, *Chem. Rev.*, 2008, **108**, 2064.

55. B. Glaser, J. Lehmann, and W. Zech, *Biol. Fertil. Soils*, 2002, **35**, 219.

56. J.-P. Korb, S. Godefroy, and M. Fleury, *Magn. Reson. Imag.*, 2003, **21**, 193.

57. A. J. Simpson, D. J. McNally, and M. J. Simpson, *Progr. Nucl. Magn. Reson. Spectrosc.*, 2011, **58**, 97.

58. A. J. Simpson, M. J. Simpson, and R. Soong, *Environ. Sci. Technol.*, 2012, **46**, 11488.

59. B. P. Hills, Y. L. Wang, and H.-R. Tang, *Mol. Phys.*, 2001, **99**, 1679.

60. G. Kassab, D. Petit, J.-P. Korb, T. Tajouri, and Levitz P., *Comptes Rendus Chimie*, 2006, **9**, 493.

61. D. J. Pusiol and E. Anoardo, *Braz. J. Phys.*, 1998, **28**, 1.

62. J.-P. Korb, L. Monteilhet, P. J. McDonald, and J. Mitchell Cement, *Concrete Res.*, 2007, **37**, 295.

63. S. Muncaci, C. Mattea, S. Stapf, and I. Ardelean, *Magn. Reson. Chem.*, 2013, **51**, 123.

64. T. Vitha, V. Kubíček, J. Kotek, P. Hermann, L. Vander Elst, R. N. Muller, I. Lukeš, and J. A. Peters, *Dalton Trans.*, 2009, **17**, 3204.

65. D. Ó. Hógáin, G. R. Davies, S. Baroni, S. Aime, and D. J. Lurie, *Phys. Med. Biol.*, 2011, **56**, 105.

66. J. Seliger and V. Žagar, *J. Magn. Reson.*, 2008, **193**, 54.

67. M. W. I. Schmidt, H. Knicker, P. G. Hatcher, and I. Kögel-Knaber, *Eur. J. Soil Sci.*, 1997, **48**, 319.

68. R. J. Smernik and J. A. Baldock, *Plant Soil*, 2005, **275**, 271.

69. Y. Yeshurun, A. P. Malozemoff, and A. Shaulov, *Rev. Modern Phys.*, 1996, **68**, 911.

Chapter 9
Mobile NMR

Ernesto Danieli, Bernhard Blümich and Federico Casanova

Institute for Technical Chemistry and Macromolecular Chemistry, RWTH Aachen University, Aachen D-52074, Germany

9.1 INTRODUCTION

Whereas the trend toward miniaturization is often observed in many technological areas, it is not the case for NMR. Although from the point of view of electronics, NMR has benefited from the development of integrated circuits and smaller and more powerful computers to considerably reduce the size of the spectrometer console, superconducting magnets have become larger to generate stronger fields, thereby improving the field homogeneity. When sensitivity and spectral resolution are the key parameters to be maximized in the magnet optimization process, the size of the magnet rapidly scales up. Today, huge magnets are installed in NMR facilities where to samples must be brought for analysis. However, there are many applications where portability is just as important as sensitivity

or resolution, simply because measurements have to be conducted at the sample site. This is the case when the objects to be studied cannot be transported to the laboratory, either because of size or safety reasons, or when the NMR information is required without time delays after sample preparation or production. This approach has become attractive to many areas such as materials testing,[1] industrial quality control,[2] art conservation,[3] food science,[4] medicine,[5,6] and chemistry.[7,8] A variety of dedicated mobile NMR instruments has become available in the last years. To achieve the compactness and robustness required for field measurements, most sensors are equipped with permanent magnets of different geometries.

In this chapter, we describe the most relevant sensors published in the area. Besides the challenging technical issues solved with the different magnet arrays, we also discuss the NMR methodologies applicable in each case. Two main groups of magnets are defined. First, open magnets generating a sensitive volume external to the magnet array are introduced. Such magnets are useful to study arbitrarily large samples that would not fit inside conventional devices. Then, closed magnets generating a sensitive volume inside the magnet array are discussed. For this case, the main goal is to generate strong and homogeneous fields over a defined volume with the smallest magnet possible.

NMR Spectroscopy: A Versatile Tool for Environmental Research
Edited by Myrna J. Simpson and André J. Simpson
© 2014 John Wiley & Sons, Ltd. ISBN: 978-1-118-61647-5

9.2 SINGLE-SIDED NMR

The concept of single-sided NMR was introduced in the 1980s when the oil industry proved that NMR signals from arbitrarily large samples can be measured by placing them in the stray field of open magnet arrays.[9] In those years, several companies engaged in developing NMR tools suitable to measure downhole the properties of fluids confined in rock formations.[9–12] To do this, the magnet and radiofrequency (RF) probe have to project magnetic and RF fields to volumes external to the sensor, where the object is located. This approach was later followed by building the NMR-MOUSE® (Mobile Universal Surface Explorer) which, in its initial design, utilized a magnet in the shape of a "U" or a horseshoe[13] to generate a magnetic field of about $1/2$ T and a gradient of about $20\,T\,m^{-1}$ in a volume located right above the magnets. Even though the development of single-sided NMR lifted the constraints on the sample size, it introduced the problem of exciting and detecting NMR signals under large static and RF-field gradients across an extended sample.

Once inhomogeneous fields were accepted for the NMR measurement, the most simple NMR sensor, i.e., the simple bar magnet, could be obtained.[14] The various magnet designs, which had been reported for single-sided NMR within the past decade, are based on the two aforementioned geometries: the U-shaped magnet and the single bar magnet. Whereas magnets based on the U-shaped geometry generate a magnetic field parallel to the surface of the sensor and are combined with spiral surface coils, those based on the bar magnet generate a field perpendicular to the sensor surface and require "figure-8"-type coils, which generate a transverse B_1 field. Besides this technical difference, both geometries can be categorized as magnets generating a magnetic field with a strong gradient. In the presence of such a field profile, the excited volume approaches a slice parallel to the sensor surface. This is convenient to obtain spatial localization within the object.

A second group of magnets pursues maximizing sensitivity but sacrificing depth resolution. Such magnets are designed to generate a so-called "sweet spot" or point in space where at least the first-order derivatives of the magnetic field magnitude are zero. By doing so, the size of the excited volume is maximized. In general, because of symmetry reasons, the first-order derivatives along the lateral directions are zero, while the main coefficient that must be cancelled is G_0 (gradient along the depth direction). Even though volume (V) maximization is achieved at the expense of magnetic field strength, the product $V\,B_0^2$, which defines the sensitivity of the magnet, is the largest for these type of magnets.[15] Figure 9.1 shows some of the magnets available for each family.

9.2.1 Magnets Generating a Uniform Gradient

The early design of the NMR-MOUSE offered a typical depth resolution of about one millimeter, while the maximum penetration depth was less than 5 mm.[13] Under these conditions, the analysis hardly implied depth profiling but rather only crude depth selection. To improve the uniformity of the static gradient, several magnet arrays had been optimized during the last decade (see Figure 9.1).[16–19] By means of numerical simulations, the size, position, shape, and polarization direction of the multiple magnet blocks forming the arrays were varied to maximize the depth range and lateral extension where a uniform gradient is generated. This last feature is very important when the sensor is used for depth profiling and slice-selective imaging.[24]

9.2.1.1 Depth Profiling with High Resolution

In the presence of an ideal uniform static gradient (G_0), a depth profile of an object can be measured using G_0 as a read gradient. However, two important technical limitations complicate implementing this technique in a straightforward way. First, the strong gradients generated, in practice, by real sensors reduce the excited depth range to a slice hardly thicker than 1 mm. Second, the efficiency of the RF-coil quickly decays with depth, thereby complicating the uniform excitation of thick slices. Thus, two approaches have been adopted to scan a large depth range. In magnets generating a uniform gradient over an extended depth, the RF-probe can be retuned to excite slices covering the desired depth range in successive experiments. Even though it is simple and fast, the main limitation of this approach is the systematic contrast variation across the sample obtained in the profiles. This problem stems from the fact that the contrast parameters, such as relaxation times or self-diffusion coefficients, depend on the B_0 and B_1 values, which are constantly changing in the retuning process. A different way of obtaining depth profiles is by changing the relative position of the sample with respect to the sensitive slice[18,25] while keeping

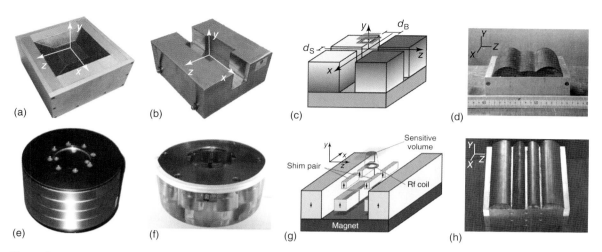

Figure 9.1. (a) Magnet optimized to obtain flat field profiles over a large depth region and extended lateral area.[16] (b) Modified U-shaped geometry designed for slice-selective imaging.[17] (c) *profile*-NMR-MOUSE. The U-shaped geometry is split in two, introducing a small gap d_S useful to adjust the gradient uniformity with high precision.[18] (d) Magnet designed using the scalar potential model.[19] (e) Cylindrical bar magnet with a hollow (barrel magnet [20]). The geometry is obtained by combining two concentric bar magnets with opposite polarization (see text). (f) The NMR-MOLE is a discrete version of the barrel magnet.[21] (g) U-shaped magnet equipped with a set of movable magnets forming a shim unit. By adjusting the position of the small magnets, a volume of high field homogeneity can be generated outside the magnet.[22] (h) Sweet-spot magnet designed using the scalar potential method. The field profile is optimized by shaping the pole pieces mounted on top of the magnets.[23]

the excitation frequency constant as in STRAy Field Imaging (STRAFI)[26] (see Figure 9.2a). It is a truly distortion-free procedure, where the whole profile is measured under exactly the same magnetic and RF field distributions. Furthermore, it also has important implications on the sensor design. Both the magnet and RF-coil can be optimized to work at only one fixed depth. Therefore, better performance and a simpler geometry are expected. An example of this is the *profile*-NMR-MOUSE,[18] which achieves micrometer resolution over a large depth range while maintaining simplicity and high sensitivity.

The most efficient strategy to profile a sample requires that one sets the RF-pulse to the shortest duration to maximize the thickness of the excited slice and then adjust the acquisition time to define the desired resolution inside the selected slice. Subsequently, a profile of the excited region is measured at each depth. By sweeping the position of the excited volume across the sample in steps equal to the thickness of the excited slice, the full depth profile is obtained as the catenation of the individual profiles. Figure 9.2(b) illustrates this procedure to measure a high-resolution profile.

One interesting application of the profiling technique described above is to characterize multilayer plastic

structures. The walls of plastic containers combine several types of polymers separated by barrier layers intended to prevent compounds from diffusing to the exterior. Examples of such a laminated material are gasoline tanks. Their walls consist of a regrind and a polyethylene (PE) layer separated by a barrier layer made from ethyl-vinylalcohol-copolymer which is attached to both sides with an adhesive. Figure 9.2(c) shows profiles measured through such a fuel tank wall before and after different times of exposure to gasoline. The profiles show how the gasoline enters from the regrind side. The solvent front moves into the sample until it meets the barrier, where it is stopped. After a sufficient time, the amplitude of the regrind layer reaches a saturated state, while the external PE layer remains unaltered.

This nondestructive profiling technique has also become a valuable tool in the field of cultural heritage. Since the transport of cultural heritage objects is often not possible or very expensive, mobile equipment is brought to the site of the object for analysis. Single-sided NMR has been applied to study different kinds of objects of cultural heritage such as frescoes,[27] mummies,[28] or ancient paper.[29–32] Old master paintings, for example, present a challenging structure and

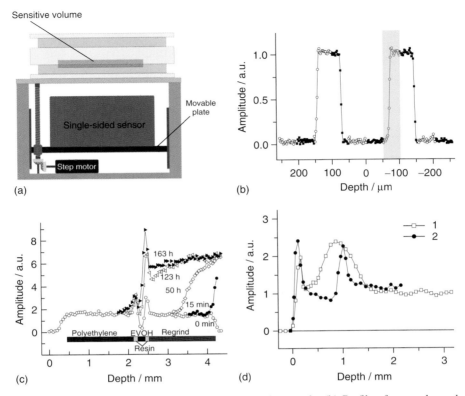

Figure 9.2. (a) Lift used to reposition the sensor with respect to the sample. (b) Profile of a sample made of two latex layers 70 μm thick separated by a glass layer 150 μm thick. The complete profile was obtained as the catenation of 50 μm profiles obtained as the FT of the echo signal measured every 50 μm by moving the sensor with a nominal resolution of about 4 μm. (c) Series of profiles measured through a fuel tank wall before and at different times after exposure to regular-grade gasoline (0, 0.25, 50, 123, 163 h). The profile taken before the exposure shows that the different layers (PE, resin, barrier, resin, regrind) can clearly be distinguished. The gasoline enters through the regrind layer and is then blocked at the barrier layer. (d) Measurements at two different spots in the *Adoration of the Magi* by Perugino (1470A.D.) show the multilayer structure of the painting consisting from left to right of paint, primer, canvas, and wood. However, the canvas layer in spot 1 is much larger than in all other cases. A possible explanation could be that several canvas layers are used for reinforcement in places where wooden pieces are joined together.

complexity. Depending on the style, paintings are made of several layers of paint/primer on a support; normally a wooden panel or canvas. Figure 9.2(d) presents two profiles measured at two spots of the *Adoration of the Magi* by Perugino. The first peak (depth 0 mm) corresponds to the paint layer, which has a thickness of about 130 μm. This is followed by a flat region that corresponds to the primer layer. The second peak at about 1 mm depth is assigned to canvas that is glued on the wood support. This single-sided NMR technique has proven to be very helpful in measuring the thickness of the different

layers, information usually obtained by destructive methods.[33]

An alternative way of measuring a large depth range is to design a sensor with reduced gradient G_0. Thus, short RF-pulses can excite larger regions in a single experiment.[19,34] Although, a sensor working according to this principle would be the ideal solution from a measurement point of view (FT of the echo signal already gives the profile), the design of such a sensor represents a real challenge. First, a reduction of the main gradient magnitude can only be achieved at the expense of field strength. Second, a reduction of

G_0 must be accompanied by a similar decrease in the lateral field variations in order to maintain the spatial resolution. Third, the requirements placed on the RF-coil are very high: good lateral selection, low B_1 gradient, and high efficiency (short RF pulses)/low Q (broadband) at the same time.

9.2.1.2 Spatial Encoding by Fourier Imaging

Spatial resolution along the lateral directions of the sensitive slice has been achieved by implementing pure phase-encoding imaging methods working in the presence of the slice-selecting static gradient. To generate the pulsed gradients required by these methods, flat gradient coils have been designed and mounted on different single-sided sensors.[35–37] Prado *et al.* implemented a single-point spin-echo imaging method[35] that combined a Hahn-echo sequence, applied to refocus the magnetization dephased due to the static background gradient, with gradient pulses applied during the free evolution periods to encode the spatial information in the phase of the echo. The technique was extended in 2002 when a prototype based on a bar magnet was used to measure 2D images of the sensitive volume. Moreover, thanks to the relatively uniform gradient generated by this magnet along the depth direction, this sensor was used to demonstrate that flat slices could be selected at different depths within the object.[36] Even though 3D spatial localization was achieved then, the poor sensitivity inherent to single-sided sensors extended the experimental times to limits far too long for practical uses.

An important advance in the applicability of imaging techniques with single-sided sensors was achieved when multiecho imaging became available. In this way, long trains of echoes generated by applying a Carr-Purcell-Meiboom-Gill (CPMG) sequence could be co-added in order to increase the sensitivity during detection. By doing so, the experimental time could be reduced by a factor proportional to the number of echoes acquired in the train. However, when the pure phase-encoding method is combined with the subsequent application of a train of refocusing pulses, the space information encoded in the phase of the signal by the gradient pulses is quickly distorted as the echo number increases. This results from the loss of the quadrature channel perpendicular to the 180° pulses during the refocusing train. The CPMG sequence preserves the component of the magnetization set by the 90° pulse along the refocusing train. Nonetheless, the component

perpendicular to it decays completely within a few echoes.[38,39]

A solution to this problem requires the combination of two experiments to acquire both quadrature components. The encoding period of the sequence is a Hahn-echo sequence where gradient pulses are applied to encode position in the phase of the echo signal. After the echo is formed, a first experiment is executed with refocusing pulses applied to preserve the real channel (the imaginary channel of this experiment is ignored). In a second experiment, the refocusing pulses are applied incrementing their phase by $\pi/2$ to preserve the imaginary channel (disregarding the real channel). By combining the real component of the first experiment with the imaginary of the second one, the full complex signal is built. Contrast in the image can be produced by using the Hahn-echo sequence as a diffusion or T_{2Hahn} filter or directly the CPMG decay to introduce T_{2CPMG} weighting.

Three-dimensional spatial resolution is achieved by combining the 2D phase encoding method with slice selection. Figure 9.3(a) shows an object with a 3D structure obtained by stacking a set of letters forming the word MOUSE cut from a sheet of 2 mm thick natural rubber. Having no spacer in between the letters, the total structure is 10 mm high. After calibrating the frequency dependence with the depth, 1 mm thick slices were selected inside each letter. The 2D image of each letter can be observed in Figure 9.3(b), demonstrating good slice selection performance and no gradient distortions across the full range.

9.2.1.3 Flow Imaging

Conventional NMR has proven to be a powerful tool to noninvasively characterize molecular motion.[1,40] Several sequences based on the use of pulsed field gradients (PFG) have been implemented and tested in homogeneous magnetic fields, however, PFG do not work well in extremely inhomogeneous fields of open sensors. In the presence of a static magnetic field gradient G_0, the echo signal generated by most PFG sequences is attenuated by diffusion.[41,42] This fact complicates and sometimes prevents the use of this method for displacement encoding in the presence of background gradients. To reduce the diffusive signal attenuation due to G_0, Cotts *et al.* developed the so-called 13-interval sequence. It includes 180° pulses applied during the coding intervals of the stimulated echo sequence (STE) to cancel the phase spread introduced by G_0 during each of these periods,

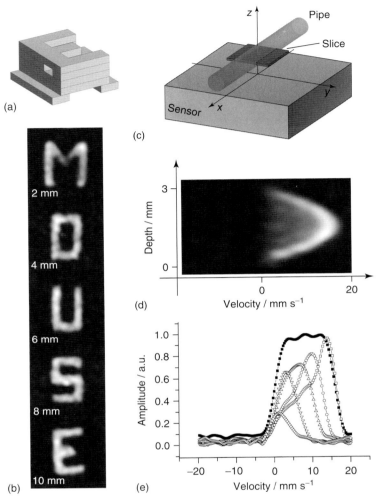

Figure 9.3. (a) Object made by stacking the letters of the word MOUSE cut from a 2 mm thick rubber sheet. (b) Images of each letter obtained by applying the slice-selective multiecho imaging method. The total time to obtain each letter was 45, 45, 90, 120, and 180 s, respectively. (c) Scheme showing the sensitive volume of the sensor selecting a slice inside a circular pipe. (d) Velocity distribution of water undergoing laminar flow in a circular pipe 3 mm in diameter, spatially resolved along the depth direction. The total experimental time to resolve the velocity distribution was about 30 min. The position $y = 0$ in the plot is at 7 mm from the surface of the sensor. (e) Velocity distribution of the different slices across the circular pipe. The total velocity distribution (black squares) obtained by integration of the velocity profiles shows the typical hat function expected for this tube geometry.

while the encoding phase is defined by means of bipolar gradient pulses. In contrast to the PFG-STE sequence, signal attenuation due to diffusion under G_0 arises only during the two encoding periods, but not during the evolution time Δ. To adapt this sequence to be used on an open sensor, where the B_0 and B_1 fields are strongly inhomogeneous, a suitable phase-cycling routine has been proposed to filter the large number of undesired coherence pathways.[43] To improve the signal-to-noise ratio, the multiecho acquisition scheme CMPG-CP with composite inversion pulses implemented for the imaging experiments was attached to the 13-interval sequence.

This velocity method was combined with a slice selection procedure to spatially resolve the velocity profile in the object along the depth direction. To illustrate the performance of the method, the profile of water flowing in a circular pipe with a 3 mm inner diameter was measured by gradation of the excitation frequency (Figure 9.3c). Figure 9.3(d) shows the spatially resolved velocity distribution where the parabolic profile is clearly visible.[37] Moreover, Figure 9.3(e) illustrates the propagators measured at different depths inside the tube. The complete propagator of the tube can be reconstructed by integrating the distributions along the depth direction. Figure 9.3(e) shows the characteristic hat function distribution expected for this geometry.

9.2.2 Magnets Generating a Volume of Homogeneous Field

9.2.2.1 Sweet-spot Magnets

For a given magnet geometry, the static gradient can be canceled by combining two units of the same type with opposite polarization and different sizes. Taking the case of the bar magnet geometry, two bar magnets with different diameters and opposite polarization can be concentrically arranged to generate at a certain distance from the sensor surface a point where the total gradient is zero. This is the basic idea behind the barrel magnet[20] (see Figure 9.1e). While the magnetic field of the big and small magnets are the same at the surface, the gradient of the smaller block is much larger and decays faster with depth. In this way, the two gradients match at a certain depth to generate a zero crossing of the total gradient (position of the sweet spot). The depth where the sweet spot is generated depends on the dimensions of the main and shim magnets. The optimization of this geometry focuses on maximizing the sensitivity of the magnet given by $B_0^2 V^*$, where V^* is the volume excited for a defined bandwidth. One possibility to introduce a certain degree of control over the field profile is to introduce a second magnet (central magnet) in the hole of the barrel and vary its position and size.[20] The two main advantages of this modification are the increments of the sweet-spot size and the field strength. Both changes contribute toward increasing the magnet sensitivity. Another possibility of introducing control on the field profile is to adjust the inclination angle of the barrel walls. This can be practically

achieved by using a discrete version of the barrel as in the case of the NMR-MOLE (MObile Lateral Explorer) shown in Figure 9.1(f).[21]

The U-shaped magnet geometry can also be modified to generate a sweet spot. To do this, a pair of small magnets arranged in antiparallel configuration has to be placed in the gap of the magnet with their polarizations opposing the one of the main magnet. As for the barrel magnet, the size, gaps, and positions of all blocks need to be optimized to maximize the sensitivity. This geometry has also been optimized using the scalar potential method.[23] Figure 9.1(h) depicts a magnet were the permanent magnet blocks are covered by a high permeability cap with a shape designed to cancel the gradient at about 10 mm.

9.2.2.2 Spectroscopy Magnets

In 2001, an ingenious technique developed by Meriles *et al.* triggered a huge advance toward recovering high-resolution spectroscopy in single-sided sensors.[44] The methodology is based on the use of an inhomogeneous RF-field with the same spatial variation of the static magnetic field. In the case when the two fields are matched, the dephasing introduced by the static field inhomogeneity can be refocused under the application of an RF-pulse leading to the formation of a so-called nutation-echo.[44] In 2005, this method was implemented on a portable sensor.[45] The key step in achieving the high matching quality required for this experiment was the use of an inverse shimming concept where the spatial dependence of $B_0(r)$ was shaped by varying the magnet geometry to copy the $B_1(r)$ profile generated by a surface RF-coil. To do this, control variables were incorporated in the magnet geometry to shape the spatial dependence of the magnetic field, and, at the same time, to reduce the gradient strength of the static field. Both goals were achieved by adding a shim unit in the gap of the main U-shaped magnet. It was made of four permanent magnet blocks arranged also in a U-shaped configuration with a polarization opposing the one of the main unit (see Figure 9.4a). Once the shim magnets were incorporated into the main magnet, the total field had a spatial dependence that could be matched to the field of the RF-coil. In this way, the broad line covering a range of about 3000 ppm measured in the presence of the static gradient of the original magnet was first reduced to 200 ppm by means of the shim unit. Finally, by implementing the nutation echo experiment, a spectral resolution of 8 ppm could be obtained (see Figure 9.4b).

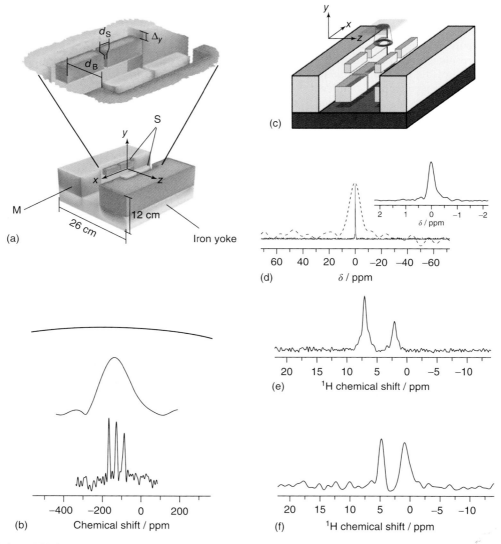

Figure 9.4. (a) U-shaped magnet equipped with a shim unit used to match the B_0 and B_1 field profiles. (b) Spectra acquired in the presence of the natural gradient of the U-shaped magnet (top), once the shim unit is mounted in the main magnet (middle), and applying the ex situ methodology (bottom). (c) U-shaped magnet equipped with a movable shim unit suitable to shim the field to high homogeneity. (d) Comparison of the best resolution achieved with the $B_1 - B_0$ matching technique and the field shimming approach. (e) Spectrum of toluene measured on a sample much larger than the sensitive volume. (f) Spectrum of crude oil mixed with water.

After the shim magnets proved to be very helpful to modify the spatial dependence of the magnetic field, the tenet that a volume of high homogeneity cannot be generated outside the magnet was questioned. A mechanical shimming approach, consisting of a suitable arrangement of small movable magnetic blocks placed within the main magnet could be implemented for improving homogeneity analogous to the resistive shim coil systems used in conventional superconducting NMR magnets.

This strategy was used to shim the field of a conventional U-shaped, single-sided magnet (Figure 9.4c). The mechanical shim unit consists of four magnet-block pairs placed inside the gap of the main magnet. Two pairs are fixed at the bottom and produce a strong gradient along the depth direction y, and two further pairs on the top are movable. Together they generate a total of eight shim components. The first-order shim components x, y, and z are generated by displacing the two movable pairs along the x, y, and z directions, respectively. The second-order terms x^2 and z^2 are adjusted by varying the distance between the pairs while keeping their centers fixed with respect to the main magnet. Moreover, cross terms xy, xz, and yz can also be generated by displacing the magnets asymmetrically.[22,46] The shim unit also includes three single-sided coils generating x, y, and z gradient fields used for fine tuning of the field (not shown in the figure). The spatial region where the required degree of homogeneity is achieved (the so-called "sweet spot") occupies only a limited volume over the magnet. Combining a 90° soft pulse for excitation with the natural lateral selection of a surface RF-coil (blue volume in Figure 9.4c) is required to excite and detect the NMR signal corresponding to this region. Figure 9.4(d) shows the spectrum of a water sample with a volume much larger than the sensitive volume (an arbitrarily large sample) placed on top of the sensor. It is compared with the best resolution achieved using the $B_1 - B_0$ matching technique. The line width is 2.2 Hz, corresponding to a spectral resolution of about 0.25 ppm. The spectral resolution has been improved by a factor of about 30 with a concomitant fivefold extension of the excited volume. Together with a sensitivity-optimized surface RF-coil, the increased size of the sensitive volume leads to an appreciably higher signal-to-noise ratio. The achieved sub parts per million resolution enabled one to resolve different molecular structures such as toluene (Figure 9.4e). These results have proven that ex situ NMR can be used to determine molecular composition by ^{1}H NMR spectroscopy. Figure 9.4(f) shows the ^{1}H spectrum of a mixture of water and crude oil. From the line integrals, the water/oil ratio can be quantified, a result of interest for well logging in the oil industry. Moreover, ex situ ^{1}H NMR spectroscopy can be utilized to noninvasively screen molecular composition, monitor chemical reactions, and identify target compounds.

9.3   CLOSED MAGNETS

Closed magnet geometries generate a magnetic field much stronger and more homogeneous than those achieved with single-sided magnets. In addition, as they are usually equipped with solenoidal RF-coils for excitation and detection of the NMR signal, the RF-pulses are highly uniform. Using permanent magnets for NMR is as old as this technique itself. In the 1960s and 1970s, high-resolution NMR spectroscopy was already available with systems built from permanent magnetic pieces like the Varian T-60. These systems were based in the so-called C-magnet geometry, where the main magnetic field is generated by two magnetic blocks located one in front of the other and attached to an iron frame closing the magnetic field lines. The surfaces of both blocks determining the magnet gap were covered by iron poles properly designed in order to homogenize the magnetic field within the gap. Such permanent magnet systems, however, were rapidly replaced by superconducting technology since superconductors of similar size achieved higher magnetic fields.

When the size of the magnet is reduced to gain mobility, it becomes difficult to control the magnetic field homogeneity over a large working volume, seriously compromising the spectral resolution. Up to a certain degree, however, control can be achieved by using resistive shim coils.[47] But due to the limited efficiency of these coils, there is a limit to the inhomogeneities that can be corrected with this approach. Nonetheless, resistive shim coils have successfully been used to shim customized desktop NMR systems based on the C-magnet geometry.[7,48,49] When the miniaturization of the system is pursued even further, the inhomogeneities become out of range. Another limitation faced when the size of the C-magnet is reduced is imposed by the gap distance between the poles, which is already small in desktop systems. For smaller systems, the RF-probe and shim coil system have to be accordingly reduced, limiting sample size, and with it, the sensitivity. In another limit, when larger volumes with modest homogeneity are required for magnetic resonance imaging (MRI), there is no alternative other than opening the gap at the expense of field strength. These examples explain the need to explore new magnet geometries providing higher magnetic fields and homogeneities in larger volumes.

9.3.1 New Magnet Designs

Among the different closed magnet designs normally used to build mobile NMR sensors, the Halbach magnet[50] is the most promising one. The ideal cylindrical Halbach magnet (Figure 9.5a) requires a polarization that varies continuously along the circumference to generate in the central region the optimum magnetic field in terms of strength and homogeneity, and no stray field outside. The advantages of this magnet are that it offers a generous volume for sample positioning (large bore-to-magnet size ratio) and, at the same time, generates relatively strong magnetic fields $(1-2\,T)$ using conventional magnetic materials such as SmCo or NdFeB. Moreover, it generates a magnetic field perpendicular to the axis of symmetry of the magnet, allowing the use of sensitive solenoid RF coils to detect NMR signals. In practice, the continuous cylinder is approximated by discrete arrays. Different ways of building such arrays with identical blocks such as trapezoids, squares, or circles have been proposed in the past (Figure 9.5b–d).[50–52] Regarding the benefits of using identical pieces with cubic or cylindrical geometry, the fact that all the magnet blocks have the same polarization direction makes it easier and cheaper to obtain them from the manufacturers. Afterward, when mounted, the polarization orientation of each piece is changed by rotating the piece depending on the angular position it occupies in the final ring. The total height of the cylinder is usually achieved by stacking several of these rings.

Another geometry used to obtain a homogeneous magnetic field with reduced magnet size is the so-called Halbach box which is built from permanent magnet blocks arranged as shown in Figure 9.5(e). This design also includes steel pole pieces and is totally enclosed by a steel flux return path. Depending on the applied magnetic material, field strengths ranging from 1 to 2 T can be achieved.

Although in theory these designs are optimized to provide sub parts per million homogeneity, the small errors in the positioning of the pieces when the magnets are built as well as the intrinsic inhomogeneity of the magnetic elements and piece-machining errors reduce the expected performance. In order to tackle these problems, different strategies were followed to obtain spatial and chemical shift information.

9.3.2 Limited Sample Volumes

For some applications where the amount of sample is limited, confining it within a reduced spatial

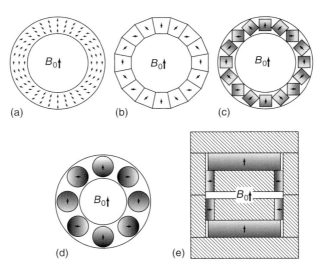

(a)   (b)   (c)

(d)   (e)

Figure 9.5. (a) Ideal Halbach magnet in which the polarization varies continuously along the ring circumference. (b) Discrete approximation of a Halbach array built from trapezoidal pieces polarized in different directions. (c) Identical cubic magnets (MANDHaLa array) and (d) cylindrical rods. (e) Halbach box built by combining permanent magnet elements polarized according to the black arrows and steel elements (pole caps and flux return paths) corresponding to the dashed regions. In all the plots, the direction of the generated B_0 field is indicated by a black arrow located in the position of the sensitive volume.

region where the variations of the magnetic field are of the order of some Hertz allows one to obtain high-resolution NMR information. Typically reported volumes are some nanoliters, which need to be detected by using delicately manufactured microcoils RF-probes in order to improve the sensitivity of the signal coming from such small volumes. To find the location of the "sweet spot", i.e., the region with the highest homogeneity, the probe is placed in the nominal center of the magnet and is then moved until a maximum amplitude of the lines is found in the spectrum. Figures 9.6(a) and (b) show a compact Halbach magnet and a Halbach box, used to detect spectra of fluorinated and protonated compounds, respectively.[53,54] These setups have been shown to be convenient for hyphenating NMR spectroscopy with other chemical techniques, such as capillary electrophoresis to improve identification of components in a complex mixture by first chemically separating the compounds and then using NMR to identify them.[55] Moreover, recent advances in nanoparticles and biotechnology combined with these NMR systems allow the development of diagnostic magnetic resonance sensors, which can be very important for the early detection of diseases.[56,57]

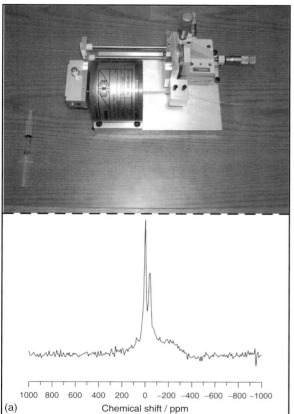

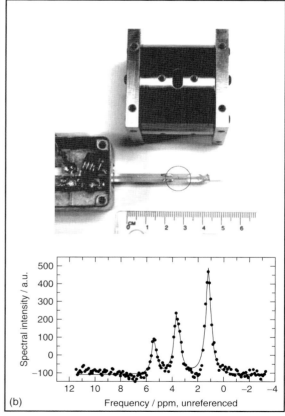

Figure 9.6. Compact magnets for capillary NMR. (a) Top, Halbach magnet composed of eight elements generating a magnetic field of 2 T. Bottom, Fourier transform of the NMR signal of Fluorinert (FC-43), acquired using a 320 μm inner diameter microcoil. The homogeneity reported for the sensitive volume is 17 ppm. (b) Top, Halbach box magnet generating a magnetic field of 1 T within the 5 mm gap region. The sub-parts per million homogeneity of the sweet spot can be appreciated in the spectrum of pure ethanol shown at the bottom.

9.3.3 Homogeneity Improvement by Passive Shimming

Different approaches based on the use of extra small permanent magnets[58-60] or ferromagnetic pieces[61] added to the main structure have been proposed to correct the field homogeneity. Once the main magnet array is built, the field is mapped in order to determine the nature of the corrections required. Afterward, the homogeneity is improved by inserting small magnetic blocks at defined positions or moving them in an iterative process.

A standard way to improve the homogeneity of superconducting MRI systems is to use small pieces of ferromagnetic material.[62] A similar approach with movable systems has also proven itself to be valid. Figure 9.7 shows a sketch of the NMR CUFF (Cut-open, Uniform, Force Free) magnet, which uses a Halbach array built from blocks of square geometry that can be opened and closed at a particular angular position using minimum force.[61] The design is attractive, since it enables nondestructive and in situ investigation of systems such as plants or fruits, which otherwise would be impossible to characterize in a nondestructive way. A modest homogeneity is required to perform spatially resolved experiments. In this case, this was achieved by adding 1 mm steel pole caps to all the magnets, thereby obtaining a homogeneity of 50 ppm over a spherical volume of 5 mm diameter. The magnet weighs 3 kg and generates a magnetic field of 0.57 T. In combination with two sets of plane-parallel gradient coils, the system allows imaging intact plants and measuring water (xylem) flow within a region of about 1 cm in diameter.

By extending the concept of mechanical shim units, based on small movable permanent magnets implemented on single-sided magnets,[22] it is possible to increase the homogeneity of a cylindrical Halbach array to values of greater than 10 ppm (more than two orders of magnitude) over a large sensitive volume $V_S \sim (L_M/10)^3$, whereby L_M denotes a characteristic length for the magnet size.[59] The key to the method is to set the shim unit to reproduce the spatial dependence of the main field with the smallest average field strength possible. To do this, the shim unit is also built following the Halbach design and is placed inside the main unit (Figure 9.8a). Then, by setting the polarization of the shim unit opposite to that of the main field, the inhomogeneities of the latter are corrected while the total field strength is maintained at an acceptable magnitude. Moreover, the

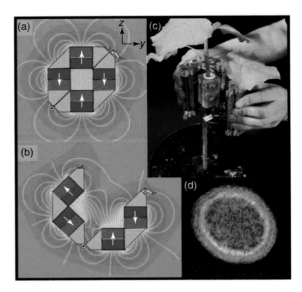

Figure 9.7. Drawings of the closed (a) and opened (b) configurations of the NMR-CUFF. (c) Picture of the sensor being mounted on a plant stem. Note that the RF-probe and gradient coils have been previously mounted. (d) NMR image of a 3 mm thick excised castor bean (*Ricinus communis*) stem slice with a diameter of 6 mm, acquired with a standard spin-echo sequence. The nominal resolution is 120×60 μm².

movement of the shim magnets around their optimum positions generates shim terms with well-defined spatial dependencies used to obtain the simulated performance. The high homogeneity achieved in the sensitive volume of the magnet allows implementation of powerful imaging sequences, such as the RARE sequence, enabling the acquisition of 3D images with in-plane resolution of 0.5 mm within approximately 10 min (Figure 9.8b and c). This performance has been achieved over a cylindrical volume 4 cm in diameter and 4 cm long, with a magnet that is 30 cm in diameter and 30 cm long. Reducing the size of the sample, sub parts per million resolution has been obtained in standard 10 mm NMR tubes, allowing the acquisition of chemical shift-resolved spectra of protonated compounds.

9.3.4 Self-shimeable Main Magnet

The passive shimming approaches described so far are based on the addition of extra elements to the main unit. In this way, the homogeneity is improved at the

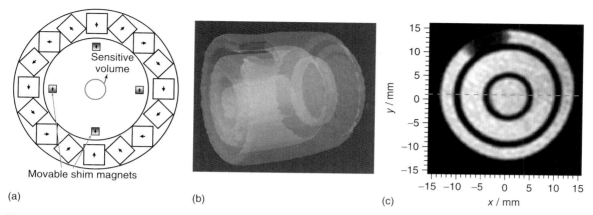

Figure 9.8. (a) In-plane view of the 16 NdFeB square elements Halbach array composing the main magnetic unit of a mobile tomograph. The four smaller permanent magnets located within the bore represent the movable shim unit. (b) 3D image of a phantom sample composed of three concentric tubes filled with water. The nominal resolution is $0.5 \times 0.5 \times 2 \, \text{mm}^3$. A slice of the 3D image can be seen in (c).

expense of magnetic field strength and the size of the magnet bore, since the shim magnets are placed in the bore of the main magnet. It is possible, however, to improve the field homogeneity by changing the polarization direction or the position of the elements that compose the main magnetic source.

Moresi and Magin,[51] pioneered reducing the magnet size for NMR applications by building a magnet using 15 cm long and 2 cm diameter cylindrical rods of SmCo magnetic material arranged in a Halbach configuration (Figure 9.5d). In addition, two steel yokes were placed within the bore close to the sensitive volume. Once the magnets were mounted in their corresponding positions, the homogeneity of the system was improved by slightly adjusting the angular orientation of each element. Using this approach, a homogeneity of 40 ppm was achieved in a volume of 3 mm diameter and 5 mm long, where the strength of the magnetic field is 0.6 T.

Recently it was demonstrated[63] that by combining magnetic blocks with different transverse geometries to build a Halbach array, some of the pieces can be moved to correct the inhomogeneities of the field without sacrificing compactness or appreciably reducing the internal bore size. Figure 9.9(a) illustrates this magnet geometry. It combines three cylindrical Halbach rings with different geometric proportions optimized to account for the field distortions along the cylinder axis due to the finite magnet length. To correct the inhomogeneities left after the assembly process, each ring is composed of fixed trapezoidal elements

forming parallel gaps between them. Rectangular magnets placed in the gaps can then be radially moved to mechanically shim the magnetic field in a highly efficient and accurate way (Figure 9.9b). By displacing the rectangular blocks in each ring with defined angular modulations and amplitudes, it is possible to independently generate the set of spherically harmonic corrections to the magnetic field up to the order $N/2$, where N is the number of rectangular blocks in the ring. As an example, Figure 9.9(b) shows the required modulation for the displacements of the rectangular magnets to generate tesseral spherical harmonic corrections of the order and degree three (top) and four (bottom) together with the corresponding two-dimensional field maps. The homogeneity attained with this magnet can be appreciated from the spectra of water and toluene acquired in a 5 mm NMR tube (Figure 9.9c). The line width at half height of 4.5 Hz (0.15 ppm) in water was achieved with the magnet alone, where only gradient coils were used for finer field adjustment of the first-order terms.

The compactness of the sensor allowed its installation in a chemical fume hood to study an atmosphere-controlled chemical reaction. The good performance of the sensor in terms of resolution and sensitivity allowed the real-time monitoring of the trimerization of propionaldehyde catalyzed by indium chloride.[8] Another example exploiting the benefits of this new design and shimming methodology is the desktop tomograph. Using these concepts, an array was built, which consisted of nine rings scaled up to

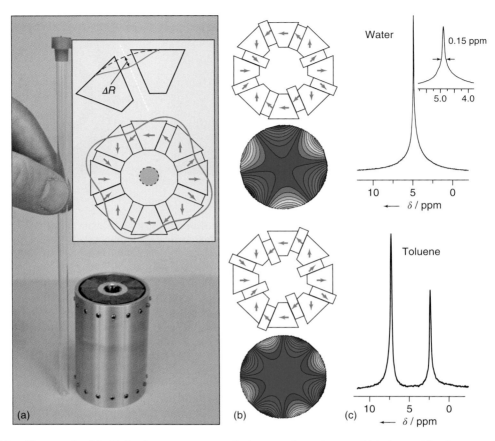

Figure 9.9. Photograph of the Halbach magnet composed of rectangular and trapezoidal pieces of SmCo generating a field of 0.7 T. It is 80 mm long and has inner and outer diameters of 15 and 35 mm, respectively. (b) Displacements of the rectangular magnets required to generate corrections of third and fourth order as illustrated by the two-dimensional field maps underneath. (c) Line shape for a water and toluene sample placed in a conventional 5 mm NMR tube. The spectra correspond to the average of 16 and 64 FIDs, respectively. The zoomed inset in (a) shows the amplitude of the movements ΔR of the rectangular magnets, which are the control variables for the mechanical shimming.

build a desktop tomograph fitting samples with diameters of up to 40 mm.[64] Improved field homogeneity was obtained by extending the number of correction terms that can be generated by the magnet.

9.3.5 Perspectives

Although high-resolution NMR using permanent magnets is already available, in cases such as biological systems, the homogeneity of the magnetic field deteriorates because of residual dipolar or chemical shift anisotropy interactions, and by local magnetic field gradients arising from variations in the bulk magnetic susceptibility of the samples. Conventionally, these problems are solved by rotating the sample at the magic angle with respect to the magnetic field direction.[65,66] However, this is not always possible since the sample might be damaged due to centrifugal forces. An alternative solution is to rotate the magnet while the sample remains at rest. Even though both approaches principally lead to the same results, the latter one is only possible using permanent magnets instead of superconducting magnets. Figure 9.10(a) illustrates a cylindrical permanent magnet design generating a magnetic field whose direction is tilted with respect to the magnet axis by the magic angle. This magnet combines a Halbach

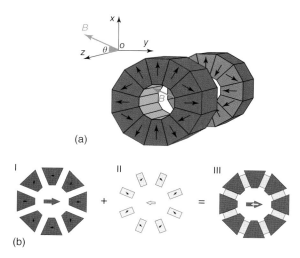

(a)

(b)

Figure 9.10. (a) Schematic diagram of a dipolar hybrid cylinder made by assembling two rings. Each ring interleaves blocks whose polarization follows a Halbach and Auber design. A magnetic field of 0.253 T tilted at an angle of 54.7° with respect to the z-axis (magic angle) is obtained combining transverse and longitudinal components of 0.207 and 0.146 T, respectively. (b) Temperature self-compensated permanent magnet design. Halbach arrays I and II are built from different magnetic materials. The orientation of the magnetic field of the arrays oppose each other. Temperature-compensated Halbach array III is built by combining the geometries of I and II.

ring, which generates a B_0 field perpendicular to the magnet axis, with an Auber geometry,[67] generating a longitudinal B_0 field. The desired tilt in the magnetic field is obtained by setting the proper proportions of each geometry. Subsequently, by rotating the magnet around its axis it is possible to average the effects of the anisotropic interactions. The outer diameter and length of the magnet are 9.5 and 7.5 cm respectively, and the field strength is 0.22 T.[68]

Another major problem inherent to permanent magnetic materials is the strong dependence of the field strength on temperature. Depending on the magnetic material, temperature variations of some degrees Celsius lead to drifts in the resonance frequency of several tens of kilohertz, thereby requiring a retuning system for the RF excitation/detection circuit. To overcome this problem, a new concept based on a combination of materials with different magnetic properties has been proposed.[22] If the field generated by the two magnetic materials oppose each other, it is possible to cancel the drift of the total field. This method requires one

to adjust the ratio of the magnetic fields generated by each unit with the inverse of the ratios of the thermal coefficients of each material. Figure 9.10(b) depicts a magnet design which combines SmCo trapezoids and NdFeB rectangles. For example, a cylindrical magnet 8 cm long with inner and outer radius of 1 and 2.5 cm, respectively, would generate a magnetic field of 0.25 T.

9.4   CONCLUSION

In this account, the most important mobile sensors reported to date have been outlined. The progress observed in the field during the last years has been accelerated by the development of powerful shimming strategies. They are useful in controlling the homogeneity of the magnetic field generated by discrete permanent magnet arrays. The possibility to correct errors introduced either during the assembly of the magnet or by the imperfection of the magnet blocks has triggered the efficient miniaturization of NMR magnets. Both open and closed magnet arrays have profited from these approaches making it possible to achieve higher and more homogeneous magnetic fields over larger volumes. These improved experimental conditions have enabled the implementation of sophisticated methodologies on mobile probes. The combination of more powerful magnets with modern techniques is making mobile NMR an alternative tool for material analysis.

ACKNOWLEDGMENTS

Different parts of this work were supported by the Deutsche Forschung Gemeinschaft grant CA660/33, the Virtual Institute for Portable NMR founded by the Helmholtz Association (HGF), and the European Community project CHARISMA (Cultural Heritage Advanced Research Infrastructures: Synergy for a Multidisciplinary Approach to Conservation/Restoration). E. D. thanks the Alexander von Humboldt foundation for a postdoctoral fellowship.

RELATED ARTICLES IN EMAGRES

Abdominal MRA

Low-Field Whole Body Systems

Resistive and Permanent Magnets for Whole Body MRI

Shimming of Superconducting Magnets

Stray-Field (STRAFI) NMR: Imaging in Large Field Gradients

Well Logging

REFERENCES

1. B. Blümich, 'NMR Imaging of Materials', Clarendon Press: Oxford, 2000.

2. G. Guthausen, H. Todt, W. Burk, D. Schmalbein, and A. Kamlowski, in 'Modern Magnetic Resonance', ed. G. A. Webb, Royal Society of Chemistry: London, Part III, 2006, pp. 1735–1738.

3. B. Blümich, F. Casanova, J. Perlo, F. Presciutti, C. Anselmi, and B. Doherty, *Acc. Chem. Res.*, 2010, **43**, 761.

4. J. P. M. van, Duynhoven, G. J. W. Goudappel, G. van, Dalen, P. C. van, Bruggen, J. C. G. Blonk, and A. P. A. M. Eijkelenboom, *Magn. Reson. Chem.*, 2002, **40**, S51.

5. T. Shirai, T. Haishi, S. Utzusawa, Y. Matsuda, and K. Kose, *Magn. Reson. Med. Sci.*, 2005, **4**, 137.

6. H. Yoshioka, S. Ito, S. Handa, S. Tomiha, K. Kose, T. Haishi, A. Tsutsumi, and T. Sumida, *J. Magn. Reson. Imaging*, 2006, **23**, 370.

7. A. Nordon, A. Diez-Lazaro, C. W. L. Wong, C. A. McGill, D. Littlejohn, M. Weerasinghe, D. A. Mamman, M. L. Hitchman, and J. Wilkie, *Analyst*, 2008, **133**, 339.

8. S. K. Küster, E. Danieli, B. Blümich, and F. Casanova, *Phys. Chem. Chem. Phys.*, 2011, **13**, 13172.

9. J. A. Jackson, L. J. Burnett, and J. F. Harmon, *J. Magn. Reson.*, 1980, **41**, 411.

10. R. K. Cooper and J. A. Jackson, *J. Magn. Reson.*, 1980, **41**, 400.

11. R. F. Paetzold, A. Delossantos, and G. A. Matzkanin, *Soil Sci. Soc. Am. J.*, 1987, **51**, 287.

12. R. F. Paetzold, G. A. Matzkanin, and A. Delossantos, *Soil Sci. Soc. Am. J.*, 1985, **49**, 537.

13. G. Eidmann, R. Savelsberg, P. Blümler, and B. Blümich, *J. Magn. Reson., Ser. A*, 1996, **122**, 104.

14. B. Blümich, V. Anferov, S. Anferova, M. Klein, R. Fechete, M. Adams, and F. Casanova, *Concepts Magn. Reson. A*, 2002, **15**, 255.

15. F. Casanova and J. Perlo, in 'Single-sided NMR', eds F. Casanova, J. Perlo, and B. Blümich, Springer: Manheim, 2011, Chapter 2.

16. P. J. Prado, *Magn. Reson. Imaging*, 2003, **21**, 397.

17. H. Popella and G. Henneberger, *COMPEL*, 2001, **20**, 269.

18. J. Perlo, F. Casanova, and B. Blümich, *J. Magn. Reson.*, 2005, **176**, 64.

19. A. E. Marble, I. V. Mastikhin, B. G. Colpitts, and B. J. Balcom, *J. Magn. Reson.*, 2006, **183**, 228.

20. E. Fukushima and J. A. Jackson, Unilateral magnet having a remote uniform field region for nuclear magnetic resonance, U.S. Pat. 6489872, 2002.

21. B. Manz, A. Coy, R. Dykstra, C. D. Eccles, M. W. Hunter, B. J. Parkinson, and P. T. Callaghan, *J. Magn. Reson.*, 2006, **183**, 25.

22. J. Perlo, F. Casanova, and B. Blümich, *Science*, 2007, **315**, 1110.

23. A. E. Marble, I. V. Mastikhin, B. G. Colpitts, and B. J. Balcom, *J. Magn. Reson.*, 2005, **174**, 78.

24. J. Perlo, F. Casanova, and B. Blümich, *J. Magn. Reson.*, 2004, **166**, 228.

25. P. J. McDonald, P. S. Aptaker, J. Mitchell, and M. Mulheron, *J. Magn. Reson.*, 2007, **185**, 1.

26. P. J. McDonald, *Prog. Nucl. Magn. Reson. Spectrosc.*, 1997, **30**, 69.

27. N. Proietti, D. Capitani, R. Lamanna, F. Presciutti, E. Rossi, and A. L. Segre, *J. Magn. Reson.*, 2005, **177**, 111.

28. F. Ruhli, T. Boni, J. Perlo, F. Casanova, E. Egarter, and B. Blümich, *Am. J. Phys. Anthropol.*, 2007, **44**, 204.

29. D. Capitani, M. C. Emanuele, J. Bella, A. L. Segre, D. Attanasio, B. Focher, and G. Capretti, *TAPPI J.*, 1999, **1**, 117.

30. B. Blümich, S. Anferova, S. Sharma, A. L. Segre, and C. Federici, *J. Magn. Reson.*, 2003, **161**, 204.

31. C. Casieri, L. Senni, M. Romagnoli, U. Santamaria, and F. De Luca, *J. Magn. Reson.*, 2004, **171**, 364.

32. I. Viola, S. Bubici, C. Casieri, and F. De Luca, *J. Cult. Herit.*, 2004, **5**, 257.

33. F. Presciutti, J. Perlo, F. Casanova, S. Gloggler, C. Miliani, B. Blümich, B. G. Brunetti, and A. Sgamellotti, *Appl. Phys. Lett.*, 2008, **93**, 033505.

34. S. Rahmatallah, Y. Li, H. C. Seton, I. S. Mackenzie, J. S. Gregory, and R. M. Aspden, *J. Magn. Reson.*, 2005, **173**, 23.

35. P. J. Prado, B. Blümich, and U. Schmitz, *J. Magn. Reson.*, 2000, **144**, 200.

36. F. Casanova and B. Blümich, *J. Magn. Reson.*, 2003, **163**, 38.

37. J. Perlo, F. Casanova, and B. Blümich, *J. Magn. Reson.*, 2005, **173**, 254.

38. S. Meiboom and D. Gill, *Rev. Sci. Instrum.*, 1958, **29**, 688.

39. M. D. Hürlimann and D. D. Griffin, *J. Magn. Reson.*, 2000, **143**, 120.

40. P. T. Callaghan, 'Principles of Nuclear Magnetic Resonance Microscopy', Clarendon Press: Oxford, 1991.

41. R. M. Cotts, M. J. R. Hoch, T. Sun, and J. T. Markert, *J. Magn. Reson.*, 1989, **83**, 252.

42. P. Z. Sun, J. G. Seland, and D. Cory, *J. Magn. Reson.*, 2003, **161**, 168.

43. F. Casanova, J. Perlo, and B. Blümich, *J. Magn. Reson.*, 2004, **171**, 124.

44. C. A. Meriles, D. Sakellariou, H. Heise, A. J. Moule, and A. Pines, *Science*, 2001, **293**, 82.

45. J. Perlo, V. Demas, F. Casanova, C. A. Meriles, J. Reimer, A. Pines, and B. Blümich, *Science*, 2005, **308**, 1279.

46. E. Danieli, in 'Single-sided NMR', eds F. Casanova, J. Perlo, and B. Blümich, Springer: Manheim, 2011, Chapter 7.

47. M. J. E. Golay, *Rev. Sci. Instrum.*, 1958, **29**, 313.

48. T. W. Skloss, A. J. Kim, and J. F. Haw, *Anal. Chem.*, 1994, **66**, 536.

49. M. A. Vargas, M. Cudaj, K. Hailu, K. Sachsenheimer, and G. Guthausen, *Macromolecules*, 2010, **43**, 5561.

50. K. Halbach, *Nucl. Instrum. Methods*, 1980, **169**, 1.

51. G. Moresi and R. Magin, *Concepts Magn. Reson. B*, 2003, **19**, 35.

52. H. Raich and P. Blümler, *Concepts Magn. Reson. B*, 2004, **23**, 16.

53. V. Demas, J. L. Herberg, V. Malba, A. Bernhardt, L. Evans, C. Harvey, S. C. Chinn, R. S. Maxwell, and J. Reimer, *J. Magn. Reson.*, 2007, **189**, 121.

54. A. McDowell and E. Fukushima, *Appl. Magn. Reson.*, 2008, **35**, 185.

55. J. Diekmann, K. L. Adams, G. L. Klunder, L. Evans, P. Steele, C. Vogt, and J. L. Herberg, *Anal. Chem.*, 2011, **83**, 1328.

56. H. Lee, E. Sun, D. Ham, and R. Weissleder, *Nat. Med.*, 2008, **14**, 869.

57. H. Lee, T.-J. Yoona, J. L. Figueiredoa, F. K. Swirskia, and R. Weissleder, *Proc. Natl. Acad. Sci. U.S.A.*, 2009, **106**, 12459.

58. R. C. Jachmann, D. R. Trease, L. S. Bouchard, D. Sakellariou, R. W. Martin, R. D. Schlueter, T. F. Budinger, and A. Pines, *Rev. Sci. Instrum.*, 2007, **78**, 035115.

59. E. Danieli, J. Mauler, J. Perlo, B. Blümich, and F. Casanova, *J. Magn. Reson.*, 2009, **198**, 80.

60. C. Hugon, F. D'Amico, G. Aubert, and D. Sakellariou, *J. Magn. Reson.*, 2010, **205**, 75.

61. C. W. Windt, H. Soltner, D. van, Dusschoten, and P. Blümler, *J. Magn. Reson.*, 2011, **2011**, 27.

62. D. I. Hoult and D. Lee, *Rev. Sci. Instrum.*, 1985, **56**, 131.

63. E. Danieli, J. Perlo, B. Blümich, and F. Casanova, *Angew. Chem. Int. Ed. Engl.*, 2010, **49**, 4133.

64. E. Danieli, K. Berdel, J. Perlo, W. Michaeli, U. Masberg, B. Blümich, and F. Casanova, *J. Magn. Reson.*, 2010, **207**, 53.

65. E. R. Andrew, A. Bradbury, and R. G. Eades, *Nature*, 1958, **182**, 1659.

66. I. J. Lowe, *Phys. Rev. Lett.*, 1959, **2**, 285.

67. G. Aubert, Cylindrical permanent magnet with longitudinal induced field, U.S. Pat. 5014032, 1991.

68. D. Sakellariou, C. Hugon, A. Guiga, G. Aubert, S. Cazaux, and P. Hardy, *Magn. Reson. Chem.*, 2010, **48**, 903.

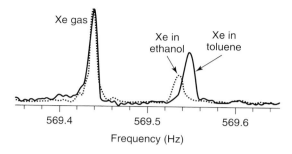

Figure 10.3. ^{129}Xe NMR Spectrum of Xe gas and Xe dissolved in ethanol and toluene acquired in the Earth's magnetic field. (Reproduced with permission from Ref. 25. © 2005 American Physical Society)

in the Earth's magnetic field, in which ^{129}Xe has a Larmor frequency of approximately 570 Hz. Figure 10.3 presents a ^{129}Xe NMR spectrum acquired in the Earth's magnetic field, which demonstrates the chemical shift between ^{129}Xe gas and ^{129}Xe dissolved in ethanol and in toluene.[25]

More recently, Appelt *et al.*[26–28] further defied the conventional wisdom to demonstrate that in cases where the magnetic equivalence between two groups of homonuclear spins is broken by the presence of unique J-couplings to a heteronucleus, homonuclear J-couplings can be observed. In these remarkable experiments, the samples were prepolarized by a 1-T Halbach permanent magnet array and then manually transported to the EFNMR probe for excitation by a ULF pulse and subsequent detection of the free precession. The use of the Halbach array for

prepolarization provides a polarization advantage that is orders of magnitude greater than the more traditional electromagnetic approach because the latter is limited to much weaker magnetic fields for the practical reasons of field switching and resistive heating. The use of the Halbach permanent magnet for prepolarization is especially advantageous because this type of array is largely self-screening and therefore can be located relatively close to the EFNMR probe without perturbing the homogeneity of the Earth's field. Figure 10.4 presents a high-resolution spectrum of tetramethylsilane (TMS) acquired in the Earth's magnetic field, which demonstrates heteronuclear J-coupling between the protons and the 4.7% natural abundance ^{28}Si, the heteronuclear J-coupling between the protons and the 1% natural abundance ^{13}C, and the homonuclear J-coupling between the ^{13}CH$_3$ and the ^{12}CH$_3$ protons.[27]

In addition to these advances in 1D spectroscopy, multidimensional spectroscopy has also been demonstrated to be possible in the Earth's field. In 2006, Robinson *et al.*[29] presented the first 2D COSY spectrum acquired in the Earth's magnetic field. In the case of multidimensional spectroscopic experiments, the polarization enhancement solution presented by the Halbach array method is limited in its applicability because of the practical difficulties associated with sample transport between the array and the EFNMR probe as well as problems with the consistency of the polarization between transients due to uncertainties associated with the sample polarization and transfer time. Therefore, for multidimensional experiments,

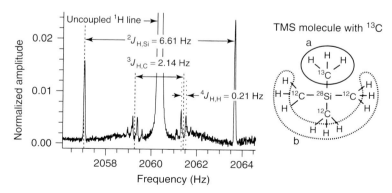

Figure 10.4. ^{1}H NMR spectrum of TMS acquired in the Earth's magnetic field, which shows the heteronuclear J-coupling between ^{1}H and natural abundance (4.7%) ^{28}Si and between ^{1}H and natural abundance (1%) ^{13}C as well as the homonuclear ^{1}H–^{1}H coupling, revealed by the presence of the ^{1}H–^{13}C heteronuclear coupling. (Reproduced with permission from Ref. 27. © 2007 Elsevier)

other sensitivity enhancement schemes have been explored, in particular, the use of dynamic nuclear polarization (DNP).

DNP is a method of transferring polarization from unpaired electron spins to target nuclear spins and is widely used for sensitivity enhancement and contrast in high-field NMR and MRI. In the high-field case, the maximum polarization enhancement is given by the ratio of the Zeeman polarizations, i.e., the ratio of magnetogyric ratios, which is approximately 658 for protons. At ultralow fields, such as the Earth's magnetic field, there is a significant sensitivity enhancement advantage to be obtained from employing nitroxide-free radicals, where the unpaired electron spin experiences a strong hyperfine coupling to a neighboring nitrogen nucleus. In the ultralow field regime, this coupling accounts for a larger portion of the electron polarization than the Zeeman interaction with the prevailing external magnetic field and so the maximum obtainable enhancement is much greater than that predicted by the ratio of magnetogyric ratios alone. At the Earth's field, the maximum theoretical sensitivity enhancement over thermal polarization is more than 3 orders of magnitude. The low-field DNP regime has been studied extensively as a means of indirectly obtaining EPR spectra of free radicals[30-35] and has also been successfully implemented for sensitivity enhancement of NMR magnetometers,[36] Earth's field MRI,[35] and multidimensional EFNMR spectroscopy.[37] Figure 10.5 presents a 2D COSY spectrum of 2,2,2-trifluoroethanol acquired in the Earth's magnetic field using DNP for signal enhancement.[37]

10.5 FUTURE DIRECTIONS

Magnetic resonance at low fields offers a number of significant advantages. For example, at low field, the spin-lattice relaxation time becomes sensitive to slow motions in macromolecules, thus making it a very effective method for contrast in biological tissue.[38] This could lead to the use of Earth's field MRI for medical diagnostics where low-field relaxation contrast may give a more definitive screening for cancer. A further advantage of MRI at low field is the ability of the ULF electromagnetic waves, associated with excitation and detection, to penetrate metal, thus allowing imaging of liquids inside metal containers.[39] The remarkable homogeneity of the Earth's magnetic field allows for very high resolution spectroscopy so that very weak hetero- and homonuclear J-couplings can,

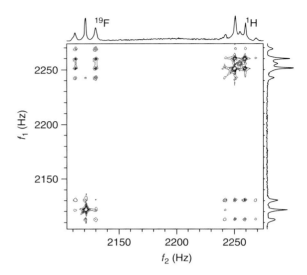

Figure 10.5. 2D COSY to 2,2,2-trifluoroethanol acquired in the Earth's magnetic field using DNP for sensitivity enhancement. (Reproduced with permission from Ref. 37. © 2008 Elsevier)

in principle, be measured using this method, and recent advances in multidimensional low-field spectroscopy make a range of correlation and exchange experiments possible. Finally, advances in the use of enhanced polarization techniques such as DNP or spin-exchange methods and advanced detection technologies such as SQUIDs and atomic magnetometers, suggest a means of considerably enhancing EFNMR and MRI sensitivity, permitting more sophisticated experiments on smaller samples over shorter time periods.

RELATED ARTICLES IN EMAGRES

Dynamic Nuclear Polarization: Applications to Liquid-State NMR Spectroscopy

Dynamic Nuclear Polarization and High-Resolution NMR of Solids

Terrestrial Magnetic Field NMR

REFERENCES

1. M. Packard and R. Varian, *Phys. Rev.*, 1954, **93**, 941.

2. D. P. Stern, *Rev. Geophys.*, 2003, **40**, 3.

3. R. J. S. Brown and B. W. Gamson, *Trans Am Inst Min Metall Eng.*, 1960, **219**, 199.

4. EFNMR Instrument, Teach Spin Inc., Buffalo, NY; http://www.teachspin.com.

5. Terranova-MRI, Magritek Limited, Wellington, New Zealand; http://www.magritek.com.

6. P. T. Callaghan and M. Legros, *Am. J. Phys.*, 1982, **50**, 709.

7. A. G. Semenev, M. D. Schirov, A. V. Legchenko, A. I. Burshtein, and A. Ju. Pusep, Device for measuring the parameters of Underground mineral deposit. GB Patent 2198540B, 1989

8. D. V. Tushkin, O. A. Shushakov, and A. V. Legchenko, *Geophys. Prosp.*, 1994, **42**, 855.

9. D. V. Tushkin, O. A. Shushakov, and A. V. Legchenko, *Geophys. Prosp.*, 1995, **43**, 623.

10. O. A. Shushakov, *Geophysics*, 1996, **61**, 998.

11. A. Legchenko, *Appl. Magn. Reson.*, 2004, **25**, 621.

12. P. T. Callaghan, C. D. Eccles, and J. D. Seymour, *Rev. Sci. Instrum.*, 1997, **68**, 4263.

13. P. T. Callaghan, C. D. Eccles, T. G. Haskell, P. J. Langhorne, and J. D. Seymour, *J. Magn. Reson.*, 1998, **133**, 148.

14. P. T. Callaghan, R. Dykstra, C. D. Eccles, T. G. Haskell, and J. D. Seymour, *Cold Reg. Sci. Technol.*, 1999, **29**, 153.

15. O. R. Mercier, M. W. Hunter, and P. T. Callaghan, *Cold Reg. Sci. Technol.*, 2005, **42**, 96.

16. D. G. Noris and J. M. Hutchison, *Magn. Reson. Imaging*, 1990, **8**, 33.

17. J. Stepisnik, V. Erzen, and M. Kos, *Magn. Reson. Med.*, 1990, **15**, 386.

18. M. E. Halse, A. Coy, R. Dykstra, C. D. Eccles, M. W. Hunter, and P. T. Callaghan, Multidimensional Earth's field NMR, in Magnetic Resonance Microscopy, eds S. Codd and J. D. Seymour, Wiley-VCH Verlag GmbH & Co.: Weinheim, 2009, 15–29.

19. A. Mohoric, G. Planinsic, M. Kos, A. Duh, and J. Stepisnik, *Instrum. Sci. Technol.*, 2004, **32**, 655.

20. M. E. Halse, A. Coy, R. Dykstra, C. D. Eccles, M. W. Hunter, R. Ward, and P. T. Callaghan, *J. Magn. Reson.*, 2006, **182**, 75.

21. A. Mohoric, J. Stepisnik, M. Kos, and G. Planinsic, *J. Magn. Reson.*, 1999, **136**, 22.

22. I. M. Savukov, *Phys. Rev. Lett.*, 2005, **94**, 123001.

23. S. Xu, C. W. Crawford, S. Rochester, V. Yashchuk, D. Budker, and A. Pines, *Phys. Rev. A*, 2008, **78**, 013404.

24. V. S. Zotev, A. N. Matlashov, P. L. Volegov, I. M. Savukov, M. A. Espy, J. C. Mosher, J. J. Gomez, and R. H. Kraus Jr., *J. Magn. Reson.*, 2008, **194**, 115.

25. S. Appelt, F. W. Häsing, H. Kühn, J. Perlo, and B. Blümich, *Phys. Rev. Lett.*, 2005, **94**, 197602.

26. S. Appelt, H. Kühn, F. W. Häsing, and B. Blümich, *Nature Phys.*, 2006, **2**, 105.

27. S. Appelt, F. W. Häsing, H. Kühn, U. Sieling, and B. Blümich, *Chem. Phys. Lett.*, 2007, **440**, 308.

28. S. Appelt, F. W. Häsing, H. Kühn, and B. Blümich, *Phys. Rev. A*, 2007, **76**, 023420.

29. J. N. Robinson, A. Coy, R. Dykstra, C. D. Eccles, M. W. Hunter, and P. T. Callaghan, *J. Magn. Reson.*, 2006, **182**, 343.

30. T. Guiberteau and D. Grucker, *J. Magn. Reson. B*, 1996, **110**, 77.

31. C. Polyon, D. J. Lurie, W. Youngdee, C. Thomas, and I. Thomas, *J. Phys. D: Appl. Phys.*, 2007, **40**, 5527.

32. T. Guiberteau and D. Grucker, *J. Magn. Reson. A*, 1993, **105**, 98.

33. D. J. Lurie, I. Nicholson, and J. R. Mallard, *J. Magn. Reson.*, 1991, **95**, 405.

34. D. J. Lurie, *Appl. Magn. Reson.*, 1992, **3**, 917.

35. G. Planinsic, T. Guiberteau, and D. Grucker, *J. Magn. Reson. B*, 1996, **110**, 205.

36. N. Kernevez and H. Glenat, *IEEE Trans. Magn.*, 1991, **27**, 5402.

37. M. E. Halse and P. T. Callaghan, *J. Magn. Reson.*, 2008, **195**, 162.

38. S.-K. Lee, M. Mößle, W. R. Myers, N. Kelso, A. H. Trabesinger, A. Pines, and J. Clarke, *Magn. Reson. Med.*, 2005, **53**, 9.

39. M. Mößle, S.-I. Han, W. R. Myers, S.-K. Lee, N. Kelso, M. Hartridge, A. Pines, and J. Clarke, *J. Magn. Reson.*, 2006, **179**, 146.

PART B
NMR for Air, Soil and Water

Chapter 11
Dissolved Organic Matter

Perry J. Mitchell, André J. Simpson and Myrna J. Simpson

Department of Chemistry and Environmental NMR Centre, University of Toronto, Toronto, Ontario, M1C 1A4, Canada

11.1 INTRODUCTION

Dissolved organic matter (DOM) is a complex mixture of organic molecules that is found ubiquitously in aquatic, terrestrial, and atmospheric environments such as oceans, lakes, rivers, soils, sediments, aerosols, and rainwater.[1–7] It is derived from several sources including decaying plants and microbes in addition to anthropogenic inputs,[8,9] and is composed primarily of polar, water-soluble molecules of varying size and composition.[9] It can also be operationally defined as soluble organic material that passes through either a

0.22 or a 0.45 μm filter.[9,10] Although often present in varying concentrations ($\sim$1–100 mg L^{-1}) in aquatic environments,[11] it is a major pool of actively cycling organic carbon and plays a key role in the global biogeochemical carbon cycle.[12,13] For example, oceans contain as much organic carbon in the form of marine DOM as there is total atmospheric carbon[14] and there is continual interchange of organic carbon between these two pools.[15] Furthermore, approximately 1% of marine organic carbon is stored in biomass, whereas over 90% is present in the form of DOM.[16] Therefore, a detailed understanding of DOM structure and reactivity is crucial for understanding global-scale environmental processes such as carbon fluxes and climate change. In addition, DOM has been implicated in the sorption and complexation of organic and metal contaminants (see Chapters 19 and 20, respectively) and may play a role in transporting these pollutants in the environment.[17–19] However, DOM is highly variable in composition and may contain thousands of compounds as estimated using high-resolution mass spectrometry techniques.[20,21] The heterogeneity of DOM has proven to be a major obstacle to fully understand the composition and subsequent reactivity of DOM in natural settings.[3,22,23] Thus, identifying the individual molecules that are present in a DOM sample is difficult and prevents scientists from better understanding how DOM interacts with other environmental components such as soil organic matter (OM) or minerals.

NMR Spectroscopy: A Versatile Tool for Environmental Research
Edited by Myrna J. Simpson and André J. Simpson
© 2014 John Wiley & Sons, Ltd. ISBN: 978-1-118-61647-5

NMR spectroscopy is a powerful technique for the characterization of DOM in natural environmental samples in both the solid- and the solution-state.[11] Solid-state ^{13}C NMR has been widely used to characterize the structure of DOM and to examine changes in DOM composition with different environmental conditions such as spatial and temporal distributions[7,24] or after sorptive fractionation by soils and minerals.[25-28] Advances in solution-state NMR techniques have provided more detailed structural information such as the identity of individual molecules and classes of compounds that are major components of DOM.[3,22,29-32] NMR has also been used to study interactions between DOM and organic contaminants (see Chapter 20) such as organofluorine compounds and glyphosate.[33-37] In addition, several NMR-active metal nuclei such as ^{51}V, ^{113}Cd, and ^{135}C have been shown to interact with DOM (see Chapter 19), which may play a role in their transport in the environment.[38-41] This chapter provides an overview of recent progress in the field of DOM characterization using both solid and solution-state NMR spectroscopy, including a discussion of the potential applications and environmental implications of these advances as well as future research directions. It should be noted that many of the examples that we highlight in this chapter have used NMR in conjunction with other analytical techniques such as mass spectrometry and infrared and fluorescence spectroscopies to study DOM structure and reactivity; however, it is primarily the contribution of NMR that is emphasized here.

11.2   CHARACTERIZATION OF DOM USING SOLID-STATE NMR SPECTROSCOPY

11.2.1   Overview

Solid-state NMR spectroscopy is an advantageous tool for DOM characterization because of the ease of analysis (see Chapter 4), which involves packing a small amount of dry sample inside a zirconium rotor sealed with a Kel-F cap.[11] However, a great amount of time is spent isolating DOM for analysis using techniques such as ultrafiltration, reverse osmosis, or lyophilization.[42-44] The presence of salts in the DOM sample must also be considered because of their potentially hygroscopic nature. Furthermore, high concentrations of inorganic components in DOM

samples that contain low organic carbon content may lower the signal-to-noise ratio of the sample. Passing DOM samples through an ion-exchange resin to remove ionic salts may partially alleviate these issues.

Solid-state ^{13}C NMR spectra of DOM are often collected using magic-angle spinning (MAS) experiments employing either direct polarization (DP) or cross polarization (CP) of the ^{13}C nucleus (see Chapter 4).[43,45] CPMAS ^{13}C NMR is more sensitive than DPMAS because of the additional polarization from protons as well as the shorter T_1 relaxation time of the ^{1}H nucleus (determines the recycle delay in CPMAS) versus the ^{13}C nucleus (determines the recycle delay in DPMAS). However, CPMAS tends to be less quantitative in general than DPMAS.[11,46] Like soil OM,[47] the solid-state ^{13}C NMR spectra of DOM samples often contain broad resonances which prevent the precise identification of individual compounds. These peaks are usually attributed to several general classes of functional groups that are known to resonate in specific regions of the ^{13}C NMR spectrum (Figure 11.1), which includes aliphatic (0–50 ppm), O-alkyl (50–95 ppm), anomeric (95–110 ppm), aromatic and phenolic (110–165 ppm), and carboxyl and carbonyl carbon (165–215 ppm).[48,49] Aliphatic DOM signals arise from components with long alkyl chains such as fatty acids, whereas the O-alkyl region is composed of labile moieties such as carbohydrates,

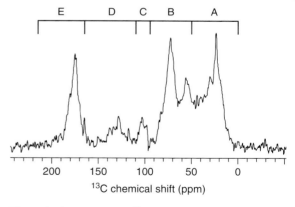

Figure 11.1.   Solid-state ^{13}C NMR spectrum of compost-derived DOM showing broad peaks due to the overlapping resonances of multiple DOM components. The major classes of molecules are indicated using braces as follows: (A) aliphatic; (B) O-alkyl; (C) anomeric; (D) aromatic and phenolic; and (E) carboxyl and carbonyl. (Reproduced from Ref. 9, © 2011 Elsevier)

amino acids, peptides, and proteins.[47] The signal of the methoxy group of lignin has also been shown to resonate in the *O*-alkyl region at 56 ppm.[49] Signals in the aromatic region are usually attributed to substituted aromatic DOM components such as phenolic species, which include lignin-derived compounds and peptides and proteins with aromatic side chains.[9,49] Strong signals in the carboxyl and carbonyl carbon regions of DOM samples suggest a high abundance of carboxylic acid functionalities.[25,27,43] This is consistent with the data obtained using Fourier transform ion cyclotron resonance mass spectrometry (FT-ICR-MS), which have also identified a large number of carboxylic components.[4,13]

11.2.2 Solid-state ^{13}C NMR for Studying Spatial and Temporal Changes in DOM Composition

Integrating solid-state ^{13}C NMR spectra can provide semiquantitative insight into the general composition and distribution of organic carbon in DOM samples, which may be used to study spatial changes in DOM composition. A comprehensive study by Mao *et al.*[43] used several different NMR techniques to study the composition of coastal and marine DOM samples. DPMAS NMR techniques were employed such that all carbon atoms in the samples were observed, not just those bound to hydrogen that can be biased during CPMAS ^{13}C NMR experiments.[46] Dipolar dephasing (see Chapter 4) was also used to observe only the signals from nonprotonated, quaternary carbons and mobile functionalities such as methyl groups bonded to oxygen and carbon. This technique revealed that the majority of aromatic carbons in coastal DOM are not protonated, whereas their marine DOM counterparts are protonated. Several spectral editing techniques were also used to derive detailed structural information about the DOM samples. For example, a chemical shift anisotropy filter was applied to distinguish between sp^3- and sp^2-hybridized carbons, which helped to clarify overlapping resonances between anomeric carbon and aromatic signals in the region between 90 and 120 ppm. Additional experiments were conducted in which spectra for only CH or CH$_2$ groups were collected, which allowed for an estimation of the proportions of methine and methylene carbon in the samples. Overall, the results demonstrated that various NMR techniques can be used to provide greater

information about DOM samples in the solid state despite the broad, overlapping resonances that are often observed.[43]

Another study of the spatial variability of DOM was conducted by Jaffé *et al.*[44] who characterized DOM from six major watersheds across North America and used several different analytical techniques including solid-state ^{13}C NMR to examine differences in DOM structure. The NMR results showed that *O*-alkyl carbon was the dominant type of carbon in all of the DOM samples (>40%) followed by alkyl moieties, whereas aromatic and carboxyl functionalities each comprised <20% of DOM. Overall, the NMR spectra of the DOM samples contained similar features. Therefore, this study demonstrated that DOM samples from a wide range of geographic locations in North America exhibit similar composition, suggesting that DOM structure does not vary greatly across freshwater environments.[44]

Solid-state ^{13}C NMR can also be used to examine temporal changes in DOM composition. For example, Sanderman and Kramer[24] used ^{13}C NMR to characterize DOM samples extracted from soils along a chronosequence in Hawaii, including samples that ranged in age from 300 to 4 100 000 years. The results showed remarkable similarity in the macromolecular composition of the different samples despite the wide range of sample ages. On the basis of this evidence, the authors proposed that DOM structure converges over time toward a relatively similar composition, which may proceed via the preferential degradation of specific structures by microbes or via photochemical decomposition of labile material while more recalcitrant components accumulate in DOM.

Compost is commonly added to soils to improve soil fertility, and solid-state ^{13}C NMR has been used to study chemical changes in the composition of compost during its maturation.[50,51] For example, Chefetz *et al.*[51] monitored the bulk structure of compost DOM at five sampling points over a 6-month period. The authors found that the *O*-alkyl content of the DOM decreased with time, likely because of increased microbial activity due to the prevalence of labile substrates which became depleted over time. A 50% increase in the concentration of aromatic and phenolic DOM components was also observed in the compost over time, implying that these structures are more recalcitrant and accumulate in mature compost.

11.2.3 Using Solid-state NMR to Study Organo-mineral Interactions Involving DOM

Interactions between DOM and soil components are of interest as they may play a key role in carbon cycling. For example, some components of DOM may sorb strongly to soil particles and become a part of stabilized soil OM, whereas other DOM constituents may remain in solution where they are bioavailable and can potentially be used as an energy source by microbes.[52,53] The primary binding sites for DOM in terrestrial environments are the reactive surfaces of exposed soil minerals.[25–28,54] For instance, sorption experiments have demonstrated that DOM binds to both crystalline clay minerals such as kaolinite, montmorillonite and vermiculite[55–58] and amorphous minerals such as iron and aluminum oxides.[25–28,54] Solid-state ^{13}C NMR spectroscopy has been used to examine changes in DOM structure upon sorptive fractionation by both pure soil minerals and whole soils of varying composition. A field study by McKnight et al.[26] characterized DOM before and after interactions with suspended colloids of iron and aluminum oxides at the confluence of two waterways in Colorado. The authors integrated the solid-state ^{13}C NMR spectra of DOM samples and found that the sorbed fraction contained greater contributions from aromatic and carboxylic carbons, suggesting that these components are preferentially retained by minerals as the DOM flows along the watershed. A subsequent laboratory study by Gu et al.[28] used batch equilibration sorption experiments and solid-state ^{13}C NMR to study interactions between DOM and hematite, a form of iron oxide, both before and after sorption. The results showed that hydroxyl-containing *O*-alkyl components and carboxyl moieties were preferentially sorbed onto the hematite surfaces, in partial agreement with the findings of McKnight et al.[26] A later study by Gu et al.[27] also noted a reduction in the signal intensity of the *O*-alkyl region upon sorption of a hydrophobic DOM fraction to hematite. Conte et al.[9] studied the sorption of compost DOM to montmorillonite and kaolinite and characterized the DOM samples before and after sorptive fractionation using solid-state ^{13}C NMR. Reduced signal intensity from aromatic- and hydroxyl-containing moieties was observed after sorption to montmorillonite, suggesting the preferential sorption of these components. Similar analysis of DOM after interaction with kaolinite revealed that only aromatic components were retained. The above studies collectively indicate that soil minerals preferentially bind specific types of functional groups in DOM, especially hydroxyl- and carboxyl-rich moieties. Thus, these organo-mineral interactions, which may result in DOM stabilization, appear to depend on both the mineralogy and the composition of the DOM.

The surfaces of soil minerals are often coated with OM,[59] thus it is also important to understand how DOM interacts with native soil OM as it leaches through the soil profile. Kaiser et al.[25] reported the first use of solid-state ^{13}C NMR to examine the sorptive fractionation of forest floor DOM by two different whole soils. The authors noted decreases in the proportion of carboxyl and aromatic carbons and corresponding increases in the amount of alkyl carbon upon sorption to the soils, which agrees with the findings of McKnight et al.[26] for DOM sorption to colloidal iron and aluminum oxides in rivers. This suggests that similar sorptive interactions occur when DOM binds to minerals in both aquatic and terrestrial environments. Sanderman et al.[42] examined the composition of DOM extracted from soil water with increasing depth in prairie and forest soils. The authors noted that concentrations of aromatic and carboxyl carbon decreased with depth, further supporting the hypothesis that these components are selectively retained by soils as they are carried downward through soil by the flow of water. However, it remains unclear whether these interactions are due to the binding of OM to available mineral surfaces or whether OM–OM interactions also result in DOM retention in soils.

11.2.4 Solid-state ^{15}N NMR Applications

Dissolved organic nitrogen (DON) refers to the portion of DOM that contains nitrogenous compounds and has been described as a poorly characterized and refractory component of natural OM because of its inherent complexity and analytical limitations.[60,61] However, several nitrogen-containing functional groups in DOM can be identified using solid-state ^{15}N NMR, including amine, amide, and heterocyclic moieties such as indole and pyrrole groups and their substituted and protonated forms.[61,62] Similar to solid-state ^{13}C NMR experiments, CPMAS is frequently used to transfer magnetization from protons to ^{15}N rather than direct polarization of the ^{15}N nucleus.[43,60,61,63] However, owing to the low natural abundance of ^{15}N, the signal-to-noise ratios are often low and a large number of scans may be needed to produce adequate resolution for spectral quantification.[61,63] Several studies have used solid-state ^{15}N NMR to characterize DON as a component of DOM samples.[43,61–63] McCarthy *et al.*[63] reported the first study in which solid-state ^{15}N NMR was applied to the analysis of DON and found that the majority of high molecular weight marine DON was in the amide form, which is biologically derived, rather than in heterocyclic forms of nitrogen, which could form abiotically. Templier *et al.*[61] fractionated riverine DOM samples from France into hydrophobic and moderately hydrophilic fractions using XAD resins and used solid-state ^{15}N NMR to investigate the nitrogenous components in the individual fractions. The authors reported that the DON exhibited varying degrees of complexity that differed by fraction and DOM source with amide being the most prevalent form of DON and lower concentrations of heterocyclic nitrogen components also present. Similarly, Mao *et al.*[43] used ^{15}N NMR as a part of an extensive solid-state NMR investigation into the structure of DOM from coastal and marine environments. The authors found that resonances from amides were most prominent, whereas the signal from heterocyclic nitrogen was weaker. These three studies collectively suggest that DON may have a similar composition across multiple environments and geographic locations with amides playing a prominent role and with lower concentrations of heterocyclic nitrogen also contributing. Amide signals may arise from peptide bonds in proteins that are considered labile DOM components as well as from more stable biopolymers such as chitin and peptidoglycan that contain amide side chains. To better understand the chemical forms of amide nitrogen in DOM and to estimate their abundance, Aluwihare *et al.*[62] subjected high molecular weight marine DOM samples to a mild acid hydrolysis reaction. This treatment destroys the peptide bonds in proteins but only hydrolyzes side chains from the biopolymers, leaving the glycosidic linkage intact and not causing any depolymerization of the macromolecules. This enabled the authors to differentiate between the two main contributors to the amide signal and to determine the relative contribution of each to the total amount of amide that was present in the DOM samples. After the hydrolysis step, the authors monitored the changes in the intensity of amide and amine signals using solid-state ^{15}N NMR. Results showed that the amount of biopolymer-derived amide nitrogen increased with depth, suggesting that more labile proteinaceous amide nitrogen is present in the surface layer, whereas more recalcitrant chitin- and peptidoglycan-derived amides

DON accumulate in the deeper ocean. This study represents an important finding as DON was a largely uncharacterized pool of DOM, and determining its spatial composition is important to understand processes such as nutrient cycling and primary productivity in large water bodies such as oceans.

11.3 CHARACTERIZATION OF DOM USING SOLUTION-STATE NMR SPECTROSCOPY

11.3.1 Overview

Overlapping resonances observed in the solid-state NMR spectra of DOM often limits the amount of detailed structural information that can be obtained.[43] An alternative technique is solution-state NMR spectroscopy, which is capable of providing greater resolution of individual components in complex samples such as DOM and soil OM.[49] For this technique, samples must be compatible with a deuterated solvent such as D_2O or DMSO-d_6.[22,31,49,64] In the case of D_2O, the solution pH can be adjusted to basic conditions using NaOD to enhance DOM component solubility,[22,29,31,65] thus increasing the amount of material in solution and potentially improving the signal-to-noise ratio. The amount of dissolved analyte can also be enhanced by isolating and redissolving DOM samples at higher concentrations, although the samples are no longer present in their natural state. Samples with high concentrations of ionic salts such as marine DOM may benefit from analysis using NMR microcoil probes (e.g., 1.7 mm diameter),[22,29,31] which are more tolerant to salts and also require less total sample. When D_2O is used as the NMR solvent or when DOM is analyzed directly, water suppression techniques are often required to provide greater resolution of components that resonate near water ($\delta = 4.7$ ppm), which may otherwise be obscured. For example, Herzog *et al.*[66] applied a technique called water attenuation by transverse relaxation (WATR) to suppress water signals in river, bog, soil leachate, and groundwater DOM samples. Another water suppression technique known as presaturation utilizing relaxation gradients and echoes (PURGE) has been shown to effectively suppress signals from residual water in DOM samples.[67] More recently, a specialized NMR pulse program known as shaped presaturation-WATER suppression by gradient-tailored excitation using optimized W5 pulse trains (SPR-W5-WATERGATE) was developed by Lam and Simpson.[32] This technique was shown to provide sufficient water suppression to enable the direct analysis of DOM from ocean, lake, and river water samples without any initial sample preconcentration even at very low DOM concentrations ($\sim$1–8 mg L^{-1}). As such, this pulse sequence represents a major advance in DOM characterization, as it simplifies sample preparation and reduces potential sample losses and experimental preparation time.[32]

A number of experiments are available for the characterization of DOM using solution-state NMR, which can provide complementary structural information (see Chapter 1).[11] Basic ^{1}H NMR spectra provide a complete overview of all protons in the DOM sample, whereas diffusion-edited ^{1}H NMR can be used to attenuate the signals from small molecules, allowing for a more detailed examination of larger, less mobile DOM constituents.[5,31] For example, attenuation of the signals from small molecules in lake water DOM allowed for the identification of peptidoglycan, a key component of the cell walls of Gram-positive bacteria.[5] Two-dimensional (2D) NMR experiments can provide information about bond connectivity and assist with DOM component identification. For example, correlation spectroscopy (COSY) and heteronuclear single quantum coherence (HSQC) experiments have been used to infer the presence of specific molecules that cannot be assigned from one-dimensional (1D) ^{1}H NMR alone.[22,31] Specific molecules in DOM are assigned based on the consistency of experimental peak splitting and chemical shift values with published spectra of standards and NMR databases.[22,30,31,68]

As with solid-state ^{13}C NMR spectroscopy, solution-state ^{1}H NMR spectra of DOM exhibit distinct regions where signals from specific components are known to resonate. Therefore, the ^{1}H NMR spectra of DOM can be integrated into several general regions to examine changes in the distribution of these constituents across samples.[31,69] The region between approximately 0.6 and 1.6 ppm has been attributed to a group of aliphatic DOM components termed material derived from linear terpenoids (MDLT).[3] Lam et al.[3] used advanced 2D NMR experiments such as heteronuclear multiple bond correlation (HMBC) and heteronuclear multiple quantum coherence (HMQC) as well as extensive spectral simulations to establish that signals in this region arise from degradation products of conjugated long-chain aliphatic compounds such as carotenoids. Signals arising specifically from

the conjugated terpenoid precursors were not detected, suggesting that these structures are degraded into oxidized products with abundant hydroxyl and carboxyl groups that comprise a significant portion of DOM. It should be noted that the MDLT region may also contain resonances from aliphatic functionalities such as methyl groups in fatty acids and peptides, which may be microbial- or plant-derived.[3,49] A second defined region of the ^{1}H NMR spectra of DOM between 1.6 and 3.2 ppm has been attributed to carboxyl-rich alicyclic molecules (CRAMs) by Hertkorn et al.[4] based on evidence from multidimensional NMR and FT-ICR-MS investigations. The authors proposed that CRAMs are composed of carboxyl-rich products of the degradation of cyclic terpenoid biomolecules such as steroids and hopanoids. Previous studies using solution-state ^{1}H NMR have found that CRAMs may comprise >40% of the total composition of DOM samples,[3,13,31,69] thus they may be one of the most abundant classes of molecules in aquatic systems. The ability to assign the resonances in the CRAM region to a specific class of molecules represented a major step forward in the structural characterization of DOM. This region also contains resonances from commonly observed short-chain carboxylic acids such as succinic acid and acetic acid, which may be microbial-derived metabolites.[30,31,68] Previous studies have established that carbohydrates and peptides resonate in the region between 3.2 and 4.5 ppm in ^{1}H NMR spectra of DOM.[3,22,30,32] These structures have a biological origin and are believed to be among the most labile and rapidly degraded DOM components, as they are a readily available energy source for microbes.[30,68] Owing to overlapping resonances from numerous carbohydrate and peptide moieties, it is often difficult to positively identify specific molecules within this region.[30,31,68] The region between 4.5 and 6.5 ppm contains few signals with the exception of water, whereas aromatic DOM components resonate between approximately 6.5 and 8.5 ppm.[69] Several studies that have used solution-state ^{1}H NMR to characterize DOM in a variety of environments have noted a low abundance of aromatic components relative to aliphatic structures such as MDLT, CRAMs, and carbohydrates and peptides.[2,22,29–31,68,70] As such, it may be difficult to reliably integrate this region if the signal-to-noise ratio is high. On the basis of the collective evidence of several solution-state NMR investigations, it appears that the majority of DOM components are aliphatic molecules of varying structure and polarity.[3,22,29–32,68,70]

11.3.2 Small Molecule Identification

Early studies analyzed DOM samples using basic ^{1}H NMR without water suppression techniques or preliminary separation steps and reported broad resonances that permitted only a few specific molecules to be identified.[71-73] For example, Wilson *et al.*[71] examined the solution-state ^{1}H NMR spectra of dissolved soil OM components including humic and fulvic acids and were able to identify simple biomolecules such as lactic acid, succinic acid, methanol, and formic acid. These compounds are likely present in high concentrations as their resonances were distinguishable within the complex mixture of DOM. With advances in water suppression that remove the intense, broad signal for water, a greater number of individual DOM components can be identified. For example, Pautler *et al.*[30] analyzed glacier meltwater using the SPR-W5-WATERGATE water suppression technique and reported abundant small molecules including several amino acids and short-chain carboxylic acids (Figure 11.2). This indicated that the majority of the glacier DOM was likely microbial derived, which may have implications for climate change such as the influx of nutrients into ecosystems upon glacial melting. Water suppression techniques are of major benefit when trying to discern the signals of small molecules in DOM. However, even with the advent of sophisticated water suppression techniques, many DOM samples that are not subjected to initial separation steps still exhibit broad peaks because of the overlap of potentially thousands of individual molecules.[20,21]

11.3.3 Molecular Aggregation and Diffusion in DOM

In addition to studying small molecules in DOM, sophisticated solution-state NMR techniques can be used to study larger molecular aggregates in solution (see Chapter 2). For example, Lam and Simpson[74] used diffusion-ordered spectroscopy techniques to study the aggregation of DOM components at varying concentrations in water. The authors reported a shift toward large molecular weight material with increasing DOM concentration, suggesting that DOM constituents aggregate when present at high concentrations. CRAMs appear to play a prominent role in aggregation, possibly because of their high abundance, whereas aliphatic MDLT constituents play a minor role. To further clarify whether these components

self-aggregate or whether they interact with other DOM components in solution, saturation transfer difference experiments were performed. The results showed correlations between major components with all other species, including strong interactions between CRAMs and aromatic components.[74] Thus, it appears that concentrated DOM solutions involve a high degree of aggregation between multiple types of components as well as self-aggregation.

Solution-state NMR can also be used to estimate the size of DOM components, which has been shown to vary from individual molecules to high molecular weight aggregates.[22,74] For example, Zheng and Price[75] applied a novel pulse sequence entitled pulsed gradient stimulated-echo (PGSTE)-WATERGATE to measure the diffusion coefficients and average hydrodynamic radii of DOM components in pond, river, and seawater samples without preconcentration as well as in an artificially concentrated pond water DOM sample. The authors divided the ^{1}H NMR spectra of DOM into aliphatic compounds, CRAMs, and carbohydrates and used a number of mathematical equations involving the experimental pulse parameters to calculate diffusion coefficients. Results showed that the three DOM components had distinctly different diffusion coefficients at dilute concentrations, whereas in the concentrated pond water DOM, the three diffusion coefficients were similar. This suggests that the components mainly self-associate in dilute solution, whereas at higher concentrations, the components begin to interact more strongly with each other as reported by Lam and Simpson.[74] These studies collectively suggest that DOM reactivity and aggregation behavior in solution is concentration-dependent and complex. Future work is needed to extract more specific information from NMR data regarding DOM organization in solution; however, this is complicated by the often overlapping resonances that are encountered for DOM samples.

11.3.4 Sorptive Interactions between DOM and Solid Phases

Both solution-state ^{1}H and ^{13}C NMR experiments have been used in addition to solid-state ^{13}C NMR to enhance our understanding of DOM reactivity with respect to sorption in soil environments. For example, Kaiser[76] used solution-state ^{13}C NMR to examine the sorption of several different types of DOM samples to iron oxide, including hydrophobic and hydrophilic DOM fractions that were separated on an

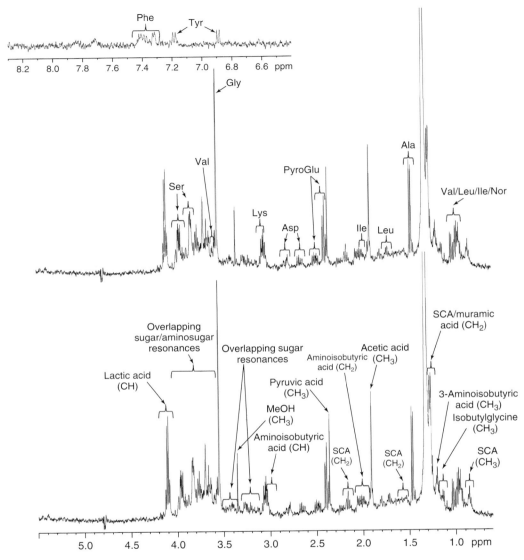

Figure 11.2. Solution-state ^1H NMR spectra of melted glacial ice obtained using the SPR-W5-WATERGATE water suppression technique. The spectra are dominated by a mixture of small molecules including carbohydrates, short-chain carboxylic acids, and amino acids, suggesting a microbial source. (Reproduced from Ref. 30, © 2011 American Chemical Society)

XAD resin as well as humic and fulvic acids. The solution-state ^{13}C NMR spectra of the samples were integrated and changes in the distribution of different forms of organic carbon were compared. It was found that fractions richer in carboxyl carbon exhibited the highest affinity for iron oxide, confirming earlier studies that noted this preferential binding based on solid-state ^{13}C NMR results.[26-28] Another study using solution-state ^{13}C NMR was conducted by Kaiser et al.[77] in which DOM samples at different depths in a forest soil were characterized. The authors reported that concentrations of carboxyl and aromatic components decreased with depth, suggesting that they are preferentially removed from solution via sorptive interactions with soil minerals and OM as they leach downward through the soil profile.

Other studies have demonstrated that solution-state ${}^1H$ NMR can also be used to study interactions involving DOM and solid phases. For example, Feng *et al.*[56] used ${}^1H$ high-resolution magic-angle spinning (HR-MAS) NMR (see Chapter 5) to study the binding of peat humic acid to the clay minerals kaolinite and montmorillonite. Using this technique, the authors found that montmorillonite preferentially sorbed more aromatic and protein components of the humic acid, whereas kaolinite sorbed more aliphatic moieties. Mitchell *et al.*[31] compared the structure of biosolids-derived DOM samples before and after sorption to three alkaline mineral soils with varying mineralogy. 1D ${}^1H$ NMR spectra were collected and integrated into the four major regions outlined previously (MDLT, CRAM, carbohydrates and peptides, and aromatic components), and 2D COSY and HSQC experiments were used to assist with structural assignments. As shown in Figure 11.3, greater resolution of small molecules was observed after sorptive fractionation, suggesting that larger DOM components were preferentially sorbed, which agreed with previous reports.[27,78–80] Aliphatic MDLT components of DOM were preferentially retained by a soil with high concentrations of montmorillonite, a high surface area clay mineral, whereas these components were not retained by the two soils that contained predominantly sand and low concentrations of clay. It was suggested that the high clay surface area provided favorable sites for van der Waals interactions with aliphatic MDLT components. CRAMs were preferentially sorbed by all three soils, likely because of favorable cation bridging or ligand-exchange interactions between abundant carboxyl and hydroxyl groups and the mineral surfaces. Conversely, carbohydrate and peptide DOM components did not exhibit any preferential sorption behavior, possibly because of their polar nature and favorable hydrogen-bonding interactions with water. Low signal intensity prevented a thorough characterization of the sorption behavior of aromatic DOM components. This study illustrated that DOM is highly reactive and that preferential fractionation of DOM occurs in lower soil horizons with exposed mineral surfaces and low concentrations of OM. Furthermore, these studies[31,56,76,77] collectively established that both

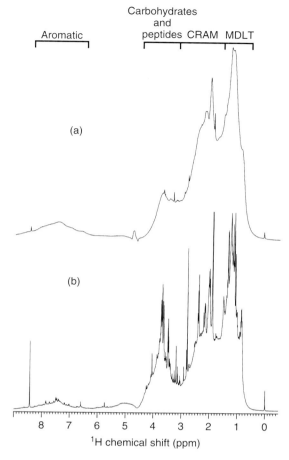

Figure 11.3. Solution-state ${}^1H$ NMR spectra of biosolids-derived DOM (a) before and (b) after sorptive fractionation by an alkaline mineral soil. Greater resolution of individual small molecules was observed after sorptive interactions. (Reproduced from Ref. 31, © 2013 CSIRO Publishing)

solution-state ${}^1H$ and ${}^{13}C$ NMR techniques can provide valuable information regarding DOM reactivity during sorption processes in soil.

11.3.5 Coupling Solution-state NMR with Chromatography and Separation Techniques

The introduction of a sample cleanup or chromatographic separation step before solution-state NMR analysis (see Chapter 3) holds the potential to reduce

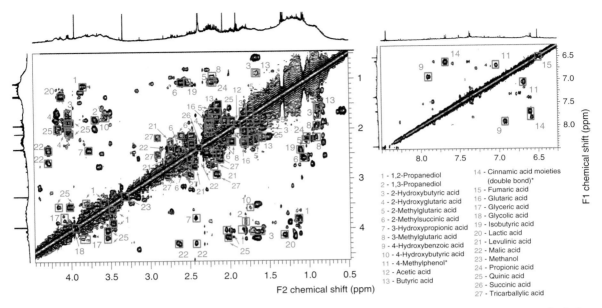

1 - 1,2-Propanediol
2 - 1,3-Propanediol
3 - 2-Hydroxybutyric acid
4 - 2-Hydroxyglutaric acid
5 - 2-Methylglutaric acid
6 - 2-Methylsuccinic acid
7 - 3-Hydroxypropionic acid
8 - 3-Methylglutaric acid
9 - 4-Hydroxybenzoic acid
10 - 4-Hydroxybutyric acid
11 - 4-Methylphenol*
12 - Acetic acid
13 - Butyric acid
14 - Cinnamic acid moieties
(double bond)*
15 - Fumaric acid
16 - Glutaric acid
17 - Glyceric acid
18 - Glycolic acid
19 - Isobutyric acid
20 - Lactic acid
21 - Levulinic acid
22 - Malic acid
23 - Methanol
24 - Propionic acid
25 - Quinic acid
26 - Succinic acid
27 - Tricarballylic acid

Figure 11.4. 2D COSY NMR spectrum of DOM fractionated using offline hydrophilic interaction chromatography before NMR analysis with identified compounds between 0.5 and 4.5 ppm labeled. The inset shows a zoomed view of the aromatic proton region between 6.3 and 8.5 ppm. (Reproduced from Ref. 22, © 2011 American Chemical Society)

the chemical complexity of DOM into individual components. Offline DOM separations are advantageous as the challenges of interfacing different experimental techniques are avoided (see Chapter 3). Examples of offline separation steps that have been used to simplify DOM samples include solid phase extraction (SPE)[23] and high-performance liquid chromatography (HPLC).[22,29] For example, Kim *et al.*[23] reported that passing DOM samples through a C_{18} SPE disk increased the number of identifiable components using 2D solution-state NMR. For one DOM sample, the SPE step was conducted at a remote site in the field, allowing for the preconcentration of DOM on site without the need to transport large quantities of water back to the laboratory for concentration. Offline chromatographic separation using HPLC techniques has also been used to separate DOM in preparation for NMR analysis.[22,29] For instance, Woods *et al.*[22] used hydrophilic interaction chromatography (HILIC) to separate Suwannee River DOM into 80 fractions which were subsequently analyzed using 1D and 2D solution-state NMR. The chromatographic step was shown to significantly improve the resolution of small molecules, which were confirmed by matching ^{1}H chemical shifts with NMR databases and also via

2D NMR experiments such as COSY and HSQC. As shown in Figure 11.4, a number of short-chain carboxylic acids were identified, which are likely biologically-derived and present in appreciable concentrations as these peaks were distinguishable from the broad background signals. This complements data obtained from FT-ICR-MS analysis of DOM which reported, which reported high levels of oxidation (O/C atomic ratios) in many DOM samples, suggesting an abundance of carboxylic acids in DOM.[4,13]

2D chromatographic separation of DOM has also been reported to improve the resolution of solution-state NMR spectra of DOM and to identify major components. Woods *et al.*[29] developed a 2D HILIC–HILIC system using two sequential HPLC columns and used this to separate Suwannee River DOM into close to 10 000 fractions, of which some were selected for subsequent analysis by 2D and three-dimensional NMR. Figure 11.5 illustrates how the chromatographic steps greatly increase the ^{1}H NMR spectral resolution of individual components. The extremely broad profile in the unfractionated DOM is simplified by 1D HPLC and many discrete resonances begin to be resolved from the broader resonances. Using 2D HPLC, the resolution increased

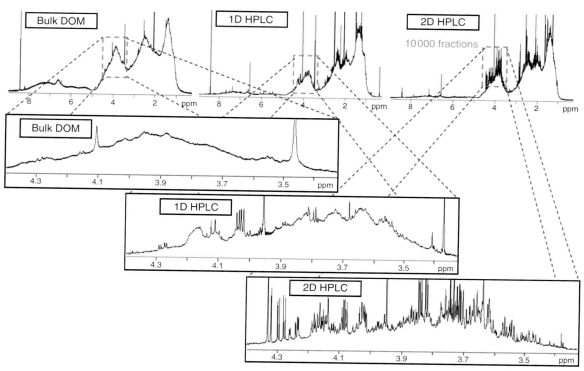

Figure 11.5. Illustration of solution-state ^{1}H NMR spectra of bulk DOM and spectra after undergoing sequential HILIC–HILIC separation using HPLC. With increasing chromatographic separation, the signals from a greater number of components in the complex mixture are resolved.

further and it is apparent that the broader resonances in the unfractionated material arise from the overlap of hundreds if not thousands of similar structures. With the enhanced spectral resolution obtained by the chromatographic separation steps and using spectral simulations, the authors reported that oxidized sterol and hopanoid-like structures such as cholic acid comprise a large proportion of DOM. These molecules are large alicyclic biomolecules, which may degrade slowly in the environment because of their molecular size and thus accumulate in DOM. This is consistent with the results of Hertkorn et al.[4] who reported abundant CRAMs which are believed to be derived from oxidized sterols and hopanoids. The results of these studies demonstrate that an initial offline separation step can provide much more structural information than the direct analysis of bulk DOM. However, broad resonances were still observed and future developments in this field such as improvements to the stationary and mobile phase combination that improve

the separation are needed to enable the comprehensive identification of all DOM components.

Online chromatographic separation before NMR analysis presents a significant analytical challenge as highly sophisticated instruments that automate multiple steps such as chromatography, sample concentration, and NMR spectroscopy must be interfaced (see Chapter 3). However, such systems are commercially available and offer greater ease of sample analysis once experimental conditions are developed and optimized. An early study by Simpson et al.[81] used liquid chromatography interfaced to NMR (LC-NMR) to study the composition of wetland- and lake-derived DOM. LC-NMR was used to separate the DOM into 150 fractions which were subsequently analyzed using ^{1}H NMR. Results showed that the separation step improved the resolution of individual peaks and the addition of tetrabutylammonium hydrogen sulfate as an ion-pair reagent assisted with the separation of aromatic DOM components. Additional experiments were carried out by introducing an online

SPE step to concentrate samples before NMR analysis (thus, LC-SPE-NMR) of DOM obtained via alkaline soil extraction. Concentration using SPE was shown to greatly improve the signal-to-noise ratio and allowed for the identification of some individual small molecules in the aromatic region based on peak splitting and chemical shifts. However, the authors noted that it may be time consuming to identify all of the peaks in a complex mixture based on this approach.[81]

Online chromatography can also be used to separate DOM components based on their molecular size before NMR analysis. For example, Woods *et al.*[65] used online high-performance size-exclusion chromatography using gel permeation in conjunction with ^{1}H NMR to characterize DOM samples from three different environments. The authors were able to separate the DOM samples into three main size fractions with distinct properties. The largest fraction contained resonances from carbohydrates and aromatic components, the intermediate fraction was rich in CRAMs and the smallest size fraction contained mainly signatures for aliphatic MDLT components. This suggests that DOM is heterogeneous not only in terms of the number of compounds present but also in the molecular size and distribution of those moieties. The presence of a large quantity of DOM in the form of macromolecular aggregates is challenging to study using NMR because of potential overlapping resonances at similar chemical shift values. Future hyphenated NMR techniques with improved preconcentration steps or the introduction of online mass spectrometry (e.g., LC-NMR-MS) may allow for more detailed analysis of DOM samples.

11.4 NMR AS A COMPLEMENTARY ANALYTICAL TECHNIQUE FOR DOM CHARACTERIZATION

NMR is frequently used in conjunction with other advanced analytical techniques to provide complementary information that further enhances our understanding of DOM composition and reactivity.[4,7,13,68,82] Optical techniques such as fluorescence spectroscopy are commonly used for the characterization of DOM to determine the source of major components,[7,68,83] which has been shown to support NMR data interpretation. For example, Pautler *et al.*[68] studied the composition of DOM from glacial ice using solution-state NMR in addition to fluorescence spectroscopy. Using NMR, the authors reported the presence of several small molecules which were hypothesized to be microbial derived. Complementary fluorescence data showed that the majority of fluorescence signals were attributed to protein-like OM components, thus the complementary results suggest that DOM from glaciers contains a large amount of small labile molecules from microbial sources. Similarly, Guéguen *et al.*[7] used solution-state NMR in conjunction with fluorescence spectroscopy to characterize DOM from the Athabasca River, Canada, an area that is undergoing land use changes with the development of the Athabasca oil sands deposit. It was hypothesized that increased industrial activity may change the composition of DOM, thus samples were collected at various points along the river, both upstream and downstream from the oil sands development. An increase in the amount of aliphatic carbon was observed using ^{1}H NMR in samples flowing from upstream to downstream. In addition, a decrease in the concentration of aromatic ring-derived components in samples along this transect was obtained from fluorescence data, as the ^{1}H NMR signals for the aromatic region were low (<7% of the total NMR signal). Thus, the two techniques were used together to suggest that riverine DOM composition is being altered by industrial activity in the Athabasca River watershed.

Peatlands are a large sink in the biogeochemical carbon cycle, thus understanding the molecular-level processes in these ecosystems may provide insight into global carbon dynamics. Tfaily *et al.*[82] used solution-state ^{1}H NMR along with specific ultraviolet absorbance (SUVA) to characterize DOM from North American peatlands, including DOM present in water obtained from deep bog and fen environments. NMR was used to examine the distribution of several functional groups, whereas the SUVA at 254 nm was used as an index of humification to assess the development of aromaticity under different wetland conditions. The results showed that aromatic and carbohydrate components comprise up to 70% of DOM from deep bogs, whereas these components are found in much smaller quantities in deep fen DOM. These results were supported by decreased SUVA at 254 nm for the fen sample relative to the deep bog DOM. The drastic differences in DOM structures may be attributable to differences in the microbial community composition, which may be influenced by oxygen availability and thus the prevalence of aerobic or anaerobic conditions in the peatlands.

FT-ICR-MS provides ultrahigh resolution accurate mass data that can be used to calculate molecular formulae of the components of a complex mixture such as DOM.[1,4,13] Atomic ratios and molecular formulae can be used to propose plausible structure and the NMR spectra of those molecules can be simulated using specialized software. Comparing the spectra of actual DOM and simulated spectra of model components may be used to infer the presence of specific molecules or classes of molecules. FT-ICR-MS data can also be used to generate van Krevelen plots in which the atomic ratios H/C are plotted against O/C, which can be used to infer the level of oxidation of DOM components,[1,4,13] again providing insights into potential chemical structures. This deduction process was used in a pioneering study by Hertkorn *et al.*[4] to identify CRAMs in marine DOM. The authors used FT-ICR-MS to acquire mass data, calculated molecular formulae and then proposed potential structures within the constraints of O/C and H/C ratios. The types of molecules that best fit the criteria resembled degradation products of sterols and hopanoids with multiple carboxyl groups. As such, FT-ICR-MS has emerged as a prominent tool for the analysis of DOM and has been used in conjunction with NMR to further our understanding of DOM structure.[13]

NMR experiments can also be used in conjunction with Fourier transform infrared (FTIR) spectroscopy to characterize DOM samples. For example, Oren and Chefetz[84] studied the sorptive fractionation of biosolids-derived DOM using FTIR, whereas Mitchell *et al.*[31] characterized the same DOM samples using solution-state ^{1}H NMR. FTIR results showed that stretching bands for carbonyl groups were more intense in the spectra of bound DOM compared to the nonfractionated DOM. Moreover, ^{1}H NMR integration values for CRAMs decreased upon sorption to the mineral soils compared with the bulk DOM sample. These results collectively suggested that carboxyl-containing DOM components were preferentially sorbed by mineral soils, thus providing further evidence for the role of these moieties as a major reactive component of DOM.[25–28]

11.5 CONCLUSIONS AND FUTURE RESEARCH DIRECTIONS

Although DOM is a complex environmental matrix, advances in NMR spectroscopy have provided the tools needed to extract detailed information about DOM structure and reactivity. Solid-state NMR has been used to study the bulk composition of DOM samples including the abundance of several classes of molecules[47,49] and how this distribution changes under different environmental conditions such as after sorption to mineral surfaces[25–28] or with increasing soil residence time.[24] Solution-state NMR offers a wider range of potential experiments which provide greater insight into DOM composition including the presence of small molecules[30,31,71] and aggregation behavior.[74,75] Coupling solution-state NMR with both offline and online chromatographies and separation steps has been shown to vastly improve spectral resolution, which permits the identification of a greater number of individual DOM components.[22,23,29,65,81] As DOM is a highly complex and heterogeneous matrix which is difficult to characterize, NMR is often used in conjunction with other sophisticated analytical techniques that provide complementary structural information which helps to further our understanding of DOM composition in various environments.[4,7,13,68,82,84]

In the future, new advances in analytical instrumentation may lead to the hyphenation of NMR with other separation techniques that enable a more detailed analysis of DOM structure. For example, Cottrell *et al.*[85] showed that a technique known as counterbalance capillary electrophoresis (CE) can be used to separate DOM into several hundred constituent components after optimizing experimental conditions such as buffer concentration and pH. Future developments may allow for the interfacing of CE and NMR such that CE fractions can be repeatedly collected until a sufficient quantity of sample is obtained for detailed NMR analysis. Although such analytical development is challenging from an engineering standpoint, this technology may have the capability to provide as yet unachievable levels of resolution of small molecules in DOM.

ACKNOWLEDGMENTS

The authors thank support from the Natural Science and Engineering Research Council of Canada.

REFERENCES

1. P. Schmitt-Kopplin, G. Liger-Belair, B. P. Koch, R. Flerus, G. Kattner, M. Harir, B. Kanawati, M. Lucio, D. Tziotis, N. Hertkorn, and I. Gebefügi, *Biogeosciences*, 2012, **9**, 1571.

2. P. J. Seaton, R. J. Kieber, J. D. Willey, G. B. Avery, and J. L. Dixon, *Atmos. Environ.*, 2013, **65**, 52.

3. B. Lam, A. Baer, M. Alaee, B. Lefebvre, A. Moser, A. Williams, and A. J. Simpson, *Environ. Sci. Technol.*, 2007, **41**, 8240.

4. N. Hertkorn, R. Benner, M. Frommberger, P. Schmitt-Kopplin, M. Witt, K. Kaiser, A. Kettrup, and J. I. Hedges, *Geochim. Cosmochim. Acta*, 2006, **70**, 2990.

5. M. V. McCaul, D. Sutton, A. J. Simpson, A. Spence, D. J. McNally, B. W. Moran, A. Goel, B. O'Connor, K. Hart, and B. P. Kelleher, *Environ. Chem.*, 2011, **8**, 146.

6. A. Nebbioso and A. Piccolo, *Anal. Bioanal. Chem.*, 2013, **405**, 109.

7. C. Guéguen, D. C. Burns, A. McDonald, and B. Ring, *Chemosphere*, 2012, **87**, 932.

8. A. Stubbins, E. Hood, P. A. Raymond, G. R. Aiken, R. L. Sleighter, P. J. Hernes, D. Butman, P. G. Hatcher, R. G. Striegl, P. Schuster, H. A. N. Abdulla, A. W. Vermilyea, D. T. Scott, and R. G. M. Spencer, *Nat. Geosci.*, 2012, **5**, 198.

9. P. Conte, C. Abbate, A. Baglieri, M. Nègre, C. De Pasquale, G. Alonzo, and M. Gennari, *Org. Geochem.*, 2011, **42**, 972.

10. P. S. M. Santos, M. Otero, E. B. H. Santos, and A. C. Duarte, *Talanta*, 2010, **82**, 1616.

11. A. J. Simpson, D. J. McNally, and M. J. Simpson, *Prog. Nucl. Magn. Reson. Spectrosc.*, 2011, **58**, 97.

12. R. M. W. Amon and R. Benner, *Nature*, 1994, **369**, 549.

13. N. Hertkorn, M. Harir, B. P. Koch, B. Michalke, and P. Schmitt-Kopplin, *Biogeosciences*, 2013, **10**, 1583.

14. J. I. Hedges, *Mar. Chem.*, 1992, **39**, 67.

15. J. Dachs, M. L. Calleja, C. M. Duarte, S.del Vento, B. Turpin, A. Polidori, G. J. Herndl, and S. Agustí, *Geophys. Res. Lett.*, 2005, **32**, 1.

16. H. Ogawa and E. Tanoue, *J. Oceanogr.*, 2003, **59**, 129.

17. S. Hernandez-Ruiz, L. Abrell, S. Wickramasekara, B. Chefetz, and J. Chorover, *Water Res.*, 2012, **46**, 943.

18. R. Navon, S. Hernandez-Ruiz, J. Chorover, and B. Chefetz, *J. Environ. Qual.*, 2011, **40**, 942.

19. N. S. Bolan, D. C. Adriano, A. Kunhikrishnan, T. James, R. McDowell, and N. Senesi, *Adv. Agron.*, 2011, **110**, 1.

20. B. P. Koch, T. Dittmar, M. Witt, and G. Kattner, *Anal. Chem.*, 2007, **79**, 1758.

21. T. Dittmar and J. Paeng, *Nat. Geosci.*, 2009, **2**, 175.

22. G. C. Woods, M. J. Simpson, P. J. Koerner, A. Napoli, and A. J. Simpson, *Environ. Sci. Technol.*, 2011, **45**, 3880.

23. S. Kim, A. J. Simpson, E. B. Kujawinski, M. A. Freitas, and P. G. Hatcher, *Org. Geochem.*, 2003, **34**, 1325.

24. J. Sanderman and M. G. Kramer, *Biogeochemistry*, 2013, **113**, 259.

25. K. Kaiser, G. Guggenberger, L. Haumaier, and W. Zech, *Eur. J. Soil Sci.*, 1997, **48**, 301.

26. D. M. McKnight, K. E. Bencala, G. W. Zellweger, G. R. Aiken, G. L. Feder, and K. A. Thorn, *Environ. Sci. Technol.*, 1992, **26**, 1388.

27. B. Gu, J. Schmitt, Z. Chen, L. Liang, and J. F. McCarthy, *Geochim. Cosmochim. Acta*, 1995, **59**, 219.

28. B. Gu, J. Schmitt, Z. Chen, L. Liang, and J. F. McCarthy, *Environ. Sci. Technol.*, 1994, **28**, 38.

29. G. C. Woods, M. J. Simpson, and A. J. Simpson, *Water Res.*, 2012, **46**, 3398.

30. B. G. Pautler, A. J. Simpson, M. J. Simpson, L. Tseng, M. Spraul, A. Dubnick, M. J. Sharp, and S. J. Fitzsimons, *Environ. Sci. Technol.*, 2011, **45**, 4710.

31. P. J. Mitchell, A. J. Simpson, R. Soong, A. Oren, B. Chefetz, and M. J. Simpson, *Environ. Chem.*, 2013, **10**, 333.

32. B. Lam and A. J. Simpson, *Analyst*, 2008, **133**, 263.

33. J. G. Longstaffe, D. Courtier-Murias, and A. J. Simpson, *Chemosphere*, 2013, **90**, 270.

34. J. G. Longstaffe and A. J. Simpson, *Environ. Toxicol. Chem.*, 2011, **30**, 1745.

35. J. G. Longstaffe, M. J. Simpson, W. Maas, and A. J. Simpson, *Environ. Sci. Technol.*, 2010, **44**, 5476.

36. A. Shirzadi, M. J. Simpson, Y. Xu, and A. J. Simpson, *Environ. Sci. Technol.*, 2008, **42**, 1084.

37. P. Mazzei and A. Piccolo, *Environ. Sci. Technol.*, 2012, **46**, 5939.

38. E. M. Perdue, N. Hertkorn, and A. Kettrup, *Appl. Geochem.*, 2007, **22**, 1612.

39. J. Li, E. M. Perdue, and L. T. Gelbaum, *Environ. Sci. Technol.*, 1998, **32**, 483.

40. X. Xu, A. G. Kalinichev, and R. James Kirkpatrick, *Geochim. Cosmochim. Acta*, 2006, **70**, 4319.

41. X. Q. Lu, W. D. Johnson, and J. Hook, *Environ. Sci. Technol.*, 1998, **32**, 2257.

42. J. Sanderman, J. A. Baldock, and R. Amundson, *Biogeochemistry*, 2008, **89**, 181.

43. J. Mao, X. Kong, K. Schmidt-Rohr, J. J. Pignatello, and E. M. Perdue, *Environ. Sci. Technol.*, 2012, **46**, 5806.

44. R. Jaffé, Y. Yamashita, N. Maie, W. T. Cooper, T. Dittmar, W. K. Dodds, J. B. Jones, T. Myoshi, J. R. Ortiz-Zayas, D. C. Podgorski, and A. Watanabe, *Geochim. Cosmochim. Acta*, 2012, **94**, 95.

45. M. G. Kramer, J. Sanderman, O. A. Chadwick, J. Chorover, and P. M. Vitousek, *Global Change Biol.*, 2012, **18**, 2594.

46. K. J. Dria, J. R. Sachleben, and P. G. Hatcher, *J. Environ. Qual.*, 2002, **31**, 393.

47. M. J. Simpson and A. J. Simpson, *J. Chem. Ecol.*, 2012, **38**, 768.

48. M. J. Simpson, A. Otto, and X. Feng, *Soil Sci. Soc. Am. J.*, 2008, **72**, 268.

49. J. S. Clemente, E. G. Gregorich, A. J. Simpson, R. Kumar, D. Courtier-Murias, and M. J. Simpson, *Environ. Chem.*, 2012, **9**, 97.

50. B. Chefetz, Y. Hadar, and Y. Chen, *Acta Hydrochim. Hydrobiol.*, 1998, **26**, 172.

51. B. Chefetz, P. G. Hatcher, Y. Hadar, and Y. Chen, *Soil Sci. Soc. Am. J.*, 1998, **62**, 326.

52. D. Courtier-Murias, A. J. Simpson, C. Marzadori, G. Baldoni, C. Ciavatta, J. M. Fernández, E. G. López-de-Sá, and C. Plaza, *Agric. Ecosyst. Environ.*, 2013, **171**, 9.

53. C. Plaza, D. Courtier-Murias, J. M. Fernández, A. Polo, and A. J. Simpson, *Soil Biol. Biochem.*, 2013, **57**, 124.

54. J. Chorover and M. K. Amistadi, *Geochim. Cosmochim. Acta*, 2001, **65**, 95.

55. G. U. Balcke, N. A. Kulikova, S. Hesse, F. D. Kopinke, I. V. Perminova, and F. H. Frimmel, *Soil Sci. Soc. Am. J.*, 2002, **66**, 1805.

56. X. Feng, A. J. Simpson, and M. J. Simpson, *Org. Geochem.*, 2005, **36**, 1553.

57. A. J. Simpson, M. J. Simpson, W. L. Kingery, B. A. Lefebvre, A. Moser, A. J. Williams, M. Kvasha, and B. P. Kelleher, *Langmuir*, 2006, **22**, 4498.

58. L. Zhang, L. Luo, and S. Zhang, *Colloids Surf. A*, 2012, **406**, 84.

59. K. Wang and B. Xing, *J. Environ. Qual.*, 2005, **34**, 342.

60. U. Lankes, H. Lüdemann, and F. H. Frimmel, *Water Res.*, 2008, **42**, 1051.

61. J. Templier, F. Miserque, N. Barré, F. Mercier, J. Croué, and S. Derenne, *J. Anal. Appl. Pyrolysis*, 2012, **97**, 62.

62. L. I. Aluwihare, D. J. Repeta, S. Pantoja, and C. G. Johnson, *Science*, 2005, **308**, 1007.

63. M. McCarthy, T. Pratum, J. Hedges, and R. Benner, *Nature*, 1997, **390**, 150.

64. B. G. Pautler, A. Dubnick, M. J. Sharp, A. J. Simpson, and M. J. Simpson, *Geochim. Cosmochim. Acta*, 2013, **104**, 1.

65. G. C. Woods, M. J. Simpson, B. P. Kelleher, M. McCaul, W. L. Kingery, and A. J. Simpson, *Environ. Sci. Technol.*, 2010, **44**, 624.

66. H. Herzog, S. Haiber, P. Burba, and J. Buddrus, *Fresenius J. Anal. Chem.*, 1997, **359**, 167.

67. A. J. Simpson and S. A. Brown, *J. Magn. Reson.*, 2005, **175**, 340.

68. B. G. Pautler, G. C. Woods, A. Dubnick, A. J. Simpson, M. J. Sharp, S. J. Fitzsimons, and M. J. Simpson, *Environ. Sci. Technol.*, 2012, **46**, 3753.

69. G. C. Woods, M. J. Simpson, B. G. Pautler, S. F. Lamoureux, M. J. Lafrenière, and A. J. Simpson, *Geochim. Cosmochim. Acta*, 2011, **75**, 7226.

70. P. S. M. Santos, E. B. H. Santos, and A. C. Duarte, *Sci. Total Environ.*, 2012, **426**, 172.

71. M. A. Wilson, P. J. Collin, R. L. Malcolm, E. M. Perdue, and P. Cresswell, *Org. Geochem.*, 1988, **12**, 7.

72. H. Ma, H. E. Allen, and Y. Yin, *Water Res.*, 2001, **35**, 985.

73. J. Peuravuori and K. Pihlaja, *Anal. Chim. Acta*, 1998, **363**, 235.

74. B. Lam and A. J. Simpson, *Environ. Toxicol. Chem.*, 2009, **28**, 931.

75. G. Zheng and W. S. Price, *Environ. Sci. Technol.*, 2012, **46**, 1675.

76. K. Kaiser, *Org. Geochem.*, 2003, **34**, 1569.

77. K. Kaiser, G. Guggenberger, and L. Haumaier, *Biogeochemistry*, 2004, **70**, 135.

78. J. Hur and M. A. Schlautman, *J. Colloid Interf. Sci.*, 2003, **264**, 313.

79. M. Ochs, B. Ćosović, and W. Stumm, *Geochim. Cosmochim. Acta*, 1994, **58**, 639.

80. Q. Zhou, P. A. Maurice, and S. E. Cabaniss, *Geochim. Cosmochim. Acta*, 2001, **65**, 803.

81. A. J. Simpson, L. Tseng, M. J. Simpson, M. Spraul, U. Braumann, W. L. Kingery, B. P. Kelleher, and M. H. B. Hayes, *Analyst*, 2004, **129**, 1216.

82. M. M. Tfaily, R. Hamdan, J. E. Corbett, J. P. Chanton, P. H. Glaser, and W. T. Cooper, *Geochim. Cosmochim. Acta*, 2013, **112**, 116.

83. O. Pisani, Y. Yamashita, and R. Jaffé, *Water Res.*, 2011, **45**, 3836.

84. A. Oren and B. Chefetz, *J. Environ. Qual.*, 2012, **41**, 526.

85. B. A. Cottrell, W. R. Cheng, B. Lam, W. J. Cooper, and A. J. Simpson, *Analyst*, 2013, **138**, 1174.

Chapter 12
Atmospheric Organic Matter

Regina M.B.O. Duarte and Armando C. Duarte

CESAM & Department of Chemistry, University of Aveiro, 3810-193 Aveiro, Portugal

12.1 INTRODUCTION

Understanding the significance of atmospheric organic matter, either organic aerosols or dissolved organic compounds in suspended clouds, precipitation, and fog droplets, to a variety of processes occurring in the atmosphere is a focus of major scientific and policy concern. Indeed, the overwhelming focus of the atmospheric research community has been on the detailed structural characterization of the atmospheric water-soluble organic matter (WSOM). In the particular case of atmospheric aerosols, as much as 10–80% of organic aerosols have been found to be water-soluble. The growing interest on the

NMR Spectroscopy: A Versatile Tool for Environmental Research
Edited by Myrna J. Simpson and André J. Simpson
© 2014 John Wiley & Sons, Ltd. ISBN: 978-1-118-61647-5

ubiquitous aerosol WSOM is fueled by the realization that this organic fraction has an important, yet highly uncertain, role on the radiative budget of the global climate system.[1-5] Wet-deposition fluxes of atmospheric WSOM can also be important as a temporal source of organic carbon to surface waters, thus having major implications on the global carbon cycle.[6] The WSOM in rainwater can also affect a variety of processes occurring in atmospheric waters, including attenuation and spectral distribution of incoming solar radiation,[7] number of cloud droplets (implications on cloud lifetime),[8,9] and trace metal complexation.[10] Fog droplets are usually viewed as an aqueous-phase reactor combining scavenged particles, water-soluble gasses, and their reaction products. As such, aqueous-phase reactions in fog droplets can contribute to secondary formation of organic aerosols, in which the participation of water-soluble organics is likely to be highly significant.[11]

The sources and formation mechanisms of atmospheric WSOM are another emerging issue, for which the atmospheric research community has only qualitative but not sufficient understanding. Biomass burning and secondary formation (involving both anthropogenic and biogenic volatile organic compounds) are considered to be major sources of atmospheric WSOM.[12] To further enhance the diversity and the complexity of its constituent molecules, the atmospheric WSOM and their precursor gasses can also evolve in the atmosphere, becoming increasingly oxidized, less volatile, more hygroscopic, and, therefore, more water-soluble.[13]

Despite the potential importance of atmospheric WSOM, details on its chemical composition, physicochemical properties, and fate are still poorly constrained at present, mostly because of its dynamic nature, which is translated into a multitude of molecular forms, sources, and reactivity. Nevertheless, determining the primary molecular composition and structure of atmospheric WSOM is warranted to increase the current understanding of its role in various atmospheric processes. This review highlights the use of NMR spectroscopy, focusing on both ^{1}H and ^{13}C nuclei, as a major breakthrough for targeting structural details of atmospheric WSOM. There is a special emphasis on the richness of information that is obtainable with solid-state and one-dimensional (1D) and two-dimensional (2D) solution-state NMR methods, and on how these data can be employed for source apportionment in different areas. The limitations of such advanced techniques, the complementarity with other sophisticated tools, and methodological aspects concerning sample preparation are also discussed. Finally, the major challenges ahead for improvement of the current knowledge about the chemical structures of atmospheric WSOM based on NMR spectroscopy are outlined.

12.2 SETTING THE SCENE

A recent critical review settled the scenario regarding the use of several different offline and online sophisticated analytical techniques for mitigating the complexity of atmospheric WSOM and to unravel the structure and composition of this organic component.[14] NMR spectroscopy fits into the group of offline techniques that allow a more complete description of the whole mass of atmospheric WSOM while providing resolution on functional groups and substructural components. It should be noted, however, that NMR spectroscopy does not allow capturing WSOM compositional changes on timescales consistent with atmospheric variability (typically between a few seconds and 1 h). In fact, a proper application of NMR techniques requires large amounts of sample, resulting in low temporal resolution (from a few hours to 7 days). Nevertheless, unfolding the chemical classes within WSOM using NMR techniques at the expense of near real-time data offers great promise in answering major unsolved questions regarding the composition and structure of atmospheric WSOM.

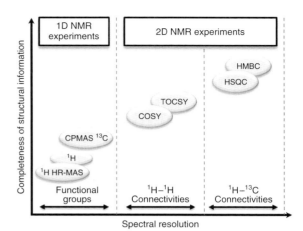

Figure 12.1. Representation of the range of NMR techniques currently employed for the characterization of atmospheric WSOM, highlighting the level of structural information and spectral resolution attained by the different methods.

Figure 12.1 provides an overview on how the different NMR methods applied so far for the structural characterization of atmospheric WSOM compare for two important parameters: completeness of structural information and the level of spectral resolution. 1D NMR methods, namely solution-state ^{1}H and solid-state cross polarization magic-angle spinning (CPMAS) ^{13}C NMR, have been undoubtedly the most widely used techniques for the past 15 years, as reported in a very recent review.[14] These two techniques provide a semiquantitative overview as to the distribution of the various functional groups present in atmospheric WSOM. As in the case of NMR spectra of aquatic and terrestrial organic matter, the structural interpretation of 1D ^{1}H and ^{13}C NMR spectra of atmospheric WSOM is hampered by the strong overlapping of chemical resonances. In such cases, the identification of specific molecular structures is impossible and only some generic structural assignments and average parameters (e.g., aliphatic, oxygenated aliphatic, aromatic, and/or carboxylic acid content) can be made on the basis of these 1D NMR methods alone.[14] Nevertheless, as discussed later, the information provided by these methods can be used for molecular modeling,[15] source apportionment,[16,17] and seasonal and temporal characterizations[18-20] of atmospheric WSOM. High-resolution (HR)-MAS NMR is a method suitable for studying semi-soluble components of a complex organic mixture (see Chapter 5). This technique provides information on structures hidden in less-soluble domains (e.g., hydrophobic moieties), which become easily accessed using penetrating solvents [e.g., dimethyl sulfoxide-d_6 (DMSO-d_6)].[21] Thus far, only one study has applied ^{1}H HR-MAS NMR spectroscopy, combined with solution- and solid-state NMR methods, for building a general overview of the organic composition of surface films originated from atmospheric deposition in urban environments.[22] In comparison to solution-state ^{1}H NMR, the use of ^{1}H HR-MAS NMR alone provides less detail on the distribution of all protons in the sample, which may lead to an incomplete description of the sample components. Used in combination with either 1D or 2D solution-state NMR methods, ^{1}H HR-MAS NMR is likely to give an enhanced contribution for achieving a detailed structural identification within atmospheric WSOM. At this point, the reader may also cogitate on the possibility of applying both solid-state CPMAS ^{13}C and ^{1}H HR-MAS NMR methods to the structural studies of the whole atmospheric particulate matter. The issues of aerosol composition (organic versus inorganic component, including paramagnetic trace elements), the low amount of the aerosol organic component (usually <35% of the total aerosol mass), the difficulties in recovering the atmospheric aerosols from the collection media (usually quartz fiber filters) without sample contamination, and the sample pretreatment procedure (e.g., the use of hydrofluoric acid to help remove the inorganic component) are major obstacles that prevent the straightforward application of such NMR methods to the structural investigation of the whole atmospheric particulate matter.

^{1}H–^{1}H homonuclear solution-state 2D NMR experiments, namely correlation spectroscopy (COSY) and total correlation spectroscopy (TOCSY), have also been applied to the structural characterization of the water-soluble fraction of rainwater[20] and organic aerosols.[22–24] Both these 2D NMR methods allow resolving and detecting chemical functionalities that may be hidden in a 1D ^{1}H NMR spectrum because of strong peak overlap. By providing connectivity information between protons in neighboring units (COSY) or regarding protons that are interacting within two to three bonds (TOCSY), these 2D NMR experiments allow producing a map of the partial structures present in the WSOM sample with a higher degree of confidence. Combining this ^{1}H–^{1}H homonuclear coupling information with that provided by the short- and long-range ^{1}H–^{13}C heteronuclear connectivities

[heteronuclear single-quantum correlation (HSQC) and heteronuclear multiple bond correlation (HMBC), respectively] will enable the identification of the H–C backbone of an organic compound and, therefore, one can start to envisage the molecular structures within the atmospheric WSOM. Indeed, this information can offer significant gains for an improved understanding of the structural composition, source attribution, and possible formation mechanisms of the complex atmospheric organic matter.[23,24]

In deciding the specific NMR experiment to be implemented, the atmospheric chemists should question the level of information needed for resolving the many open questions related to the origin and formation mechanisms of atmospheric WSOM and its effects on atmospheric processes and climate.[25] If one is developing a model for linking the molecular structure of WSOM to its primary and/or secondary origins, then the implementation of a comprehensive analytical approach based on 2D solution-state NMR techniques is highly desirable. However, if one is more interested in understanding the composition of the atmospheric WSOM and its variation with respect to sources, then the typical 1D NMR approach for targeting structural average parameters may be all that are warranted. As demonstrated in the following sections, the synergetic application of NMR techniques (solid-state and 1D and 2D solution-state NMR) may be an advantageous approach for unraveling the composition and structure of atmospheric WSOM, as well as for identifying molecular fingerprints of its different sources, which is valuable for source apportionment studies.

12.3 SOLID-STATE NMR STUDIES OF ATMOSPHERIC ORGANIC MATTER

Solid-state CPMAS ^{13}C NMR spectroscopy presents itself as a valuable technique for investigating the distribution of carbon functional groups in atmospheric organic matter. However, thus far, there has been only five studies reporting the use of solid-state CPMAS ^{13}C NMR spectroscopy for the characterization of organic matter, including the water-soluble fraction, in atmospheric aerosols[18,26–28] and fog water.[29] A very important concern regarding the application of this technique to the analysis of atmospheric organic matter, is the acquisition of resolved CPMAS ^{13}C NMR spectra with a high signal-to-noise ratio. Unfortunately, this technique is insufficiently sensitive to be applied to the analysis of WSOM at the concentration levels usually found in the atmosphere. Typically, this NMR technique requires about 20–100 mg of sample (depending on the size of the NMR probe), which means that it is possible to acquire a CPMAS ^{13}C NMR spectrum with an adequate signal-to-noise ratio, provided that the sample contains about 10–50 mg of carbon.[18,27] Being aware of this constraint, most researchers usually combine samples according to similar ambient conditions, yielding representative composite samples of a given sampling period and/or seasonal event.[18,26–29] As previously mentioned, this particular experimental detail may be considered a weakness of this offline analytical approach as it hampers any investigation of real-time changes in aerosol chemistry.

Nevertheless, if one decides to use solid-state CPMAS ^{13}C NMR spectroscopy, it is critical to make the atmospheric WSOM samples amenable to analysis to take full advantage of the potential of this NMR technique. In this sense, atmospheric WSOM must be isolated from the inorganic matrix (including the naturally occurring paramagnetic trace elements) and prepared as a solid (e.g., freeze drying) before analysis. The isolation procedure is of particular concern, as it must allow the recovery of an unbiased and uncontaminated fraction of the WSOM from the original atmospheric sample (aerosols, fog water, and rainwater). A similar concern of sample preparation (i.e., extraction/desalting procedure) is also critical for enhancing the resolution capability of both 1D and 2D solution-state NMR techniques. Solid-phase extraction (e.g., hydrophobic bonded-phase silica sorbents and polymer-based packing materials), ion-exchange chromatography, and size-exclusion chromatography have been employed to isolate and simultaneously fractionate, the atmospheric WSOM. Of the available isolation procedures, solid-phase extraction and ion-exchange chromatography are currently the most commonly used methods.[14,20,29–32] It must be emphasized, however, that depending on the applied procedure, the isolated WSOM is likely to be enriched in those organic species targeted by the physicochemical mechanisms governing their extraction.[14,30] Isolation procedures based on the use of ion-exchange resins exploit the acidic character of the WSOM and allow fractionating the WSOM mixture into (i) neutral compounds, (ii) carboxylic acids (mono and di), and (iii) polycarboxylic acids. On the other hand, the use of hydrophobic bonded-phase silica and polymer-based sorbents allows fractionating the WSOM samples at preadjusted acidic conditions

(pH ≈ 2) into operationally defined hydrophobic and hydrophilic fractions.[14] Indeed, the lack of a unified approach for the quantitative isolation of a representative mixture of a given atmospheric WSOM sample is a key issue still requiring further systematic studies. A complete survey of the different methods established for the isolation of atmospheric WSOM is well beyond the scope of this overview. Readers are encouraged to consult the reviews of Graber and Rudich[30] and Duarte and Duarte[2,14] and references therein, as well as the research chapters of Herckes *et al.*,[29] Mazzoleni *et al.*,[31] Santos *et al.*,[32] and Seaton *et al.*[20] to obtain additional information.

Despite the aforementioned challenges, the CPMAS ^{13}C NMR technique exhibits important advantages, especially when compared to 1D solution-state ^{1}H NMR spectroscopy in the analysis of atmospheric organic matter: (i) it is nondestructive and, after analysis, the WSOM samples can be used for other complementary structural investigations[18,26,29]; (ii) it is not prone to solvent effects that may alter chemical shifts of the functional groups, mask some of the chemical resonances because of solvent signals, or even cause the loss of some peaks (especially those of the carboxylic acids because of the presence of rapidly exchangeable protons)[28,33]; and (iii) the limited solubility of the organic material in the selected solvent may result in lower resolution and sensitivity in the solution-state ^{1}H NMR spectra.[33] It should also be mentioned that these features holds true for 1D solution-state ^{13}C NMR spectroscopy of atmospheric WSOM. Indeed, in deciding whether to use solution-state ^{13}C or solid-state CPMAS ^{13}C NMR spectroscopy for the structural studies of atmospheric WSOM, the latter is usually preferred over the former. Although one may intuitively consider that solution-state NMR is of better choice to investigate the soluble WSOM, the low natural abundance of ^{13}C (1.1%) and the low concentrations of atmospheric WSOM are likely to produce NMR spectra with low signal-to-noise ratios and poor sensitivity. These two important aspects and the aforementioned advantages of solid-state CPMAS ^{13}C NMR make this technique a powerful alternative to its solution-state counterpart. Another interesting motivation for using the CPMAS ^{13}C NMR technique is the possibility of obtaining a semiquantitative measure of the relative contribution of the different functional groups to the organic carbon present in the atmospheric WSOM. This information has been used for investigating aerosol WSOM composition and its structural variations with changes in parameters such as aerosol

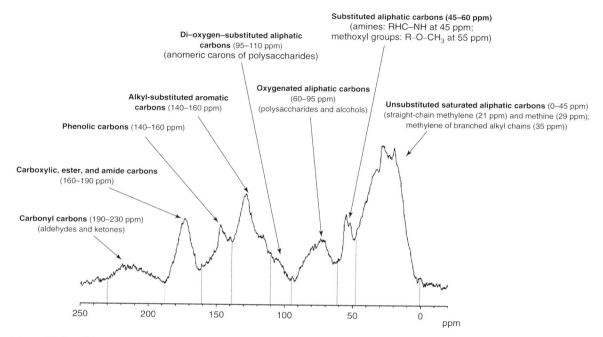

Figure 12.2. ^{13}C chemical-shift assignments of functional groups identified in a solid-state CPMAS ^{13}C NMR spectrum of WSOM from atmospheric aerosols. [Spectrum acquired at 125.77 MHz on a Bruker Avance-500 NMR, transients (3188) recorded with a contact time of 1.5 ms, spinning rate of 9 kHz, recycle delay of 5 s, and length of the proton 90° pulse of 3.5 μs. Chemical shifts are quoted in ppm from the external calibrant tetramethylsilane].

sources and meteorological conditions.[18,26,28] It should be pointed out, however, that this CPMAS ^{13}C NMR semiquantitative approach via data integration has yet to be applied in studies of fog water and rainwater organic matters. Atmospheric chemists should also be aware that the absolute quantification of structural groups from CPMAS ^{13}C NMR spectra is questionable, and the distribution of carbon derived from a particular spectrum must be interpreted with caution. As discussed further in Section 12.5, the acquisition of semiquantitative CPMAS ^{13}C NMR spectra of atmospheric WSOM have benefited from the knowledge produced so far in the studies of natural organic matter from aquatic and terrestrial environments. All these samples are complex mixtures of organic structures, which mean that they share the same analytical challenges with respect to the chemical characterization by means of CPMAS ^{13}C NMR spectroscopy, both as a whole or as fractions separated from the whole.

The ^{13}C chemical-shift ranges used to identify WSOM constituents in aerosols and fog water are thoroughly described in the literature,[2,14,29] and structural assignment is based on those found for terrestrial and aquatic natural organic matter.[21,34] Figure 12.2 shows an example of a solid-state CPMAS ^{13}C NMR spectrum of WSOM from atmospheric aerosols. The question is now focused on the level of compositional information that can be withdrawn from the CPMAS ^{13}C NMR spectra of WSOM, and whether it is well worth to expend the effort in applying this powerful tool to advance the understanding on atmospheric WSOM. The works reported thus far demonstrate that almost all CPMAS ^{13}C NMR spectra, including that shown in Figure 12.2, are very broad with overlapping peaks, just allowing the identification of typically five to eight types of functional groups. On average, the atmospheric WSOM is mostly aliphatic, also having contributions from oxygenated aliphatics (e.g., carbohydrate-type structures and alcohols), aromatics, and carboxylic acid functional groups. The published CPMAS ^{13}C NMR data also shows evidence that the WSOM from aerosols (rural, urban,

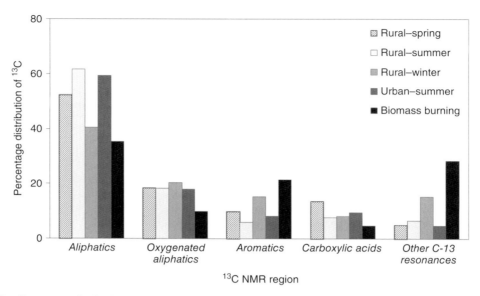

Figure 12.3. Percentage distribution of the main carbon functional groups in WSOM samples extracted from atmospheric aerosols collected in a rural location during three different seasons,[18] in an urban area in summer, and during a biomass-burning event[28].

and biomass burning)[18,28] and fog waters[29] exhibit the same main carbon functional groups, but their relative abundances are quite different. Figure 12.3 illustrates the relative abundance (as percentage of total NMR peak area) of the main carbon functional groups identified in aerosol WSOM samples collected by Duarte *et al.*[18] at a rural location, and by Sannigrahi *et al.*[28] at an urban area and during a biomass-burning event. In addition to the comparison between WSOM samples from different locations and sources, this semiquantitative assessment of the NMR data also enables one to envisage a possible conceptual model of the organic structures that are likely to be representative of a given atmospheric WSOM sample. Duarte *et al.*[18] also concluded that the high aromaticity of the aerosol WSOM collected in colder periods, together with resonances attributable to methoxyl groups and oxygen-substituted aromatic ring carbons, was consistent with the notion that wood-burning processes in domestic fireplaces has a major contribution to the bulk chemical properties of atmospheric organic matter. It is with no surprise that the resonances assigned to aromatic moieties are more prominent in WSOM samples from biomass burning.[28] The CPMAS ^{13}C NMR results of Sannigrahi *et al.*[28] also suggested that the urban aerosol WSOM sample is likely to be produced by direct emissions from

motor vehicle or biomass combustion and secondary organic aerosol formation. The CPMAS ^{13}C NMR studies of Herckes *et al.*[29] indicated the contribution of biological sources to organic matter in atmospheric fog droplets. The presence of resonances likely attributable to nitrogen-containing organic species, namely aliphatic amines (resonance at 40–55 ppm of the C–N linkage) and aromatic amines (shoulder at 145 ppm), also suggested the need for future studies by other analytical methods.

From these examples, it is apparent that much new and valuable information about atmospheric WSOM is obtainable by implementing solid-state CPMAS ^{13}C NMR experiments. However, atmospheric chemists can give a step forward into the detailed structural characterization of WSOM if they are willing to embrace more sophisticated solid-state NMR techniques. Indeed, during these last 12 years, there has been an enormous effort toward the development and use of advanced solid-state NMR spectral editing techniques for targeting and quantifying specific functional groups of sp[2]- and sp[3]-hybridized carbons (e.g., CH, CH_2, mobile CH_3, OCH_3, CN, nonprotonated aromatic C–C, and anomeric O–C–O and nonprotonated anomeric O–C(R,R′)–O groups) in humic substances from various origins.[35] It has been demonstrated that the structural information provided by these advanced solid-state NMR techniques combined with that provided by 2D ^{1}H–^{13}C heteronuclear correlation (HETCOR) NMR is promising for achieving a new and deeper understanding of the structure, heterogeneity, and domains of such complex organic mixtures.[35] Likewise, no studies have been found in the literature reporting the use of solid-state NMR of nuclei other than ^{13}C in atmospheric WSOM. As previously mentioned, nitrogen-containing organic species in the WSOM of both aerosols and fog waters has been observed.[14] These structures are usually difficult to identify by the conventional CPMAS ^{15}N NMR spectroscopy, mostly because of the very low natural abundance of the ^{15}N isotope (0.36%), its magnetic properties (low and negative gyromagnetic ratio), and the very low content of nitrogen-containing organic species, all contribute to produce CPMAS ^{15}N NMR spectra with a sensitivity 50 times lower than those of CPMAS ^{13}C NMR spectra.[36] The use of spectral editing NMR techniques, such as ^{13}C–{^{14}N} saturation-pulse-induced dipolar exchange with recoupling (SPIDER), can be a promising tool for determining the nature of structures to which nitrogen is bonded. With this technique, one can determine how nitrogen is incorporated in the water-soluble organic matrix, and, therefore, achieve a qualitative characterization of the nitrogen-containing functional groups. The latest work of Mao and coworkers[35] is a good starting point for those who wish to apply a systematic protocol based on these advanced solid-state NMR techniques in order to accomplish an in-depth understanding of the structural features of the complex atmospheric WSOM.

12.4 SOLUTION-STATE NMR STUDIES OF ATMOSPHERIC ORGANIC MATTER

12.4.1 Application of One-dimensional Solution-state NMR Spectroscopy

Solution-state ^{1}H NMR seems to be the method of choice for rapid screening and characterization of WSOM in aerosols, bulk deposition, and fog water samples, as well as to quantify the distribution of different functional groups with C–H bonds.[14,17,19,20,23,30,37–39] In most of these studies, D$_2$O is the solvent of choice for the ^{1}H NMR measurements. However, the use of this solvent makes it impossible to determine carboxyl and hydroxyl

groups, because the hydrogen atoms of these groups rapidly exchange with a deuterium of D$_2$O rendering them invisible by ^{1}H NMR. DMSO-d_6 or methanol-d_4 (MeOH-d_4) can be used instead as solvents for the ^{1}H NMR analyses of atmospheric WSOM. The redissolution of the organic matter increases generally in the order of increasing polarity index of the solvent: D$_2$O > DMSO-d_6 > MeOH-d_4. Therefore, those researchers who decide to use an organic solvent may experience some difficulties in ensuring the complete dissolution of the WSOM sample and, consequently, may observe a loss of sensitivity of the ^{1}H NMR analyses.[23] Regardless of the solvent employed in the analyses, if the sample is not thoroughly dried, the dissolution of the WSOM will always produce a ^{1}H NMR spectrum with a clear HOD (deuterium hydrogen oxide) signal at $\delta_H = 4.8$ ppm, in addition to the organic solvent signal at $\delta_H = 2.5$ ppm/$\delta_C = 39.5$ ppm for DMSO-d_6, and $\delta_H = 3.31$ ppm/$\delta_C = 49.1$ ppm for MeOH-d_4 (the chemical shifts observed for the ^{13}C NMR spectrum are also indicated, as they are particularly important when implementing short- and long-range ^{1}H–^{13}C heteronuclear connectivity NMR experiments). To suppress the residual HOD peak, there are a multitude of NMR water suppression approaches available. Readers are referred to the review work of Simpson *et al.*[40] and references therein, which addresses different water suppression sequences on a ^{1}H NMR spectrum of a dissolved organic matter sample. However, when using DMSO-d_6 or MeOH-d_4, the additional organic solvent signal could still mask important ^{1}H NMR signals from functional groups with chemical shifts within the solvent signal area,[23,41] thus hindering any attempt to identify or even quantify those functional groups. The sensitivity achieved in an NMR spectrum can also be influenced by the magnetic field strength employed, and the number of scans accumulated in the spectrum (i.e., overall time of scanning).[42] There are a few studies on atmospheric WSOM that already employ high-field instrumentation (500–600 MHz) for achieving higher sensitivity and to unveil specific molecular assignments in WSOM, which are likely to be undistinguishable in low-field instruments (300–400 MHz). In either case, it is advisable to collect a sufficient number of scans (up to 1000 scans, depending on sample concentration) in order to obtain high-quality ^{1}H NMR spectra. Tagliavini *et al.*[42] estimated that 800 scans, corresponding to an overall time of scanning of about 55 min with a 600 MHz instrument, are adequate to detect levoglucosan at an air concentration of $0.04 \, \mu g \, m^{-3}$.

Despite the great effort employed in the acquisition of high-quality ^1H NMR spectra, their interpretation can be further hindered by the complex overlapping profile. A typical ^1H NMR spectrum of atmospheric WSOM displays broad bands (arising from structural assemblies of organic molecules), superimposed by a relatively small number of sharp peaks (arising from either small molecules or highly mobile segments of the structural assemblies) dispersed over a small chemical-shift range (0–10 ppm). Usually, only four main categories of functional groups carrying C–H bonds are identified in atmospheric WSOM: (i) Ar–H: aromatic protons (6.5–8.3 ppm); (ii) H–C–O: protons bound to oxygenated aliphatic carbons atoms, such as aliphatic alcohols, ethers, and esters (3.3–4.1 ppm); (iii) H–C–C=: protons bound to aliphatic carbon atoms adjacent to unsaturated groups, such as alkenes, carbonyl, imino, or aromatic groups (1.9–3.2 ppm); and (iv) H–C: aliphatic protons in extended alkyl chains (0.5–1.9 ppm). There are a few studies also reporting the presence of terminal CH_3 groups (0.5–1.1 ppm) and CH_2 in alkyl chains (1.1–1.9 ppm) in WSOM from cloud water[39] and bulk deposition,[43] as well as acetalic (O–CH–O) and vinylic (=C–H) protons (5.0–5.5 ppm) in WSOM from atmospheric aerosols.[17,42] Using quantitative integration of each spectral region, it has been concluded that protons in aliphatic structures are the dominant moieties in atmospheric WSOM, followed by oxygenated aliphatic compounds and unsaturated aliphatic groups (including H–C–C=O and H–C–C=C), and only a minor contribution from aromatic groups.[14,17,19,20,39,42,43] This overall set of data provided by ^1H NMR analysis has been applied for source apportionment, and seasonal and temporal characterizations of atmospheric WSOM. In 2007, Decesari *et al.*[16] pioneered the technique to source apportion organic aerosols based on the ^1H NMR signature and its NMR integrals. Since then, this method has been applied to establish a new fingerprint for aerosol WSOM from a highly urbanized and industrially influenced location,[17] as well as to assess the influence of season events and air-mass trajectories on the NMR signature of WSOM in rainwater.[20] The basics and disadvantages of this methodology are discussed further in Section 12.5.

Given the complex nature of atmospheric WSOM and its influence on the obtained NMR profile, the use of a chromatographic separation procedure before ^1H

NMR analysis is likely to have considerable promise for investigating the structural features of WSOM. Probably, the most important chromatographic technique employed in the chemistry of complex organic mixtures, such as those of natural organic matter, is size-exclusion chromatography (SEC). Besides reducing the heterogeneity of atmospheric WSOM and, thus, generating more homogeneous fractions, SEC hyphenated with NMR will allow the determination of size-distinguished fractions and how they differ in composition. Recently, Duarte and coworkers[43] developed a comprehensive two-dimensional liquid chromatographic protocol for resolving the chemical heterogeneity of natural organic matter. The authors concluded that the combination of two independent separation mechanisms is promising for extending the range of natural organic matter separation. These findings suggest that the synergy between a multidimensional separation protocol and NMR analysis may hold great promise for structural identification and molecular fingerprint of atmospheric WSOM.

12.4.2 Application of Two-dimensional Solution-state NMR Spectroscopy

Till date, only four studies have applied 2D solution-state NMR techniques to investigate atmospheric organic matter.[20,22–24] The combined use of the information provided by ^1H–^1H homonuclear (COSY and TOCSY) and ^1H–^{13}C heteronuclear (HSQC and HMBC) connectivities allows a higher spectral resolution and, therefore, greater detail on the C–H backbone of the substructures present in the complex atmospheric organic matter mixture. Atmospheric chemists who are not completely fluent in these sophisticated 2D NMR experiments may experience some difficulties in selecting the most appropriate 2D experiments for answering their specific question. Readers are referred to the excellent reviews of Simpson and Simpson,[21] and Simpson and coworkers[40] and references therein, where they can find suggestions as to the key 2D NMR experiments and pulse programs that may be useful for studying very complex natural mixtures. As for 1D solution-state ^1H NMR spectroscopy, here it is also important to consider (i) the amount of sample used (high sample concentrations could promote molecular aggregation), (ii) the sample solubility (the amount of dissolved organic carbon could be too low for the application of solution-state ^{13}C NMR techniques),[23] and (iii) the interference from the solvent signals. Of these three important considerations, the molecular size (or aggregates) of sample constituents is a key issue that can limit the acquisition and interpretation of 2D NMR spectra. Small molecules (or aggregates) often have long spin–spin relaxation times (T_2), which enables the detection of highly resolved 2D NMR spectra. On the other hand, if T_2 is very short, such as in the case of large molecules (or aggregates), the signal may be difficult or even impossible to detect. Taking into account that the molecular size distribution of atmospheric WSOM is at the lower molecular size range (<600 Da),[45] it is very likely that 2D NMR would be most effective for structural elucidation of these complex organic mixtures. For the interpretation of 2D NMR data of atmospheric organic matter, it is advisable to use NMR databases containing chemical-shift assignments (^1H and ^{13}C nuclei) of both complex natural organic mixtures (e.g., humic and fulvic acids) and standard organic compounds, which are both likely to be present in atmospheric organic matter.[23]

The studies reported so far in the literature for atmospheric organic matter share a common feature: they all exhibit high-quality 2D NMR spectra.[20,22–24] Simpson and coworkers[22] applied a combination of solid-state CPMAS ^{13}C NMR with 2D HR-MAS and solution-state NMR (TOCSY, HSQC) spectroscopy to provide a general overview of the structural components in atmospheric urban deposits. The authors concluded that these deposits represent a complex mixture of anthropogenically and biogenically derived materials, encompassing carbohydrates, various aliphatic groups (including acids, alcohols, alkanes, alkenes, and esters), polybutadiene, and functionalized polyaromatic species. Duarte and coworkers[23] combined solution-state COSY, HSQC, and HMBC 2D NMR techniques to deliver qualitative information on the substructures present in WSOM from fine atmospheric aerosols collected during winter and spring/summer seasons at a rural-coastal location with high agricultural activity. The authors concluded that the aliphatic material of both samples consists of long-chain (carbons >3 or 4) and branched mono- and dicarboxylic acids, carbonyl, and ester structural types. The presence of such structural fragments was associated to secondary organic aerosol formation. Spectral signatures typical of anhydrosugars from cellulose and methoxyphenols from lignin were also clearly identified among the carbohydrate and aromatic moieties of the winter sample. Their presence was linked

to the occurrence of wood-burning processes in domestic fireplaces during the colder period. In a recent review, Duarte and Duarte[14] further demonstrated how these 2D solution-state NMR methods combined with solid-state CPMAS ^{13}C NMR spectroscopy can offer great promise in unraveling the composition and structure of the whole aerosol WSOM sample. Besides a comprehensive description of the substructures present in the WSOM, their exercise clearly demonstrates that these high-resolution NMR techniques, when considered together, can be successfully used to identify molecular fingerprints of the different aerosol sources. Schmitt-Kopplin and coworkers[24] combined solution-state 2D NMR techniques (COSY, TOCSY, HSQC, HMBC, and distortionless enhancement by polarization transfer (DEPT)-HSQC) with high-resolution mass spectrometry to investigate the molecular signatures of the water-soluble fraction of secondary organic aerosols. The typical aliphatic chemical environment within the studied samples was heteroatom-substituted functional groups adjacent to highly branched aliphatics, likely in the form of strongly coupled fused alicyclic ring spin systems (e.g., terpenoid-like molecules). Aromatics were found to be highly substituted, and the presence of electron withdrawing groups and (O)NO$_x$ substitution was considerably more common than the presence of electron-donating oxygen-containing functional groups and neutral substitution (aliphatic carbon). The obtained solution-state 2D NMR dataset allowed the authors to improve the current knowledge on secondary organic aerosol formation by suggesting possible chemical reaction pathways involving CHO precursor molecules and sulfuric acid in gas-phase photoreactions. Using a COSY NMR experiment, Seaton and coworkers[20] were also able to identify correlations consistent with fatty acids and C5 or higher di-acids or ox-acids in WSOM from rainwater.

At this point, it is clear that the application of solution-state 2D NMR to atmospheric organic matter is still in its infancy. It should also be mentioned that the acquisition time required for collecting a complete solution-state 2D NMR dataset may require several days, thus suggesting that not all atmospheric chemists will consider these techniques as a routine tool for the characterization of atmospheric organic matter. However, the prospect of mining the molecular structural constituents of the complex atmospheric organic matter with possible identification of its sources and formation mechanisms is a truly exciting challenge that justifies the efforts for implementing such sophisticated NMR approaches.

12.5 THE QUEST FOR QUANTIFICATION AND SOURCE APPORTIONMENT

The enormous impact that NMR spectroscopy has made in the area of atmospheric organic matter research is a direct consequence of its utility for structural elucidation, as well as a quantitative tool. Notwithstanding these unique capabilities, the quantitative reliability of some of these techniques, namely of solid-state CPMAS ^{13}C NMR spectroscopy in the analysis of complex natural organic mixtures, has always been a topic of much debate. The CP process is known to be semiquantitative and it is usually recommended for comparisons between samples whose spectra have been obtained under similar NMR conditions.[21] To improve the quantitative reliability of solid-state CPMAS ^{13}C NMR spectra of complex organic mixtures, a number of different approaches have been recommended, usually through optimization of NMR parameters, such as the rate of MAS, contact time, and recycle delay, so that ^{13}C nuclei in all carbon functional groups will be equally detected. Readers are encouraged to consult the comprehensive reviews of Simpson and Simpson,[21] Cook,[33] and Conte and coworkers,[46] and references therein to obtain a more complete understanding of the complexities, issues, and recommended procedures to obtain the most possible quantitatively correct solid-state CPMAS ^{13}C NMR spectra of natural organic matter. Nevertheless, the readers should also consider the recent developments made on the use of advanced solid-state NMR editing techniques for the characterization of natural organic matter (see Section 12.3). These techniques offer new tools that enable one to identify and quantify the relative abundances of specific functional groups whose resonances are usually overlapped in the traditional solid-state CPMAS ^{13}C NMR technique.[35]

In the works published thus far focusing on solid-state CPMAS ^{13}C NMR analysis of atmospheric WSOM, MAS frequencies range from 5 to 10 kHz, contact times are generally between 1 and 5 ms, and recycle delay times range from 1 to 5 s. In order to ensure a satisfactory assessment of all carbon functional groups in a CPMAS ^{13}C NMR spectrum of atmospheric WSOM, it is advisable to use a MAS rate of 10 kHz in order to reduce the number and intensity of spinning-side bands arising from the aromatic and

carbonyl carbon moieties.[46] A contact time of 1.0 ms has been reported as giving the highest values of NMR peak area for all functional groups.[28] Regarding the most adequate recycle delay time, its settings depend on the nature of the sample under analysis and should be adjusted until the maximum spectral intensity is obtained. A recycle delay of 5 s has been reported as suitable for the acquisition CPMAS ^{13}C NMR spectrum of atmospheric WSOM.[28] To ensure complete relaxation between scans, the recycle delay should be set at five times the T_1 of protons in a particular sample. Even if all these NMR conditions are properly set, the reader should be aware that the CP technique may still underestimate ^{13}C nuclei remote from ^{1}H, such as those of quaternary aliphatic carbons, carbonyl groups, and condensed polycyclic aromatic structures, because of weak ^{1}H–^{13}C coupling. This outcome must be considered when interpreting the NMR data because the resulting lower signal of H-poor components may underestimate their relative abundance using the traditional integration approach. However, the obtained results, such as those shown in Figure 12.3, can be used to investigate the composition of different atmospheric WSOM samples, and to deliver new valuable information regarding their relative variation with changes in their sources, sampling locations (rural, urban, and pristine areas), season events, and meteorological conditions.

The substantial improvements made thus far in the acquisition of high-quality solution-state ^{1}H NMR spectra (see Section 12.4.1) enable the quantitative integration of the ^{1}H spectral regions, after correction for possible baseline drift and calibration to an internal standard [typically, 3-trimethylsilylpropane sulfonic acid (DSS), sodium 3-trimethylsilyl-2,2,3,3-d_4-propanoate (TSP-d_4), or dibromomethane (CH_2Br_2)]. These NMR integrals provide information on the relative abundance of the major ^{1}H functional groups present in atmospheric WSOM samples, being this important to define their chemical characteristics. Using the NMR integrals and ^{1}H NMR signatures, Decesari and coworkers[16] suggested a new approach for source attribution of WSOM from atmospheric aerosols. In this method, the measured hydrogen contents of the ^{1}H NMR functional groups are converted into organic carbon contents considering (i) the average stoichiometry of aliphatic groups (H/C ratios of 1.1 and 0.4 for oxygenated aliphatics (H–C–O) and aromatic functional groups, respectively) and (ii) the hypothesis of highly substituted aromatic rings. The obtained results are then used to estimate and plot two indexes related to the aliphatic composition: (i) the aliphatic carbon fraction accounted for by carbonylic/carboxylic groups (H–C–C=O/total aliphatic carbon) versus (ii) the fraction of aliphatic carbon accounted for by hydroxyl groups (O–C–H/total aliphatic carbon). The total amount of aliphatic carbonyl and carboxylic (H–C–C=O) groups is indirectly estimated from the amount of hydrogen atoms adjacent to unsaturated carbon atoms (H–C–C=), after subtraction of the contribution from aromatic protons. The total aliphatic carbon includes the saturated (H–C–O) and unsaturated (H–C–C=O) oxygenated functional groups, nonoxygenated groups [i.e., the benzylic groups (H–C–Ar)], and the unfunctionalized alkyls (H–C). By applying this procedure to aerosol WSOM samples from multiple sites and influences for which a source apportionment has already been performed (namely, marine organic aerosols, secondary organic aerosols, and biomass-burning aerosols), Decesari and coworkers[16] were able to delineate and plot different source areas on the basis of their ¹H functional group distributions, as schematically illustrated in Figure 12.4. Source attribution for the new samples (e.g., urban organic aerosols[17]) is then performed by comparing their ¹H functional group distributions with the NMR fingerprints of the sources of WSOM previously established. Atmospheric researchers should be aware, however, that solution-state ¹H NMR spectroscopy has low sensitivity for detecting functional groups that do not carry protons (e.g., substituted aromatic compounds) or contain acidic functions with rapidly exchangeable protons (e.g., carboxylic acids).[33] In fact, Decesari and coworkers[16] reported a reconstructed carbon content of 86% of the total water-soluble organic carbon measured by a total organic carbon analysis method, which confirms the assumption that some structures of the WSOM samples remain unidentified through solution-state ¹H NMR spectroscopy. Indeed, the most highly arguable aspect of the method developed by them is the indirect estimation of the H–C–C=O groups. To overcome this problem and to achieve a more direct determination of carbonyl and carboxylic functional groups, Tagliavini *et al.*[42] and Moretti *et al.*[47] suggested a time-consuming chemical derivatization method coupled to offline solution-state ¹H NMR analysis. Although progress has been made using these indirect protocols for elaborating empirical functional group composition models of atmospheric WSOM and discussion of its possible sources, it is questionable

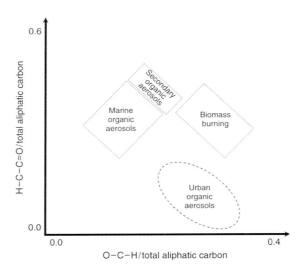

Figure 12.4. Schematic representation of functional group distribution for urban atmospheric aerosol WSOM samples[17] based on the analysis of Decesari and coworkers[16] with specific well-established aerosol sources (marine organic aerosols, secondary organic aerosols, and biomass-burning organic aerosols).

whether the obtained results truly represent the chemical nature of these samples, particularly because a highly variable fraction (14–40%) of the water-soluble organic carbon still eludes solution-state ¹H NMR detection.[16,42,47]

12.6 CONCLUSIONS

Undoubtedly, NMR spectroscopy is an indispensable tool in atmospheric organic matter research. Perhaps, the main disadvantage of using this advanced offline technique is the long time resolution required for obtaining sufficient amount of sample, with the consequent loss of information regarding the very own atmospheric variability of the organic matter composition. However, the structural identification, of the type described here for its WSOM component, is definitely key observation that assists those interested in assessing the sources, formation pathways, seasonal, and regional characterization of atmospheric organic matter. The semiquantitative solid-state CPMAS ¹³C NMR spectroscopy has the potential for looking at the relative changes in carbon distribution across different atmospheric WSOM samples, as long as their NMR spectra are acquired under similar experimental conditions. The recent improvements made in the use of advanced spectral editing techniques for solid-state NMR have revealed the promise to provide deeper insights into the molecular connectivity and functional group identification and quantification of complex natural organic mixtures. These sophisticated techniques have yet to be applied to atmospheric WSOM, but their application will greatly contribute to the identification and quantification of specific functional groups (including those containing nitrogen), which are otherwise hidden in conventional solid-state CPMAS ^{13}C NMR spectra. The lack of analytical expertise in the use of these solid-state NMR spectral editing techniques, particularly in selecting the most appropriate pulse sequence, as well as handling and interpretation of the acquired NMR data are major challenges that prevent the application of these sophisticated solid-state techniques for studies of atmospheric WSOM.

While the semiquantitative solid-state CPMAS ^{13}C NMR gives an indication of the overall carbon composition, the level of structural detail revealed in the studies reported in this review is only possible through high-resolution solution-state NMR spectroscopy. The vast dataset provided by the combined application of 2D NMR techniques (e.g., COSY, HSQC, and HMBC) enables the visualization of the H–C backbone of molecules, and, therefore, it allows for detailed structural assignments of components within atmospheric WSOM. The synergistic application of both solid- and solution-state 2D NMRs is an invaluable strategy for the study of WSOM composition and structure, as well as for identifying molecular fingerprints related to the sources and formation mechanisms of atmospheric WSOM. In both solid- and solution-state NMRs, it is critical to make the atmospheric WSOM samples amenable to analysis to enhance the resolution capability of these techniques. The implementation of a robust and standardized analytical methodology for WSOM extraction and isolation is obviously a prerequisite to realize the full potential of these NMR techniques. Additionally, a few words of concern are important in regard to the sample solubilization and the solvent employed for solution-state NMR studies.

Of course, NMR spectroscopy cannot stand alone as the single tool for all structural queries within atmospheric WSOM. The combination of high-resolution NMR with other advanced techniques (e.g., high-resolution mass spectrometry) will enable unequivocal elemental and composition assignment of structural components within the complex atmospheric WSOM. Future research should also consider the chromatographic separation of WSOM before NMR analysis. It is clear that chemical complexity of the atmospheric WSOM limits the quality of NMR data, thus making their interpretation difficult. Reducing the heterogeneity of WSOM, using a separation protocol based on polarity, molecular weight/size, or even better the combination of two independent separation mechanisms, before offline NMR characterization, can be of particular value for targeting unique molecular structures within the complex atmospheric WSOM.

With the current availability of a huge number of pulse sequences and the increasingly sophisticated features of NMR spectral editing techniques, there is no doubt that NMR spectroscopy will continue to play a key role in structural studies of atmospheric WSOM. Although never applied to the structural investigations of atmospheric WSOM, microcoil and/or cryogenic probe technology also presents itself as a very valuable and promising tool to enhance sensitivity and improve the signal-to-noise ratio of solution-state NMR measurements of the 'mass limited' atmospheric WSOM. As these sophisticated tools evolve, newer and deeper understanding of the structural composition of atmospheric WSOM becomes readily obtained, which will allow suggesting fruitful directions for assessing the effects of atmospheric organic matter in the climate system, atmospheric processes, and human health.

ACKNOWLEDGMENTS

Centre for Environmental and Marine Studies (PEsT-c/MAR/LA0017/2011, University of Aveiro, Portugal) and the Portuguese Science and Technology Foundation, through the European Social Fund (ESF) and 'Programa Operacional Potencial Humano–POPH', are acknowledged for financial support. This work was also funded by FEDER under the Operational Program for Competitiveness Factors – COMPETE and by National funds via FCT within the framework of research project ORGANOSOL (FCOMP-01-0124-FEDER-019913; PTDC/CTE-ATM/118551/2010).

RELATED ARTICLES IN EMAGRES

Environmental Comprehensive Multiphase NMR

Environmental NMR: High-resolution Magic-angle Spinning

REFERENCES

1. P. Forster, V. Ramaswamy, P. Artaxo, T. Berntsen, R. Betts, D. W. Fahey, J. Haywood, J. Lean, D. C. Lowe, G. Myhre, J. Nganga, R. Prinn, G. Raga, M. Schulz, and R. Van Dorland, in 'Climate Change 2007: The Physical Science Basis', Contribution of Working Group I to the Fourth Assessment Report of the Intergovernmental Panel on Climate Change, eds S. Solomon, D. Qin, M. Manning, Z. Chen, M. Marquis, K. B. Averyt, M. Tignor, and H. L. Miller, Cambridge University Press: Cambridge, UK; New York, NY, 2007, Chapter 2.

2. A. C. Duarte and R. M. B. O. Duarte, in Biophysico-Chemical Processes Involving Natural Nonliving Organic Matter in Environmental Systems, eds N. Senesi, B. Xing and P. M. Huang, John Wiley & Sons: Hoboken, NJ, 2009, Chapter 12.

3. N. Mladenov, I. Reche, F. J. Olmo, H. Lyamani, and L. Alados-Arboledas, *J. Geophys. Res.*, 2010, **115**, G00F11. DOI: 10.1029/2009JG000991.

4. L. T. Padró, D. Tkacik, T. Lathem, C. J. Hennigan, A. P. Sullivan, R. J. Weber, L. G. Huey, and A. Nenes, *J. Geophys. Res.*, 2010, **115**, D09204. DOI: 10.1029/2009JD013195.

5. E. O. Fors, J. Rissler, A. Massling, B. Svenningsson, M. O. Andreae, U. Dusek, G. P. Frank, A. Hoffer, M. Bilde, G. Kiss, S. Janitsek, S. Henning, M. C. Facchini, S. Decesari, and E. Swietlicki, *Atmos. Chem. Phys.*, 2010, **10**, 5625.

6. E. Jurado, J. Dachs, C. M. Duarte, and R. Simó, *Atmos. Environ.*, 2008, **42**, 7931.

7. R. J. Kieber, R. F. Whitehead, S. N. Reid, J. D. Willey, and P. J. Seaton, *J. Atmos. Chem.*, 2006, **54**, 21.

8. S. Decesari, M. C. Facchini, M. Mircea, F. Cavalli, and S. Fuzzi, *J. Geophys. Res.*, 2003, **108**, 4685. DOI: 10.1029/2003JD003566.

9. S. Decesari, M. C. Facchini, S. Fuzzi, G. B. McFiggans, H. Coe, and K. N. Bower, *Atmos. Environ.*, 2005, **39**, 211.

10. R. J. Kieber, S. A. Skrabal, B. Smith, and J. D. Willey, *Environ. Sci. Technol.*, 2005, **39**, 1576.

11. J. D. Blando and B. J. Turpin, *Atmos. Environ.*, 2000, **34**, 1623.

12. M. Hallquist, J. C. Wenger, U. Baltensperger, Y. Rudich, D. Simpson, M. Claeys, J. Dommen, N. M. Donahue, C. George, A. H. Goldstein, J. F. Hamilton, H. Herrmann, T. Hoffmann, Y. Iinuma, M. Jang, M. E. Jenkin, J. L. Jimenez, A. Kiendler-Scharr, W. Maenhaut, G. McFiggans, Th. F. Mentel, A. Monod, A. S. H. Prévôt, J. H. Seinfeld, J. D. Surratt, R. Szmigielski, and J. Wildt, *Atmos. Chem. Phys.*, 2009, **9**, 5155.

13. J. L. Jimenez, M. R. Canagaratna, N. M. Donahue, A. S. H. Prevot, Q. Zhang, J. H. Kroll, P. F. DeCarlo, J. D. Allan, H. Coe, N. L. Ng, A. C. Aiken, K. S. Docherty, I. M. Ulbrich, A. P. Grieshop, A. L. Robinson, J. Duplissy, J. D. Smith, K. R. Wilson, V. A. Lanz, C. Hueglin, Y. L. Sun, J. Tian, A. Laaksonen, T. Raatikainen, J. Rautiainen, P. Vaattovaara, M. Ehn, M. Kulmala, J. M. Tomlinson, D. R. Collins, M. J. Cubison, E. J. Dunlea, J. A. Huffman, T. B. Onasch, M. R. Alfarra, P. I. Williams, K. Bower, Y. Kondo, J. Schneider, F. Drewnick, S. Borrmann, S. Weimer, K. Demerjian, D. Salcedo, L. Cottrell, R. Griffin, A. Takami, T. Miyoshi, S. Hatakeyama, A. Shimono, J. Y. Sun, Y. M. Zhang, K. Dzepina, J. R. Kimmel, D. Sueper, J. T. Jayne, S. C. Herndon, A. M. Trimborn, L. R. Williams, E. C. Wood, A. M. Middlebrook, C. E. Kolb, U. Baltensperger, and D. R. Worsnop, *Science*, 2009, **326**, 1525.

14. R. M. B. O. Duarte and A. C. Duarte, *Trends Anal. Chem.*, 2011, **10**, 1659.

15. S. Fuzzi, S. Decesari, M. C. Facchini, E. Matta, M. Mircea, and E. Tagliavini, *Geophys. Res. Lett.*, 2001, **20**, 4079.

16. S. Decesari, M. Mircea, F. Cavalli, S. Fuzzi, F. Moretti, E. Tagliavini, and M. C. Facchini, *Environ. Sci. Technol.*, 2007, **41**, 2479.

17. M. J. Cleveland, L. D. Ziemba, R. J. Griffin, J. E. Dibb, C. H. Anderson, B. Lefer, and B. Rappenglück, *Atmos. Environ.*, 2012, **54**, 511.

18. R. M. B. O. Duarte, E. B. H. Santos, C. A. Pio, and A. C. Duarte, *Atmos. Environ.*, 2007, **41**, 8100.

19. P. S. M. Santos, E. B. H. Santos, and A. C. Duarte, *Sci. Total Environ.*, 2012, **426**, 172.

20. P. J. Seaton, R. J. Kieber, J. D. Willey, G. B. Avery Jr., and J. L. Dixon, *Atmos. Environ.*, 2013, **65**, 52.

21. A. J. Simpson and M. J. Simpson, in Biophysico-Chemical Processes Involving Natural Nonliving Organic Matter in Environmental Systems, eds N. Senesi, B. Xing and P. M. Huang, John Wiley & Sons: Hoboken, NJ, 2009 Chapter 15.

22. A. J. Simpson, B. Lam, M. L. Diamond, D. J. Donaldson, B. A. Lefebvre, A. Q. Moser, A. J. Williams, N. I. Larin, and M. P. Kvasha, *Chemosphere*, 2006, **63**, 142.

23. R. M. B. O. Duarte, A. M. S. Silva, and A. C. Duarte, *Environ. Sci. Technol.*, 2008, **42**, 8224.

24. P. Schmitt-Kopplin, A. Gelencsér, E. Dabek-Zlotorzynska, G. Kiss, N. Hertkorn, M. Harir, Y. Hong, and I. Gebefügi, *Anal. Chem.*, 2010, **82**, 8017.

25. S. Fuzzi, M. O. Andreae, B. J. Huebert, M. Kulmala, T. C. Bond, M. Boy, S. J. Doherty, A. Guenther, M. Kanakidou, K. Kawamura, V.-M. Kerminen, U. Lohmann, L. M. Russell, and U. Pöschl, *Atmos. Chem. Phys.*, 2006, **6**, 2017.

26. Y. Subbalakshmi, A. F. Patti, G. S. H. Lee, and M. A. Hooper, *J. Environ. Monit.*, 2000, **2**, 561.

27. R. M. B. O. Duarte, C. A. Pio, and A. C. Duarte, *Anal. Chim. Acta*, 2005, **530**, 7.

28. P. Sannigrahi, A. P. Sullivan, R. J. Weber, and E. D. Ingall, *Environ. Sci. Technol.*, 2006, **40**, 666.

29. P. Herckes, J. A. Leenheer, and J. L. Collett, *Environ. Sci. Technol.*, 2007, **41**, 393.

30. E. R. Graber and Y. Rudich, *Atmos. Chem. Phys.*, 2006, **6**, 729.

31. L. R. Mazzoleni, B. M. Ehrmann, X. Shen, A. G. Marshall, and J. L.Collett Jr., *Environ. Sci. Technol.*, 2010, **44**, 3690.

32. P. S. M. Santos, M. Otero, O. M. S. Filipe, E. B. H. Santos, and A. C. Duarte, *Talanta*, 2010, **83**, 505.

33. R. L. Cook, *Anal. Bioanal. Chem.*, 2004, **378**, 1484.

34. G. Abbt-Braun, U. Lankes, and F. H. Frimmel, *Aquat. Sci.*, 2004, **66**, 151.

35. J. Mao, N. Chen, and X. Cao, *Organ. Geochem.*, 2011, **42**, 891.

36. I. Kögel-Knabner, *Geoderma*, 1997, **80**, 243–270.

37. N. Havers, P. Burba, J. Lambert, and D. Klockow, *J. Atmos. Chem.*, 1998, **29**, 45.

38. Y. Suzuki, M. Kawakami, and K. Akasaka, *Environ. Sci. Technol.*, 2001, **35**, 2656.

39. G. J. Reyes-Rodríguez, A. Gioda, O. L. Mayol-Bracero, and J.Collett Jr., *Atmos. Environ.*, 2009, **43**, 4171.

40. A. J. Simpson, D. J. McNally, and M. J. Simpson, *Prog. Nucl. Magn. Reson. Spectrosc.*, 2011, **58**, 97.

41. V. Samburova, T. Didenko, E. Kunenkov, C. Emmenegger, R. Zenobia, and M. Kalberer, *Atmos. Environ.*, 2007, **41**, 4703.

42. E. Tagliavini, F. Moretti, S. Decesari, M. C. Facchini, S. Fuzzi, and W. Maenhaut, *Atmos. Chem. Phys.*, 2006, **6**, 1003.

43. C. Miller, K. G. Gordon, R. J. Kieber, J. D. Willey, and P. J. Seaton, *Atmos. Environ.*, 2009, **43**, 2497.

44. R. M. B. O. Duarte, A. C. Barros, and A. C. Duarte, *J. Chromatogr. A*, 2012, **1249**, 138.

45. R. M. B. O. Duarte and A. C. Duarte, *Anal. Chim. Acta*, 2011, **688**, 90.

46. P. Conte, R. Spaccini, and A. Piccolo, *Prog. Nucl. Magn. Reson. Spectrosc.*, 2004, **44**, 215.

47. F. Moretti, E. Tagliavini, S. Decesari, M. C. Facchini, M. Rinaldi, and S. Fuzzi, *Environ. Sci. Technol.*, 2008, **42**, 4844.

Chapter 13
Soil Organic Matter

Anne E. Berns[1] and Heike Knicker[2]

[1]*Forschungszentrum Jülich GmbH, Jülich 52425, Germany*
[2]*Department of Geoecology and Biogeochemistry – Organic Biogeochemistry, Instituto de Recursos Naturales y Agrobiología de Sevilla (IRNAS-CSIC), Sevilla 41012, Spain*

13.1 INTRODUCTION

Over the past three decades, solid- and solution-state NMR spectroscopies have become an indispensable tool in soil science. This chapter focuses on the contribution of these techniques to the understanding of the structure and dynamics of soil organic matter (SOM). Therefore, we present the changing views of SOM with time. This chapter starts with a short history on the developments of NMR spectroscopy to set a rough time frame of the availability of the different NMR technologies. It then shortly summarizes views before the first NMR spectra of SOM were recorded and continues stepwise through the decades pointing out the gain in knowledge achieved by this technique.

NMR Spectroscopy: A Versatile Tool for Environmental Research
Edited by Myrna J. Simpson and André J. Simpson
© 2014 John Wiley & Sons, Ltd. ISBN: 978-1-118-61647-5

13.2 A FEW MILESTONES IN THE HISTORY OF NUCLEAR MAGNETIC RESONANCE SPECTROSCOPY

The birth year of NMR spectroscopy is commonly considered to be 1946, when the groups of Felix Bloch[1–3] and Edward Purcell[4,5] independently recorded NMR signals of protons at room temperature in ordinary liquid (water) and solid (paraffin) materials, respectively. Unlike earlier researchers, both groups were aware of relaxation processes and avoided saturation, a necessary prerequisite for the acquisition of signals. Bloch and coworkers placed a spherical water sample in a strong static magnetic field (z-direction), surrounded by a transmitter coil (x-direction) and a receiver coil (y-direction), and observed the oscillating voltage in the receiver coil induced by the applied oscillating radiofrequency (RF) field in the transmitter coil. Although NMR spectrometers became very sophisticated over time, the basic arrangement of the classical NMR experiment essentially has not changed much since then.

In 1949/1950, Hahn[6,7] presented pulsed NMR experiments, allowing the accurate determination of relaxation times through observation of the free evolution of the Larmor precession after application of short RF pulses. While originally, with the continuous wave (CW) technique, each frequency was excited independently, the development of more powerful computers in the 1960s enabled the use of pulsed Fourier transformation NMR spectroscopy presented by Ernst and Anderson,[8] where a frequency band was exploited with one short pulse. By the end of the 1950s, the main principles of nuclear magnetism were already clarified and explained by quantum mechanical approaches.[9]

In 1958, Andrew and Newing[10] suggested to reduce the anisotropic dipole–dipole interactions, responsible for the broad linewidths in spectra of solids, by their very fast rotation. In 1959, Lowe[11] found that the broad signal of CaF_2 could be split into a much narrower center line with satellite signals at a distance of the spinning frequency or multiples thereof, when the sample was spun at 54.7°, the so-called magic-angle. This technique soon became indispensable for the recording of high-resolution NMR spectra in the solid state.

In 1966, Ernst[12] introduced broadband decoupling as a mean to eliminate signal splittings caused by heteronuclear couplings, which simplified complicated high-resolution spectra. In 1962, Hartmann and Hahn[13] presented the cross polarization (CP) technique to transfer magnetization from abundant protons to less sensitive and dilute nuclei such as ^{13}C and ^{15}N. Later, this technique was combined with high proton decoupling by Pines *et al.*[14,15] This resulted not only in an increase in the sensitivity of the experiment by a factor determined by the ratio of the gyromagnetic ratios of the involved nuclei but also in a considerable decrease in the measurement time. The latter was possible because the short spin–lattice relaxation times of protons rather than those of the dilute spins had to be considered to avoid saturation. In 1976, Schaefer and Stejskal[16,17] combined the CP technique with magic-angle spinning (MAS), which nowadays represents the routine experiment in solid-state NMR (see Chapter 4). In the same year, Ernst *et al.*[18] published an essay on two-dimensional (2D) NMR spectroscopy, the experimental application of which was yet constrained by the incapacity of the spectrometer to deal with large data sets. Nevertheless, this approach opened the door to a new class of experiments with huge possibilities. With the development of superconducting magnets in the mid-1950s,[19] at the end of the 1970s, all the essential requirements for the successful application of solid-state NMR to SOM were in place.

13.3 CHANGING PERCEPTION OF SOIL ORGANIC MATTER THROUGH TIME

13.3.1 Pre-NMR Times

An early and very thorough review on the nature of SOM or soil 'humus' was presented by Waksman in 1926.[20–24] Apart from a complete lack of uniformity in terminology, a number of controversial hypotheses on the origin and nature of SOM prevailed. It was believed that humic acids (HAs), extracted from the soil with strong base and precipitated by strong acid, consisted of a few definite chemical compounds and that HAs originating from different soils shared the same chemical composition. Nitrogen found in this precipitate was considered as an impurity or as being absorbed by HAs from the atmosphere. However, various researches already viewed SOM as it is largely seen today, a decaying mass of residues from plants, animals, and microorganisms being composed of a large number of compounds belonging to different but distinct classes. The study of SOM at these times was dominated by the examination of humic substances extracted from soils. Although this isolation, based on their solubility in aqueous solutions at different pH values, was known to be far from perfect, it was considered to be the best available method at the time.

One of the favored theories of the formation of humus was its genesis from lignin. This was based on similarity in physical and certain chemical properties of lignin and HAs and the assumption that lignin is not easily degraded by microorganisms and, hence, could accumulate in soil. However, the theory failed to explain the nitrogen content of humus. At the end of the 1930s, the ligno-protein theory was introduced to fill this gap,[25] explaining humus formation by accumulating lignin that reacts with nitrogenous compounds such as proteins derived from microbial activities. In 1951 and 1954, Bremner[25,26] published two reviews on SOM. Although the ligno-protein theory was still popular, a number of researchers had shown that after entering the soil, lignin is altered and Flaig suggested in 1964[27] that with ongoing decomposition of the lignin, its side chains are degraded and demethylation occurs, which reduces the linking of the monomers. The latter repolymerize by reacting with nitrogen (polyphenol theory). Nitrogen was, hence, viewed as an integral part of SOM, with heterocyclic nitrogen playing a major role.[28]

13.3.2 Modern Times: 1970s – The First NMR Spectra of Soil Components

13.3.2.1 The Situation at the Beginning of the Decade

In 1974, the 50th anniversary of the International Society of Soil Science, HAs and fulvic acids (FAs) were considered as complex polymers or heteropoly-condensates containing phenolic and quinone units, which originated from lignin or other substances.[29] Amino acids, peptides, and other organic constituents were believed to be linked – mostly by adsorption – to this polymer framework. FAs were believed to have smaller ranges of molecular weights and to contain more polysaccharides than HAs.[30] It was generally agreed that no two molecules of HA or FA were identical. On the nitrogen front, the hypothesis of a continuous internal nitrogen cycle was formulated.[29] The fact that a part of the nitrogen was sequestered from cycling was explained by its incorporation into the humic network after ammonification of quinones and subsequent further transformation into hete-rocyclic aromatic N. Alternatively, the formation of refractory soil organic N was explained by the so-called Maillard reaction,[31,32] during which car-bonyl groups react with amino groups to form yellow to dark-brown melanoidins containing among others N-heterocyclic aromatic structures.[33,34]

Application of radioisotopes, spectrofluorometry, ultraviolet, visible, and IR spectroscopy, electron spin resonance, and X-ray analysis were new techniques at the hands of soil scientists during this period. However, NMR spectroscopy was the emerging star among them.

13.3.2.2 The NMR Contribution

Although the very first solution-state ^{1}H NMR data of methylated FAs were already published in 1963 by Barton and Schnitzer,[35] a decade passed before this expensive technique was applied by others. Ishiwatari,[36] Neyroud and Schnitzer,[37] and Grant[38] were among the first to publish solution-state ^{1}H and ^{13}C NMR spectra of HAs or FAs. They were recorded on either methylated humic substances or organic extracts to increase the solubility in deuterated organic solvents for solution-state ^{1}H NMR. Both CW ^{1}H and Fourier transform (FT) ^{13}C NMR spectra revealed the presence of aliphatic carbon, thus far believed to be a minor component of humic substances. They, however, failed to confirm the importance of aromatic

structures. This was attributed to either low content of aromatic protons, caused by high substitution, or ex-tremely fast relaxation, owing to unpaired electrons in stable aromatic radicals. The first well-resolved solution-state FT ^{13}C spectra of HAs and FAs were published by Gonzalez-Vila *et al.*[39] They specifically avoided solution-state ^{1}H NMR methods as H/D ex-change between the analyte and the deuterated sol-vents limited the recording of the spectra to nonex-changeable protons. Their solution-state ^{13}C NMR spectra exhibited the expected aromatic structures as well as aliphatic structures (Figure 13.1). However, as

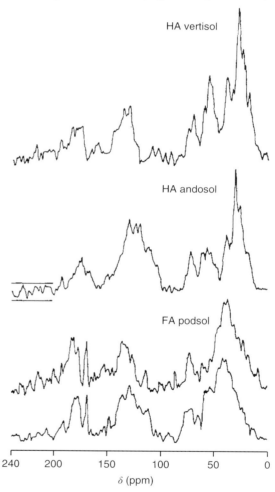

Figure 13.1. Solution-state FT ^{13}C NMR spectra of humic acids (HAs) and fulvic acid (FA) extracted from different soils. (Reprinted with permission from Ref. 39. © 1976 Elsevier)

different relaxation behavior of rigid and mobile regions inside the complex humic molecules may have caused saturation, quantification of the spectra was not performed. The first solution-state FT ^{1}H spectrum of humic material in aqueous NaOH was recorded by Wilson *et al.*[40] and still contained a prominent water signal at around 5 ppm, but the aliphatic and aromatic regions were sufficiently well resolved for a detailed signal assignment. On the basis of the low contribution of aromatic protons and the high contribution of aromatic and carboxylic carbon detected by ^{13}C NMR spectroscopy, the authors assigned the aromatic signals to either polycyclics or heavily substituted monocyclics. Furthermore, a large proportion of the aliphatic protons were found to be associated with carbon bound to oxygen, indicating sugarlike, polyhydroxy, or polyether components.

13.3.2.3 Gain in Knowledge

The most important insight from the first NMR spectra was that aliphatic structures were a more important building block of humic substances than formerly assumed and that the highly aromatic polyphenol theory had to be reviewed.

13.3.3 Modern Times: 1980s – The First In Situ Measurements in the Solid State

13.3.3.1 The NMR Contribution

Solution-state NMR spectroscopy requires that samples are fully dissolved, which results in some disadvantages with respect to SOM analysis. Apart from the problem of analyzing only the fully soluble portion of the organic matter (OM), oxidation reactions may alter the humic substances before analysis. Furthermore, solvent effects, especially in solution-state ^{1}H NMR, may produce artifacts. For a detailed account on solution-state NMR on dissolved OM, the reader is kindly referred to Chapters 4 and 11. Solid-state NMR techniques are nondestructive and allow SOM analysis without the need of prior extraction, provided that carbon content is high enough and that the presence of paramagnetic iron is low (see Chapter 4).

Barron *et al.*[41] published first solid-state ^{13}C NMR spectra on bulk soils in 1980. However, as sample spinning was not applied, the low resolution resulted in the observation of only an aliphatic and an aromatic region. Nevertheless, the variable nature of the SOM

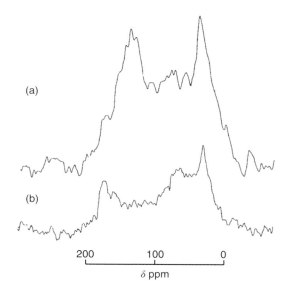

Figure 13.2. Solid-state ^{13}C NMR spectra of (a) Hokitika soil and (b) Tirau soil. (Reprinted with permission from Ref. 43. © 1981 Nature Publishing Group)

of different sources from highly aliphatic to moderately aromatic was revealed. A year later, Barron and Wilson[42] and Wilson *et al.*[43] published better resolved solid-state ^{13}C CPMAS NMR spectra of bulk soils, confirming their high variability (Figure 13.2). Using a 300 MHz instrument, they faced challenges with the chemical shift anisotropy that was only partly removed because of insufficient high sample spinning. This causes echoes, the so-called spinning side bands, which appear with a distance of the spinning frequency at both sides of the parent signal and can obscure signal assignment (see Chapter 4). With the customary achievable spinning speeds of the time, ^{13}C CPMAS NMR spectra obtained with 100 MHz spectrometer were, hence, easier to interpret. One of the most comprehensive ^{13}C CPMAS NMR studies with over 20 different samples of sedimentary deposits, peat, and humic isolates from soils was published by Hatcher *et al.*[44]

In 1982, Preston and Ripmeester[45] applied solid-state ^{13}C NMR spectroscopy to four soils and their humic substances and included one of the first dipolar dephasing (DD) experiments in soil science highlighting nonprotonated carbons. The HA and humin fractions had carbon distributions comparable to those of the respective soils, but the FAs were dominated by carbohydrates. Plotting results of DD experiments as a pseudo-2D spectrum, in which one coordinate gives the chemical shift information and the other the magnitude of the dipolar interaction, Wilson[46] was able to distinguish highly substituted, condensed from protonated aromatic carbons. In the same year, Preston *et al.*[47] described the impact of copper on the CP dynamics in organic soils. Although a contact time of 1 ms was recommended for recording quantitative NMR spectra, they advised to still be careful if soils rich in paramagnetic metals are examined. They revealed that paramagnetic metals can selectively decrease spin–lattice relaxation time in the rotating frame ($T_{1\rho}^{H}$) of protons associated with carbohydrates because of the preferential interaction of copper with the hydroxyl groups. In 1985, Wilson *et al.*[48] examined humus samples from ornithogenic soils of the Antarctic with standard CP and DD solid-state experiments and demonstrated that lignin is not a prerequisite for aromatic structures in soil. Two years later, using DD experiments on lignin samples, Hatcher[49] indicated that this technique can be used to estimate the degree of substitution on the phenyl-propane structural units. It became quickly clear that well-designed pulse sequences allowed the extraction of more information from soil samples than the mere bulk chemical composition.[50,51]

While ^{13}C NMR spectroscopy turned out to be a powerful alternative to more traditional analytical techniques, ^{15}N NMR spectroscopy was widely unused, mostly because it poses particular problems that are not encountered in ^{13}C or ^{1}H NMR spectroscopy. Owing to the quadrupolar properties of the most abundant ^{14}N isotope, it is impossible to acquire well-resolved ^{14}N spectra. The low natural abundance of 0.37% and the low and negative gyromagnetic ratio of the ^{15}N isotope equates to an NMR sensitivity of approximately 50 times lower than that of ^{13}C. These difficulties are even more pronounced for soils, as N concentrations are usually lower than 1%. Thus, the early applications of ^{15}N NMR spectroscopy in soil science were performed with ^{15}N-enriched materials.

The first ^{15}N CPMAS NMR spectrum of an ^{15}N-labeled HA was published in 1983 by Benzing-Purdie *et al.*[52] demonstrating with the additional use of ^{13}C NMR its close relationship with synthetic melanoidins. Secondary amides were identified as the main nitrogen form and the presence of sterically hindered secondary amides, which are very resistant to acid hydrolysis, was suggested to partly explain the difficulties encountered in the elucidation of nitrogen structures in soil. In a further study,[53]

investigating the fate of ^{15}N-labeled glycine in sterile and nonsterile peat samples after 6 months, the authors detected aliphatic amines and secondary amide structures in varying intensities in all samples. In the nonsterile samples, their presence was assumed to result from microbial activity, extracellular enzymatic activity, or chemical reaction. In the sterile samples, they were explained as derived from the Maillard reaction.

13.3.3.2 Gain in Knowledge

By the end of the decade, solid-state NMR spectroscopy had demonstrated its usefulness in soil science. Along with pyrolysis–gas chromatography/mass spectrometry (GC/MS), it was one of the few methods that allowed analysis of bulk soil samples. The solid-state ^{13}C CPMAS NMR spectra confirmed the results obtained using solution-state NMR methods on humic substances for the bulk soil, namely that aliphatic structures were as abundant, or even more abundant, than aromatic components. The application of ^{15}N CPMAS NMR spectroscopy, however, generated unexpected results. While the presence of secondary amides and some aliphatic amines was not surprising, the massive dominance of the first in nonsterile samples did raise questions, as pyrolysis–GC/MS data seemed to support the presence of heterocyclic structures. The CP technique was suspected to underestimate structures, either because recording parameters were wrongly set or because high contents of paramagnetic compounds could obscure the intensity distribution because of selective suppression of signals. As pyrolysis–GC/MS was prone to produce heterocyclics from amino acids during pyrolysis, the highly aromatic structure and the involvement of heterocyclic nitrogen remained a matter of debate.

13.3.4 Modern Times: 1990s – The First Natural Abundance ^{15}N CPMAS NMR Spectra

13.3.4.1 Exemplary Views of SOM at the Beginning of the Decade

In the first half of the decade, several models or concepts on HAs and SOM structure were developed and discussed. The model proposed by Schulten and Schnitzer[54] was based on pyrolysis–GC/MS, pyrolysis–field ionization mass spectrometry (FIMS), ^{13}C NMR, chemical, oxidative, and reductive degradation, colloid-chemical, and electron microscope data and consisted of a carbon skeleton with alkylbenzene structures as major components associated with carboxyl, phenolic and hydroxyl, carboxylic ester, and ether groups. Nitrogen was integrated as heterocyclics and nitriles. To match the elemental composition of natural HAs, the authors assumed that 10% carbohydrates and 10% proteinaceous materials were either adsorbed on or bonded covalently to the HAs.

On the basis of NMR and chemical degradation studies, Wershaw[55] proposed a membrane-micelle model, where amphiphilic molecules, originating from plant degradation products, and lignin–carbohydrate complexes interacted with mineral grains and formed membrane-like coatings. The author considered these coatings to constitute the humus in soils. In addition, plant degradation products were also assumed to exist as micelles and monomeric units in solution. A comparable concept, but focusing rather on physical aspects than on the chemical composition, was presented by Ladd *et al.*[56] after reviewing a large set of literature supporting the concept of physical protection of substrate (SOM) and microbes. The association of SOM with clay particles conferred stability even to otherwise easily decomposable substrates.

13.3.4.2 *^{13}C CPMAS NMR Spectroscopy*

In the mid-1990s, Preston[57] published a comprehensive summary· on NMR applications to SOM. By then solution- and solid-state NMR had been applied not only to simply characterize SOM but also to understand decomposition processes, to characterize biopolymers, to study the binding of metal ions (see Chapter 19) and xenobiotics (see Chapter 20), and to investigate the effect of soil cultivation, the importance of fire events, and the influence of climatic factors.[58] In general, the decomposition of SOM was characterized by an initial decrease in the *O*-alkyl/alkyl C ratio because of the loss of easily metabolizable carbohydrates. The accumulation of alkyl C was attributed to a combination of selective preservation of plant biopolymers such as suberin and cutin and de novo synthesis of microbial biomass. However, with advancing decomposition, the carbohydrate contents determined from wet chemical analysis and NMR started to diverge. This was ascribed to the formation of a range of ill-defined 'pseudo-polysaccharides', which contained the typical *O*- and di-*O*-alkyl C

signatures but were no longer part of recognizable carbohydrates. With prolonged decomposition generally, the alkyl C and aromatic C regions both became broader and contained less characteristic signals.

The DD sequence, differentiating between carbons with strong and weak dipolar interactions, was used to identify overlapping signals. For example, in the di-*O*-alkyl C region around 100–105 ppm, the signal of anomeric carbon of carbohydrates is rapidly lost in a DD spectrum while the nonprotonated carbons of condensed tannins remain. The technique of proton spin-relaxation edited (PSRE) subspectra was applied to generate subspectra with long or short T_1^H values and revealed plant-like (cellulose and hemicellulose; long T_1^H), humified (high content of *O*-containing groups; short T_1^H), and 'recalcitrant' (high content of polymethylene C; longest T_1^H) structures.[59] While the presence of paramagnetic ions is usually not welcomed, the binding of these ions can be used to probe the chemical structure of SOM. By measuring the reductions in the $T_{1\rho}^H$ values of the different regions, the binding sites of these ions can be identified.[60] A thorough review on metals in the environment is given in Chapter 19.

In the 1990s, both solution- and solid-state NMR were started to be used in pesticide studies (see Chapter 20). The labeling with NMR-active nuclei such as ^{13}C of specific reactive positions in xenobiotics permitted the monitoring of changes in chemical shift of these positions and to draw conclusions on the type of interaction with the soil matrix. From sorption studies with hydrophobic organic compounds, the concept of a dual reactive domain model with glassy and rubbery domains in SOM also emerged (see Chapter 20).[61,62]

13.3.4.3 *^{15}N CPMAS NMR Spectroscopy*

The first natural abundance ^{15}N CPMAS NMR spectrum of a peat was published by Preston *et al.*[63] in 1986. First spectra of a bulk soil and soil fractions extracted from various soils were presented by Knicker *et al.*[64] in 1993. In opposition to structural models of the time, still assigning a significant part of the nitrogen to heteroaromatic structures, the ^{15}N NMR spectra showed that over 80% of the signal intensity was in the amide/peptide region. The intensity contribution in the region typical for heteroaromatic nitrogen was <10%, which is too low to justify their importance for N sequestration in fire-unaffected soils. The spectra furthermore revealed small amounts of secondary amines and free NH_2 groups (Figure 13.3).

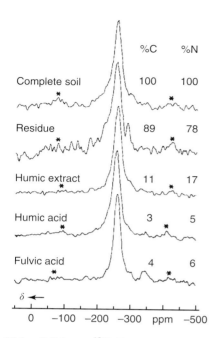

%C %N

Complete soil   100   100

Residue   89   78

Humic extract   11   17

Humic acid   3   5

Fulvic acid   4   6

$\delta \leftarrow$

0   −100   −200   −300   ppm   −500

Figure 13.3. Solid-state ^{15}N CPMAS spectra of a 'black' calcaric Regosol from Ismaning and its humic fractions (* = spinning side bands). (Reprinted with permission from Ref. 64. © 1993 Springer)

Studies on ^{15}N-enriched plant residues during prolonged biological degradation[65,66] resulted in comparable spectra. Applying ^{13}C CPMAS NMR spectroscopy, the authors found that during initial degradation carbohydrates were preferentially metabolized but that after approximately 100 days, all compound classes decomposed at comparable rates. Neither the ^{13}C nor the ^{15}N NMR spectra indicated the synthesis of new chemical structures. Especially, the ^{15}N NMR spectra were remarkably uniform during the entire incubation period. The authors hypothesized that the major pathway of nitrogen metabolism during degradation involved the breakdown of the proteinaceous plant material with a simultaneous de novo synthesis of bacterial and fungal proteins. Such an internal recycling of nitrogen from one amide structure to another would result in similar spectra. The authors concluded that, under the applied conditions, the formation of nitrogen-containing heterocycles can only play a very minor role.

Kelley and Stevenson[67] discussed literature results from acid hydrolysis, NMR experiments, and analytical pyrolysis. Acid hydrolysis with its subsequent

wet chemical determination methods identified only one-third to one-half of the organic nitrogen compounds. About 25–35% could not be hydrolyzed by the acid treatment and 10–20% was found in the hydrolyzable unknown-N fraction. In the hydrolyzable fraction ammonia, amino acids and amino sugars were identified. Ammonia was assumed to derive from indigenous fixed NH_4^+ and from the partial destruction of amino sugars. Part of the amino acid-N was presumed to originate from peptides or proteins. The revised ^{15}N CPMAS NMR literature on ^{15}N-labeling experiments indicated that 80–87% of N in humic substances was present in amide forms. Amino N (9–12%) and pyrrole-type (4–9%) compounds were identified as minor components. All of the pyrrole-type N was accounted for in the hydrolyzable unknown-N fraction. Pyrolysis–FIMS identified amines, nitriles, and heterocyclic N fragments.

With the years, solid-state ^{15}N NMR spectroscopy developed into a routine approach in soil science and soils of various environments were studied by this technique. They confirmed that, in fire-unaffected soils, almost all of the organic N is assignable to amide-N but that, in soil with fire history, heterocyclic N can occur in higher amounts and even dominate the present N-forms. They were hypothesized to be derived from cyclization and dehydration of peptides in plant residues during the charring process.[68] The solid-state ^{15}N NMR spectrum of a soil from Antarctica showed clear signals in the region of heterocyclic N, which were assignable to uric acid being a major component of Penguin guano.[69]

13.3.4.4   The Dispute on the Reliability of Solid-state NMR of Soils

The initial enthusiasm for solid-state NMR spectroscopy among soil scientists was quickly abated by the upcoming discussion whether integration of the different chemical shift regions yields in quantitative information on chemical structures or not (see Chapter 4). The debate on quantification had already started in the 1980s. Some groups[43,44] referred to contact time and relaxation experiments done on coal to argue in favor of quantification. They stated that owing to the much lower aromatic content in humic substances, these samples were less prone to intensity distortions resulting from the isolation of carbons from protons. This approach was countered[70] with the consideration that owing to the larger variety of structural components in humic substances, greater differences in relaxation times and CP efficiencies could occur and that, hence, results from coal samples could not necessarily be applied to humic substances. Others[47] pointed out that paramagentic ions present in soil samples could cause line broadening or signal quenching.

Starting in the 1990s, first comprehensive studies on the quantification problems in solid-sate NMR were performed.[71–74] The CPMAS NMR experiment quickly turned to the standard method in soil science as it is much easier to avoid signal saturation in CP experiments, while still being able to acquire large numbers of transients to obtain a well-resolved spectrum. However, the CP technique itself can be a source of quantification problems. As the $T_{1\rho}^{H}$ starts during the process of polarization transfer (contact time), the complete magnetization I_0 of the X nuclei can never be achieved. An intensity maximum is reached at an optimal contact time, where gain of signal intensity due to polarization transfer is still higher than loss due to relaxation. However, the CP depends on molecular properties that are specific for each chemical group in a heterogeneous mixture and optimum contact times can greatly differ. As only one contact time can be chosen, this can result in a high variability of the intensity losses of the signals of the various groups. Conte *et al.*[75] performed a series of variable contact time (VCT) experiments on a number of humic substances from volcanic soils. For the humic materials under analysis, the authors found optimal contact times ranging from 250 to 800 μs. They concluded that for quantification purposes VCT experiments are a necessary prerequisite as an a priori choice of a recommended average contact time may easily lead to erroneous quantification. However, earlier studies[45,50] found that in humic material comparable intensity losses are observed for all carbon classes using a contact time of 1 ms.

Considerable problems can occur in soils containing higher amounts of paramagnetic material (mainly Fe^{3+}, also Cu^{2+} or Mn^{2+}), which may interact preferentially with one functional group and selectively reducing their intensity due to very fast $T_{1\rho}^{H}$.[47] Differences in the spin–lattice relaxation constant of the unpaired electron (T_{1e}) are a source for varying effects of paramagnetic cations on SOM spectra.[60]

In 1994, Skjemstad *et al.*[76] recommended the demineralization of soil samples using 2% hydrofluoric (HF) acid as the most effective way to remove paramagnetic components and increase the concentration of SOM. Schmidt *et al.*[77] proposed the use of 10% HF

with less extraction steps. Gonçalves *et al.*[78] found that after the second extraction, the quality of the NMR spectra decreased, mostly because the HF had dissolved the mineral phase and released paramagnetic ions that were previously sequestered in the mineral network. In subsoils, with very low organic material, which is mostly adsorbed to the mineral phase, the percentual C losses can be considerable, because of the dissolution of this OM after destruction of the mineral phase. However, if those losses are nonselective, the respective NMR data can still be used to elucidate the SOM composition of the bulk soils. In order to assess possible alteration of the chemical composition, the C/N (carbon-to-nitrogen) ratios before and after treatment can be compared. In most cases, no major alteration of the C/N ratio was evidenced, which is in line with the fact that cold HF solution does not hydrolyze or destroy organic compounds.[79]

In the 1990s, high-field NMR instruments with MAS probes allowing spinning rates of up to 20 kHz became commercially available allowing fast sample rotation that avoids spinning side bands even at higher field strengths. However, high-speed spinning interferes with the Hartmann–Hahn matching condition. When the spinning speed matches or exceeds the order of magnitude of the dipolar decoupling, the Hartmann–Hahn profile breaks down into a series of matching side bands, the maxima of which are resonance-offset dependent, which impedes or even prevents the optimization of the Hartmann–Hahn match for all functional groups. To overcome these motional modulations of carbon–proton couplings at larger spinning speeds, variable amplitude sequences, which increment one of the polarization amplitudes during the contact time, were developed. Cook *et al.*[80] applied an increasing ramp on the ^1H radio frequency field during the spin locking to restore a flat Hartmann–Hahn matching profile at high spinning speeds, which generated better resolved spectra. The ramped CP sequence with high spinning rates was quickly accepted as a necessary prerequisite to record quantifiable spectra.

13.3.4.5 Gain in Knowledge

In the 1990s, solid-state NMR spectroscopy had established itself as essential though not easily available analytical method in soil science. It became clear that SOM could have very diverging compositions depending on input material and environmental conditions and that an all-encompassing structural model could not exist. In 1997, a book[81] and a special issue on NMR in environmental chemistry and soil science were published.[82] Along with contributions on solution- and solid-state NMR, these publications also covered the application of quadrupolar nuclei, imaging, ^{29}Si, ^{27}Al, ^{31}P, ^{19}F, and ^{129}Xe NMR spectroscopy in soil science. The community was well aware that the enormous potential of NMR was far from being used to its full extent. This was partly due to the fact that only few experienced operators were available in the field of soil science and partly due to the constraints imposed by the analyte, namely generally low concentrations of NMR active nuclei and the presence of paramagnetic components. Nevertheless, more sophisticated techniques such as DD and proton spin relaxation editing (PSRE) experiments were applied, revealing overlapping signals and probing the mobility of the different chemical components. Soil models, hence, started to focus more on the spatial arrangement and domain mobility of SOM. Furthermore, attempts were made to link the identified structures to functions through the combination with physical fractionation techniques aimed at the discrimination of pools with identifiable functions.

13.3.5 Modern Times: 2000s – Advanced NMR Techniques

13.3.5.1 Exemplary Views at the Beginning of the Decade

Commonly, SOM is defined as being composed of partially decayed plant residues, microbes, and by-products of decomposition undergoing humification.[83] Hatcher *et al.*[84] considered humic substances to be closely related to the biomolecules from which they derive. The authors described them as complex substances containing a lot of the polar functional groups found in biomolecules while displaying some hydrophobicity at the same time. Burdon[85] revised a number of traditional concepts related to the structures of humic substances from the 1970s, 1980s, and 1990s and concluded that there was no strong evidence for any of them. The author instead argued in favor of mixtures of plant and microbial constituents and their respective microbial degradation products as main components of humic substances. Abiotic pathways such as the Maillard reaction were ruled out as major sources of humic substances, although they were considered to be present in soil. In general, there

was a tendency to refrain from drawing structural models of humic substances and to focus on the spatial conformation of SOM and its interaction with the mineral soil components.

13.3.5.2 Advanced Solution- and Solid-state Techniques

Advances in spectrometer design in the 2000s and the use of increasing magnetic field strengths made it possible to apply advanced solution-state NMR techniques ranging from 1D to very complex 6D experiments. While the latter have limited application in environmental sciences, three-dimensional (3D) NMR, combining heteronuclear multiple quantum coherence (HMQC) with two-dimensional total correlation spectroscopy (TOCSY), has already been successfully applied to unlabeled humic substances,[86] allowing the further separation of overlapping signals in 2D spectra (Figure 13.4). The authors concluded from these 3D-spectra that the major part of the aliphatic components in a forest FA was most likely derived from the leaf cuticles.

In solid-state NMR, 2D experiments are still limited. One of the first two-dimensional solid-state experiments on SOM-like material was double CP $^1H \rightarrow {}^{15}N \rightarrow {}^{13}C$ NMR spectroscopy applied to study

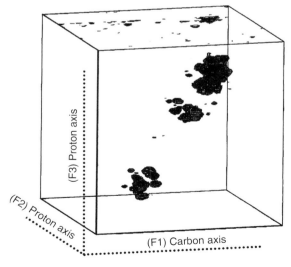

Figure 13.4. 3D HMQC-TOCSY spectrum of pine forest fulvic acid. The F2–F3 plane contains TOCSY information. (Reprinted with permission from Ref. 86. © 2003 American Chemical Society)

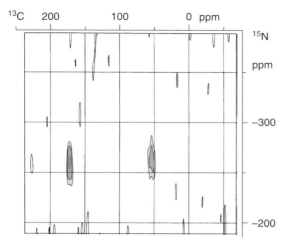

Figure 13.5. Solid-state 2D double CPMAS ^{15}N ^{13}C NMR spectrum of an acid-insoluble fraction obtained from ^{13}C- and ^{15}N-labeled plant residues. (Reprinted with permission from Ref. 88. © 2002 Elsevier)

the immobilization of nitrogen during humification.[87] With this sequence, the authors were able to selectively observe which carbon is directly bound to nitrogen. No evidence for the formation of heterocyclic N was found, whereas proteinaceous N remained the major organic N component (Figure 13.5).[88]

Mao *et al.*[89] published the first heteronuclear correlation (HETCOR) experiments of a HA using proton line-narrowing techniques and total suppression of spinning side bands to obtain greatly improved resolutions. The HETCOR spectra revealed which protons were attached to which carbons and with the additional application of variable DD times the authors were able to determine which functional groups were in close proximity to each other. C-bonded methyl groups were found to be close to both aliphatic and *O*-alkyl but not aromatic groups. Most methoxyl groups were attached directly to aromatic rings, as in lignin, and they made up about a third of the aromatic C–O groups. Both protonated and unprotonated anomeric O–C–O carbons were identified. Carboxyl groups were mostly found in OCH$_n$–COO environments. Some COO groups were also connected to aromatic rings and aliphatic groups. The authors contrasted their results with eight structural humic substance models from the literature and rejected most of them as they lacked most of the identified structures. The model considered to fit best with their results was the FA model of Leenheer *et al.*,[90] who focused on the acidity of carboxyl groups and

presented three possible structural models containing carboxyl groups with pK_a values below 3.0. All three of them, however, were free of nitrogen.

Schmidt-Rohr and Mao[91] developed an advanced spectral-editing pulse sequence, called *dipolar distortionless enhancement by proton transfer (DEPT)*, able to reliably isolate CH groups while suppressing CH$_2$ and quaternary carbons. The authors were able to show that the major fraction of the unsubstituted aliphatic carbons in a HA must be from CH$_2$ and CH$_3$ groups as only a small fraction of the region between 0 and 45 ppm was retained in the CH-only spectra. The combination with a chemical shift filter enabled the authors to differentiate NCH from CCH(C,C) sites in the CH-only spectrum.

Solid-state NMR is rarely applied to ^{1}H in soil science as the spectra are unstructured because of very large peak widths. However, Jäger *et al.*[92] demonstrated that, through mathematical analysis of the Lorentzian and Gaussian contributions in an ^{1}H-wideline spectrum, the amounts of mobile and rigid protons can be quantified (see Chapter 18). In the past decade, fast-field cycling (FFC) relaxometry at low fields also found its way into environmental and soil science (see Chapter 8).[93,94]

13.3.5.3 *Soil Organic Nitrogen*

In 2011, Knicker[95] published a thorough review on the role of soil organic nitrogen (SON). From both 1D and 2D solid-state ^{15}N NMR experiments on plant incubates and OM, which were derived from fire-unaffected soils, it was demonstrated that amide N dominated the SON pool. Heterocyclic N, aromatic amines, or nitroaromatic N were not detected in amounts high enough to justify the abiotic recondensation pathways as the main N immobilization mechanisms in biologically active soils. Those studies rather confirmed that N-stabilization within proteinaceous material represents one of the main N-sequestration mechanisms, although proteinaceous material is often assumed to be biochemically labile. On the other hand, recalcitrant peptide-like compounds were evidenced to be ubiquitous and to exist even in fossilized OM-rich deposits. The predominance of these peptide N forms in soils is assumed to be caused by continuous recycling of biomass residues for buildup of new cell material and the accumulation of N-containing residues, which cannot be enzymatically attacked due to steric hindrance or encapsulation into hydrophobic macromolecules of the cell wall.

In addition, peptide-like material may be physically protected by incorporation within aggregates or sequestration in nanopores that are too small for entering and functioning of soil enzymes. On the basis of the solid-state ^{15}N NMR results presented in the literature up to then, the author suggested that peptides must represent more important players within the SOM sequestration process than commonly assumed. Along the same line, a diffusion-edited solution-state ^{1}H NMR study by Simpson *et al.*[96] concluded that the previously reported estimations of microbial contributions to SOM were much too low and that over 80% of the organic nitrogen in humic fractions is potentially from microbial sources. The majority of the organic nitrogen was detected as amide form.

On the other hand, in ecosystems that were exposed to natural and prescribed fires, a shift of the chemical SON composition toward heterocyclic N was revealed using solid-state ^{15}N NMR spectroscopy.[97] The frequent appearance of their signals in ^{15}N NMR spectra of such soils indicates that their lack in fire-unaffected environments is unlikely to be caused by the failure of ^{15}N CPMAS NMR to detect them. Those spectra together with solid-state spectra of plant charring experiments demonstrated further that organic N has to be considered as an integral part of charcoal derived from N-rich sources.[98] Furthermore, recent experiments revealed that compared to charcoal from wood, such N-rich pyrogenic OM show a lower stability toward biochemical[99,100] and chemical oxidation.[98]

13.3.5.4 *Impact of Environmental Factors on SOM Composition*

Considering the important role of SOM with respect to soil functions and as C and N sink within the global elemental cycles, the factors that define its stability and quality are of high interest in environmental research. Hence, ^{13}C CPMAS NMR spectroscopy became a routine application in soil studies investigating the influence of environmental factors, such as temperature or humidity, agricultural practices, compost application, liming, or fire. In soil studies investigating the effect of increasing temperatures, it was found that decomposition slows down with time. Therefore, fresh litter should react more sensitive to increased temperatures than older SOM.[101,102] Studies along a climosequence in southern Brazil confirmed that SOM contents increased with enhanced soil humidity and decreasing temperature.[103] It was suggested that SOM accumulation is enhanced by organo-mineral

interactions with iron oxides contributing to the stabilization of SOM.

Solid-state NMR is frequently used to monitor changes induced by differing agricultural practices. The comparison of soils with various crops to a control soil under grassland showed lower amounts of alkyl and higher contributions of aromatic C.[104] This was attributed to differences in the microbial community and therefore in the quality of its biochemical products. Nitrogen fertilization did not affect the particulate OM and silt-size fraction but increased the alkyl C content in the clay fraction. On the basis of the NMR data, it was concluded that in the studied soil the SOM was affected by land use but not by variations in the arable crops. Studies on grazing practices in semiarid steppe ecosystems[105] indicated that management changes in steppe ecosystems do not necessarily increase carbon sequestration and their assumed potential to act as carbon sinks was questioned.

Several studies found that compost application did not contribute to the mid-term accumulation of stable forms of OM in soil.[106] Results suggested that decomposition of the added composts in soil is an ongoing humification process of the composts themselves and affects the overall composition of the soil only to a minor degree.[107] In contrast, liming of a mature Norway spruce forest induced major changes in the composition of the organic surface layer. The proportion of alkyl C, the alkyl C/*O*-alkyl C ratio, and the content of extractable lipids detected by GC/MS decreased. These changes were explained by increased fine root formation in the organic layer after liming. The decrease in the C/N ratio was attributed to the higher microbial activity.[108]

Considerable changes in the composition of SOM were detected after wild fires.[109] Owing to the incorporation of incompletely combusted vegetation residues mainly, an increase in the relative contribution of aromatic C and heteroaromatic N was found.[110] Often, a harsh chemical oxidation procedure is used to identify charcoal residues in soils.[111–113] However, recent NMR studies demonstrated that this treatment is inefficient in removing noncharred hydrophobic compounds such as paraffinic structures.[114] Consequently, quantification of charcoal residues only by the determination of the amount of chemical oxidation resistant C may lead to biased data, and, hence, further spectroscopic characterization of the oxidation residue is obligatory.[115] The extent of the fire-induced alteration of SOM was shown to depend on the intensity of the fire and the source material. During high intensity forest fires (see Chapter 15) and prescribed grass land fires, most of the organic material is volatilized. Increasing amounts of charcoal usually accumulate with decreasing fire intensity. [109] N-rich source material was shown to produce charcoal that increases both the aromatic and the alkyl C contents of the soil.[98] With increasing recovery time after the fire, the charcoal residues experience oxidation.[116] Although the formation of carboxylic groups may accelerate its loss because of leaching with the soil solution into deeper soil horizons or to increased accessibility to microbial degradation, charcoal residues from Neolithic times and older are frequently found.[117–119] Possibly, oxidation of pyrogenic OM enhanced the potential to be adsorbed to the mineral phase, which subsequently led to its efficient sequestration.[120]

13.3.5.5   *NMR Analysis of Subsoils*

With increasing improvement of NMR instrumentation, both with respect to sensitivity and data acquisition it became possible to analyze bulk soils with low OM content. Subsoils so far were mostly excluded from in-depth characterization not only with NMR but also with most other analytical tools. The demineralization with HF opened the door to investigate subsoil SOM,[78] even though considerable OM losses of up to 80% and sometimes even more can be encountered in subsoil samples. As a common pattern in OM composition with soil depths, a decrease in the relative contribution of *O*-alkyl C and an increase in the chemical shift region of carboxyl C were detected.[121] In many fire-unaffected soils, no distinct trend with respect to aromaticity could be identified. It was, hence, concluded that organic C input into subsoils occurs in dissolved form (see Chapter 11) as aboveground or root litter and exudates along root channels and/or through bioturbation.[122] One of the most important factors leading to the survival of SOM in subsoils is assumed to be the spatial separation of SOM, microbes, and extracellular enzyme activity. In contrast, in soils with a recurrent fire history, the aromatic C contribution to the total organic C increased with depths,[74,103,104] owing to accumulation of leached pyrogenic OM.[97,120,123] Fire was, hence, identified as an important factor also in subsoil formation.[120]

13.3.5.6   *Gain in Knowledge*

In the past decade, several reviews on NMR in environmental sciences[95,124–126] and three special issues

on soil NMR were published.[127–129] At the end of the preceding decade, it became more and more accepted that SOM represents a mixture of partly degraded biogenic residues that are protected physically, either by entrapment in aggregates[130] or hydrophobic organic networks[131] or by adsorption to the mineral phase,[132] and chemically, either through steric obstruction or other molecular properties decreasing the accessibility of enzymes for complete mineralization. Furthermore, reduced microbial activity, owing to unfavorable soil conditions such as oxygen depletion, high and low pH, extreme dryness, or high humidity, is also discussed to contribute to the survival of SOM.[133,134]

Moreover, a new concept, in opposition to the long-lived polymer or macromolecules model, suggested that SOM represents a supramolecular association of relatively small and diverse organic molecules, which form clusters linked by hydrogen bonds and hydrophobic interactions[135–138] or by complexation with metal cations.[139] However, although NMR studies revealed support for most of those pathways, there is still no common agreement on which mechanisms are responsible for SOM sequestration, possibly because all of them can occur concomitantly. A very thorough review in two parts on the different concepts of SOM was published by Schaumann.[140,141] The author concluded that the very different experimental conditions and molecular composition of the investigated humic materials were the main reasons for the diverging concepts. The author considered SOM and humic substances as highly amorphous materials with a certain degree of microcrystalline regions such as polymethylene crystallites.

13.4   CONCLUSIONS AND FUTURE PROSPECTS

During the past century, the view on SOM has changed dramatically. At the beginning of the twentieth century, the thesis of formation of humus from lignin failed to explain the nitrogen content of humus. The ligno-protein theory then explained humus formation by accumulating lignin, which reacts with nitrogenous compounds such as proteins derived from microbial activities. Later on, the polyphenol theory suggested that with ongoing decomposition of the lignin, its side chains are degraded and demethylation occurs, which reduces the linking of the monomers. The latter were thought to repolymerize by reacting with nitrogen, which was viewed as an integral part of SOM, with heterocyclic nitrogen playing a major role. The first solution-state NMR spectra of dissolved humic substances then clearly demonstrated that the aromatic content of humic substances was overestimated and that aliphatic components were as important or could even outrange the aromatic structures. This was confirmed for the whole SOM in solid-state NMR studies. Solid-state ^{15}N NMR spectra on whole soil samples revealed that the major part of nitrogen in SOM was present in amide form and that heterocyclic nitrogen played only a minor role in fire-unaffected soils. Modern concepts on SOM now recognize the survival of partially degraded biopolymers because of physical and chemical protection mechanisms and SOM is nowadays viewed as a mixture of partly degraded biogenic residues. With the introduction of highly sophisticated NMR sequences, the focus turned to a more detailed elucidation of the spatial conformation of SOM and the interactions of its functional groups either with one another or with other soil components.

The joint analysis of samples in multidimensional solution-state NMR and Fourier transform ion cyclotron resonance mass spectrometry (FT-ICR-MS)[142] offers the possibility to get complementary structural information. The coupling of multidimensional solution-state NMR with liquid chromatographic techniques [LC-NMR (liquid chromatography nuclear magnetic resonance), see Chapter 3], so far successfully applied to dissolved SOM samples,[143] is a further development that allows to obtain very detailed information on the composition of dissolved OM or soluble soil components such as humic substances. The isolation and identification of specific biomarkers will allow a more precise understanding of the interlinking of nutrient cycles in soil.[144] Furthermore, new techniques such as dynamic nuclear polarization (DNP) NMR[145] promise to push the limits of detection in solid-state NMR.

RELATED ARTICLES IN EMAGRES

Environmental NMR: Solid-state Methods

Environmental NMR: Solution-state Methods

Environmental NMR: Fast Field Cycling Relaxometry

Soil-Water Interactions

Dissolved Organic Matter

Organic Pollutants in the Environment

Metals in the Environment

Forest Ecology and Soils

REFERENCES

1. F. Bloch, W. W. Hansen, and M. Packard, *Phys. Rev.*, 1946, **69**, 127.

2. F. Bloch, *Phys. Rev.*, 1946, **70**, 460.

3. F. Bloch, W. W. Hansen, and M. Packard, *Phys. Rev.*, 1946, **70**, 474.

4. E. M. Purcell, H. C. Torrey, and R. V. Pound, *Phys. Rev.*, 1946, **69**, 37.

5. N. Bloembergen, E. M. Purcell, and R. V. Pound, *Phys. Rev.*, 1948, **73**, 679.

6. E. L. Hahn, *Phys. Rev.*, 1949, **76**, 145.

7. E. L. Hahn, *Phys. Rev.*, 1950, **80**, 580.

8. R. R. Ernst and W. A. Anderson, *Rev. Sci. Instrum.*, 1966, **37**, 93.

9. A. Abragam, 'The Principles of Nuclear Magnetism', 1st edn, Oxford University Press: London, UK, 1961.

10. E. R. Andrew and R. A. Newing, *Proc. Phys. Soc. Lond.*, 1958, **72**, 959.

11. I. J. Lowe, *Phys. Rev. Lett.*, 1959, **2**, 285.

12. R. R. Ernst, *J. Chem. Phys.*, 1966, **45**, 3845.

13. S. R. Hartmann and E. L. Hahn, *Phys. Rev.*, 1962, **128**, 2042.

14. A. Pines, M. G. Gibby, and J. S. Waugh, *J. Chem. Phys.*, 1972, **56**, 1776.

15. A. Pines, M. G. Gibby, and J. S. Waugh, *J. Chem. Phys.*, 1973, **59**, 569.

16. J. Schaefer and E. O. Stejskal, *J. Am. Chem. Soc.*, 1976, **98**, 1031.

17. E. O. Stejskal, J. Schaefer, and J. S. Waugh, *J. Magn. Reson.*, 1977, **28**, 105.

18. W. P. Aue, E. Bartholdi, and R. R. Ernst, *J. Chem. Phys.*, 1976, **64**, 2229.

19. G. B. Yntema, *Phys. Rev.*, 1955, **98**, 1197.

20. S. A. Waksman, *Soil Sci.*, 1926, **22**, 123.

21. S. A. Waksman, *Soil Sci.*, 1926, **22**, 221.

22. S. A. Waksman, *Soil Sci.*, 1926, **22**, 323.

23. S. A. Waksman and F. G. Tenney, *Soil Sci.*, 1926, **22**, 395.

24. S. A. Waksman, *Soil Sci.*, 1926, **22**, 421.

25. J. M. Bremner, *J. Soil Sci.*, 1954, **5**, 214.

26. J. M. Bremner, *J. Soil Sci.*, 1951, **2**, 67.

27. W. Flaig, *Geochim. Cosmochim. Acta*, 1964, **28**, 1523.

28. E. Griffiths, *Biol. Rev.*, 1965, **40**, 129.

29. J. Macura, *Geoderma*, 1974, **12**, 311.

30. F. J. Stevenson and K. M. Goh, *Geochim. Cosmochim. Acta*, 1971, **35**, 471.

31. L. C. Maillard, *C. R. Soc. Biol.*, 1912, **72**, 599.

32. L. C. Maillard, *Ann. Chim. France*, 1916, **5**, 258.

33. A. Nissenbaum and I. R. Kaplan, *Limnol. Oceanogr.*, 1972, **17**, 570.

34. J. I. Hedges, in 'Humic Substances and Their Role in the Environment', eds F. H. Frimmel and R. F. Christman, John Wiley & Sons Ltd.: Chichester, 1988.

35. D. H. R. Barton and M. Schnitzer, *Nature*, 1963, **198**, 217.

36. R. Ishiwatari, *Chem. Geol.*, 1973, **12**, 113.

37. J. A. Neyroud and M. Schnitzer, *Can. J. Chem.*, 1974, **52**, 4123.

38. D. Grant, *Nature*, 1977, **270**, 709.

39. F. J. Gonzalez-Vila, H. Lentz, and H.-D. Lüdemann, *Biochem. Bioph. Res. Commun.*, 1976, **72**, 1063.

40. M. A. Wilson, A. J. Jones, and B. Williamson, *Nature*, 1978, **276**, 487.

41. P. F. Barron, M. A. Wilson, J. F. Stephens, B. A. Cornell, and K. R. Tate, *Nature*, 1980, **286**, 585.

42. P. F. Barron and M. A. Wilson, *Nature*, 1981, **289**, 275.

43. M. A. Wilson, R. J. Pugmire, K. W. Zilm, K. M. Goh, S. Heng, and D. M. Grant, *Nature*, 1981, **294**, 648.

44. P. G. Hatcher, I. A. Breger, L. W. Dennis, and G. E. Maciel, in 'Aquatic and Terrestrial Humic Materials', eds R. F. Christman and E. T. Gjessing, Ann Arbor Science Publishers: Michigan, 1983, Chapter 3.

45. C. M. Preston and J. A. Ripmeester, *Can. J. Spectrosc.*, 1982, **27**, 99.

46. M. A. Wilson, *J. Soil Sci.*, 1984, **35**, 209.

47. C. M. Preston, R. L. Dudley, C. A. Fyfe, and S. P. Mathur, *Geoderma*, 1984, **33**, 245.

48. M. A. Wilson, K. M. Goh, P. J. Collin, and L. G. Greenfield, *Org. Geochem.*, 1986, **9**, 225.

49. P. G. Hatcher, *Org. Geochem.*, 1987, **11**, 31.

50. R. Fründ and H.-D. Lüdemann, *Sci. Total Environ.*, 1989, **81/82**, 157.

51. M. A. Wilson, 'NMR Techniques and Applications in Geochemistry and Soil Chemistry', 1st edn, Pergamon Press: London, UK, 1987.

52. L. Benzing-Purdie, J. A. Ripmeester, and C. M. Preston, *J. Agr. Food Chem.*, 1983, **31**, 913.

53. L. M. Benzing-Purdie, M. V. Cheshire, B. L. Williams, G. P. Sparling, C. I. Ratcliffe, and J. A. Ripmeester, *J. Agr. Food Chem.*, 1986, **34**, 170.

54. H. R. Schulten and M. Schnitzer, *Naturwissenschaften*, 1993, **80**, 29.

55. R. L. Wershaw, *Environ. Sci. Technol.*, 1993, **27**, 814.

56. J. N. Ladd, R. C. Foster, and J. O. Skjemstad, *Geoderma*, 1993, **56**, 401.

57. C. M. Preston, *Soil Sci.*, 1996, **161**, 144.

58. M. A. Nanny, in 'Nuclear Magnetic Resonance Spectroscopy in Environmental Chemistry', eds M. A. Nanny, R. A. Minear and J. A. Lenheer, Oxford University Press: London, 1997.

59. I. Kögel-Knabner, *Geoderma*, 1997, **80**, 243.

60. R. J. Smernik and J. M. Oades, *Geoderma*, 1999, **89**, 219.

61. B. Xing, J. J. Pignatello, and B. Gigliotti, *Environ. Sci. Technol.*, 1996, **30**, 2432.

62. E. J. Leboeuf and W. J.Weber Jr, *Environ. Sci. Technol.*, 1997, **31**, 1697.

63. C. M. Preston, J. A. Ripmeester, S. P. Mathur, and M. Levesque, *Can. J. Spectrosc.*, 1986, **31**, 63.

64. H. Knicker, R. Fründ, and H. D. Lüdemann, *Naturwissenschaften*, 1993, **80**, 219.

65. G. Almendros, R. Fründ, F. J. González-Vila, K. M. Haider, H. Knicker, and H. D. Lüdemann, *Febs Lett.*, 1991, **282**, 119.

66. H. Knicker and H. D. Lüdemann, *Org. Geochem.*, 1995, **23**, 329.

67. K. R. Kelley and F. J. Stevenson, *Fert. Res.*, 1995, **42**, 1.

68. H. Knicker and J. O. Skjemstad, *Aust. J. Soil Res.*, 2000, **38**, 113.

69. H. Knicker, R. Fründ, and H. D. Lüdemann, in ''Nuclear Magnetic Resonance Spectroscopy in Environmental Chemistry', eds M. A. Nanny, R. A. Minear

and J. A. Lenheer, Oxford University Press: London, 1997.

70. R. L. Wershaw, in 'Humic Substances in Soil, Sediment, and Water', eds G. R. Aiken, D. M. McKnight and R. L. Wershaw, John Wiley & Sons: New York, 1985, Chapter 22.

71. P. G. Hatcher and M. A. Wilson, *Org. Geochem.*, 1991, **17**, 293.

72. M. A. Wilson, in 'Soil Analysis. Modern Instrumental Techniques', ed. K. A. Smith, Marcel Dekker, Inc.: New York, USA, 1991.

73. R. J. Smernik, J. A. Baldock, and J. M. Oades, *Solid State Nucl. Mag.*, 2002, **22**, 71.

74. D. P. Dick, C. N. Gonçalves, R. S. D. Dalmolin, H. Knicker, E. Klamt, I. Kögel-Knabner, M. L. Simoes, and L. Martin-Neto, *Geoderma*, 2005, **124**, 319.

75. P. Conte, A. Piccolo, B. van Lagen, P. Buurman, and P. A. de Jager, *Geoderma*, 1997, **80**, 327.

76. J. O. Skjemstad, P. Clarke, J. A. Taylor, J. M. Oades, and R. H. Newman, *Aust. J. Soil Res.*, 1994, **32**, 1215.

77. M. W. I. Schmidt, H. Knicker, P. G. Hatcher, and I. Kögel-Knabner, *Eur. J. Soil Sci.*, 1997, **48**, 319.

78. C. N. Gonçalves, R. S. D. Dalmolin, D. P. Dick, H. Knicker, E. Klamt, and I. Kögel-Knabner, *Geoderma*, 2003, **116**, 373.

79. I. Schöning, H. Knicker, and I. Kögel-Knabner, *Org. Geochem.*, 2005, **36**, 1378.

80. R. L. Cook, C. H. Langford, R. Yamdagni, and C. M. Preston, *Anal. Chem.*, 1996, **68**, 3979.

81. M. A. Nanny, R. A. Minear, and J. A. Lenheer (eds), 'Nuclear Magnetic Resonance Spectroscopy in Environmental Chemistry', 1st edn, Oxford University Press: London, UK, 1997.

82. M. A. Hemminga and P. Buurman, *Geoderma*, 1997, **80**, 221.

83. N. J. Mathers, X. A. Mao, Z. H. Xu, P. G. Saffigna, S. J. Berners-Price, and M. C. S. Perera, *Aust. J. Soil Res.*, 2000, **38**, 769.

84. P. G. Hatcher, K. J. Dria, S. Kim, and S. W. Frazier, *Soil Sci.*, 2001, **166**, 770.

85. J. Burdon, *Soil Sci.*, 2001, **166**, 752.

86. A. J. Simpson, W. L. Kingery, and P. G. Hatcher, *Environ. Sci. Technol.*, 2003, **37**, 337.

87. H. Knicker, P. G. Hatcher, and F. J. González-Vila, *J. Environ. Qual.*, 2002, **31**, 444.

88. H. Knicker, *Org. Geochem.*, 2002, **33**, 237.

89. J. D. Mao, B. Xing, and K. Schmidt-Rohr, *Environ. Sci. Technol.*, 2001, **35**, 1928.

90. J. A. Leenheer, R. L. Wershaw, and M. M. Reddy, *Environ. Sci. Technol.*, 1995, **29**, 399.

91. K. Schmidt-Rohr and J. D. Mao, *J. Am. Chem. Soc.*, 2002, **124**, 13938.

92. A. Jäger, G. E. Schaumann, and M. Bertmer, *Org. Geochem.*, 2011, **42**, 917.

93. R. Kimmich and E. Anoardo, *Prog. Nucl. Magn. Reson. Spectrosc.*, 2004, **44**, 257.

94. P. Conte, C. Abbate, A. Baglieri, M. Nègre, C. De Pasquale, G. Alonzo, and M. Gennari, *Org. Geochem.*, 2011, **42**, 972.

95. H. Knicker, *Soil Biol. Biochem.*, 2011, **43**, 1118.

96. A. J. Simpson, M. J. Simpson, E. Smith, and B. P. Kelleher, *Environ. Sci. Technol.*, 2007, **41**, 8070.

97. H. Knicker, *Biogeochemistry*, 2007, **85**, 91.

98. H. Knicker, *Org. Geochem.*, 2010, **41**, 947.

99. A. Hilscher and H. Knicker, *Org. Geochem.*, 2011, **42**, 42.

100. J. M.de la Rosa and H. Knicker, *Soil Biol. Biochem.*, 2011, **43**, 2368.

101. J. A. Trofymow, T. R. Moore, B. Titus, C. Prescott, I. Morrison, M. Siltanen, S. Smith, J. Fyles, R. Wein, C. CamirT, L. Duschene, L. Kozak, M. Kranabetter, and S. Visser, *Can. J. Forest Res.*, 2002, **32**, 789.

102. C. M. Preston, J. R. Nault, and J. A. Trofymow, *Ecosystems*, 2009, **12**, 1078.

103. R. S. D. Dalmolin, C. N. Gonçalves, D. P. Dick, H. Knicker, E. Klamt, and I. Kögel-Knabner, *Eur. J. Soil Sci.*, 2006, **57**, 644.

104. J. Dieckow, J. Mielniczuk, H. Knicker, C. Bayer, D. P. Dick, and I. Kögel-Knabner, *Eur. J. Soil Sci.*, 2005, **56**, 705.

105. M. Steffens, A. Kölbl, and I. Kögel-Knabner, *Eur. J. Soil Sci.*, 2009, **60**, 198.

106. F. J. González-Vila, G. Almendros, and F. Madrid, *Sci. Total Environ.*, 1999, **236**, 215.

107. J. Leifeld, S. Siebert, and I. Kögel-Knabner, *Eur. J. Soil Sci.*, 2002, **53**, 299.

108. W. Rosenberg, K. G. J. Nierop, H. Knicker, P. A.de Jager, K. Kreutzer, and T. Weiss, *Soil Biol. Biochem.*, 2003, **35**, 155.

109. H. Knicker, G. Almendros, F. J. González-Vila, J. A. González-Pérez, and O. Polvillo, *Eur. J. Soil Sci.*, 2006, **57**, 558.

110. H. Knicker, F. J. González-Vila, O. Polvillo, J. A. González, and G. Almendros, *Soil Biol. Biochem.*, 2005, **37**, 701.

111. C. Rumpel, M. Alexis, A. Chabbi, V. Chaplot, D. P. Rasse, C. Valentin, and A. Mariotti, *Geoderma*, 2006, **130**, 35.

112. M. J. Simpson and P. G. Hatcher, *Org. Geochem.*, 2004, **35**, 923.

113. E. S. Krull, C. W. Swanston, J. O. Skjemstad, and J. A. McGowan, *J. Geophys. Res. Biogeo.*, 2006, **111**, G04001.

114. H. Knicker, P. Müller, and A. Hilscher, *Geoderma*, 2007, **142**, 178.

115. H. Knicker, M. Wiesmeier, and D. R. Dick, *Geoderma*, 2008, **147**, 69.

116. C. H. Cheng, J. Lehmann, and M. H. Engelhard, *Geochim. Cosmochim. Acta*, 2008, **72**, 1598.

117. J. O. Skjemstad, J. A. Taylor, and R. J. Smernik, *Commun. Soil Sci. Plan.*, 1999, **30**, 2283.

118. B. Glaser, L. Haumaier, G. Guggenberger, and W. Zech, *Org. Geochem.*, 1998, **29**, 811.

119. E. Eckmeier, R. Gerlach, E. Gehrt, and M. W. I. Schmidt, *Geoderma*, 2007, **139**, 288.

120. H. Knicker, *Quatern. Int.*, 2011, **243**, 251.

121. C. Rumpel, I. Kögel-Knabner, and F. Bruhn, *Org. Geochem.*, 2002, **33**, 1131.

122. C. Rumpel and I. Kögel-Knabner, *Plant Soil*, 2011, **338**, 143.

123. M. Egli, G. Mastrolonardo, R. Seiler, S. Raimondi, F. Favilli, V. Crimi, R. Krebs, P. Cherubini, and G. Certini, *Catena*, 2012, **88**, 14.

124. P. Conte, R. Spaccini, and A. Piccolo, *Prog. Nucl. Magn. Reson. Spectrosc.*, 2004, **44**, 215.

125. R. L. Cook, *Anal. Bioanal. Chem.*, 2004, **378**, 1484.

126. L. A. Cardoza, A. K. Korir, W. H. Otto, C. J. Wurrey, and C. K. Larive, *Prog. Nucl. Magn. Reson. Spectrosc.*, 2004, **45**, 209.

127. P. Conte, A. E. Berns, A. Pohlmeier, and G. Alonzo, *Open Magn. Res. J.*, 2010, **3**, 14.

128. A. E. Berns, P. Conte, A. Pohlmeier, and G. Alonzo, *Org. Geochem.*, 2011, **42**, 865.

129. *J. Environ. Qual.*, 2002, **31**(2), 369. doi:10.2134/jeq2002.3690

130. J. O. Skjemstad, L. J. Janik, M. J. Head, and S. G. McClure, *J. Soil Sci.*, 1993, **44**, 485.

131. H. Knicker and P. G. Hatcher, *Naturwissenschaften*, 1997, **84**, 231.

132. P. Sollins, C. Swanston, M. Kleber, T. Filley, M. Kramer, S. Crow, B. A. Caldwell, K. Lajtha, and R. Bowden, *Soil Biol. Biochem.*, 2006, **38**, 3313.

133. J. A. Baldock, C. A. Masiello, Y. Gelinas, and J. I. Hedges, *Mar. Chem.*, 2004, **92**, 39.

134. M. Kleber, *Environ. Chem.*, 2010, **7**, 320.

135. R. L. Wershaw, *Soil Sci.*, 1999, **164**, 803.

136. P. Conte and A. Piccolo, *Chemosphere*, 1999, **38**, 517.

137. P. Conte and A. Piccolo, *Environ. Sci. Technol.*, 1999, **33**, 1682.

138. A. Piccolo, *Soil Sci.*, 2001, **166**, 810.

139. A. J. Simpson, W. L. Kingery, M. H. B. Hayes, M. Spraul, E. Humpfer, P. Dvortsak, R. Kerssebaum, M. Godejohann, and M. Hofmann, *Naturwissenschaften*, 2002, **89**, 84.

140. G. E. Schaumann, *J. Plant Nutr. Soil Sci.*, 2006, **169**, 145.

141. G. E. Schaumann, *J. Plant Nutr. Soil Sci.*, 2006, **169**, 157.

142. N. Hertkorn, M. Harir, B. P. Koch, B. Michalke, and P. Schmitt-Kopplin, *Biogeosciences*, 2013, **10**, 1583.

143. G. C. Woods, M. J. Simpson, P. J. Koerner, A. Napoli, and A. J. Simpson, *Environ. Sci. Technol.*, 2011, **45**, 3880.

144. W. Amelung, S. Brodowski, A. Sandhage-Hofmann, and R. Bol, in 'Advances in Agronomy', ed. D. L. Sparks, Elsevier Academic Press Inc.: San Diego, USA, 2008, vol. 100.

145. H. Müller, M. Etzkorn, and H. Heise, *Top. Curr. Chem.*, 2013, **335**, 121.

Chapter 14
Chemical Ecology

Bernd Schneider

Max Planck Institute for Chemical Ecology, Jena 07745, Germany

14.1 INTRODUCTION

Chemical ecology investigates natural product-based interactions between different organisms. Unlike classical natural product chemistry, which mostly aims to identify natural products to be used for pharmacological purposes, chemical ecology considers natural products from the viewpoint of the producing and/or receiving organisms.[1] However, natural products involved in plant–plant, plant–insect, plant–microbial, insect–microbial, and other interspecies interactions are studied not only for their specific activities in the producer and their role for the emitting and receiving organisms but also to investigate their structure, stereochemistry, localization, biogenetic and metabolic pathways, and the quantitative composition of their mixtures. Many analytical methods are necessary to retrieve the chemical information needed in chemical ecology, with NMR spectroscopy and mass spectrometry (MS) being the most important. Since the 1970s, NMR has revolutionized the structural elucidation of new natural products[2-4] and, through the use of stable isotope labeling, biosynthetic and metabolic pathway studies.[5,6] Structure elucidation and biosynthetic studies remain challenging tasks in natural products chemistry and chemical ecology, as many ecologically relevant and pharmacologically interesting compounds are yet to be discovered and their biosynthesis is yet to be identified. NMR spectroscopy is especially powerful in assigning stereochemical properties of natural products. Knowledge of the spatial orientation of atoms and functional groups of a molecule is crucial for estimating its biological function. Assigning a compound's relative and absolute configurations is also an important prerequisite for studying the mechanistic aspects of metabolic conversions. Coupling separation methods [liquid chromatography (LC) and capillary electrophoresis] with NMR analysis is another trend in natural products chemistry[7-9] (see Chapter 3) and has been employed successfully in ecological studies. In recent years, improvements in NMR instrumentation and methodology have contributed to sensitivity enhancement, making low-concentrated metabolites accessible to NMR analysis. On the

NMR Spectroscopy: A Versatile Tool for Environmental Research
Edited by Myrna J. Simpson and André J. Simpson
© 2014 John Wiley & Sons, Ltd. ISBN: 978-1-118-61647-5

basis of sensitivity-enhancement techniques, appealing applications of NMR have been emerging in chemical ecology. For example, by combining NMR with microscopic sampling methods such as laser microdissection (LMD),[10] scientists are able to study the cell- and tissue-specific occurrences of specialized metabolites. Information about the specific occurrence and distribution on a microscopic scale within specialized cells of an organism is important for understanding the ecological function of metabolites. For micrometabolomic studies,[11] metabolites do not necessarily have to be separated, but methods have been developed to analyze compounds in mixtures such as crude extracts or partially purified fractions. The high dynamic range of modern spectrometers allows quantitative NMR analysis of very low concentrations of components occurring in mixtures along with abundant concentrations of other components. NMR-based metabolomics and corresponding studies based on MS and other techniques are used to identify as many compounds as possible in a sample simultaneously. However, the goal, namely, the unbiased identification and quantification of all metabolites in a biological sample,[12,13] will continue to be a challenge for NMR and probably other methods as well. As chemoecological investigations frequently focus on metabolites, which change in concentration in response to genotypic or phenotypic alterations, and are affected by an organism's pathophysiological history and during development, methods detecting differences between samples obtained from an unchallenged and a challenged organism are of special interest. The number of ecological studies applying NMR-based comparative approaches and other NMR methods for the identification, formation, and distribution of metabolites has been constantly increasing, as the sensitivity of NMR has been enhanced.[14–16] By presenting selected examples, this chapter aims to demonstrate various applications of NMR in chemical ecology.

14.2 NMR-BASED IDENTIFICATION AND STRUCTURE ELUCIDATION OF ECOLOGICALLY RELEVANT COMPOUNDS

14.2.1 General Remarks

Natural products from plants, insects, and other organisms play multiple roles in ecological interactions.[1]

Examples from plants are defense compounds, which are formed in response to tissue damage by herbivores, after microbial infection (phytoalexins), or compounds that occur constitutively in plants (phytoanticipins). Plant defense compounds can be repellent or even toxic for herbivores or microbes. Similar scenarios occur in insects and other organisms. In addition to defense, natural products are involved in intra- or interspecies communication. For example, specialized metabolites in flowers (nectar components, flower pigments, and volatile scent components) attract pollinators by their taste, color, or smell, guiding insects to the nectar.[17,18] Other signaling molecules are involved in social behavior (aggregation, attraction, or selection of mating partners, alarming, etc.).[19–22]

Many ecologically relevant compounds are readily accessible to NMR and other tools of natural product analysis. However, most compounds are of very low natural abundance, and in many cases, it has been necessary to synthesize the compounds not only to confirm the activity in bioassays but also because the accessible amount is insufficient for de novo structure elucidation by two-dimensional (2D) NMR methods. Recent advances in NMR such as cryogenic probes, miniaturization, and special sampling techniques have greatly improved sensitivity, enabling current NMR-based identification of natural products to occur down to the nanogram scale. Numerous studies in the literature report structural elucidation of natural products involved in various ecological interactions.[20,22–24] Several further examples are summarized and discussed in the following paragraphs.

14.2.2 Identification of Ecologically Relevant Compounds from Plants, Insects, and Microorganisms

Flower constituents such as petal pigments affect the interaction of plants with pollinators. In this context, a problem in natural products chemistry that has persisted for more than 70 years[25] has been solved recently by revising the previously proposed structure of nudicaulins,[26] the unique yellow flower pigments of *Papaver nudicaule*. Extensive NMR studies of underivatized nudicaulins I and II, of the partially hydrogenated compounds, and of the *N,O*-permethyl derivatives made it possible to revise the previously proposed structure. [1]H–[13]C heteronuclear multiple bond correlations (HMBCs) of the key

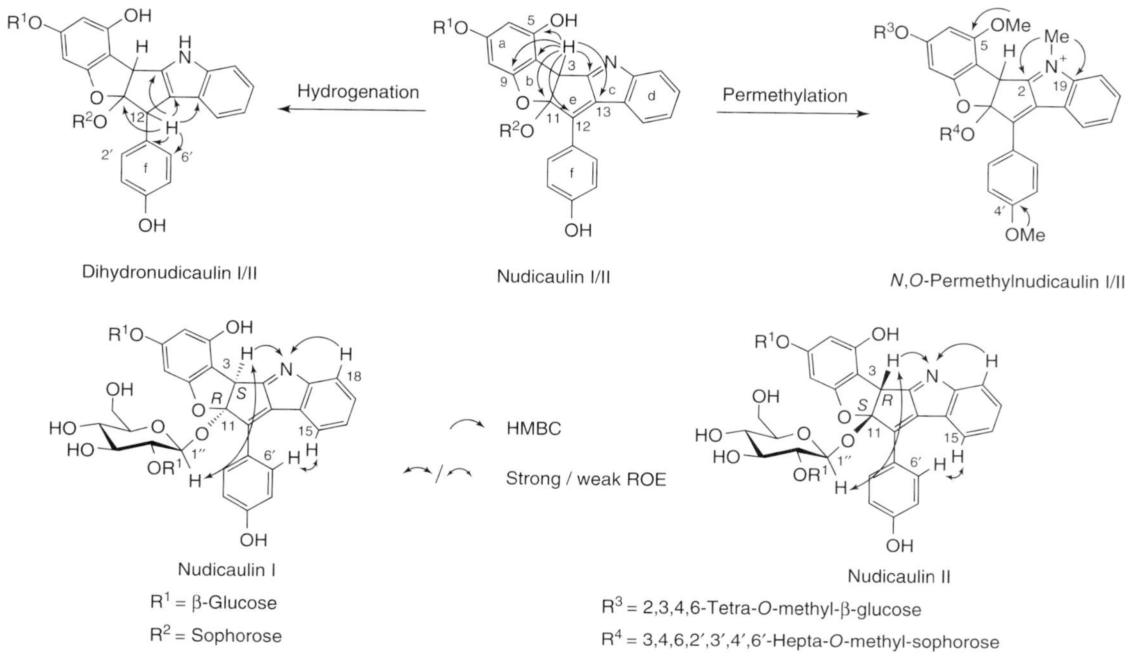

Figure 14.1. 2D NMR correlations established the structure of the nudicaulin aglycone. Nudicaulins are flower pigments of Island poppy, *Papaver nudicaule*. $^1H–^{13}C$ HMBC cross signals of H-3 in nudicaulins I and II and of H-12 of the dihydro derivatives are important keys with which to elucidate the planar structure. $^1H–^{15}N$ HMBC and ROESY signals between H-2'/6' and H-15 supported annelation of the indole ring to the central cyclopentene ring. $^1H–^{13}C$ HMBC correlations of the *N*- and *O*-methyl signals in the spectra of permethylnudicaulins I and II assigned the positions that correspond to the free amino and hydroxy groups of the parent nudicaulins. The different intensity of cross signals between H-3 and H-1'' in the ROESY spectra of nudicaulins I and II assigned the relative configuration at the two stereogenic centers of the aglycones at C-3 and C-11[27].

proton H-3 of underivatized nudicaulin and H-12 in dihydronudicaulin established the annelation of the pentacyclic carbon skeleton rings (Figure 14.1). The structure was further supported by $^1H–^{13}C$ HMBC correlations of the *N*- and *O*-methyl signals in the spectra of nudicaulins I and II permethyl derivatives. These signals were used to assign the methyl groups to previously unsubstituted oxygen and nitrogen functionalities of underivatized nudicaulins. Assignment of the relative configuration of the two stereogenic centers in the aglycon was also performed by NMR. A strong cross signal in the Rotating-frame Overhauser Effect Spectroscopy (ROESY) NMR spectrum between H-3 and H-1'' (Figure 14.1) of nudicaulin I showed that H-3 and the glucose attached to C-11 were oriented to the same side of the aglycon (cis). The corresponding ROE signal of nudicaulin II was far less intense than that of nudicaulin I and was also

in agreement with cis-configuration. The absolute configuration of the two diastereomers, nudicaulins I and II, has been assigned by interpreting the experimental electronic circular dichroism (ECD) spectra using quantum-chemical ECD calculations.[27]

Floral nectar is another plant resource containing ecologically relevant natural products. The major pigment responsible for the dark brown nectar of *Leucosceptrum canum* (Labiatae), the 'birds' Coca Cola tree', has recently been isolated and identified as a unique symmetrical proline–quinone conjugate, 2,5-di-(*N*-(−)-prolyl)-*para*-benzoquinone ((−)DPBQ).[28] On the basis of the molecular formula and NMR measurements, a structure composed of two proline units and a benzoquinone was suggested. An unusually strong HMBC coupling through four bonds from H-3 to C-6 and from H-6 to C-3 was observed in the $^1H–^{13}C$ HMBC

2,5-Di-(*N*-prolyl)-*ortho*-benzoquinone
(hypothetical alternative structure)

2,5-Di-(*N*-(−)-prolyl)-*para*-benzoquinone
(−)-DPBQ

Figure 14.2. Structures of 2,5-di-(*N*-(−)-prolyl)-*para*-benzoquinone ((−)-DPBQ), which is identical in all respects with the natural bird-attracting pigment from the colored floral nectar of *Leucosceptrum canum*, and a hypothetical alternative planar structure, 2,5-di-(*N*-prolyl)-*ortho*-benzoquinone. The symmetrical structure of (−)-DPBQ is indicated by an unusually strong HMBC coupling through four bonds from H-3 to C-6 and from H-6 to C-3. The arrows indicate ^{1}H–^{13}C HMBC correlations[28].

NMR spectrum (Figure 14.2), and indicated a symmetrical structure for the compound, which could be either 2,5-di-(*N*-prolyl)-*para*-benzoquinone or 2,5-di-(*N*-prolyl)-*ortho*-benzoquinone. The two stereoisomeric 2,5-di-(*N*-prolyl)-*para*-benzoquinones ((−)-DPBQ and (+)-DPBQ), synthesized from benzoquinone and (*S*)-L-(−)-proline and (*R*)-D-(+)-proline, respectively, showed ^{1}H and ^{13}C NMR spectra that were identical to the isolated pigment. Measurement of the circular dichroism (CD) spectrum and optical rotation showed that (−)-DPBQ was identical with the natural product and ruled out (+)-DPBQ. Behavioral experiments with both isolated and synthetic samples indicated that DPBQ functions mainly as a color attractant to bird pollinators.

Phytoalexins and phytoanticipins belong to structurally diverse classes of phytochemicals such as phenolics, terpenoids, and glucosinolates. Phenylphenalenones, a group of polycyclic phenolics, were identified as phytoalexins in the Musaceae (banana family), which is susceptible to various pathogens such as fungi and nematodes. However, some *Musa* varieties, e.g., 'Yangambi km5' show resistance to several pathogens. New phenalenones were identified from 'Yangambi km5' by ^{1}H and 2D NMR spectroscopic methods such as ^{1}H–^{1}H correlation spectroscopy (^{1}H–^{1}H COSY), ^{1}H–^{13}C HMBC, and ^{1}H–^{13}C heteronuclear multiple quantum coherence (HMQC) spectra.[29] In vitro, the new compounds displayed strong activity against *Mycosphaerella fijiensis*, the causative agent of the Black sigatoka, a devastative disease of banana. Challenging in vitro plants of *Musa acuminata* with the nonpathogenic yeast strain *Sporobolomyces salmonicolor* resulted in the induction of phenylphenalenone-type compounds, which again were identified by one-dimensional (1D) and 2D NMR spectroscopic methods.[30]

The model plant *Arabidopsis thaliana* has been extensively employed in ecological plant–pathogen interaction studies. Using wild-type and mutant root cultures of *A. thaliana* as an experimental system, and the root-pathogenic oomycete, *Pythium sylvaticum*, for infections, the aromatic metabolite profiles in extracts from uninfected and infected roots, as well as from the surrounding medium, were determined. A number of indolic, heterocyclic, and phenylpropanoid compounds were structurally identified by NMR and MS.[31] Indol-3-ylmethylamine and other intermediates of a new glucosinolate metabolism pathway in living plant cells[32] were identified by NMR spectroscopy and other analytical tools. The pathway mediates the broad-spectrum antifungal defense in *A. thaliana* and differs from the glucosinolate metabolism pathway activated by chewing insects.

Insect-associated microorganisms are another source of ecologically relevant natural products. Beewolf digger wasps (*Philanthus triangulum*) cultivate specific symbiotic bacteria (*Streptomyces* spp.) that are incorporated into the larval cocoon for protection against pathogens. The bacteria produce a cocktail of nine antibiotic substances. Acting complementarily, the cocktail defends the wasp larvae against all tested microbes in a way that parallels the combination prophylaxis known from human medicine.[33] NMR, together with MS, readily identified streptochlorine, a substance previously isolated from a marine *Streptomyces* species,[34] and several piericidin derivatives also known from Streptomyceta.

14.2.3 Enzyme Product Analysis

The identification of small molecules, which are produced by recombinant proteins or enzymes isolated (or partially purified) from plant or insect tissue, is an important objective in chemical ecology. Owing to the limited quantity of product formed in enzyme assays, in many cases, only ¹H NMR using a cryoprobe has been sufficiently sensitive to detect such compounds. For example, to study the mechanism of the biosynthetic conversion of the furanocoumarin (+)-columbianetin to angelicin, specifically deuterated *syn*-[30-²H]columbianetin was incubated with the angelicin synthase CYP71AJ4 functionally expressed in yeast cells. Liquid chromatography-solid-phase extraction (LC-SPE) ¹H NMR analysis of the product in a cryoprobe (500 MHz) showed data that are consistent with the enzyme-derived product such as *syn*-[30-²H](*anti*-30-hydroxycolumbianetin).[35] The LC-NMR study established the mechanistic details of the enzyme reaction and closed a gap in furanocoumarin biosynthesis.

The ecological function of cocaine appears to be to protect the coca plant (*Erythroxylum coca*) against herbivores and pathogens. However, limited information is available about the biosynthesis of this tropane alkaloid. The reduction of the 3-keto function of 2β-carbomethoxy-3-tropinone (methylecgonone) is a key step in the biosynthetic pathway. Methylecgonone reductase (MecgoR), purified from young coca leaves, was shown by ¹H NMR spectroscopy to stereospecifically reduce methylecgonone to 2β-carbomethoxy-3β-tropine (methylecgonine).[36] The ¹H NMR spectrum of the product obtained from incubating MecgoR with methylecgonone was compared with the spectrum of an authentic sample of methylecgonine. Chemical shifts and multiplicities of ¹H NMR signals of authentic methylecgonine (Figure 14.3a) matched closely with corresponding signals of enzyme product (Figure 14.3b), which, therefore, was assigned 3β-configuration. ¹H NMR spectra were recorded at 500 MHz in MeOH-d_4 in capillary tubes (2 mm) using a triple-resonance inverse (TCI) cryoprobe.

Metabolic enzymes can be involved in the detoxification of active natural products or synthetic pesticides. Joußen *et al.*[37] have shown that the resistance toward the insecticide fenvalerate in an Australian strain of *Helicoverpa armigera* is due to a unique P450 enzyme. This chimeric protein, CYP337B3, arose from the unequal crossover between two parental P450 genes. CYP337B3 was capable of converting fenvalerate into a metabolite whose mass spectrum suggested an additional oxygen atom in the molecule. NMR spectroscopy

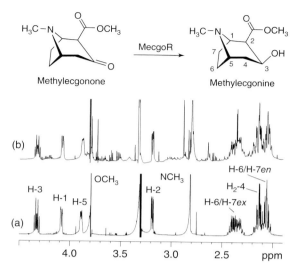

Figure 14.3. Stereospecificity of methylecgonone reductase (MecgoR)-catalyzed reduction demonstrated by ¹H NMR spectroscopy (500 MHz, MeOH-d_4, 2 mm capillary tubes, 5 mm TCI cryoprobe). The spectrum of the product of methylecgonone reduction (b) matches closely with the spectrum of authentic methylecgonine (3β-product) (a).[36]

was used to establish the structure, especially the position of the hypothetical new hydroxyl group. As the available amount of material was insufficient for acquiring heterocorrelation experiments, the position of hydroxylation was inferred from comparing ¹H NMR spectra of the metabolite with the spectrum of the (2S,αS)-enantiomer, esfenvalerate. Instead of the signals of the monosubstituted phenyl ring of the phenoxybenzyl alcohol moiety, an AA'MM' spin system characteristic of a *p*-substituted aromatic ring was found in the spectrum of the metabolite. Thus, the new hydroxyl group was assigned to position 4' of the phenyl ring. The coupling pattern of H-2'/6' and H-3'/5' can be clearly distinguished from another AA'MM' spin system assignable to the chlorophenyl ring of the aromatic acid moiety by its low-frequency chemical shift. Moreover, the presence of two sets of ¹H NMR signals in a 1:1 ratio indicated the racemic character of the metabolite 4'-hydroxyfenvalerate with two pairs of enantiomers, likely (2S,αS)/(2R,αR) and (2S,αR)/(2R,αS). Further examples of applying NMR methods to identify ecologically relevant or otherwise interesting products produced by recombinant or isolated enzymes have been reviewed recently.[38]

14.3 NMR-BASED MIXTURE ANALYSIS

14.3.1 One-dimensional (1D) NMR

The NMR-based analysis of mixtures of natural products in biological samples has traditionally been performed using 1D NMR methods, namely ^{1}H and ^{13}C NMR spectroscopies. The major advantages of ^{1}H NMR are its relatively high sensitivity and its quantitative character. A simple example of how ^{1}H NMR was used to determine components in a mixture is shown in Figure 14.4. Buch *et al.*[39] have identified two isomeric naphthoquinones, plumbagin and 7-methyljuglone, by ^{1}H NMR in pitcher tissue of *Nepenthes* species. Signals of the two compounds were readily distinguished by their different integral intensity, chemical shifts, multiplicity, and coupling constants. The naphthoquinones are secreted into the fluid of the pitfall, which contains debris from formerly trapped insects. During the digestion process, the naphthoquinones are supposed to preserve the digestive fluids and trapped insects from microbial infestation.

Because of the frequently limited availability of ecologically relevant compounds, methods for improving sensitivity are in high demand. Miniaturization is one of the strategies to improve the signal-to-noise ratio of NMR.[40] A microcoil probe operating at 600 MHz with a 5 μl sample volume was used to identify 13 new steroids in a limited amount of firefly extract (*Lucidota atra*).[41] Owing to the extreme signal overlap of the multiple similar steroids, partial fractionation via high-performance liquid chromatography (HPLC) was necessary. Fractions containing one to three components were subjected to offline NMR analysis.

The introduction of cryogenically cooled probes further enhanced NMR sensitivity[42] and has enabled many of the studies discussed in this chapter. A high-temperature superconducting 1 mm probe has been constructed, combining microcoil and cryogenic technologies, which drastically improves the sensitivity of NMR.[43] Using this probe facilitated analysis of samples that were milked from single insect, *Anisomorpha buprestoides* (a 'walking stick' species), by ^{1}H NMR spectroscopy.[44] The spectra of samples of insect individuals revealed the occurrence of a heterogeneous mixture of defensive dolichodial-like stereoisomers. The composition of the mixture varied among *A. buprestoides* individuals and over time.

14.3.2 Combining Different Magnetic Resonance Methods

For complex mixtures, the limited signal dispersion of ^{1}H NMR spectra may create detrimental signal overlap, which adversely affects the feasibility of using ^{1}H NMR to identify individual components.[45] To increase the signal resolution, high-frequency magnets are very useful but often too expensive for natural product laboratories. Chemometric methods can be used to resolve overlapping ^{1}H NMR signals, and 1D spectra devoid of proton–proton coupling can be reconstructed from 2D J-resolved spectra.[46] Signal overlap in ^{13}C NMR is not problematic, but sensitivity and the nonquantitative characters are issues. Considering the advantages and drawbacks, a combination of different NMR methods can help identify and quantify the components of a mixture.

The composition of floral nectar, especially the concentration and relative proportion of the carbohydrates (glucose, fructose, and sucrose), amino acids, and special secondary metabolites, plays an important role in plant–pollinator and plant–herbivore interactions and evolutions.[47] A combination of magnetic resonance methods have been used, for example, to characterize the chemical composition and aspects of formation of

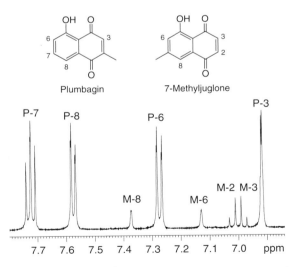

Plumbagin   7-Methyljuglone

Figure 14.4. Partial ^{1}H NMR spectrum of an extract of *Nepenthes alata* pitcher tissue showing methine signals of plumbagin (P) and 7-methyl-juglone (M). The spectrum was used to identify low concentrations of the naphthoquinones in the pitfall fluids of carnivorous *Nepenthes* species[39].

floral nectar of Australian *Anigozanthos* species (kangaroo paw, family Haemodoraceae) with a minimum of sample preparation and without derivatization.[48,49] Minor components such as amino acids, organic acids, and other components were detected and quantified by ^{1}H NMR and major carbohydrates by ^{13}C NMR. A ratio of glucose : fructose of 1.2 : 1, which is inconsistent with the simple hydrolysis of phloem-derived sucrose by invertases, was found in the floral nectar of *Anigozanthos*. The phenomenon of an unbalanced glucose : fructose ratio, which is known from many other nectars as well, deserved detailed investigation. A combination of ^{13}C NMR and magnetic resonance imaging (MRI) methods, applied to *Anigozanthos* inflorescences, and considerations about the labeling position of the nectar hexoses, finally resulted in a hypothesis, which explained the unbalanced glucose : fructose ratio found in the nectar of this plant.[49] [1]H gradient- and ^{1}H spin-echo images visualized vascular bundles of *A. flavidus* peduncles. [1]H spin-echo images of flowers were recorded to study the architecture of nectaries. After ^{13}C-labeled carbohydrates were hydroponically administered to excized peduncles, chemical shift imaging (CSI) and cross polarization (CP) experiments using a cryogenic microimaging probe (400 MHz) were carried out to locate the joint [1]H–[13]C cross signal of C-1 of β-D-glucopyranose and the glucose moiety of sucrose in the vascular bundles. The positions of ^{13}C in the nectar hexoses and the carbohydrate composition in extracts of peduncle segments taken from different positions below the flower were determined by ^{13}C NMR. The results clearly indicated that the sucrose after uptake from the vascular bundles must be hydrolyzed, so that the hexoses can enter into the glycolysis, gluconeogenesis, and pentose phosphate pathways, finally resulting in the observed unbalanced glucose : fructose ratio.

14.3.3 Two-dimensional (2D) NMR

2D NMR spectra in their homo- or heteronuclear shift-correlation versions improve resolution by spreading the resonances in a second spectroscopic dimension. Overlapping signals of one nucleus (i.e., ^{1}H) can be resolved along the dimension of the second nucleus, which in natural products mostly is ^{13}C or ^{15}N but can also be ^{31}P or another nucleus. For example, the analysis of the crude content of pygidial glands of the rove beetle *Stenus solutus* using various 2D NMR methods resulted in the structural

elucidation of a new pyridine alkaloid.[50] The signals of the high-frequency part of the spectrum are readily assignable to the pyridine ring and a double bond. However, the region of carbohydrate and aliphatic signals was crowded because of a variety of matrix components. Nevertheless, standard ^{1}H–^{1}H COSY, [1]H–[13]C heteronuclear single quantum coherence (HSQC), and ^{1}H–^{13}C HMBC spectra recorded with a cryoprobe at 500 MHz ^{1}H resonance frequency allowed the allocation of the signals of the aliphatic part of the side chain and, thereby, elucidating the structure of the new alkaloid from the crude mixture as (*E*)-3-(2-methyl-1-butenyl)-pyridine. This alkaloid has antimicrobial properties and is potentially involved in skimming, a method of locomotion of the beetles on water surfaces.

Although 2D NMR methods enormously improve the performance of mixture analyses and provide atomic correlations, further increases in sensitivity and resolution are needed to achieve the ultimate goal of metabolomics, namely, the identification of all the possible metabolites of a biological system (see Chapter 27). Therefore, recording heteronuclear 2D NMR experiments of samples from plants labeled with stable isotopes has been proposed.[51,52] Such spectra, when compared to the spectra of unlabeled samples, display cross signals of enhanced signal-to-noise ratio. Furthermore, the subtraction of 2D spectra obtained from mutant or stimuli-treated samples and untreated control samples revealed metabolites that were either increased or decreased in concentration among those samples. As quantitative and qualitative differences between samples from, for example, organisms with different genetic backgrounds or organisms that have been exposed to different stimuli are most relevant in chemical ecology, methods for differential analysis of 2D NMR spectra have been developed.

Double-quantum-filtered correlation spectroscopy (DQF-COSY) was used to analyze ethyl acetate extracts of the broths of insect–pathogenic fungal cultures.[53] A two-step protocol for a differential analysis of the DQF-COSY NMR spectra, later designated as differential analysis by two-dimensional nuclear magnetic resonance spectroscopy (DANS),[54] was developed for this study. The first step consisted of a graphic analysis based on the multiplicative stacking of bitmaps derived from magnitude mode versions of the DQF-COSY NMR spectra. DQF-COSY-derived bitmaps were prepared using commonly available image editing software. This technique clearly distinguished signals present in one spectrum from signals present in another spectrum or common to several spectra. The second step of the DANS protocol consists of a detailed analysis of the signals representing unique or unusual metabolites in a specific extract. The corresponding spin systems were characterized based on the phase-sensitive originals of the DQF-COSY NMR spectra. If required, additional 2D NMR experiments such as $^1H-^{13}C$ HSQC and HMBC NMR spectra were carried out for signal assignment.

The DANS approach was applied to compare the pheromone composition of a signaling-deficient mutant, daf-22, of the nematode *Caenorhabditis elegans*, which is unable to enter the dauer stage, with that of wild-type nematodes.[54] For a differential analysis of the DQF-COSY NMR spectra, the spectrum derived from the mutant sample was superimposed onto the spectrum derived from the wild-type pheromone (Figure 14.5), using a specific algorithm that suppressed signals present in both spectra. The study revealed that three previously identified ascarosides were missing in the mutant but were present in the wild type. In addition, four new ascarosides were identified by the DANS approach in the wild-type nematode. Biological evaluation revealed that some of the ascarosides identified via DANS strongly attracted male nematodes and showed significant dauer induction. Hence, the DANS approach

enabled metabolites to be linked with their biological function.

Recently, Izrayelit *et al.*[55] combined DANS with principal component analysis (PCA) to develop multivariate differential analysis by 2D NMR spectroscopy (dvaDANS). In this approach, coefficients from the PCA loadings were back-projected onto the DQF-COSY NMR spectra, which revealed a large number of cross peaks that are up- or downregulated in daf-22 mutants relative to wild-type *C. elegans* nematodes. The methods for identifying metabolites in complex mixtures, especially such compounds that correlate with changes in genotype and phenotype, have been reviewed.[56]

Covariance NMR[57] is another approach to enhance the resolution of homonuclear 2D NMR spectra. By applying covariance processing instead of Fourier transformation along the indirect dimension, the 2D correlation information is reconstructed by statistical means resulting in fully symmetrical spectra with identical resolution along the two dimensions. Covariance total correlation spectroscopy (TOCSY) NMR was used to analyze the unpurified venom mixture of an individual adult female walking stick insect (*A. buprestoides*). In addition to glucose, the venom mixture contained two stereoisomeric dialdehydes, each in slow chemical exchange with a diol.[58] The 1 mm high-temperature superconducting NMR probe[43] was

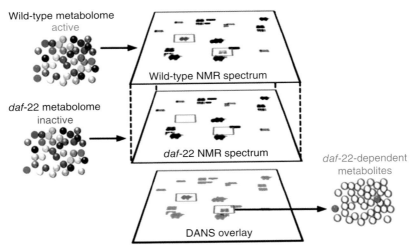

Figure 14.5. Schematic representation of differential analysis by 2D NMR spectroscopy (DANS).[54] Differences between the spectra of extracts obtained from *Caenorhabditis elegans* wild-type and daf-22 mutant were used to identify candidate molecules with dauer-inducing and male-attracting activities. (Reproduced with permission from Ref. 54. © 2009 National Academy of Sciences)

used as described earlier for a single insect NMR study.[44] The covariance TOCSY NMR spectrum displayed spin–spin connectivity information across each molecule of the mixture and allowed scientists to deconvolute the spectrum by DemixC cluster analysis into subspectra of the individual mixture components.[59] Finally, the chemical components were identified by screening the subspectra against an NMR database.

Metabolomic studies make use of statistical techniques such as PCA, partial least squares (PLS), and discriminant analysis (DA) to search for correlations between spectral signals and the biological states of the analyzed samples.[60,61] In metabolomic practice, chemical shift variation due to fluctuations in experimental conditions such as pH and sample temperature is an issue that counteracts the application of high-resolution NMR, including 2D techniques, and hampers pattern-recognition algorithms. Bucketing (binning) is used to overcome this drawback but causes significant loss of valuable information from spectral resolution, complicating subsequent interpretation and the assignment of relevant metabolites. Alignment methods are, therefore, highly demanded. A peak alignment algorithm, termed 'hierarchical alignment of two-dimensional spectra – pattern recognition' (HATS-PR) has been developed.[62] PCA and PLS regressions of full-resolution TOCSY NMR spectra aid the assignment and interpretation of statistical pattern-recognition results by producing back-scaled loading plots that look like traditional TOCSY NMR spectra but incorporate qualitative and quantitative biological information. Using HATS-PR, sets of TOCSY NMR spectra obtained from exudate (exometabolome) samples of two nematode species, *Pristionchus pacificus* and *Panagrellus redivivus*, were compared.[62] The method showed that the two species produce significantly different sets of polar small molecules. *P. redivivus* released a less diverse set of polar molecules compared to *P. pacificus* but produced large amounts of the disaccharides trehalose and sucrose, even so no glucose was observed, whereas *P. pacificus* released a very diverse set of polar small molecules including glucose but the two disaccharides, sucrose and trehalose, were not detected. Spectra and principal component loadings suggest that the two nematodes produce different mixtures of ascaroside-like compounds.

14.3.4 Observing Molecular Interactions by NMR

As mentioned earlier, the ecological function of cocaine and its derivative, cinnamoylcocain, appears to protect the coca plant (*E. coca*) against herbivores and pathogens. Plants that synthesize cocaine derivatives (or other toxic natural products) may store these compounds in their vacuoles in order to avoid autotoxicity. To keep toxins trapped within the vacuoles, complexation with polyphenols such as chlorogenic acid can be an efficient mechanism. A physical interaction between cocaine and chlorogenic acid was observed and quantified in vitro using NMR and absorption spectroscopic methods.[63] Chemical shift differences ($\Delta\delta$) were extracted from the ^{1}H NMR spectrum of chlorogenic acid in the presence of excess amounts of cocaine and from the ^{1}H NMR spectrum of cocaine in the presence of excess amounts of chlorogenic acid (Figure 14.6). The extent of the $\Delta\delta$ values was interpreted as a measure of molecular interaction. All chemical shift differences detected were toward lower frequency, which means greater nuclear shielding caused by increased magnetic anisotropy because of the induced field from the overlap of the p orbitals of the two compounds, as proposed for the interaction of caffeine with chlorogenic acid.[64] The results of this study suggested that storing cocaine and related coca alkaloids in *E. coca* vacuoles involves hydroxycinnamoyl quinate esters as complexation partners.

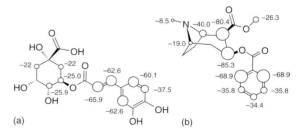

Figure 14.6. Chemical shift differences ($\Delta\delta$) in the ^{1}H NMR spectra of chlorogenic acid (a) and cocaine (b) observed in the presence of the other compound as compared to the pure reference. Values of ^{1}H NMR chemical shift differences ($\Delta\delta$ in Hertz) are shown next to the circles that differ in size according to the magnitude of the shift. (Reproduced with permission from Ref. 63. © 2012 Elsevier)

14.4   NMR-BASED CELL-TYPE-SPECIFIC LOCALIZATION

Isolating and identifying individual compounds from a plant or other organisms in conventional natural products chemistry involves the preparation of a homogeneous extract followed by the tedious separation by liquid–liquid distribution, solid-phase extraction, and/or chromatographic methods before the properties can be explored using bioassays. Unlike studies of whole organisms or organs in classical natural products chemistry, chemical ecology requires micrometabolic profiling analyses to gain information about the cell- or cell-type-specific distribution and localization of metabolites in order to make inferences about their role in interactions with other organisms. For micrometabolomics and micrometabolic profiling studies, metabolites do not necessarily have to be separated, but methods have been developed to analyze compounds in a mixture such as crude extracts or partially purified fractions (see Section 14.3). Magnetic resonance is able to detect metabolites in specific cells or tissues by means of two methods, MRI and magnetic resonance spectroscopy (MRS), usually called NMR. MRI is typically executed in vivo, i.e., it is nondestructive not only with respect to the detected compounds but also with respect to the studied tissue. MRI[65,66] has had many uses as a tool with which to observe developmental processes, water status and dynamics, stress response, host–pathogen interaction, gene expression and function, and metabolism, most of which are beyond the scope of this chapter. However, the detection of secondary metabolites in specific plant cells or tissue, which is of special interest in chemical ecology, often suffers from low concentrations of these compounds and, therefore, MRI has been applied less frequently for that purpose. Hence, only abundant metabolites such as carbohydrates and some amino acids have been localized by CSI and other MRI techniques.[67] Stable isotope labeling of substrates can enhance the sensitivity of ^{13}C NMR detection. For example, a combination of NMR spectroscopy and CP imaging localized β-D-glucopyranose and sucrose in vascular bundles of *A. flavidus* and showed the conversion of administered ^{13}C-labeled glucose to sucrose during transport to the nectaries (see Section 14.3.2).[49]

NMR spectroscopic analysis of extracts obtained from cells or tissue pieces of interest represents another approach to investigate the spatial distribution of small molecules on a microscopic scale. This approach requires separating the target material from the surrounding tissue of the organism under investigation. As microscopic tissue pieces usually are subject to such studies, several methods for harvesting single cells or specimens of microscopic size have been developed among which LMD probably is the most useful.[10,11]

LMD has been applied to harvest secretory cavities from leaves and petals of *Dilatris pillansii* (Haemodoraceae) and a herbarium specimen of *Dilatris corymbosa* flowers.[68] Extracts of a few of these relatively large secretory cells were sufficient to obtain ^{1}H NMR spectra of good signal-to-noise ratio and resolution using a cryogenically cooled 5 mm NMR probe and to identify *O*-methyl derivatives of phenylphenalenones by comparing their spectra with those of previously isolated reference compounds (Figure 14.7). For example, a signal-to-noise ratio of 310 for a typical one-proton doublet was obtained from only 10 dissected cavities when the ^{1}H NMR spectrum was measured at 500 MHz for 35 min (800 transients). The ratio of the three major compounds of the sample from *D. pillansii* was readily assessed from representative integrals. It is suggested that the *O*-methylphenylphenalenones accumulating in the secretory cavities are involved in the defense of this plant against herbivores.

The combination of LMD with NMR was also applied to investigate the contents of specialized cells in the bark of Norway spruce (*Picea abies*). First, the so-called stone cells (sklereids) which, because of their lignified structure, are thought to be involved in mechanical defense against chewing insects, were dissected using the laser microscope, followed by NMR analysis. However, the identification of astringin, a stilbene glucoside and a dihydroxyquercetin glucoside indicated that sklereids are not completely lignified; rather, although somehow depleted, they also contain low-molecular-weight defense metabolites. A sample obtained from less than 1 mg laser-microdissected tissue was sufficient to record not only ^{1}H NMR but also ^{1}H–^{1}H COSY and ^{1}H–^{13}C HSQC spectra at 500 MHz proton resonance frequency using a 5 mm cryoprobe.[69]

Continuing the study of specialized cells of Norway spruce, phloem parenchyma cells that swell and change their contents upon attack by the bark beetle *Ips typographus* and its accompanying fungus *Ceratocystis polonica* were analyzed with LMD and ^{1}H NMR spectroscopy.[70] Phloem parenchyma cells and material from adjacent sieve cells were dissected from 130 to 230 cross sections (thickness 40 mm), corresponding to an area of $1–2$ mm^2. The

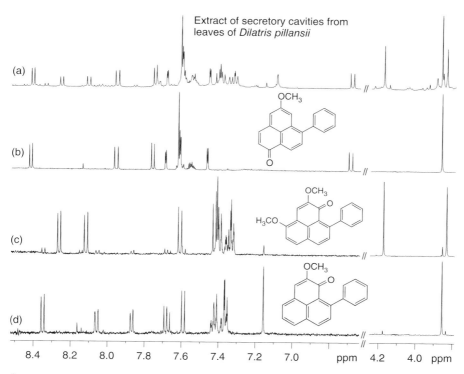

Figure 14.7. ¹H NMR spectroscopic comparison of an (a) extract of secretory cavities dissected from leaves of *Dilatris pillansii* with spectra of phenylphenalenones previously isolated from this plant. (b) 5-Methoxy-7-phenylphenalen-1-one; (c) 2,4-dimethoxy-9-phenylphenalen-1-one; and (d) methoxyanigorufone.[68]

dissected cell material was directly extracted with CD₃OD (200 µl) and the extracts were transferred to 3 mm NMR tubes for ¹H NMR measurements at 500 MHz in a cryoprobe (5 mm). ¹H NMR spectra typically were recorded with 2K transients to obtain spectra with reasonable signal-to-noise ratio. The residual water signal was suppressed using the pre-saturation utilizing relaxation gradients and echoes (PURGE) sequence.[71] As in the sklereids of Norway spruce bark tissue, the stilbene glucoside astringin was also detected in the phloem parenchyma cells. Contamination from the plastic tube used to collect the samples and from other sources is an issue in long-term ¹H NMR measurements. However, in the case of phloem parenchyma cells, signals of contamination were used as a reference to estimate the astringin level in the different samples (Figure 14.8). Interestingly, the spectra revealed that the concentration of astringin declined in the phloem parenchyma cells after infection, whereas (+)-catechin became detectable.

Trichomes are well known as a major site of the production of defense compounds in plants. For example, a mutant of *Nicotiana attenuata* possesses enlarged trichomes on the leaf surface, which allowed the scientist to harvest sufficient cell material for NMR-based metabolite profiling. LMD was not needed to sample segments (up to four cells), and the heads of two different trichome types but tweezers and a glass capillary, were used instead. ¹H NMR spectra (500 MHz; 5 mm cryoprobe) of trichome segment extracts displayed signals of nicotine, phaseoloidin, and acyl sugars.[72] The deshielded NMR signals of nicotine and phaseoloidin were used to determine the relative levels of the two substances in the different samples. It was assumed that all trichome cells have the same water content and so the integral of the water signal was used as a reference.

The surface of *Cannabis sativa* leaves is densely settled with different types of trichomes, which produce various cannabinoids, presumably to defend the plant against herbivores and pathogens. Again,

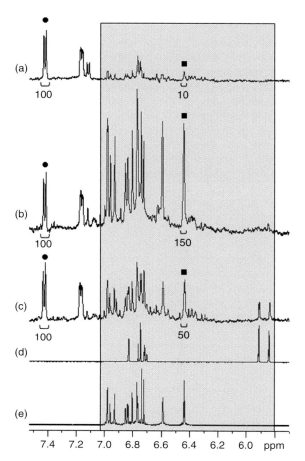

Figure 14.8. Partial ^{1}H NMR spectra from MeOH extracts of laser-microdissected Norway spruce bark cells. (a) Sieve cells; (b) phloem parenchyma cells; (c) phloem parenchyma cells 7 days after inoculation with *Ceratocystis polonica*; (d) (+)-catechin reference; and (e) astringin reference.[70] Signals outside the box originate from contaminants of the sample collection device. ■: Signal of H-12 of astringin with the integral value for quantification listed below the spectra; ●: signal used as a reference to estimate the astringin level in spectra (a), (b), and (c). (Reproduced with permission from Ref. 70. © 2012 Wiley)

^{1}H NMR using a cryoprobe confirmed the occurrence of some major cannabinoids in capitate-stalked and capitate-sessile trichomes, both of which had been detected earlier by LC-MS. In addition, ^{1}H NMR using a cryoprobe (500 MHz) enabled the identification of cannabichromenic acid in extracts of

laser-dissected trichomes. In contrast to other cannabinoids, cannabichromenic acid was not identified by LC-MS, as a standard was not available.[73]

14.5 NMR-BASED BIOSYNTHETIC AND METABOLIC LABELING STUDIES

NMR methods have been used extensively to establish biosynthetic and metabolic pathways and to identify reaction mechanisms. Nuclei used for stable isotope labeling are directly detectable by NMR (e.g., ^{13}C, ^{15}N, and ^{2}H) or through their isotope-induced effect on the chemical shift of adjacent nuclei (e.g., ^{16}O).[5,6,74]

Glucosinolates are a group of plant natural products mainly occurring in the Brassicaceae plant family. Glucosinolates and their breakdown products – for example, isothiocyanates – play an important role in plant–herbivore and plant–pathogen interactions. The structure of glucosinolates comprises a sulfonated oxime, a β-thioglucose moiety, and a variable aliphatic, indolic, or benzylic side chain. Biogenetically, the side chain originates from amino acids.[75,76] In many cases, the biosynthetic pathway involves elongation cycles, which insert one or more methylene groups into the amino acid side chain. NMR-based stable isotope labeling studies contributed much to elucidate the mechanism of the side-chain elongation. For example, ^{13}C NMR studies using the leaves of *Eruca sativa* revealed that ^{13}C-labeled methionine, with the exception of the COOH carbon, is uniformly incorporated as a unit into 4-methylthiobutylglucosinolate (4MTB).[77] The ^{13}C NMR spectrum of the desulfated 4-methylthiobutylglucosinolate (ds-4MTB) (Figure 14.9a) exhibited enhanced signals of C-2, C-3, C-4, and the S-methyl group. Moreover, the incorporation of C-2, C-3, and C-4 as a unit from methionine was inferred from the ^{13}C–^{13}C couplings $J_{C-2–C-3} = J_{C-3–C-4} = 34.7$ Hz of the signals at $\delta_C = 37.0$ (C-4), 31.7 (C-3), and 30.4 (C-2). The central signals of C-2, C-3, and C-4 representing ^{13}C$_1$-isotopologs were not enhanced, confirming that the C–C bonds among C-2, C-3, and C-4 of administered [U-^{13}C]methionin remained uncleaved during the biosynthesis. The ^{13}C NMR spectrum of ds-4MTB (Figure 14.9b) obtained from a [2-^{13}C]acetate feeding experiment showed central signal lines of the pseudotriplets of C=N ($\delta_C = 160.8$) and C-1 ($\delta_C = 35.7$), which compared with the signals of the nonenriched (natural abundance, Figure 14.9c) C-2–C-4 were significantly enhanced.

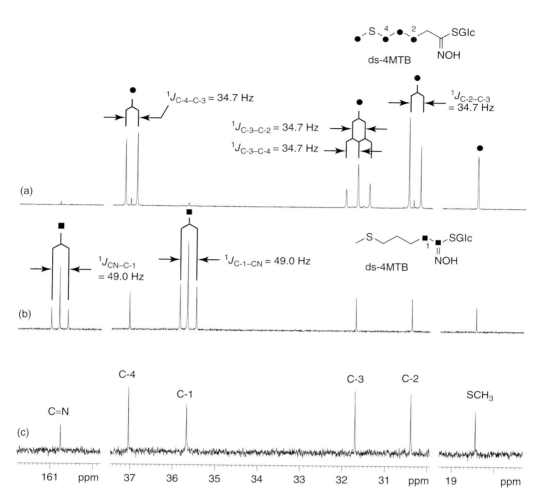

Figure 14.9. Aglycon region of ^{13}C NMR spectra (125 MHz, D$_2$O) of 4-methylthiobutylglucosinolate (isolated in the desulfated form, ds-4MTB) obtained by biosynthetic labeling in *Eruca sativa*.[77] (a) The spectrum of ds-4MTB obtained after administration of [U-^{13}C]methionine shows enhancement of signals at C-2, C-3, C-4, and SCH$_3$. The extensive ^{13}C–^{13}C coupling among C-2, C-3, and C-4 suggests the incorporation of this fragment as a unit. (b) The spectrum of ds-4MTB obtained after the administration of [2-^{13}C]acetate shows the enhancement of the signals at C=N and C-1. (c) Spectrum of unlabeled ds-4MTB. ●, ■ = ^{13}C. The data proved a chain-elongation mechanism involved in the biosynthesis of aliphatic glucosinolates.

The enlarged central ^{13}C NMR signal lines indicate high levels of [1-^{13}C$_1$]- and [2-^{13}C$_1$]ds-4MTB isotopologs; these are formed by incorporating two ^{13}C-labeled methyl groups of [2-^{13}C]acetate into C=N and C-1 of the 4MTB. The formation of C=N/C-1 doubly labeled isotopologs from two ^{13}C-labeled methyl groups of [2-^{13}C]acetate was inferred from the C–C coupling ($J_{C-1–C-2} = 49.0$ Hz). It was concluded from these data that two chain-elongation cycles must

have occurred in the biosynthesis of 4MTB. This finding is in agreement with the mechanism, which was later substantiated by biochemical and molecular studies.[78,79]

Understanding specific plant–herbivore interactions is an important subject in chemical ecology. Larvae of *Pieris rapae* (cabbage white butterfly) feed exclusively on glucosinolate-producing plants of Brassicaceae and related families. In the intact plant, glucosinolates are stored separately from their hydrolytic enzymes, the thioglucosidases (myrosinases). Upon tissue damage, for example, by herbivory, glucosinolates and myrosinases come into contact, and the glucosinolates are hydrolyzed to their aglycones, which are unstable and rearrange themselves into isothiocyanates and other breakdown products. *P. rapae* larvae express the so-called nitrile-specifier protein that redirects the hydrolysis of glucosinolates toward the formation of nitriles instead of toxic isothiocyanates. In order to study the detoxification of benzylglucosinolates in the insect gut and identify metabolites downstream in the pathway, larvae of *P. rapae* were allowed to feed on *A. thaliana* leaves, which had taken up [1-^{13}C]benzylglucosinolate.[80] The extracts of larvae feces were collected and separated by HPLC, and the chromatographic fractions were analyzed by 1D and 2D NMR experiments using a cryoprobe (500 MHz). Phenylacetylglycine was identified as a product, and the enrichment of ^{13}C in the benzylic methylene group (C-5) was substantiated by three NMR measurements. First, the ^{1}H NMR signal of 5-^{13}CH$_2$ displayed intense ^{1}H–^{13}C coupling satellites ($^{1}J_{C-H} = 129$ Hz) (Figure 14.10, F2 projection);[81] second, the enhanced cross signal of C-5/H$_2$-5 in the HSQC spectrum was detected at 500 MHz in less than 10 min, whereas the methylene signal of C-2/H$_2$-2 became visible only after 3 h under the same conditions; third, the enlargement of the ^{13}C NMR signal of C-5 at $\delta_C = 43.6$ (Figure 14.10, F1 projection) was observed.[6] Feeding of [2-^{13}C]phenylacetonitrile and [2-^{13}C]phenylacetic acid to the insect larvae (*P. rapae*) also resulted in the formation of [5-^{13}C]phenylacetylglycine. From these data, the metabolic pathway shown in Figure 14.10 was finally established.[80]

Leaf beetle larvae (Chrysomelinae) defend themselves against predators by releasing secretions from dorsal glandular reservoirs containing the iridoid chrysomelidial. Isotope labeling and in vitro isotope exchange experiments revealed stereodivergent cyclization of the ultimate precursor 8-oxogeranial to chrysomelidial in *Gastrophysa* and *Phaedon* species.[82] ^{1}H NMR spectroscopy, in addition to MS, was used to assess the ^{2}H-labeling positions of chrysomelidial after an in vitro isotope exchange experiment. The ^{1}H NMR spectra measured 0 and 24 h after adding ^{2}H$_2$O to the larval secretions of *G. viridula* revealed the position of ^{1}H/^{2}H exchange in chrysomelidial (Figure 14.11). The protons of the exocyclic methyl group were clearly exchanged as shown by the largely

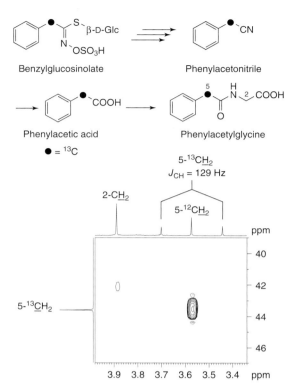

Figure 14.10. Conversion of benzylglucosinolate by the concerted action of myrosinase and nitrilase specifier protein in the gut of *Pieris rapae* (the cabbage white butterfly larvae).[80] Phenylacetylglycine was isolated from the feces of caterpillars feeding on plants treated with [1-^{13}C]benzylglucosinolate. The origin of phenylacetylglycine from benzylglucosinolate was established by ^{13}C labeling in the benzylic position. The intensities of the ^{1}H NMR signals (F2 projection), the 2D HSCQ cross peaks (500 MHz, acetone-d_6, cryoprobe) as well as ^{13}C NMR signals (125 MHz; F1 projection) indicated ^{13}C enrichment of phenylacetylglycine in the methylene group at position 5 but not at position 2.

decreased intensity of the singlet at $\delta_H = 2.1$. Moreover, the doublet at $\delta_H = 0.8$, assignable to the methyl group next to the aldehyde in the C$_3$ side chain, collapsed to a singlet, thus indicating that the acidic H-8 was exchanged. One of the diastereotopic allylic ring methylene protons, namely H-3β, which is oriented above the ring, was also selectively exchanged as shown by the decreased complex multiplet signal at $\delta_H = 2.5$.

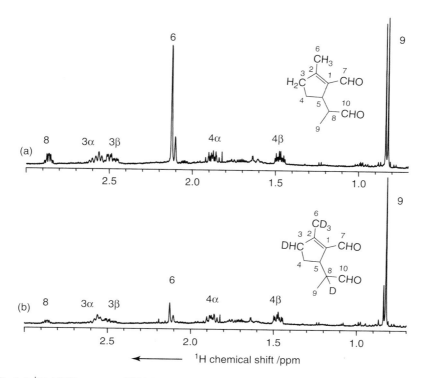

Figure 14.11. Partial ^{1}H NMR spectra (500 MHz, cryoprobe) of *Gastrophysa viridula* leaf beetle larvae secretions in ^{2}H$_2$O.[82] (a) Recorded directly after adding ^{2}H$_2$O and (b) after 24 h. Chrysomelidial was formed by incubation of unlabeled 8-hydroxygeraniol with larval secretions in ^{2}H$_2$O. (Reproduced with permission Ref. 82. © 2013 Wiley)

14.6 NMR DETECTION OF THE METABOLIC MODIFICATION OF PLANT POLYMERS

In addition to small molecules, biological polymers may also play a specific role in ecological interactions. While NMR studies of proteins, polysaccharides, lignins, chitins, etc. are beyond the scope of this chapter, the low-molecular-weight components released from biological polymers or breakdown products are accessible by high-resolution solution-state and high-resolution magic-angle spinning (HR-MAS) NMR methods. Two examples with ecological implication using solution-state NMR are discussed here.

Pectin polymers account for 35% of the cell wall mass of dicotyledonous plants and consist of α-D-galacturonic acid units with variable numbers of methyl ester and *O*-acetyl groups. Herbivore attack and the treatment of puncture wounds with

larval oral secretions increased the activity of leaf pectin methylesterases and decreased the degree of pectin methylation, as determined by solution-state ^{1}H NMR. Demethylation of the pectin polymers by methylesterases releases one molecule of methanol per α-D-galacturonic acid unit. Therefore, it was suggested that an essential portion of methanol released from plants to the atmosphere originates from the activation of pectin methylesterases by herbivore attack.[83] Solution-state ^{1}H NMR was used to rapidly measure free methanol both before and after the saponification of pectin extracts to estimate the level of their methyl ester residues (Figure 14.12). Alkaline conditions cleaved the methyl ester as well as the acetyl linkages of the galacturonic acid residues, allowing the detection of the previously undetected ^{1}H NMR signals of the saponified pectin.[84]

Environmental stresses play an essential role in regulating lignin biosynthesis in lignin-deficient plants. In addition to other altered properties, *N. attenuata*

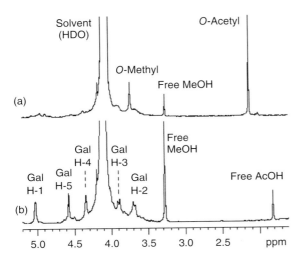

Figure 14.12. ^{1}H NMR spectra (400 MHz, 353 K, D_2O, TMSP-d_4 as internal standard) of pectin from *Nicotiana attenuata* wounded leaf extracts before and after saponification.[83] NMR signals are labeled according to reported assignments.[84] After the first measurement, (a) NaOD in D_2O was added to the sample in the NMR tube and (b) spectrum was recorded to detect the loss of methyl ester and *O*-acetyl groups upon saponification. Note that the signals of methyl ester and *O*-acetyl groups disappear after saponification and that the signal of free MeOH increases.

plants silenced in two cinnamyl alcohol dehydrogenase genes (ir-CAD plants) showed low lignin contents and rubbery, structurally unstable stems when grown in the glasshouse.[85] However, when planted into their native desert habitat, ir-CAD plants exhibited lignin contents comparable to those of wild-type plants and produced similarly robust stems that survived wind storms. Hence, it was suggested that growth under unprotected field conditions alters lignin composition of ir-CAD plants. Solution-state NMR was used to better understand the differences in lignin composition between wild-type and ir-CAD plants grown in different environments. Samples obtained by cellulase treatment of the cell walls isolated from *N. attenuata* stems were acetylated and analyzed by ^{1}H–^{13}C HSQC NMR.[86,87] The NMR data showed, for example, the incorporation of atypical phenolic units (syringaldehyde and sinapaldehyde moieties) into the lignin of ir-CAD plants (Figure 14.13) and the high abundance of feruloyl tyramine in plants in the field. The findings of this study reflect different lignification pathways in stress-exposed plants grown in the natural environment

compared to plants grown under protected conditions in the glasshouse.

14.7 SUMMARY AND CONCLUSIONS

Chemical ecology, the science of chemically mediated intra- and interspecies interactions between organisms, requires powerful analytical tools to identify relevant chemical structures. To discover benefits for the producing organism and the function of natural products in the ecological interaction, biosynthetic and metabolic pathways as well as the tissue- and cell-specific localizations of the compounds have to be elaborated, in addition to identification and structure elucidation. NMR spectroscopy, as the most powerful and informative analytical method in natural product chemistry, has the potential to accomplish all these challenging tasks or, at least, to contribute to solving questions related to chemical ecology. In addition, a number of analytical methods such as MS, absorption spectroscopy, and vibrational spectroscopy are required to retrieve chemical information that complements data obtained from NMR. The examples presented in this overview aim to demonstrate how NMR has been used in the identification and de novo structure elucidation of ecologically interesting natural products and metabolites isolated from plants, insects, and microorganisms. Examples illustrate how the identification of products produced by isolated or recombinant enzymes is used to prove an enzyme's metabolic function.

Although the separation of natural products from a biological source long preceded the structure elucidation, and methods have been developed to couple separation methods such as LC with NMR, recently a trend can be observed toward identifying ecologically active components in crude mixtures without or before separation. A variety of examples of mixture analysis, ranging from 1D ^{1}H NMR spectroscopy to 2D shift-correlation NMR methods and further to novel metabolomic strategies that combine NMR with statistical methods, are reviewed here. Using statistical methods[88] for processing and interpreting, spectroscopic data facilitates linkage to biological properties[56] and will advance our understanding of chemoecological interactions. Graphic analysis based on the multiplicative stacking of bitmaps and statistical methods, such as DANS[54] and the covariance processing[57] of TOCSY NMR spectra followed by deconvolution using DemixC cluster analysis,[59]

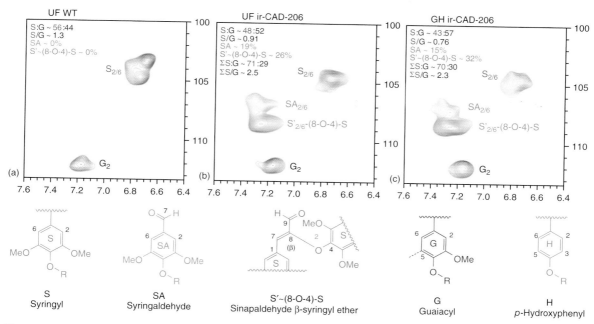

Figure 14.13. Partial ^{1}H–^{13}C HSQC spectra (500 MHz, cryoprobe, acetone-d_6) of acetylated lignin samples from stems of *Nicotiana attenuata*. (a) Wild-type plants grown under unprotected field (UF) conditions. (b) Plants silenced in cinnamyl alcohol dehydrogenase genes (ir-CAD) grown under UF conditions. (c) ir-CAD plants grown under glasshouse (GH) conditions. (Reproduced with permission from Ref. 85. © 2012 American Society of Plant Biologists)

among other approaches, exemplify some recent applications in chemical ecology. Statistical total correlation spectroscopy (STOCSY) is another approach to analyzing metabolomic data to identify bioactive metabolites in mixtures.[89] STOCSY has been used extensively in biomedical studies but, despite its potential, has so far rarely been applied in chemical ecology.

Metabolic profiling of single cells based on MS and other methods has been recently reviewed.[90–92] NMR spectroscopy, although much less sensitive than mass spectrometry, has also been employed in micrometabolic profiling studies[11] to analyze extracts from cells of the same type (cell-type-specific analysis). LMD has been introduced in microanalytical studies to obtain sufficient amounts of specific cells or cell populations for subsequent NMR analysis.[10] Examples of combined LMD/^{1}H NMR studies from *Dilatris* species, Norway spruce, and cannabis have been reviewed in this overview. Spatiotemporal resolution studies of metabolite accumulation in plants and other organisms are considered trend-setting and a methodical extension of the NMR-based localization

studies is reviewed here. Such investigations seem to be feasible by NMR, especially when considering recent and future sensitivity enhancements.

NMR- and MS-based biosynthetic labeling studies, in addition to molecular techniques, mutant studies, and biochemical methods, extend our understanding of the evolutionary shaped ecological network of species. Corresponding metabolic investigations advance our knowledge of the mechanisms of detoxification and deactivation, and the sequestration of ecologically important natural products. NMR-based labeling studies and the underlying methods used to detect the incorporation of heavy isotopes, e.g., ^{13}C, to establish precursor–product relationships and metabolic pathways have been reviewed previously.[5,6] In the present overview, a few ecologically relevant biosynthetic and metabolic investigations using ^{13}C and a mechanistic study using ^{2}H labeling demonstrate the potential of the NMR spectroscopy.

In summary, the interest in disclosing ecological relationships is growing. Understanding the chemical processes underlying the interactions between species and individuals requires the continual development and constructive application of analytical methods. Magnetic resonance techniques will certainly be in the forefront of future progress in this area.

ACKNOWLEDGMENTS

The author thanks the Max Planck Society, the Deutsche Forschungsgemeinschaft, the European Union, and the Alexander von Humboldt Foundation for the financial support of our work in this field. Emily Wheeler is gratefully acknowledged for editorial assistance.

RELATED ARTICLES IN EMAGRES

Biosynthesis and Metabolic Pathways: Carbon-13 and Nitrogen-15 NMR

Plant Physiology

Plant Metabolomics

Cryogenic NMR Probes: Applications

REFERENCES

1. J. Meinwald, *J. Nat. Prod.*, 2011, **74**, 305.

2. I. H. Sadler, *Nat. Prod. Rep.*, 1988, **5**, 101.

3. A. E. Derome, *Nat. Prod. Rep.*, 1989, **6**, 111.

4. S. Berger and D. Sicker, Classics in Spectroscopy, Wiley-VCH: Weinheim, 2009.

5. J. C. Vederas, *Nat. Prod. Rep.*, 1987, **4**, 277.

6. B. Schneider, *Prog. Nucl. Magn. Reson. Spectr.*, 2007, **51**, 155.

7. J. W. Jaroszewski, *Planta Med.*, 2005, **71**, 691.

8. J. W. Jaroszewski, *Planta Med.*, 2005, **71**, 795.

9. C. Daolio and B. Schneider, in Hyphenated and Alternative Methods of Detection in Chromatography, ed. R. A. Shalliker, Taylor & Francis: CRC Press, 2012, 61, Vol. 104.

10. D. Hölscher and B. Schneider, in Progress in Botany, eds E. U. Lüttge, W. Beyschlag and J. Murata, Springer: Berlin, 2008, 141, Vol. 69.

11. S. Moco, B. Schneider, and J. Vervoort, *J. Proteome Res.*, 2009, **8**, 1694.

12. J. K. Nicholson, J. C. Lindon, and E. Holmes, *Xenobiotica*, 1999, **29**, 1181.

13. O. Fiehn, *Comp. Funct. Genom.*, 2001, **2**, 155.

14. F. C. Schroeder and M. Gronquist, *Angew. Chem. Int. Ed.*, 2006, **45**, 7122.

15. T. F. Molinski, *Nat. Prod. Rep.*, 2010, **27**, 321.

16. S. Nojima, D. J. Kiemle, F. X. Webster, C. S. Apperson, and C. Schal, *PLoS One*, 2011, **6**, e18178.

17. W. R. Thompson, J. Meinwald, D. Aneshansley, and T. Eisner, *Science*, 1972, **177**, 528.

18. G. A. Wright, D. D. Baker, M. J. Palmer, D. Stabler, J. A. Mustard, E. F. Power, A. M. Borland, and P. C. Stevenson, *Science*, 2013, **339**, 1202.

19. S. F. Cummins, C. H. Schein, Y. Xu, W. Braun, and G. T. Nagle, *Peptides*, 2005, **26**, 121.

20. B. D. Morris, R. R. Smyth, S. P. Foster, M. P. Hoffmann, W. L. Roelofs, S. Franke, and W. Francke, *J. Nat. Prod.*, 2005, **68**, 26.

21. S. Vandermoten, M. C. Mescher, F. Francis, E. Haubruge, and F. J. Verheggen, *Insect Biochem. Mol. Biol.*, 2012, **42**, 155.

22. M. Hilker, C. Häberlein, U. Trauer, M. Bünnige, M.-O. Vicentini, and S. Schulz, *ChemBioChem*, 2010, **11**, 1720.

23. A. T. Dossey, S. S. Walse, O. V. Conle, and A. S. Edison, *J. Nat. Prod.*, 2007, **70**, 1335.

24. S. H.von Reuß, M. Kai, B. Piechulla, and W. Francke, *Angew. Chem. Int. Ed.*, 2010, **49**, 2009.

25. J. R. Price, R. Robinson, and R. J. Scott-Moncrieff, *Chem. Soc.*, 1939, 1465.

26. W. Schliemann, B. Schneider, V. Wray, J. Schmidt, M. Nimtz, A. Porzel, and H. Böhm, *Phytochemistry*, 2007, **67**, 191.

27. E. C. Tatsis, A. Schaumlöffel, A. C. Warskulat, G. Massiot, B. Schneider, and G. Bringmann, *Org. Lett.*, 2013, **15**, 156.

28. S.-H. Luo, Y. Liu, J. Hua, X.-M. Niu, S.-X. Jing, X. Zhao, B. Schneider, J. Gershenzon, and S.-H. Li, *Org. Lett.*, 2012, **14**, 4146.

29. F. Otálvaro, J. Nanclares, L. E. Vasquez, W. Quinones, F. Echeverri, R. Arango, and B. Schneider, *J. Nat. Prod.*, 2007, **70**, 887.

30. K. Jitsaeng and B. Schneider, *Phytochem. Lett.*, 2010, **3**, 84.

31. P. Bednarek, B. Schneider, A. Svatoš, N. J. Oldham, and K. Hahlbrock, *Plant Physiol.*, 2005, **138**, 1058.

32. P. Bednarek, M. Piślewska-Bednarek, A. Svatoš, B. Schneider, J. Doubský, M. Mansourova, M. Humphry, C. Consonni, R. Panstruga, A. Sanchez-Vallet, A. Molina, and P. Schulze-Lefert, *Science*, 2009, **323**, 95.

33. J. Kroiss, M. Kaltenpoth, B. Schneider, M. G. Schwinger, C. Hertweck, R. K. Maddula, E. Strohm, and A. Svatoš, *Nat. Chem. Biol.*, 2010, **6**, 261.

34. H. J. Shin, H. S. Jeong, H. S. Lee, S. K. Park, H. M. Kim, and H. J. Kwon, *J. Microbiol. Biotechnol.*, 2007, **17**, 1403.

35. R. Larbat, A. Hehn, J. Hans, S. Schneider, H. Judgé, B. Schneider, U. Matern, and F. Bourgaud, *J. Biol. Chem.*, 2009, **284**, 4776.

36. J. Jirschitzka, M. Reichelt, G. W. Schmidt, B. Schneider, J. Gershenzon, and J. C. D'Auria, *Proc. Natl. Acad. Sci. U. S. A.*, 2012, **109**, 10304.

37. N. Joußen, S. Agnolet, S. Lorenz, S. E. Schöne, R. Ellinger, B. Schneider, and D. G. Heckel, *Proc. Natl. Acad. Sci. U. S. A.*, 2012, **109**, 15206.

38. B. Schneider, in Progress in Botany', eds U. Lüttge, W. Beyschlag, B. Büdel and D. Francis, Springer: Berlin, 2011, 183, Vol. 72.

39. F. Buch, M. Rott, S. Rottloff, C. Paetz, I. Hilke, M. Raessler, and A. Mithöfer, *Ann. Bot.*, 2013, **111**, 375.

40. D. L. Olson, T. L. Peck, A. G. Webb, R. L. Magin, and J. V. Sweedler, *Science*, 1995, **270**, 1967.

41. M. Gronquist, J. Meinwald, T. Eisner, and F. C. Schroeder, *J. Am. Chem. Soc.*, 2005, **127**, 10810.

42. H. Kovacs, D. Moskau, and M. Spraul, *Prog. Nucl. Magn. Reson. Spectrosc.*, 2005, **46**, 131.

43. W. W. Brey, A. S. Edison, R. E. Nast, J. R. Rocca, S. Saha, and R. S. Withers, *J. Magn. Reson.*, 2006, **179**, 290.

44. A. T. Dossey, S. S. Walse, J. R. Rocca, and A. S. Edison, *ACS Chem. Biol.*, 2006, **1**, 511.

45. J. L. Ward and M. H. Beale, in Biotechnology in Agriculture and Forestry, eds K. Saito, R. A. Dixon and L. Willmitzer, Springer: Berlin, 2006, 81, Vol. 57.

46. M. R. Viant, *Biochem. Biophys. Res. Commun.*, 2003, **310**, 943.

47. D. Kessler and I. T. Baldwin, *Plant J.*, 2006, **49**, 840.

48. D. Hölscher, S. Brand, M. Wenzler, and B. Schneider, *J. Nat. Prod.*, 2008, **71**, 251.

49. M. Wenzler, D. Hölscher, T. Oerther, and B. Schneider, *J. Exp. Bot.*, 2008, **59**, 3425.

50. I. Lusebrink, K. Dettner, A. Schierling, T. Müller, C. Daolio, B. Schneider, J. Schmidt, and K. Seifert, *Z. Naturforsch.*, 2009, **64c**, 271.

51. J. Kikuchi, K. Shinozaki, and T. Hirayama, *Plant Cell Physiol.*, 2004, **45**, 1099.

52. E. Chikayama, M. Suto, T. Nishihara, K. Shinozaki, T. Hirayama, and J. Kikuchi, *PLoS One*, 2008, e3805.

53. F. C. Schroeder, D. F. Gibson, A. C. L. Churchill, P. Sojikul, E. J. Wursthorn, S. B. Krasnoff, and J. Clardy, *Angew. Chem. Int. Ed.*, 2007, **46**, 901.

54. C. Pungaliya, J. Srinivasan, B. W. Fox, R. U. Malik, A. H. Ludewig, P. W. Sternberg, and F. C. Schroeder, *Proc. Natl. Acad. Sci. U. S. A.*, 2009, **106**, 7708.

55. Y. Izrayelit, S. L. Robinette, N. Bose, S. H.von Reuss, and F. C. Schroeder, *ACS Chem. Biol.*, 2013, **8**, 314.

56. R. Forseth and F. C. Schroeder, *Curr. Opin. Chem. Biol.*, 2011, **15**, 38.

57. R. Brüschweiler and F. Zhang, *J. Chem. Phys.*, 2004, **120**, 5253.

58. F. Zhang, A. T. Dossey, C. Zachariah, A. S. Edison, and R. Brüschweiler, *Anal. Chem.*, 2007, **79**, 7748.

59. F. Zhang, S. L. Robinette, L. Bruschweiler-Li, and R. Brüschweiler, *Magn. Reson. Chem.*, 2009, **47**, 118.

60. R. Madsen, T. Lundstedt, and J. Trygg, *Anal. Chim. Acta*, 2010, **659**, 23.

61. S. L. Robinette, R. Brüschweiler, F. C. Schroeder, and A. S. Edison, *Acc. Chem. Res.*, 2012, **45**, 288.

62. S. L. Robinette, R. Ajredini, H. Rasheed, A. Zeinomar, F. C. Schroeder, A. T. Dossey, and A. S. Edison, *Anal. Chem.*, 2011, **83**, 1649.

63. J. C. Pardo-Torre, G. W. Schmidt, C. Paetz, M. Reichelt, B. Schneider, J. Gershenzon, and J. C. D'Auria, *Phytochemistry*, 2013, **91**, 177.

64. N. D'Amelio, L. Fontanive, F. Uggeri, F. Suggi-Liverani, and L. Navarini, *Food Biophys.*, 2009, **4**, 321.

65. L. Borisjuk, H. Rolletschek, and T. Neuberger, *Plant J.*, 2012, **70**, 129.

66. A. G. Hart, R. W. Bowtell, W. Köckenberger, T. Wenseleer, and F. L. W. Ratnieks, *J. Insect Sci.*, 2003, **3**, 1.

67. M. Heidenreich, W. Köckenberger, R. Kimmich, N. Chandrakumar, and R. Bowtell, *J. Magn. Reson.*, 1998, **132**, 109.

68. D. Hölscher and B. Schneider, *Planta*, 2007, **225**, 767.

69. S.-H. Li, B. Schneider, and J. Gershenzon, *Planta*, 2007, **225**, 771.

70. S.-H. Li, N. E. Nagy, A. Hammerbacher, P. Krokene, X.-M. Niu, J. Gershenzon, and B. Schneider, *ChemBioChem*, 2012, **13**, 2702.

71. A. J. Simpson and S. A. Brown, *J. Magn. Reson.*, 2005, **175**, 340.

72. A. Weinhold, K. Shaker, M. Wenzler, B. Schneider, and I. T. Baldwin, *J. Chem. Ecol.*, 2011, **37**, 1091.

73. N. Happyana, S. Agnolet, R. Muntendam, A. Van Dam, B. Schneider, and O. Kayser, *Phytochemistry*, 2013, **87**, 51.

74. J. M. Risley and R. L. Van Etten, in NMR. Basic Principles and Progress, eds P. Diehl, E. Fluck, H. Günther, R. Kosfeld and J. Selig, Springer: Berlin, 1990, 81, Vol. 22.

75. C. D. Grubb and S. Abel, *Trends Plant Sci.*, 2006, **11**, 89.

76. B. A. Halkier and J. Gershenzon, *Annu. Rev. Plant Biol.*, 2006, **57**, 302.

77. G. Graser, B. Schneider, N. J. Oldham, and J. Gershenzon, *Arch. Biochem. Biophys.*, 2000, **378**, 411.

78. K. L. Falk, C. Vogel, S. Textor, S. Bartram, A. Hick, J. A. Pickett, and J. Gershenzon, *Phytochemistry*, 2004, **65**, 1073.

79. S. Textor, J.-W. de Kraker, B. Hause, J. Gershenzon, and J. G. Tokuhisa, *Plant Physiol.*, 2007, **144**, 60.

80. F. Vergara, A. Svatoš, B. Schneider, M. Reichelt, J. Gershenzon, and U. Wittstock, *ChemBioChem*, 2006, **7**, 1982.

81. B. Schneider, J. Gershenzon, G. Graser, D. Hölscher, and B. Schmitt, *Phytochem. Rev.*, 2003, **2**, 31.

82. M. Kunert, P. Rahfeld, K. H. Shaker, B. Schneider, A. David, K. Dettner, J. M. Pasteels, and W. Boland, *ChemBioChem*, 2013, **14**, 353.

83. C. C. von Dahl, M. Hävecker, R. Schlögl, and I. T. Baldwin, *Plant J.*, 2006, **46**, 948.

84. L. Bedóuet, B. Courtois, and J. Courtois, *Carbohydr. Res.*, 2003, **338**, 379.

85. H. Kaur, K. Shaker, N. Heinzel, J. Ralph, I. Gális, and I. T. Baldwin, *Plant Physiol.*, 2012, **159**, 1545.

86. J. Ralph, R. D. Hatfield, J. Piquemal, N. Yahiaoui, M. Pean, C. Lapierre, and A. M. Boudet, *Proc. Natl. Acad. Sci. U. S. A.*, 1998, **95**, 12803.

87. M. Bunzel and J. Ralph, *J. Agric. Food Chem.*, 2006, **54**, 8352.

88. T. M. D. Ebbels and R. Cavill, *Prog. Nucl. Magn. Reson. Spectrosc.*, 2009, **55**, 361.

89. O. Cloarec, M.-E. Dumas, A. Craig, R. H. Barton, J. Trygg, J. Hudson, C. Blancher, D. Gauguier, J. C. Lindon, E. Holmes, and J. Nicholson, *Anal. Chem.*, 2005, **77**, 1282.

90. M. Heinemann and R. Zenobi, *Curr. Opin. Biotechnol.*, 2011, **22**, 26.

91. A. Svatoš, *Anal. Chem.*, 2011, **83**, 5037.

92. A. Oikawa and K. Saito, *Plant J.*, 2012, **70**, 30.

FURTHER READINGS

H. K. Kim, Y. H. Choi, and R. Verpoorte, *Trends Biotechnol.*, 2011, **29**, 267.

F. A. Macías, J. L. G. Galindo, and J. C. G. Galindo, *Phytochemistry*, 2007, **68**, 2912.

E. K. Prince and G. Pohnert, *Anal. Bioanal. Chem.*, 2010, **396**, 193.

O. Serra, S. Chatterjee, W. Huang, and R. E. Stark, *Plant Sci.*, 2012, **195**, 120.

Chapter 15
Forest Ecology and Soils

Sylvie A. Quideau[1], Charlotte E. Norris[1], Laure N. Soucémarianadin[1] and Roderick E. Wasylishen[2]

[1]*Department of Renewable Resources, University of Alberta, Edmonton, Alberta T6G 2E3, Canada*
[2]*Gunning-Lemieux Chemistry Centre, Department of Chemistry, University of Alberta, Edmonton, Alberta T6G 2G2, Canada*

15.1 INTRODUCTION

Humans have long relied on forested ecosystems for their livelihoods. Direct benefits provided by forests include wood for timber, fiber, or fuel and understory plants for food and medicinal uses. More recently, a longer list of services provided by forests has emerged. These services, which typically pertain to the larger landscape or even global scale, acknowledge the important role that forests play in the functioning of our planet. For instance, forests regulate and filter our valuable water resources, maintain soil quality through erosion control, preserve soil fertility by generating, storing, and cycling nutrients, and provide a habitat for many plants and animals, hence contributing to global biodiversity.

Forest management, which used to solely consider timber and fiber production, has evolved over the past two decades to mature as a broader, ecosystem-based approach. The concept of sustainable development has been adopted as the key concept against which the success of forest management practices is assessed. This approach also acknowledges the dependence of forest sustainability on the interactions and feedback mechanisms among soils, water, the biotic components (vegetation and wildlife), and human activities. Essential to these resources is the soil, where most of the recycling of nutrients, filtering of water, and detoxification of pollutants occur. Indeed, achieving the right balance between increasing demands for forest products and environmental constraints is central to achieving the long-term sustainability of forest ecosystems. In addition, recent discussions of climate change mitigation and carbon trading have generated much interest in the long-term potential of forest soils to sequester carbon. The two most important carbon reserves occur at low latitude in tropical forest aboveground biomass

NMR Spectroscopy: A Versatile Tool for Environmental Research
Edited by Myrna J. Simpson and André J. Simpson

and high latitude in the huge belowground reservoirs of boreal ecosystems. Carbon stocks in boreal forest soils alone are estimated at 470 Gt (10^{12} kg), which is equivalent to more than half the carbon in the atmosphere. However, the efficacy of strategies to sequester carbon in forests necessitates a better understanding of carbon dynamics in these ecosystems.

Many key questions related to the biogeochemical functioning of forest ecosystems have been answered by contributions from NMR studies.[1–4] Early studies, which were conducted as far back as three decades ago, may have been more motivated by understanding the link among management practices, soil fertility, and forest productivity.[5,6] More recent work has been directed at characterizing distinct organic matter pools in forest ecosystems, as these relate to issues of carbon sequestration and turnover rates.[7,8] By providing a unique tool to characterize organic matter in soils, NMR spectroscopy can be credited for many of the recent insights into processes regulating carbon storage in forest ecosystems.

Early NMR work in forest biogeochemistry aimed to validate the methodology. Solution-state NMR targeted humic substances that could be extracted from soils. However, soluble carbon present in extracts may not be representative of the entire soil organic matter pool. It was recognized early on that solid-state NMR held considerable promise for characterizing organic matter of whole soils (see Chapter 4).[9] Spinning these samples rapidly at the 'magic-angle' of 54.74° reduced ^{13}C linewidths to levels that were comparable to those obtained through solution-state NMR. Solid-state NMR hence rapidly emerged as the tool of choice for characterization of the chemical composition of soil organic matter in forest ecosystems.[10] In forest ecology and soils, solid-state NMR has been widely used to determine the distribution of ^{13}C and ^{15}N into key chemical classes,[10–12] whereas solution-state NMR has also been applied to ^{1}H and ^{31}P.[10,13]

Traditionally, forest soil carbon was characterized based on its solubility in alkaline and acid solutions to isolate the following three fractions: fulvic acids, soluble in both solutions; humic acids, soluble in alkaline but not in acidic solutions; and humin fractions, insoluble in both. These three fractions were found to differ in some of their key chemical characteristics, with, for instance, greater oxygen content, greater concentration of acidic functional groups, but lower carbon content and molecular weight for fulvic acids compared to humic acids and humin. Seminal work on forest pedogenesis used this fractionation scheme

to define the key influence of forest vegetation in controlling the composition and properties of soil organic matter.[14] In particular, soluble fulvic acids, which are produced in large quantities by coniferous vegetation, were found to be key players in the process of podzolization, and were responsible for the subsoil accumulation of organic matter, as well as iron and aluminum oxides.[15] Although traditional extraction methods and fractionation into humin, fulvic acid, and humic acid are still used nowadays in the study of soil humus quality in forest ecosystems, mostly because they are inexpensive and require little in terms of equipment, solid-state NMR has slowly become the reference method (see Chapter 4). Consequently, carbon composition in forest ecosystems is now referred to in terms of carbon types, including alkyl, *O*-alkyl, aryl, and carboxyl carbons.

Several excellent reviews of NMR applications to forests and soils exist in the literature.[4,10–12,16–21] Earlier reviews described the appropriate techniques for the most effective use of solid-state NMR to the study of soil organic matter and the inherent limitations of the technique, in particular, as these are linked to its quantitative reliability.[4,11,16,17,22] Some particular issues to be aware of include spectrometer background, spinning sidebands, and cross polarization (CP) dynamics as well as high moisture content or paramagnetics, or both (see Chapter 4). Introduction of sample pretreatment and the development of techniques such as two-dimensional (2D) NMR can be applied to soluble fractions for improved quantification of organic matter dynamics.[10,17,23] A recent review outlines the advantages of combining molecular-level analytical approaches to NMR techniques to elucidate organic matter source and stability within soils, and its structure shifts in response to environmental changes.[20]

The purpose of this contribution is to outline some of the key questions in forest biogeochemistry that NMR techniques have helped tackle successfully, including the characterization of carbon inputs to forest soils, their decomposition processes, and organic matter distribution into different soil pools. This chapter also covers some of the more recent advances in tracing organic matter fluxes through forest ecosystems and provides an overview of how NMR can be used to quantify changes in forest soil quality linked to natural and anthropogenic disturbances. Several reviews concerning solid-state NMR methodology are available in the literature and described elsewhere in this eMagRes (see Chapters 4 and 13), so our discussion of experimental considerations is brief. Some key considerations to keep in mind when using NMR to study forest ecology and forest soils are presented in the following sections.

15.2 ORGANIC MATTER INPUTS TO FOREST SOILS

Solid-state NMR is able to provide a fingerprint of plant litter materials, and as such, has been widely used to help define a useful index of forest litter quality. Litter chemical composition directly determines the quality and quantity of microbial substrates in forest floors, the organic-rich layers that characteristically accumulate on top of mineral soils in forest ecosystems. Forest floor accumulation increases with increasing latitude, as lower temperatures inhibit decomposition processes. In boreal forests, where decomposition rates are notably low, forest floor layers may reach >1 m in thickness and constitute the largest carbon pool in the ecosystem.[24] In addition to climatic effects, litter chemical composition exerts a controlling influence on rates of decomposition. Litter with lower quality substrates, such as coniferous litter, usually decomposes slower than the litter derived from broad-leaved trees.[25] Consequently, characterization of litter quality is most often based on chemical indices, including elemental concentrations and ratios, as well as proximate organic fractions.[26] Examples of chemical parameters that have been used as litter quality indices include nitrogen concentration, lignin content, carbon-to-nitrogen ratio, lignin-to-nitrogen ratio, or the lignocellulose index, defined as the lignin-to-(lignin + cellulose) ratio.[27] In particular, litter decomposition rates are often found negatively correlated to their lignin content.[28] Lignin content is typically quantified using proximate analysis, which is based on the partitioning of plant fractions depending on their solubility in different solvents. Following the extraction of water-soluble constituents and litter components that are soluble in neutral organic solvents, a sulfuric acid digestion effectively hydrolyses crystalline (cellulose) and noncrystalline polysaccharides such as plant hemicelluloses. The residue remaining after digestion is assumed to be recalcitrant to microbial degradation. NMR research in this area has helped clarify the chemical structure of the litter fractions isolated by proximate analysis. In particular, the so-called Klason lignin, which is the residue obtained following acid hydrolysis, has been shown through solid-state ^{13}C NMR analysis to not only include lignin but also contain nonnegligible amounts of nonlignin components such as fatty acids, cutin, suberin, and condensed tannins.[7] Similarly, analysis using solid-state ^{13}C NMR spectroscopy demonstrated that different extraction methods may isolate cellulose with varying amounts of lipids, waxes, and lignins.[29] These studies clearly demonstrate the interest of using NMR to elucidate the composition of isolated plant fractions. While no method may be fully efficient at isolating purified biopolymers, it is nonetheless important to understand their limitations, so that results from both litter decomposition and environmental reconstruction studies may be appropriately interpreted.

Carbon inputs to forest soils include various aboveground vegetative parts, such as leaves, branches, bark, and stems, as well as belowground inputs, namely roots and root exudates. Solid-state ^{13}C NMR has been extensively employed to research decomposition processes in forest ecosystems, as it can readily follow chemical changes occurring during litter degradation.[26] Spectra are typically divided into regions that correspond to different carbon types and are assigned based on local minima of the spectra. For instance, based on the solid-state ^{13}C NMR spectra shown in Figure 15.1, the following regions were used for integration: 0–45 ppm attributed to alkyl carbons, 45–92 ppm attributed to methoxyl- and O-alkyl carbons, 92–125 ppm attributed to di-O-alkyl and some aromatic carbons, 125–140 ppm attributed to aromatic carbons, 140–164 ppm attributed to phenolic carbons, and 165–200 ppm attributed to carboxylic and carbonyl carbons. In addition to values calculated from integrations, valuable qualitative information can be obtained by examining specific peaks in the ^{13}C NMR spectra.[26] For instance, the first peak in the alkyl region that arises at 19–20 ppm corresponds to methyl carbons, and the second one at 29–30 ppm corresponds to alkyl methylene chains. The additional peak present around 38 ppm can be assigned to quaternary alkyl carbons. The O-alkyl region, as is most often the case for spectra obtained from forest litter, is dominated by overlapping peaks at approximately 73 ppm, which are characteristic of the C-2, C-3, and C-5 carbons of pyranoside rings in cellulose and hemicelluloses. The shoulder at 65 ppm can be similarly assigned to the C-6 carbons in carbohydrates. In the di-O-alkyl region of the spectra, the anomeric carbons of carbohydrates are expected to give peaks around 105 ppm. However, this peak may be obscured by aromatic carbon peaks arising from lignins and tannins. The assignment is less ambiguous for the peak at 118 ppm, which corresponds

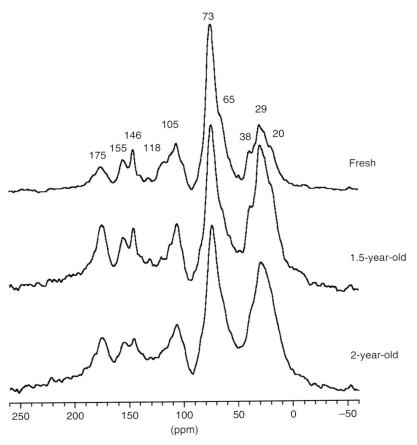

Figure 15.1. ^{13}C CPMAS NMR spectra of fresh and decomposing manzanita leaf litter. Spectra were obtained on a Varian Chemagnetics CMX Infinity 200 ($B_0 = 4.7$ T, $v_L(^{13}C) = 50.3$ MHz) NMR spectrometer using a 7.5 mm double-resonance MAS probe. All ^{13}C NMR spectra were referenced to TMS ($\delta_{iso} = 0.0$ ppm) by setting the high-frequency isotropic peak of solid adamantane to 38.56 ppm; a [1]H 90° pulse duration of 4.5 μs, a pulse delay of 5.0 s, a contact time of 1.0 ms, an acquisition time of 17.1 ms, and a spinning frequency of 6.5 kHz were used. One thousand transients were collected for each sample analyzed. A Gaussian line broadening of 100 Hz was used to process all spectra. (Reprinted with permission from Ref. 37. © 2005 Elsevier)

to C-substituted aromatic carbons, whereas phenolic carbons appear at 146 and 155 ppm. Finally, the peak at 175 ppm is indicative of the carbonyl carbon in acetyl and ester moieties. Although the integration regions described earlier are typical, readers should be aware that there is some variation in the selection of regions, depending on the samples and resulting spectra;[12,30] chemical shift ranges are discussed in several texts.[31,32]

From the numerous NMR studies of organic matter decomposition conducted in various types of forest ecosystems over the years, the alkyl/O-alkyl carbon ratio emerges as the most commonly used index of litter

quality.[1,33] With chemical degradation, the O-alkyl C, corresponding to carbohydrates that are preferentially decomposed, tends to decrease, whereas the alkyl C corresponding to more recalcitrant moieties usually increases. Consequently, the alkyl/O-alkyl C ratio typically increases with increasing decomposition. The Canadian Intersite Decomposition Experiment (CIDET) study, an extensive investigation of forest litter decomposition that involved the placement of >10 000 litter bags across the major ecoclimatic regions of Canada, consistently showed an increase in alkyl carbon and a decrease in O-alkyl carbon during long-term decomposition of foliar litters.[7] This is also apparent from Figure 15.1 for foliage litter decomposing in a Mediterranean-type ecosystem, where the alkyl carbon increased by 36%, while the O-alkyl carbon decreased by 28% after 1.5 years in the field; consequently, the alkyl/O-alkyl carbon ratio increased from 0.6 in the fresh litter to 1.1 after 1.5 years.[26] As opposed to rapid changes that may occur during foliage decomposition, wood seems to undergo limited changes in composition during degradation.[34,35] Because of the preferential loss of O-alkyl carbon from carbohydrates, foliage litters from different trees tend to show increasing similarity with decomposition and an eventual convergence to a composition rich in alkyl, phenolic, and carbonyl carbons.[7,34] Wood litters, on the other hand, may exhibit nonselective decomposition,[35] at least in the presence of white-rot fungi that can degrade both lignin and carbohydrates.[36]

Results from NMR spectral signatures are useful for the semiquantitative evaluation of chemical alterations during litter decomposition. However, while the chemical composition of litter obviously influences its rate of decomposition, researchers should not forget about other potentially important factors, such as abiotic environmental conditions, as well as the physical characteristics, including surface area and toughness (e.g., cuticle thickness). Similarly, for forest floors, various physical characteristics affect moisture retention, aeration, and temperature regimes that all have been shown to influence decomposition rates. Consequently, the importance of an integrative approach cannot be overstated when aiming to define a predictive index of litter decomposition.

15.3 ORGANIC MATTER POOLS IN FOREST SOILS

CP is the most commonly used solid-state ^{13}C NMR technique (see Chapter 4) in forest ecology, as it is the fastest method, and presents useful semiquantitative values to compare the chemical composition of distinct forest carbon pools. Factors affecting the quality of the NMR signal (i.e., the signal-to-noise ratio) include the concentration of carbon and the presence of paramagnetic species (e.g., Mn^{2+}, Fe^{2+}, Fe^{3+}, Cu^{2+}, and Ni^{2+}) in samples. Litter and forest floor materials, which contain both high carbon levels and relatively low concentrations of paramagnetic materials, are particularly attractive for NMR analysis. Indeed, the

majority of studies in forest ecosystems have focused on characterizing the chemical structure of these two carbon pools. Characteristic peaks typically occur at approximately the same chemical shifts in all spectra derived from either litter or forest floor samples. As illustrated in Figure 15.2, spectra are typically

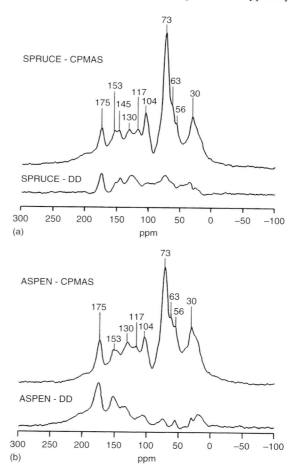

Figure 15.2. (a,b) ^{13}C CPMAS and DD NMR spectra of the Oe + Oa horizon of a white-spruce-dominated (SPRUCE) and a trembling-aspen-dominated (ASPEN) stand. CP spectra were obtained as described in the caption to Figure 15.1; the spectra were obtained with a ^{1}H decoupling field of 56 kHz. A Gaussian line broadening of 100 Hz was used to process all spectra. The contribution of the background signal was subtracted from all Fourier-transformed ^{13}C NMR spectra before analysis. Corrections for spinning sidebands were not applied because they did not appear to have a strong effect on any of the spectra. (Reproduced with permission from Ref. 37. © 2004 ASA, CSSA, SSSA)

dominated by *O*-alkyl carbons, followed by alkyl, aromatic, and finally, carbonyl carbons.[37] What does vary among them, however, is the relative intensity of these peaks. Forest floors have unique NMR signatures that differ depending on their macromolecular chemistry. In the case of the NMR spectra for boreal forest floor samples shown in Figure 15.2, integration of the spectral areas revealed a greater concentration of aromatic carbon in the forest floor formed under spruce compared to the aspen forest floor. In addition to standard CP magic-angle spinning (MAS) analysis, a particularly useful technique is to produce dipolar-dephased (DD) ^{13}C NMR spectra (see Chapter 4) by inserting a delay period between the CP and the acquisition parts of the CPMAS pulse sequence.[38] The peaks present in DD spectra arise from quaternary carbons, or from carbons that belong to mobile molecular fragments (e.g., methyl groups). In forest ecology, this is particularly suitable to distinguish between lignin and tannin. While both polymers may yield a comparable signal in the aromatic region of CPMAS spectra, in DD spectra, lignin produces a characteristic methoxyl peak around 56 ppm, whereas two clearly separated peaks at approximately 145 and 153 ppm are characteristic of materials with high tannin content. In Figure 15.2, the DD spectrum from the aspen forest floor shows a distinct peak at 56 ppm consistent with high lignin content, whereas the presence of tannins is evident in the spectrum of the spruce forest floor.

Plant nutrient concentrations such as nitrogen are greater in litter and forest floors than in the underlying mineral soil, which may explain why most studies of soil fertility focus on these surficial soil layers. Yet, it is within the mineral soil that organic matter is stabilized for longer periods.[39] Stabilization of organic matter is achieved by chemical binding to mineral edges, or by physical protection inside aggregates that may form within the mineral soil profile. The formation of aggregates, or peds, during pedogenesis, is favored in fine-textured soils dominated by clay, and soil texture directly affects the potential of forest mineral soils to accumulate and store organic matter.[39] Hence, because it may be desirable to characterize the chemistry of organic matter within the mineral soil matrix itself, pretreatment methods have been developed to remove ferromagnetic particles, reduce paramagnetic ions, and concentrate carbon in the sample of interest.[40–43] One method of removing the interfering ferromagnetic particles is the use of a magnetic bar.[40] As well, dithionite and dithionite/citrate extractions have been found useful in removing iron from SOM.[40,43] Another method, digestion with HF, is a preferred technique, as it dissolves mineral particles, hence reducing paramagnetic species, and concentrates the organic matter while not substantially altering the composition of biopolymers.[40–42] Of course, extraction with HF may also be successful in improving ^{15}N NMR analyses.[41] As a rule of thumb, it can be assumed that an acceptable NMR spectrum can be obtained if the total carbon concentration exceeds 5% in the absence of paramagnetics.

Treatment with HF, which necessitates a high level of safety precautions in view of its extremely corrosive nature and potentially severe health hazards, can be replaced if appropriate with the much gentler physical fractionation approach. Physical fractionation was not originally developed for the purpose of NMR but for isolating soil organic matter pools of different turnover rates. Yet, even for mineral soils that have a low carbon concentration when considered as bulk samples, some of their isolated fractions typically show sufficient carbon to acquire a reasonable NMR spectrum.[44] Experimentally, fractionation involves density and size fractionation of primary organo-mineral particles in whole soil. Density separation is most useful in concentrating organic matter turning over rapidly in the low-density fraction. In practical terms, low-density organic matter is isolated by flotation in liquids of specific density $1–2\,g\,cm^{-3}$. This labile light fraction plays a central role in short-term nutrient cycling, and has been suggested to provide a better and earlier indication of the consequences of land use changes than total soil carbon content. Because it mostly comprises plant and animal residues and very little mineral matter, the light fraction is naturally rich in carbon, and particularly suitable for analysis by NMR. The soil heavy fraction, which includes organo-mineral complexes, may be further divided into different size fractions such as sand (5–2000 µm), silt (2–5 µm), and clay (<2 µm). In addition to the separation of primary organo-mineral particles in whole soil, physical separation of secondary organo-mineral particles (i.e., aggregates) has been extensively utilized. For this, four levels or divisions are typically considered in the aggregate hierarchy: <20, 20–90, 90–250, and >250 mm. Organo-mineral complexes found in the coarse clay may be the longest lived fractions in the soil, and the most relevant to long-term carbon sequestration. Carbon associated with clay may also favor the stability of microaggregates, which is essential for aeration, and storage of water available to plants. The fine fractions, clay- and silt-sized, are typically much more concentrated in carbon than bulk soil, and as such will yield a much improved ^{13}C NMR response.

NMR spectroscopy applied to physical separates of forest soils has repeatedly shown a decrease in the content of *O*-alkyl carbon (cellulose and hemicelluloses) with decreasing particle size.[4,33,45] At the same time, aromatic carbon (lignin and tannins), alkyl carbon (aliphatic compounds such as waxes), and carbonyl carbon are more abundant in the clay and silt separates compared to the sands. These NMR results have allowed the clarification of litter decomposition and transformation pathways in forest soils. Changes in chemical composition among fractions must be explained not only based on the recalcitrant nature of some plant constituents but also considering the synthesis of new microbial products during biodegradation. Another interesting application of NMR to forest soil fractions has included investigations of vegetation and climatic influences on both labile and stabilized organic matter pools.[8,46] Key findings to date have revealed the increasing contribution of roots to low-density organic matter with increasing depth in the profiles,[8] and have indicated the key influence of vegetation in controlling composition of the organic matter associated with the fine silt fractions.[46]

15.4 TRACING ORGANIC MATTER FLUXES THROUGH FOREST ECOSYSTEMS

The first step in tracing organic matter fluxes through forests is the ability to identify inputs to the system; i.e., the ability to trace the source of the different soil organic matter pools. The chemical makeup of soil organic matter may provide the best clue to separate among different source materials, including different plant species and faunal and microbial residues (see Chapter 13). However, while solid-state ^{13}C NMR analysis can provide an overview of organic matter composition of these pools, additional investigations may be required for a more accurate assessment of organic matter inputs and necessitates extraction and separation of soil organic matter into smaller fragments that are then analyzed by colorimetric or chromatographic methods.[47] Techniques that strive to separate fractions representative of plant or microbe biopolymers are the most promising. For instance, the alkaline cupric oxide oxidation has been applied to the study of lignin decomposition and preservation in forest humus layers. Cupric oxide oxidation

yields a suite of lignin-derived phenols whose relative distribution can distinguish between angiosperm and gymnosperm, as well as between woody and nonwoody origins.[48,49] Because pentoses are not synthesized to any great extent by microorganisms, pentose/hexose ratios have been used to characterize the contribution of microbial synthesis to soil carbohydrates, whereas the ratio of glucosamine to galactosamine can indicate the relative contribution of fungi. Finally, the dominance of particular *n*-alkane fragments in soil organic matter can provide chemotaxonomic and quantitative information on local plant inputs.[50] In all cases, combining solid-state ^{13}C NMR fingerprints with other molecular-level analyses is the most promising approach to characterize the mechanistic interactions between soil organic matter chemistry and source inputs.

By characterizing biopolymers within organic matter inputs, forest floor, and soils, carbon fluxes of forest ecosystems may begin to be elucidated. For instance, NMR has proven to be a key tool in determining primary succession along forest developmental pathways. With succession, and associated changes in organic matter inputs, the chemical composition of the forest floor and soil organic matter will slowly evolve to reflect these changes. A more immediate indicator of forest ecosystem status may be soil-dissolved organic carbon (DOC). It is a highly mobile pool of carbon that is readily influenced by both biotic and abiotic conditions (see Chapter 11). Concentrated bulk soil solutions can be freeze-dried and analyzed using solid-state ^{13}C NMR.[51,52] This provides a much more direct assessment and indication of DOC aromaticity than more traditional UV adsorption techniques. Results to date have demonstrated seasonal changes in DOC composition,[51] as well as chemical alteration with increasing depth in the soil profiles, namely a decrease in *O*-alkyl carbons but an increase in aromatic carbons.[52] On the basis of more recent work (see Chapter 11), it appears that multidimensional solution-state NMR, combined with chromatographic techniques, possesses the best potential to evaluate this pool.[53]

The ability to characterize the different organic matter pools and their persistence in forest soils is key to a better understanding and prediction of the entire ecosystem performance. Hence, how to probe into the underlying processes controlling carbon sequestration and turnover rates of distinct soil organic matter pools has been a key issue driving recent research work in forest ecosystems. To move beyond the simple chemical characterization of these pools such as what can be provided by NMR, it is necessary to use complementary tools that can provide precise estimates of turnover rates. In situations where archived soils (i.e., samples taken before the detonation of thermonuclear weapons in the 1960s) are available for comparison with modern soils, radiocarbon (^{14}C) enrichment can be used as a sensitive tracer to estimate soil carbon turnover rates.[54] In addition, measuring $^{13}C/^{12}C$ ratios in soil fractions can indicate the extent of microbial processing. The concentration in ^{13}C typically increases with decreasing particle size, and microbial discrimination against ^{13}C during decomposition has been proposed as a possible cause.[55] Respired CO_2 is comparatively depleted in ^{13}C compared to bulk soil organic matter, whereas microbial-derived products become enriched in ^{13}C. However, interpretation of ^{13}C data is complicated by the fact that the isotopic composition of the starting litter material is not homogeneous. Both cellulose and hemicellulose typically are enriched in ^{13}C by 4–7% relative to lignins and waxes. Following the evolution of $^{13}C/^{12}C$ in fractions in combination with NMR analyses may prove very useful to identify the relative contributions of plant- and microbe-derived substances to soil organic matter. The full potential of this combination of techniques is yet to be realized.

Determining, and refining, carbon fluxes between forest ecosystems and the atmosphere is becoming increasingly important to accurately model carbon at the global scale. Critical work in resolving turnover rates of organic matter is being pursued using stable isotope-enriched (or depleted) organic matter, in particular involving carbon (^{13}C, ^{12}C) and nitrogen (^{15}N, ^{14}N) stable isotopes. Tracing the movement of isotopes through their respective soil biogeochemical cycles allows for a more accurate determination of turnover rates than is allowed by following total carbon fluxes. Isotopic tracers from the enriched material (see Chapter 27) can be tracked through the forest ecosystem, as they are transferred among different pools, for instance, from leaf litter to soil microbial biomass, or as they are respired back to the atmosphere. In addition, by adding ^{13}C-enriched compounds to soils, it is possible to use solid-state ^{13}C NMR to directly follow their fate against the background soil organic matter. This approach has been exploited to follow the transformations of simple compounds such as ^{13}C-glucose and ^{13}C-glycine.[56,57] More recently, NMR analysis has

been used to determine specific biopolymer enrichment during a whole plant labeling experiment.[58] A direct, added benefit of enriching organic matter in ^{13}C or ^{15}N is an enhancement of signal-to-noise (S/N) ratios and improvement of macromolecular characterization in the NMR spectra. After a pulse labeling of $^{13}CO_2$ and $K^{15}NO_3$ to tree seedlings, leaf enrichment in ^{13}C was originally concentrated within the O-alkyl carbons (carbohydrates); ^{13}C isotopes redistributed to more complex biopolymers and other parts of the seedlings after 1 week (Figure 15.3). Within the roots, there was an initial increase in amine nitrogen after the addition of ^{15}N, whereas ^{13}C enrichment was uniform across the biopolymers. In this particular study, direct polarization (DP) or Bloch decay (BD) NMR spectroscopy was used (see Chapter 4). Using this technique, ^{13}C (or ^{15}N) nuclei are polarized directly by the applied magnetic field, and is, therefore, more quantitative in its assessment of organic matter chemical composition. However, owing to the typically long ^{13}C and ^{15}N spin–lattice relaxation times, acquiring DPMAS spectra may take a very long time (>24 h per spectrum); this may not be a feasible approach if spectrometer availability is limited. Hence, the DP technique is particularly suited for the characterization of isotopically enriched organic matter where the increased abundance of ^{13}C or ^{15}N decreases the necessary acquisition times. In the case of nitrogen, the use of ^{15}N-enriched materials is particularly important, as the very low natural ^{15}N abundance ($\approx 0.365\%$) and relatively small nuclear magnetic moment make it difficult to obtain NMR spectra with reasonable signal-to-noise ratios.

15.5 FOREST RESPONSE TO DISTURBANCE

New approaches to forest management, such as variable retention harvesting, attempt to emulate natural disturbance regimes in preserving forest biodiversity and functional integrity. Experiments aimed at testing the efficiency of these approaches typically compare processes in managed ecosystems to those in natural settings. NMR analyses are well adapted to follow the response of soil quality to harvesting, and in particular changes in soil organic matter composition. Short-term changes following harvesting may include an increase in aromatic carbon concentration, indicating a decreased degree of decomposition in the absence of fresh tree litter inputs.[59–61] In the longer term,

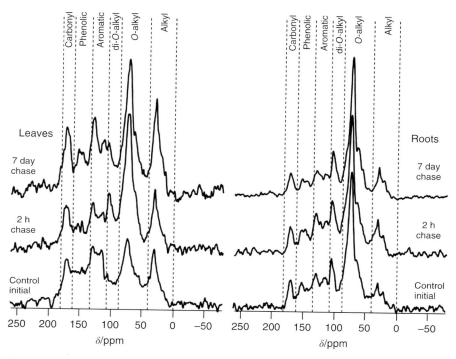

Figure 15.3. ^{13}C DPMAS NMR spectra of ^{13}C-enriched seedling leaves and roots with control-initial, one pulse, and 2 h chase and one pulse 7 day chase. Samples for ^{13}C were analyzed using a CMX Infinity 200 spectrometer as described in the caption to Figure 15.1, but using a spin-echo pulse to eliminate a background signal. A total of 500 transients (2000 for the control initial leaves) were acquired for each sample with a 5.0 kHz spinning rate, a 4.5 μs 90° pulse duration, and 100 s recycle time. Processing the spectra included zero filling to 1 K, line broadening of 100 Hz with phase and baseline correction. (Reproduced with permission from Ref. 58. © 2012 Nature Publishing Group)

changes may also reflect an increase in alkyl carbon. Another common ecological disturbance in forests is fire, which may have a lasting effect on soil organic matter composition; again, this can be readily followed using solid-state ^{13}C NMR spectroscopy. Following a fire, the forest floor is typically enriched in aromatic forms of carbon, which can be attributed to the formation of char or black carbon.

Examining the molecular composition of forest soils led to the realization that black carbon is an important carbon pool in these ecosystems.[62,63] Recent efforts have only begun to determine its quantity, chemistry, and role in forest ecology. The initial concept of black carbon as a stable and inert stock of soil carbon is beginning to be challenged through results based on a variety of techniques including NMR spectroscopy. Usually, solid-state studies focusing on organic matter tend to use the CP technique, as it allows acquisition of a ^{13}C NMR spectrum in a shorter time period. However, CP observability of aromatic ^{13}C nuclei in chars is relatively low because of the low concentration of nearby ^{1}H nuclei (see Chapter 4). Consequently, these carbons are underrepresented in CPMAS NMR spectra relative to those that undergo efficient CP (Figure 15.4). The alternative is to use the time-consuming but more quantitative DP technique.[63] In principle, this technique detects the whole combustion continuum, which ranges from slightly charred biomass to graphite black carbon.[64] Indeed, ^{13}C NMR is an unparalleled technique that allows both the characterization and the quantification of black carbon moieties in forest floors. Spin-counting experiments (see Chapter 4) can also be performed to precisely evaluate the quantity of black carbon present in organic samples.[65,66] Briefly, spin counting relies on comparing the NMR-integrated signal intensity of a sample for which the mass and carbon content are known with that of a standard, which is known

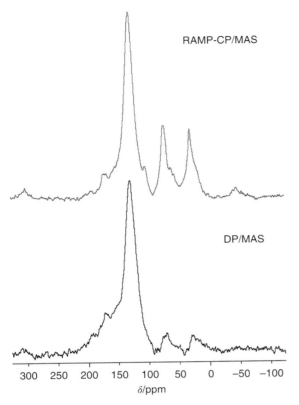

Figure 15.4. Ramped-amplitude (RAMP) CPMAS and DPMAS ^{13}C NMR spectra of an historical black carbon layer from a boreal black spruce forest floor. Spectra were obtained on a Bruker Avance 300 (B_0 = 7.05 T, $v_L(^{13}C)$ = 75.5 MHz) spectrometer using a 4.0 mm Bruker double-resonance MAS probe with a spinning frequency of 13 kHz ($\pm$10 Hz). All ^{13}C NMR spectra were referenced to TMS as discussed in the caption to Figure 15.1. The ^{13}C RAMP CP NMR spectrum was acquired using a ^{1}H 90° pulse width of 4.5 μs duration, a pulse delay of 10.0 s, and a contact time of 1.0 ms. A total of 7548 transients were obtained. For the ^{13}C DP NMR spectrum, 1800 transients were acquired using a 90° ^{13}C pulse of 4.5 μs duration. To allow for complete relaxation of all carbon nuclei, the pulse delay was set to 100 s. DEPTH pulse sequences were used to remove the background (probe) signal. Processing the spectra included zero filling to 8 K, line broadening of 100 Hz with phase, and baseline correction.

to provide a quantitative signal. The carbon observability (C_{obs}) further corresponds to the fraction of the expected NMR intensity that is actually observed. It is assumed that the intensities of the NMR resonances quantitatively reflect 'the distribution of ^{13}C

environments present in the sample' if C_{obs} = 100%.[65] Inversely, a C_{obs} less than 100% implies that part of the potential signal is not detected.

Many NMR studies have followed changes in soil organic matter composition during natural primary succession and forest ecosystem development. Equally interesting is the use of NMR to monitor changes after anthropogenic initiation of primary succession such as what happens during land reclamation following open-pit mining. In that case, organic matter amendments are often added to the soil surface, and planting of a variety of trees and shrubs is conducted to accelerate the re-establishment of self-sustaining, target ecosystems. In other anthropogenic chronosequences, such as harvested sites, characteristic shifts in alkyl/O-alkyl C ratios reflect the initially decreased vegetation inputs and changes in decomposition rates.[60] However, after reclamation, composition of soil organic matter initially reflects the type of amendment used during soil reconstruction. For instance, the aromatic carbon spectral intensity has been used to infer the contribution of coal residues to the overall organic matter signature.[67] In the case of land reclamation following oil sands mining, the alkyl/O-alkyl C ratio was used to follow the evolution of the peat amendment that is used when re-establishing a boreal forest cover.[68] Indeed, as vegetation grows on these rehabilitated sites, composition of soil organic matter begins to reflect inputs from the new forest vegetation. A stronger carbohydrate signal intensity has been reported in older reclaimed sites where labile carbon inputs from forest litter increased.[67] On the other hand, the alkyl/O-alkyl C ratio was seen to decrease following oil sands mining, as forest litter slowly replaces the peat amendment as a source of soil organic matter.[69] In all cases, in reclamation studies where soil quality is deeply disturbed, ^{13}C NMR techniques, in combination with other soil analytical methods such as the determination of isotopic enrichment values, can serve to establish monitoring indicators of developing forest ecosystem health. Research in novel forest ecosystems also benefits from investigations at the molecular level using plant biomarkers.[50]

15.6 OVERALL EXPERIMENTAL CONSIDERATIONS

Although many modern solid-state NMR techniques have been applied to investigate soil organic matter (see Chapter 13), the favored techniques remain the acquisition of one-dimensional ^{13}C NMR spectra of MAS samples, either with direct excitation of the ^{13}C nuclei (DP) or indirectly through a transfer of magnetization from ^{1}H (CP). A brief overview of experimental factors to be considered for these experiments is presented later. However, a detailed discussion of the theory and experimental considerations for such experiments is beyond the scope of this book; readers are referred to Chapter 4, recent textbooks,[70–72] and chapters,[10,16,17,19,23,73,74] including those highlighting applications to soil organic matter (see Chapter 13).

Probably, the two most important questions to be asked before undertaking a solid-state NMR investigation of soil organic matter are (i) which instrument should one use and (ii) is quantitative information needed? Of course, the answer to the first question is often dictated by the availability of instruments at the facility. However, if presented with a variety of options, this is a factor that must be considered carefully.[75–77] For most NMR applications, the greater sensitivity afforded by the higher field-strength instruments make them the preferred choice. However, gains in spectral resolution are minimal; interpretation of ^{13}C spectra of SOM obtained at high-field strengths may be difficult if one cannot remove the effects of spinning sidebands from the region of interest. Techniques such as total sideband suppression (TOSS)[78] have been developed to eliminate spinning sidebands (see Chapter 4); however, it is important to realize that the technique does not refocus the intensity of the spinning sidebands to the isotropic peak, so this method does not yield quantitative spectra. Alternatively, one may avoid the problem of spinning sidebands by very fast sample spinning. However, fast spinning probes usually only work with smaller sample volumes (4 and 3.2 mm rotors) and may not be suitable for samples with dilute carbon concentrations. It is important to note that CP efficiency generally decreases with MAS frequency.[77] Smernik[77] has developed a technique (RESTORE) to improve CP quantification (see Chapter 4).

Selecting the proper experiment to obtain the desired information is probably the most important decision for the spectroscopist. If quantitative information is not critical, then the typical experiment is CPMAS ^{13}C NMR spectroscopy. A major advantage of this technique is that the recycle time of the experiment is usually dictated by the ^{1}H spin–lattice relaxation time, which is typically one or two orders of magnitude less than that for ^{13}C. Combined with the potential CP enhancement of up to a factor of 4, the technique allows the spectroscopist to obtain spectra in a relatively

short period of time. In addition to typical acquisition parameters set by the spectroscopist (e.g., 90° pulse duration and acquisition time), the experiment is sensitive to the contact time, which generally is on the order of microseconds. It is important for the spectroscopist to understand that the resulting spectra are not quantitative, as the various carbon nuclei in the sample may not be enhanced equally for a given contact time. Techniques such as ramped-amplitude (RAMP) CP[79] reduce the sensitivity of the experiment to the contact time (see Chapter 4). Water content and the presence of paramagnetic ions have also been shown to influence ^{13}C NMR quantification and nuclear spin relaxation rates in soil organic matter.[12,40,80] The spin-counting technique discussed earlier can be used to assess the extent to which the carbon nuclei are detected.[65] If quantitative data are required, then the DP method is generally preferred; however, it is essential that the recycle delay be selected such that all nuclei fully relax.[12] Thus, the time required to obtain acceptable spectra may be an order of magnitude or greater more than that required to obtain spectra with CP. Another difficulty with the DP experiment is that the resulting spectra may contain a significant background signal.[81] One method of correcting for this factor is to obtain a spectrum of an empty rotor under identical conditions as that for the sample of interest, and to subtract it from the latter. Pulse sequences have also been developed to remove the background signal.[81,82]

When practical, it is desirable to quantify the reproducibility of the NMR data obtained. One easy approach is to obtain a spectrum for a given sample several times to calculate the coefficient of variation (%) of each carbon type by dividing the standard deviation by the mean value. Reproducibility is generally quite good, and the variation is typically less than 5% for any given carbon type.

Finally, in a rapidly developing field such as solid-state NMR, it is not surprising that researchers are investigating the potential for other solid-state NMR techniques to investigate soil organic matter. For example, ^{31}P CPMAS and DPMAS NMR spectra of samples have been used to investigate soil organic matter.[13] Solid-state ^{113}Cd CPMAS NMR combined with ^{13}C NMR, potentiometry, and voltammetry were used to investigate the Cd binding sites of some exudates in soil organic matter (see Chapter 19).[83] Thermally polarized and hyperpolarized ^{129}Xe NMR spectroscopy has been used to evaluate the potential of these techniques in characterizing porous materials in natural soil materials.[84] Pore size distribution has been studied by ^{1}H NMR relaxometry (see Chapter 18).[85] The recent introduction of PFGs for MAS probes has allowed the investigation of the mobility of organic contaminants in soil organic matter (see Chapter 20).[86]

15.7   CONCLUSIONS

Early NMR studies in forest ecology focused on establishing the link among management practices, soil fertility, and forest productivity. More recent work is motivated by issues of carbon sequestration and turnover rates. As outlined in Section 15.2, a particular area in forest ecology that has benefited from NMR analysis is the characterization of plant litter materials. Chemical changes during decomposition of tree litters, including both foliar and woody materials, were determined, and a sensitive index of organic matter decomposition in forest ecosystems was defined based on these analyses. In addition, NMR has allowed a better insight into litter and organic matter degradation processes. Changes in litters during decomposition appear universal across all climates, from boreal to tropical ecosystems. Another important paradigm that NMR has helped to clarify is the debate around lignin degradation. Originally thought to be highly recalcitrant, this biopolymer is now known to degrade, where the appropriate environmental conditions are present.

Section 15.3 describes how insights into the composition of key soil carbon pools have profited from NMR work. These include litter and forest floor pools and also organic matter associated with deeper mineral soil horizons. Similarly, as outlined in Section 15.4, NMR spectroscopy has proven to be a key tool in characterizing labile carbon such as that present in water percolating through soil horizons. Finally, as illustrated in Section 15.5, NMR spectroscopy has demonstrated that degraded organic matter in forest soils (including charred organic matter) is not as aromatic as originally thought. For all of these possible NMR applications to the study of forest ecosystems, combining NMR to isotopic and other molecular-level tools should always be the preferred approach.

ACKNOWLEDGMENTS

S.A. Quideau and R.E. Wasylishen thank NSERC of Canada for supporting their research programs and Dr. Guy M. Bernard for several helpful suggestions.

RELATED ARTICLE IN EMAGRES

Agriculture and Soils

REFERENCES

1. W. Zech, F. Ziegler, I. Kögel-Knabner, and L. Haumaier, *Sci. Total Environ.*, 1992, **117/118**, 155.

2. I. Kögel-Knabner, W. Zech, and P. G. Hatcher, *Z. Pflanzenernähr. Bodenk.*, 1988, **151**, 331.

3. J. A. Baldock, J. M. Oades, P. N. Nelson, T. M. Skene, A. Golchin, and P. Clarke, *Aust. J. Soil Res.*, 1997, **35**, 1061.

4. C. M. Preston, *Soil Sci.*, 1996, **161**, 144.

5. C. M. Preston, *Sci. Total Environ.*, 1992, **113**, 107.

6. C. M. Preston and R. H. Newman, *Geoderma*, 1995, **68**, 229.

7. C. M. Preston, J. R. Nault, and J. A. Trofymow, *Ecosystems*, 2009, **12**, 1078.

8. C. E. Norris, S. A. Quideau, J. S. Bhatti, and R. E. Wasylishen, *Global Change Biol.*, 2011, **17**, 480.

9. M. A. Wilson, R. J. Pugmire, K. W. Zilm, K. M. Goh, S. Heng, and D. M. Grant, *Nature*, 1981, **294**, 648.

10. A. J. Simpson, D. J. McNally, and M. J. Simpson, *Prog. Nucl. Magn. Reson. Spectrosc.*, 2011, **58**, 97.

11. I. Kögel-Knabner, *Geoderma*, 1997, **80**, 243.

12. H. Knicker, *Org. Geochem.*, 2011, **42**, 867.

13. A. L. Doolette and R. J. Smernik, in Phosphorus in Action: Biological Processes in Soil Phosphorus Cycling, eds E. K. Bünemann, A. Oberson and E. Frossard, Springer: Berlin, 2011, Vol. 26, Chap. 1.

14. M. M. Kononova, Soil Organic Matter: Its Nature, its Role in Soil Formation and in Soil Fertility, Pergamon Press: Oxford, 1961.

15. F. C. Ugolini and R. A. Dahlgren, in Podzols et Podzolisation, eds D. Righi and A. Chauvel, AFES et INRA: Plaisir et Paris, 1987, 195.

16. C. M. Preston, *Can. J. Soil Sci.*, 2001, **81**, 255.

17. N. J. Mathers, X. A. Mao, Z. H. Xu, P. G. Saffigna, S. J. Berners-Price, and M. C. S. Perera, *Aust. J. Soil. Res.*, 2000, **38**, 769.

18. M. J. Simpson, *Soil Sci. Soc. Am. J.*, 2006, **70**, 995.

19. M. J. Simpson and C. Preston, in Soil Sampling and Methods of Analysis, 2nd edn, eds M. R. Carter and E. G. Gregorich, CRC Press: Boca Raton, 2008.

20. X. Feng and M. J. Simpson, *J. Environ. Monit.*, 2011, **13**, 1246.

21. M. J. Simpson and A. J. Simpson, *J. Chem. Ecol.*, 2012, **38**, 768.

22. F. Ziarelli, S. Viel, S. Sanchez, D. Cross, and S. Caldarelli, *J. Magn. Reson.*, 2007, **188**, 260.

23. J. Mao, N. Chen, and X. Cao, *Org. Geochem.*, 2011, **42**, 891.

24. J. W. Harden, K. P. O'Neill, S. E. Trumbore, H. Veldhuis, and B. J. Stocks, *J. Geophys. Res.*, 1997, **102**, 28805.

25. P. W. Flanagan and K. Van Cleve, *Can. J. Forest Res.*, 1983, **13**, 795.

26. S. A. Quideau, R. C. Graham, S.-W. Oh, P. F. Hendrix, and R. E. Wasylishen, *Soil Biol. Biochem.*, 2005, **37**, 1988.

27. J. M. Melillo, J. D. Aber, A. E. Linkins, A. Ricca, B. Fry, and K. J. Nadelhoffer, *Plant Soil*, 1989, **115**, 189.

28. C. M. Preston and J. A. Trofymow, *Can. J. Bot.*, 2000, **78**, 1269.

29. J. B. Gaudinski, T. E. Dawson, S. Quideau, E. A. G. Schuur, J. S. Roden, S. E. Trumbore, D. R. Sandquist, S.-W. Oh, and R. E. Wasylishen, *Anal. Chem.*, 2005, **77**, 7212.

30. J.-D. Mao, W. G. Hu, K. Schmidt-Rohr, G. Davies, E. A. Ghabbour, and B. Xing, *Soil Sci. Soc. Am. J.*, 2000, **64**, 873.

31. E. Breitmaier and W. Voelter, Carbon-13 NMR Spectroscopy. High-Resolution Methods and Applications in Organic Chemistry and Biochemistry, VCH Publishers: New York, 1987.

32. H.-O. Kalinowski, S. Berger, and S. Braun, Carbon-13 NMR Spectroscopy, John Wiley & Sons: Chichester, 1988.

33. J. A. Baldock, J. M. Oades, A. G. Waters, X. Peng, A. M. Vassallo, and M. A. Wilson, *Biogeochemistry*, 1992, **16**, 1.

34. M. Strukelj, S. Brais, S. A. Quideau, and S.-W. Oh, *Can. J. Forest Res.*, 2012, **42**, 772.

35. C. E. Johnson, T. J. Blumfield, S. Boyd, and Z. Xu, *J. Soil. Sediment.*, 2013, **13**, 854.

36. J. A. Baldock and C. M. Preston, in Carbon Forms and Functions in Forest Soils, eds W. W. McFee and J. M. Kelly, Soil Science Society of America: Madison, WI, 1995, 89.

37. K. D. Hannam, S. A. Quideau, S.-W. Oh, B. E. Kishchuk, and R. E. Wasylishen, *Soil Sci. Soc. Am. J.*, 2004, **68**, 1735.

38. S. J. Opella and M. H. Frey, *J. Am. Chem. Soc.*, 1979, **101**, 5854.

39. A. Chabbi, I. Kögel-Knabner, and C. Rumpel, *Soil Biol. Biochem.*, 2009, **41**, 256.

40. J. O. Skjemstad, P. Clarke, J. A. Taylor, J. M. Oades, and R. H. Newman, *Aust. J. Soil Res.*, 1994, **32**, 1215.

41. M. W. I. Schmidt, H. Knicker, P. G. Hatcher, and I. Kögel-Knabner, *Eur. J. Soil Sci.*, 1997, **48**, 319.

42. X. Fang, T. Chua, K. Schmidt-Rohr, and M. L. Thompson, *Geochim. Cosmochim. Acta*, 2010, **74**, 584.

43. C. M. Preston, R. H. Newman, and P. Rother, *Soil Sci.*, 1994, **157**, 26.

44. S. A. Quideau, M. A. Anderson, R. C. Graham, O. A. Chadwick, and S. E. Trumbore, *For. Ecol. Manage.*, 2000, **138**, 19.

45. N. Mahieu, D. S. Powlson, and E. W. Randall, *Soil Sci. Soc. Am. J.*, 1999, **63**, 307.

46. S. A. Quideau, O. A. Chadwick, A. Benesi, R. C. Graham, and M. A. Anderson, *Geoderma*, 2001, **104**, 41.

47. I. Kögel-Knabner, *Org. Geochem.*, 2000, **31**, 609.

48. J. I. Hedges and J. R. Ertel, *Anal. Chem.*, 1982, **54**, 174.

49. S. M. Tareq, N. Tanaka, and K. Ohta, *Sci. Total Environ.*, 2004, **324**, 91.

50. C. E. Norris, J. A. Dungait, A. Jones, and S. A. Quideau, *Org. Geochem.*, 2013, **64**, 9.

51. J. Sanderman, J. A. Baldock, and R. Amundson, *Biogeochemistry*, 2008, **89**, 181.

52. K. H. Dai, C. E. Johnson, and C. T. Driscoll, *Biogeochemistry*, 2001, **54**, 51.

53. G. C. Woods, M. J. Simpson, P. J. Koerner, A. Napoli, and A. J. Simpson, *Environ. Sci. Technol.*, 2011, **45**, 3880.

54. S. A. Quideau, O. A. Chadwick, S. E. Trumbore, J. L. Johnson-Maynard, R. C. Graham, and M. A. Anderson, *Org. Geochem.*, 2001, **32**, 247.

55. S. A. Quideau, R. C. Graham, X. Feng, and O. A. Chadwick, *Soil Sci. Soc. Am. J.*, 2003, **67**, 1544.

56. A. Golchin, P. Clarke, and J. M. Oades, *Biogeochemistry*, 1996, **34**, 71.

57. E. A. Webster, J. A. Chudek, and D. W. Hopkins, *Biol. Fertil. Soils*, 1997, **25**, 389.

58. C. E. Norris, S. A. Quideau, S. M. Landhäusser, G. M. Bernard, and R. E. Wasylishen, *Sci. Rep.*, 2012, **2**, 719. DOI: 10.1038/srep00719.

59. K. D. Hannam, S. A. Quideau, B. E. Kishchuk, S.-W. Oh, and R. E. Wasylishen, *Can. J. Forest Res.*, 2005, **35**, 2457.

60. C. E. Norris, S. A. Quideau, J. S. Bhatti, R. E. Wasylishen, and M. D. MacKenzie, *Can. J. Forest Res.*, 2009, **39**, 642.

61. E. Thiffault, K. D. Hannam, S. A. Quideau, D. Paré, N. Bélanger, S.-W. Oh, and A. D. Munson, *Plant Soil*, 2008, **308**, 37.

62. H. Knicker, F. J. González-Vila, O. Polvillo, J. A. González, and G. Almendros, *Soil Biol. Biochem.*, 2005, **37**, 701.

63. C. M. Preston and M. W. I. Schmidt, *Biogeosciences*, 2006, **3**, 397.

64. C. A. Masiello, *Mar. Chem.*, 2004, **92**, 201.

65. R. J. Smernik and J. M. Oades, *Geoderma*, 2000, **96**, 101.

66. R. J. Smernik and J. M. Oades, *Geoderma*, 2000, **96**, 159.

67. D. P. Dick, H. Knicker, L. G. Ávila, A. V. Inda Jr, E. Giasson, and C. Bissani, *Org. Geochem.*, 2006, **37**, 1537.

68. I. Turcotte, S. A. Quideau, and S.-W. Oh, *Org. Geochem.*, 2009, **40**, 510.

69. P. T. Sorenson, S. A. Quideau, M. D. MacKenzie, S. M. Landhäusser, and S. W. Oh, *Appl. Soil Ecol.*, 2011, **49**, 139.

70. D. E. Axelson, Solid State Nuclear Magnetic Resonance: A Practical Introduction, Natural Resources Canada: Ottawa, 2012.

71. D. C. Apperley, R. K. Harris, and P. Hodgkinson, Solid State NMR: Basic Principles & Practice, Momentum Press: New York, 2012.

72. M. H. Levitt, Spin Dynamics. Basics of Nuclear Magnetic Resonance, 2nd edn, Wiley: Chichester, 2008.

73. D. L. Bryce, G. M. Bernard, M. Gee, M. D. Lumsden, K. Eichele, and R. E. Wasylishen, *Can. J. Anal. Sci. Spectrosc.*, 2001, **46**, 46.

74. P. Conte, R. Spaccini, and A. Piccolo, *Prog. Nucl. Magn. Reson. Spectrosc.*, 2004, **44**, 215.

75. K. J. Dria, J. R. Sachleben, and P. G. Hatcher, *J. Environ. Qual.*, 2002, **31**, 393.

76. R. J. Smernik, E. Eckmeier, and M. W. I. Schmidt, *Aust. J. Soil Res.*, 2008, **46**, 122.

77. R. J. Smernik, *Geoderma*, 2005, **125**, 249.

78. W. T. Dixon, J. Schaefer, M. D. Sefcik, E. O. Stejskal, and R. A. McKay, *J. Magn. Reson.*, 1982, **49**, 341.

79. G. Metz, X. Wu, and S. O. Smith, *J. Magn. Reson., Ser A*, 1994, **110**, 219.

80. R. J. Smernik, *Eur. J. Soil Sci.*, 2006, **57**, 665.

81. R. J. Smernik and J. M. Oades, *Solid State Nucl. Magn. Reson.*, 2001, **20**, 74.

82. D. G. Cory and W. M. Ritchey, *J. Magn. Reson.*, 1988, **80**, 128.

83. V. Lenoble, C. Garnier, A. Masion, F. Ziarelli, and J. M. Garnier, *Anal. Bioanal. Chem.*, 2008, **390**, 749.

84. S. Filimonova, A. Nossov, A. Dümig, A. Gédéon, I. Kögel-Knabner, and H. Knicker, *Geoderma*, 2011, **162**, 96.

85. J. V. Bayer, F. Jaeger, and G. E. Schaumann, *Open Magn. Reson. J.*, 2010, **3**, 15.

86. K. W. Fomba, P. Galvosas, U. Roland, J. Kärger, and F.-D. Kopinke, *Environ. Sci. Technol.*, 2009, **43**, 8264.

Chapter 16
Biofuels

Antonio G. Ferreira[1], Luciano M. Lião[2] and Marcos R. Monteiro[1]

[1]*Department of Chemistry, Federal University of São Carlos, São Carlos, 13565-905, Brazil*
[2]*Federal University of Goiás, Goiânia, 74001-970, Brazil*

16.1 INTRODUCTION

While nonrenewable energy sources (i.e., coal, petroleum, and natural gas) are being depleted, the global demand for energy has grown, increasing the urgency for biofuels. Finding alternative renewable energy sources has been a major technological challenge around the world. The need for technical developments in the energy sector is amplified by the following problems: the security of the international energy supply; the growth in global demand; and the negative environmental effects, including global warming, caused by burning fossil fuels and atmospheric emissions of greenhouse gases. Therefore, developments and innovations in biofuels, such as bioethanol, biogas, biobutanol, and biohydrogen, have emerged as important goals for global energy.

For the production and use of biofuels on a large scale, we highlight the following experiences: producing ethanol from sugarcane in Brazil and producing biodiesel from vegetable oils and fats in Europe. The global consumption of these biofuels was estimated to be approximately 2 billion barrels per day in 2011, which was 80% higher than the consumption in 2007. The International Energy Agency (IEA) estimates an average annual growth rate of 7%, which means that by 2035, biofuels will account for approximately 8% of the total road transportation fuel demand, whereas biofuels only represent approximately 3% of the current demand.[1]

Production processes have advanced, especially those regarding technological alternatives and different ways of thinking about the energy matrix. Accordingly, we can classify biofuel production into four generations:[2]

1. *First-generation biofuels*: biofuels made from sugar, starch, vegetable oil, or animal fats using conventional technology (i.e., microbiological fermentation and transesterification by alkaline catalyst). The basic feedstock for this production is often either grass (e.g., sugarcane, corn, and sorghum) that is fermented into bioethanol or seeds (e.g., soybean, sunflower, and peanuts)

NMR Spectroscopy: A Versatile Tool for Environmental Research
Edited by Myrna J. Simpson and André J. Simpson
© 2014 John Wiley & Sons, Ltd. ISBN: 978-1-118-61647-5

that yields vegetable oil, which can be used for biodiesel production.

2. *Second-generation biofuel*: biofuels produced from a variety of nonfood crops. The feedstock for second-generation biofuels is lignocellulosic biomass, such as wood (e.g., *Pinus* and *Eucalyptus*), organic waste, switchgrass, and agricultural residue from corn, wheat, sugarcane, and similar plants. Many of the resulting biofuels, such as biohydrogen, biomethanol, and biohydrogen diesel, are under development.

3. *Third-generation biofuel*: while the bioconversion process changed for second-generation biofuels, third-generation biofuels are based on creating biomass sources with high specific productivity (e.g., using genetic engineering to design oil plants that are more productive). The types of feedstock for third-generation biofuels are those that require further research and development to become commercially feasible. These feedstocks include perennial grasses, fast-growing trees, and large-scale algae production. These feedstocks are designed exclusively for fuel production and are commonly referred to as 'energy crops'.

4. *Fourth-generation biofuel*: biofuels combining the development of genetically optimized plants with the capacity to capture large amounts of carbon, thus removing CO_2 from the atmosphere and for storage in the branches, trunks, and leaves of the plant. The produced biomass is then converted into fuel and gas using second-generation biofuel techniques. The produced carbon dioxide can then be sequestered geologically by storing the carbon dioxide in old oil and gas fields or in saline aquifers.[2,3]

The emissions of the principal pollutants are much lower when combusting biofuels than when combusting conventional petroleum derivatives. Regulated emissions (i.e., total unburned hydrocarbons, carbon monoxide, and particulate matter) and unregulated emissions (i.e., sulfates, polycyclic aromatic hydrocarbon, and others) from 100% biodiesel (B100) and biodiesel/diesel blends are significantly lower than those from diesel, except for nitrogen oxide.[4] The use of 20% biodiesel and 80% diesel B20 and B100 biofuel reduces the emission of the total unburned hydrocarbons by 20% and 67%, respectively, relative to that of conventional diesel and reduces sulfate emission by 100%.[5] Using 20% bioethanol in gasoline reduces the total unburned hydrocarbon by approximately 28% and the carbon monoxide by 21%, but the aldehyde and nitrogen oxides levels may be higher, depending on the engine characteristics.[6]

The expansion and consolidation of biofuels in global energy involve a wide range of issues, including the following: the need to increase productivity and competitiveness in the production chain of biofuels, the need to guarantee quality control in the final product, and the use of conversion routes that are more productive and less harmful to the environment. Developing biofuel requires standardization of analytical techniques, especially those for raw material characterization, process monitoring, product quality control, production technology optimization, and analysis of the methods for producing biofuels and their blends. These steps are important for the inclusion of biofuels in the world market. Therefore, NMR, combined with other spectroscopic and hyphenated spectrometric techniques, e.g., high performance liquid chromatography-mass spectrometry (HPLC-MS), gas chromatography-mass spectrometry (GC-MS), Fourier transform infrared spectroscopy (FTIR), physicochemical analysis, and statistical tools, provides an important set of techniques for the development of biofuels in the world market.

16.2 NMR IN PROSPECTING FOR RAW MATERIALS

Several sources of raw materials can be used to produce biofuels, and lignocellulose that is produced from biomass through hydrolysis, pyrolysis, or enzymatic reactions is undoubtedly one of them.[7] Other very important sources include vegetable oils (e.g., palm, soybean, castor, jatropha, corn, sunflower, linseed, rapeseed, cottonseed, canola, peanut, and macadamia), animal fats, and residual oil.[8,9] More recently, algae have also been used as a raw material.[10,11] However, the important criteria for choosing a raw material are the energy potential, the cost to produce the material, and the environmental effects from its commercial exploitation. With these criteria, several sources of raw materials can be evaluated using NMR as an analytical tool.

When using lignocellulosic biomass as a raw material (e.g., *Pinus*, *Eucalyptus*, bamboo— *Neosinocalamus affinis*, sugarcane bagasse, and rice straw), it is necessary to promote the depolymerization of these materials to their more basic constituents (i.e., cellulose, hemicellulose, and lignin). After this

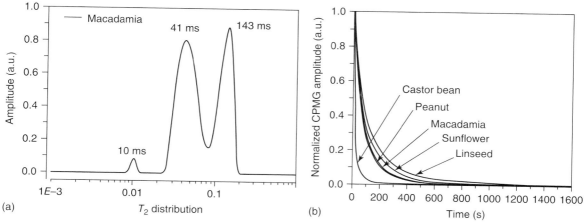

Figure 16.1. Plot of T_2 decay using the CPMG pulse sequence in a 10 MHz spectrometer with a 100 mm diameter probe. (a) Measurement of the water and saturated and unsaturated fatty acid contents and (b) differentiation of seeds based on total fatty acid content.

depolymerization, NMR can be used to evaluate the amount of material and consequently its biofuel conversion potential. Using NMR measurements, we can evaluate the content of lignin and sugar extracts in the raw material using ^1H, ^{13}C, and ^{31}P (in the solution state) and ^{13}C nuclei (in the solid state).[12]

For vegetable oils, an important step is measuring the content of water and fatty acid (saturated and unsaturated) in the seeds of the plant that will be used as the raw material. The water content should ideally be in the range from 27% to 38% to avoid seed germination. Otherwise, the material that must be stored for days, weeks, or months before processing can change the total fatty acid content during storage. Furthermore, the total fatty acid content is important for estimating the energy conversion capacity for methanolic or ethanolic esterification into biofuel. The polyunsaturated fatty acid content is also important because these levels may influence the stability over time for both the starting material and the resultant biofuel. Polyunsaturated compounds are less stable over time and are more susceptible to oxidative processes. In contrast, monounsaturated or saturated fatty acids tend to solidify with decreasing temperature, and this solidification can have serious practical consequences for biofuel use.[13–15]

A fast way to quantify all these parameters is to use a low-cost and efficient technique, such as low frequency NMR, which directly measures the ^1H spin–spin or transverse relaxation time (T_2) of intact

seeds, typically using a Carr-Purcell-Meiboom-Gill (CPMG) pulse sequence. This approach does not require any sample preparation, does not destroy the sample, is less expensive, is environmentally friendly, and produces results similar to those of conventional physicochemical analysis (i.e., gravimetric water content measurements, total fatty acids content measurements using extraction with a mixture of organic solvents, and unsaturation degree measurements using an iodine reaction). The measurements take a few seconds and result in a plot of signal intensity versus time decay (Figure 16.1). For these signals, each exponential decay curve represents a different relaxation time, which can be correlated to the various water and fatty acid (saturated and unsaturated) content in the sample. This measurement allows not only the discrimination but also the quantification of this content when these data are compared with a previously obtained standard calibration curve.[16,17] Figure 16.1(a) and (b) shows an example of this measurement. In Figure 16.1(a), the differentiation and possibly also the quantification of intracellular water (10 ms) and saturated (41 ms), unsaturated, and/or polyunsaturated (142 ms) fatty acids are shown for intact macadamia seeds. In Figure 16.1(b), the differentiation of total fatty acids content in several intact seeds (e.g., castor, peanut, macadamia, sunflower, and linseed oil) is shown.

In addition to low frequency NMR, high frequency NMR can also be used to determine the unsaturated fatty acid content, which is related to the iodine value

(IV). The IV is the amount of iodine that reacts with 100 g of oil under the analysis conditions. However, before measuring IV, it is necessary to extract the oil from the seed, and solution-state NMR measurements should be performed using a deuterated solvent as an internal or external reference. The oil can be extracted by pressing the seeds and/or by an extraction using an organic solvent and an extractor, such as Soxhlet extraction. Specific regions in a quantitative ^{1}H NMR spectrum can be integrated to determine which fatty acids are present in a specific oil or animal fat sample. The fatty acid content can be calculated by summing the areas of the hydrogens that are common to all fatty acids (U) (i.e., α-, β-carbonyl and methyl hydrogens) and dividing by the respective number of areas that represent specific hydrogens. Subsequently, this area is divided by three to obtain the value proportional to that hydrogen. Applying the corresponding formulas given in equations (16.1–16.4), we can calculate the content of monounsaturated, diunsaturated, and triunsaturated fatty acids. The saturated fatty acid content is obtained by excluding the levels of unsaturated acids. Figure 16.2 shows an example of a solution-state ^{1}H NMR spectrum from linseed oil in CDCl$_3$ (80:20 v/v).

$$U = \frac{1}{3}\left(\frac{A}{2} + \frac{B}{2} + \frac{C+D}{3}\right) \qquad (16.1)$$

$$X\ (\%\ \text{of monounsaturated}) = \frac{\left(\frac{C}{3} \times 100\right)}{U} \qquad (16.2)$$

$$Y\ (\%\ \text{of diunsaturated}) = \frac{\left(\frac{E}{2} - 2\frac{C}{3} \times 100\right)}{U} \qquad (16.3)$$

$$Z\ (\%\ \text{of triunsaturated}) = \frac{\left(\frac{F}{4} \times 100\right)}{U} - (X + Y) \qquad (16.4)$$

Table 16.1 lists the fatty acid contents of the various vegetable oils and animal fats that are most commonly used as raw materials for biofuel production. These data were partially in agreement with the data obtained from solution-state ^{1}H NMR of several of these oils.

Quantitative measurements of the unsaturation degree can also be obtained using solution-state ^{13}C NMR spectra. The advantage of observing the ^{13}C nucleus is the greater dispersion of the spectral signals, which consequently decreases the possibility for signal overlap from the sample and/or from impurities in the sample. However, as for solution-state ^{1}H NMR spectra, a quantitative measurement is required. For ^{13}C, the spin–lattice or longitudinal relaxation time (T_1) is considerably longer than that for ^{1}H nuclei, and, therefore, ^{13}C measurements are much more time consuming, possibly requiring several hours for each sample. However, only minimal sample treatment, consisting

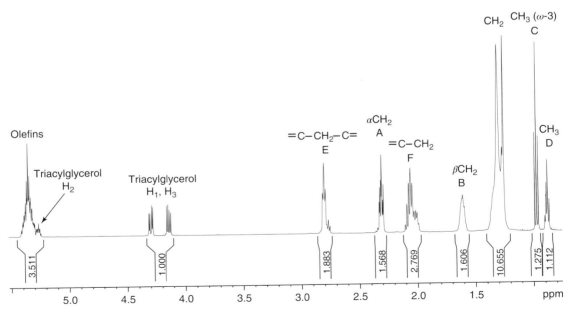

Figure 16.2. ^{1}H NMR spectrum of linseed oil in CDCl$_3$, using a 500 MHz spectrometer.

Table 16.1. Composition of fatty acid in vegetable oil and in animal fat (compiled from references 18,19).

Oil or Fat	Fatty Acid Composition (wt. %)						
	Lauric $C_{12}H_{24}O_2$	Myristic $C_{14}H_{28}O_2$	Palmitic $C_{16}H_{32}O_2$	Stearic $C_{18}H_{36}O_2$	Oleic $C_{18}H_{34}O_2$	Linoleic $C_{18}H_{32}O_2$	Linolenic $C_{18}H_{30}O_2$
Babassu	40–45	11–27	5.2–11	1.8–7.4	9–20	1.4–6.6	nd
Canola	nd	nd	1.5–6	1–2.5	52–66.9	16.1–31	6.4–14.1
Coconut	44–51	13–20.6	7.5–10.5	1–3.5	5–8.2	1–2.6	0–0.2
Corn	nd	0–0.3	7–16.5	1–3.3	20–43	39–62.5	0.5–1.5
Cottonseed	nd	0.6–1.5	21.4–26.4	2.1–5	14.7–21.7	46.7–58.2	nd
Linseed	nd	nd	6–7	3.2–5	13–37	5–23.0	26–60
Olive	0–1.3	7–20	0.5–5	55–84.5	3.5–21	nd	nd
Palm	0.5–2.4	32–47.5	3.5–6.3	36–53	6–12	nd	nd
Peanut	nd	0–0.5	6–14	1.9–6	36.4–67.1	13–43	nd
Rapeseed	nd	0–1.5	1.6	0.5–3.5	8–60	9.5–23	1–13
Safflower	nd	nd	5.3–8	1.9–2.9	8.4–23.1	67.8–83.2	nd
Sesame	nd	nd	7.2–9.2	5.8–7.7	35–46	35–48	nd
Soybean	nd	nd	2.3–13.3	2.4–6	17.7–30.8	49–57.1	2–10.5
Sunflower	nd	nd	3.5–7.6	1.3–6.5	14–43	44–74	nd
Tallow	nd	2.1–6.9	25–37	9.5–34.2	14–50	26–50	nd

nd—not detected.

of sample dilution in a deuterated solvent (e.g., $CDCl_3$ 1 : 3 v/v) or using a capillary with a deuterated solvent (i.e., external reference), is required. By integrating the total peak areas for the four groups of peaks (i.e., β-carbonyl, divinyl methylene, *cis*-allylic carbons, and mid-chain methylene) in the region from $\delta = 24.5$ to 30.0 ppm, one can calculate the number of double bonds in triacylglycerols; consequently, the IV can be calculated, and the average molecular weight can be estimated. This procedure has shown good agreement with the data obtained using conventional methods.[20]

Using ^{1}H NMR in the semi-solid state (^{1}H HR-MAS NMR; see Chapter 5), it is possible to measure the same information as that measured using low- and high-field NMR techniques. In this technique, the material is inserted into the rotor (typically 12–50 μl volume) without being subjected to any extraction or purification process (see Chapter 5). Small fragments of material are inserted into the rotor with the addition of one drop of deuterated water to provide a lock signal for adjust the local field homogeneity. This technique is convenient for measuring intact vegetable or animal samples and also for running a number of routine one-dimensional (1D) and two-dimensional (2D) NMR experiments.[21] Interested readers should also see Chapter 5.

16.3 NMR IN PRODUCTION CONTROL

The main step during biodiesel production is the transesterification of triacylglycerol from animal fats (e.g., tallow, lard, poultry, and fish wastes) or oils (e.g., vegetable, waste cooking, and algae) in the presence of an alcohol (i.e., methanol, ethanol, and higher alcohols) and a catalyst (usually a base, e.g., NaOH, $NaOCH_3$, and KOH). Figure 16.3 shows an example of a typical reaction using triacylglycerol as a source of fatty acid. Biodiesel can also be produced by direct esterification of free fatty acids. However, in this reaction, the alkaline catalysis should be avoided to avoid soap formation that will consume the catalyst as well as the feedstock, decreasing the reaction yield and making difficult to separate glycerol from biodiesel. It is also possible to use various catalysts such as acids (e.g., sulfuric acid), enzymes, and heterogeneous (e.g., calcium/barium acetates, zinc oxide supported on aluminum, and hydrotalcites).

Besides the chemical and biochemical developments to optimize the reaction, it is important to monitor the level of products and by-products that arise from the transesterification reaction (i.e., glycerol, free fatty acids, monoacylglycerols, diacylglycerols, alcohols, and catalysts) to ensure the biodiesel quality. In this context, GC and HPLC are the techniques typically employed. However, these methods require

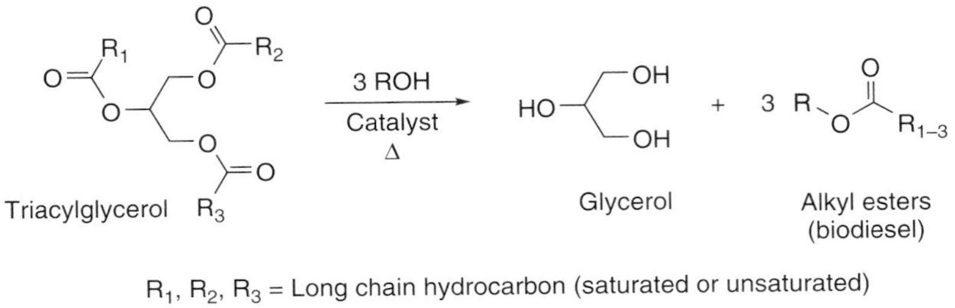

Figure 16.3. Transesterification reaction of triacylglycerol into biodiesel.

time-consuming analysis, are environmentally unfriendly because of large consumption of solvents, and, in the case of GC, typically require a derivatization reaction to make the compounds of interest volatile. Moreover, NMR can be used to analyze many samples, quantifying various compounds in a single analytical protocol, working quickly, and consuming very little laboratory supplies. A quantitative solution-state ^{1}H NMR spectrum can be used for monitoring the yield of transesterification reaction and to characterize the by-products formed. The peaks of methylene group adjacent to ester moiety in triacylglycerols (α-CH$_2$, $\delta = 2.3$ ppm, t) and the methoxy group in the esters (OCH$_3$, $\delta = 3.6$ ppm, s) are used to follow the reaction progress (Figure 16.4). The percentage of conversion can be calculated from the areas of those peaks, using the equation 16.5:

$$C = 100 \times \frac{2AOCH_3}{3A \, \alpha \, CH_2} \quad (16.5)$$

Alternatively, the integration of methylene glycerol hydrogens ($\delta = 4.10–4.40$ ppm) from feedstock can be included and used for monitoring the reaction progress.[22] Transesterification can also be monitored by solution-state ^{13}C NMR using the signal at $\delta = 14.5$ ppm, which is from a methyl group that is not altered by the reaction and can be used as internal standard. The glycerol carbons at $\delta = 62–71$ ppm along with methoxy carbon of fatty esters at $\delta = 51$ ppm are used to determine the conversion rate.[23]

Methyl and α-carbonyl methylene hydrogens can be related to biodiesel as well as to mono-, di-, and triacylglycerols from uncompleted transesterification reaction and free fatty acids from raw material. A ratio 3 : 2 for methyl and α-carbonyl methylene, respectively, is expected for pure biodiesel, and this is a quality indicator for obtained product. 1-Monoacylglycerols, 2-monoacylglycerols,

1,2-Diacylglycerols, 1,3-diacylglycerols, and triacylglycerols can be differentiated through chemical shift variations of H-1, H-2, and H-3 attached to glycerol carbons. When one or two acyl groups are migrated from triacylglycerols in transesterification process, H-1, H-2, and H-3 would shift toward to lower chemical shift values, because of the loss of the high electron density of the acyl group, whereas the hydrogens on the acyl group could not shift. This analysis can also be done by solution-state ^{13}C NMR that minimizes the signal overlap.[24]

Another alternative approach for characterizing mixtures of multisubstituted acylglycerols involves phosphitylation of hydroxyl groups from substituted glycerol with 2-chloro-4,4,5,5-tetramethyl-1,3,2-Dioxaphospholane (TMDP) or 2-chloro-1,3,2-Dioxaphospholane (DOP), followed by a quantitative solution-state ^{31}P NMR analysis. Using cyclohexanol derivatized with TMDP as internal standard 1-monoleoylglycerol presents signals at $\delta = 146.3$ and 147.8 ppm. For 2-monoleoylglycerol, only one signal is observed at $\delta = 147.9$ ppm. The ^{31}P signals for 1,2- and 1,3-dioleoylglycerols are, respectively, observed at $\delta = 147.7$ and 146.4 ppm. Phosphitylation results in signals for free glycerol at $\delta = 147.1$ and 146.1 ppm, methanol at $\delta = 147.9$ ppm, ethanol at $\delta = 146.3$ ppm, and isopropanol at $\delta = 146.4$ ppm, which can be used to differentiate the compounds. Quantification of these compounds can be obtained by integration of respective signal area.[25]

Diffusion ordered spectroscopy (DOSY) (see Chapter 2) can also be applied to resolve mixtures of unreacted mono-, di-, and triacylglycerols along with their methyl esters and residual methanol. In this 2D experiment, higher molecular-weight compounds diffuse more slowly than those of lower molecular weight and, therefore, can be resolved the spectra without the physical compounds separation.[26]

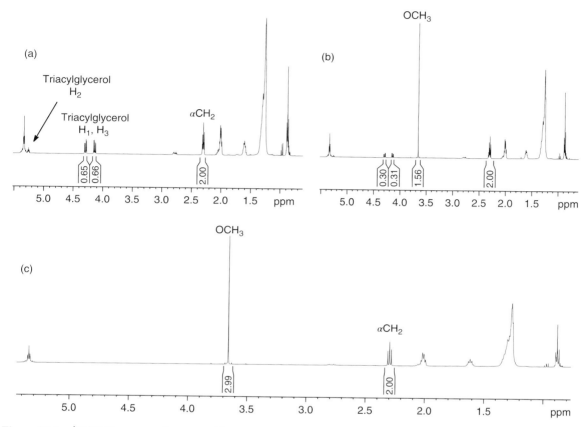

Figure 16.4. ^{1}H NMR spectra of transesterification reaction in CDCl$_3$, using a 500 MHz spectrometer. (a) Raw material (vegetable oil), (b) incomplete reaction: 48% raw material and 52% product, and (c) only product (biodiesel).

Traditionally, the monitoring of the conversion of vegetable oil to biodiesel has been accomplished by high-resolution NMR. However, quantitative measurements based on T_2 values can be performed in a low-resolution NMR for this type of monitoring. High-resolution NMR monitoring is the most sensible technique, nevertheless more expensive, and uses solvents that have limited applications in an industrial environment. On the other hand, low-resolution NMR spectrometers are more suitable for monitoring transesterification reactions in industrial settings.

The use of solution-state ^{1}H NMR is also a practical and fast method to monitor kinetics and product distributions in transesterification reactions.[27,28] In real time, a small aliquot from reaction flask can be used to quantify the reactant consumption and product formation. Thereby, the conversion rate can be determined and used as an important tool in the transesterification

reaction optimization. Kinetic studies can also be performed and monitored in situ with solution-state ^{1}H NMR spectroscopy, as exemplified in palmitic acid esterification with deuterated methanol, in the presence of the *p*-sulfonic acid calix[4]arene and sulfuric acid.[29] ^{1}H NMR can also be used to determine the average degree of unsaturation on fatty acid methyl esters by comparing integrations of the ^{1}H resonances from the olefinic ($\delta = 5.20-5.45$ ppm) and methoxy ($\delta = 3.60$ ppm) groups. Special attention should be given not to integrate the methine hydrogen of triacylglycerol together with the olefinic. Moreover, integrations of olefinic and ester ethoxy methylene hydrogens ($\delta = 4.05-4.20$ ppm) could be used for fatty acid ethyl esters unsaturation degree. On the other hand, polyunsaturated fatty acids can be detected and quantified by the signal at $\delta = 2.70$ ppm from the bisallylic hydrogen (methylene group between two double bonds).[30]

^{1}H NMR methodology to monitor oil ethanolysis as well as to quantify fatty acid ethyl esters in blends of biodiesel and vegetable oil is also available. The region of $\delta = 4.05-4.40$ ppm (ester ethoxy and triacylglycerol methylene hydrogens) is used for quantification in this case. Two doublets of doublets signals referring to methylene hydrogens from triacylglycerol are observed in this region for pure oil. Otherwise, only one quartet signal from ester ethoxy is observed when ethanolysis is complete and triacylglycerol is not present. During the reaction, overlapped signals are observed as shown in Figure 16.5. The overlapped resonances represent one methylene group from triacylglycerol and one from fatty acid ethyl ester. This signal region compared to unoverlapped methylene signal region can be used to monitor and quantify blend contents for the ethanolysis reaction.[31] In the same way,

syntheses of propyl, butyl, and higher biodiesels that have a triplet signal from OCH_2 group could be monitored and quantified by solution-state ^{1}H NMR.

Bioethanol, an important renewable energy resource, can be obtained through fermentation of sugar biomasses but can also be obtained by fermentation of cellulosic wastes generated by industrial and agricultural activities.[32] An important stage in this process is cellulose solubilization that involves hydrolysis of carbohydrates of the raw material to low-molecular-weight sugars and subsequent fermentation into ethanol. Solubilization can be done in several solvent mixtures, which include phosphoric acid, currently the most used to cellulose pretreatment before its degradation to glucose units. This solubilization mechanism was elucidated using high- and low-field NMR spectroscopies. High-frequency ^{13}C and ^{31}P NMR spectra showed formation of direct bonding between phosphoric acid and dissolved

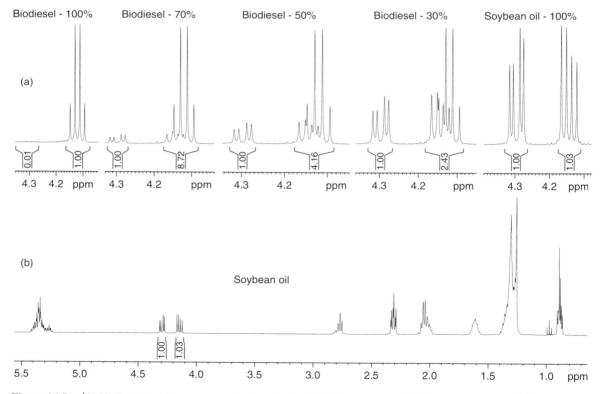

Figure 16.5. ^{1}H NMR to monitor soybean oil ethanolysis in $CDCl_3$, using a 400 MHz spectrometer. (a) Expansion of overlapping signals and integrals that were obtained to determine the ester ethoxy and triacylglycerol methylene hydrogen contents and (b) full spectra of soybean oil.

cellulose. On the other hand, molecular dynamics studies by low-frequency NMR revealed two different H_3PO_4 relaxing components. The fastest spin–lattice relaxation rate (R_1) was assigned to the H_3PO_4 molecules bound to the biopolymer and the slowest R_1 was attributed to the bulk solvent.[32] NMR can also be applied to determine molecular composition of lignocellulosic biomass and bio-oils; pyrolysis of this material produces very complex mixtures of oxygenated organic compounds by depolymerization of cellulose, hemicellulose, and lignin. During storage, these organic compounds can react to form polymers causing changes in physical properties of bio-oils, such as an increase in viscosity. Moreover, the composition can vary significantly with both feedstock and reactor conditions. In this context, the determination of the molecular composition is critical to process development. It is also possible to use the ^{13}C solid-state NMR (see Chapter 4) through cross polarization magic-angle spinning (CPMAS) to determine the crystallinity index of cellulose, an important parameter on the utilization of this compound as a feedstock for biofuels production.

16.4 NMR IN QUALITY CONTROL OF THE FINISHED PRODUCT

Physical chemistry properties of biodiesel are very similar to those of diesel, and consequently this biofuel can be used as an alternative fuel successfully in a diesel engine without any modifications. However, important properties of biofuel, such as cetane number (CN), melting point, degree of saturation, density, cloud point, pour point, viscosity, and nitrogen oxides exhaust emission (NOx), vary significantly with the composition of the fatty acid presents in raw material used in biodiesel production.[33]

CN is a parameter for the ignition quality of a diesel fuel and an important fuel quality indicator.[34] The CN is a measure of the fuels ignition delay, that is, the time between the start of injection and the start of ignition of the fuel and it is used to correlate the delay time of ignition to the time of entry of fuel into the combustion chamber measuring how fast self-ignition occurs in conditions similar to diesel engines. It is a dimensionless number and reflects the quality of fuel used and serves as a comparison between different types of fuels. Cetane is a linear hydrocarbon chain with 16 carbon atoms (*n*-hexadecane) that becomes readily combustible when compressed

and, for this reason, has been assigned the CN of 100, whereas the α-methylnaphthalene was given the number between 0 and 15 to heptamethylnonane.[35,] All other hydrocarbon fuels can be compared to CN, however, the experimental determination of this number according to ASTM D-613[36] requires very specific conditions of analysis, besides being costly and presents problems of repeatability and reproductibility.[36,37] The solution-state ^1H NMR spectrum can be used to correlate CN and fatty acid ester contents establishing a parallel between C–H bond strength and its chemical shift. The spectrum is divided into seven different regions according to the hydrogen chemical environment in the fatty esters: methylic $(\delta = 0.85–1.28 \, ppm)$, paraffinic $(\delta = 1.29–1.84 \, ppm)$, allylic $(\delta = 1.85–2.14 \, ppm)$, α-carbonylic $(\delta = 2.15–2.49 \, ppm)$, bisallylic $(\delta = 2.50–3.49 \, ppm)$, carbinolic $(\delta = 3.50–4.49 \, ppm)$, and vinylic $(\delta = 4.50–7.00 \, ppm)$. The chemical shift and number of hydrogen obtained by NMR for different fatty esters can be used as input data for principal component analysis (PCA) and fuzzy clustering analysis (FCA) and correlated to CN values from the literature. PCA and FCA pointed two major characteristics as determinants for the CN values: the number of carbon–carbon double bonds and the structure of alcohol moiety in each fatty ester. Neural networks can also be used to quantify CN.[38]

Viscosity is another important parameter that affects the utilization of biodiesel fuels, and is significantly influenced by fatty acid esters structure.[38] High viscosity can cause excessive fuel injection pressures when the engine is in the warm-up stage and, low pressure when the temperature is low; consequently, the engine moves slowly because the fuel moves slowly into the filters and pipes. Kinematic viscosity increases with chain length, and depends on the nature and number of double bonds.[39] The unsaturated fatty acid ester contents can be determined through integration of specific regions from the solution-state ^1H NMR spectra, similarly to what is shown in Figure 16.2, using the equations (16.1–16.4) and the saturated ones can be obtained by excluding the unsaturated contents.

Other important quality parameter related to degree of unsaturation is IV, which is directly related with unsaturated fatty acid ester contents and is useful as a guide to prevent problems in engines. A simplified method is to use 1,4-dioxane as internal standard in the quantitative ^1H NMR spectrum. In this case, only olefinic and 1,4-dioxane hydrogen signals need to be integrated and correlated. For ethylic biodiesel, the IV

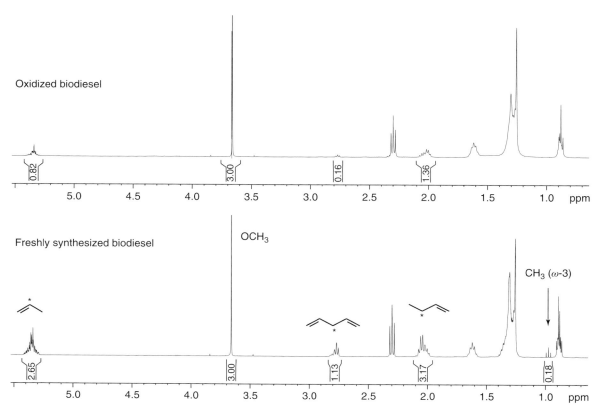

Figure 16.6. ^{1}H NMR spectrum of a new and an oxidized biodiesel in CDCl$_3$ at 345 K for 600 h and bubbling compressed air, using a 500 MHz spectrometer. The highlighted and integrated areas indicate the signals used to monitor the oxidation.

determination could be performed replacing the integral value of the methoxy group by integral value of the ester ethoxy methylene hydrogens.[40] Alternatively, determination of the IV could be related to the molecular weight and number of double bond of the fatty acid methyl esters, which is described in equation (16.6).[41]

$$IV = \frac{253.808}{DBs} \times MW \times 100 \qquad (16.6)$$

where DBs is the number of double bonds and MW average molecular weight.

Oxidative stability is a parameter that is based on polyunsaturated fatty ester contents. Oxidized biodiesel can affect kinematic viscosity, acidity, and CN, and produce insoluble high molecular weight polymers that block fuel lines and filters, leading to incomplete combustion and engine damage. This oxidation can be monitored through integral area of the hydrogen signals at $\delta = 0.9$ ppm (ω-3 CH$_3$), $\delta = 2.8$ ppm (bisallylic CH$_2$), $\delta = 3.6$ ppm (OCH$_3$),

and $\delta = 5.4$ ppm (olefinic) in a quantitative ^{1}H NMR spectrum (Figure 16.6). For ethyl, propyl, and higher chain biodiesels, the hydrogen signal at $\delta = 4.2$ ppm region (OCH$_2$) can be used. Signals of OCH$_3$ and OCH$_2$ groups of the esters remained constant throughout the oxidation reaction, whereas other signals area decrease.[42]

Biodiesel properties and its specifications are directly related to its fatty ester contents and consequently to raw material used in biodiesel preparation. One typical problem is the high viscosity of some biodiesel, e.g., castor oil, which can be met when mixed with lower viscosity biodiesel from the other sources. These properties also give the biodiesel different commercial values. In this context, ^{1}H NMR associated with PCA can be used to identify pure biodiesel from different sources (e.g., castor and soybean) but also a blend of different

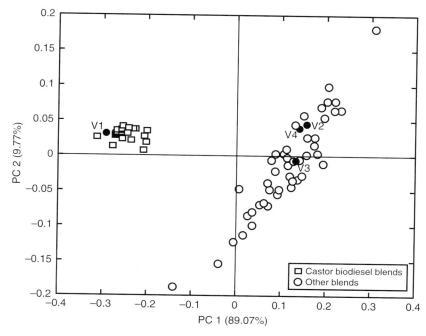

Figure 16.7. A PCA scores plot (PC1 89.07%) × PC2 (9.77%) obtained from the $\delta = 2.18-2.24$ ppm region of ^{1}H NMR spectral data shows pure and blended biodiesel samples. V1 to V4 are the samples that were used for test validation.

biodiesels. In this case, specific regions of the spectra from each source are selected and evaluated in PCA. The region for methylene group in allylic and α-hydroxylic position ($\delta = 2.18-2.24$ ppm) is used to identify samples containing biodiesel from castor oil [43] (Figure 16.7). For soybean-containing biodiesel, the regions for methylic and bisallylic hydrogens in linolenic esters ($\delta = 0.94-1.00$ and 2.79–2.83 ppm, respectively) are used. If other linolenic ester-containing biodiesels are present, integration of the vinylic and allylic hydrogens can contribute to its evaluation. The signals from methylene groups ($\delta = 1.20-1.40$ ppm) in the hydrocarbon chain of saturated fatty acids are the major discriminant signals for tallow-containing biodiesel samples. The less intense signals in region ranging from $\delta = 1.40$ to 1.50 ppm are characteristic of cotton-containing biodiesel samples. If castor-containing biodiesel is present, the highest signal in this region refers to hydrogen beta to the hydroxyl group.[43]

The blend level of biodiesel in diesel can also be determined through the use of quantitative ^{1}H NMR

where the integral area of the methyl ester hydrogens ($\delta = 3.55-3.65$ ppm), methylene and methyl hydrogens ($\delta = 0.42-3.30$ ppm), and the aromatic hydrogens ($\delta = 6.50-8.80$ ppm) can be related and employed to construct calibration curves used for determining blend biodiesel/diesel levels.[44] Another approach to quantify biodiesel and vegetable oil in mixtures with mineral diesel is the association of partial least squares (PLS) regression and principal components regression (PCR) algorithms with ^{1}H NMR data. For this purpose, aliphatic ($\delta = 0.40-3.30$ ppm), methoxy ($\delta = 3.60$ ppm), glycerol ($\delta = 4.00-4.40$ ppm), and olefin hydrogen regions ($\delta = 5.20-5.40$ ppm) are selected to design PLS and PCR models to evaluate blends of oil/diesel, biodiesel/diesel, and oil/biodiesel/diesel.[45]

Ethanol is another important biofuel used in vehicles in pure or in blend forms. Commercial gasoline (e.g., gasoline C in Brazil) has a content of anhydrous bioethanol between 20% and 25%. Ethanol and ether-like methyl *tert*-butyl (MTBE) and *tert*-amyl methyl (TAME) have an important role as octane boosters to minimize harmful exhaust emissions.

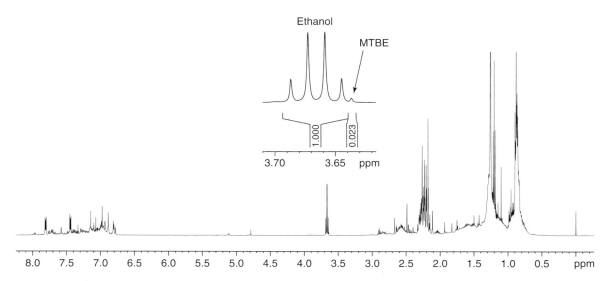

Figure 16.8. ¹H NMR spectrum of gasoline containing ethanol and MTBE in CDCl₃ using a 500 MHz spectrometer.

Through quantitative ¹H NMR, quantification of oxygenated additive present in gasoline can be calculated by dividing its molecular weight and integrals of the alpha carbon group (OCH₂–ethanol, OCH₃–MTBE, etc.) by number of hydrogen in the alpha carbon group of oxygenates, as showed in Figure 16.8.[46]

16.5 CONCLUSIONS AND FUTURE CHALLENGES

Because of worsening environmental conditions, limitations of natural resources energy, sustainability of the current consumption, instability of the supply chain, and other factors, many countries have sought to reduce their dependence on petroleum-based energy because global demand is expected to grow while the supply of fossil fuel decreases. Additional factors make biofuels an important alternative to petroleum-based fuels: the urgent need to control and reduce the emission of greenhouse gases and to improve the air quality in cities; the large dependence of global energy on nonrenewable fuels; and the generation of income and jobs.

In some countries, biofuels are used to replace all or part of conventional fuels, such as bioethanol–gasoline blends and diesel–biodiesel, which are used in different proportions, and the usage varies according to the laws of each country.

Although biofuels are already a reality, there are many limitations to the potential of biofuels as a strategic and sustainable energy option. Several challenges must be overcome for biofuels to become viable on a global scale. First, a substantial increase in biofuel production worldwide, with competitive prices, is required to supply stable products of sufficient quality. New crops that are better able to compete with petroleum must be introduced to increase production. Second, although biofuels are known for their ability to reduce the emission of greenhouse gases compared with that from fossil fuels, the environmental effects on local production cannot be neglected. It is important to stimulate the production and use of biofuels in the world global energy market, which mainly includes developing countries. Third, government policies for the promotion, development, and commercialization of biofuels in international trade must be implemented, which requires reduced barriers to trade biofuels in the world with the assurance of product quality. Last, technology solutions must be found for the challenges in the biofuel production chain, and analytical methods must be developed that allow analysis and quality control of the raw materials, process, products, and blends (i.e., diesel/biodiesel and gasoline/bioethanol) while guaranteeing traceability.

The consolidation and expansion of biofuels along with standardization methods are important factors for biofuel success, and NMR can contribute significantly because it can be used for quality control in all stages of the production process as highlighted in this overview. NMR techniques are very fast and noninvasive, and multiple results can be obtained in a single analysis with minimal environmental impact. Despite the versatility of NMR and its diversity of use, there are challenges that must be overcome to increase its application with biofuel matrices. We can highlight a better understanding of the NMR data correlation with others techniques such as spectroscopy (e.g., FT-IR, Raman, UV–vis, and EM), chromatography (e.g., GC and HPLC), and physicochemical measurements (e.g., viscosity index, CN, MON, RON, distillation residue, flash point, and total sulfur content), where through simple measures (either spectroscopic or physical-chemistry) and using chemometric/statistical tools to predict with a high degree of reliability the biofuel physicochemical properties and also in their blends with petroleum products. This methodology has been used for others matrices using NMR (e.g., food NMR and body fluids analysis) but, still under explored in biofuels. Another important point for NMR technique consolidation in biofuel is undoubtedly makes it more accessible throughout the production chain, possibly using low-field NMR (e.g., measurements of T_2 and relaxometry) using benchtop and portables spectrometers, which are low cost and with easy operation. This is most suitable for fieldwork where it is necessary for quick analysis and has been used very often in the petroleum industry exploration.

RELATED ARTICLES IN EMAGRES

Fossil Fuels

Relaxation: An Introduction

Tycko, Robert: The Multidisciplinarity of Solid-State NMR

REFERENCES

1. EIA, International Energy Statistics, Biofuel Consumption, 2013. http://www.eia.gov/renewable/ (accessed April 26, 2013).

2. University of Edinburgh, Generations of Biofuels, 2012. http://energyfromwasteandwood.weebly.com/ generations-of-biofuels.html (accessed April 29, 2013).

3. World Energy Council, Biofuel: Policies, Standards and Technologies, 2010. http://www.worldenergy.org/ publications/2010/biofuels-policies-standards-and-technologies (accessed April 29, 2013).

4. R. L. McCormick and T. I. Alleman, in The Biodiesel Handbook, eds G. Knothe, J. V. Gerpen and J. Krahl, AOCS Press: Illinois, 2005, Chap. 7.

5. United States Environmental Protection Agency, A Comprehensive Analysis of Biodiesel Impacts on Exhaust Emissions, 2002. Draft Technical Report, EPA420-P-02-001.

6. M. Schulz and S. Clark, *J. ASTM Int.*, 2011, STP, **1477**, 577.

7. V. Menon and M. Rao, *Prog. Energy Combust. Sci.*, 2012, **38**, 522.

8. B. L. Salvi and N. L. Panwar, *Renew. Sustain. Energy Rev.*, 2012, **16**, 3680.

9. Y. G. Linck, M. H. M. Killner, E. Danieli, and B. Bluemich, *Appl. Magn. Reson.*, 2013, **44**, 41.

10. N. K. Singh and D. W. Dhar, *Agron. Sustain. Dev.*, 2011, **31**, 605.

11. V. Vasudevan, R. W. Stratton, M. N. Pearlson, G. R. Jersey, A. G. Beyene, J. C. Weissman, M. Rubino, and J. I. Hileman, *Environ. Sci. Technol.*, 2012, **46**, 2451.

12. Y. Pu, S. Cao, and A. J. Ragauskas, *Energy Environ. Sci.*, 2011, **4**, 3154.

13. M. Al-Sabawi and J. Chen, *Energy Fuels*, 2012, **26**, 5373.

14. L. A. Colnago, M. Engelsberg, A. A. Souza, and L. L. Barbosa, *Anal. Chem.*, 2007, **79**, 1271.

15. E. L. Bakota, J. K. Winkler-Moser, and D. E. Palmquist, *J. Am. Oil Chem. Soc.*, 2012, **89**, 2135.

16. R. A. Prestes, L. A. Colnago, L. A. Forato, L. Vizzotto, E. H. Novotny, and E. Carrilho, *Anal. Chim. Acta*, 2007, **596**, 325.

17. G. P. Mambrini, C. Ribeiro, and L. A. Colnago, *Magn. Reson. Chem.*, 2012, **50**, 1.

18. F. D. Gunstone, J. L. Harwood, and F. B. Padley, The Lipid Handbook, 2nd edn, Chapman & Hall: London, 1994, 1273.

19. T. H. Applewhite, in Kirk-Othmer, Encyclopedia of Chemical Technology, 3rd edn, eds M. Grayson and D. Eckroth, John Wiley & Sons: New York, 1980, Vol. 9, Chap. Fats and Fatty Oils, 795.

20. S. Ng and P. T. Gee, *Eur. J. Lipid Sci. Technol.*, 2001, **103**, 223.

21. N. Merkley and R. T. Syvitski, *J. Appl. Phycol.*, 2012, **24**, 535.

22. G. Knothe, *J. Am. Oil Chem. Soc.*, 2000, **77**, 489.

23. T. Dimmig, W. Radig, C. Knoll, and T. Dittmar, *Chem. Technol. (Leipzig)*, 1999, **51**, 326.

24. F. Jin, K. Kawasaki, H. Kishida, K. Tohji, T. Moriya, and H. Enomoto, *Fuel*, 2007, **86**, 1201.

25. M. Nagy, B. J. Kerr, C. J. Ziemer, and A. J. Ragauskas, *Fuel*, 2009, **88**, 1793.

26. A. M. Socha, G. Kagan, W. Li, R. Hopson, J. K. Sello, and P. G. Williard, *Energy Fuels*, 2010, **24**, 4518.

27. S. A. Fernandes, A. L. Cardoso, and M. J. Silva, *Fuel Process. Technol.*, 2012, **96**, 98.

28. R. Guzatto, T. L. Martini, and D. Samios, *Fuel Process. Technol.*, 2011, **92**, 2083.

29. S. A. Fernandes, R. Natalino, M. J. Silva, and C. F. Lima, *Catal. Commun.*, 2012, **26**, 127.

30. G. J. Suppes, K. Bockwinkel, S. Lucas, J. B. Botts, M. H. Mason, and J. A. Heppert, *J. Am. Oil Chem. Soc.*, 2001, **78**, 139.

31. P. R. Costa-Neto, M. S. B. Caro, L. M. Mazzuco, and M. G. Nascimento, *J. Am. Oil Chem. Soc.*, 2004, **81**, 1111.

32. P. Conte, A. Maccota, C. Pasquale, S. Bubici, and G. Alonzo, *J. Agric. Food Chem.*, 2009, **57**, 8748.

33. F. Ma and M. A. Hanna, *Bioresour. Technol.*, 1999, **70**, 1.

34. L. F. Ramírez-Verduzco, J. E. Rodríguez-Rodríguez, and A. R. Jaramillo-Jacob, *Fuel*, 2012, **91**, 102.

35. H. Wang, S. J. Warner, M. A. Oehlschlaeger, R. Bounaceur, J. Biet, P. A. Glaude, and F. Battin-Leclerc, *Combust. Flame*, 2010, **157**, 1976.

36. ASTM D-613. Test method for ignition quality of diesel fuel by cetane methods. Annu. Book ASTM stand. 1988. Vol. 5, 4.

37. O. L. Gulder and B. Glavincesvsk, *Ind. Eng. Chem. Prod. Res. Dev.*, 1986, **25**, 153.

38. D. V. Nadai, J. B. Simões, C. E. N. Gatts, and P. C. M. L. Miranda, *Fuel*, 2013, **105**, 325.

39. G. Knothe and K. R. Steidley, *Fuel*, 2005, **84**, 1059.

40. M. Oromí-Farrús, G. Villorbina, J. Eras, F. Gatius, M. Torres, and R. Canela, *Fuel*, 2010, **89**, 3489.

41. R. Kumar, V. Bansal, M. B. Patel, and A. S. Sarpal, *Energy Fuels*, 2012, **26**, 7005.

42. C. J. Chuck, C. D. Bannister, R. W. Jenkins, J. P. Lowe, and M. G. Davidson, *Fuel*, 2012, **96**, 426.

43. I. S. Flores, M. S. Godinho, A. E. Oliveira, G. B. Alcantara, M. R. Monteiro, S. M. C. Menezes, and L. M. Lião, *Fuel*, 2012, **99**, 40.

44. M. R. Monteiro, A. R. P. Ambrozin, L. M. Lião, and A. G. Ferreira, *Fuel*, 2009, **88**, 691.

45. M. R. Monteiro, A. R. P. Ambrozin, M. S. Santos, E. F. Boffo, E. R. Pereira-Filho, L. M. Lião, and A. G. Ferreira, *Talanta*, 2009, **78**, 660.

46. W. R. Kalsi, A. S. Sarpal, S. K. Jain, S. P. Srivastava, and A. K. Bhatnagar, *Energy Fuels*, 1995, **9**, 574.

FURTHER READINGS

A. Murugesan, C. Umarani, R. Subramanian, and N. Nedunchezhian, *Renew. Sustain. Energy Rev.*, 2009, **13**, 653.

D. J. Roddy, in *Comprehensive Renewable Energy*, ed. A. Sayigh, Elsevier: Amsterdam, 2012, Vol. 5, Chap. Biomass and Biofuels—Introduction, 1.

EPA, A Comprehensive Analysis of Biodiesel Impacts on Exhaust, 2001. http://www.epa.gov/oms/models/analysis/biodsl/p02001.pdf (accessed April 29, 2013).

ETC/ACC, Effect of Biodiesel and Bioethanol on Exhaust Emission, Technical Paper, 2008. http://acm.eionet.europa.eu/reports/ETCACC_TP_2008_5_biofuels_emissions (accessed April 29, 2013).

A. K. Agarwal, *Prog. Energy Combust. Sci.*, 2007, **33**, 233.

A. Demirbas, Biodiesel: A Realistic Fuel Alternative for Diesel Engines, Springer-Verlag: London, 2008, 218.

L. F. Cabeça, L. V. Marconcini, G. P. Mambrini, R. B. V. Azeredo, and L. A. Colnago, *Energy Fuels*, 2011, **25**, 2696.

S. Y. Reda, B. Costa, and R. J. S. Freitas, *Ann. Magn. Reson.*, 2007, **6**, 69.

S. Park, D. K. Johnson, C. I. Ishizawa, P. A. Parilla, and M. F. Davis, *Cellulose*, 2009, **16**, 641.

L. Ingram, D. Mohan, M. Bricka, P. Steele, D. Strobel, D. Crocker, B. Mitchell, J. Mohammad, K. Cantrell, and C. U. Pittman, *Energy Fuels*, 2008, **22**, 614.

R. Freeman, 'Spin Choreography: Basic Steps in High Resolution NMR', Oxford University Press, Oxford, 1988, p. 408.

M. R. Monteiro, A. R. P. Ambrozin, L. M. Lião, A. G. Ferreira, *Talanta*, 2008, **77**, 593.

R. G. Brereton, 'Chemometrics—Data Analysis for the Laboratory and Chemical Plant', John Wiley & Sons Ltd. Chichester, 2003, p. 504.

G. Knothe, J. Krahl, and J. Van Gerpen, 'The Biodiesel Handbook', 2nd edn., AOCS Press, Champaing, IL, 2010, p. 516.

L. G. Schumacher, J. Van Gerpen, and B. Adams, 'Biodiesel Fuels', in *Encyclopedia of Energy*, ed. C. J. Cleveland, Elsevier, New York, 2004, Vol. 1, p. 151–162.

Chapter 17
Clay Minerals

Karl T. Mueller[1,2], Rebecca L. Sanders[3] and Nancy M. Washton[2]

[1]*Department of Chemistry, Pennsylvania State University, University Park, PA 16802, USA*
[2]*Pacific Northwest National Laboratory, Environmental Molecular Sciences Laboratory, Richland, WA 99352, USA*
[3]*Department of Geosciences, Princeton University, Princeton, NJ 08544, USA*

17.1 INTRODUCTION

Clay minerals are common secondary minerals formed during weathering and sedimentation processes. These micron-sized hydrous aluminum phyllosilicates (sheet silicates) have large surface-area-to-volume ratios, high cation exchange capacities, plasticity when wet, and unique catalytic properties. These properties result in structural durability, high sorptivity, and low permeability, making clay minerals useful for a variety of applications including utilization as drilling mud,

NMR Spectroscopy: A Versatile Tool for Environmental Research
Edited by Myrna J. Simpson and André J. Simpson
© 2014 John Wiley & Sons, Ltd. ISBN: 978-1-118-61647-5

contaminant sorbents, and filters and as components in pharmaceuticals, cosmetics, and paints.[1] In nature, clay minerals can have both beneficial and harmful properties. Weathering of clay minerals, such as micas, releases potassium (an important plant nutrient) into soils. Smectites can swell to retain water and prevent erosion in surface soils. However, in the subsurface, the shrink/swell properties of smectites can reduce the shear properties of the soil, resulting in the possibility of building damage or creation of landslides.[2,3] The structures, chemistries, and dynamics of clay minerals and their substituents are industrially and environmentally significant, and physicochemical characterization tools play a large role in understanding their properties and behaviors. Solid-state NMR is an especially useful tool for elucidating the static and dynamic properties of clay minerals, and this chapter provides a brief historical perspective of NMR applications to these systems as well as a prospectus on new experiments aimed at understanding the chemical reactivities of clay minerals.

17.2 CLAY MINERAL STRUCTURE

17.2.1 Bulk Structures of Clay Minerals

Clay minerals are composed of layers formed from the association of two or three tetrahedral and octahedral

Table 17.1. Classification of clay minerals

Layer type	Charge per formula unit	Group	Subgroup	Example species
1 : 1	0	Serpentine–kaolin	Serpentines	Chrysotile
			Kaolins	Kaolinite, dickite, and nacrite
2 : 1	0	Talc–pyrophyllite	Dioctahedral	Pyrophyllite
			Trioctahedral	Talc
2 : 1	0.2–0.6	Smectite	Dioctahedral	Montmorillonite and beidellite
			Trioctahedral	Saponite and hectorite
2 : 1	0.6–0.9	Vermiculite	Dioctahedral	Dioctahedral and vermiculite
			Trioctahedral	Trioctahedral and vermiculite
2 : 1	1.0	Mica	Dioctahedral	Muscovite
			Trioctahedral	Biotite
2 : 1	2.0	Brittle mica	Dioctahedral	Margarite
			Trioctahedral	Clintonite
2 : 1	Variable	Chlorite	Dioctahedral	Donbassite
			Trioctahedral	Clinochlore
			Di,trioctahedral	Cookeite

Clay minerals are classified by layer type, charge per formula unit, group, subgroup, and species

oxide sheets.[1,3,4] Tetrahedral sheets are composed of central atoms (e.g., Si^{4+} and Al^{3+}) coordinated by four oxygen atoms (three basal and one apical). Similarly, octahedral sheets include central cations (e.g., Al^{3+}, Mg^{2+}, Fe^{3+}, and Fe^{2+}) coordinated with six bridging oxygens or hydroxyls. Common clay mineral layer types are 1 : 1 clay minerals, which are composed of layers containing one tetrahedral sheet bound through the apical oxygens to one octahedral sheet, and 2 : 1 clay minerals, having layers with an octahedral sheet sandwiched between two tetrahedral sheets. Layers associate to form crystallites and an interlayer space is located between the layers.

Classification of clay minerals is based on layer type, charge per formula unit, group, subgroup, and species.[5,6] While ideal 1 : 1 clay minerals do not have a formal charge, 2 : 1 clay minerals are classified by the charge per formula unit, with a negative charge resulting from the degree of isomorphic substitution present in each layer (e.g., Al^{3+} for Si^{4+} in tetrahedral sheets or Mg^{2+} for Al^{3+} in octahedral sheets). The clay minerals are further divided into subgroups based on the composition of the octahedral sheets. Divalent cations form trioctahedral sheets, whereas trivalent cations form dioctahedral sheets with every third octahedral site vacant, because the positive cation charge compensates for the vacancies. Finally, clay minerals are

designated by species. As an aid for the reader, while both examining this chapter and perusing the vast clay mineral literature, a clay mineral classification table is included (Table 17.1).

Idealized 1 : 1 clay minerals (e.g., kaolins and serpentines) and 2 : 1 clay minerals (e.g., talc and pyrophyllite) have little to no isomorphic substitution in their layers and, therefore, little to no charge per formula unit. The layers are held together only by weak electrostatic forces.[3,7] All other clay minerals have some degree of isomorphic substitution in their layers, creating a permanent charge per formula unit. Cations are incorporated into the interlayers of the clay structure to balance the negative charge of the layers created by isomorphic substitution. When the charge per formula unit of the clay layers is large, such as 1.0 for micas, alkali cations with large radii, such as K^+ and Cs^+, easily lose water from their hydration spheres and then interact directly with basal oxygens, forming inner-sphere complexes.[3,4] Clay minerals with lower charges per formula unit such as vermiculites and smectites have interlayers filled with smaller cations such as Na^+, Mg^{2+}, and Ca^{2+}. These cations retain their hydration spheres and are incorporated into the interlayers by hydrogen bonding to basal oxygens, forming outer-sphere complexes. The intercalation of hydrated ions results in the swelling behavior of clays and is limited to clay minerals with small charge per formula unit.[6]

17.2.2 Surface Structures of Clay Minerals

Clay minerals have two distinct types of surfaces, denoted as basal planes and edges, as depicted in Figure 17.1. The ratio of basal planes to edges is not consistent or constant for each clay mineral species and can vary within the species. The two different surfaces can have quite different chemical properties (i.e., chemical reactivities). For instance, dissolution experiments performed on clay minerals under mildly acidic conditions indicate that edge sites react preferentially compared to basal planes during dissolution.[8–13] The basal planes can obtain a permanent charge because of isomorphic substitution in the tetrahedral or octahedral sheets. Clays, such as micas, that have large charge per formula unit, balance the charge with tightly bound interlayer cations such as K^+. Clay minerals with lower charges per formula unit, such as smectites and vermiculites, have interlayer cations that can be readily exchanged when placed in solutions with different solutes or concentrations through cation exchange. The extent to which this swap occurs is related to the cation exchange capacity (CEC) of a clay mineral, which depends on both permanent charge on the layers and variable charge on the edges that change as a function of pH.[7] Montmorillonite and vermiculite have large CECs, whereas kaolinite, talc, and micas such as biotite and muscovite have small CECs.[7]

As shown schematically in Figure 17.2, the charge on edge surfaces is variable and often a function of pH. Layers are terminated by chemisorption of water species with a variable number of protons based on the pH of the surrounding environment.[15] For instance, dioctahedral phyllosilicates have edge sites that are fully protonated in extremely acidic conditions ($\equiv SiOH_2^{+1}$ and $\equiv AlOH_2^{+1/2}$), creating positive charge on the edge sites. At pH values of 3–4.5, protons are lost on the Si sites ($\equiv SiOH$), leaving the sites neutral. At pH 6.5 and 7.5, the Al sites have lost protons ($\equiv AlOH^{-1/2}$ and $\equiv AlO^{-3/2}$, respectively). Finally, at pH 9, the tetrahedral edge sites are deprotonated ($\equiv SiO^{-1}$).[14]

17.3 NMR STUDIES OF CLAY MINERALS

17.3.1 General Considerations

There are a number of structural, dynamical, and chemical questions and problems that involve clay minerals and their properties, and many of these studies require advanced spectral characterization. Solid-state NMR spectroscopy has proven to be a powerful tool for the study of complex crystalline solids, poorly crystalline materials, and even amorphous samples because of its isotopic selectivity and the dependence of solid-state NMR spectral parameters on the local electromagnetic environment.[16,17] A primary discriminator between different local environments is the isotropic chemical shift, and advances in magnet technology and spectral resolution methods (especially, magic-angle spinning (MAS) capabilities) have driven subsequent studies and applications.[18] Of particular importance in clay mineral systems are ^{29}Si, ^{27}Al, and ^{17}O nuclides, as well as the NMR-active nuclides of cations such as Cs^+ and Na^+. Studies of local silicon environments suffer somewhat from the lower natural abundance of ^{29}Si (4.7%) as well as long spin–lattice relaxation times for this spin-1/2 nuclide, whereas ^{27}Al is a quadrupolar nuclide (with $I = 5/2$) whose spectra often display a decrease in resolution because of the second-order quadrupolar broadening.[16] However, the ^{27}Al nuclide is 100% abundant in nature and typical ^{27}Al spin–lattice

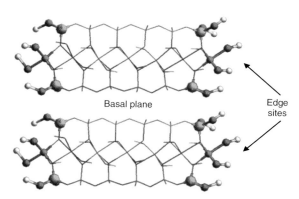

Figure 17.1. A schematic of the structure of montmorillonite (a 2 : 1 layered aluminosilicate) illustrates the presence of both basal planes and edge surfaces, each of which may differ in their specific chemical reactivities. (Reproduced with permission from Ref 54. © 2010 American Chemical Society)

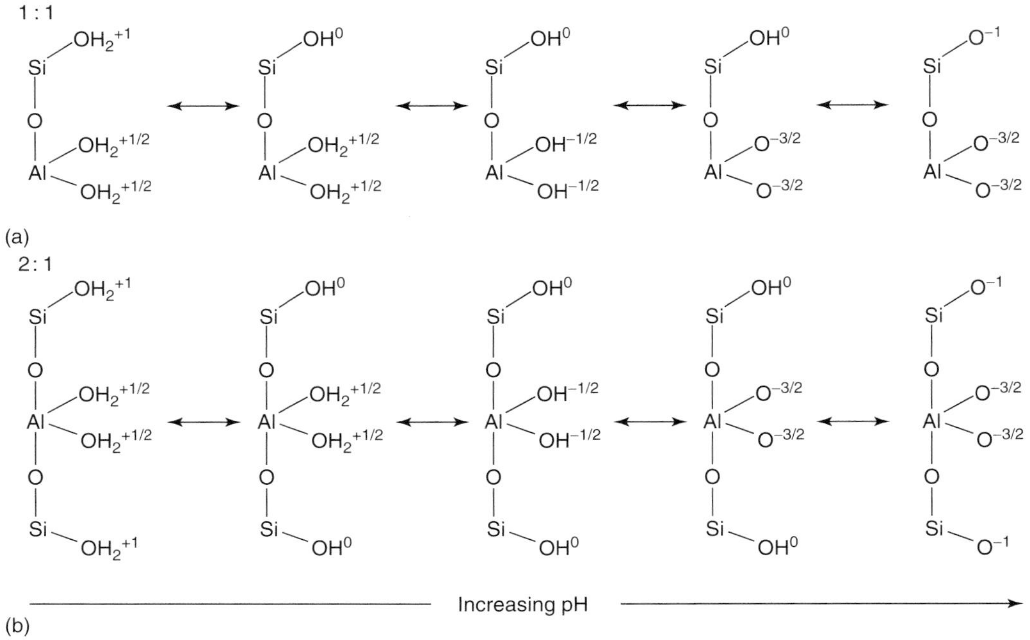

Figure 17.2. The charges and chemical functional group change on edges of (a) 1 : 1 and (b) 2 : 1 dioctahedral phyllosilicates as a function of pH, affecting the chemistries of the edge sites. (Reproduced with permission from Ref. 14. © 1988 The Clay Minerals Society)

relaxation times are short in many clay mineral systems. The ^{17}O nuclide is both quadrupolar ($I = 5/2$) and low in natural abundance (only 0.037%), and typically, the quadrupolar coupling constants (and, therefore, the widths of the ^{17}O central transition resonances) are large enough to cause spectral overlap of resonances, as the nuclides typically observed in aluminosilicate minerals have quite similar isotropic shifts. Some of these limitations can be circumvented by isotopic enrichment of ^{29}Si or ^{17}O for synthetic or laboratory-modified samples, and by collection of spectra at higher magnetic field strengths for quadrupolar nuclei, as spectral lines are narrowed as the field strength increases because of the dependence of the linewidth and the second-order quadrupolar shift on the inverse of the square of the magnetic field strength.[16] However, the spin–lattice relaxation times are also field dependent, and experiments carried out at ultrahigh magnetic field strengths can result in quite long (>60 s, and often much longer) relaxation time constants for ^{29}Si nuclei. Additional methods for obtaining purely isotropic (i.e., unbroadened) high-field NMR resonances from quadrupolar nuclides have

been introduced, including dynamic angle spinning (DAS),[19,20] double rotation (DOR),[21,22] multiple quantum magic-angle spinning (MQMAS),[23,24] and satellite transition magic-angle spinning (STMAS).[25] As discussed later in this section, line-narrowing capabilities for quadrupolar species and other advanced techniques are finding wide-ranging applications in studies of clay minerals.

Natural clay minerals may also contain a range of paramagnetic species, primarily as Fe^{3+} and Mn^{2+}, that are substituted into the clay structure. These paramagnetic ions can not only reduce relaxation times of nearby nuclides but may also introduce line broadening into NMR spectra and, therefore, decrease spectral resolution.[26] While interlayer paramagnetic ions could be removed through cation exchange to improve spectral resolution, most paramagnetic ions such as Fe^{3+} are incorporated into the clay layers and cannot be removed without altering the layer composition. Natural materials where clays are present may also contain ferromagnetic or other magnetic impurities, causing additional line shifts and broadening or, as is often the case, the complete loss of signals.[27]

17.3.2 Historical and Simple Studies of Clay Minerals Using Solid-State NMR

Solid-state NMR spectroscopy has been well utilized in studies of clay mineral systems since the 1980s, as described in the comprehensive text by Engelhardt and Michel published in 1987.[27] Lippmaa *et al.*[28] obtained some of the earliest ^{29}Si NMR spectra of clay minerals, and well-resolved resonances were obtained for both ^{29}Si and ^{27}Al species in early MAS experiments.[29] For many clay mineral samples, the numbers and intensities of the resonances in the ^{29}Si MAS NMR spectra are determined by the local ordering of atoms around a tetrahedrally coordinated silicon. In layered phyllosilicates, resonances from Q^3 silicon species (denoting that they contain a central tetrahedrally coordinated Si atom with one nonbridging oxygen atom) are shifted depending on the number of aluminum atoms in the second coordination sphere of the silicon. Such sites are usually denoted with the $Q^3[m\text{Al}]$ notation, where *m* is the number of next-nearest-neighbor aluminums (here, ≤ 3) in the tetrahedral layer. If aluminum avoidance in the silicate layer is assumed (Lowenstein's avoidance rule), then the Si/Al ratio can be calculated from the intensities of the $Q^3[m\text{Al}]$ resonances.[27] The interested reader is referred to the text by Engelhardt and Michel[27] as well as the seminal chapter by Kirkpatrick in *Reviews in Mineralogy*[30] for a more complete description of early ^{29}Si MAS NMR studies of layered silicates and related structures.

In clay minerals, ^{27}Al can be either present in octahedral alumina sheets or contained within tetrahedral sheets as a result of isomorphic substitution. At moderate and higher magnetic field strengths (generally, 7.0 T and greater), resonances from these two species (denoted Al^{VI} and Al^{IV}, respectively) appear in the MAS spectra with isotropic shifts that are separated by much greater spectral frequencies than the intrinsic linewidths. If quantitative excitation conditions are applied for the central ($-1/2 \leftrightarrow +1/2$) transition – which is the transition typically observed in routine characterization experiments – then the fractions of each of the two species can be determined. Additional excitation and detection of satellite transitions is possible[31] and useful for determining quadrupolar parameters and providing spectral separation of resonances from very similar aluminum sites. As described in the Section 17.3.3, MQMAS methods may also be applied to investigate subtle structural differences or changes.[32–34]

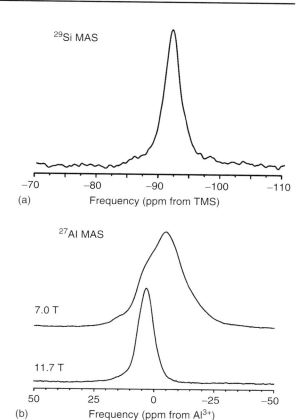

Figure 17.3. (a) The ^{29}Si MAS NMR spectrum of kaolinite obtained at a magnetic field strength of 9.4 T reveals one resonance at approximately −92 ppm. (b) The ^{27}Al MAS NMR spectrum of kaolinite changes as a function of magnetic field strength, shifting to higher frequency and narrowing, as the field is raised. (Reprinted with permission from Ref. 64. © 2006 American Chemical Society)

A number of researchers have presented structural studies of simple (or specimen) clay minerals such as kaolinite (with ideal structural formula of $\text{Al}_2\text{Si}_2\text{O}_5(\text{OH})_4$). Figure 17.3 shows the ^{29}Si MAS and ^{27}Al MAS NMR spectra from a specimen kaolinite sample (KGa-1b, obtained from the Source Clays Repository at Purdue University). The single ^{29}Si MAS NMR resonance appears at −92 ppm, arising from the $Q^3[0\text{Al}]$ sites in the sample. Other natural and synthetic kaolinite samples display a 'split' resonance that researchers have attributed to the presence of crystallographically nonequivalent Si sites.[27,35] A comprehensive description of the ^{29}Si, ^{27}Al, and ^{1}H MAS NMR spectra and spin–lattice relaxation times for kaolinite was presented by Hayashi *et al.* in 1992.[35,36] Aside from minor impurity phases, the ^{27}Al MAS NMR spectrum in Figure 17.3 displays a single resonance from Al in octahedral coordination. The resonance is broadened and shifted from the true isotropic chemical shift for the [27]Al nuclei in this sample because of the second-order quadrupolar interactions. The resonance also shifts and narrows as a function of magnetic field strength. From the shift in the center of gravity of the line, both the isotropic chemical shift and the quadrupolar product (describing the size of the second-order quadrupolar interaction scaled by a symmetry term) may be determined. A recent study by Zhou *et al.*[37] addressed conclusions reached by diffraction methods that indicated two crystallographic aluminum sites in kaolinite with differing site symmetries. The study discusses the trade-off between obtaining spectra at high magnetic field strengths, where chemical shift anisotropy and dipolar interactions overwhelm spectral effects of quadrupolar couplings, and obtaining the same spectra at lower field strengths, where species with similar local environments (and hence, similar isotropic chemical shifts and quadrupolar parameters) will be indistinguishable. Although a suite of high-resolution experiments (MAS at multiple fields, satellite transition MAS, and MQMAS) failed to resolve the two Al sites, results from *ab initio* quantum mechanical modeling were used to aid in obtaining and constraining the NMR parameters (chemical shift and quadrupolar parameters) for the aluminum sites in kaolinite (reported as $\delta_{iso} = 6.25$ ppm, $C_q = 2.6$ MHz, and $\eta = 0.75$).

17.3.3   Solid-State NMR Studies with Advanced Methods

Oxygen atoms play a central role in the structure and chemistry of aluminosilicate minerals. Ultrahigh-field MAS NMR is useful for resolving resonances from apical, basal, and hydroxyl oxygen species in clay minerals, as Lee and Stebbins[33] demonstrated in studies of kaolinite at 18.8 T. The MQMAS experiment is also useful for gaining increased information and understanding from complex clay mineral systems. A combination of [17]O and/or [27]Al MQMAS studies have been performed by Lee and Stebbins[33] on natural kaolinite and muscovite systems as well as mixed synthetic clay minerals.[34] For the natural samples, parent materials were sealed in gold tubes and allowed

to react hydrothermally with 46% [17]O-enriched water. Synthetic sample preparations, originally with compositions mimicking those of pyrophyllite and kaolinite, produced mixed composition samples. The latter hydrothermal preparations were accomplished in the presence of excess [17]O-enriched water. The triple-quantum MQMAS spectra of [17]O, acquired at a field strength of 14.1 T, are shown in Figure 17.4 for enriched natural kaolinite and muscovite as well as a synthetic clay sample (with an original composition target matching that of kaolinite). The absence of paramagnetic impurities in the synthetic sample results in increased spectral resolution and the observation of multiple resonances arising from the presence of a smectite phase in this sample. In these studies, two resonances from the kaolinite sample arising from basal oxygen atoms are resolved, and additional resonances are identified from apical oxygens as well as oxygen atoms in hydroxyl groups. For muscovite, the improvement gained by employing the MQMAS method also allows for resolution of several basal oxygen and hydroxyl species.

Both thermal and mechanical transformations of clay minerals have been followed with solid-state NMR methods, focusing primarily on [29]Si MAS and a combination of [27]Al MAS and advanced resolution-enhancement and/or heteronuclear correlation methods. A study by Ashbrook *et al.*[32] examined the structural outcomes of mechanical treatment of mixtures of kaolinite with gibbsite (γ-Al(OH)$_3$), which are the starting materials for the commercially important synthesis of mullite. The synthesis of mullite is achieved via thermal processing of kaolinite and gibbsite starting materials; however, the kaolinite and gibbsite are also often ground together at ambient temperatures to change the required processing conditions (presumably through decreasing particle size). However, other chemical precursors containing Si–O–Al bonds could also form during mechanical grinding of kaolinite and gibbsite, and it was postulated that such species could be identified using advanced solid-state NMR methods. Ashbrook *et al.*[32] applied [27]Al triple-quantum MQMAS for the enhancement of spectral resolution, as well as a number of cross polarization (CP) experiments between ^{1}H and [27]Al spins to establish spatial proximity of these two nuclei. Overlapping peaks were studied by a combination of polarization transfer and MQMAS methods to obtain spectral editing based on radiofrequency field strengths and quadrupolar parameters. They report that at short times, mechanical treatment

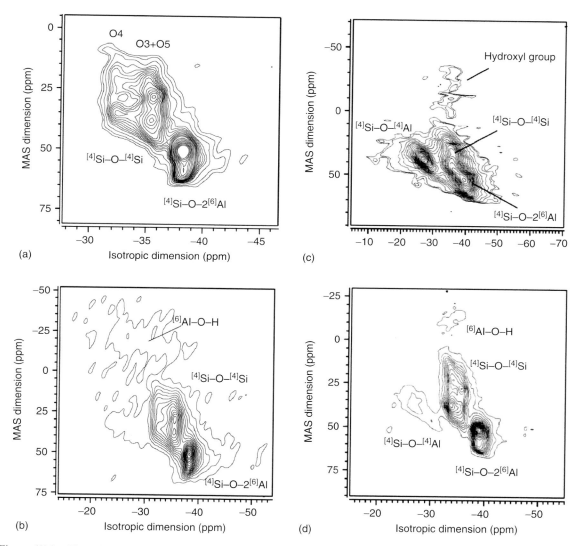

Figure 17.4. The triple-quantum magic-angle spinning (3QMAS) spectra from ^{17}O in kaolinite at 14.1 T [spectra (a) and (b)] display differences based on the use of different apodization functions that allow for isolation or augmentation of different resonances. Processing for spectrum (b) utilized more smoothing and allows observation of ^{17}O resonances from aluminol groups. Spectrum (c) is from a 3QMAS experiment performed on a sample of ^{17}O-enriched muscovite, also obtained at 14.1 T, whereas spectrum (d) is the 3QMAS spectrum from a synthetic sample containing kaolinite and smectite. (Reprinted with permission from Ref. 34. © 2003 American Chemical Society)

did not produce measurable structural changes other than a possible increase in amorphous character. At longer grinding time, new resonances appeared in the spectra, arising from multiple aluminum sites with differences in local coordination numbers with oxygen.

The thermal transformations of kaolinite into metakaolinite and mullite have also been the subject of NMR investigations. One recent study[38] conducted at ultrahigh magnetic field strength (18.8 T; 800 MHz for ^1H) demonstrates the usefulness of NMR for identification of phases that are not observable with standard diffraction methods. For example, the X-ray diffraction (XRD) pattern from a kaolinite sample treated at 600 °C loses the structural detail observed in patterns from unheated samples because of the dehydroxylation of the sample and structural change into metakaolinite, which lacks long-range order. Mullite, along with cristobalite, is identified in the XRD patterns only after treatments that reach 1200 °C and above. ^{29}Si and ^{27}Al MAS data display a variety of resonances from intermediate phases between 600 and 1200 °C. Two-dimensional double-quantum-filtered satellite transition magic-angle spinning (DQF-STMAS) experiments provide a detailed picture of the evolving aluminum structural environments as a function of treatment temperature, as new structures with different local coordination environments are formed as a function of treatment temperature (Figure 17.5).

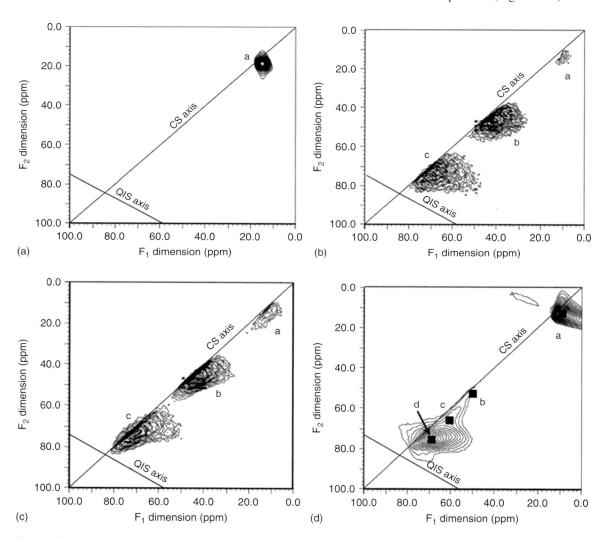

Figure 17.5. Double-quantum-filtered satellite transition magic-angle spinning (DQF-STMAS) results from kaolinite that had been thermally processed at (a) 300°C, (b) 800°C, (c) 1000°C, and (d) 1600°C. The new sites formed display a range of isotropic chemical shifts and quadrupolar parameters that also change as a function of treatment temperature. (Reprinted with permission from Ref. 38. © 2013 Springer)

17.4 STUDIES OF SURFACES, INTERLAYERS, AND ADSORBED SPECIES

Beyond the structural details of atomic surroundings in the bulk of a clay mineral, additional detailed investigations have described the structures of surfaces, interlayers, edge sites, charge-balancing cations, and molecular or polymeric intercalates in clay mineral systems. Two reports describing the study of modified clays and polymer/clay nanocomposites were published in 2004 and 2006 by Grandjean[39,40] and include a comprehensive and important description of NMR studies of inorganic cations in clay interlayer spaces.

17.4.1 Examples of NMR Investigations of Adsorbed or Sequestered Species

17.4.1.1 *Uptake of Boron*

The study of species sorbed onto clay surfaces is important for understanding many environmental processes, including the fate of pollutants in the subsurface. The sorption of boron onto an illite clay mineral was studied by Kim and Kirkpatrick[41] as part of a larger study of boric acid interacting with the surfaces of silica gel, boehmite, and illite. Motivated by the need to understand the transport of boron in the environment, each sample was prepared by reacting it with boric acid solutions (0.1 or 0.01 M) at a range of pH values. The illite sample had a structural formula of $(K_{0.82}Na_{0.04}Ca_{0.01})(Si_{3.12}Al_{0.88})(Al_{1.98}Mg_{0.02})O_{10}(OH)_2$ and no expandable layers. The speciation of boron on the surfaces of these materials was followed with ^{11}B MAS NMR (Figure 17.6), which readily differentiates between boron in tetrahedral and trigonal bonding environments. While infrared (IR) spectroscopy can also reveal details of boron speciation on surfaces with increased sensitivity compared to NMR, the measured NMR spectra do not display any overlapping signals from the clay mineral itself (which is not always the case with IR methods). For the illite sample in this study, ^{11}B MAS NMR spectra from samples prepared at lower pH display resonances from B(3) and B(4) species adsorbed as outer-sphere complexes. An additional B(4) resonance appears at pH 5 and above, growing in intensity as the pH is increased up to a value of 11. Additional changes also occur within the B(3) region of the spectrum, and both of these additional sites are assigned to inner-sphere

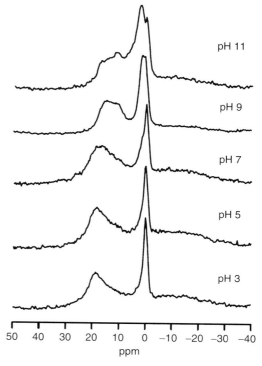

Figure 17.6. Resonances from three- and four-coordinate boron species (at approximately 10–20 and 0 ppm, respectively) are evident in the ^{11}B MAS NMR spectra of illite clay samples that have been reacted for 24 h with 0.1 M boric acid solutions. At high pH values, resonances also appear that may arise from B–O–Al linkages. (Reprinted with permission from Ref. 41. © 2006 Elsevier)

complexes or new phases containing B–O–Al bonds (similar species were also detected in parallel studies of boron sorption onto boehmite). These B–O–Al resonances are proposed to arise from species formed during ligand-exchange reactions, where broken edges of illite crystallites expose reactive aluminol groups.

17.4.1.2 *Uptake of Metal Cations*

Kim and coworkers have also investigated alkali metal uptake on clay mineral surfaces using NMR, including ^{23}Na and ^{133}Cs studies of cation uptake and competitive binding on illite and kaolinite clay minerals.[42–45] Both NMR-active species have 100% natural abundance, but Na and Cs can behave very differently in their binding to surfaces because of the higher hydration energy of Na, which tends to favor the formation of outer-sphere complexes (whereas Cs can also be found in inner-sphere complexes). Importantly, the relative humidity of the samples can alter surface speciation, as demonstrated for illite where the amount of surface-adsorbed water changes dramatically. For illite and kaolinite samples that were reacted with 0.1 M CsCl at 25% relative humidity for 5 days, ^{133}Cs NMR resonances appear from inner sphere and outer sphere Cs$^+$-containing complexes, as well as CsCl that has precipitated onto the kaolinite surface. Changes in the spectra as a function of relative humidity also occur because of the changing dynamics/mobility of the cations as a function of surface hydration.

Cadmium is an important element with environmental impact (see Chapter 19), and solid-state NMR has been used to monitor cadmium interactions with a number of clay minerals.[46] Cadmium has two NMR-active isotopes, and both ^{111}Cd and ^{113}Cd are spin-1/2 nuclides with moderate natural abundance and sensitivity, as well as a wide chemical shift range (>900 ppm) reported for ^{113}Cd. The shift depends strongly on the local coordination of the cation, and, therefore, is an excellent reporter of cadmium environments in clay mineral systems.

17.4.1.3 Interlayer Cation Sequestration for Environmental Remediation

Advanced pulsed NMR techniques such as quadrupolar Carr–Purcell–Meiboom–Gill (QCPMG) sequences[47] are valuable tools for investigating low-sensitivity nuclides such as ^{87}Sr (a spin-9/2 nuclide with a natural abundance of approximately 7%) in clay mineral systems. Coupled with preparatory schemes such as a double-frequency sweep (DFS),[48] which enhances the population difference across the observed central transition, the QCPMG method provides added sensitivity that is critical for observing low-γ, low-abundance nuclei. An aluminosilicate micaceous mineral with nominal composition of Na$_4$Mg$_6$Al$_4$Si$_4$O$_{20}$F$_4$ was studied with solid-state NMR methods,[49] and this mica undergoes selective exchange of strontium for sodium, making it suitable as a strontium remediation material. Strontium loading of about 468 meq per 100 g of mica was achieved by aqueous exchange from a saturated solution, and then the sample was collapsed to trap the strontium via heat treatment and water removal at 500 °C for 4 h. The ^{87}Sr DFS-QCPMG spectrum from the heat-treated sample is shown in Figure 17.7. A single strontium

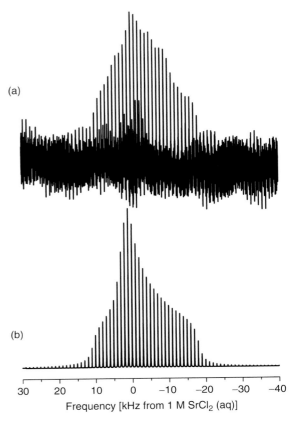

Figure 17.7. (a) The double-frequency sweep quadrupolar Carr–Purcell–Meiboom–Gill (DFS-QCPMG) spectrum from ^{87}Sr entrapped in a collapsed mica clay mineral displays one resonance. The best-fit spectrum is displayed in (b), from which the isotropic chemical shift and quadrupolar parameters could be determined. (Reprinted with permission from Ref. 49. © 2006 American Chemical Society)

resonance is observed, although the acquisition period is quite long (67 h when using an ultrahigh magnetic field strength of 21.14 T). An estimated 1800 days of spectrometer time would have been required to obtain similar signal-to-noise ratios if using a lower magnetic field strength of 11.7 T. The observed ^{87}Sr resonance is associated with the strontium cations bound in the proton-free interlayer, and Figure 17.7 also displays a best-fit simulated QCPMG pattern from a single quadrupolar line shape, corresponding to a quadrupolar coupling constant of 9.02 MHz and an asymmetry parameter of 1.0.

17.4.1.4 Reactions with Silanes

Alkoxysilanes are typically used as coupling agents to make the hydrophilic surfaces of clays organophilic.[50,51] Most trifunctional organosilanes react with surfaces by forming both covalent bonds to surface hydroxyl groups and cross-links to other silane species. However, as monofunctional silanes react only with single hydroxyl groups on a surface, these species have also been shown to be an effective tool for quantifying reactive hydroxyl sites on aluminosilicate surfaces including glass fibers, natural volcanic glasses, and clay minerals.[52–56] In these experiments, the reactive probe molecule that is attached to aluminosilicate surfaces is (3,3,3-trifluoropropyl)dimethylchlorosilane (TFS). Quantification of the number of attached TFS molecules is accomplished with ^{19}F MAS NMR of the treated solid sample. It has been established that monochlorosilanes attach selectively to nonhydrogen-bonded Q^3 silanol groups,[56–60] and ^{19}F is a very accessible NMR nuclide with high gyromagnetic ratio and low natural abundance in most clay systems. When studying ^{19}F, sensitivity is enhanced approximately threefold (compared to direct detection of, say, the disappearance of reactive hydroxyls), as each TFS molecule contains three fluorine atoms. A comparison of integrated peak intensities of TFS-treated samples with a reference containing a known amount of ^{19}F spins provides the number of reactive surface hydroxyl groups per gram of sample (referred to as the *specific hydroxyl number*), which may be converted to a measurement on a per unit area basis if the specific surface area of the sample is known. However, as the reactivity of basal and edge sites on a clay will differ, such conversions must be considered carefully while analyzing the chemical properties of a clay mineral.

The first examinations of the reactions of clay minerals with TFS probe molecules were reported by Washton *et al.*[56] during studies of the number of reactive sites on a chronosequence of volcanic glasses obtained from Kozushima, Japan. Some samples (especially one that had weathered for 52 000 years – the longest aged in the chronosequence) had considerable surface coverage by weathering products (primarily, halloysite, a clay mineral with empirical formula $Al_2Si_2O_5(OH)_4$). In order to quantify the contribution made by weathering products to reactive surface site measurements on this sample, a similar specimen clay sample (the specimen kaolinite KGa-2) was first investigated. A measurement of the density of reactive sites

on the kaolinite for attachment of TFS was normalized to the edge-site surface area based on atomic force microscopy (AFM) measurements,[61] giving a result of 0.44 reactive hydroxyl sites per square nanometer.

In these same studies, a pretreatment comprising extensive sonication was necessary to uncover the weathered rhyolite surface. The rhyolite contains only tetrahedrally coordinated aluminum (Al^{IV}) species, whereas the clay contains octahedrally coordinated aluminum (Al^{VI}) species. The Al^{VI} signal from the clay appears at approximately 3 ppm, fully separated and resolved from all resonances arising from Al^{IV} species (Figure 17.8). After 10 h of treatment, a small amount of clay (approximately 0.5 wt%) remained on or within the sample, which is consistent with small amounts of weathering products located in vesicles or micropores in these samples. By tracking the changes in reactivity to TFS as a function of surface cleaning, a correction to overall surface reactivity of the glass samples was accomplished.

17.5 NMR STUDIES OF SURFACE REACTIVITY OF CLAY MINERALS

A number of NMR studies have focused on the chemical reactivity of the surfaces of clay minerals, a subject vital for understanding such varied phenomena within the environment as dissolution, precipitation, and pollutant transport. Important in these studies are either exploiting NMR-active nuclides within the clay mineral structure itself (e.g., ^{29}Si, ^{27}Al, or ^{17}O) or using 'reporter' molecules that adsorb either physically or chemically to the clay surface (as shown earlier, for both boron-containing species and monochlorosilane probe molecule TFS). It is also possible to use NMR-active nuclides in a molecule that reacts and degrades on the surfaces of a clay mineral, sometimes forming new species that are detectable with high-resolution methods. If the reaction of interest depends on the total area available from all surfaces of the clays, then physical adsorption of a probe molecule is the preferred adsorption route [which is also the basis for such techniques as the BET (Brunauer–Emmett–Teller) isotherm, used to measure specific surface area]. However, if reactions take place at specific sites on a clay surface (such as specific functional groups on edge sites or basal planes, or both), then chemical differentiation is required, and a particularly fruitful advance is to use specific sites for chemisorption as a proxy for the

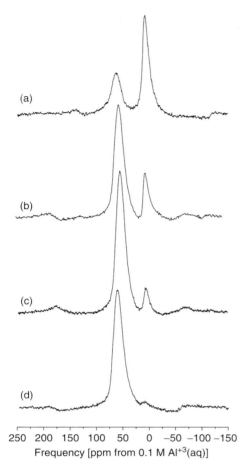

250 200 150 100 50 0 −50 −100 −150
Frequency [ppm from 0.1 M Al⁺³(aq)]

Figure 17.8. (a–d) The ^{27}Al MAS NMR spectra of a 52 ka rhyolitic glass sample display a clay-weathering product (halloysite) through the resonance at approximately 3 ppm. As a function of increasing sonication time, this clay mineral is removed from the sample to enable quantitative reactive surface area measurements. (Reprinted with permission from Ref. 56. © 2008 Elsevier)

so-called reactive surface area. As noted earlier, clay mineral surfaces can be categorized as basal planes or edge sites, and the chemical functionalities present on these surfaces change as a function of humidity and/or solution hydronium activity. We now discuss a number of examples of important environmental reactions at clay surfaces that have been studied with solid-state NMR, including the degradation of pollutants, the formation of new phases under caustic weathering, and the dissolution of clay minerals under low pH conditions.

17.5.1 The Degradation of Perfluorinated Compounds on Clay Mineral Surfaces

The presence of perfluorinated compounds in the environment is a persistent issue because of the high resistance to degradation that is common for materials such as perfluoropolyethers (PFPEs). The degradation of these materials is typically slow because of their strong C–F bonds. A solid-state NMR study by Denkenberberger *et al.* followed the thermal degradation of Krytox® 1506, a common PFPE, adsorbed on to the surfaces of γ-Al_2O_3 and a kaolinite clay mineral (KGa-1b). The degradation process for the PFPE was followed with ^{19}F MAS, ^{27}Al MAS, and $^{19}F \rightarrow ^{27}Al$ CP MAS NMR studies. During degradation, a set of new ^{19}F MAS NMR resonances were observed from (oxy)fluorine species that were selectively associated with octahedrally coordinated aluminum atoms on the surfaces. After thermal treatment at 300°C for times up to 20 h, results from the kaolinite system displayed additional resonances (most likely arising from direct Al–F bonds) in ^{27}Al MAS and $^{19}F \rightarrow ^{27}Al$ CP-MAS spectra. As the reaction progresses in time, more AlF_3-like products were observed, which could ultimately lead to removal of Al from the kaolinite lattice and the production of additional reactive sites for the degradation of the PFPE.

17.5.2 Clay Mineral Reactions and Formation of Neophases under Caustic Conditions

During World War II and the Cold War, the Hanford Site in Washington State was the main production facility for plutonium in the United States. During the production of plutonium at this site, millions of gallons of nuclear waste were generated. To safely dispose of this immense volume of nuclear waste, 177 stainless steel tanks were buried and filled at Hanford. Over time, leaks in these tanks have released over 1 million gallons of the initial 54 million gallons of waste into the surrounding soil. Tank remediation attempts were undertaken, and as a result, the remaining sludge is defined by high pH (~13), OH⁻, Na, Al⁺, and NO_3^- and radionuclide contamination including Cs⁺, Sr²⁺, and I⁻. Solid-state NMR has been utilized to investigate the reactivity of clay minerals with simulated tank waste leachate (STWL) solutions that have high pH and ionic strengths to aid in prediction of the effects of further radioactive tank waste leachates on near-field soils and sediments at the Hanford Site. In these studies, ^{29}Si and ^{27}Al MAS NMR spectroscopies have provided structural and kinetic information for kaolinite and montmorillonite specimen clay systems weathered with high pH and high ionic strength solutions that contain varying amounts of Cs and Sr co-contaminants to mimic the compositions of highly radioactive tank waste leachates that have entered the environment.[62–64] By performing MAS NMR experiments at a range of magnetic field strengths, full sets of relative populations were determined for tetrahedrally coordinated aluminum sites in four new phases formed by the weathering of kaolinite in a STWL for up to 369 days.[64] The kinetics of secondary phase formation can then be mapped based on the separate formation kinetics of each of the neophases, which also changed as a function of co-contaminant Sr and Cs concentrations in the STWL (Figure 17.9). In all cases, the data indicate incipient formation of the feldspathoid mineral sodalite, followed by transformation to more stable cancrinite. Therefore, the clay minerals are transformed through the caustic dissolution and precipitation process to form mineral phases that are actually capable of sequestering radioactive Cs and Sr, as they leak into the environment.

17.5.3 Clay Mineral Dissolution under Acidic Conditions

One proposed rate-limiting step for acid-mediated dissolution for aluminosilicates involves attachment of hydronium species to a nonhydrogen-bonded silanol, followed by the breaking of the bridging oxygen–metal bond.[56,65,66] This fact couples strongly with the proposal that under a certain range of acidic pH conditions, the dissolution of clay minerals proceeds from particle edge sites, resulting in a slight increase in total specific surface area over time while decreasing the number of edge sites overall, as more are dissolved away.[13] Such behavior would lead to a decrease in reactive surface area (on a per gram basis) as a function of the total reaction time. As discussed earlier, it is possible to selectively count reactive hydroxyl species with the TFS probe molecule and follow their changing populations, and hence, changing reactive surface area, with solid-state ^{19}F MAS NMR.

In clay systems, the reactivities and selectivities of TFS probe molecules were studied systematically for a set of kaolinite and bentonite clay samples.[54]

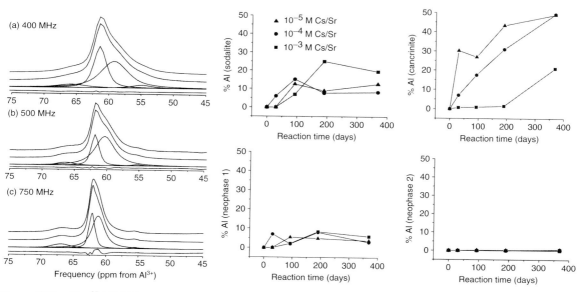

Figure 17.9. The ^{27}Al MAS NMR spectra from kaolinite reacted with a simulated tank waste leachate (with varying initial concentrations of Sr and Cs co-contaminants) provide a quantitative measure of the formation of four new phases. Two of the phases are identified as well-known feldspathoid minerals (sodalite and cancrinite), whereas the other two neophases are most likely an amorphous alumina phase and a second poorly crystalline aluminosilicate phase. (Reprinted with permission from Ref. 64. © 2006 American Chemical Society)

The kaolinites included a low-defect kaolin (KGa-1b) and a high-defect kaolin (KGa-2) and the two bentonite samples were a calcium-rich sample (STx-1b) and a sodium-rich sample (SWy-2). Silicon-29 NMR experiments sensitive to the bulk of the sample (such as one-pulse ^{29}Si MAS NMR studies) do not detect the TFS molecule attachment to the surface; however, more surface-sensitive measurements (such as ^{1}H-^{29}Si CPMAS NMR, see Figure 17.10) demonstrate how the functionalization of the surface can be assessed. As the signal generated by the CP experiment is not easily quantified for samples containing such different types of Si species, and the background ^{1}H signal from the sample overwhelms the ^{1}H MAS NMR signals, the use of ^{19}F signal as a quantification tool is crucial for correct determination of site-specific reactivities.

It is important to note that ^{19}F MAS NMR experiments have shown the presence of surface-adsorbed fluoride in kaolinite KGa-1b.[67] In the ^{19}F MAS NMR spectrum reported by Cochiara and Phillips, this fluorine signal is observed in the range approximately -130 to -150 ppm. ^{19}F(^{27}Al) transfer of populations in double resonance (TRAPDOR) experiments indicated that the naturally occurring fluorine was from Al−F species and ruled out the possibility of significant signal contribution from Si−F species. However, such species were not readily detectable in TFS-treated KGa-1b samples, possibly because of the pretreatment of the samples that is intended to remove surface defects and impurities.

As described earlier, results from initial TFS studies demonstrated that the quantification of ^{19}F spins in the TFS-treated kaolinite samples provided a robust and sensitive measure of the number of reactive hydroxyl sites on a mass-normalized (per gram) basis. The reactive surface site densities for the kaolinite samples were also found to be proportional to edge-site fractions determined by AFM. After acid treatment (initial pH of 2.9) for 10 days, essentially no change was found in results from BET measurements of specific surface area for kaolinites. However, a significant decrease in reactive surface site density was measured via NMR after a single, fixed period of acid-mediated dissolution.

With a measured decrease in the reactive surface area, direct observation of a concomitant loss in reactivity should show strong correlation between reactive site measurement on clay surfaces and the

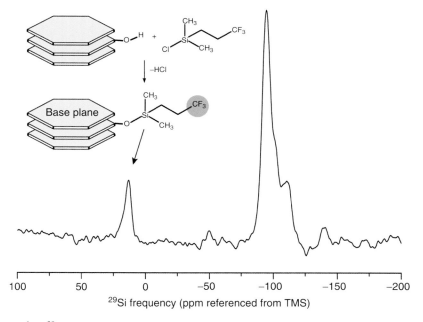

Figure 17.10. The ^{1}H–^{29}Si CPMAS NMR spectrum of bentonite sample STx-1b after treatment with (3,3,3-trifluoropropyl)dimethylchlorosilane (TFS) displays a peak at 13 ppm from Si atoms within the TFS that is attached to the edge surface of the clay (see illustration). Peaks from -80 to -120 ppm are from the silicon species in the bulk clay as well as an opal-CT intergrowth. (Reproduced with permission from Ref 54. © 2010 American Chemical Society)

rates of release of cations into solution during a dissolution event. Therefore, subsequent studies of kaolinite and montmorillonite samples investigated how changes in reactive surface area are tied to decreases in rates of silicon and aluminum releases into solution.[55] Again using TFS treatment and ^{19}F MAS NMR as a solid-state NMR proxy, changes in reactive surface area were monitored for a series of batch dissolution experiments of KGa-1b (low-defect kaolinite) and STx-1b (Ca-rich bentonite, which is a montmorillonite-rich clay containing an opal-CT impurity). In the reported experiments, dissolution took place at 21°C at an initial pH of 3. The specific surface area of kaolinite samples did not change during 80 days of dissolution, whereas bentonite-specific surface area did decrease quickly (to about 50% of the original value) because of the changes in interlayer cation concentrations. Solid-state ^{19}F MAS NMR was used to measure the decrease in the number of reactive surface sites per gram as a function of dissolution time (Figure 17.11). The depletion of reactive edge sites was shown to be proportional to the measured decrease in the rates of release of Si and Al into solution. Overall, these results demonstrate the need to quantify the number of reactive sites present on a per gram basis and measure their changes in order to develop and utilize dissolution rate models for clay

minerals and other heterogeneous materials in the environment.

17.6   CONCLUSIONS

NMR spectroscopy serves an important role in the study of clay minerals for both laboratory-generated and natural samples. Ranging from simpler one-pulse studies of local atomic ordering to more complex multidimensional studies of evolving site identities and symmetries, NMR provides an unprecedented atomic-level understanding of the structure and reactivities of these important systems. Newer studies of reactive sites and chemistries on the surfaces of clay minerals have also been presented and hold promise for increased understanding of the coupling of geochemical cycles with clay mineral behavior. In the future, advanced molecular-level understandings of chemical cycling and reactions in the environment will benefit from the discovery and development of an array of probe molecules that can be used to predict and describe important chemistries. Further progress in these areas of study will be accelerated by advancements in NMR techniques and instrumentation, including the use of polarization enhancement mechanisms (such as high-field dynamic nuclear polarization)[68] and multidimensional solid-state heteronuclear correlation NMR methods.

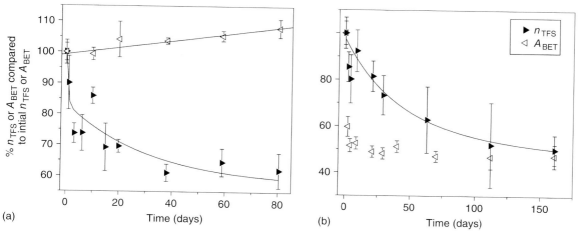

Figure 17.11.   For (a) kaolinite and (b) bentonite solids that are dissolving in solutions with an initial pH of 3, the change in the number of sorbed (3,3,3-trifluoropropyl)dimethylchlorosilane (TFS) molecules per gram (n_{TFS}) is measured with ^{19}F MAS NMR. This change is found to correlate with the dissolution rate of the clay mineral as a function of time. (Reprinted with permission from Ref. 55. © 2012 Elsevier)

ACKNOWLEDGMENTS

We wish to thank Pennsylvania State University and the NSF-funded Center for Environmental Kinetics Analysis (CEKA) at Pennsylvania State (NSF CHE-0431328) for financial support in our studies of reactive surface area in the environment. Close collaborations with Susan Brantley, James Kubicki, and others within the CEKA program were greatly appreciated. We also acknowledge the collaborators Jon Chorover (Arizona) and Peggy O'Day (UC Merced) who have worked with us on the analysis of Hanford waste problems for many years, supported by the Department of Energy Biological and Environmental Research Program. KTM and NMW now work at the Environmental Molecular Research Laboratory, a national scientific user facility sponsored by the US Department of Energy's Office of Biological and Environmental Research located at Pacific Northwest National Laboratory and operated for DOE by Battelle.

REFERENCES

1. D. M. Moore and R. C. Reynolds, X-ray Diffraction and the Identification and Analysis of Clay Minerals, 1st edn, Oxford University Press: New York, 1989, 332 pages.

2. G. Borchardt, in Minerals in Soil Environments, eds J. B. Dixon and S. B. Weed, Soil Science Society of America: Madison, WI, 1989, 675.

3. C. Klein and C. S. Hurlbut, Manual of Mineralogy, 21st edn, John Wiley & Sons, Inc.: New York, 1993, 704 pages.

4. D. G. Schulze, in Minerals in Soil Environments, eds J. B. Dixon and S. B. Weed, Soil Science Society of America: Madison, WI, 1989, 1.

5. R. A. Schoonheydt, in Mineral Surfaces, eds D. J. Vaughan and R. A. D. Pattrick, Chapman & Hall: London, 1995, 303.

6. B. Velde and A. Meunier, The Origin of Clay Minerals in Soils and Weathered Rocks, 1st edn, Springer: Berlin, 2008, 406 pages.

7. D. L. Sparks, Environmental Soil Chemistry, 2nd edn, John Wiley and Sons, Inc.: New York, 2003, 352 pages.

8. E. Wieland and W. Stumm, *Geochim. Cosmochim. Acta*, 1992, **56**, 3339.

9. J. Ganor, J. L. Mogollon, and A. C. Lasaga, *Geochim. Cosmochim. Acta*, 1995, **59**, 1037.

10. D. Bosbach, L. Charlet, B. Bickmore, and M. F. Hochella, *Am. Mineral.*, 2000, **85**, 1209.

11. B. R. Bickmore, D. Bosbach, M. F. Hochella, L. Charlet, and E. Rufe, *Am. Mineral.*, 2001, **86**, 411.

12. M. P. Turpault and L. Trotignon, *Geochim. Cosmochim. Acta*, 1994, **58**, 2761.

13. S. J. Kohler, D. Bosbach, and E. H. Oelkers, *Geochim. Cosmochim. Acta*, 2005, **69**, 1997.

14. G. N. White and L. W. Zelazny, *Clays Clay Miner.*, 1988, **36**, 141.

15. W. F. Bleam, G. J. Welhouse, and M. A. Janowiak, *Clays Clay Miner.*, 1993, **41**, 305.

16. M. J. Duer, Introduction to Solid-State NMR Spectroscopy, 1st edn, Blackwell Publishing: Oxford, UK, 2004, 349 pages.

17. B. C. Gerstein and C. R. Dybowski, Transient Techniques in NMR of Solids: An Introduction to Theory and Practice, Academic Press, Inc.: New York, 1985, 295 pages.

18. C. A. Fyfe, Solid State NMR for Chemists, CFC Press: Guelph, ON, 1983, 593 pages.

19. A. Llor and J. Virlet, *Chem. Phys. Lett.*, 1988, **152**, 248.

20. K. T. Mueller, B. Q. Sun, G. C. Chingas, J. W. Zwanziger, T. Terao, and A. Pines, *J. Magn. Reson.*, 1990, **86**, 470.

21. A. Samoson, E. Lippmaa, and A. Pines, *Mol. Phys.*, 1988, **65**, 1013.

22. Y. Wu, B. Q. Sun, A. Pines, A. Samoson, and E. Lippmaa, *J. Magn. Reson.*, 1990, **89**, 297.

23. L. Frydman and J. S. Harwood, *J. Am. Chem. Soc.*, 1995, **117**, 5367.

24. A. Medek, J. S. Harwood, and L. Frydman, *J. Am. Chem. Soc.*, 1995, **117**, 12779.

25. Z. Gan, *J. Am. Chem. Soc.*, 2000, **122**, 3242.

26. A. R. Grimmer, F. v. Lampe, M. Mägi, and E. Lippmaa, *Z. Chem.*, 1983, **23**, 343.

27. G. Engelhardt and D. Michel, High-Resolution Solid-State NMR of Silicates and Zeolites, 1st edn, John Wiley & Sons: New York, 1987, 485 pages.

28. E. Lippmaa, M. Mägi, A. Samoson, G. Engelhardt, and A.-R. Grimmer, *J. Am. Chem. Soc.*, 1980, **102**, 4889.

29. J. Sanz and J. M. Serratosa, *J. Am. Chem. Soc.*, 1984, **106**, 4790.

30. R. J. Kirkpatrick, in Reviews in Mineralogy, ed F. C. Hawthorne, Mineralogical Society of America: Chelsea, Michigan, 1988, 341, Vol. 18.

31. A. Samoson, *Chem. Phys. Lett.*, 1985, **119**, 29.

32. S. E. Ashbrook, J. McManus, K. J. D. MacKenzie, and S. Wimperis, *J. Phys. Chem. B*, 2000, **104**, 6408.

33. S. K. Lee and J. F. Stebbins, *Am. Mineral.*, 2003, **88**, 493.

34. S. K. Lee, J. F. Stebbins, C. A. Weiss, and R. J. Kirkpatrick, *Chem. Mater.*, 2003, **15**, 2605.

35. S. Hayashi, T. Ueda, K. Hayamizu, and E. Akiba, *J. Phys. Chem.*, 1992, **96**, 10922.

36. S. Hayashi, T. Ueda, K. Hayamizu, and E. Akiba, *J. Phys. Chem.*, 1992, **96**, 10928.

37. B. Zhou, B. L. Sherriff, and T. Wang, *Am. Mineral.*, 2009, **94**, 865.

38. X. Lin, K. Ideta, J. Miyawaki, Y. Wang, I. Mochida, and S.-H. Yoon, *Appl. Magn. Reson.*, 2013, **44**, 1081.

39. J. Grandjean, in Clay Surfaces: Fundamentals and Applications, eds E. Wypch and K. G. Satyanarayana, Elsevier: Amsterdam, The Netherlands, 2004, 216.

40. J. Grandjean, *Clay Miner.*, 2006, **41**, 567.

41. Y. Kim and R. J. Kirkpatrick, *Geochim. Cosmochim. Acta*, 2006, **70**, 3231.

42. Y. Kim, R. T. Cygan, and R. J. Kirkpatrick, *Geochim. Cosmochim. Acta*, 1996, **60**, 1041.

43. Y. Kim and R. J. Kirkpatrick, *Geochim. Cosmochim. Acta*, 1997, **61**, 5199.

44. Y. Kim and R. J. Kirkpatrick, *Am. Mineral.*, 1998, **83**, 661.

45. Y. Kim, R. J. Kirkpatrick, and R. T. Cygan, *Geochim. Cosmochim. Acta*, 1996, **60**, 4059.

46. P. D. Leo and J. Cuadros, *Clays Clay Miner.*, 2003, **51**, 403.

47. A. N. Garroway, *J. Magn. Reson.*, 1977, **28**, 365.

48. D. Iuga, H. Schafer, R. Verhagen, and A. P. M. Kentgens, *J. Magn. Reson.*, 2000, **147**, 192.

49. G. M. Bowers, R. Ravella, S. Komarneni, and K. T. Mueller, *J. Phys. Chem. B*, 2006, **110**, 7159.

50. N. N. Herrera, J.-M. Letoffe, J.-P. Reymond, and E. Bourgeat-Lami, *J. Mater. Chem.*, 2005, **15**, 863.

51. E. Ruiz-Hitzky and J. J. Fripiat, *Clays Clay Miner.*, 1976, **24**, 24.

52. R. Fry, C. G. Pantano, and K. T. Mueller, *Phys. Chem. Glasses*, 2003, **44**, 64.

53. R. A. Fry, N. Tsomaia, C. G. Pantano, and K. T. Mueller, *J. Am. Chem. Soc.*, 2003, **125**, 2378.

54. R. L. Sanders, N. M. Washton, and K. T. Mueller, *J. Phys. Chem. C*, 2010, **114**, 5491.

55. R. L. Sanders, N. M. Washton, and K. T. Mueller, *Geochim. Cosmochim. Acta*, 2012, **92**, 100.

56. N. M. Washton, S. L. Brantley, and K. T. Mueller, *Geochim. Cosmochim. Acta*, 2008, **72**, 5949.

57. T. Kawai and K. Tsutsumi, *Coll. Polym. Sci.*, 1998, **276**, 992.

58. C. H. Lochmuller and M. T. Kersey, *Langmuir*, 1988, **4**, 572.

59. P. Vandervoort, I. Gillisdhamers, and E. F. Vansant, *J. Chem. Soc. Faraday Trans.*, 1990, **86**, 3751.

60. X. S. Zhao and G. Q. Lu, *J. Phys. Chem. B*, 1998, **102**, 1556.

61. B. R. Bickmore, K. L. Nagy, P. E. Sandlin, and T. S. Crater, *Am. Mineral.*, 2002, **87**, 780.

62. J. Chorover, S. Choi, M. K. Amistadi, K. G. Karthikeyan, G. Crosson, and K. T. Mueller, *Environ. Sci. Tech.*, 2003, **37**, 2200.

63. S. Choi, G. Crosson, K. T. Mueller, S. Seraphin, and J. Chorover, *Geochim. Cosmochim. Acta*, 2005, **69**, 4437.

64. G. Crosson, S. Choi, J. Chorover, M. K. Amistadi, P. A. O'Day, and K. T. Mueller, *J. Phys. Chem. B*, 2006, **110**, 723.

65. W. H. Casey, A. C. Lasaga, and G. V. Gibbs, *Geochim. Cosmochim. Acta*, 1990, **54**, 3369.

66. Y. T. Xiao and A. C. Lasaga, *Geochim. Cosmochim. Acta*, 1994, **58**, 5379.

67. S. G. Cochiara and B. L. Phillips, *Clays Clay Miner.*, 2008, **56**, 90.

68. R. G. Griffin and T. F. Prisner, *Phys. Chem. Chem. Phys.*, 2010, **12**, 5737.

Chapter 18
Soil–Water Interactions

Gabriele E. Schaumann[1] and Marko Bertmer[2]

[1]*Department of Environmental and Soil Chemistry, University Koblenz-Landau, Landau D-76829, Germany*
[2]*Institute of Experimental Physics II, Leipzig University, Faculty of Physics and Earth Sciences, Leipzig D-04103, Germany*

18.1 INTRODUCTION

Among others, soil maintains ecological functions in the water cycle, including water purification and provision of water supply for plants, animals, and microorganisms in terrestrial ecosystems.[1–3] Water in soil can occur as pore water, and the transport, mobility, and the availability of water for plants are directly related to the pore size. Water can also be incorporated in swollen

NMR Spectroscopy: A Versatile Tool for Environmental Research
Edited by Myrna J. Simpson and André J. Simpson
© 2014 John Wiley & Sons, Ltd. ISBN: 978-1-118-61647-5

organic matrices such as biofilms, polysaccharides, and soil organic matter (SOM). Depending on the surface characteristics of the soil particles, soils can be water repellent,[4–6] resulting in modifications in water transport and water supply patterns in ecosystems.[7–10] In addition, water can also play a significant role in the stabilization of supramolecular structures in SOM[11–13] and, therefore, has direct impact on binding and entrapment of organic chemicals and availability of nutrients. In contrast to the importance of water in soil, many aspects of soil–water interactions, such as their mechanisms and implications for soil quality and soil functioning, have up to now not been fully explored at the molecular level.[11,14]

Low-field NMR[15] and low-resolution NMR[16] represent powerful techniques to overcome this lack of knowledge. They can be used for nondestructive investigation of soil–water interactions both on the molecular/supramolecular level and on the microscopic and macroscopic level in moist, wet, and dry soils,[17–19] and recent mobile NMR techniques can be applied in the field (Chapter 9). NMR relaxometry has been primarily used for studies in geosciences but needs to be further developed for studying soils, because the pore system differs from that found in rock formations.[15] The main challenge for application to soil is, in this context, its huge complexity and heterogeneity and its up to now only scarcely understood chemistry of SOM (see Chapter 13).[14,20] Nevertheless, efforts have been made to use NMR relaxometry to describe pore size distributions in soils,[18] as well as processes occurring during water uptake, such as wetting, swelling of organic matter, and redistribution of water.[17,19,21]

NMR studies of soil–water interactions are commonly done by observing the ^{1}H nucleus. In ^{1}H NMR relaxometry, spins are excited by a radiofrequency (RF) pulse with the characteristic Larmor frequency of the protons, causing a detectable magnetization. Depending on the pulse sequence, the macroscopic magnetization is deflected in 90° or 180° direction. After excitation, spins relax to their equilibrium orientation; the transverse relaxation of magnetization reflects in the free induction decay (FID).[22] Relaxation occurs as spin–lattice (longitudinal) and spin–spin (transverse) relaxation[22], and is generally a first-order process characterized by the relaxation time under the limit of Bloch equations, that is, for uniform (isotropic) systems.

The spin–lattice relaxation time (T_1) describes how effective interactions between the spin system and the environment are in dissipating magnetic energy from spins to the lattice. The stronger the interactions are, the faster the energy is dissipated and the shorter is T_1. Measuring T_1 can be very time consuming and is, so far, not often used in soil science applications, although it may be the more appropriate measure than the spin–spin relaxation time (T_2) in many cases.[22,23] In contrast, T_2 normally originates from a dephasing of precessing spins, for example, because of slight differences in Larmor frequency from local field inhomogeneities.[22] Variations in the magnetic field caused by neighboring nuclei are weakest in systems where spins can move freely and inhomogeneities due to neighboring spins are small. As the dephasing of the spins can only take place in the presence of a longitudinal magnetization, T_2 can be smaller than or equal to T_1, but it can never be longer.[15,23] Therefore, limitation of mobility can generally reduce T_2. In contrast, T_1 can be either increased or reduced by a reduction in mobility, depending on the Larmor frequency and the correlation time for the relaxation-relevant interaction[22]. Molecular diffusion in field gradients affects T_2 but not T_1, because no energy exchange is involved in this relaxation mechanism.[24] The relaxation rate due to diffusion in field gradients is proportional to the diffusion coefficient and the square of local field gradients.[22] The local field gradients increase with increasing external field strength. Therefore, measurements in systems such as soils, where large local field gradients are generally omnipresent, are to be carried out preferentially in low fields up to 10–50 MHz.[15]

Field-cycling NMR explicitly investigates the field and frequency dependencies of T_1 and T_2 at proton Larmor frequencies between 10 kHz and 40 MHz or higher and is, therefore, a promising tool to study dynamic molecular interactions and to distinguish between the molecular effects and effects of local field gradients or sample heterogeneity[25] (see Chapter 8).

18.2 RELAXATION TIMES (T_1 AND T_2) IN POROUS SYSTEMS

T_1 and T_2 of protons in bulk water generally are in the range 1–3 s.[17,26] If confined in porous media, relaxation is accelerated because of solid–fluid interactions at pore surfaces with fixed spins, paramagnetic ions, or paramagnetic crystal defects.[18,26] The surface relaxation process is superimposed by diffusion of water molecules between bulk water and the surface. Further, spin–spin relaxation occurs via diffusion in local field gradients. The macroscopically measurable relaxation rate is a result of the superposition of these subprocesses:[23]

$$\left(\frac{1}{T_1}\right)_{\text{total}} = \left(\frac{1}{T_1}\right)_{\text{B}} + \left(\frac{1}{T_1}\right)_{\text{S}}$$
$$\left(\frac{1}{T_1}\right)_{\text{total}} = \left(\frac{1}{T_2}\right)_{\text{B}} + \left(\frac{1}{T_2}\right)_{\text{S}} + \left(\frac{1}{T_2}\right)_{\text{diff-FG}}$$

$$(18.1)$$

where 'B' stands for 'bulk', 'S' for 'surface', and 'diff-FG' for diffusion in field gradients. The surface relaxation term contains information of the pore system and is, therefore, of special relevance. Its contribution depends on the dominance of the respective subprocess, where fast-diffusion (surface-limited) and slow-diffusion (diffusion-limited) regimes have to be distinguished.[27] In the surface-limited regime, surface relaxation depends on the internal surface area S, internal pore volume V, and the surface relaxivity ρ.[23] The latter is strongly influenced by paramagnetic ions on the surface such as Mn^{2+} or Fe^{3+}:

$$\text{Surface-limited}: \frac{1}{T_{1,2S}} = \rho_{1,2}\frac{S}{V} \approx \rho_{1,2}\frac{\alpha}{r} \quad (18.2)$$

where r is the pore radius and α the shape factor (1, 2, and 3 for planar, cylindrical, and spherical pore geometries, respectively).[28]

The diffusion-limited regime requires large pores or fast surface relaxation, for example, because of the presence of paramagnetic centers.

Diffusion-limited : $\dfrac{1}{T_{1S}} = \dfrac{1}{T_{2S}} = D\dfrac{c}{r^2}$ \hspace{1em} (18.3)

where D is the diffusion coefficient and c a shape-dependent factor. For a more detailed summary, please refer to the review by Bayer *et al.*[15] For analysis of the relaxation decay curves, several different algorithms and software are applied.[15] They convert relaxation decay curves from the time domain into the relaxation time domain, resulting in relaxation time distributions.

Magnetic resonance imaging (MRI) studies provide insights into soil—water interactions on the spatial scale. As they provide qualitative and quantitative information about local water distribution, they help understanding water transport in soils and porous media, especially the infiltration patterns and development of preferential flow pathways of water and hydrocarbons.[29,30] For more details, see Chapter 7.

18.3 NMR SPECTROSCOPY TO STUDY WATER MOBILITY AND WATER BINDING

NMR spectroscopy is mainly used to determine the chemical nature of soil components (see Chapters 1, 4, and 5). For example, in solid-state NMR, functional groups can be identified using ^{13}C NMR, which exhibits a larger shift range and, therefore, provides better resolution compared to ^1H NMR. In solution state, the linewidth is lower compared to the solid state (even using magic-angle spinning (MAS)). However, the question arises whether the total soil can be analyzed or only the soluble parts (see Chapter 6).

Only a few publications use nonspinning, static wideline solid-state NMR to obtain information on soils.[13,16,31–33] Although no chemical information can be directly obtained, the mobility of proton-containing material, such as SOM segments or especially water in soil, is directly reflected by the linewidth of the signals (Figure 18.1). Liquid-like signals show a narrow Lorentzian line, whereas rigid solid-like signals are characterized by a broad Gaussian line. This method is especially useful for medium to low soil water contents. The direct relation between linewidth and mobility is hindered by paramagnetic agents in soils, mostly iron or manganese ions in clay-containing samples. Nevertheless, signals with small linewidth and, therefore, high mobility are less affected by this,

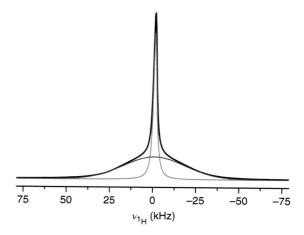

Figure 18.1. ^1H wideline NMR spectrum of an air-dried sapric peat (black line). The red and blue lines derive from a deconvolution of the spectrum corresponding to the Lorentzian and Gaussian contributions, preferably water and soil organic matter, respectively.

as the interaction time with the paramagnetic agents is small.

For samples with high water content, the objective is mostly to study the diffusion of water through pores in the soil, and the SOM signals cannot be detected because of a too long probe deadtime of the resonators used for these measurements. As the water content is high, the water—water interactions dominate and a very small linewidth of almost that of free water is obtained that does not allow for mobility studies. However, it can be used for MRI to document phenomena such as water transport (see Chapter 7).

For very low water contents, water binding to specific soil segments becomes more important to investigate. The ^1H chemical shift range of water incorporated in crystalline materials or confined water in porous materials is observed to be rather large (at least, between 3 and 12 ppm)[34,35] and thus would, in principle, enable detection of bound water. However, the strong $^1H–^1H$ dipole–dipole coupling together with the small shift range for protons leads to low resolution. Furthermore, soil is a very heterogeneous system, which leads to a strong overlap of signals and usually an unresolved spectrum is obtained. The recently introduced ultrafast MAS probes reaching spinning frequencies of up to 70 kHz might, therefore, not even bring any improvement. To our knowledge, there are no studies of water-binding signals in soils using NMR.

18.4 WATER IN WET SOILS

18.4.1 Water in Soil Pores and Pore Size Distribution

Water uptake and redistribution in soils are of great importance for nearly all soil ecological and soil chemical processes.[3,4,6,14] They are determined by the water content, the matric potential (ψ), and the water retention curve of a soil.[3] Measurement of water retention curves is time consuming and cannot be carried out *in situ*. ^{1}H NMR relaxometry can be a helpful tool for studying water distribution in soil pores,[15,18] as the ^{1}H relaxation time of protons in pore water is determined by the pore size distribution. Surface relaxation in soil pores is strongly influenced by the presence of paramagnetic centers including paramagnetic ions (Fe^{3+} and Mn^{2+}) and crystal defects.[36–38] In addition, bulk relaxation is accelerated by the presence of dissolved paramagnetic ions and depends on the speciation of the ion.[37,38] Several ^{1}H NMR relaxometry studies have addressed water uptake into soils or clays.[17,19,39–43] As soils generally have a pore system with a defined pore size distribution, the ^{1}H NMR relaxation times will reveal the water distribution[19] with three or four separate peaks representing different water molecule populations in the pore system.[15] NMR relaxometry distinguishes: (i) bound water at small relaxation times ($T_2 < 60$ ms), (ii) water in mesopores or loosely bound water at 60 ms $< T_2 <$ 300 ms, and (iii) macropore water ($T_2 > 300$ ms) and bulk water ($T_2 \approx 2000$ ms).[15]

The first study assessing the pore size distribution of soils by ^{1}H NMR relaxometry was conducted by Hinedi *et al*.[41] Several later studies agree that relaxation time distributions of soil samples are related to pore sizes, but quantitative assessment is restricted because of high but unknown influences of paramagnetic centers.[39,44–46] In 2005, NMR relaxometry was recommended only as an additional method to characterize pore connectivity.[46] Verification of pore size distribution determined via ^{1}H NMR relaxometry by conventionally obtained pore size distribution has been performed for rocks[47] and only recently for soils.[18] Four central problems occur in soils and rocks: (i) iron concentration is generally unevenly distributed, which necessarily results in a distribution of surface relaxivities,[28,37] (ii) relaxation times correlate with contents of different soil constituents depending on the pore size,[18] (iii) the diffusion is surface limited only in small pores resulting in relaxation times below 10 ms, and (iv) most pores cause situations in the intermediate and slow-diffusion regime; the latter mainly for sandy soil samples.[18] As soon as these restrictions are considered and two different surface relaxivities are assumed, pore size distribution determined by conventional soil water retention measurements correlated well with that obtained by ^{1}H NMR relaxometry ($R^2 = 0.98$).[18]

18.4.2 Water Uptake in Soils: Wetting and Swelling

In the course of water uptake in soils, relaxation time distributions often shifted toward smaller relaxation times and peaks at shorter relaxation times increased in size,[17,19,40,41,46] suggesting a decrease in pore size and/or successive water movement into smaller pores, respectively. The hypothesis that these changes are solely due to changes in the paramagnetic environment of the sample was excluded.[19] The increase of peak intensity at shorter relaxation times is most strikingly observed when water-repellent soil samples were investigated,[19,39] where large pores are filled first and small pores are filled only successively, when pore walls become increasingly hydrophilic.[19,42] Another explanation may be due to the swelling processes.[39] The time constants of these processes varied from hours to days/weeks, and their activation energies ranged between 5 and 50 kJ mol^{-1}.[15]

For a deeper understanding, we will distinguish between wetting and swelling as two separate and more or less independent processes.[40] Wetting is fast in wettable soils, whereas hydrophobic surfaces of soil particles can become hydrophilic slowly, resulting in a slower wetting rate. The hydrophobicity of a surface can affect relaxation time by exclusion of surface–water contact in small pores, resulting in relaxation times up to ≈ 1 s[48,49] and by successive hydrophilization of hydrophobic surfaces by chemical or physicochemical processes.[19,50] ^{1}H NMR relaxometry in combination with time-resolved contact angle measurement is a promising approach to increase our understanding on the currently almost completely unraveled processes rendering hydrophobic soil surfaces hydrophilic[4] and vice versa.

In contrast, water uptake into the solid phase, for example, of SOM, is referred to as swelling.[19]

Swelling processes are relevant for the understanding of binding and entrapment of contaminants in SOM by the formation of new sorption domains or changes in rigidity[13,40], or by opening up new binding sites.[51] During swelling, the properties of water change, as discussed in Section 18.4.3.

18.4.3 Properties of Water in Hydrated Peat

McBrierty *et al.*[43] were among the first to study the state of water in peats using NMR relaxometry at stronger field strengths 7.05 T in combination with thermal analysis. McBrierty *et al.*[43] distinguished four different water states, with two forms of loosely bound water, bulk water, and tightly bound water that did not freeze at temperatures as low as 160 K. The loosely bound water froze around 210 K and bulk water at 273 K, indicating also the temperature range above which each water form became mobile. Drying and rewetting of peat samples did shrink and swell the peat matrix and with that, changed the amount of loosely bound water, but the amount of strongly surface-associated water was similar after each change in moisture status.[52] Nonfreezing water was associated with hydration water, that is, water in a gel-like layer at the solid surface or water that chemically interacts with the hydrophilic moieties on the surface.[43] Therefore, the amount of nonfreezable water could be an indicator for surface–water interactions and gel water.[15,52,53]

A detailed study of water uptake of a peat sample over 9 months followed by ^{1}H NMR relaxometry, thermal analysis, and T_1-T_2 and T_2-D (diffusion coefficient) correlation measurements revealed dramatic physical and physicochemical changes in two peats and enabled for the first time to develop a mechanistic model for peat swelling.[17] For this, the authors combined NMR data of frozen and unfrozen samples with results from differential scanning calorimetry (DSC) analysis to assess the freezability of water. The following types of water were distinguished by NMR[17]: (i) water unfreezable above $-34\,°C$, (ii) freezable bound water freezing between -5 and $-34\,°C$, (iii) physically immobilized water freezing between 0 and $-5\,°C$, and (iv) free water. Amount and characteristics of these water types are summarized in Table 18.1. With DSC, a similar classification was possible, leading to three different water types[17] (Table 18.1): (1) water unfreezable above $-90\,°C$ and two types of freezable water: (2) water melting in a sharp peak around $0\,°C$, and (3)

water melting in a broad peak above $-5\,°C$, the position and width of which depend on the heating rate and swelling time.

Figure 18.2 visualizes the distribution of the water among these types identified by NMR and DSC. It shows that the amount of water melting below $-5\,°C$ largely corresponds between NMR (i) and (ii) and DSC (1). Water type (2) was only identified by DSC; its amount corresponds to part of the physically immobilized water (iii), while the freezable water (3) melting in a broad peak in DSC comprises both free (iv) and physically immobilized water (iii) identified by NMR.

18.4.4 Swelling of Peat and Water Status in Swollen Peats

Re-evaluation of these data[17] and results discussed earlier show that combination of qualitative and quantitative information from DSC and NMR analyses distinguishes five types of waters present in swollen peats (A–E; Figure 18.3).

Jähnert *et al.*[54] demonstrated that silica nanopore walls are surrounded by a layer of unfreezable water enclosing freezable water. Assuming that in organic materials the border between freezable and unfreezable waters is more continuous, the nanopore water could be separated into three fractions with different mobility, representing the water populations A (nanopore surface water), B (nanopore layer water), and C (nanopore bulk water), which are located within organic matter particles. Swollen peats would then contain 1.1 g–1.4 g intraparticular nanopore water per gram swollen peat. The swollen peat particle could, therefore, be regarded as a traditional gel phase. Water types D and E are expected to be located in larger pores (e.g., interparticular micropores or macropores). The diffusional restriction of type D could be due to the presence of dissolved and swollen organic substances on particle surfaces penetrating into the water (swellable micropores), whereas type E represents water in unswellable macropores. The type D water would indicate regions in the peat where particles are glued to each other by swellable organic substances, which successively reorientate in the course of swelling. The microgels originating from this process are expected to stabilize the organic matter aggregates in the peat. This could represent one of the mechanisms responsible for aggregate stabilization by organic substances.[59]

Table 18.1. Types of water distinguished by NMR and differential scanning calorimetry (DSC) in a sapric histosol sample[17]

	NMR
(i) Unfreezable water (−34 °C)	• did not freeze above −34 °C • 3–10% of the total water, relaxation time (0.4 ± 0.1) ms • during swelling, the amount increased slightly • potentially located close to functional groups or nanopore walls[54]
(ii) Freezable bound water	• freezes between −5 °C and −34 °C • 1% of total water, relaxation time (1.0 ± 0.1) ms • still strongly restricted in mobility, most probably via hydrogen bonds or in gel phases[17] • swelling increased relaxation time, thus mobility or pore diameter increased.
(iii) Physically immobilized water	• freezes between 0 °C and −5 °C • 12%–35% of total water, relaxation time 10 ms–800 ms, diffusivity 35%–60% of pure water[17] • partly in nanopores or in gel phases, but indicated by the high T_1/T_2 ratios; at least, part is located in micropores, but restricted in mobility because of incorporation in gel phases[55,56] • during swelling, amount increased from 12%–16% to 22%–35%
(iv) Free water	• freezes around 0 °C • 65%–84% of total water, relaxation time 300 ms–800 ms, diffusivity corresponds to that of pure water • retained by the peat, but behaves like free bulk water

	DSC
(1) Unfreezable water (−90 °C)	• did not reveal melting above −90 °C • amount corresponds to that of water (i) and (ii) identified by NMR • amount did not change in amount upon swelling
(2) Freezable water I	• melts in a very sharp temperature interval (−3 °C to 0 °C) • occurs only in swollen gels and cannot contain salts or solutes • potentially free water entrapped in nanopores[43,57,58] not allowing hydrated solutes to enter[17] • amount increased from 3%–6% of the total water to 4%–11%
(3) Freezable water II	• melts in a broad peak superimposing the sharp melting (2) • 88%–93% of total water; melting point strongly depends on heating rate and swelling time

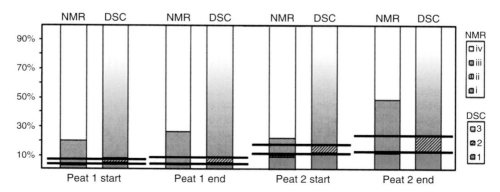

Figure 18.2. Water types identified by Jaeger *et al.*[17] with NMR (i, ii, iii, and iv) and DSC (1, 2, and 3) before swelling ('start') and after swelling ('end'). Blue regions (i, ii, and 1) indicate water that does not freeze below −5 °C. Orange and white (ii, iv, 2, and 3) indicate freezable water with different melting behavior and diffusivity. Horizontal lines visualize the border between unfreezable and freezable waters (bottom line) and the freezable water I with the sharp melting peak.

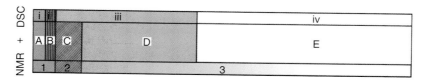

Figure 18.3. Types of water found in swollen peat. A: Water unfreezable above −34 °C; B: water freezable between −5 °C and −34 °C; C: water restricted in diffusivity, but melting in a sharp peak around −2 ± 1 °C: fast melting but diffusion-limited water; D: slowly melting water restricted in diffusivity; and E: slowly melting, mobile water.

Swelling is defined as hydration of SOM, which increases the thickness of the SOM coating or SOM particle. Peat swelling was observed by changes in pore size distribution, volumetric expansion, and redistribution of water, increasing amounts of unfreezable and loosely bound waters, and formation of gel phases and reduction of the translational and rotational mobilities of water molecules.[17] Swelling in peat and soil samples leads to a decrease in interparticular pore size[17,19,40] during which nanopores and interparticular microgel pores increased in quantity on the cost of interaggregate macropores (Table 18.2 and Figure 18.4).

The time dependence of the swelling process at three different temperatures differentiated three subprocesses having activation energies of (5–15) kJ mol^{-1} for fast processes occurring within minutes, (15–25) kJ mol^{-1} for processes with time constants of hours, and (25–50) kJ mol^{-1} for the slowest processes lasting weeks to months.[17] These activation energies point to physical and physico-chemical processes. In a similar, but less detailed and shorter study on other soils, Todoruk *et al.*[39] even found activation energies of >80 kJ mol^{-1} in an organic soil sample, indicating chemical transformations such as ester hydrolysis or more complex rearrangements of SOM components.[39]

18.4.5 Microbial Influences on Soil–Water Interactions

Microorganisms can form biofilms in order to relieve water stress and use nutrients more efficiently, consisting of networks of EPS, which bind water very effectively and form highly hydrated gels.[60] Biofilms[60–62] or small biofilm-like structural units can also be formed in soils.[5,21,63–65]

Like swollen peat, biofilms and EPS networks can affect relaxation times.[21,66,67] For example, the monomodal relaxation time distribution of water became bimodal in the presence of a biofilm, but peak resolution declined in the porous system, because water in part of the mineral pores has similar relaxation times as the water entrapped in biofilms.[66] In soil samples, the detection of biofilm growth is challenging. Biofilms are expressed in significantly lower concentrations than in pure culture systems. Microorganisms in soils are mainly attached via EPS to particle surfaces and primarily found in pores with diameters of (1–30) μm.[60] In a biofilm growth experiment with soil samples, enhancement of microorganism activity resulted in a stronger reduction of T_2 than in untreated samples, which could be due to the increased production of EPS and origination

Table 18.2. Classification of water in swollen peats, re-evaluated from data by Jaeger *et al.*[17]

(A) Nanopore surface water	Does not freeze below −34 °C, located at the pore walls of nanopores, strongly restricted in mobility
(B) Nanopore layer water	Freezes between −5 °C and −34 °C, restricted mobility and freezability because of neighborhood of nanopore surface water
(C) Nanopore bulk water	Fast melting water restricted in mobility, located in gel nanopores where reorientation due to freezing is possible
(D) Interparticular microgel pore water	Slowly melting interparticular pore water, restricted in mobility, probably because of organic molecules and molecule parts penetrating from the swelling particle surfaces into the interparticular pore space
(E) Interaggregate pore water	Slowly melting but mobile water located in macropores, mainly between aggregates

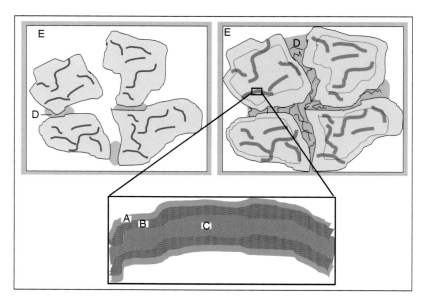

Figure 18.4. Conceptual model of the water distribution in swollen peat. The classification of the five water types A–E is given in Table 18.2 and Figure 18.3.

of gel phases.[21] A 50 times higher spin–spin than spin–lattice rate was determined for agar gels at 30 MHz.[17] This finding can be interpreted in terms of a reduced rotational mobility of the water molecules because of water structuring of the polymer.[68] Thus, a combination of T_1 and T_2 measurements is recommended for a more detailed study of biofilm or other gel phases,[17] such as inside the SOM matrix.[40] This may be helpful to determine different water states and to discriminate between the effects of water mobility and pore size distribution in biofilm- or gel-containing porous media.

18.4.6 Potential and Limitations of ^{1}H NMR Relaxometry

The potential of ^{1}H NMR relaxometry to study soil–water interactions is still far from being fully exploited. For this, the method needs to be further adapted to the specific complexity and heterogeneity of soils, and targeted studies focusing on soil-forming processes have to be conducted. The standard ^{1}H NMR relaxometry can be extended by the more intensive use of pulsed field gradient (PFG) NMR for the determination of diffusion coefficients in combination with NMR relaxometry[46,69] and the development of two-dimensional (2D) correlation maps (T_1–T_2 and T_2–D).[17] Mobile NMR devices (see Chapter 9) will help to investigate porosity and water distribution in samples *in situ*: the NMR MOUSE (mobile universal surface explorer) a unilateral scanner[70,71] and in-side-out NMR devices such as the Halbach scanner for bore hole applications.[22,72] Field-cycling NMR techniques are promising to study interaction dynamics (see Chapter 8).

The potential of NMR relaxometry lies in its versatility. This, however, hides the danger of severe misinterpretations especially in the complex and heterogeneous soil systems. Conclusions on soil processes, therefore, have to be drawn with care and on the basis of a detailed targeted process analysis.[15]

18.5 SOIL–WATER INTERACTIONS AT LOW WATER CONTENTS: SUPRAMOLECULAR WATER NETWORKS

The supramolecular structure of SOM plays a central role for soil wettability and for the availability of nutrients, water, and organic chemicals. Release and degradation of compounds can be controlled by physical entrapment in aggregate structures[73] or by kinetic barriers in the SOM matrix.[14,74] Rigid matrices generally slow down the transport of ions and molecules.[75] Among others, glassy phases in natural organic matter are made responsible for sequestration, sorption hysteresis, and increased desorption resistance of organic chemicals.[14,76] Changes in temperature or moisture can soften them,[14,77] which can suddenly increase kinetic availability of any compound entrapped in the matrix. Matrix stiffness can be increased by the presence of multivalent cations and by water molecules, which can form cation bridges (CaB) and water molecule bridges (WaMB), respectively, between molecular segments of organic matter.[11,13] Matrices stiffened by WaMB undergo slow physicochemical aging, in which the redistribution of water molecules increases the number and strength of WaMB.[13,78,79] The stability of WaMB is characterized by the temperature (T^*) at which they are disrupted.[13,80] Such WaMB transitions have been observed in SOM by DSC and thermomechanical analysis (TMA).[79–81] Currently, only DSC is able to give a quantitative information on T^*.[82]

To identify the supramolecular structure of SOM, especially WaMB, on a microscopic level, solid-state wideline NMR spectroscopy seems to be the method of choice. In this respect, not the WaMB itself but their effects on the lineshape after a heating event hint at the existence of WaMB. Treatment of SOM at high temperature (e.g., 110 °C for 30 min) in a closed container is assumed to destroy WaMB. By comparing the wideline spectra before and directly after heating (both measurements at room temperature), an increase in the fraction of the narrow Lorentzian line is observed (Figure 18.5), which could be assigned to mobilizable, now mobile water molecules originating from WaMB with more rigid, motionally restricted water molecules.[16] In this context, it is assumed that other proton-containing material does not change its lineshape because of this heat treatment. These proton-containing materials can be rigid protons of SOM giving rise to a broad Gaussian line and mobile protons from water molecules not in WaMB, showing a Lorentzian line or possibly mobile methyl groups in SOM. The increase in Lorentzian line ratio directly reflects the amount of water in WaMB. Absolute quantification, however, is complicated, as the number of protons is not directly related to the organic content in soil. On the basis of the wideline analysis, the applicability of the WaMB model has been demonstrated on various peats and soils.[13,16,31]

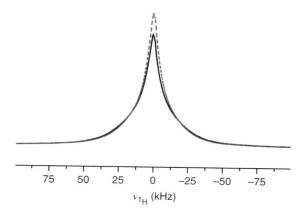

Figure 18.5. ^{1}H wideline NMR spectra of a gleyic podzol soil, recorded before (solid black line) and after (dashed gray line) the heating event. Both spectra were recorded at room temperature. A decrease in broad Gaussian intensity and a concomitant increase in narrow Lorentzian intensity are observed because of the heating event.

In addition, cations can form bridges between functional groups of molecular SOM segments. Despite fundamental theoretical knowledge, the nature of CaB in SOM is still not fully understood.[83] Not only direct CaB but also associations of CaB and WaMB are present in SOM[31,84] (Figure 18.6), especially if functional groups are too far in distance from each other to interact with one cation directly.[83] Involvement of WaMB in CaB is highly relevant, as it will lead to higher dynamics than direct CaB and may even overbalance individual cation effects. It is, therefore, essential to know to which extent direct cross-links, WaMB and CaB control matrix rigidity and surface properties, and how they interact with each other.[82,85]

Wideline spectroscopy opens direct evidence for different mobilities of proton populations. However, it is clear that the assumption of only two mobilities is somewhat too simplified though the simplicity is one of the striking benefits of this method. In an approach to understand the origin of linewidth in SOM, different model compounds have been tested and lead to a more complicated analysis though, in general, does not contradict the simple model.

Another important point is the connection between linewidth and chemical composition. In this respect, NMR offers at least two methods to address this question. The first is to prepare soil samples with D_2O instead of H_2O. If full exchange is possible, and disregarding exchangeable protons at

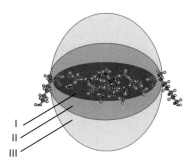

Figure 18.6. Conceptual model of a cation bridge (CaB)–water molecule bridge (WaMB) association between two segments of soil organic matter (SOM) developed from a combined ^{1}H NMR relaxometry and ^{1}H NMR wideline analysis.[31] The dark blue region (I) marks the range in which the central multivalent cation is surrounded by a solid-like supramolecular network of WaMB. The two lighter areas (II and III) highlight regions where water molecules are still influenced by the CaB–WaMB association, but have clearly higher mobility. The latter is indicated by the detectability by low-field ^{1}H NMR relaxometry and relaxation times of 5–30 ms (^{1}H resonance frequency 20 MHz). (Modified with friendly permission from the Institute of Hydrology and Hydromechanics, Ref. 31. © 2013 Walter de Gruyter GmbH)

hydroxyl or carboxyl functionalities in SOM, this enables the separation of lineshapes from water and SOM by measuring two different spectra, that is, both ^{1}H and ^{2}H NMR spectra. While the replacement causes some problems, first results suggest that the narrow Lorentzian line originates purely from water.[16] The second possibility involves the wideline separation (WISE) sequence.[86] Here, ^{1}H mobility can be separated via ^{13}C chemical shifts at slow-spinning MAS. While this has proven to be very efficient for heterogeneous polymer systems, for SOM, this is not straightforward. First, ^{13}C detection often requires a very high number of accumulations, and there are often no significant differences in linewidth for the different components in SOM. Second, water mobility cannot be seen directly, possibly only when efficient spin diffusion is present.

The detection of water-binding signals is probably impossible in air-dried soils, as these materials still contain a significant number of water molecules, so that not isolated water molecules near SOM are present but fast exchange between binding sites and more mobile water molecules is observed.

18.6 CONCLUSIONS AND OUTLOOK

Using low magnetic fields and low resolution NMR spectroscopy proves to be a powerful method to study labile interactions between water and soil constituents, as well as providing insights into the nature of supramolecular structures in SOM. ^{1}H NMR relaxometry is capable to study molecular mobility of water molecules in wet and moist soil samples, and with this, the development of pore systems and gel phases can be used as a simple, fast, and nondestructive method to determine pore size distribution in soil. In combination with thermal analysis and other NMR applications, a detailed study of the state and properties of the enclosed water molecules is possible, which will help to study the formation of biofilms in soil as well as studying water binding in the rhizosphere under the influence of mucilage. ^{1}H NMR wideline analysis gives detailed insights into an even lower level of organization, which is the supramolecular structure of SOM and which links the molecular structure of SOM with its functions. The detection and investigation of WaMB is only one field of application. In contrast to the low field NMR relaxometry, NMR wideline analysis also probes protons in the solid matter. Using D_2O as solvent instead of H_2O opens the possibility to study SOM–water interactions in more detail by analyzing individually the water mobility and identify possible effects on the SOM mobility.

ACKNOWLEDGMENT

This study has been a part of the projects SCHA849/5 and SCHA849/8 funded by the German Research Foundation (DFG).

RELATED ARTICLES IN EMAGRES

Environmental NMR: Solution-state Methods

Environmental NMR: Diffusion Ordered Spectroscopy Methods

Environmental NMR: Magnetic Resonance Imaging

Environmental NMR: Fast Field Cycling Relaxometry

Soil Organic Matter

REFERENCES

1. K. U. Totsche, T. Rennert, M. H. Gerzabek, I. Kögel-Knabner, K. Smalla, M. Spiteller, and H.-J. Vogel, *J. Plant Nutr. Soil Sci.*, 2009, **173**, 88. DOI: 10.1002/jpln.200900105.

2. H. D. Foth, *Fundamentals of Soil Science*, John Wiley & Sons, Inc.: New York, 1990.

3. K. H. Tan, *Environmental Soil Science*, 3rd edn, CRC Press: Boca Raton, 2009.

4. D. Diehl, *Colloid Surf. A Physicochem. Eng. Asp.*, 2013, **432**, 8. DOI: 10.1016/j.colsurfa.2013.05.011.

5. G. E. Schaumann, B. Braun, D. Kirchner, W. Rotard, U. Szewzyk, and E. Grohmann, *Hydrol. Process.*, 2007, **21**, 2276. DOI: 10.1002/hyp.6746.

6. D. Diehl and G. E. Schaumann, *Hydrol. Process.*, 2007, **21**, 2255. DOI: 10.1002/hyp.6745.

7. M.-O. Goebel, J. Bachmann, M. Reichstein, I. A. Janssens, and G. Guggenberger, *Global Change Biol.*, 2011, **17**, 2640. DOI: 10.1111/j.1365-2486.2011.02414.x.

8. E. Gimeno-García, J. A. Pascual, and J. Llovet, *CATENA*, 2011, **85**, 48. DOI: 10.1016/j.catena.2010.12.001.

9. S. H. Doerr, S. H. Shakesby, and R. P. D. Walsh, *Earth Sci. Rev.*, 2000, **51**, 33. DOI: 10.1016/S0012-8252(00)00011-8.

10. S. H. Doerr and C. J. Ritsema, *eMagRes of Hydrological Sciences*, John Wiley & Sons, Ltd: Chichester, 2006.

11. G. E. Schaumann and S. Thiele-Bruhn, *Geoderma*, 2011, **169**, 55. DOI: 10.1016/j.geoderma.2011.04.024.

12. A. J. A. Aquino, D. Tunega, G. E. Schaumann, G. Haberhauer, M. H. Gerzabek, and H. Lischka, *J. Phys. Chem. C*, 2009, **113**, 16468. DOI: 10.1021/jp9054796.

13. G. E. Schaumann and M. Bertmer, *Eur. J. Soil Sci.*, 2008, **59**, 423. DOI: 10.1111/j.1365-2389.2007.00959.x.

14. G. E. Schaumann, *J. Plant Nutr. Soil Sci.*, 2006, **169**, 157. DOI: 10.1002/jpln.200521791.

15. J. V. Bayer, F. Jaeger, and G. E. Schaumann, *Open Magn. Reson. J.*, 2010, **3**, 15. DOI: 10.2174/1874769801003010015.

16. A. Jäger, G. E. Schaumann, and M. Bertmer, *Org. Geochem.*, 2011, **42**, 917. DOI: 10.1016/j.orggeochem.2011.03.021.

17. F. Jaeger, A. Shchegolikhina, H. Van As, and G. E. Schaumann, *Open Magn. Reson. J.*, 2010, **3**, 27. DOI: 10.2174/1874769801003010027.

18. F. Jaeger, S. Bowe, and G. E. Schaumann, *Eur. J. Soil Sci.*, 2009, **60**, 1052. DOI: 10.1111/j.1365-2389.2009.01192.x.

19. G. E. Schaumann, E. Hobley, J. Hurraß, and W. Rotard, *Plant Soil*, 2005, **275**, 1. DOI: 10.1007/s11104-005-1708-7.

20. G. E. Schaumann, *J. Plant Nutr. Soil Sci.*, 2006, **169**, 145. DOI: 10.1002/jpln.200521785.

21. F. Jaeger, E. Grohmann, and G. E. Schaumann, *Plant Soil*, 2006, **280**, 209. DOI: 10.1007/s11104-005-3035-4.

22. K.-J. Dunn, D. J. Bergman, and G. A. Latorraca, *Handbook of Geographical Exploration-Seismic Exploration: Nuclear Magnetic Resonance-Petrophysical and Logging Applications*, Pergamon: Amsterdam, 2002.

23. R. L. Kleinberg, *Methods in the Physics of Porous Media*, Academic Press: San Diego, 1999, 337.

24. R. L. Kleinberg, C. Straley, W. E. Kenyon, R. Akkurt, and S. A. Farooqui, In 68th Annual Technical Conference and Exhibition of the Society of Petroleum Engineers, Houston 1993, 553.

25. R. Kimmich and E. Anoardo, *Prog. Nucl. Magn. Reson. Spectrosc.*, 2004, **44**, 257.

26. R. Kimmich, *NMR: Tomography, Diffusometry, Relaxometry*, Springer: Berlin, 1997.

27. K. R. Brownstein and C. E. Tarr, *Phys. Rev. A*, 1979, **19**, 2446.

28. S. Godefroy, J. P. Korb, M. Fleury, and R. G. Bryant, *Phys. Rev. E Stat. Nonlinear Soft Matter Phys.*, 2001, **64**, 021605/1.

29. S. Davies, A. Hardwick, D. Roberts, K. Spowage, and K. J. Packer, *Magn. Reson. Imaging*, 1994, **12**, 349.

30. T. Baumann, R. Petsch, and R. Niessner, *Environ. Sci. Technol.*, 2000, **34**, 4242.

31. G. E. Schaumann, D. Diehl, M. Bertmer, A. Jaeger, P. Conte, G. Alonzo, and J. Bachmann, *J. Hydrol. Hydromech.*, 2013, **61**, 50. DOI: 10.2478/johh-2013-0007.

32. J. Mao, G. Ding, and B. Xing, *Commun. Soil Sci. Plant Anal.*, 2002, **33**, 1679.

33. P. Chatakananda, L. C. Dickinson, and P. Chinachoti, *J. Agric. Food Chem.*, 2003, **51**, 7445.

34. T. Tsukahara, A. Hibara, Y. Ikeda, and T. Kitamori, *Angew. Chem. Int. Ed.*, 2007, **46**, 1180.

35. A. Vyalikh, T. Emmler, B. Grunberg, Y. Xu, I. Shenderovich, G. H. Findenegg, H. H. Limbach, and G. Buntkowsky, *ZPC*, 2007, **221**, 155. DOI: 10.1524/zpch.2007.221.1.155.

36. R. L. Kleinberg, W. E. Kenyon, and P. P. Mitra, *J. Magn. Reson. Ser. A*, 1994, **108**, 206.

37. T. R. Bryar, C. J. Daughney, and R. J. Knight, *J. Magn. Reson.*, 2000, **142**, 74.

38. F. Jaeger, N. Rudolph, F. Lang, and G. E. Schaumann, *Soil Sci. Soc. Am. J.*, 2008, **72**, 1694. DOI: 10.2136/sssaj2007.0427.

39. T. R. Todoruk, M. Litvina, A. Kantzas, and C. H. Langford, *Environ. Sci. Technol.*, 2003, **37**, 2878.

40. G. E. Schaumann, J. Hurraß, M. Müller, and W. Rotard, *Humic Substances: Nature's most Versatile Materials*, Taylor and Francis, Inc.: New York, 2004, 101.

41. Z. R. Hinedi, Z. J. Kabala, T. H. Skaggs, D. B. Borchardt, R. W. K. Lee, and A. C. Chang, *Water Resour. Res.*, 1993, **29**, 3861.

42. J. Hurraß and G. E. Schaumann, *Soil Sci. Soc. Am. J.*, 2007, **71**, 280. DOI: 10.2136/sssaj2006.0142.

43. V. J. Mcbrierty, G. E. Wardell, C. M. Keely, E. P. O'neill, and M. Prasad, *Soil Sci. Soc. Am. J.*, 1996, **60**, 991.

44. Z. R. Hinedi, A. C. Chang, and M. A. Anderson, *Water Resour. Res.*, 1997, **31**, 2687.

45. Z. R. Hinedi, A. C. Chang, M. A. Anderson, and D. B. Borchardt, *Water Resour. Res.*, 1997, **33**, 2697.

46. N. R. A. Bird, A. R. Preston, E. W. Randall, W. R. Whalley, and A. P. Whitmore, *Eur. J. Soil Sci.*, 2005, **56**, 135. DOI: 10.1111/j.1351-0754.2004.00658.x.

47. H.-K. Liaw, R. Kulkarni, S. Chen, and A. T. Watson, *AIChE J.*, 1996, **42**, 538.

48. F. P. Manalo, A. Kantzas, and C. H. Langford, *Environ. Sci. Technol.*, 2003, **37**, 2701. DOI: 10.1021/es0259685.

49. F. M. Hum and A. Kantzas, *J. Can. Petroleum Technol.*, 2006, **45**, 23.

50. D. Diehl and G. E. Schaumann, *Geophys. Res. Abstr.*, 2005, **7** EGU05-A-00414; SSS8-1FR5P-0160.

51. M. Borisover and E. R. Graber, *Langmuir*, 2002, **18**, 4775.

52. V. J. Mcbrierty, S. J. Martin, and F. E. Karasz, *J. Mol. Liq.*, 1999, **80**, 179.

53. A. Prusova, F. J. Vergeldt, and J. Kucerik, *Carbohydr. Polym.*, 2013, **95**, 515. DOI: 10.1016/j.carbpol.2013.03.031.

54. S. Jähnert, F. Vaca Chavez, G. E. Schaumann, A. Schreiber, M. Schönhoff, and G. H. Findenegg, *Phys. Chem. Chem. Phys.*, 2008, **39**, 6039. DOI: 10.1039/b809438c.

55. M. Holz, S. R. Heila, and A. Saccob, *Phys. Chem. Chem. Phys.*, 2000, **2**, 4740.

56. P. P. Mitra, P. N. Sen, and L. M. Schwartz, *Phys. Rev. B*, 1993, **47**, 8565.

57. S. Radosta and F. Schierbaum, *Starch Stärke*, 1990, **42**, 142.

58. M. S. Jhon and J. D. Andrade, *J. Biomed. Mater. Res.*, 1973, **7**, 509.

59. P. Puget, D. A. Angers, and C. Chenu, *Soil Biol. Biochem.*, 1999, **31**, 55.

60. H.-C. Flemming and J. Wingender, *Vom Wasser*, 2000, **94**, 245.

61. H. C. Flemming, Forces that keep biofilms together. (1996). In: *DECHEMA Monographien: Biodeterioration and Biodegradation*. Verlag Chemie, Weinheim, 1996, 311.

62. J. Wingender, H. C. Flemming, and T. R. Neu, *Microbial Extracellular Polymeric Substances*, Springer: Berlin, 1999, 1.

63. E. J. Bouwer, H. H. M. Rijnaarts, A. B. Cunningham, and R. Gerlach, in *Biofilms II Process Analysis and Applications*, ed J. D. Bryers, John Wiley & Sons, Inc: New York, 2000, 123.

64. I. M. Young, J. W. Crawford, N. Nunan, W. Otten, and A. Spiers, *Advances in Agronomy*, Elsevier Academic Press Inc: San Diego, 2008, Vol. 100, 81.

65. R. Nazir, J. A. Warmink, H. Boersma, and J. D. Van Elsas, *Fems Microbiol. Ecol.*, 2010, **71**, 169. DOI: 10.1111/j.1574-6941.2009.00807.x.

66. B. C. Hoskins, L. Fevang, P. D. Majors, M. M. Sharma, and G. Georgiou, *J. Magn. Reson.*, 1999, **139**, 67.

67. P. N. L. Lens and H. Van As, *Biofilms in Medicine, Industry and Environmental Biotechnology: Characteristics, Analysis and Control*, IWA Publishing: London, UK, 2003, 285.

68. S. Radosta, F. Schierbaum, and W. P. Yuriev, *Starch Stärke*, 1989, **41**, 428.

69. P. Kinchesh, A. A. Samoilenko, A. R. Preston, and E. W. Randall, *J. Environ. Qual.*, 2002, **31**, 494.

70. B. Bluemich, F. Casanova, J. Perlo, S. Anferova, V. Anferov, K. Kremer, N. Goga, K. Kupferschläger, and M. Adams, *Magn. Reson. Imaging*, 2005, **23**, 197.

71. H. Kühn, M. Klein, A. Wiesmath, D. E. Demco, B. Blümich, J. Kelm, and P. W. Gold, *Magn. Reson. Imaging*, 2001, **19**, 497.

72. S. Anferova, E. Talnishnikh, V. Anferov, B. Blümich, J. Arnold, H. Pape, and C. Clauser, *Magn. Reson. Imaging*, 2007, **25**, 547.

73. I. Virto, C. Moni, C. Swanston, and C. Chenu, *Geoderma*, 2011, **156**, 1. DOI: 10.1016/j.geoderma.2009.12.028.

74. G. E. Schaumann, C. Siewert, and B. Marschner, *J. Plant Nutr. Soil Sci.*, 2000, **163**, 1. DOI: 10.1002/(SICI)1522-2624(200002)163:1<1::AID-JPLN1>3.0.CO;2-N.

75. A. I. Suvorova, I. S. Tjukova, and E. I. Trufanova, *J. Environ. Polym. Degrad.*, 1999, **7**, 35.

76. E. J. Leboeuf and W. J. Weber Jr, *Environ. Sci. Technol.*, 2000, **34**, 3632.

77. E. J. Leboeuf and W. J. Weber, *Environ. Sci. Technol.*, 2000, **34**, 3623.

78. G. E. Schaumann, *Colloids Surf. A: Physicochem. Eng. Aspects*, 2005, **265**, 163. DOI: 10.1016/j.colsurfa.2005.02.040.

79. J. Hurraß and G. E. Schaumann, *Geochim. Cosmochim. Acta*, 2007, **71**, 691. DOI: 10.1016/j.gca.2006.09.022.

80. G. E. Schaumann and E. J. Leboeuf, *Environ. Sci. Technol.*, 2005, **39**, 800. DOI: 10.1021/es0490931.

81. G. E. Schaumann, E. J. Leboeuf, R. C. Delapp, and J. Hurraß, *Thermochimica Acta*, 2005, **436**, 83. DOI: 10.1016/j.tca.2005.07.009.

82. G. E. Schaumann, D. Gildemeister, D. Diehl, Y. K. Mouvenchery, and S. Spielvogel. *J. Soil Sediments* 2013, DOI: 10.1007/s11368-013-0746-7; DOI: 10.1007/s11368-013-0746-7.

83. Y. Kunhi Mouvenchery, J. Kučerík, D. Diehl, and G. E. Schaumann, *Rev. Environ. Sci. Biotechnol.*, 2012, **11**, 41. DOI: 10.1007/s11157-011-9258-3.

84. A. J. A. Aquino, D. Tunega, G. E. Schaumann, G. Haberhauer, M. H. Gerzabek, and H. Lischka, *Int. J. Quantum Chem.*, 2011, **111**, 1531. DOI: 10.1002/qua.22693.

85. Y. Kunhi Mouvenchery, A. Jaeger, A. J. A. Aquino, D. Tunega, D. Diehl, M. Bertmer, and G. E. Schaumann, *PLoS ONE*, 2013, **8**, e65359. DOI: 10.1371/journal.pone.0065359.

86. K. Schmidt-Rohr, J. Clauss, and H. W. Spiess, *Macromolecules*, 1992, **25**, 3273. DOI: 10.1021/ma00038a037.

FURTHER READINGS

G. E. Schaumann, D. Diehl, M. Bertmer, A. Jaeger, P. Conte, G. Alonzo, and J. Bachmann, (2013): Combined proton NMR wideline and NMR relaxometry to study SOM-water interactions of cation-treated soils. Journal of Hydrology and Hydromechanics 61(1), 50–63.

A. Jäger, G. E. Schaumann, and M. Bertmer, (2011): Optimized NMR spectroscopic strategy to characterize water dynamics in soil samples. Organic Geochemistry 42(8), 917–925.

J. V. Bayer, F. Jaeger, and G. E. Schaumann, (2010): Proton nuclear magnetic resonance (NMR) relaxometry in soil science applications. The Open Magnetic Resonance Journal 3, 15–26.

F. Jaeger, A. Shchegolikhina, H. van As, and G. E. Schaumann, (2010): Proton NMR relaxometry as a useful tool to evaluate swelling processes in peat soils. The Open Magnetic Resonance Journal 3, 27–45.

F. Jaeger, S. Bowe, and G. E. Schaumann, (2009): Evaluation of ^{1}H NMR relaxometry for the assessment of pore size distribution in soil samples. European Journal of Soil Science 60(6), 1052–1064.

F. Jaeger, N. Rudolph, F. Lang, and G. E. Schaumann, (2008): Effects of soil solution's constituents on proton NMR relaxometry of soil samples. Soil Science Society of America Journal 72(12), 1694–1707.

J. Hurraß, and G. E. Schaumann, (2007): Hydration kinetics of wettable and water repellent soil samples. Soil Science Society of America Journal 71(2), 280–288.

F. Jaeger, E. Grohmann, and G. E. Schaumann, (2006): ^{1}H NMR relaxometry in natural humous soil samples: Insights in microbial effects on relaxation time distributions. Plant and Soil 280(1–2), 209–222.

G. E. Schaumann, E. Hobley, J. Hurraß, and W. Rotard, (2005): H-NMR relaxometry to monitor wetting and swelling kinetics in high organic matter soils. Plant and Soil 275(1–2), 1–20.

Chapter 19
Metals in the Environment

Andre Sutrisno and André J. Simpson

Department of Chemistry and Environmental NMR Centre, University of Toronto, Toronto, Ontario, M1C 1A4, Canada

19.1 INTRODUCTION

Heavy metals are one of the most problematic classes of contaminants because of their high toxicity at low concentrations, abundance, and accumulation by plants and animals.[1,2] They constitute a large group of inorganic chemicals that are hazardous such as arsenic (As), cadmium (Cd), chromium (Cr), copper (Cu), mercury (Hg), manganese (Mn), nickel (Ni), lead (Pb), selenium (Se), and zinc (Zn). Unlike organic contaminants, most heavy metals do not undergo biological or chemical degradation.[3] Hence, they tend to accumulate in solids and sediments, posing risks and health hazards through direct contact (e.g., absorption via the skin), inhalation, or ingestion of contaminated soil (and vegetation grown therein), water, and air.[4]

In addition to their bulk metal counterparts, quantum dots and nanoparticles are now also emerging as major environmental concerns.[5–7] Their applications include electronics, solar energy generation, biomedical imaging, and nanomedicine tools. The unique physical and chemical properties of nanoparticles, having by far the greatest surface area to volume ratio, lead to very different reactivity compared to their bulk counterparts. Quantum dots typically consist of a metalloid crystalline core (e.g., CdTe or CdSe), shielded by a shell coating (e.g., ZnS) that makes them bioavailable.[8] By tailoring their coatings, quantum dots can have highly specific bioactivities making them useful for targeted sorption. Adversely, however, if for some reason the metalloid cores are revealed, they may exhibit considerable toxicity for organisms in the environment.

As heavy metals are of widespread environmental concern, there are numerous methods employed to try to remove and recover metals from our environment. These may include redox reactions, separations/extractions, adsorption, and ion exchange. Interested readers are encouraged to refer to reviews on the topic,[1–4,6] as remediation of heavy metal-contaminated sites is beyond the scope of this chapter. Understanding the chemistry of a particular metal and its interaction with various environment matrices is of considerable importance to remediation and risk assessment. Metal speciation (i.e., the occurrence of a metal in a variety of different chemical forms: free metal ions, metal complexes dissolved in solution and/or sorbed onto solid surfaces, and metal

NMR Spectroscopy: A Versatile Tool for Environmental Research
Edited by Myrna J. Simpson and André J. Simpson

species in solids) greatly determines the behavior and toxicity of metals in the environment. Hence, metal complexes play an important role in controlling the availability and fate of metals in environment, as they generally increase the solubility and mobility of metals in surface and groundwaters.

As previously mentioned, it is very important to understand where and how metal contaminants bind in soils, sediments, and living species. Such insights will improve our understanding of their bioavailability and how to best proceed with remediation. Much of the environmental research has focused on trying to explain the uptake, distribution, excretion, and bioaccumulation of these heavy metals, in order to predict more subtle long-term effects of exposure in living species.[2,9] Accurate and sensitive analytical tools are required, such that one can monitor metal contaminants from the source, through an environment (e.g., soil/sediment), and into a living organism at the molecular level. Despite being a challenging task, such studies are required to fully understand and predict true environmental fate, toxicity, and long-term risk.

This chapter aims to provide a brief overview on the application of direct metal NMR spectroscopy to environmental samples. It is by no means to be comprehensive, but hopefully, will act a starting point for scientists exploring the field. For more information and reviews on NMR studies of environmental systems, the readers are encouraged to refer to these chapters[10–12] (also see Chapters 11, 12, 13 and 17).

19.2 GENERAL METAL NMR SPECTROSCOPY

NMR spectroscopy is arguably one of the single most powerful analytical tools for the analysis of structure and dynamics at the molecular level, as it probes short-range ordering (as opposed to long-range ordering in diffraction methods) and local structure around the nucleus of interest. Nearly, all elements in the periodic table have at least one or more NMR-active nucleus, which means that most materials can be studied directly by some form of NMR spectroscopy (refer to Table 19.1 for the list of the nuclear properties of all the NMR-active nuclei of most common heavy metal contaminants[13,14]).

As illustrated in Table 19.1, some of these heavy metal nuclei have spin number of $1/2$, and they often have large chemical shift anisotropies and correspondingly broad powder patterns in solids. In solution NMR spectroscopy, the fast and isotropic molecular tumbling averages out most of the anisotropic NMR interactions, leaving only chemical shift information. Cross polarization magic-angle spinning (CPMAS) NMR experiments with high-power proton decoupling are common for acquiring high-resolution NMR spectra of solids.[15–17] In cases where they are limited by weak dipolar couplings and poor excitation bandwidth, Carr-Purcell-Meiboom-Gill (CPMG) experiments can provide another alternative, as they were reintroduced for the piecewise acquisition of wideline NMR spectra.[18,19] They have been particularly useful

Table 19.1. Nuclear properties of NMR-active nuclei for common heavy metal contaminants

Heavy metal	NMR-active nucleus	Spin (*I*)	Natural abundance (%)	*Q* (millibarn)	NMR frequency at 11.75 T (MHz)	R^C (receptivity relative to ^{13}C)
Arsenic	^{75}As	3/2	100.0	314	85.6	149
Cadmium	^{111}Cd	1/2	12.80	—	106.1	7.27
	^{113}Cd	1/2	12.22	—	111.0	7.94
Chromium	^{53}Cr	3/2	9.50	−150	28.3	0.51
Copper	^{63}Cu	3/2	69.15	−220	132.6	382
	^{65}Cu	3/2	30.85	−204	142.1	208
Mercury	^{199}Hg	1/2	16.87	—	89.6	5.89
	^{201}Hg	3/2	13.18	387	33.1	1.16
Manganese	^{55}Mn	5/2	100.0	330	124.0	1050
Nickel	^{61}Ni	3/2	1.14	162	44.7	0.24
Lead	^{207}Pb	1/2	22.10	—	104.6	11.8
Selenium	^{77}Se	1/2	7.63	—	95.4	3.15
Zinc	^{67}Zn	5/2	4.10	150	31.3	0.69

All values were taken from Bruker "NMR Frequency Table" (www.bruker-biospin.com), except for the Q values, that were adapted from Ref. 26.

in enhancing the signal-to-noise (S/N) ratios and reducing the total experimental times. For very broad patterns (>300–400 kHz), wide, uniform rate, and smooth truncation (WURST)–CPMG experiments have been shown to further reduce acquisition times because of their larger excitation bandwidths compared to standard high-power rectangular pulses, as one would not need to acquire as many subspectra as they would have to use rectangular pulses.[14,20]

Aside from spin $1/2$ nuclei, the majority of the other NMR-active heavy metals are quadrupolar nuclei ($I > 1/2$), which means that the nuclear charge distribution within the nuclei is asymmetric (as opposed to spherical charge distribution in spin $1/2$). Excellent reviews on NMR spectroscopy of quadrupolar nuclei can be found here.[21,22] The acquisition of NMR spectra of quadrupolar nuclei can be challenging, as the S/N ratios of such spectra are generally low (because the total integrated signal intensity is usually spread over very broad regions) and often require large, uniform excitation bandwidths. Furthermore, environmental samples generally have very low metal concentration (see also Chapters 1, 4, 5 and 6 for different methods used in environmental research). Consequently, metal NMR studies of these heavy metal nuclei have not been used routinely to directly characterize many important environmental samples such as natural organic matter (NOM), soils, plants, rocks, and clays.

As mentioned previously, to determine the best strategy for remediation, a good understanding of how and where the metal binds or interacts with the systems (i.e., protein, lignin, carbohydrate, aliphatics, or minerals in soils) is required.[23–25] In general, solution-state NMR studies can characterize the metal-binding sites in soil or organic matter, and monitor the changes in pH and concentration, as the chemical shift is very sensitive to these changes. Furthermore, NMR parameters such as spin–lattice (T_1, or longitudinal) and spin–spin (T_2, or transverse) relaxation times provide information regarding structural characteristics and molecular dynamics (i.e., correlation times and reorientational motion).[27] If the relative mobility of each species is known, then this could be very useful in understanding the fate and transport of metals in soil. Solid-state NMR experiments could potentially be used to better understand the sequestration in soils by monitoring the changes of specific functional groups both quantitatively and qualitatively (see Section 19.4).

Despite the fact that some information can be obtained indirectly by acquiring the more common ^{1}H,

^{13}C, ^{31}P, and ^{15}N NMR spectra (interested readers are referred to *excellent reviews* by Preston,[28] Lens and Hemminga,[29] Larive *et al.*,[30] and Simpson *et al.*[23,31]), direct observation of metal nuclei is a more sensitive approach for probing the local environment around the metal-binding sites. To the best of our knowledge, most of the metal NMR studies on environmentally relevant samples have focused mainly on ^{113}Cd,[32–51] both in solution and solid, with some exceptions that include ^{51}V[52] and ^{207}Pb.[53] Other exchangeable metal cations (such as ^{7}Li, ^{23}Na, ^{87}Rb, and ^{133}Cs) or transition metal (e.g., ^{27}Al) have also been utilized.

The next section discusses several literature examples regarding the applications of metal NMR spectroscopy to environmental samples. There are many complimentary reports in the literature, as such only a few selected examples will be discussed in detail that demonstrate a variety of different metal-based NMR approaches.

19.3 SELECTED LITERATURE EXAMPLES

As mentioned earlier in this chapter, ^{113}Cd NMR spectroscopy is by far the most commonly used approach to directly probe the metal sites in environmental research.[32–51] The ability of cadmium to form complexes with many different conformations and ligand numbers is the main impetus for using it as a model metal. A broad chemical shift range (~900 ppm) of ^{113}Cd also results in a high sensitivity of the chemical shifts because of small variations in the local chemical environments around the cadmium nuclei. These chemical shifts have been shown to be dependent on the nature, number, and geometry of the coordinating atoms.[54] In general, any ligands that are bound or coordinated to the Cd through oxygen donor atoms will increase its shielding (-180 ppm < δ_{iso} < 100 ppm). On the other hand, nitrogen-containing ligands tends to decrease the shielding (30 ppm < δ_{iso} < 300 ppm), whereas sulfur donor atoms further decrease it (400 ppm < δ_{iso} < 800 ppm).

19.3.1 Characterization of Different Metal-binding Sites, Peak Assignments, and CS Tensor Analysis

Solid-state ^{113}Cd NMR spectroscopy has been used to characterize the metal-binding sites in Cd-exchanged montmorillonites.[32,34,42,43] Montmorillonite has a crystalline-layered clay structure with a high surface area and a high internal negative charge that attracts hydrated cations. Metal ions such as Cd^{2+} can either substitute for Si or Al sites within the sheets or intercalate between the layers and occupy edge positions.

Bank *et al.*[32] reported the first solid-state ^{113}Cd NMR study of Cd-exchanged montmorillonite samples. They showed that the ^{113}Cd MAS NMR spectra contained different Cd oxoanion sites with distinct chemical shifts and linewidths, mainly because of metal ion dynamics and paramagnetic effects. The observed linewidths in both MAS and static spectra within the same sample are comparable; this lack of linewidth-narrowing effect of the MAS spectra indicates that the broad features are governed by homogeneous (dynamic) process. This is further confirmed by the echo experiments in which the spectra are dependent on the various echo delays. At least, two types of ion motion can be identified, which included an interlayer ion motion of intermediate rate (broad component) and a faster motion at sheet or edge sites (narrow component). Paramagnetic broadening due to Fe metal or Fe^{3+} ion in the octahedral layers is very common (though the iron can be removed); thus, the spectra are very sensitive to sample preparation. For this reason, the authors also looked at the effects of pH variation, exchanged clay form, pretreatment with alginic and humic acids, and air- and freeze-drying. Air-drying, oven-drying, hydration, and dithionite treatment produce essentially no change in the spectra, whereas pH variation (Figure 19.1), pretreatment with alginic and humic acids, and freeze-drying have distinct or observable changes. In summary, this study provides some insight to the migration of metal contaminants in environmental samples. The results indicate that ion mobility in clays is independent of the anion or the solubility of the metal salt.

Tinet *et al.*[34] built on this work via chemical shift tensor analysis on oriented samples of Cd-exchanged montmorillonite and concluded that the observed ^{113}Cd NMR signal is not necessarily due to multiple Cd sites. The experimental line shape was attributed to the interaction between interlayer Cd and hydroxyls in the hexagonal cavities. It is also possible to differentiate between Cd exchanged outside and inside the clay layers in the case of clay gels by varying the level of water in the sample. This is because the interactions between Cd with water and Cd and clay are sufficiently different. Finally, NMR analysis can

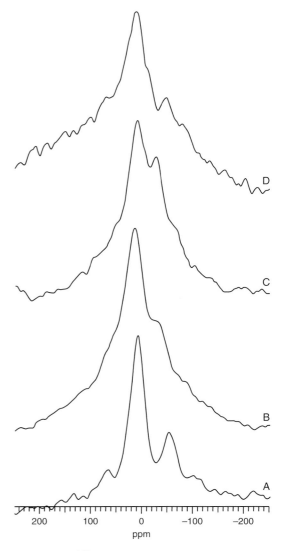

Figure 19.1. ^{113}Cd MAS NMR spectra of Cd absorbed on montmorillonite at pH 3 (A), 5 (B), 7 (C), and 8 (D). (Reprinted with permission from Ref. 32. © 1989 American Chemical Society)

provide an estimate of the number of clay sheets involved in the formation of clay tactoids (stacks of parallel, equally spaced clay platelets at ~10 Å separations).

Additional solid-state ^{113}Cd NMR studies on Cd-exchanged montmorillonite were conducted by Di Leo[42,43] in 1999 and 2000. They found that Cd^{2+}

ions are localized in two different sites: (i) in the interlayer sites as hydrated Cd^{2+} and $CdCl^+$ ions and (ii) on the external surface sites as edge Cd^{2+} with only a few H_2O molecules hydrating it. The two monolayers of adsorbed H_2O present in the interlayer prevent motional averaging between Cd^{2+} and $CdCl^+$ environments, hence two resonances depicting two unique interlayer species.

19.3.2 Chemical Shift Changes with Different Cd Concentration and pH

Solution-state ^{113}Cd NMR spectroscopy also offers great means for characterizing humic substances binding sites by examining the reactivity of humic substances and Cd metal ions.[37–39,41] Humic substances are heterogeneous mixtures of decomposition products of plant and animal residues, which are ubiquitous in soils and water. Fulvic acid (FA) is a fraction of humic substances that are soluble at all pH values in aqueous solvents. All the early ^{113}Cd NMR observations with humic substances are fairly consistent, implying that oxygenated structures (via the carboxylate ligand) are the predominant binding groups in FA for Cd^{2+}.

Figure 19.2 illustrates how ^{113}Cd NMR chemical shift changes with varying Cd concentration bound to a soil FA.[38] The apparent difference in the chemical shift may be caused by the presence of different functional groups of FA, such as carboxylate, phenolic −OH, alcoholic −OH, and minor amounts of N, hence creating a multiligand environment for the Cd^{2+} ions. The single resonance observed reflects the rapidly exchanging conditions in the aqueous solution between the free and the bound Cd species. This is consistent with the fast exchange model, where the observed peak is just the average of the two species. Such fast exchange contrasts sharply with many ^{113}Cd NMR studies of ligand (small molecules) binding to protein and humic acid[37] where two distinct peaks are usually observed.

The fact that such a fast exchange occurs in FA suggests that a competitive binding study with any metal ions (such as Ca^{2+}) could yield additional equilibrium constants. Indeed, Otto *et al.*[46] determined an average association equilibrium constant for Cd (K_{Cd}) from NMR spectra measured for the titration of FA with Cd^{2+}, and ranges between 1.2 and 3.5×10^3 M^{-1}. Competitive binding between Ca^{2+} and Cd^{2+} was then used to indirectly calculate the equilibrium constant for Ca (K_{Ca}), and the values range from 4.6 to 7.8 $\times$ 10^2 M^{-1}.

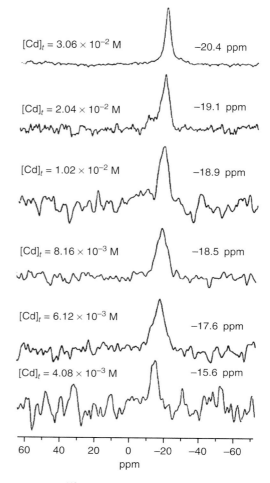

Figure 19.2. ^{113}Cd NMR spectra of varying Cd^{2+} concentration bound on fulvic acid (FA) in aqueous solution: $[FA]_t = 0.24$ equiv. mol^{-1} and pH = 6. (Reproduced with permission from Ref. 38. © 1996 NRC Research Press)

^{113}Cd NMR was also used to examine the role of pH on the complexation of Suwannee River NOM.[41] The study provides direct evidence regarding the nature of Cd-binding sites in NOM. At pH 3–9 range, Cd^{2+} is primarily complexed via O donor (−COOH), whereas, at alkaline pH values, Cd^{2+} ions are also present in N atom-containing binding sites. The Cd–NOM exchange rates are also affected by the concentration of free Cd^{2+} ions. At acidic pH, Cd–NOM exchange rates are relatively fast because of higher concentration of free Cd^{2+} ions, indicating only weak binding to the organic matter, hence only a single weighted-average peak is observed in the NMR spectra. On the other hand, at alkaline pH, slower exchange rate caused by a stronger interaction results in broader, more complex observed spectra that contain poorly resolved peaks.

A recent and systematic study by Hertkorn *et al.*[48,50] examined Cd complexation to NOM using [113]Cd NMR spectroscopy at two different magnetic field strengths. They investigated the binding of Cd to Suwannee River NOM over a broad range of Cd/C ratios (0.00056–0.0056) and pH values (3.5–11). Figure 19.3 reveals several general trends: as pH is increased, (i) the linewidth of the NMR signal increases steadily; (ii) at pH range of 4–7, the observed chemical shift becomes smaller and smaller

until it reaches the smallest at neutral pH and then, at pH 7–10, the shift becomes larger again; and (iii) line shape gradually deviates from a Lorentzian shape. In addition, at any given pH, the linewidth also increases as the Cd/C ratio decreases. Two general effects, slower chemical exchange rate at lower Cd ion concentration and increased heterogeneity of exchanging species at higher pH, are believed to be responsible for the observed line broadening. The NMR data suggest that (i) oxygen-derived functional groups (aliphatic and aromatic carbonyl and hydroxyl groups) are the prevailing ligands in which Cd is bound to NOM and (ii) at higher pH and lower Cd/C ratio, Cd^{2+} preferentially binds to N-donor ligands.

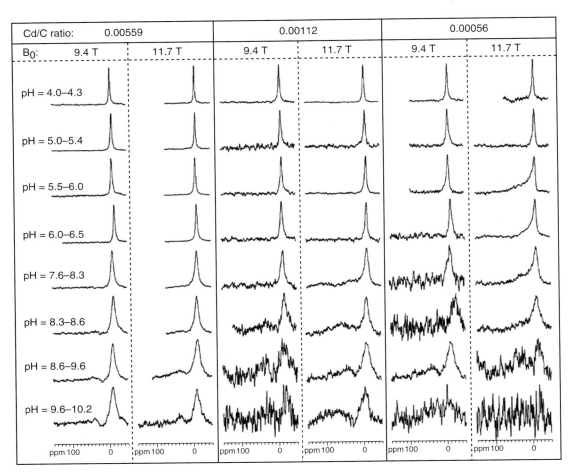

Figure 19.3. Selected [113]Cd NMR spectra at 303 K of [113]Cd(ClO$_4$)$_2$ bound to Suwannee River organic matter at various Cd/C ratios, [113]Cd NMR frequencies, and pH. (Reprinted with permission from Ref. 48. © 2004 American Chemical Society)

The effect of temperature on observed chemical shift and linewidth was also briefly examined.[48,50] As temperature increases, the observed chemical shift does not significantly change; however, the linewidth decreases significantly. The authors concluded at the end of this chapter that ^{113}Cd NMR spectra of cadmium ions interacting with NOM will always reflect a superposition of slow, intermediate, and fast exchange characteristics. However, the relative proportion of slow chemical exchange generally increases with increasing pH, higher B_0, decreasing Cd/C ratio, and lower temperature. Hence, an NMR analysis that is solely based on chemical shift information should be avoided.

19.3.3 Relaxation Measurement

NMR parameters, such as T_1 and T_2 relaxation times, provide information regarding structural characteristics and molecular dynamics (correlation times) in solution.[27] The complexation of Cd^{2+} by Suwannee River FA was investigated using ^{113}Cd NMR relaxation measurements.[45] A series of ^{113}Cd inversion-recovery experiments (Figure 19.4) were performed in order to determine the T_1 relaxation times, whereas standard spin–echo experiments were utilized to determine the T_2 relaxation times. They found that the ratio between two relaxation rates (R_2/R_1 or T_1/T_2) increases with the increasing rotational correlation time (τ_c).

These ratios, with the aid of model ligands, can then be used to map relative motion of a given molecule or molecular complex. Four model ligands, (i) acetate (single carboxylate), (ii) salicylate (bidentate carboxylate), (iii) nitrilotriacetate (carboxylate with N donor), and (iv) cyclopentanetetracarboxylic acid (polydentate carboxylate) were used to represent a series of different carboxylate-binding scenarios. The relaxation rates for all these model ligands were then measured to provide a more complete understanding of the effect of different binding motifs on Cd relaxation rates.

In summary, the ^{113}Cd NMR relaxation measurements data showed that there are two distinct types of Cd binding sites on FA: first, a strong Cd^{2+} binding site per molecule that is mainly due to a polydentate carboxylate ligand or its derivatives (this corresponds to its fast relaxation rate arising from efficient mechanisms provided by sterically hindered binding site); and second, a weaker binding site that relaxes slower.

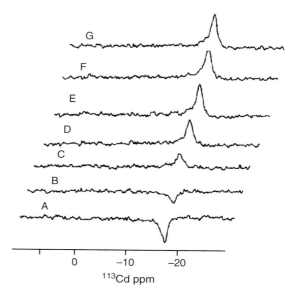

Figure 19.4. ^{113}Cd inversion-recovery spectra measured for a solution containing 3 mg mL Suwannee River fulvic acid and 2 mM ^{113}Cd(II). The relaxation delays (in ms) are (A) 0.001, (B) 50, (C) 250, (D) 450, (E) 650, (F) 900, and (G) 1400. (Reprinted with permission from Ref. 45. © 2001 American Chemical Society)

19.3.4 Diffusion

Diffusion ordered spectroscopy (DOSY) NMR can be utilized to examine the chemical and physical properties of a complex mixture (see also Chapter 2). A series of spin–echo spectra is measured with different PFG strengths, with the goal to separate the NMR signals of different species according to their diffusion coefficients. It is often being referred to as *NMR chromatography*, because it takes away the need for physical separation of the sample (e.g., chromatography and recrystallization).

The study of cadmium binding to an oak forest soil NOM sample using ^{113}Cd DOSY NMR was reported by Simpson.[47] The average downfield shift and a considerable broadening of the ^{113}Cd resonance from the water-soluble forest soil fulvic acid (WS-FSFA) sample compared to the reference CdCl$_2$ result primarily from the interaction between Cd and oxygen-containing functional groups (Figure 19.5a and b). ^{113}Cd DOSY NMR experiments confirm this observation. Figure 19.5d illustrates the ^{113}Cd DOSY spectrum of CdCl$_2$ after the addition of WS-FSFA.

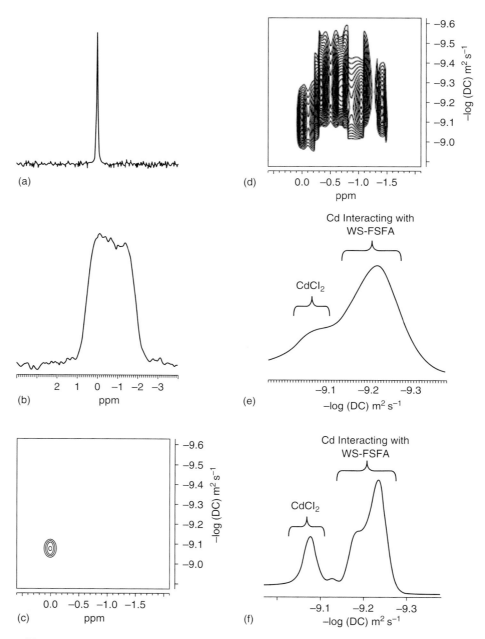

Figure 19.5. [113]Cd NMR spectra of CdCl$_2$ at a concentration of 50 mg ml^{-1} metal ions (a, 8 scans); after the addition of 5 mg of WS-FSFA (b, 2048 scans); DOSY of CDCl$_2$ (linewidth factor 1) (c); DOSY after the addition of the WS-FSFA (linewidth factor 2.5) (d); F_1 diffusion projection (e); and F_1 diffusion projection with linewidth factor 0.3 (f). (Reproduced with permission from Ref. 47. © 2002 Wiley)

The signal from free $CdCl_2$ at 0 ppm remains, but signal from Cd interacting with the WS-FSFA (-0.5 to -1.5 ppm) exhibits diffusivities between that of free $CdCl_2$ (log DC $\sim$9.1) and of WS-FSFA (log DC $\sim$9.5). This result supports the fast exchange model proposed earlier by Moon *et al.*[38] and the signal reflects the equilibrium between free and bound states. Such findings imply that although Cd has a strong affinity for NOM, it binds reversibly and may be mobile under different environmental conditions.

19.4 POTENTIAL EXPERIMENTS

NMR spectroscopy, although incredibly powerful, is still an inherently insensitive analytical technique, and this is especially true for most heavy metal nuclei. Thus, almost all technical advances in NMR spectroscopy have always been revolved around increasing the S/N ratio. In general, a few important parameters that affect the sensitivity of NMR signals are as follows: (i) static magnetic field (B_0) and the gyromagnetic ratio of the nuclei of interest, (ii) sample concentration and natural abundance of the nuclei of interest, (iii) temperature, and (iv) pulse sequence design. In the last couple of decades, a number of technologies, such as cryogenically cooled NMR probes, microcoil NMR probes, and DNP, show promise to improve the detection limits and/or reduce experimental time and sample size.

19.4.1 Cryogenic and Microcoil Probes

The use of cryoprobes (cryogenically cooled probes, mostly for solution-state NMR spectroscopy) has great potential in environmental research for sensitivity enhancement, as the population difference between spin levels gets larger at lower temperature. In theory, a three- to fourfold sensitivity increase can be achieved using cryogenic probes, which means up to 9–16-fold of reduction in experiment time.[55] They have been found useful in applications such as biomolecular NMR, NMR screening and ligand binding, metabonomics, and structure verification of low-molecular-weight compounds where the sample concentration is often very low. However, the main drawback of a cryoprobe system is likely the high associate maintenance cost because of the use of helium compressors and cryogens.

Microcoil probes (coils of inner diameter <1.0 mm) are much less expensive than cryoprobes. They have also been shown to significantly improve S/N ratios. Smaller coil diameter generally results in a more homogeneous and much larger RF fields for excitation, albeit with smaller sample volumes.[56] In addition, the S/N ratio of microcoils with a fixed length-to-diameter ratio improves as the coil decreases (i.e., the smaller the coil is, the greater will be the transverse magnetic field strength per unit volume). The application of microcoils to directly observe metal nucleus (^{91}Zr) in inorganic salts has been reported.[57] However, to the extent of our knowledge, there has not been any reports on direct metal NMR studies on environmentally related samples using microcoils.

19.4.2 Polarization Enhancement Using Dynamic Nuclear Polarization (DNP) and Optics

While cryoprobes and microcoils are great in increasing NMR sensitivity, DNP likely holds the greatest potential for environmental research. DNP involves polarization transfer from unpaired electron to the nuclei of interest. The maximum signal enhancement, in theory, is proportional to the gyromagnetic ratio between electron and polarized nucleus ($\sim$650 in the case of ^{1}H). In terms of experimental approach, most of the modern DNP experiments are now performed at high-field ($\sim$3–5 T), high-power ($\sim$10 W) microwave (terahertz/gigahertz) source, low temperature (<90 K) with the use of sapphire rotors (rather than zirconia rotors, in order to permit light to penetrate through the rotor wall), and suitable polarizing agent (e.g., TOTAPOL, TEMPO, TRITYL, BDPA, and BTnE).[58,59] A few relevant and recent examples include ^{27}Al[60] (as an example of a metal) and ^{17}O[61] (as an example of nucleus with very low inherent sensitivity). Figure 19.6 shows ^{27}Al MAS spectra of mesoporous alumina at 103 K ($\sim$15$\times$ enhancement with DNP),[60] whereas ^{17}O detected DNP NMR spectra of water/glycerol/glass enabled $\sim$80$\times$ enhancement (Figure 19.7).[61]

Optics, such as laser, is one of other possible ways to increase sensitivity. A few key advantages include sensitivity (up to 10 orders of magnitude), selectivity (only certain region in space, such as surfaces), and speed (rather an instantaneous process that depends only on laser intensity and not on temperature). Although it has not been used much, a ^{113}Cd NMR study of CdS single crystal irradiated with circularly polarized light

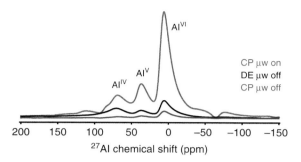

Figure 19.6. ^{27}Al MAS NMR of mesoporous alumina sample at ~103 K. A DNP enhancement of ~15 was measured by comparing the signal intensities of CPMAS experiments with (red line) and without (blue line) microwave radiation. (Reprinted with permission from Ref. 60. © 2012 American Chemical Society)

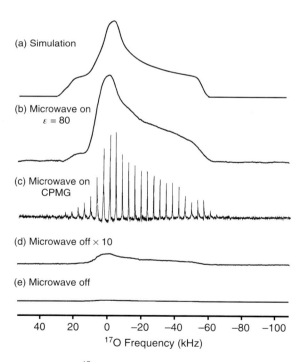

Figure 19.7. ^{17}O static CP NMR spectra of glycerol/D$_2$O/H$_2$O at 82 K. Comparing (b) (microwave on, 128 scans) and (e) (microwave off, 4864 scans), DNP provides a signal enhancement of ~80. (Reprinted with permission from Ref. 61. © 2012 American Chemical Society)

(at 2.60 eV for 600 s) was reported,[62] and has shown to be able to provide ~4× signal enhancement.

19.4.3 Indirect Methods

In addition to the direct observation methods described earlier, another alternative to study the metal-binding interactions is through the use of indirect methods. To our knowledge, indirect methods have yet to be applied to understand interactions of metals in an environmental context. In light of this, brief examples relating to the interaction of organic contaminants are provided in order demonstrate the future potential of the experiments for metal studies.

Saturation transfer difference (STD) is an experiment based on different cross-relaxation mechanisms and can be used to identify which organic components specifically interact or bind to the various systems (e.g., soil). In simple terms, a heteronucleus is saturated and this saturation propagates onto any surrounding organic moieties in close proximity. Most commonly, the signals from ^{1}H nuclei are observed. This approach has shown great promise over the last decade for the use in drug screening and discovery, as it is utilized frequently to investigate ligand binding in proteins. In the context of environmental research, it has also been used to study the molecular interactions of pesticides in either whole peat soil or humic acid.[63,64] Recently, Longstaffe and Simpson[65] reported ^{1}H{^{19}F} reverse heteronuclear saturation transfer difference (RH-STD) NMR spectroscopy of peat humic acid mixed with different aromatic organofluorides to study the noncovalent interactions between xenobiotics and NOM. This experiment opens up the possibility of similar approach in the form of ^{1}H{M} RH-STD NMR experiments, where M could be a heavy metal such as ^{113}Cd or ^{207}Pb. The idea here is to saturate the metal and look for changes in the ^{1}H NMR spectrum. Such experiments, to the extent of our knowledge, have yet to be applied in an environmental context.

CPMAS experiment is another commonly used technique, not only for acquiring high-resolution NMR spectra, but also as a spectral editing method.[66] A ^{19}F → ^{1}H CPMAS NMR study performed on perfluorooctanoic acid and heptafluoronaphthol mixed with peat, lignin, or albumin took advantage of the latter.[67] In this study, magnetization was passed from a perfluorinated contaminant (^{19}F) onto ^{1}H from the surrounding soil organic matter to identify binding motifs. Several preliminary studies involving ^{19}F → ^{1}H → ^{13}C double CPMAS experiments have also been reported (see Chapter 6, for example, for $^{19}F \rightarrow {}^1H \rightarrow {}^{13}C$ double CPMAS NMR study of perfluorooctanesulfonic acid mixed with broccoli leaves and/or a whole peat soil). This experiment utilizes ^{13}C rather than ^1H detection and as such provides much more information as to the binding sites considering the higher spectral dispersion and improved resolution of ^{13}C versus ^1H in the solid state.

In theory, metal $\rightarrow {}^1H$ CPMAS and/or metal $\rightarrow {}^1H \rightarrow {}^{13}C$ double CPMAS can be applied to detect which organic components in a soil (i.e., protein, lignin, lipids, or cellulose) the metals bind to. This is a key for understanding the fundamental fate of metals in soil and sediment. However, to the best of our knowledge, such experiments have never been reported in the literature in the context of environmental research. Single CP, for example, metal $\rightarrow {}^1H$, should be feasible with current technology. Double CPMAS NMR experiments are feasible in soil at natural abundance using ^{19}F as a source of magnetization (see Chapter 6), but the largest hurdle for these experiments utilizing metals will be the limited sensitivity, given that metal nuclei have a much lower intrinsic NMR sensitivity than ^{19}F. However, double CP experiments involving metals may find more widespread application in the future, as NMR technology with better detection limits is continually being developed (as discussed in Section 4.1). Such experiments hold great potential because of their ability to directly identify metal-binding sites in complex environmental matrices.

19.5   CONCLUSIONS AND OUTLOOK

In summary, there is still a considerable potential for metal NMR spectroscopy in environmental research. Metals are ubiquitous in our environment and are found in the atmosphere and aquatic and terrestrial ecosystems. Within each system, metals are present in association with water (freely dissolved metal or as organic and inorganic metal complexes), particles (sorbed, precipitated, or incorporated within a mineral phase), and air. Thus, a molecular-level understanding of different processes is central to predict and model their mobility and potential impacts on human health.

NMR spectroscopy is often the only tool that can provide molecular-level insights regarding chemical structure and their interactions in their natural state. It is a very versatile tool suited for analyzing very

complex and/or multiphase samples in environmental research. Considering that NMR spectroscopy can be used in (i) characterizing the metal-binding sites, (ii) monitoring the changes in pH and concentration, and (iii) providing information regarding structural characteristics and molecular dynamics, it has a profound role in the field of environmental science. Hopefully, a better understanding of distribution, fate, and transport of metals in the environment within and among the various ecosystems can be achieved in the future with more collaboration between NMR spectroscopists and environmental scientists.

ACKNOWLEDGMENTS

We would like to thank the Natural Sciences and Engineering Research Council of Canada (NSERC, Strategic and Discovery Grants Programs), the Canada Foundation for Innovation (CFI), the Ontario Ministry of Research and Innovation (MRI), and the Krembil Foundation for providing funding. Andre Simpson would like to thank the Government of Ontario for an Early Researcher Award. We would like to thank everyone at Bruker BioSpin (Germany, USA, and Canada) for providing continued support and collaboration that made this work possible.

RELATED ARTICLES IN EMAGRES

Acquisition of Wideline Solid-State NMR Spectra of Quadrupolar Nuclei

Agriculture and Soils

Atmospheric Organic Matter

Coal Structure from Solid State NMR

Clay Minerals

Dissolved Organic Matter

Diffusion-Ordered Spectroscopy

Environmental NMR: Solution-state Methods

Environmental NMR: Solid-state Methods

Environmental NMR: High-resolution Magic-angle Spinning

Environmental Comprehensive Multiphase NMR

Environmental NMR: Diffusion Ordered Spectroscopy Methods

Fossil Fuels

Optically Enhanced Magnetic Resonance

Quadrupolar Interactions

Quadrupolar Nuclei in Solids: Influence of Different Interactions on Spectra

Quadrupolar NMR in Earth Sciences

Soil Organic Matter

REFERENCES

1. T. Kikuchi and S. Tanaka, *Crit. Rev. Env. Sci. Technol.*, 2012, **42**, 1007.

2. R. A. Wuana and F. E. Okieimen, *ISRN Ecol.*, 2011, **2011** Article ID 402647.

3. I. K. Wernick and N. J. Themelis, *Annu. Rev. Energy Env.*, 1998, **23**, 465.

4. F. Fu and Q. Wang, *J. Environ. Manage.*, 2011, **92**, 407.

5. M. F. Hochella and A. S. Madden, *Elements*, 2005, **1**, 199.

6. N. C. Mueller and B. Nowack, *Elements*, 2010, **6**, 395.

7. M. J. D. Clift and V. Stone, *Theranostics*, 2012, **2**, 668.

8. R. Hardman, *Environ. Health Perspect.*, 2006, **114**, 165.

9. A. Kabata-Pendias, *Trace Elements in Soils and Plants*, 4th edn, CRC Press: Boca Raton, 2010.

10. M. A. Nanny, R. A. Minear, and J. A. Leenheer, *Nuclear Magnetic Resonance Spectroscopy in Environmental Chemistry (Topics in Environmental Chemistry)*, Oxford University Press: USA, 1997, 344 pp.

11. Special Issue: NMR in Soil Science; ed. M. A. Hemminga and P. Buurman, *Geoderma*, 1997, **80**, 221.

12. Special Issue: NMR and MRI Research in Environmental Soil Science; ed. Y. Chen, *J. Environ. Qual.*, 2002, **31**, 375.

13. R. K. Harris and A. Sebald, *Magn. Reson. Chem.*, 1987, **25**, 1058.

14. A. W. MacGregor, L. A. O'Dell, and R. W. Schurko, *J. Magn. Reson.*, 2011, **208**, 103.

15. S. R. Hartmann and E. L. Hahn, *Phys. Rev.*, 1962, **128**, 2042.

16. G. Sinning, M. Mehring, and A. Pines, *Chem. Phys. Lett.*, 1976, **43**, 382.

17. M. Mehring and G. Sinning, *Phys. Rev. B*, 1977, **15**, 2519.

18. H. Y. Carr and E. M. Purcell, *Phys. Rev.*, 1954, **94**, 630.

19. S. Meiboom and D. Gill, *Rev. Sci. Instrum.*, 1958, **29**, 688.

20. L. A. O'Dell and R. W. Schurko, *Chem. Phys. Lett.*, 2008, **464**, 97.

21. S. E. Ashbrook, *Phys. Chem. Chem. Phys.*, 2009, **11**, 6892.

22. R. E. Wasylishen, S. E. Ashbrook, and S. Wimperis, *NMR of Quadrupolar Nuclei in Solid Materials*, John Wiley & Sons, Ltd: Chichester, 2012, 550 pp.

23. A. J. Simpson, D. J. McNally, and M. J. Simpson, *Prog. Nucl. Magn. Reson. Spectrosc.*, 2011, **58**, 97.

24. W. L. Kingery, A. J. Simpson, F. Han, and B. Xing, in *Humic Substances and Chemical Contaminants*, eds. C. E. Clapp, M. H. B. Hayes, N. Senesi, P. R. Bloom and P. M. Jardine, Soil Science Society of America: Madison, 2001, 397.

25. A. J. Simpson, W. L. Kingery, and M. H. B. Hayes, in *Heavy Metals Release in Soils*, eds. H. M. Selim and D. L. Sparks, CRC Press: Boca Raton, 2001, 237.

26. P. Pyykoo, *Mol. Phys*, 2008, **106**, 1965.

27. W. R. Carper, *Concepts Magn. Reson.*, 1999, **11**, 51.

28. C. M. Preston, *Soil Sci.*, 1996, **161**, 144.

29. P. N. L. Lens and M. A. Hemminga, *Biodegradation*, 1998, **9**, 393.

30. L. A. Cardoza, A. K. Korir, W. H. Otto, C. J. Wurrey, and C. K. Larive, *Prog. Nucl. Magn. Reson. Spectrosc.*, 2004, **45**, 209.

31. A. J. Simpson, M. J. Simpson, and R. Soong, *Environ. Sci. Technol.*, 2012, **46**, 11488.

32. S. Bank, J. F. Bank, and P. D. Ellis, *J. Phys. Chem.*, 1989, **93**, 4847.

33. V. Majidi, D. A. Laude, and J. A. Holcombe, *Environ. Sci. Technol.*, 1990, **24**, 1309.

34. D. Tinet, A. M. Faugere, and R. Prost, *J. Phys. Chem.*, 1991, **95**, 8804.

35. H.-Y. D. Ke and G. D. Rayson, *Environ. Sci. Technol.*, 1992, **26**, 1202.

36. W. Zhang and V. Majidi, *Appl. Spectrosc.*, 1993, **47**, 2151.

37. K. H. Chung and C. H. Moon, *Environ. Technol.*, 1994, **15**, 795.

38. K. H. Chung, S. W. Rhee, H. S. Shin, and C. H. Moon, *Can. J. Chem.*, 1996, **74**, 1360.

39. C. K. Larive, A. Rogers, M. Morton, and W. R. Carper, *Environ. Sci. Technol.*, 1996, **30**, 2828.

40. M. Grassi and G. Gatti, *Ann. Chim. (Rome)*, 1998, **88**, 537.

41. J. Li, E. M. Perdue, and L. T. Gelbaum, *Environ. Sci. Technol.*, 1998, **32**, 483.

42. P. Di Leo and P. O'Brien, *Clays Clay Miner.*, 1999, **47**, 761.

43. P. Di Leo, *Clays Clay Miner.*, 2000, **48**, 495.

44. H. Xia and G. D. Rayson, *Adv. Environ. Res.*, 2000, **4**, 67.

45. W. H. Otto, S. D. Burton, W. R. Carper, and C. K. Larive, *Environ. Sci. Technol.*, 2001, **35**, 4900.

46. W. H. Otto, W. R. Carper, and C. K. Larive, *Environ. Sci. Technol.*, 2001, **35**, 1463.

47. A. J. Simpson, *Magn. Reson. Chem.*, 2002, **40**, S72.

48. N. Hertkorn, E. M. Perdue, and A. Kettrup, *Anal. Chem.*, 2004, **76**, 6327.

49. M. Grassi and V. Daquino, *Ann. Chim. (Rome)*, 2005, **95**, 579.

50. E. M. Perdue, N. Hertkorn, and A. Kettrup, *Appl. Geochem.*, 2007, **22**, 1612.

51. V. Lenoble, C. Garnier, A. Masion, F. Ziarelli, and J. M. Garnier, *Anal. Bioanal. Chem.*, 2008, **390**, 749.

52. X. Lu, W. D. Johnson, and J. Hook, *Environ. Sci. Technol.*, 1998, **32**, 2257.

53. M. Grassi, E. Oldani, and G. Gatti, *Ann. Chim. (Rome)*, 1997, **87**, 353.

54. M. F. Summers, *Coord. Chem. Rev.*, 1988, **86**, 43.

55. H. Kovacs, D. Moskau, and M. Spraul, *Prog. Nucl. Magn. Reson. Spectrosc.*, 2005, **46**, 131.

56. K. Takeda, *Solid State Nucl. Magn. Reson.*, 2012, **47–48**, 1.

57. J. A. Tang, L. A. O'Dell, P. M. Aguiar, B. E. G. Lucier, D. Sakellariou, and R. W. Schurko, *Chem. Phys. Lett.*, 2008, **466**, 227.

58. T. Maly, G. T. Debelouchina, V. S. Bajaj, K.-N. Hu, C.-G. Joo, M. L. MakJurkauskas, J. R. Sirigiri, P. C. A. van der Wel, J. Herzfeld, R. J. Temkin, and R. G. Griffin, *J. Chem. Phys.*, 2008, **128**, 052211.

59. M. D. Lingwood, S. Han, and A. W. Graham, in *Annual Reports on NMR Spectroscopy*, ed. G. A. Webb, Academic Press, Elsevier: USA, 2011, Vol. 73, 83.

60. D. Lee, H. Takahashi, A. S. L. Thankamony, J.-P. Dacquin, M. Bardet, O. Lafon, and G. D. Paëpe, *J. Am. Chem. Soc.*, 2012, **134**, 18491.

61. V. K. Michaelis, E. Markhasin, E. Daviso, J. Herzfeld, and R. G. Griffin, *J. Phys. Chem. Lett.*, 2012, **3**, 2030.

62. T. Pietraß and M. Tomaselli, *Phys. Rev. B*, 1999, **59**, 1986.

63. A. Shirzadi, M. J. Simpson, R. Kumar, A. J. Baer, Y. Xu, and A. J. Simpson, *Environ. Sci. Technol.*, 2008, **42**, 5514.

64. A. Shirzadi, M. J. Simpson, Y. Xu, and A. J. Simpson, *Environ. Sci. Technol.*, 2008, **42**, 1084.

65. J. G. Longstaffe and A. J. Simpson, *Environ. Toxicol. Chem.*, 2011, **30**, 1745.

66. X. L. Wu, S. T. Burns, and K. W. Zilm, *J. Magn. Reson. Ser. A*, 1994, **111**, 29.

67. J. G. Longstaffe, D. Courtier-Murias, R. Soong, M. J. Simpson, W. E. Maas, M. Fey, H. Hutchins, S. Krishnamurthy, J. Struppe, M. Alaee, R. Kumar, M. Monette, H. J. Stronks, and A. J. Simpson, *Environ. Sci. Technol.*, 2012, **46**, 10508.

Chapter 20
Organic Pollutants in the Environment

Gabriela Chilom and James A. Rice

Department of Chemistry and Biochemistry, South Dakota State University, Brookings, SD 57007-0896, USA

20.1 INTRODUCTION

NMR spectroscopy is one of the most useful tools for studying specific and nonspecific (i.e., covalent and noncovalent) molecular interactions and the chemical nature of chemical substances. It is well suited to study both specific and nonspecific interactions in complex environmental systems. Perhaps no material is more complex than natural organic matter (NOM), so it is not surprising that NMR has found wide use in its characterization (see Chapters 1, 4, 5, 11, 12, and 13). A significant portion of the application of NMR has been to fingerprint NOM in order to ascertain the relative proportions of aliphatic, anomeric, aromatic, and carboxyl/carbonyl carbons in a sample (see Chapters 4 and 13). However, the application of NMR to study chemical interactions within NOM, between NOM and environmental contaminants, the fate of environmental contaminants is slowly becoming more widespread.

This trend can be seen in the four reviews that have summarized this work over the last 10 years; the most recent of these reviews was in 2011.[1-4] This chapter provides a broad overview of the literature with emphasis on recent work with reference to that older work, which helps place the newer work into context. This overview is organized into "mechanistic studies", which focuses on the interactions of NOM and organic contaminants, and "analytical studies", which reviews the application of NMR in the study of organic contaminant fate.

20.2 MECHANISTIC STUDIES OF CONTAMINANT FATE

Development of efficient decontamination methods and estimation of long-term environmental risks of organic contaminants require a better understanding of the mechanism of their interactions with environmental matrices. Among the numerous methods developed, the NMR techniques have been shown to make significant contributions in this area and, therefore, have been used more frequently.[2,5] NMR can be used to study the fate of organic contaminants in the environment directly and/or indirectly. The indirect approach focuses on establishing correlations between contaminant fate and NMR-based structural descriptors of environmental samples.[1] The direct approach focuses on the use of NMR to probe the interactions between specific atoms within molecules of the contaminated environmental samples.[1,3]

NMR Spectroscopy: A Versatile Tool for Environmental Research
Edited by Myrna J. Simpson and André J. Simpson
© 2014 John Wiley & Sons, Ltd. ISBN: 978-1-118-61647-5

20.2.1 Indirect Approach Using Correlation of Structure and Binding Coefficients

The indirect approach is based on establishing correlations between data from analyses that are performed independently. One analysis refers to quantification of contaminant fate and is generally performed as sorption experiments in the laboratory.[6] These experiments enable the calculation of the distribution coefficient (K_d) and the organic carbon normalized sorption coefficient (K_{oc}), two commonly used parameters for assessing the fate of organic pollutants in the environment. The use of K_{oc} is supported by the finding and currently accepted paradigm that NOM from soils/sediments plays a predominant role in the sorption process even.[7] The other analysis refers to the determination of structural descriptors and involves NMR analysis of the environmental samples; mostly solid-state ^{13}C NMR experiments (see Chapter 4). From these NMR spectra, the relative distribution of carbon type including alkyl, O-alkyl, aromatic, and carboxyl carbons, or a combination of them is usually reported.[8]

Sorption studies have been applied to materials of different diagenetic origin, such as soils and coals, and to various components of soil organic matter such as humic acid and humin. For example, Abelmann et al.[9] studied the sorption of five hydrophobic organic contaminants (HOCs) on a total of 18 geosorbents, including seven coals, three peat soils, and nine mineral soils. The authors reported a positive correlation between K_{OC} and aromaticity (defined by the C/H-aryl content) and a negative correlation with polarity (defined by the ratio between carbonyl, O-alkyl, O-aryl carbons, and C/H aryl and alkyl C). On the basis of these correlations, the authors suggested that nonpolar aromatic domains in NOM represent the main sorbents for HOCs. Similar findings were reported by Ahmad et al.,[10] who used solid-state ^{13}C NMR data and elemental composition in a molecular mixing model to study the sorption of carbaryl and phosalone on 24 soils from various agroecological regions. The authors suggested that aromatic structures such as those from lignin and charcoal contributed the most to sorption of the two pesticides. In the context of this study, it is worth noting that because of NOM's chemical heterogeneity, its ^{13}C solid-state NMR spectra are complex with broad resonances whose linewidths are typically measured in "10s" of ppms; "pure" lignin or charcoal samples display a similar spectra.

In a study of pyrene sorption on NOM samples of various aromaticity, Chefetz et al.[11] showed that humin (aromaticity 8%) had significantly higher K_{OC} values than humic acid, lignin, and lignite samples (aromaticity between 30 and 55%). Several other studies emphasized the role of aliphatic carbon for sorption of HOCs, and positive correlations between K_{OC} and aliphaticity were reported for aliphatic-rich NOM samples.[12,13] However, these compound-specific correlations with aromaticity and aliphaticity of organic matter samples are limited by the relatively small number and variety of systems considered. In a recent review, Chefetz and Xing[14] compiled a large and diverse literature dataset of phenanthrene sorption on natural and engineered sorbents and found no significant correlations between either NOM aromaticity or aliphaticity, as determined by solid-state ^{13}C NMR, and sorption affinity. The authors suggested that correlations with molecular descriptors as aromaticity and aliphaticity are valid for relatively homogeneous and chemically similar sorbents and neither one can be used to predict the sorption affinity of diverse sorbents.

Other NOM structural characteristics are also considered including the contaminant accessibility and physical conformation of organic matter.[15] Mao et al.[16] found a positive correlation between phenanthrene sorption capacity and the amount of amorphous nonpolar aliphatic poly(methylene) domains for seven geosorbents. They used ^1H inversion recovery with ^{13}C NMR detection to identify domains of poly(methylene) and of branched nonpolar aliphatic segments and suggested that the rubber-like mobility of these domains play a major role in controlling the sorption of HOCs. The role of rigid, condensed, and flexible, expanded, domains was considered also by Gunasekara et al.[17] to explain the sorption behavior of the structurally altered humic acid samples. The authors used proton spin–spin (T_2^H) relaxation measurements by solid-state NMR to identify separate rigid (condensed) and flexible (expanded) ^1H domains and determine their distribution. They noticed that the samples subjected to bleaching, a process that selectively removes aromatic moieties, had a more partition-like sorption behavior and showed an increase of expanded protons indicating a more expanded structure.

Correlation studies contributed significantly in establishing relationships between sorption behavior and NOM characteristics as determined by NMR; however, they were not able to identify the sorption mechanisms. The complexity and heterogeneity of NOM make it difficult to determine specific structural details. This limitation constrains the inferences that can be drawn from the indirect observation of contaminant–NOM interactions; studying the interactions with a bulk observation usually provides only weak inferences on interaction and behavior because of this heterogeneity.

20.2.2 Direct Interactions between Organic Contaminants and Environmental Matrices

20.2.2.1 Covalent Interactions

It is estimated that between 20 and 70% of an organic contaminant entering a terrestrial system becomes irreversibly bound to soil and cannot be extracted by nondestructive methods.[18] These bound contaminants may have reduced toxicity because of limited bioavailability, and, therefore, immobilization can be an important decontamination method for reducing the risk associated with contaminated soils. One of the mechanisms that contributes to immobilization of a contaminant is the covalent binding to the environmental matrix, binding that can be mediated by either enzymatic or abiotic catalysts.[18] NMR spectroscopy can detect covalent binding by monitoring changes in chemical shift of the contaminant. Isotopically enriched contaminants are often used for these experiments, such that the signal from contaminant is appreciably higher above the background signal from complex material such as NOM.[1] Labeling of specific nuclei within contaminant molecule allows a detailed description of the type of interactions occurring between organic pollutants and environmental samples. For example, Berns *et al.*[19] used solid-state ^{15}N cross polarization magic-angle spinning (CPMAS) NMR to study the binding of simazine, a triazine herbicide, to soil. In order to improve the detectability of simazine by ^{15}N NMR analysis, they incubated ^{15}N-labeled simazine, at both core and sidechain (Figure 20.1), on a ^{15}N-depleted artificial compost obtained from maize and wheat plants grown on sand, with ^{15}N-depleted NH_4NO_3 as sole nitrogen source. Their data indicated that degradation products resulting from *N*-dealkylation and triazine ring destruction were binding to soil matrix.

Thorn *et al.*[20] provided direct spectroscopic evidence of covalent binding of aromatic amines to soil humic materials using solution- and solid-state ^{15}N NMRs. The authors showed that aniline underwent

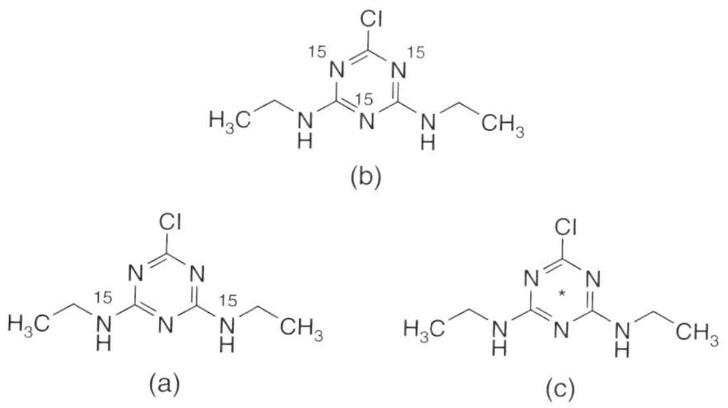

Figure 20.1. Simazine (a) ^{15}N-side-chain-labeled; (b) ^{15}N-core-labeled; and (c) ^{14}C-[U]-core-labeled. (Reprinted with permission form Ref. 19. © 2005 Elsevier)

nucleophilic addition reactions with the quinone and other carbonyl groups of the humic samples and became incorporated in the form of anilinohydroquinone, anilinoquinone, anilide, imine, and heterocyclic nitrogen. The role of anilic nitrogen for covalent binding has also been the focus of recent studies of sulfonamide antimicrobial fate in soils.[21,22] Bialk and Pedersen[21] used solution-state ^{1}H–^{15}N heteronuclear multiple bond correlation (HMBC) NMR to study the phenoloxidase-mediated reaction of ^{15}N-enriched sulfapyridine with both catechol and humic acid. The NMR spectra revealed the presence of Michael adducts, whereas no covalent binding was observed in the absence of phenoloxidase. A similar NMR technique was used by Gulkowska *et al.*[22] to explore other aspects of sulfonamide fate in soils such as the covalent binding to other carbonyls without oxidants and the involvement of radical reactions in the covalent bonding in the presence of oxidants.

^{15}N NMR has been often used to study the binding of 2,4,6-trinitrotoluene (TNT) to various soil matrices under aerobic and anaerobic conditions.[23,24] Using solid-state ^{15}N CPMAS NMR, Bruns-Nagel *et al.*[23] found 58% of the ^{15}N residues bound to the anaerobic/aerobic composted ^{15}N TNT-spiked soils, from which 23% of ^{15}N was present in mostly heterocyclic structures, 15% was bound in a covalent manner, 15% could be detected as amino functions, and 2% as nitro functions. While most of the NMR studies reporting on TNT binding to soil use one-dimensional (1D) solid-state ^{15}N NMR, a two-dimensional (2D) NMR approach was used by Knicker.[24] The authors employed solid-state double cross polarization magic-angle spinning (DCPMAS) ^{15}N–^{13}C NMR to

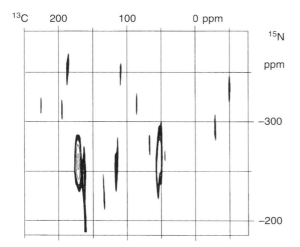

Figure 20.2. Two-dimensional solid-state DCPMAS $^{15}N-^{13}C$ NMR spectra of the acid-insoluble fraction obtained from ^{13}C-enriched plant residues that were incubated after addition of $^{15}N_3$-TNT for 11 months. (Reprinted with permission from Ref. 24. © 2003 Elsevier)

investigate the incorporation of ^{15}N-enriched TNT into ^{13}C-enriched plant material under aerobic conditions. The authors found that the covalent binding of TNT transformation products to plant-derived organic matter is mediated by alkylation and acetylation reactions rather than by 1,4 addition of TNT-derived nitrogenous groups to quinones in the plant material. The DCPMAS spectra (Figure 20.2) contained only the signals of ^{13}C that couple with ^{15}N and a crosspeak correlated the amide N to *N*-alkyl C, demonstrating that most of the amide N were alkylated.

Covalent binding of contaminants containing phenolic moieties, such as chlorophenols, has been studied by solution-state ^{13}C NMR. Using ^{13}C-labeled 2,4-dichlorophenol (DCP) at C-2 and C-6 positions, Hatcher *et al.*[25] observed changes in chemical shifts of the DCP carbons as the humic acid covalently binds to the 2,4-DCP in the presence of oxidoreductive enzymes. The authors suggested that the various new peaks observed (Figure 20.3) represent sites at or near the labeled carbons that had formed covalent bonds with the 2,4-DCP itself, or with the humic acid through carbon–oxygen ester, carbon–oxygen ether, and carbon–carbon linkages. Solution-state ^{13}C NMR was also used by Fukushima *et al.*[26] to study the enzymatic oxidation of pentachlorophenol (PCP) in the presence of humic substances. The authors

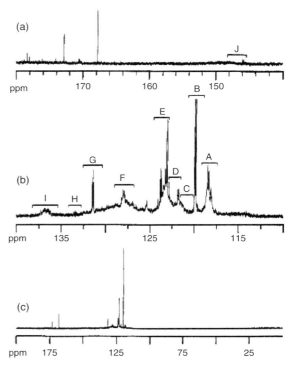

Figure 20.3. ^{13}C NMR spectrum of ^{13}C-2,4-DCP enzymatically bound to Minnesota peat humic acid after treatment with horseradish peroxidase. (a) Expanded region, 140–180 ppm; (b) expanded region, 110–140 ppm; and (c) full spectrum, 0–200 ppm. (Reprinted with permission from Ref. 25. © 1993 American Chemical Society)

found that the presence of humic substances reduced the formation of the more toxic dimers formed when chlorophenols are oxidized by peroxidase, and ligninase that contain porphyrin–Fe^{3+} acts as an active center. The NMR data showed that the chlorinated intermediates from PCP were covalently incorporated into humic substances mainly in the form of ether C–O linkages.[26]

Solid-state ^{13}C CPMAS and direct polarization magic-angle spinning (DPMAS) NMRs were used by Tao and Maciel[27] to describe the chemical behavior of methyl bromide. The solid-state ^{13}C NMR study of $^{13}CH_3Br$-treated soil and soil components such as humic materials and clays showed the formation of the methoxy ($^{13}CH_3$–O–) linkages suggesting that methylation of soil organic matter may be the major pathway for degradation of CH_3Br in soils. Chemical transformation of polyaromatic hydrocarbons in soil systems was demonstrated by chemical shift changes in the solid-state ^{13}C NMR spectrum of the humic acid isolated from phenanthrene-spiked soils.[28] Ester- and ether-bound phenanthrene derivatives were identified based on structural assignment of the chemical shifts and with the help of mass spectrometry (MS) analysis of the products obtained after alkaline hydrolysis of contaminated samples.

20.2.2.2 Noncovalent Interactions

The majority of the NMR investigations of noncovalent interactions between contaminants and NOM can be classified as either studying the existence and nature of contaminant sorption domains in NOM or using the contaminants themselves as probes of the chemistry taking place within the domains. For example, Kohl et al.[29] used solid-state ^{19}F NMR and hexafluorobenzene (HFB) to provide spectroscopic evidence for the existence of two domains consistent with the dual-mode sorption model[30–32] of soil and sediment (i.e., solid phase) NOM–contaminant interactions. Lipid components were found to compete with HFB for high-energy sorption sites. Kinetic studies demonstrated that sorption was rapid and occurred first in the partitioning (or rubbery) domain with the transfer of HFB to the rigid (or glassy) domain with time. Subsequently, Cornelissen et al.[33] reported similar observations of dual-mode sorption domains in peaty sediments using HFB and solid-state ^{19}F NMR. Similarly, Khalaf et al.[34] used this solid-state ^{19}F NMR to obtain evidence for the existence of multiple sorption domains in low-molecular-weight NOM fractions but only two domains were observed in higher molecular weight fractions. Their data indicated that the rigid sorption domains were more aliphatic than aromatic in nature.

Golding et al.[35] used solid-state ^{13}C NMR with proton spin relaxation editing (PSRE) and phenanthrene ^{13}C-labeled in the 9 and 10 positions to show that NOM has a rapidly relaxing component and a slowly relaxing component consistent with a dual-mode sorption model. Aromatic carbon in their samples was present in both rapidly and slowly relaxing domains emphasizing the heterogeneity of NOM. Smernik[36] used solid-state ^{13}C CPMAS NMR, PSRE, and ^{13}C-labeled benzoic acid, benzophenone, naphthalene, phenanthrene, or palmitic acid to study sorption to NOM. The chemical shifts of the individually sorbed species were shifted upfield by as much as 3 ppm, and the resonances broadened indicating a range

of different chemical environments. PSRE identified a rapidly relaxing, charcoal-rich domain and more slowly relaxing domain in humic materials as well. The charcoal-rich domain had a greater affinity for the organic compounds, especially for the more hydrophobic compounds (benzoic acid ≈ benzophenone < naphthalene < phenanthrene). Mao et al.[16] used solid-state ^{1}H inversion-recovery NMR experiments to show that NOM contained ~10 nm diameter rubbery domains formed by branched aliphatic carbon compounds, and that these domains were responsible for the sorption of hydrophobic organic compounds. Subsequently, Mao and Schmidt-Rohr[37] used ^{1}H–^{13}C HETCOR (heteronuclear correlation) experiments to conclude that these saturated carbon domains did not have significant contributions from carbohydrates. Using toluene as a probe of these domains and ^{1}H MAS pulsed-field gradient NMR, Fomba et al.[38] reached a similar conclusion. Mao et al.[39] used ^{1}H–^{13}C 2D wide-line separation (WISE) solid-state NMR to show that a continuum of sorption domain mobility exists in NOM between the rigid and the mobile domain end-members. Using quadrupole-echo solid-state ^{2}H NMR and benzene-d_6 as a probe, Eastman et al.[40] observed more than one sorption environment in a series of humic and fulvic acids. They suggested that the thermodynamic driving force was a transition from an entropic effect to an enthalpic effect, as the organic matter structure deforms on interaction with the benzene-d_6.

Chien et al.[41] used solution-state ^{19}F NMR and the hydrophilic paramagnetic species, 2,2,6,6-tetramethyl-1-piperidinyloxy (TEMPO), as a probe of atrazine's interaction with concentrated alkaline solutions of humic acid. The absence of any observable paramagnetic relaxation of atrazine in the presence of TEMPO indicated that atrazine had been solubilized by the humic acid micelles and confirmed the existence of hydrophobic domains in these assemblies. On the basis of ^{2}H and ^{19}F NMR studies of the interaction of acid and a suite of labeled fluorobenzenes, Herbert and Bertsch[42] postulated that the sorption process in humic acid may be inherently different than partitioning into a micelle. On the basis of small changes in the observed chemical shifts and large correlation times that were on the same order of magnitude as those observed for large-molecular-weight polymers, the authors hypothesized that substituted fluorobenzenes in humic acid solutions exist within a three-dimensional organic matter cage. Smejkalova and Piccolo[43] used

^{1}H diffusion ordered spectroscopy (DOSY) NMR and measurement of T_1^H and T_2^H relaxation times to investigate the association of DCP with humic and fulvic acids isolated from soil. Stronger DCP binding to humic acid was observed and attributed to a greater tendency to form hydrophobic domains that formed a host–guest complex with DCP. This is similar to the conclusions reached by Nanny *et al.*,[44] using ^{13}C-labeled acenaphthenone and $^1T_1^C$ relaxation measurements, and Chefetz *et al.*[11]

Deuterium NMR studies[45,46] of the interaction of dissolved humic materials and monoaromatic compounds indicate that noncovalent binding among these materials are the result of interactions including aromatic–aromatic interactions.

When applied to NOM–contaminant interactions, saturation transfer double difference (STDD) NMR spectroscopy[47] can selectively observe resonances from only the bound fraction of a contaminant present in a system permitting the contaminant to function as a direct probe of its binding environment.[48,49] In this application of STDD, NOM is selectively saturated using a weak radio frequency field and any bound contaminant receives saturation from NOM. With STDD, the interaction at each NMR-active atom in the contaminant with NOM can be quantified and the mechanism of interaction described. Shirzadi *et al.*[48,49] applied this technique to the study of the interaction of common pesticides with humic acid in aqueous media. They concluded that at neutral pH values, the F and Cl atoms in diflufenzopyr, acifluorfen, and chlorsulfuron and the resulting polar interactions with NOM were the primary influence on binding in these contaminants.[48] When STDD was combined with high-resolution magic-angle spinning solid-state ^{1}H NMR to study the interaction of trifluralin, acifluorfen, (4-nitro-3-(trifluoromethyl) phenol and 1-naphthol, and humic acid dipolar interactions, H-bonding, hydrophobic associations, and possibly π–π interactions were the predominant binding mechanisms.[49]

The complexity of NOM as a sorbent is reflected not only in its molecular heterogeneity and polydispersity but in the coexistence of multiple NOM phases in a single natural system, a reflection of NOM's tendency to self-assemble.[50] It is not unusual to find NOM present in the dissolved (i.e., in solution) and solid phases. In many systems, a third, gel-like (e.g., colloidal) phase exists as well. Fractionation of NOM to separate these phases almost certainly modifies the organization of these components and consequently NOM's ability to interact with contaminants.

Courtier-Murias *et al.*[51] developed a comprehensive multiphase-nuclear magnetic resonance (CMP-NMR) probe and ^{1}H, ^{13}C, or ^{19}F HR-MAS (high-resolution magic-angle spinning) approaches that allow these phases to be individually observed without NOM sample fractionation (see Chapter 6). Using ^{19}F HR-MAS, this technique was applied to study the interaction of pentadecafluoro-octanoic acid (PFOA) with a peat. PFOA appeared to preferentially interact with the peat's aliphatic organic components.

20.3   ANALYTICAL STUDIES

In addition to probing mechanisms of contaminant interactions with environmental matrices, NMR is also a powerful tool for elucidating the structure of organic compounds and has been used to monitor the transformation of organic pollutants in various environmental samples.[52] Identification and quantification of the organic pollutant, either the parent compound or the metabolite, are important for assessing its toxicity and persistence in the environment. NMR is a nondestructive method and requires little or no sample pretreatment. However, NMR sensitivity can limit its application for monitoring the fate of contaminants in the environment. The low concentration of organic pollutants and their metabolites may make them difficult to detect, and more so in the presence of environmental matrices such as soils or sediments. Therefore, sometimes steps such as isolation, separation, and concentration are necessary before NMR analysis, and sometimes coupling NMR with other analytical techniques such as high-performance liquid chromatography (HPLC) or MS is needed (see Chapter 3). Significant gains in the sensitivity of NMR, either direct or hyphenated techniques, can be achieved by implementing recent developments in NMR such as pulse sequences, multidimensional data analysis capabilities, high-field instruments, and cryogenic probes.

20.3.1   NMR Analysis of Organic Contaminant and Metabolite Structure

The use of NMR as an analytical tool is determined by the type of organic pollutant and by the nuclei present in the contaminated environmental sample (Table 20.1).

Table 20.1. Characteristics of the nuclei used to study the fate of common organic pollutants

Type	Nuclei Natural abundance (%)	Receptivity versus ^{1}H	Chemical shift (ppm)	Common organic pollutants
^{1}H	99.9844	1	−25 to 12	Polyaromatic hydrocarbons
^{13}C	1.108	1.76×10^{-4}	−25 to 250	Chlorinated solvents
				Polychlorobiphenyls
				Trihalomethanes
^{15}N	0.365	3.85×10^{-6}	−400 to 800	Explosives and propellants
^{19}F	100	0.8328	−500 to 900	Fluorophenols, fluorinated
				surfactants, and repellents
^{31}P	100	0.0663	−500 to 600	Organophosphorous pesticides

The ^{19}F nucleus has the advantage of 100% natural abundance, high sensitivity, and large chemical shift and has little to no interference with the environmental matrices, as natural fluorine-containing organic molecules are rare in soils and sediments. Bondar *et al.*[53] used solution-state ^{19}F NMR to directly monitor the microbial degradation of fluorinated environmental pollutants such as fluorophenols by various microorganisms. The authors were able to identify and quantify the previously unidentified metabolites as well as provide information about specificity of enzymes generally considered unstable. The fate of fluorinated analogs of chlorinated phenols in aquatic plants was studied by ^{19}F NMR, and fluorinated-chlorophenol conjugates were identified as metabolites of 4-chloro-2-fluorophenol.[54] The soluble metabolites were identified by solution-state ^{19}F NMR of the plant extracts, whereas the metabolites bound to the cell wall or other plant materials were identified by solid-state ^{19}F NMR. Perfluorinated surfactants are fluorinated contaminants that also benefit from the application of ^{19}F NMR. Moody *et al.*[55] developed a solution-state ^{19}F NMR-based analytical method for the determination of total perfluorinated surfactant concentrations in aqueous samples that has a detection limit of 10 μg L^{-1} for a 100 mL sample. The method was successfully applied to determine the concentration of perfluorinated surfactants in fish tissue extracts and surface water samples. The number of fluorine-containing chemicals increased lately and the trend is expected to continue in the future, as more compounds are produced as either intermediates or end products in the synthesis of pharmaceutical, industrial, or agricultural chemicals, so the use of ^{19}F NMR as an important analytical tool in the field of contaminant fate is also expected to increase.[56]

Organophosphorous pesticides, containing the ^{31}P nucleus, are widely used and constitute major classes of insecticide, herbicide, and fungicide applications.[57] Similar to ^{19}F, the ^{31}P nuclide has 100% natural abundance, high sensitivity, and also a broad chemical shift range. There are numerous studies that describe the use of ^{31}P NMR for the analysis of the purity and stability of organophosphorous pesticides and identification of their degradation products.[58] The number of studies though that address the identification and quantification of these products in complex matrices is significantly smaller and even more limited in case of the environmental samples or systems that resembled them. Hong *et al.*[59] studied the hydrolysis of phorate under simulated natural water conditions. On the basis of solution-state ^{31}P NMR data, the authors suggested a dominant pathway reaction as a new mechanism for the phorate hydrolysis. Seger and Maciel[60] used both solid- and solution-state ^{31}P NMRs to study the decomposition of chlorpyrifos sorbed on soil components such as humic acid and clays. While the concentration range of the pesticides used by the authors exceeds the one found in the environment, the study shows that the phosphorous-containing chemicals exist at the surface of soil components in a variety of chemical and physical states. Krolski *et al.*[61] studied the degradation of sulprofos in a soil samples for 90 days. The authors used solution-state ^{31}P NMR to identify and quantify the two major metabolites and confirm their presence by HPLC and liquid scintillation counting. While the use of ^{31}P NMR as a quantitative analytical tool for monitoring the fate of pollutants in complex matrices is rather limited, the method is considered to have great potential because it is a non-invasive method.[62]

^{1}H NMR is often used for studying the contaminant fate in environmental samples because of the high concentration and sensitivity of protons although its chemical shit range is narrow in comparisons with the ^{19}F and ^{31}P and the contaminant signals can be masked by the environmental matrices. Depending on the complexity of the environmental sample, either ^{1}H NMR spectra or 2D correlation spectra are acquired for studying the biotransformation of organic contaminants.[63] The study of microbial degradation of pesticides belonging to benzothiazole family by two strains of *Rhodococcus* involved both solution-state ^{1}H NMR and long-range ^{1}H–^{15}N HMBC to identify the two metabolites 2-hydroxybenzothiazole and 2,6-dihydroxybenzothiazole. The use of 2D ^{1}H–^{15}N HMBC also allowed the identification of the metabolites for the degradation of methabenthiazuron herbicide by the fungus *Aspergillus niger*. Using solution-state ^{1}H NMR, Besse *et al.*[64] established the biodegradation pathway of solvents such as morpholine and its derivatives, whereas no other methods could establish their metabolism. Trace amounts of chemical warfare agents and their degradation products were determined in complex matrices using ^{1}H and ^{31}P NMRs.[65] NMR spectroscopy provided identification of six of the seven spiking compounds, bis-(2-chloroethyl)ethylamine, methylphosphonic acid, cyclopentyl methylphosphonate, ethyldiethanolamine, diethylaminoethanol, and bis(*N,N*-diethylaminoethyl)disulfide, in an organic extract and in samples of canal water. Application of inverse [1]H–[31]P experiments increased the sensitivity of determining trace amounts of organophosporous compounds. The 1D nonselective [1]H–[31]P heteronuclear single quantum coherence (HSQC) and heteronuclear single quantum multiple bond correlation (HSQMBC) experiments were shown to be the most sensitive NMR experiments to selectively detect constituents of chemical weapons.[66]

^{1}H NMR has been also applied for the determination of unknown contaminants in water because of its minimal sample preparation that often implies adding small amount of deuterated water. Charlton *et al.*[67] showed that using solution-state ^{1}H NMR with a cryoprobe, a wide range of contaminants can be detected and characterized at microgram per liter level in potable water with minimal sample treatment. The use of cryogenic probe increases the sensitivity of measurement by reducing the electrical noise of recorded NMR signals, as the probe is being cooled down to 25 K.

The ^{13}C nucleus does not have as high a natural abundance or sensitivity as ^{19}F or ^{31}P which create a strong interference with the matrix background because they are present in all natural environmental samples. Isotopically enriched contaminants have been used to investigate the fate of organic contaminants.[44,50] Guthrie-Nichols *et al.*[68] used [4,9-[13]C_2] pyrene to study the effect of aging on pyrene transformation in sediments. Using solution-state ^{13}C NMR, the authors were able to identify, for the first time, pyrene-4,5-quinone in lipid extracts of aged sediments. The use of isotopically enriched samples though is challenging, as concentration and specificity of the isotopic label as well as the cost make these experiments difficult.[65] Introduction of a separation step is, therefore, recommended in these cases, as it not only increases the concentration of the desired analyte but also decreases the complexity of the mixture analyzed.

20.3.2 Hyphenated NMR Analysis

Hyphenated techniques use NMR directly coupled with other analytical techniques, and they address some of the analytical restraints of direct NMR analysis such as concentration and complexity of the environmental matrices (see Chapter 3). They also avoid the potential pitfalls of offline NMR analysis such as sample contamination by avoiding the fraction collection step.[2] There are numerous studies that apply liquid chromatography-nuclear magnetic resonance (LC-NMR) for the identification of organic compounds in complex matrices, including environmental samples.[52] Explosives and related compounds and organic acids were determined in leachate samples from industrial landfills, and more recently, polar compounds were measured from more highly complex samples from municipal solid waste.[69,70] A detailed description of the principles and use of LC-NMR can be found in Cardoza *et al.*[2] and in Chapter 3. Hyphenated techniques facilitate the use of NMR for the identification of known and unknown organic pollutant transformation products in complex matrices (see examples in Chapter 3). In some cases, analyte concentrations may be too low, but this is overcome by the incorporation of solid-phase extraction (SPE) with LC-NMR techniques as well as hyphenation with MS (see Chapter 3). LC-SPE-NMR has been used to identify a number of new metabolites from organic chemicals in the environment.[71,72] Accordingly,

LC-SPE-NMR/MS shows great promise for future environmental fate studies of organic pollutants.

20.4 CONCLUSIONS

This chapter highlights how NMR spectroscopy enables the study of chemical interactions within NOM, between NOM and environmental contaminants, and the fate of environmental contaminants. In the spirit of reviewing the literature on these topics, it is worth noting that this has not been a topic that has been received much attention until recently.[4] With its broad potential for answering a variety of questions relevant to the environmental chemistry and geochemistry of contaminants and NOM–contaminant interactions, this is somewhat surprising.

Given the heterogeneity of these processes, and environmental matrices in general, the use of molecules labeled with specific NMR-active nuclei that can serve as probes of the reactions taking place in these systems is an experimental approach that could be more fully exploited. The ^{13}C background from NOM itself limits the use of this nucleus even as a probe. Nuclei that can serve as surrogates for elements contained in contaminants but that are not abundant in NOM or the environment being studied (e.g., ^{19}F in soil organic matter) have the potential to be more broadly utilized for studying contaminant–NOM interactions by NMR.[29]

Increasingly, sophisticated NMR techniques, especially solution-state methods, are being developed that allow selective excitation and or detection of NMR-active nuclei, which can also help to hide the background created by NOM.[48] Multidimensional methods[73] that reveal connectivity between components of complex mixtures also hold promise for understanding how contaminants interact with NOM. These methods explicitly acknowledge the complex, multicomponent nature of NOM. They focus on specific NOM components, or NOM–contaminant interactions, through the use of selective pulse sequences or detection methods that can effectively "ignore" components of interactions that are not of interest.[48,49,51] NMR is one of the few techniques that can do this (another being small-angle neutron scattering where it is possible to "contrast match" out components that do not contribute to the chemistry being studied[74]). This is a potentially transformative approach for dealing with the complexity of NOM

and the complexity of its interactions with organic contaminants.

REFERENCES

1. M. J. Simpson, *Soil Sci. Soc. Am. J.*, 2006, **70**, 995.

2. L. A. Cardoza, A. K. Korir, W. H. Otto, C. J. Wurrey, and C. K. Larive, *Prog. NMR Spectrosc.*, 2004, **45**, 209.

3. A.-M. Delort and M. Combourieu, *Environ. Chem. Lett.*, 2004, **1**, 209.

4. A. J. Simpson, D. J. McNally, and M. J. Simpson, *Prog. Nucl. Mag. Res. Spec.*, 2011, **58**, 97.

5. B. Pan, P. Ning, and B. Xing, *Environ. Sci. Pollut. Res. Int.*, 2008, **15**, 554.

6. A. Delle Sitea, *J. Phys. Chem. Ref. Data*, 2001, **30**, 187.

7. M. Rebhun, R. Kalabo, L. Grossman, J. Manka, and C. H. Rav-Acha, *Wat. Res.*, 1992, **26**, 79.

8. J.-D. Mao, W.-G. Hu, K. Schmidt-Rohr, G. Davies, E. A. Ghabbour, and B. Xing, *Soil Sci. Soc. Am. J.*, 2000, **64**, 873.

9. K. Abelmann, S. Kleineidam, H. Knicker, P. Grathwohl, I. Kögel-Knabner, and J. Plant, *Nutr. Soil Sci.*, 2005, **168**, 293.

10. R. Ahmad, P. N. Nelson, and R. S. Kookana, *Eur. J. Soil Sci.*, 2006, **57**, 883.

11. B. Chefetz, A. P. Deshmukh, and P. G. Hatcher, *Environ. Sci. Technol.*, 2000, **34**, 2925.

12. M. J. Salloum, B. Chefetz, and P. G. Hatcher, *Environ. Sci. Technol.*, 2002, **36**, 1953.

13. Y. Ran, K. Sun, Y. Yang, B. Xing, and E. Zeng, *Environ. Sci. Technol.*, 2007, **41**, 3952.

14. B. Chefetz and B. Xing, *Environ. Sci. Technol.*, 2009, **43**, 1680.

15. P. J. Mitchell and M. J. Simpson, *Environ. Sci. Technol.*, 2013, **47**, 412.

16. J.-D. Mao, L. S. Hundal, M. L. Thompson, and K. Schmidt-Rohr, *Environ. Sci. Technol.*, 2002, **36**, 929.

17. A. Gunasekara, M. J. Simpson, and B. Xing, *Environ. Sci. Technol.*, 2003, **37**, 852.

18. J.-M. Bollag, in *Biotechnology for the Environment: Soil Remediation*, eds S. N. Agathos and W. Reineke, Kluwer Academic Publishers: Dordrecht, 2002, 93.

19. A. Berns, R. Vinken, M. Bertmer, A. Breitschwerdt, and A. Schaeffer, *Chemosphere*, 2005, **59**, 649.

20. K. A. Thorn, P. J. Pettigrew, W. S. Goldenberg, and E. J. Weber, *Environ. Sci. Technol.*, 1996, **30**, 2764.

21. H. M. Bialk and J. A. Pedersen, *Environ. Sci. Technol.*, 2008, **42**, 106.

22. A. Gulkowska, M. Krauss, D. Rentsch, and J. Hollender, *Environ. Sci. Technol.*, 2012, **46**, 2102.

23. D. Bruns-Nagel, H. Knicker, O. Drzyga, U. Butehorn, K. Steinbach, D. Gemsa, and E. Von Low, *Environ. Sci. Technol.*, 2000, **34**, 1549.

24. H. Knicker, *Sci. Total Environ.*, 2003, **308**, 211.

25. P. G. Hatcher, J. M. Bortiatynski, R. D. Minard, J. Dec, and J.-M. Bollag, *Environ. Sci. Technol.*, 1993, **27**, 2098.

26. M. Fukushima, H. Ichikawa, M. Kawasaki, A. Sawada, K. Morimoto, and K. Tatsumi, *Environ. Sci. Technol.*, 2003, **37**, 386.

27. T. Tao and G. E. Maciel, *Environ. Sci. Technol.*, 2002, **36**, 603.

28. T. Kacker, E. T. Haupt, C. Garms, W. Francke, and H. Steinhart, *Chemosphere*, 2002, **48**, 117.

29. S. D. Kohl, P. J. Toscano, W. Hou, and J. A. Rice, *Environ. Sci. Technol.*, 2000, **34**, 204.

30. W. J. Weber, P. M. McGinley, and L. E. Katz, *Environ. Sci. Technol.*, 1992, **26**, 1955.

31. B. Xing and J. J. Pignatello, *Environ. Sci. Technol.*, 1997, **31**, 792.

32. E. J. Leboeuf and W. J. Weber, *Environ. Sci. Technol.*, 1997, **31**, 1697. M. J. Simpson, A. J. Simpson, D. Gross, M. Spraul, and W. L. Kingery, *Environ. Toxicol. Chem.*, 2007, **26**, 1340.

33. G. Cornelissen, P. C. M. Van Noort, G. Nachtegaal, and A. P. M. Kentgens, *Environ. Sci. Technol.*, 2000, **34**, 645.

34. M. Khalaf, S. D. Scott, E. Klumpp, J. A. Rice, and E. Tombacz, *Environ. Sci. Technol.*, 2003, **37**, 2855.

35. C. J. Golding, R. J. Smernik, and G. F. Gavin, *Environ. Sci. Technol.*, 2005, **39**, 3925.

36. R. J. Smernik, *J. Environ. Qual.*, 2005, **34**, 21194.

37. J.-D. Mao and K. Schmidt-Rohr, *Environ. Sci. Technol.*, 2006, **40**, 1751.

38. K. W. Fomba, P. Galvosas, U. Roland, J. Kaerger, and F.-D. Kopinke, *Environ. Sci. Technol.*, 2011, **45**, 5164.

39. J. Mao, G. Ding, and B. Xing, *Commun. Soil Sci. Plant Anal.*, 2002, **33**, 1679.

40. M. A. Eastman, L. A. Brothers, and M. A. Nanny, *J. Phys. Chem. A*, 2011, **115**, 4359.

41. Y.-Y. Chien, E.-G. Kim, and W. F. Bleam, *Environ. Sci. Technol.*, 1997, **31**, 3204.

42. B. E. Herbert and P. M. Bertsch, in *Nuclear Magnetic Resonance Spectroscopy in Environmental Chemistry*, eds M. A. Nanny, R. A. Minear and J. A. Leenheer, Oxford University Press: New York, 1997, 73.

43. D. Smejkalova and A. Piccolo, *Environ. Sci. Technol.*, 2008, **42**, 8440.

44. M. A. Nanny, J. M. Bortiatynski, and P. G. Hatcher, *Environ. Sci. Technol.*, 1997, **31**, 530.

45. M. A. Nanny and J. P. Maza, *Environ. Sci. Technol.*, 2001, **35**, 379.

46. D. Zhu, B. E. Herbert, and M. A. Schlautman, *J. Environ. Qual.*, 2003, **32**, 232.

47. B. Claasen, M. Axmann, R. Meinecke, and B. Meyer, *J. Am. Chem. Soc.*, 2005, **127**, 916.

48. A. Shirzadi, M. J. Simpson, Y. Xu, and A. J. Simpson, *Environ. Sci. Technol.*, 2008, **42**, 1084.

49. A. Shirzadi, M. J. Simpson, R. Kumar, A. J. Baer, Y. Xu, and A. J. Simpson, *Environ. Sci. Technol.*, 2008, **42**, 5514.

50. A. Piccolo, *Soil Sci.*, 2001, **166**, 810.

51. D. Courtier-Murias, H. Farooq, H. Masoom, A. Botana, R. Soong, J. G. Longstaffe, M. J. Simpson, W. E. Maas, M. Fey, B. Andrew, J. Struppe, H. Hutchins, S. Krishnamurthy, R. Kumar, M. Monette, H. J. Stronks, A. Hume, and A. J. Simpson, *J. Mag. Res.*, 2012, **217**, 61.

52. L. A. Cardoza, V. K. Almeida, A. Carr, C. K. Larive, and D. W. Graham, *TrAC*, 2003, **22**, 766.

53. V. S. Bondar, M. G. Boersma, E. L. Golovlev, J. Vervoort, W. J. H. Van Berkel, Z. L. I. Finkelstein, I. P. Solyanikova, L. A. Golovleva, and I. M. C. M. Rietjens, *Biodegradation*, 1998, **9**, 475.

54. J. M. Tront and F. M. Saunders, *Environ. Pollut.*, 2007, **145**, 708.

55. C. A. Moody, W. C. Kwan, J. W. Martin, D. C. G. Muir, and S. A. Mabury, *Anal. Chem.*, 2001, **73**, 2200.

56. B. Key, R. D. Howell, and C. S. Criddle, *Environ. Sci. Technol.*, 1997, **31**, 2445.

57. M. Sarkouhi, M. Shamsipur, and J. Hassan, *Environ. Monit. Assess.*, 2012, **184**, 7383.

58. R. Greenhalgh, B. A. Blackwell, C. M. Preston, and W. J. Murray, *J. Agric. Food Chem.*, 1983, **31**, 710.

59. F. Hong, S. O. Pehkonen, and E. Brooks, *J. Agric. Food Chem.*, 2000, **48**, 3013.

60. M. Seger and G. E. Maciel, *Environ. Sci. Technol.*, 2006, **40**, 797.

61. M. Krolski, L. L. Bosnak, and J. J. Murphy, *J. Agric. Food Chem.*, 1992, **40**, 458.

62. L. M. Condron, E. Frossard, R. H. Newman, P. Tekely, and J.-L. Morel, in Nuclear Magnetic Resonance Spectroscopy in Environmental Chemistry, eds M. A. Nanny, R. A. Minear and J. A. Leenheer, Oxford University Press: New York, 1997, 247.

63. J.-P. Grivet and A.-M. Delort, *Prog. NMR Spectrosc.*, 2009, **54**, 1.

64. P. Besse, B. Combourieu, P. Poupin, M. Sancelme, N. Truffaut, H. Veschambre, and A.-M. Delort, *J. Mol. Catal. B Enzym.*, 1998, **5**, 403.

65. M. T. Mesilaakso, *Environ. Sci. Technol.*, 1997, **31**, 518.

66. U. C. Meier, *Anal. Chem.*, 2004, **76**, 392.

67. A. J. Charlton, J. A. Donarski, B. D. May, and K. C. Thompson, *Spec. Publ. R. Soc. Chem.*, 2009, **317**, 245.

68. E. Guthrie-Nichols, A. Grasham, C. Kazunga, R. Sangaiah, A. Gold, J. Bortiatynski, M. Salloum, and P. Hatcher, *Environ. Toxicol. Chem.*, 2003, **22**, 40.

69. A. Preiss, M. Elend, S. Gerling, E. Berger-Preiss, A.-K. Reineke, and J. Hollender, *WIT Trans. Ecol. Environ.*, 2008, **111**, 127.

70. A. Preiss, E. Berger-Preiss, M. Elend, S. Gerling, S. Kühn, and S. Schuchardt, *Anal. Bioanal. Chem.*, 2012, **403**, 2553.

71. M. Godejohann, L. Heintz, C. Daolio, J.-D. Berset, and D. Muff, *Environ. Sci. Technol.*, 2009, **43**, 7055.

72. M. Godejohann, J.-D. Berset, and D. Muff, *J. Chromatogr. A*, 2011, **1218**, 9202.

73. B. Lam, A. Baer, M. Alaee, B. Lefebvre, A. Moser, A. Williams, and A. J. Simpson, *Environ. Sci. Technol.*, 2007, **41**, 8240.

74. B. T. M. Willis and C. J. Carlile, *Experimental Neutron Scattering*, Oxford University Press: Oxford, UK, 2009.

Chapter 21

Soil–Plant–Atmosphere Continuum Studied by MRI[*]

Henk Van As[1], Natalia Homan[1], Frank J. Vergeldt[1] and Carel W. Windt[2]

[1]*Laboratory of Biophysics and Wageningen NMR Centre, Wageningen University, Dreijenlaan 3, 6703 HA Wageningen, The Netherlands*
[2]*Forschungszentrum, Jülich GmbH, Jülich, 52425, Germany*

21.1 INTRODUCTION

The availability of water is one of the major factors that affect plant production, yield, and reproductive success. Water controls, at least in part, the distribution of plants over the Earth's surface. For growth, plants need to take up CO_2 for photosynthesis, and

[*]Chapter modified from H. Van As, N. Homan, F. J. Vergeldt, C. W. Windt. *MRI of Water Transport in the Soil–Plant–Atmosphere Continuum* in *Magnetic Resonance Microscopy: Spatially Resolved NMR Techniques and Applications*, S. L. Codd & J. D. Seymour (Eds). Pages 315–330. 2009. © Wiley-VCH Verlag GmbH & Co. KGaA. Reproduced with permission.

NMR Spectroscopy: A Versatile Tool for Environmental Research
Edited by Myrna J. Simpson and André J. Simpson
© 2014 John Wiley & Sons, Ltd. ISBN: 978-1-118-61647-5

evaporative water loss is an inevitable consequence of this uptake. Transpiration in higher plants accounts for about one-eighth of all of the water that evaporates to the atmosphere over the entire globe, and for about three-quarters of all water that evaporates from the land. Long-distance transport in plants thus directly affects the global water cycle, and with that, global climate.[1]

Changes in global climate are expected, which may include increased CO_2, global warming, and periods of drought and flooding. These changes will affect (the composition of) ecosystems and agricultural production in many regions of the planet, and an understanding of the short- and long-term responses of plants to climate change is therefore crucial. A key parameter in this understanding is the hydraulic conductance in the soil–plant–atmosphere continuum (SPAC).[1–3]

Water and soluble compounds are passively transported inside plant xylem conduits (vessels and tracheids) in the continuum between soil and atmosphere along a water potential gradient. When this gradient becomes too steep, it causes damage either by dehydration of living cells or by cavitation due to tensions (negative pressures) in the water columns of the xylem being too high.[2,3] Cavitations result in air embolisms that inhibit water transport in the plant and affect the plants' ability to take up CO_2 and to grow. Therefore, mechanisms are needed to maintain this gradient within a nondamaging range. The most important mechanism is regulation of the stomatal aperture in the leaves by increasing the resistance for water vapor leaving the leaves into the atmosphere with a lower water content. The hydraulic conductivity of the root and stem, together with the plants' stomatal regulation, defines the water potential gradients that exist between leaf and root.

In its path from the soil to the atmosphere, water must be transported through the soil, pass living root tissue (radial transport) to reach the root xylem, move inside the xylem conduits up to the leaves, pass living tissue in the leaves, evaporate at the cell wall surfaces in the leaves, and then diffuse through the intercellular air spaces and stomata to the atmosphere. In analyzing the components of the SPAC, it is useful to distinguish the hydraulic conductivity (K, conductance normalized to the length) of the different components. K is defined as

$$K = -\frac{Q}{(d\Psi/dx)} \tag{21.1}$$

where Q is the volume flow rate, Ψ the water potential driving the flow (pressure in soil and xylem, pressure and osmotic components in tissues such as roots and leaves), and x is the distance along the flow path. Equation (21.1) is very similar to Darcy's law. For ease of comparison, K is usually expressed relative to an area transverse to the flow path. In doing so, K-values for soil (K_s), root (K_r), xylem (K_x), and leaves (K_l) can be compared.[2]

K_s and the different plant K-values are not constant, but depend on a number of factors, including the driving force itself. Plant K-values may be subjected to short-term changes. K_r and K_l values, for instance, depend on the presence of aquaporins (water channel-forming membrane proteins), which can be opened or closed depending on internal (e.g., pH) and external (e.g., mechanical stress, temperature) factors.[4,5]

There are important similarities between the flow of water in soils and in the xylem.[2] Bulk flow in both media occurs through pores and is driven by pressure differences (including matrix potential for soil and tension in xylem). Depending on the plant species, the conduits in the xylem consist of relatively wide xylem vessels and/or narrower tracheids, which are interconnected by much narrower channels of the connecting pits. The permeability of the pits largely determines the total xylem conductance and depends

on the K^+-concentration of the xylem sap.[6] A second path of water transport is in the phloem; this is responsible for the transport of photosynthates such as sucrose from the leaves to the rest of the plant. The conductance of this phloem pathway depends on the radii and length of its sieve elements and the characteristics of the pores in the sieve plates between the elements.[7]

There is a strong need for eco-biophysical plant models to predict the dynamics in plant evaporation in relation to photosynthesis, growth, and production under variable climate and soil conditions, including stress conditions. The cohesion–tension theory[8] and the resistance flow model[9] describe a plant or tree as an integrated hydraulic system. However, reality is more complex. Several types of eco-biophysical plant models have been drawn up (e.g., Refs 1 and 10). Some models are already in use in horticultural practice, for example to estimate crop yield or to control greenhouse climate, whereas others are in use as parts of models used to predict the global water cycle and global climate. One of the most complete tree models currently available includes xylem hydraulics, leaf microclimatic factors with feedback signals from tree and soil water content, and the important interaction between xylem, phloem, and sugar content/transport.[10] However, the dynamic behavior of phloem and xylem characteristics (e.g., K_x) and of radial transport has not yet been included. One reason for this is that little is known about the dynamic behavior of phloem and xylem transport and hydraulic conductance in the living plant, in relation to soil and plant water status (storage pools), photosynthetic activity, and sugar content. Although the basic principles that underlie xylem and phloem transport have been known for about a century, their translation into accurate models has remained exceedingly difficult. The problem is that we lack methodological means to produce integrative empirical data that allow testing of explicit hypotheses and models noninvasively. The pressure gradients that drive translocation in both systems are easily disturbed by cutting or puncturing.[11,12]

Among many others, leaf water potential, leaf evaporation, stem water potential, and trunk diameter fluctuations are in use as indicators of plant water status.[10] Most of these methods are invasive, lack automation, or are difficult to interpret. On the cell level, the cell pressure probe has been proved to be very valuable to measure water (and solute) membrane permeability, either diffusional (no driving force, P_d) or under a driving force (hydrostatic or osmotic pressure gradients, P_f).[11,13] On the tissue level, the pressure bomb and root pressure probe allow water potential and tissue hydraulics to be measured.[14] However, while these techniques can be applied to excised roots, leaves, and other parts of a plant, and are clearly very informative, they are destructive. Hydraulic conductivity has been studied by the use of the high-pressure flow meter,[15] which can also only be applied to excised plant parts. Xylem pressure probes have been used to measure xylem tension in intact plants, but the practical applicability of the technique is limited.[11]

For several decades, heat tracer methods have been used to measure mass flow in the xylem. Here, the placement of the sensor itself is a source of error, particularly for the heat pulse method.[16] The calculation of mass flow rates from sap velocities obtained by heat pulse techniques requires a reliable estimate of the actual sap-conducting surface area. However, an accurate estimate of the actual flow-conducting area is extremely difficult to obtain, as changes in flow-conducting area occur under changing environmental conditions,[3,17] even without the occurrence of embolisms. These factors can result in substantial errors of actual sap flow measurement based on heat pulse techniques.

Thus, a truly noninvasive measurement of sap flow and active flow-conducting area in intact plants in an integrative relation to water content in storage pools and radial transport parameters is required that will allow one to study the dynamics (day/night, stress responses) therein. In the following, we summarize several magnetic resonance imaging (MRI) methods that are currently available for this task. In addition, the water content in soils, stem tissue, and leaves can be monitored; however, the measurement of transport in intact root and leaves remains a challenge.

21.2 MAPPING TRANSPORT AND RELATED PARAMETERS IN THE SPAC BY MRI APPROACHES

The application of in vivo NMR and MRI to plants has brought significant contributions across a wide range of topics. For some recent reviews, the reader is referred to Refs 18–23.

21.2.1 The Soil–Root System

21.2.1.1 Soil Water Content and Root Anatomy

Root water uptake mechanisms belong to the most challenging topics in soil science. Transport in soils is quite slow and can best be approached by temporally resolved water content imaging.[23–25] For a thorough understanding, 3D monitoring techniques for determining water content changes must be combined with 3D root anatomy measurements. The determination of water content in soils is complicated by the presence of paramagnetic/ferromagnetic impurities and small and partially filled pores, both of which may cause significant increases in $1/T_2$ and $1/T_1$.[26] The MRI approaches are very similar to those used to map moisture migration in food materials, as well as the problems of calibration in terms of water content.[27] At low water content, the T_2 values are rather short and single-point imaging (SPI)-type experiments are needed.[28] At higher water contents (and depending on the type of soil), multiple spin echo (MSE), turbo spin echo (TSE), or RARE (rapid acquisition with relaxation enhancement) methods can be used.[29] Root anatomy can be imaged using 3D RARE with sufficient resolution; some combined results of multi-slice, multi-echo imaging for soil water content, and 3D RARE for root anatomy are presented in Figure 21.1.

21.2.1.2 Transport in Roots

Water transport in roots is complex, and has been described in terms of a composite root transport model.[11] This model considers three pathways for water transport: the extracellular or apoplastic pathway; the symplastic pathway; and the transcellular pathway. The latter two pathways involve cell-to-cell transport and cannot be discriminated by present techniques. Apoplastic barriers occur in the endodermis and exodermis, and therefore cell-to-cell transport—including transport over membranes and via plasmodesmata—is very important. It is for this reason that membrane water permeability plays such an important role. Because of a difference in the reflection coefficient for nutrients in the apoplast and the membranes, the cotransport of water and nutrients is different in the apoplast and the cell-to-cell paths, and this results in different driving forces being employed—pressure (tension) in the apoplast and osmotic potentials in the cell-to-cell path.

Membrane permeability in tissue-containing vacuolated plant cells can be obtained from the observed

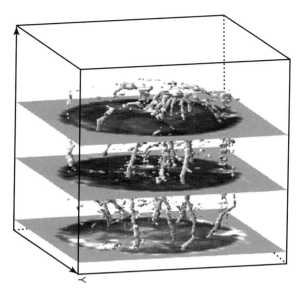

Figure 21.1. Water content changes in a sandy soil around the root system of a *Ricinus* plant, using a 3 T vertical bore MRI. The root anatomy was obtained by 3D RARE, and soil water content by multi-slice 2D multi-echo imaging ($T_E = 6.76$ ms; $N_E = 128$ with an isotropic resolution of 3.1 mm). The (extrapolated) signal intensity (S_0) was correlated to the water content. Field of view: 10×10 cm^2. (Illustration courtesy of A. Pohlmeier, F.J. Vergeldt, E. Gerkema, J. Vanderborght, M. Javaux, H. Vereecken, H. Van As, M.I. Menzel, D. van Dusschoten.)

T_2 value of water in these cells. Vacuolar $T_{2, obs}$ can be described as a function of the bulk T_2 ($T_{2, bulk}$) of the water and the probability to reach the surrounding membrane and the net loss of magnetization at that surrounding membrane. The probability to reach the membrane is defined by the diffusion time and is thus directly related to the compartment radii.[30] No evidence has been found that membranes themselves act as a relaxation sink.[31,32] The net loss of magnetization at the vacuole surrounding membrane depends on the probability to pass the membrane (membrane permeability) and the (much shorter) relaxation time of the compartment that is entered, the cytoplasm. As a result, the observed relaxation time depends, in addition to $T_{2, bulk}$, on the surface-to-volume ratio (S/V) and the net loss of magnetization at the compartment boundary, the so-called magnetization sink strength (H; units m s^{-1})[30,32]:

$$\frac{1}{T_{2,obs}} = H\left(\frac{S}{V}\right) + \frac{1}{T_{2,bulk}} \qquad (21.2)$$

where H is linearly related to the actual membrane permeability.[33] Equation (21.2) is valid only if the diffusion time to traverse the cell is shorter than the bulk relaxation time in that compartment ($R^2/2D$) < $T_{2, bulk}$. Care must be taken into account using $T_{2, obs}$ from images, as the observed T_2 value in images with respect to its value in nonimaging NMR depends on a number of contributions that relate to details of the imaging experiment, as well as to the characteristics of the plant tissues under observation. These include diffusive attenuation in the position-encoding gradient and in local field gradients originating from susceptibility artifacts, in combination with longer inter-echo time (T_E) values.[34]

For a proper interpretation of $T_{2, obs}$ in terms of membrane permeability, it is essential that the value of S/V is known [cf. equation (21.2)]. This information can be obtained by measuring the apparent diffusion coefficient (D_{app}) as a function of the diffusion-labeling time, Δ. For diffusion in a confined compartment, free diffusion is observed at short diffusion times. However, at increasing Δ the diffusion becomes restricted, but the averaging of local properties over a large enough distance does not yet occur. In that regime, D_{app} depends linearly on the square root of Δ, and the slope is determined by S/V of the compartment, irrespective of whether this compartment is connected or disconnected.[35] This S/V value can be used directly to obtain H from $T_{2, obs}$ [equation (21.2)]. At long diffusion times, hindered diffusion is observed (cf. Figure 21.4), which is determined by the permeability, and for plants reflects cell-to-cell transport, defined here as tissue permeability. The effective tissue permeability P can now be estimated. A number of approaches for different systems has been presented,[35] but all improperly ignore the effect of differences in relaxation time in the different compartments.[33]

For diffusion through a geometry consisting of a series of semipermeable membranes (thin walls) separated by a distance d, Crick obtained[36]:

$$P = \frac{(D_{inf} \, D_0)}{(D_0 - D_{inf})d} \qquad (21.3)$$

where D_0 and D_{inf} are the D-values in the limit of Δ to zero or to infinity, respectively.

If D_0, D_{inf}, and d are known, P can be obtained. In practice, the experimental range of Δ-values is limited by the relaxation times T_2 and T_1, $D(\Delta)$ is obtained over a smaller range of Δ-values, and D_{inf} is difficult to obtain. A fitting procedure to extract out

d, D_0, and D_{inf} based on a limited range of $D(\Delta)$ values has been presented.[37–39] This approach has been applied on carrot taproot in imaging mode, resulting in tissue permeability maps.[39] Alternatively, D_0 can be estimated from the initial slope at short Δ values, if experimentally available. P represents cell-to-cell transport and includes the permeability of the tonoplast, plasmalemma, cell walls, and plasmodesmata. Therefore, P is not identical to H as obtained from T_2 measurements [equation (21.2)].

In order to discriminate between the different transport pathways/compartments, the NMR signals must be further unravelled. Water in different cell compartments can best be discriminated on the basis of differences in relaxation behavior (T_2) and (restricted) diffusion behavior. By combined relaxation and diffusion measurements, together with analysis methods such as inverse Laplace transform (ILT)[40,41] or alternatives,[42] correlated D and T_2 values can be generated, which greatly enhance the discrimination of different water pools in subcellular compartments. In this way, an unambiguous correlation between relaxation time and compartment size can be obtained, resulting in a general approach to quantify water in the different cell compartments. This 2D approach is very promising in nonspatially resolved measurements,[43–45] and has been shown to allow subpixel information in imaging mode also to be obtained.[46,47]

Some preliminary results of this approach are now available. By combining the results of T_2 and D measurements on water in apple parenchyma (Granny Smith), H was found to be $\sim 1 \times 10^{-5}$ m s^{-1} ($T_{2, obs} = 1.25$ s; $R = 86$ μm). Under the assumption of parallel planes, $P = 2.9 \times 10^{-6}$ m s^{-1}.[44] For Cox apple parenchyma cells, a tonoplast water membrane permeability $P_d = 2.44 \times 10^{-5}$ m s^{-1} was reported on the basis of the Conlon–Outhred method.[48] In maize roots, a higher value of P of $\sim 5 \times 10^{-5}$ m s^{-1} was found.[37] Values of P were obtained in excised roots of normal and osmotically stressed maize and pearl millet plants: here, P was $\sim 3 \times 10^{-5}$ m s^{-1} for both normal and stressed maize, but was about 9×10^{-5} m s^{-1} for normal pearl millet plants and 3×10^{-5} m s^{-1} for stressed plants (T.A. Sibgatullin and H. Van As, unpublished results). Ionenko *et al.* reported P-values of 3×10^{-5} m s^{-1} for water in roots of maize seedlings, but these decreased by a factor of 1.7 due to water stress or HgCl$_2$ treatment.[49] The latter clearly demonstrates the contribution of aquaporin function toward P.

21.2.2 The Stem

21.2.2.1 Water Content

The stem normally has a relatively high water content that can easily be measured quantitatively using MSE, which results in an amplitude map and a T_2 map. The amplitude in each pixel represents the amount of water in the pixel volume (equal to water content × tissue density). Nonquantitative SE methods have been used to study the refilling of embolized vessels by monitoring changes in signal intensity.[50–53] However, this approach proved to be insufficient because refilling does not automatically result in a restoration of flow; and because vessels that do not exhibit cavitation may stop conducting flow.[54]

In addition to the water content, T_2 maps are obtained by MSE. As discussed in Section 21.2, $T_{2, obs}$ in combination with information on the vacuole S/V ratio provides access to information on membrane water permeability [cf. equation (21.2)]. In intact pearl millet plants, H was shown to change during osmotic stress experiments; however, in maize plants no changes were observed,[30,32] which suggested that the response to stress was most likely to be related to changes in membrane permeability due to the presence and functioning of aquaporins (cf. Section 21.2).

In general, a monoexponential fit is used to construct the amplitude and T_2 maps from MSE experiments. More detailed information on water balance in the different cell compartments is available, and can be extracted by summing the signal decay of tissue types, selected on the basis of the T_2 maps, to improve the signal-to-noise ratio (SNR).[55]

21.2.2.2 Axial Transport in Xylem and Phloem

Both nonimaging and imaging NMR methods have been applied to study water transport in the xylem and phloem of intact plants (for recent overviews, see Refs 20, 22, and 23). In these investigations, the research groups have used a variety of techniques, most of which were based on (modified) pulsed field gradient (PFG) methods, using either a limited number of PFG steps or (difference) propagator approaches (see below). Flow measurements based on the uptake and transport of (paramagnetic) tracers have also been used.

The first (nonimaging) method to measure xylem water transport in plants, presented some 20 years ago,[56,57] is based on a series of equidistant identical radiofrequency (RF) pulses (in the range $30° – 180°$) applied in the presence of a static magnetic field gradient in the direction of the flow. This method allows the averaged flow velocity and volume flow to be obtained, and also the ratio of the effective flow-conducting area. However, in order to interpret the data acquired in terms of averaged flow velocity, a calibration is required. The results depend on the actual flow profile which, within the total cross-section of a stem, is not known a priori. Moreover, neither xylem nor phloem flow can be discriminated using this method; rather, the sum of the two is observed.

Flow profile and direction are best obtained using PFG q-space or propagator techniques, whereby discrimination can be made simultaneously between nonflowing water molecules (stationary water in cells and nonactive vessels) and flowing water. The propagator $P(R, \Delta)$ represents the probability that a spin at any initial position is displaced by a distance R in time Δ (cf. Figure 21.2a). This type of measurement has been applied to measure flow in many porous systems (e.g., Refs 58 and 59).

One crucial step when quantifying the flow and flow-conducting area is to separate the contributions of nonflowing and flowing water, both of which are present in a single pixel in the plant situation. The propagator for free, unhindered, diffusing water has a Gaussian shape, centered around $R = 0$ (Figure 21.2a). The root-mean-square displacement of diffusing protons, as observed by NMR, is proportional to $\sqrt{(\Delta \times D)}$, and is directly related to the width of the Gaussian distribution (Figure 21.2a). In contrast, the mean displacement R of flowing protons is linearly proportional to $(\Delta \times v_{av})$, where v_{av} is the average flow velocity of the flowing protons. The fact that the propagator for stationary water is symmetric around displacement zero is used to separate the stationary from the flowing water. The signal in the nonflow direction is mirrored around zero and subtracted from that in the flow direction, to produce the displacement distribution of flowing and stationary water (Figure 21.2a). The probability (amplitude) of the propagator of flowing water at $R = 0$ then becomes zero. For laminar flow in small capillaries ($r < 100 \, \mu m$) at actual values of Δ, this is a correct assumption (Figure 21.2b).[60]

At increasing Δ-values, the diffusional averaging results in a shift of the lower velocities (or stationary water at the wall) toward higher velocities. A further argument for this assumption can be found in combined PFG propagator-T_2 measurements.[47] If the T_2

decay for each step in the propagator is analyzed, then stationary water has a different T_2 from flowing water, which reflects the different cell and vessel environments for water. At zero displacement, T_2 relates to that of stationary water only (Figure 21.2c). It is clear, however, that water in the vessels cannot be considered as totally isolated from water in the surrounding tissues, as exchange or radial transport has been shown to take place and may have consequences for flow measurements[61] (see also Section 21.2.2.3).

Because the signal amplitude is proportional to the density of the mobile protons, the integral of the propagator provides a measure of the amount of water. The average velocity of the flowing water can be calculated by taking the amplitude-weighted average of the velocity distribution. The volume flow rate or flux is calculated by taking the integral of (amplitude × displacement of the velocity distribution). The value of $P(R,0)$ has no effect on calculating the volume flow rate, but will affect the calculated average velocity and the flow-conducting area.[62] A propagator flow-imaging method based on RARE was developed that allowed the flow profile of every pixel in an image to be recorded quantitatively, with a relatively high spatial resolution, while keeping measurement times down to $15 – 30$ min.[55,63]

The accuracy of this approach for quantifying the sap flow and effective flow-conducting area is deduced from a comparison of the flow characteristics of water obtained by MRI with results for root water uptake and optical microscopy of the vessel dimensions. When the correct MRI settings are used, the flow results agree closely with those obtained by monitoring root water uptake.[54,55] However, care must be taken to cover the full dynamic range of the sap velocity by a proper choice of the q-step.[17,54] The successful discrimination of stationary and flowing water in single pixels containing a single vessel has been demonstrated by the agreement between the calculated cross-sectional area of some large vessels based on the MRI results and on microscopic inspection.[54] Parameters obtained by MRI range within 10% of those from microscopy. To the best of our knowledge, no other method is available at present to obtain this type of data in intact plants.

In MRI, the relaxation behavior of water in porous materials is known to be influenced by pore diameter. It has been investigated whether the T_2-relaxation behavior of xylem sap is affected by the xylem conduit diameter, and whether this must be corrected for in quantitative MR flow imaging.[64] T_2-resolved flow

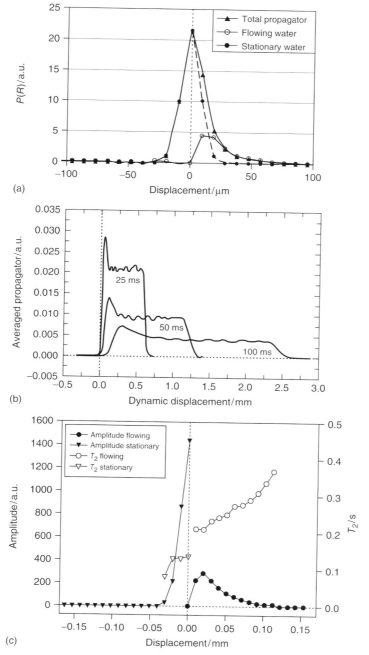

Figure 21.2. (a) Total propagator and the propagators of flowing and nonflowing water (deduced) in the xylem region in the stem of a plant; (b) propagator of laminar flow in a capillary of 100 μm radius as a function of Δ; and (c) propagator of flowing and nonflowing water in the stem of a tomato plant and the correlated T_2 values per displacement step.

imaging[47] has been performed on stem pieces with different conduit sizes through which water was pumped. At 0.7 T, the average weighted T_2 of the flowing water decreased with conduit size, which confirmed that, in the xylem, a relationship between T_2 and conduit diameter does indeed exist. When the T_2 effects were not corrected for, the largest quantification error was ~30% (volume flow) in the sample with the shortest T_2. It was concluded that a T_2 correction only becomes critical when the flow-labeling time approaches the T_2 of the flowing water. Surprisingly, in all cases the flow-conducting area (MR flow imaging) was significantly smaller than the total conduit lumen cross-sectional area (by microscopy). In samples with the widest conduits, the lowest percentage of xylem lumen cross-sectional area was found to conduct water (down to 31% of the total). In samples with narrower conduits, the highest percentage of the total xylem area was found to be active, conducting water (up to 86%).

Windt *et al.* further optimized the propagator fast-imaging method when quantitatively measuring, for the first time, the detailed flow profiles of phloem flow in large and fully developed plants.[17] This approach allowed a straightforward assessment not only of the dynamics of phloem and xylem flow, but also of the flow-conducting area. Most studies have been conducted measuring flow in the stem of a variety of plants, including fully developed maize, cucumber, tomato, castor bean, and tobacco plants, and small trees such as poplar, oak, laurel, pine, and mangrove. In a recent study, for the first time, MRI velocimetry was applied on a truss stalk to determine the ratio of xylem and phloem water transport toward a developing tomato truss.[65] In these studies, a number of totally new and unexpected results have been observed that did not confirm current hypotheses, but questioned the fundamentals of xylem and phloem transport as observed before by invasive or nonselective methods. Examples are as follows:

- effective flow-conducting area and the dynamics in flow-conducting area[17,23,54,66];
- mechanism of phloem transport: the relation between observed velocity and conductance[7];
- xylem and phloem flow in truss stalk of developing tomatoes, ratio of phloem/xylem import, backflow, and (re-)circulation of water even without fruits[65];
- xylem cavitation, repair, and neighboring vessel functioning[54];
- the relation between water content in different tissues and water potential.[23]

All these results demonstrate the potential of this MRI method for studying hydraulics in intact plants under both normal and stress conditions and pose the question whether MRI is rewriting the story of plant water hydraulics.

A very important step will be the further development of MRI method(s) to measure sugar content and transport in addition to the water content and transport.[20] Water and sugar transport are strongly interrelated and essential to sustain growth and development.[10]

21.2.2.3 Radial Transport and Exchange

As stated above (Section 21.1), xylem and phloem hydraulics and plant water content/plant water potential are important parameters that determine evaporation, growth, and stress responses. Water in the xylem and in the surrounding tissues is coupled by a resistance that is determined by the bordered pits in the xylem and the resistance of the (radial) cell-to-cell pathway in the stem. Recently, we have used two MRI approaches in an attempt to characterize these resistances.

In the first approach, the amount of flowing (and stationary) water is measured in a propagator as a function of Δ, in analogy to the approach for studying mass transfer between water in porous beads and the flowing water in chromatography columns.[67] An example of propagators in the xylem region of the stem of a poplar tree, as a function of Δ, is shown in Figure 21.3(a). Owing to exchange of water in the flow-conducting xylem vessels and stagnant water in the surrounding parenchyma cells, the amplitude of the propagator of flowing water increases and the maximum velocity decreases. Diffusional averaging within the vessels then results in a decrease in the width of velocity distribution, while the amplitude of the nonflowing water decreases at increasing Δ. In Figure 21.3(b), the amplitude of the stationary and flowing water is plotted as a function of Δ, with the results showing a clear exchange between the two water pools, mainly by the passage of water over the bordered pits in the vessel walls. However, in contrast to the chromatography columns, in these realistic (bio-)systems such as (woody) plants we have to deal with surface relaxation, internal field gradients due to susceptibility differences, and differences in relaxation times between exchanging flowing and stagnant water pools. All these effects will differently affect the propagator and signal amplitudes.[61] Water in the vessels and in parenchyma cells can be characterized on the basis of differences

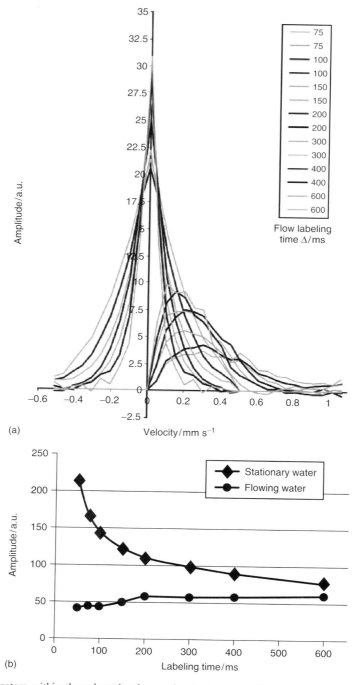

Figure 21.3. (a) Propagators within the selected xylem region in the stem of a poplar tree as a function of Δ. The total propagator has been split into a propagator for nonflowing and flowing water. (b) Integrated amplitude of the propagators of flowing and nonflowing water as a function of Δ.

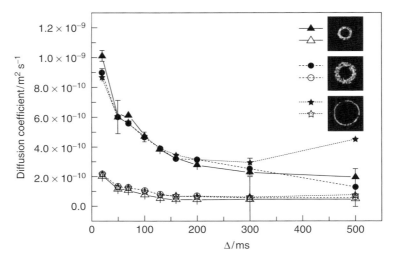

Figure 21.4. D_{app} as a function of the diffusion time Δ in three different xylem regions (see masks) in the stem of a *Viburnum* tree. The D_{app} values have been resolved on the basis of T_2 values by combined D-T_2 measurements. Hindered diffusion is observed at long diffusion times, and can be used to define the cell-to-cell or vessel-to-vessel transport and tissue permeability.

in T_2, T_1, and diffusion coefficients by correlated 2D T_1–T_2, diffusion-T_2, and flow-T_2 MRI measurements. If no exchange is present, the T_1-corrected amplitude of the stagnant fraction should be independent of Δ.[61] Here, we observed that the amplitude of stagnant water after correction for apparent T_1 ($T_{1,\,app}$) strongly increased at increasing Δ. Exchange of water from the stagnant pool to the flowing pool will shorten the residence time in the stagnant pool. The observed T_1 is then given by Homan *et al.*[61]

$$\frac{1}{T_{1,app}} = \frac{1}{T_{1,real}} + \frac{1}{\tau} \qquad (21.4)$$

where τ is the mean residence time of a water molecule in the stagnant water fraction. The use of this $T_{1,\,app}$ to correct the amplitude results in an over correction if exchange is present. $T_{1,\,real}$ is the T_1 value without exchange, which results in a corrected amplitude of the stagnant water pool that is independent of Δ. The calculated τ turned out to be defined by exchange over the barriers between the flowing and stagnant water pools, which is related to the (radial) hydraulic conductivity of this xylem tissue.

In the second approach, $D_{app}(\Delta)$ is measured in the radial direction perpendicular to the flow and analyzed according to equation (21.3), in order to determine tissue permeability (cf. Section 21.2.1.2).[39] Combined D-T_2 or flow-T_2 measurements have also been performed to resolve the water in vessels and in

tracheids/parenchyma cells. The highest T_2 values observed relate to the highest D-value (Figure 21.4) and correspond to the T_2 of flowing water in xylem vessels. The lower T_2 values represent stationary water, most probably in tracheids and (ray) parenchyma cells. The behavior of the D-values was comparable to that of water in confined geometries (see Section 21.2.1.2), and an effective P-value could be calculated using equation (21.3).

Recently, we have demonstrated the integrated approach of long-distance axial and (short distance) radial transport and the effect of water storage pools on the dynamics in the transport pathways.[68]

21.2.3   Leaves

Leaf hydraulic conductance represents a crucial factor in our understanding of water transport and transpiration. As almost all water flux to and within the leaf is lost by transpiration, measurements of such flux should allow leaf transpiration to be mapped at either the plant or the leaf level. However, to the best of our knowledge neither NMR nor MRI flow measurements in leaves have yet been reported.

An alternative approach would be to map the water content. The proton spin density—that is, the amount of water per unit volume—can be used in its own right as a marker of water content in leaf tissues.[69] Although

MRI has been used to monitor (changes in) the water content of leaves, the disadvantage of this approach is that water transport is not measured directly; rather, it is the change in water content that is observed.

In leaves, as with all other tissues, multi-exponential T_2 analyses may provide valuable information with regard to leaf water status and water compartments (see Section 21.2.1.2). Previously, NMR has shown an ability to measure changes in chloroplast water content, in combination with measurements of photosynthetic activity.[31] Chloroplast volume regulation represents a process by which chloroplasts import or export osmolytes to regulate the volume within a certain range of leaf water potential. Hence, as photosynthetic activity rates are directly coupled to changes in chloroplast volumes, these studies are of special interest for monitoring plant performance under stress conditions.[70]

After determining combined relaxation and diffusion measurements, 2D correlation plots between T_1 and T_2 or D and T_2 or T_2 and T_2 exchange[43,71] can be generated, and this greatly enhances the ability to discriminate different pools of water in (sub)cellular compartments; it also reveals the timescale of exchange of water between the different compartments. This approach shows great promise for nonspatially resolved measurements,[43] and may even be carried out using portable unilateral NMR devices (see Section 21.3).

21.3 HARDWARE FOR IN SITU PLANT STUDIES

Hardware solutions for laboratory-bound, dedicated intact plant NMR and MRI have been discussed elsewhere.[22,66] In most cases, only parts of the plant (e.g., stem, leaf, petioles, seed pods, and fruit stalk) will be selected for study, but not the plant as a whole. In this case, an optimal SNR would be obtained by optimizing the radius of the RF coil with respect to that part of the plant to be measured; the smaller the coil radius, the higher the SNR. In fact, the best approach would be to construct RF detector coils that closely fit that part of the plant or tree that is to be imaged.[17,54,55]

In many cases, dedicated hardware is required in order to image intact (woody) plants. For example, Van As *et al.*[71] applied a purposely built low-field (trans)portable NMR instrument to monitor water content and water flux in intact plants in climate rooms and in greenhouses. Relatively cheap imaging setups based on permanent magnet systems start to become available,[71–73] while for quantitative, nonspatially resolved, or imaging methods, specifically designed magnets and gradient coils are currently under development.[74,75] Very recently, a transportable MRI system for tree inspections,[76] a portable MRI system based on a hinged magnet (NMR cuff),[77] a mobile soil NMR sensor (NMR endoscope),[78] and a number of (inhomogeneous) unilateral (single-sided) NMR magnets have been presented. The last-mentioned type of magnet may be of particular interest for leaf studies,[70] and might even be combined with fluorescence-based methods for monitoring photosynthetic activity in relation to leaf water content/chloroplast volume, perhaps even in imaging mode.[79]

RELATED ARTICLES IN EMAGRES

Anisotropically Restricted Diffusion in MRI

Diffusion and Flow in Fluids

Diffusion Measurements by Magnetic Field Gradient Methods

Diffusion in Porous Media

NMR Microscopy: Resolution

Plant Physiology

Plants, Seeds, Roots, and Soils as Applications of Magnetic Resonance Microscopy

Susceptibility and Diffusion Effects in NMR Microscopy

Whole Body Magnetic Resonance Angiography

REFERENCES

1. P. J. Sellers, R. E. Dickinson, D. A. Randall, A. K. Betts, F. G. Hall, J. A. Berry, G. J. Collatz, A. S. Denning, H. A. Mooney, C. A. Nobre, N. Sato, C. B. Field, and A. Henderson-Sellers, *Science*, 1997, **275**, 502–509.

2. J. S. Sperry, U. G. Hacke, and J. P. Comstock, *Plant Cell Environ.*, 2002, **25**, 251–263.

3. M. Mencuccini, *Plant Cell Environ.*, 2003, **26**, 163–182.

4. H. Javot and C. Maurel, *J. Exp. Bot.*, 2002, **54**, 2035–2043.

5. S. H. Lee, G. C. Chung, and E. Steudle, *Plant Cell Environ.*, 2005, **28**, 1191–1202.

6. A. Nardini, S. Salleo, and S. Jansen, *J. Exp. Bot.*, 2011. [Online]. DOI: 10.1093/jxb/err208.

7. D. L. Mullendore, C. W. Windt, H. Van As, and M. Knoblauch, *Plant Cell*, 0210, **22**, 579–593.

8. H. H. Dixon and J. Joly, *Philos. Trans. R. Soc. London, Ser. B*, 1895, **186**, 563–576.

9. T. H. van den Honert, *Faraday Discuss.*, 1948, **3**, 146–153.

10. V. De Schepper and K. Steppe, *J. Exp. Bot.*, 2010, **61**, 2083–2099.

11. E. Steudle, *Annu. Rev. Plant Physiol.*, 2001, **52**, 847–875.

12. G. W. Koch, S. C. Sillet, G. M. Jennings, and S. D. Davis, *Nature*, 2004, **428**, 851–854.

13. A. D. Tomos and R. A. Leigh, *Annu. Rev. Plant Physiol.*, 1999, **50**, 447–472.

14. T. Henzler, R. N. Waterhouse, A. J. Smyth, M. Carvajal, D. T. Cooke, A. R. Schäffner, E. Steudle, and D. T. Clarkson, *Planta*, 1999, **210**, 50–60.

15. M. T. Tyree, S. Patino, J. Bennink, and J. Alexander, *J. Exp. Bot.*, 1995, **46**, 83–94.

16. M. J. Clearwater, F. C. Meinzer, J. L. Andrade, G. Goldstein, and M. Holbrook, *Tree Physiol.*, 1999, **19**, 681–687.

17. C. W. Windt, F. J. Vergeldt, P. A. de, Jager, and H. Van As, *Plant Cell Environ.*, 2006, **29**, 1715–1729.

18. Y. Shachar-Hill and P. E. Pfeffer, eds., 'Nuclear Magnetic Resonance in Plant Biology', American Society of Plant Physiologists: Rockville, MD, 1996.

19. J. A. Chudek and G. Hunter, *Prog. Nucl. Magn. Reson. Spectrosc.*, 1997, **31**, 43–62.

20. W. Köckenberger, *TRENDS Plant Sci.*, 2001, **6**, 286–292.

21. R. G. Ratcliffe, A. Roscher, and Shachar.-Hill. Y, *Prog. Nucl. Magn. Reson. Spectrosc.*, 2001, **39**, 267–300.

22. H. Van As, *J. Exp. Bot.*, 2007, **58**, 743–756.

23. H. Van As, T. Scheenen, and F. J. Vergeldt, *Photosynth. Res.*, 2009, **102**, 213–222.

24. P. A. Bottemley, H. H. Rogers, and T. H. Foster, *Proc. Natl. Acad. Sci. U.S.A.*, 1986, **83**, 87–89.

25. H. Van As and D. van, Dusschoten, *Geoderma*, 1997, **80**, 405–416.

26. F. Jäger, N. Rudolph, F. Long, and G. E. Schaumann, *Soil Sci. Soc. Am. J.*, 2008, **72**, 1694–1707.

27. J. P. M. van, Duynhoven, G.-J. W. Goudappel, W. Weglarz, C. W. Windt, P. Ramos Cabrer, A. Mohoric, and H. Van As, in 'Magnetic Resonance Microscopy: Spatially Resolved NMR Techniques and Applications', eds S. L. Codd and J. D. Seymour, Wiley-VCH Verlag GmbH: Weinheim, 2009, Chapter 21.

28. A. Pohlmeier, A. M. Oros-Peusquens, M. Javaux, M. I. Menzel, V. Vereecken, and N. J. Shah, *Magn. Reson. Imaging*, 2007, **25**, 579–580.

29. A. Pohlmeier, F. J. Vergeldt, E. Gerkema, H. Van As, D. van, Dusschoten, and H. Vereecken, *Open Magn. Reson. J.*, 2010, **3**, 69–74.

30. L. van der Weerd, M. M. A. E. Claessens, T. Ruttink, F. J. Vergeldt, T. J. Schaafsma, and H. Van As, *J. Exp. Bot.*, 2001, **52**, 2333–2343.

31. D. McCain, *Biophys. J.*, 1995, **69**, 1111–1116.

32. L. van der Weerd, M. M. A. E. Claessens, C. Efdé, and H. Van As, *Plant Cell Environ.*, 2002, **25**, 1538–1549.

33. L. van der Weerd, S. M. Melnikov, F. J. Vergeldt, E. G. Novikov, and H. Van As, *J. Magn. Reson.*, 2002, **156**, 213–221.

34. H. T. Edzes, D. van, Dusschoten, and H. Van As, *Magn. Reson. Imaging*, 1998, **16**, 185–196.

35. P. N. Sen, *Concept Magn. Reson. A*, 2004, **23**, 1–21.

36. F. Crick, *Nature*, 1970, **225**, 420–422.

37. A. V. Anisimov, N. Y. Sorokina, and N. R. Dautova, *Magn. Reson. Imaging*, 1998, **16**, 565–568.

38. R. Valiullin and V. Skirda, *J. Chem. Phys.*, 2001, **114**, 452–458.

39. T. Sibgatullin, F. J. Vergeldt, E. Gerkema, and H. Van As, *Eur. Biophys. J.*, 2010, **39**, 699–710.

40. L. Venkataramanan, Y. Q. Song, and M. D. Hurlimann, *IEEE T. Signal Proces.*, 2002, **50**, 1017–1026.

41. M. D. Hürlimann, L. Venkataramanan, and C. Flaum, *J. Chem. Phys.*, 2002, **117**, 10223–10232.

42. D. van, Dusschoten, P. A. de, Jager, and H. Van As, *J. Magn. Reson., Ser. A*, 1995, **116**, 22–28.

43. Y. Qiao, P. Galvosas, and P. T. Callaghan, *Biophys. J.*, 2005, **89**, 2899–2905.

44. T. A. Sibgatullin, P. A. de, Jager, F. J. Vergeldt, E. Gerkema, A. V. Anisimov, and H. Van As, *Biophysics*, 2007, **52**, 196–203.

45. K. E. Washburn and P. T. Callaghan, *Phys. Rev. Lett.*, 2006, **97**, 175502.

46. D. van, Dusschoten, C. T. Moonen, P. A. de, Jager, and H. Van As, *Magn. Reson. Med.*, 1996, **36**, 907–913.

47. C. W. Windt, F. J. Vergeldt, and H. Van As, *J. Magn. Reson.*, 2007, **185**, 230–239.

48. J. E. M. Snaar and H. Van As, *Biophys. J.*, 1992, **63**, 1654–1658.

49. I. F. Ionenko, A. V. Anisimov, and F. G. Karimova, *Biol. Plantarum*, 2006, **50**, 74–80.

50. N. M. Holbrook, E. T. Ahrens, M. J. Burns, and M. A. Zwieniecki, *Plant Physiol.*, 2001, **126**, 27–31.

51. M. J. Clearwater and C. J. Clark, *Plant Cell Environ.*, 2003, **26**, 1205–1214.

52. K. Kuroda, Y. Kambara, T. Inoue, and A. Agawa, *IAWA J.*, 2006, **27**, 3–17.

53. K. Fukuda, S. Utsuzawa, and D. Sakaue, *Tree Physiol.*, 2007, **27**, 969–976.

54. T. W. J. Scheenen, F. J. Vergeldt, A. M. Heemskerk, and H. Van As, *Plant Physiol.*, 2007, **144**, 1157–1165.

55. T. W. J. Scheenen, A. M. Heemskerk, P. A. de, Jager, F. J. Vergeldt, and H. Van As, *Biophys. J.*, 2002, **82**, 481–492.

56. H. Van As and T. J. Schaafsma, *Biophys. J.*, 1984, **45**, 496–472.

57. J. E. A. Reinders, H. Van As, T. J. Schaafsma, P. A. de, Jager, and D. W. Sheriff, *J. Exp. Bot.*, 1988, **39**, 1199–1210 and 1211–1220.

58. M. D. Mantle and A. J. Sederman, *Prog. Nucl. Magn. Reson. Spectrosc.*, 2003, **43**, 3–60.

59. S. Stapf and S.-I. Han, eds., 'NMR Imaging in Chemical Engineering', Wiley-VCH Verlag GmbH: Weinheim, 2005.

60. U. Tallarek, E. Rapp, T. Scheenen, E. Bayer, and H. Van As, *Anal. Chem.*, 2000, **72**, 2292–2301.

61. N. M. Homan, B. Venne, and H. Van As, *Phys. Rev. E*, 2010, **82**, 026310-1–026310-9.

62. T. W. J. Scheenen, D. van, Dusschoten, P. A. de, Jager, and H. Van As, *J. Exp. Bot.*, 2000, **51**, 1751–1759.

63. T. W. J. Scheenen, F. J. Vergeldt, C. W. Windt, P. A. de, Jager, and H. Van As, *J. Magn. Reson.*, 2001, **151**, 94–100.

64. C. W. Windt, 'NMR imaging of sap flow in plants', PhD Thesis, Wageningen University, 2007, http://edepot.wur.nl/40876 (accessed 2007).

65. C. W. Windt, E. Gerkema, and H. Van As, *Plant Physiol.*, 2009, **151**, 830–842.

66. N. M. Homan, C. W. Windt, F. J. Vergeldt, E. Gerkema, and H. Van As, *Appl. Magn. Reson.*, 2007, **32**, 157–170.

67. U. Tallarek, F. J. Vergeldt, and H. Van As, *J. Phys. Chem. B*, 1999, **103**, 7654–7664.

68. N. Homan, 'Functional magnetic resonance microscopy of long- and short-distance water transport in trees', PhD Thesis, Wageningen University, 2009, http://edepot.wur.nl/13609 (accessed 2009).

69. J. S. Veres, G. P. Cofer, and G. A. Johnson, *New Phytol.*, 1993, **123**, 769–774.

70. D. Capitani, F. Brilli, L. Mannina, N. Proietti, and F. Loreto, *Plant Physiol.*, 2009, **149**, 1638–1647.

71. H. Van As, J. E. A. Reinders, P. A. de, Jager, P. A. C. M. van der Sanden, and T. J. Schaafsma, *J. Exp. Bot.*, 1994, **45**, 61–67.

72. M. Rokitta, E. Rommel, U. Zimmermann, and A. Haase, *Rev. Sci. Instrum.*, 2000, **71**, 4257–4262.

73. T. Haishi, T. Uematsu, Y. Matsuda, and K. Kose, *Magn. Reson. Imaging*, 2001, **19**, 875–880.

74. H. Raich and P. Blümler, *Prog. Nucl. Magn. Reson. Spectrosc. B*, 2004, **23**, 16–25.

75. B. Blümich, V. Anferov, S. Anferova, M. Klein, R. Fechete, M. Adams, and F. Casanova, *Concepts Magn. Reson. B*, 2002, **15**, 255–261.

76. T. Kimura, Y. Geya, Y. Terada, K. Kose, T. Haishi, H. Gemma, and Y. Sekozawa, *Rev. Sci. Instrum.*, 2011, **82**, 053704.

77. C. W. Windt, H. Solter, D. van, Dusschoten, and P. Blümler, *J. Magn. Reson.*, 2011, **208**, 27–33.

78. O. Sucre, A. Pohlmeier, A. Miniere, and B. Blümich, *J. Hydrol.*, 2011, **406**, 30–38.

79. J. Perlo, F. Casanova, and B. Blümich, *J. Magn. Reson.*, 2004, **166**, 228–235.

PART C
NMR and Environmental Metabolomics

Chapter 22

Environmental Metabolomics

Daniel W. Bearden

Analytical Chemistry Division, National Institute of Standards and Technology, Hollings Marine Laboratory, Charleston, SC 29412, USA

22.1 INTRODUCTION

The application of metabolomics in the field of environmental science or ecology, which has developed based substantially on NMR spectroscopic approaches, is a fast-paced, rapidly developing field which seems to be poised to help reframe the discussion of environmental effects on organisms. Because of the nature of metabolomics, where experiments are based on ensembles of individuals, one is led to observations that are pertinent to population-, community-, and ecosystem-scale issues. This is in contrast to human-health-related metabolomics where one often wishes to observe or diagnose the condition of a single individual from a population. The environmental metabolomics literature is expanding and the field is maturing at a rapid pace, in part not only because of the

NMR Spectroscopy: A Versatile Tool for Environmental Research
Edited by Myrna J. Simpson and André J. Simpson
© 2014 John Wiley & Sons, Ltd. ISBN: 978-1-118-61647-5

advances in human-health metabolomics research but also because of the unique insight that this approach brings to an important area with global implications.

This chapter focuses on the study of environmental factors that impact the health and well-being of "non-model" organisms in the environment, in an effort to demonstrate that the application of NMR-based metabolomics can enhance traditional approaches to environmental science. In this view, non-model organisms represent a unique realm, distinct from the realm of direct human-health-related organisms, although these realms do interact in important ways. This realm includes some members that may serve as early warning sentinels for ecological issues, some which are commercially valuable for tourism, food, or sport, and some which deserve attention because there may be some species which, if negatively impacted by an unanticipated response to pollution, may cause widespread effects on the ecosystem structure, possibly affecting mankind.

One of the opportunities in environmental metabolomics is the number of relevant species about which little specific biochemical information is known. There are species that are relevant based on geography (diatoms in the Antarctic, or plants and animals near point pollution sources, for example) and other species that are seen as the basis for a complex, interconnected food web which may be perturbed owing to environmental change or contamination. Some populations need to be studied because of the need to preserve diversity and conserve protected species. For many of these organisms, little is known at the genomic or proteomic level, and sometimes even basic characteristics such as diet, range, or reproductive patterns are poorly characterized.

Many anthropogenic contaminants have been well characterized in terms of temporospatial distribution and toxicological impacts on relevant species, including humans. While concerns about historical contaminants such as polychlorinated biphenyls (PCBs), polycyclic aromatic hydrocarbons (PAHs), trace heavy metals, and pesticides have been addressed since the 1960s through the development of robust analytical techniques and systematic toxicological protocols, a rising awareness of less studied pollutants is raising questions that may not be as amenable to the established approaches. New materials for the production of consumer and industrial products or new drugs find their way into the environment through manufacturing processes, usage, and in the waste stream, where, for example, wastewater treatment has not been designed to properly treat the waste stream for the new contaminants.

The need to develop assessments of sublethal stressors such as those related to climate change, personal pharmaceutical products in the waste stream, new generations of pesticides, and new consumer-related chemicals entering the home and workplace continues to grow.[1] Some of these new chemical stressors may act as endocrine mimics, causing subtle effects in the reproductive biology of organisms. Others may have very species-specific interactions that are undetected in the established regulatory processes, which typically have a limited suite of biological tests, resulting in impacts on non-target organisms that only become apparent once the chemicals have been in use and their distribution is widespread.

This onslaught of new chemical and physical stressors, and the awareness of the importance of environmental services linked to non-model organisms, can overwhelm traditional approaches to environmental research (see Chapters 24, 25, 26, and 27). Environmental metabolomics provides new tools to link environmental stressors to specific biological responses, in a discovery mode where the biochemistry of the organisms can be illuminated and also in a quantitative, hypothesis-driven mode where specific questions can be addressed.

22.2 BACKGROUND

While the role of NMR in general environmental research has been growing, techniques such as

chromatography-based light spectroscopy or mass spectrometry continue to be the analytical workhorses in the area of chemical environmental research, where most pollutants are measured at trace levels. NMR spectroscopy has been shown to be useful in a number of important areas of environmental research.[2,3] For example, NMR is a primary technique in the purity assessment of compounds and is key in the identification and quantification of compounds.[4] In the general chemical sciences where, for example, new synthetic products are created, NMR has had a significant role in structure determination, including stereochemical assignments.

Since the explicit proposal that "…the thorough quantitative analysis of body fluids might permit differential diagnosis of many diseases in a more effective way than is possible at the present time" by Linus Pauling *et al.* in 1971,[5] the ability to quantitatively assess the complement of small molecule, endogenous metabolites in living organisms has shown practical results for human health through disease research, dietary studies, and numerous other health-related endeavors.

The concept was clarified and expanded beyond just disease diagnosis in 1999 in a seminal work.[6] A systems approach, where the overall variation of metabolite concentrations is considered comprehensively, is fundamentally different from the approach where a few specific metabolites are individually assessed. While the map of the important metabolic pathways was painstakingly developed through skillful classical chemical experimentation on a reaction-by-reaction basis, the concept behind the field of metabolomics is the simultaneous direct observation of as many endogenous metabolites as possible in a "snapshot" of the instantaneous physiological condition of an organism. In the last two or three decades, these ideas have been expanded, especially in relation to human health research, where the utility of NMR-based measurements has been shown to address effectively the concepts of metabolome assessment.

Given the successes in human-health metabolomics research, it is only natural to apply these approaches to environmentally relevant, non-model organisms.[7–13]

22.3 RATIONALE

NMR is an excellent tool for the assessment of the complex constitution of biomaterials because it is an unbiased detector, absent some well-documented systematic pitfalls, of the organic compounds in multi-component samples (see Chapter 23). The NMR signal is a superposition of the spectra of all the components in the sample, although there can be some nonadditive effects because of physical/chemical interactions between compounds which complicate spectral analysis. Samples can be prepared in relatively simple ways, especially for biofluids such as urine, plasma, serum, or cerebrospinal fluid (CSF), often with no need for any chromatographic separation or cleanup, avoiding the quantitative complications associated with chromatography and/or chemical derivatization. The signals of individual chemical constituents are intrinsically proportional to concentration, simplifying quantitative analysis. Most metabolites have spectra that exhibit multiple resonances which allow identification of the compound from simple ^{1}H spectra, and when correlation spectra such as correlation spectroscopy (COSY), total correlation spectroscopy (TOCSY), or heteronuclear single-quantum correlation spectroscopy (HSQC) are used, the identification of compounds becomes even more specific.

Metabolomics has been shown to be very sensitive to external effects on organisms. For chemical exposures, effects of exposure can often be detected at environmentally relevant concentrations, avoiding the difficulties involved in extrapolating from a high-exposure experiment to much lower levels of exposure.[14] However, this sensitivity is a two-edged sword in that experiments must be designed as carefully as possible to eliminate erroneous observations. In a classic example of good practice gone bad,[15] a laboratory exposure involving rats was found to be problematic because of a change in feed between the supplier and the pharmaceutical research laboratory; the rats had not been equilibrated sufficiently on the new diet before the experiments were run. The implications for environmental research are severe, since often relatively uncharacterized organisms are used in such studies. For example, one classic approach to obtain a working population for study involves field collection of organisms and equilibration in the laboratory. However, Hines *et al.*[16] contrasted laboratory-equilibrated mussels with field-frozen mussels and found significant differences in the metabolite profiles. Their recommendation was to only use field-sampled tissues and fluids to avoid increasing the metabolic variability that may mask the effect being studied. In situations that seem very amenable to laboratory studies, for example studies with microorganisms such as

bacteria,[17] small changes in sample history can cause apparent metabolomic shifts that may confuse the interpretation of results. The sensitivity of metabolomics to phenotypic variation must be appreciated and controlled as the work in this field advances.

One great advantage of using metabolomics for environmental research is the ability to distinguish different modes of action[18] due to different toxicants in sentinel organisms. Potentially, organisms from the field can be assessed for metabolic fingerprints of the different modes of action of various physical and chemical stressors, so that an effective assessment of community health can be made. These metabolic responses will be time- and dose-dependent, so that in well-modeled systems, a complete dynamic picture of ecosystem health can be developed.

While many of the organisms of interest in environmental science have not been well characterized proteomically or genetically, it is still possible to understand the stress responses from a metabolomics viewpoint. In fact, knowing the idiosyncrasies of the metabolic response may point to areas where genetic/proteomic studies should be pressed. Discovery of disproportionate metabolite signals or hitherto ignored compounds[19] may indicate novel genetic mechanisms which need investigation.

NMR has a dynamic range of several orders of magnitude, which can be increased through longer data acquisition or other approaches that increase the signal-to-noise ratio (SNR). The absolute sensitivity of NMR can be easily exceeded by the use of techniques such as fluorescence or mass spectrometry, but the trade-off in selectivity for these other techniques, coupled with the requisite need for some sort of chromatography, is often compensated by the broad-based, nondestructive, nonselective detection afforded by NMR. In terms of the natures of the compounds detected, the use of NMR affords the widest range of detection of chemical moieties in a single analysis. Sugars, organic acids, lipids, amino acids, and so on are easily detected and quantified in a single sample in a single experiment. While it is important to identify as many metabolites as possible in a sample, the mere nonquantitative detection of a metabolite using mass spectrometry, for example, does not necessarily give insight into the metabolic response to a stressor, especially for subacute responses. Given the extremely complex and correlated nature of the metabolome, one must carefully draw the line between extremely sensitive detection of every compound in a sample, and the need for quantitative or semiquantitative assessment of "important" metabolites that can help with the problem at hand. However, as the field matures, there is a building consensus that the use of multiple modalities of compound detection and quantitation provides significant advantages in understanding the metabolomic system response.

NMR fits comfortably in the continuum of measurement techniques because of the ability to obtain quantitative metabolite "patterns" while also providing quantitative chemical-specific information for a wide variety of organic compounds. These features give NMR-based metabolomics a role in the discovery of new metabolomic insight and in classical hypothesis-driven investigations that link organism biochemistry to environmental stressors.

22.4 TECHNIQUES

22.4.1 Experiment Design

Robust experiment design is key to meaningful metabolomics results. Good experimental design requires careful communication and the ability to work with people from other specializations, developing a dialog with a common vocabulary, perhaps even developing a formal ontology.[20] There is a trend in the literature showing that experiments are improving from an experimental design standpoint.[21] For example, in many published reports, the number of samples analyzed is large enough to develop meaningful statistical inferences, and the need for repeating trials is being recognized. Since many of the practitioners have ties to the environmental research community, the need for standard practices and quality control (QC) is recognized as important in improving the confidence in the reported results. Ideas for robust analytical measurements can be borrowed from the environmental analytical community, such as the use of certified reference materials, project-specific control materials, measures of analytical repeatability,[22] and interlaboratory comparison exercises.[23] In terms of the biological component of the experimental design, husbandry of the organisms must be considered in terms of effects on subsequent NMR experiments. In handling the organisms for sample collection, stress induction must be minimized and rapid quenching of metabolic processes should be of paramount importance, especially in tissues that are metabolically active such as the liver. The

effects of feeding, infection (both bacterial and viral), species misidentification, or silent phenotypes must all be considered in the biological design of the experiment. Because of the trueness and precision of NMR experiments, the repeatability of sample preparation and the robustness of the statistical tools used for data analysis, most practitioners end up confronting biological variability as the most challenging aspect of environmental metabolomics. Time spent developing a well-designed biological study will return rewards in high-quality, repeatable results with a significant impact in the field.

22.4.2 Sample Extraction and Cleanup

The broad appeal of NMR-based metabolomics is that one is able to garner meaningful metabolomics results without the additional complications of chromatographic fractionation. However, sample extraction or cleanup is a critically important factor, because there is no recognized method to isolate quantitatively all the organic metabolites from biological samples; each extraction or cleanup protocol introduces some bias in the quantitative extraction of metabolites. This has not proven to be a major stumbling block, because it is possible to make valid inferences based on well-extracted samples. Perhaps this robustness is due to the fact that the measurements sample a network of metabolites, and as long as the extractions are analytically consistent and reasonably robust, the network responses can be detected in a meaningful manner.

Because of the various organisms and matrices considered in environmental metabolomics, samples are processed in ways that are tissue-dependent, and in most cases, different species require variations in extraction protocols. The most convenient matrix to work with would be the one with the least bench workup required, such as a body fluid. However, the different constituents in fluids such as plasma or serum warrant some effort in validation of the sample workup.[24,25] For tissues, the extraction process can cause a bias if the extraction efficiency is not examined systematically.[26] Depending on the matrix, some extraction schemes are better than others, much as in environmental analytical chemistry. Since no real chromatography is performed, one may view this as a "cleanup" process whereby, for example, large molecular weight molecules such as proteins, DNA/RNA, polysaccharides, and other macromolecules are removed, leaving behind the small molecule (<500 Da) metabolites.

Various extraction schemes have been systematically optimized for some environmentally relevant matrices.[26,27] The optimizations proposed in these systematic studies are selected on the basis of the observed repeatability of the method and some measure of the amount or number of metabolites extracted. A common extraction technique is based on resolving polar and nonpolar metabolites into separate fractions, for example using a modified Bligh and Dyer scheme.[26,28] This is often desirable since some experiments may be rationalized based on more polar components (such as amino acids, tricarboxylic acid (TCA) cycle metabolites, organic acids, and aromatic compounds) in a polar solvent such as water-based buffer, while some compounds associated with lipids or cholesterol synthetic pathways would be in a nonpolar solvent such as chloroform. For high lipid-content tissues, the spectral simplification after polar/nonpolar resolution is significant, leading to clearer interpretation of the results. Some schemes have been optimized to use size exclusion cleanup based on size exclusion filter technology, either by itself or on various fractions previously separated using other techniques.[29] For blood or hemolymph studies, numerous efforts are reported for optimization of metabolite extractions in plasma and serum, often based on filtering techniques.

Well-executed projects invariably have invested suitable effort in validating the extraction/workup procedures to optimize the metabolite fingerprints and sensitivity of the experiments.

22.4.3 NMR Data Acquisition

Metabolomics analysis depends greatly on the type and quality of NMR data collected. Most environmental metabolomics projects depend on one-dimensional (1D) data collection of polar solutes in D_2O. The best spectra have very flat baselines (so that subsequent baseline corrections are easy), good phase characteristics, reasonable resolution, minimal spectral artifacts, and very good SNR. In addition, spectrum-to-spectrum consistency is very important because most of the numerical pattern detection techniques will select for features that vary between spectra.

It is good practice to develop a consistent protocol for data acquisition that provides consistently good results. For example, it is more desirable to set up the instrument carefully and then run all the samples in a project in one "session" than to run samples in multiple sessions over a few days. For very large projects,

it may be impractical to run all the samples in one session, so the protocols should involve measures that enforce and verify consistency between sessions, such as repeat runs of select samples or measurements of line widths or SNR. Protocols could cover factors such as temperature stability, temperature measurement,[30] shimming protocols, standard parameters for the pulse sequences, pulse width calibration, and standardized processing parameters. While some projects require deviations from standard protocols, having a consistent starting place for making those decisions is good practice.

For water-based samples, most laboratories use water suppression pulse sequences. Depending on the sample preparation protocols, samples may be in 90% H_2O or neat D_2O or somewhere in between, so the exact water suppression technique must be optimized for that class of sample and for the particular instrument being used. Because of the spectral artifacts that can be introduced and the need for high-quality semiquantitative spectra, the optimization of the water suppression technique is critical. Various suppression schemes have been optimized for water suppression in systematic studies,[31–33] and sequences based on a three-pulse Nuclear Overhauser effect spectroscopy (NOESY)-type sequence[32] are often used on samples prepared in D_2O. However, more rigorous water suppression techniques are not uncommon. Optimization of suppression often considers baseline distortions, intensity perturbations near the water resonance at the theoretical lobes of the suppression sequence, and difficulty of calibration and setup.[34,35] Trade-offs between these factors often come into play, and local optimization of water suppression is crucial to meaningful, consistent results.

For samples that contain residual proteins or high molecular weight lipids, such as plasma or serum, the use of spin-echo Carr–Purcell–Meiboom–Gill (CPMG) sequences can act as T_2-weighting filters that reduce the contribution of broadline signals from high molecular weight species. The optimization of these sequences balances the duration of the effective spin-echo delay against the phase distortion due to homonuclear couplings and the loss of intensity due to pulse imperfections and relaxation effects. Reports of cumulative spin-echo delays in the range of 100 ms have given satisfactory results. Combining the spin-echo sequence with strong water suppression in high H_2O content samples can also be challenging,[36–38] and must be weighed against the improvement gained by processing the samples to remove high molecular weight components.

Some reports show that 2D spectroscopy can lead to superior results in pattern recognition and compound identification. One 2D experiment that seems useful is 2D J-resolved spectroscopy (2D-JRES).[39] The tilt-corrected data can be used for identification of compounds, and a skyline projection along the direct dimension results in a homonuclear decoupled spectrum that significantly reduces the spectral complexity by collapsing the homonuclear multiplets. This projection can then be used in pattern recognition approaches in the very same manner as direct 1D spectra. Other 2D homonuclear experiments such as COSY and TOCSY can also be useful, especially for compound identification. For more reliable compound identification, heteronuclear experiments such as ^{13}C-HSQC and ^{13}C-heteronuclear multiple bond correlation (HMBC) provide nearly unambiguous compound identification in natural abundance samples, although at a somewhat higher cost than the 1D experiments because of the need for longer acquisition.

22.4.4 Data Processing and Analysis

There are many algorithmic approaches to discerning the systematic variation in the spectra from a metabolomics experiment, including many of the tools developed for fields such as functional genomics. Once patterns are detected, it is important to carefully evaluate whether the patterns are statistically significant or a result of systematic error. In some experiment designs, the number of samples is too small in an experimental group, and a determination must be made on whether the data is representative of the populations or whether outliers have a significant influence. This assessment is no different from metabolomics in other fields. Interpretation of the results depends on the robustness of the experimental design, where phenotypic variability of the sample pool was assessed, for example, and should be evaluated in the light of the results on QC samples, which help quantify the variation due to sample preparation or NMR spectral quality.

Once there is confidence that a true pattern exists, the compounds that contribute to the separation in the pattern must be identified so that linkages to metabolic pathways can be established. Therefore, pattern recognition techniques should be chosen based on their ability to provide both pattern detection and chemically relevant compound identification. In the simplest of cases, a simple univariate approach based on comparison of group-averaged spectra, for example, may lead to an understanding of the biochemical basis for differences between the treatment groups.[40–42] In these cases, there are probably very small numbers of compounds that are significantly different between treatment groups. However, metabolic response to a stressor may be more subtle, and spread over a wide range of metabolic compounds or pathways, since the whole metabolome responds to the stressors in the experiment. In this case, multivariate analysis techniques that are sensitive to coherent variation in numerous chemical signals simultaneously are most informative for detecting this coherent variation. Because of the nature of NMR spectra, where an individual compound often has numerous peaks, there is considerable correlation in the data, so that each spectral point or bin does not necessarily represent a fully independent variable. This results in a reduction of rank, and numerical methods that are robust to this nonindependence should be the most trustworthy.

The preprocessing of the data also affects the pattern recognition process. For example, baseline correction is necessary because of imperfections in instrumentation, including background signals from the probe or preamplifiers.[43,44] Sometimes, there may be effects from the receiver/digitizer system that cause rolling baselines. Baselines are corrected with any of a myriad of techniques, ranging from simple polynomial subtractions to more sophisticated algorithms. It is best to avoid severe baseline corrections by having a well-designed and maintained NMR instrument that produces flat baselines on good test samples. Good laboratory practice involves careful setup, testing and execution of experiments, in a way that detects spectral quality issues as soon as possible.

Most tools for data analysis organize the individual spectra into the rows of a matrix in which the columns then represent the chemical shift of the spectra. The number of columns of the matrix may be reduced by "binning" the spectra in a systematic manner as part of the data pretreatment.[45,46] The simplest form of binning involves dividing the spectrum into a fixed number of fixed width bins, summing the individual points that fall into a bin. The selection of the appropriate bin width is often dependent on the experience of the analyst, but using a bin width that is too large means the selectivity of the analysis is reduced, since multiple compounds may contribute to a bin, while selecting a bin width that is too small means the results may be overly sensitive to spectral features, such as the line width, so that the effect of shimming, for example, gets exaggerated. Other considerations include the possibility of inappropriate alignment of bin edges on spectral features, separation of spectral multiplets into individual bins, or peaks that shift slightly from sample to sample due to pH effects or ionic strength issues causing the spectral feature to jump between bins in different rows of the matrix. These considerations led to the development of more sophisticated binning algorithms and to the development of more robust spectral alignment tools.[47] This is an active area of development across all of NMR-based metabolomics.[48]

Parts of the NMR spectrum, such as the water region, can be excluded a priori from the analysis, and this is often done by simply deleting the data columns associated with certain chemical shift regions. For example, in a chemical dosing experiment, the toxicant may show up in the samples, and including this in the subsequent analysis might be inappropriate. Another example would be that residual solvent or inadvertent contamination from the sample extraction process remains in the sample and would contribute to the variance in the data set.

Since the effect of spectral noise on some pattern recognition techniques is not well determined, it is best to collect data in a way that keeps the SNR consistent throughout data acquisition. In cases where sample concentration cannot be controlled, due to the sampling techniques for example, experiments may need to be run with an appropriate adjustment of the number of spectral scans. Most often, spectra are normalized so that the total spectral area is constant, but there may be a reason to use a single metabolite or spectral region for data normalization because of the nature of the samples. For example, creatinine is often used in urine-based experiments because of the historical clinical practice of normalization to the creatinine level.[49]

In some experiments, there is a relatively small number of spectral peaks that dominate the spectrum, and unless those compounds are the particular ones of interest, these peaks can be scaled so that the variance of less intense peaks is detected. In extreme cases, the samples may need to be treated differently during sample preparation in order to mitigate the intense signals that may actually obscure smaller signals that convey the important information.[21] Alternatively, bins may be normalized to the variance of the data in each column, and this gives equal importance to each column variable. Other schemes, such as Pareto weighting, where each column is normalized by the square root of the variance, can reduce the influence of large peaks while keeping scale information for the other spectral regions with less overall variation. Some more complex transforms are also possible, such as log transformations.[50] Since metabolomics is evolving rapidly, numerous variance stabilizing transforms are being proposed and tested.

22.4.5   Principal Components Analysis

The workhorse of multivariate analysis in NMR-based metabolomics is principal components analysis (PCA), where the preprocessed data matrix is resolved into a "scores" matrix, which represents each sample spectrum as a point in a high-dimensional space, and an accompanying loadings matrix, which describes the optimal axes for this new space in terms of the spectral bins. These new axes are determined based on the criteria of maximizing the explained variance (EV) along each orthogonal axis. The scores are sorted by decreasing eigenvalue, since the smaller eigenvalues correspond to less explanatory power, and the overall dimensionality is reduced by considering the first few components corresponding to the largest eigenvalues. There are numerous ways to decide how many principal components (PCs) are sufficient to model the data, but seldom are more than two or three considered. Systematic investigation of higher PCs, however, is good practice. The decrease in EV for successive PCs is an indicator of the quality of the model, and often there is a significant explanatory power in the first few PCs. A very gradual increase in the cumulative EV with PC number may be an indication of a less definitive model. Auxiliary information from the PCA analysis, for example, plots of Hotelling's T^2, can be used to identify potential outliers in the data set.

Practically, the scores plots are examined for grouping or trends according to treatment group. If there is apparent grouping in scores plots, univariate testing can be done on the score values to determine the significance of the separation, even if there is significant scatter in the individual treatment groups.

Loadings plots contain information about which bins contribute to the EV for the corresponding PCs. Loadings plots are also often plotted as two-dimensional plots corresponding to the PCs in the scores plot. A loadings plot shows which compounds are correlated or anti-correlated to separations in the corresponding scores plots. In a 2D loadings plot, there is one data point for each variable or bin in the data matrix.

Although PCA scores and loadings are a powerful, unbiased way to examine the data, the interpretation of PCA scores and loadings plots is somewhat difficult because there is no constraint on the algorithm to present a linear combination of pure NMR spectra. Sometimes, there is a strong effect in the data set which is tied to a few compounds and these spectral features dominate the loadings plots, making it straightforward to identify these important compounds and progress to a biological interpretation. Unfortunately, it is difficult for people trained in the thought processes of "single response, single variable" to conceptualize a system-wide response vector. There is a natural tendency to revert to univariate thinking in discussing results, and seldom are the data treated as a multicomponent, coherent effect. If there are a few strong signals in a loadings plot, libraries of spectra or peak data tables can be used to identify the relevant compounds. In some cases, there is no small number of intense peaks, so one is faced with a much more difficult interpretation of the loadings vectors. The interpretation of loadings vectors is often more difficult, as well, when higher PCs are examined. Also, translating a coherent change in metabolite levels to a metabolic pathway interpretation is difficult, especially given the dearth of specific knowledge of metabolic pathways in non-model organisms. Metabolomics results presented in a "network" topology, often correlated with established metabolic pathways, can be useful for conveying the multivariate response that is observed in metabolomics experiments.[51] The classification capability of the analysis is often assessed using receiver operating characteristic (ROC) curves, and the associated area under the curve (AUC) parameter.[48,52]

22.4.6 PLS and PLS-DA

Often, experiments are designed to have an independent variable or classification of the treatments as the basis for determining an effect. This information is most often brought into the multivariate analysis through techniques based on partial least squares (PLS) projections, especially when the class separation is not as apparent in a PCA analysis. The classification information is incorporated into the analysis through a vector or matrix relating the experimental variable for each sample to the sample spectrum. If the experimental variable is not a continuous variable such as temperature, length, or pH, for example, but a discrete parameter such as male/female, this "Y" matrix is constructed as a discriminant matrix where the class is assigned a numerical value, leading to PLS with discriminant analysis (PLS-DA). For cases where there are multiple discriminators, the Y matrix is constructed with one column for each discriminant value and a numerical value, such as 1 or 0, is used to denote class membership. The difficulty with this approach is that the algorithm can blindly find correlations in the variables that satisfy the constraints, even if those variables are really just incidentally correlated noise. Therefore, one must very carefully test the results of PLS analysis for accidental correlations and bias, and numerous robust techniques have been developed to assess the "trueness" of detected correlations.[53–56]

PLS analysis also leads to scores and loading plots that can be used to tie the systematic variation between treatment groups to the specific chemical variation that distinguishes the groups. These identified compounds can then be linked to metabolic pathways, indicating the systematic response to the treatment variables.

22.4.7 Other Pattern Recognition Techniques

Numerous other pattern recognition tools can be and have been used for metabolomics studies. These range from the previously mentioned significant difference spectra (SDS) analysis[42] to artificial neural networks (ANN)[18] and support vector machines (SVM).[57] Each of these techniques has particular strengths, but none has found as widespread applicability in environmental metabolomics as PCA and PLS techniques.

22.4.8 Spectral Libraries

Compound identification is key to tying the spectral information to biochemical pathways and biological interpretation. In most experiments, reports indicate the assignment of approximately 50 and up to 100 compounds from NMR data. By carefully matching sample spectra to spectra of pure compounds in libraries, collected under similar sample and experimental conditions, confidence in identification and quantitation of the peaks in the mixture spectra grows. Many spectral libraries are freely publicly available[58–61] and some are commercially available. Matching can be accomplished through manual peak enumeration and comparison of chemical shift tables, or through interactive library searches or interactive peak alignment. Care must be exercised in matching because there are several chemical shift standards in use, and peaks may not match well if the chemical shift standard is not unambiguously identified.[62] For the non-model organisms of interest in environmental metabolomics, one has the potential problem that peaks from existing databases may not include metabolites of importance,[19,63–65] even though most of the libraries contain several hundred compounds. Most of the libraries are focused on more polar compounds; however, it is often desirable to consider the metabolomics of nonpolar compounds, and these libraries are less developed at this time. Under optimum conditions, the quantification of metabolites can be accomplished based on libraries, and these quantified metabolites can then be used for subsequent data analysis, rather than binned spectra. This process has been named "targeted profiling".[32] As always, ambiguities can always be resolved using analytical techniques such as authentic compound standard additions or chromatography-based purification and structure elucidation.

22.4.9 Quality Control

Analysis of individual spectra for quality parameters such as lineshape, baseline distortions, and instrumental artifacts (quadrature images, spurious radiofrequency signals) is essential for generating quality metabolomics results. Statistical analysis of QC samples that were processed with the experimental samples[22] can be used to classify the reliability of the overall experiment, although this is rather rare in the published literature at this time.

Intercomparison exercises, where participants analyze identical samples according to a specific protocol and the results are compared for consistency, have shown that at least the technical analysis of samples using NMR spectroscopy can have a high level of consistency across laboratories.[23] Even with different magnetic field strengths, instruments from different vendors, and analysis with different software packages, substantial agreement is feasible in NMR-based metabolomics. This contrasts sharply with mass-spectrometry-based metabolomics and with efforts in other -omics fields to show analytical consistency.[66–69] Because of the complexity of the biological models developed and the need for larger studies that may involve instrumental analysis across different laboratories, the ability to consistently, quantitatively analyze metabolomics samples with a high degree of interlaboratory reproducibility is crucial.

22.4.10 Data and Reporting Standards

The advancement of the field of environmental metabolomics depends on laboratory exchanges of data and consistent descriptions of data treatment and interpretation. Efforts to standardize the mechanism of data exchange are ongoing and can piggyback on efforts such as the metabolomics standards initiative (MSI).[70] For NMR data, the actual exchange and storage of raw spectral data is straightforward; however, the accompanying metadata, which describes the experimental design, the sample collection and handling, NMR data collection and processing, and subsequent multivariate analysis, is still evolving.[71,72] Once practical data standards are established and put into widespread use, data can be placed into repositories in meaningful ways. Data repositories will prove useful for future analysis with new algorithms, for long-term environmental studies, and for development of species-specific "stressor libraries" compiled from numerous independent research efforts.

22.5 CURRENT APPLICATIONS

As mentioned in the introduction, the application of NMR-based metabolomics can enhance traditional approaches to environmental science, and can address environmental factors that impact the health and well-being of "non-model" organisms in the environment. These organisms are important as functioning members of the ecosystem, forming the basis of the food web, providing important ecological services and providing us with sustenance, besides having important societal and cultural value.

22.5.1 Laboratory Exposures/Treatments

An essential element of environmental toxicology is the laboratory-based experiment. In these experiments, an organism is maintained in an artificial environment where conditions such as temperature, water conditions, or nutrition are under control. In well-designed experiments with appropriate control organisms, the response to chemical toxicants or physical stressors is measured in a way that should allow extrapolation to a "real-world" exposure or stress. However, as useful and essential as these experiments are, the laboratory environment often does not mimic every factor that may be found in the field.

The problem of linking field-collected samples to laboratory studies has not been addressed in general. The equilibration of field-collected organisms to the laboratory may cause a bias in the results and predictions. For example, the organism selection process (capture, transport, shock survival, etc.) can lead to a bias based on organism survival, biased phenotype selection, or limited gene pools. In microorganism culture experiments, only a subset of the population may be cultivable in the laboratory, so that only a small part of the representative organisms can survive to the laboratory environment. For organisms with gut or symbiotic microorganisms, change from a wild environment to laboratory environment may cause alterations in the microflora, impacting the metabolome in important ways. In addition, other factors such as adaption to consistent feeding, lack of predation, lack of temperature or physical variability, or lack of multispecies signaling may lead to confounding factors which impact the applicability of laboratory-based assessments to field observations. Factors such as full-spectrum sunlight, diurnal cycles, tidal cycles, predation, and competition for food are difficult to replicate in a consistent manner. These effects are often observed in metabolomics experiments, while for other measurement modalities, these effects may not be considered important in interpreting the experimental outcome, perhaps hiding major contributing factors to the experiment.

Naturally, laboratory-based metabolomics measurements do have some advantages. Laboratory exposures make diet, temperature, and other environmental factors controllable so that experiments can be done with reasonable sample sizes, keeping the logistics manageable. Also, single captive organisms may be followed over time as the individual responds to treatment. Often the protocols, while not "perfect," are well defined, and therefore replicable to a great degree in other laboratories. A consistent protocol allows at least for a systematic framework for comparing the toxicity of widely disparate chemical exposures and stressors.

22.5.2   Field Collections

To connect laboratory data to true environmental problems or monitoring, one has to move toward analysis of field-collected samples. Ecological aspects are probably best answered by field collections. However, in terms of interpreting environmental metabolomics results, one must consider the problem of uncontrolled

variables such as diet, temperature, predators and factors such as pollution; these may have to be explored through careful laboratory exposures. This also implies that single individual sampling for a "quick" environmental assessment will be problematic. Again, just because metabolomics is influenced by the effects of uncontrolled variables does not mean that the technique is flawed; it probably means that more of the factors influencing the results can come into play, resulting in a more robust population-level analysis.

22.5.3   Case Studies

Many environmental problems are being addressed through "case studies" where an organism is subjected to a relevant stressor and the metabolomic response is rationalized. These types of studies are important because they are the building blocks that can be used to design more comprehensive studies, better understand the biology of the non-model organisms, and develop expertise in understanding multistressor, multiorganism ecological models based on the biochemical response to stressors. Two groups of organisms, earthworms and bivalves, have been the objects of numerous studies and the expanding body of knowledge may prove to be very valuable.

As an integrated assessor of environmental processes in soil, worms are garnering a lot of attention.[21,73–88] Studies based on several species of field-collected and laboratory-dosed worms indicate that the organism has a robust metabolic response to soil contamination, including organic compounds and heavy metals. These studies show a splendid progression of interest and report on a number of sample preparation schemes, exposure routes, and elucidation of different modes of action. As an Organization for Economic Cooperation and Development (OECD) recommended test species,[89] *Eisenia fetida* is the focus of a majority of the studies, although other ecologically important species are being studied.

Mussels and clams have also been investigated using metabolomics.[7,16,90–100] Given the stationary nature of these mollusks and their aquatic environment, they may prove to be an important monitoring organism as an early warning sentinel for incipient pollution issues. Studies involving organic and inorganic pollutants have shown that bivalves are metabolically sensitive and different modes of action are apparent in their metabolic fingerprints. In one study,[92] the sex of mussels was determined using NMR-based metabolomics, and, while not as accurate as reverse transcriptase polymerase chain reaction (RT-PCR) for sex determination, metabolomics was a better indicator of functional reproductive status in both ripe and spent mussels.

Numerous case studies, which may be part of longer term investigations, have been reported since two reviews of the environmental metabolomics field were published.[10,101] Research involved NMR-based metabolomics studies of coral-associated bacteria,[17] Atlantic blue crabs,[102] and fish.[14,103–108] Environmental stressors varied from hypoxia, to microbial challenge, to temperature, to oil exposure, to heavy metals contamination.

22.5.4 Comprehensive Approaches

A recent report demonstrated the potential of environmental metabolomics to address the full range of linkages in environmental assessment from ecosystem-scale measurements to specific modes of action from environmental stressors.[95] The study illustrated the linkage of metabolomic biomarkers to an accepted assessment of organismal health based on the scope for growth (SFG), a well-defined biological index of the fitness of an organism for growth, reproduction, and survival. In a review of this work,[94] Robertson stated "The elegance in the work … is that they not only generated the models in the laboratory environment but they further field tested them in a real-world application" and indicated that the field of environmental metabolomics had reached another level of expectation and performance. Not all current studies are as comprehensive at this point, but the mark has been set and the potential is tremendous.

22.6 THE FUTURE

The opportunities for impacting the field of environmental research seem to be growing, based on the increasing number of publications and increasing scope of study. Perhaps the future holds exciting biological discoveries as more non-model organisms come under the "NMR-metaboscope" and more specific metabolic pathway maps are refined. The possibility of linking established biological indexes to metabolomic information means that there may be effective ways to assess environmental impacts and set public policy based on specific biochemical interactions, leading to better science-based management decisions.

Improving the tools of NMR-based metabolomics means that more consistent analysis protocols and metabolome mining techniques will appear, either based on specific needs of the analysis of non-model organisms or as an adaptation of approaches in human-health metabolomics.[109,110] Consistent ways of reporting, archiving, and sharing data are emerging which will allow groups to confidently leverage existing data and analysis.

As progress continues, there will be examples of long-term, regional monitoring which may help provide an early warning of encroaching environmental issues. NMR-based environmental metabolomics can develop systematic descriptions of mode of action responses to specific stressors in sensitive organisms, leading to new biological insights and better toxicological understanding of the complex, multispecies, multistressor environment in which we and our fellow creatures live.

REFERENCES

1. S. D. Richardson and T. A. Ternes, *Anal. Chem.*, 2011, **83**, 4614–4648.

2. A. J. Simpson, D. J. McNally, and M. J. Simpson, *Prog. Nucl. Magn. Reson. Spectrosc.*, 2011, **58**, 97–175.

3. L. A. Cardoza, A. K. Korir, W. H. Otto, C. J. Wurrey, and C. K. Larive, *Prog. Nucl. Magn. Reson. Spectrosc.*, 2004, **45**, 209–238.

4. F. Malz and H. Jancke, *J. Pharm. Biomed. Anal.*, 2005, **38**, 813–823.

5. L. Pauling, A. B. Robinson, R. Teranishi, and P. Cary, *Proc. Natl. Acad. Sci. U.S.A.*, 1971, **68**, 2374–2376.

6. J. K. Nicholson, J. C. Lindon, and E. Holmes, *Xenobiotica*, 1999, **29**, 1181–1189.

7. M. R. Viant, J. H. Walton, P. L. TenBrook, and R. S. Tjeerdema, *Aquat. Toxicol.*, 2002, **57**, 139–151.

8. M. R. Viant, E. S. Rosenblum, and R. S. Tieerdema, *Environ. Sci. Technol.*, 2003, **37**, 4982–4989.

9. M. R. Viant, *Mar. Ecol.: Prog. Ser.*, 2007, **332**, 301–306.

10. J. Bundy, M. Davey, and M. Viant, *Metabolomics*, 2009, **5**, 3–21.

11. M. Viant, *Metabolomics*, 2009, **5**, 1–2.

12. C. Y. Lin, M. R. Viant, and R. S. Tjeerdema, *J. Pestic. Sci.*, 2006, **31**, 245–251.

13. M. Macel, N. M. Van Dam, and J. J. Keurentjes, *Mol. Ecol. Resour.*, 2010, **10**, 583–593.

14. A. D. Southam, A. Lange, A. Hines, E. M. Hill, Y. Katsu, T. Iguchi, C. R. Tyler, and M. R. Viant, *Environ. Sci. Technol.*, 2011, **45**, 3759–3767.

15. D. G. Robertson, *Chem. Res. Toxicol.*, 2008, **21**, 1917–1922.

16. A. Hines, G. S. Oladiran, J. P. Bignell, G. D. Stentiford, and M. R. Viant, *Environ. Sci. Technol.*, 2007, **41**, 3375–3381.

17. A. F. Boroujerdi, M. I. Vizcaino, A. Meyers, E. C. Pollock, S. L. Huynh, T. B. Schock, P. J. Morris, and D. W. Bearden, *Environ. Sci. Technol.*, 2009, **43**, 7658–7664.

18. N. Aranibar, B. K. Singh, G. W. Stockton, and K. H. Ott, *Biochem. Biophys. Res. Commun.*, 2001, **286**, 150–155.

19. J. G. Bundy, E. M. Lenz, N. J. Bailey, C. L. Gavaghan, C. Svendsen, D. Spurgeon, P. K. Hankard, D. Osborn, J. M. Weeks, S. A. Trauger, P. Speir, I. Sanders, J. C. Lindon, J. K. Nicholson, and H. Tang, *Environ. Toxicol. Chem.*, 2002, **21**, 1966–1972.

20. S.-A. Sansone, D. Schober, H. Atherton, O. Fiehn, H. Jenkins, P. Rocca-Serra, D. Rubtsov, I. Spasic, L. Soldatova, C. Taylor, A. Tseng, and M. Viant, Ontology Working Group M, *Metabolomics* 2007, **3**, 249–256.

21. Q. Guo, J. Sidhu, T. Ebbels, F. Rana, D. Spurgeon, C. Svendsen, S. Stürzenbaum, P. Kille, A. Morgan, and J. Bundy, *Metabolomics*, 2009, **5**, 72–83.

22. H. M. Parsons, D. R. Ekman, T. W. Collette, and M. R. Viant, *Analyst*, 2009, **134**, 478–485.

23. M. R. Viant, D. W. Bearden, J. G. Bundy, I. W. Burton, T. W. Collette, D. R. Ekman, V. Ezernieks, T. K. Karakach, C. Y. Lin, S. Rochfort, J. Sd. Ropp, Q. Teng, R. S. Tjeerdema, J. A. Walter, and H. Wu, *Environ. Sci. Technol.*, 2009, **43**, 219–225.

24. S. Tiziani, A. H. Emwas, A. Lodi, C. Ludwig, C. M. Bunce, M. R. Viant, and U. L. Gunther, *Anal. Biochem.*, 2008, **377**, 16–23.

25. P. Bernini, I. Bertini, C. Luchinat, P. Nincheri, S. Staderini, and P. Turano, *J. Biomol. NMR*, 2011, **49**, 231–243.

26. C. Y. Lin, H. Wu, R. S. Tjeerdema, and M. R. Viant, *Metabolomics*, 2007, **3**, 55–67.

27. S. A. E. Brown, A. J. Simpson, and M. J. Simpson, *Environ. Toxicol. Chem.*, 2008, **27**, 828–836.

28. E. G. Bligh and W. J. Dyer, *Can. J. Biochem. Physiol.*, 1959, **37**, 911–917.

29. S. Stolzenburg, M. B. Lauridsen, H. Toft, P. A. Zalloua, and D. Baunsgaard, *Metabolomics*, 2010, **7**, 270–277.

30. M. Findeisen, T. Brand, and S. Berger, *Magn. Reson. Chem.*, 2007, **45**, 175–178.

31. E. Saude, C. Slupsky, and B. Sykes, *Metabolomics*, 2006, **2**, 113–123.

32. A. M. Weljie, J. Newton, P. Mercier, E. Carlson, and C. M. Slupsky, *Anal. Chem.*, 2006, **78**, 4430–4442.

33. A. J. Simpson and S. A. Brown, *J. Magn. Reson.*, 2005, **175**, 340–346.

34. N. Aranibar, K. H. Ott, V. Roongta, and L. Mueller, *Anal. Biochem.*, 2006, **355**, 62–70.

35. G. Zheng and W. S. Price, *Prog. Nucl. Magn. Reson. Spectrosc.*, 2010, **56**, 267–288.

36. Q. N. Van, G. N. Chmurny, and T. D. Veenstra, *Biochem. Biophys. Res. Commun.*, 2003, **301**, 952–959.

37. M. Kriat, S. Confort-Gouny, J. Vion-Dury, P. Viout, and P. J. Cozzone, *Biochimie*, 1992, **74**, 913–918.

38. L. H. Lucas, C. K. Larive, P. S. Wilkinson, and S. Huhn, *J. Pharm. Biomed. Anal.*, 2005, **39**, 156–163.

39. C. Ludwig and M. R. Viant, *Phytochem. Anal.*, 2009, **21**, 22–32.

40. D. R. Ekman, Q. Teng, D. L. Villeneuve, M. D. Kahl, K. M. Jensen, E. J. Durhan, G. T. Ankley, and T. W. Collette, *Environ. Sci. Technol.*, 2008, **42**, 4188–4194.

41. I. Katsiadaki, T. D. Williams, J. S. Ball, T. P. Bean, M. B. Sanders, H. Wu, E. M. Santos, M. M. Brown, P. Baker, F. Ortega, F. Falciani, J. A. Craft, C. R. Tyler, M. R. Viant, and J. K. Chipman, *Aquat. Toxicol.*, 2010, **97**, 174–187.

42. T. B. Schock, S. Newton, K. Brenkert, J. Leffler, and D. W. Bearden, An NMR-based metabolomic assessment of cultured cobia health in response to dietary manipulation', *Food Chemistry*, 2012, http://dw.doi.org/10.1016/j.foodchem.2011.12.077, in press.

43. Y. Xi and D. M. Rocke, *BMC Bioinformatics*, 2008, **9**, 324. http://dw.doi.org/10.1186/1471-2105-9-324.

44. J. C. Cobas, M. A. Bernstein, M. Martin-Pastor, and P. G. Tahoces, *J. Magn. Reson.*, 2006, **183**, 145–151.

45. E. Holmes, P. J. Foxall, J. K. Nicholson, G. H. Neild, S. M. Brown, C. R. Beddell, B. C. Sweatman,

E. Rahr, J. C. Lindon, M. Spraul, and P. Neidig, *Anal. Biochem.*, 1994, **220**, 284–296.

46. M. Spraul, P. Neidig, U. Klauck, P. Kessler, E. Holmes, J. K. Nicholson, B. C. Sweatman, S. R. Salman, R. D. Farrant, E. Rahr, C. R. Beddell, and J. C. Lindon, *J. Pharm. Biomed. Anal.*, 1994, **12**, 1215–1225.

47. P. E. Anderson, D. A. Mahle, T. E. Doom, N. V. Reo, N. J. DelRaso, and M. L. Raymer, *Metabolomics*, 2010, **7**, 179–190.

48. S. M. Kohl, M. S. Klein, J. Hochrein, P. J. Oefner, R. Spang, and W. Gronwald, *Metabolomics*, 2011. http://dw.doi.org/ 10.1007/s11306-011-0350-z.

49. S. Zhang, L. Liu, D. Steffen, T. Ye, and D. Raftery, *Metabolomics*, 2011. http://dx.doi.org/10.1007/s11306-011-0315-2.

50. P. V. Purohit, D. M. Rocke, M. R. Viant, and D. L. Woodruff, *Omics-a J. Integr. Biol.*, 2004, **8**, 118–130.

51. A. D. Southam, J. M. Easton, G. D. Stentiford, C. Ludwig, T. N. Arvanitis, and M. R. Viant, *J. Proteome Res.*, 2008, **7**, 5277–5285.

52. G. X. Xie, T. L. Chen, Y. P. Qiu, P. Shi, X. J. Zheng, M. M. Su, A. H. Zhao, Z. T. Zhou, and W. Jia, *Metabolomics*, 2012, http://dx.doi.org/ 10.1007/s11306-011-0302-7, in press.

53. E. Szymańska, E. Saccenti, A. K. Smilde, and J. A. Westerhuis, *Metabolomics*, 2011. http://dx.doi.org/ 10.1007/s11306-011-0330-3.

54. J. Westerhuis, H. Hoefsloot, S. Smit, D. Vis, A. Smilde, E. van, Velzen, J. van, Duijnhoven, and F. van, Dorsten, *Metabolomics*, 2008, **4**, 81–89.

55. S. Wiklund, D. Nilsson, L. Eriksson, M. Sjöström, S. Wold, and K. Faber, *J. Chemom.*, 2007, **21**, 427–439.

56. L. Eriksson, J. Trygg, and S. Wold, *J. Chemom.*, 2008, **22**, 594–600.

57. S. Mahadevan, S. L. Shah, T. J. Marrie, and C. M. Slupsky, *Anal. Chem.*, 2008, **80**, 7562–7570.

58. Q. Cui, I. A. Lewis, A. D. Hegeman, M. E. Anderson, J. Li, C. F. Schulte, W. M. Westler, H. R. Eghbalnia, M. R. Sussman, and J. L. Markley, *Nat. Biotechnol.*, 2008, **26**, 162–164.

59. D. S. Wishart, D. Tzur, C. Knox, R. Eisner, A. C. Guo, N. Young, D. Cheng, K. Jewell, D. Arndt, S. Sawhney, C. Fung, L. Nikolai, M. Lewis, M. A. Coutouly, I. Forsythe, P. Tang, S. Shrivastava, K. Jeroncic, P. Stothard, G. Amegbey, D. Block, D. D. Hau, J. Wagner, J. Miniaci, M. Clements, M. Gebremedhin, N. Guo, Y. Zhang, G. E. Duggan,

G. D. Macinnis, A. M. Weljie, R. Dowlatabadi, F. Bamforth, D. Clive, R. Greiner, L. Li, T. Marrie, B. D. Sykes, H. J. Vogel, and L. Querengesser, *Nucleic Acids Res.*, 2007, **35**, D521–D526.

60. D. S. Wishart, C. Knox, A. C. Guo, R. Eisner, N. Young, B. Gautam, D. D. Hau, N. Psychogios, E. Dong, S. Bouatra, R. Mandal, I. Sinelnikov, J. Xia, L. Jia, J. A. Cruz, E. Lim, C. A. Sobsey, S. Shrivastava, P. Huang, P. Liu, L. Fang, J. Peng, R. Fradette, D. Cheng, D. Tzur, M. Clements, A. Lewis, A. De Souza, A. Zuniga, M. Dawe, Y. Xiong, D. Clive, R. Greiner, A. Nazyrova, R. Shaykhutdinov, L. Li, H. J. Vogel, and I. Forsythe, *Nucleic Acids Res.*, 2009, **37**, D603–D610.

61. C. Ludwig, J. M. Easton, A. Lodi, S. Tiziani, S. E. Manzoor, A. D. Southam, J. J. Byrne, L. M. Bishop, S. He, T. N. Arvanitis, U. L. Günther, and M. R. Viant, *Metabolomics*, 2012. **8**, 8–18.

62. M. Alum, P. Shaw, B. Sweatman, B. Ubhi, J. Haselden, and S. Connor, *Metabolomics*, 2008, **4**, 122–127.

63. R. A. Kleps, T. C. Myers, R. N. Lipcius, and T. O. Henderson, *PLoS ONE*, 2007, **2**, e780.

64. S. Moura, A. Ultramari Mde, D. M. de, Paula, M. Yonamine, and E. Pinto, *Toxicon*, 2009, **53**, 578–583.

65. P. Polychronopoulos, P. Magiatis, A. L. Skaltsounis, F. Tillequin, E. Vardala-Theodorou, and A. Tsarbopoulos, *Nat. Prod. Lett.*, 2001, **15**, 411–418.

66. J. J. Chen, H. M. Hsueh, R. R. Delongchamp, C. J. Lin, and C. A. Tsai, *BMC Bioinformatics*, 2007, **8**, 412. http://dw.doi.org/10.1186/1471-2105-8-412.

67. A. W. Bell, E. W. Deutsch, C. E. Au, R. E. Kearney, R. Beavis, S. Sechi, T. Nilsson, J. J. M. Bergeron, and H. T. S. W. Group, *Nat. Methods*, 2009, **6**, 423–430.

68. J. P. A. Ioannidis, D. B. Allison, C. A. Ball, I. Coulibaly, C. Xiangqin, An. C. Culhane, M. Falchi, C. Furlanello, L. Game, G. Jurman, J. Mangion, T. Mehta, M. Nitzberg, G. P. Page, E. Petretto, and V. van, Noort, *Nat. Genet.*, 2009, **41**, 149–155.

69. M. Mann, *Nat. Methods*, 2009, **6**, 717–719.

70. O. Fiehn, D. Robertson, J. Griffin, M. Werf, B. Nikolau, N. Morrison, L. W. Sumner, R. Goodacre, N. W. Hardy, C. Taylor, J. Fostel, B. Kristal, R. Kaddurah-Daouk, P. Mendes, B. Ommen, J. C. Lindon, and S.-A. Sansone, *Metabolomics*, 2007, **3**, 175–178.

71. D. V. Rubtsov, H. Jenkins, C. Ludwig, J. Easton, M. R. Viant, U. Günther, J. L. Griffin, and N. Hardy, *Metabolomics*, 2007, **3**, 223–229.

72. N. Morrison, D. Bearden, J. G. Bundy, T. Collette, F. Currie, M. P. Davey, N. S. Haigh, D. Hancock, O. A. H. Jones, S. Rochfort, S.-A. Sansone, D. Štys, Q. Teng, D. Field, and M. R. Viant, *Metabolomics*, 2007, **3**, 203–210.

73. M. A. Warne, E. M. Lenz, D. Osborn, J. M. Weeks, and J. K. Nicholson, *Biomarkers*, 2000, **5**, 56–72.

74. J. G. Bundy, J. C. Osborn, J. M. Weeks, J. C. Lindon, and J. K. Nicholson, *FEBS Lett.*, 2001, **500**, 31–35.

75. J. G. Bundy, D. J. Spurgeon, C. Svendsen, P. K. Hankard, D. Osborn, J. C. Lindon, and J. K. Nicholson, *FEBS Lett.*, 2002, **521**, 115–120.

76. J. G. Bundy, H. Ramlov, and M. Holmstrup, *Cryo Lett.*, 2003, **24**, 347–358.

77. J. G. Bundy, H. C. Keun, J. K. Sidhu, D. J. Spurgeon, C. Svendsen, P. Kille, and A. J. Morgan, *Environ. Sci. Technol.*, 2007, **41**, 4458–4464.

78. J. G. Bundy, J. K. Sidhu, F. Rana, D. J. Spurgeon, C. Svendsen, J. F. Wren, S. R. Sturzenbaum, A. J. Morgan, and P. Kille, *BMC Biol.*, 2008, **6**, 1–25.

79. S. L. Hughes, J. G. Bundy, E. J. Want, P. Kille, and S. R. Sturzenbaum, *J. Proteome Res.*, 2009, **8**, 3512–3519.

80. S. Rochfort, V. Ezernieks, and A. Yen, *Metabolomics*, 2009, **5**, 95–107.

81. M. J. Simpson and J. R. McKelvie, *Anal. Bioanal. Chem.*, 2009. **394**, 137–149, http://dw.doi.org/10.1007/s00216-009-2612-4.

82. J. McKelvie, J. Yuk, Y. Xu, A. Simpson, and M. Simpson, *Metabolomics*, 2009, **5**, 84–94.

83. J. R. McKelvie, D. M. Wolfe, M. Celejewski, A. J. Simpson, and M. J. Simpson, *Environ. Pollut.*, 2010, **158**, 2150–2157.

84. S. A. E. Brown, J. R. McKelvie, A. J. Simpson, and M. J. Simpson, *Environ. Pollut.*, 2010, **158**, 2117–2123.

85. M. L. Whitfield Aslund, A. J. Simpson, and M. J. Simpson, *Ecotoxicology*, 2011. **20**, 836–846. http://dx.doi.org/10.1007/s10646-011-0638-9.

86. B. P. Lankadurai, D. M. Wolfe, A. J. Simpson, and M. J. Simpson, *Environ. Pollut.*, 2011. **159** (10), 2845–2851. http://dx.doi.org/10.1016/j.envpol.2011.04.044.

87. J. R. McKelvie, D. M. Wolfe, M. A. Celejewski, M. Alaee, A. J. Simpson, and M. J. Simpson, *Environ. Pollut.*, 2011. **159** (12), 3620–3626. http://dx.doi.org/10.1016/j.envpol.2011.08.002.

88. M. Whitfield Åslund, M. Celejewski, B. P. Lankadurai, A. J. Simpson, and M. J. Simpson, *Chemosphere*, 2011, **83**, 1096–1101.

89. OECD, 'OECD Guideline for Testing of Chemicals, OECD Guideline 207', Organisation for Economic Co-operation and Development (OECD): Paris, France, 1984.

90. D. R. Livingstone, J. K. Chipman, D. M. Lowe, C. Minier, C. L. Mitchelmore, M. N. Moore, L. D. Peters, and R. K. Pipe, *Int. J. Environ. Pollut.*, 2000, **13**, 56–91.

91. M. R. Viant, J. H. Walton, and R. S. Tjeerdema, *Pestic. Biochem. Physiol.*, 2001, **71**, 40–47.

92. A. Hines, W. H. Yeung, J. Craft, M. Brown, J. Kennedy, J. Bignell, G. D. Stentiford, and M. R. Viant, *Anal. Biochem.*, 2007, **369**, 175–186.

93. W. Tuffnail, G. Mills, P. Cary, and R. Greenwood, *Metabolomics*, 2009, **5**, 33–43.

94. D. G. Robertson, *Toxicol. Sci.*, 2010, **115**, 305–306.

95. A. Hines, F. J. Staff, J. Widdows, R. M. Compton, F. Falciani, and M. R. Viant, *Toxicol. Sci.*, 2010, **115**, 369–378.

96. H. Wu and W. X. Wang, *Aquat. Toxicol.*, 2010, **100**, 339–345.

97. H. Wu and W. X. Wang, *Environ. Toxicol. Chem.*, 2010, **30**, 806–812.

98. L. Zhang, X. Liu, L. You, D. Zhou, H. Wu, L. Li, J. Zhao, J. Feng, and J. Yu, *Mar. Environ. Res.*, 2011, **72**, 33–39.

99. L. Zhang, X. Liu, L. You, D. Zhou, Q. Wang, F. Li, M. Cong, L. Li, J. Zhao, D. Liu, J. Yu, and H. Wu, *Environ. Toxicol. Pharmacol.*, 2011, **32**, 218–225.

100. X. Liu, L. Zhang, L. You, M. Cong, J. Zhao, H. Wu, C. Li, D. Liu, and J. Yu, *Environ. Toxicol. Pharmacol.*, 2011, **31**, 323–332.

101. M. R. Viant, *Mol. Biosyst.*, 2008, **4**, 980–986.

102. T. B. Schock, D. A. Stancyk, L. Thibodeaux, K. G. Burnett, L. E. Burnett, A. F. B. Boroujerdi, and D. W. Bearden, *Metabolomics*, 2010, **6**, 250–262.

103. C. Y. Lin, B. S. Anderson, B. M. Phillips, A. C. Peng, S. Clark, J. Voorhees, H. D. Wu, M. J. Martin, J. McCall, C. R. Todd, F. Hsieh, D. Crane, M. R. Viant, M. L. Sowby, and R. S. Tjeerdema, *Aquat. Toxicol.*, 2009, **95**, 230–238.

104. A. R. Van Scoy, C. Yu Lin, B. S. Anderson, B. M. Philips, M. J. Martin, J. McCall, C. R. Todd, D. Crane,

M. L. Sowby, M. R. Viant, and R. S. Tjeerdema, *Ecotoxicol. Environ. Saf.*, 2010, **73**, 710–717.

105. E. Kokushi, S. Uno, T. Harada, and J. Koyama, *Environ. Toxicol.*, 2010. http://dx.doi.org/10.1002/tox.20653.

106. L. M. Samuelsson, B. Bjorlenius, L. Forlin, and D. G. Larsson, *Environ. Sci. Technol.*, 2011, **45**, 1703–1710.

107. K. H. Chiu, S. Ding, Y. W. Chen, C. H. Lee, and H. K. Mok, *Fish Physiol. Biochem.*, 2011, **37**, 701–707.

108. S. Hayashi, M. Yoshida, T. Fujiwara, S. Maegawa, and E. Fukusaki, *Z. Naturforsch., C*, 2011, **66**, 191–198.

109. G. Gibson, C. Gieger, L. Geistlinger, E. Altmaier, M. Hrabé de Angelis, F. Kronenberg, T. Meitinger, H.-W. Mewes, H. E. Wichmann, K. M. Weinberger, J. Adamski, T. Illig, and K. Suhre, *PLoS Genet.*, 2008, **4**, e1000282.

110. K. Suhre, H. Wallaschofski, J. Raffler, N. Friedrich, R. Haring, K. Michael, C. Wasner, A. Krebs, F. Kronenberg, D. Chang, C. Meisinger, H. E. Wichmann, W. Hoffmann, H. Volzke, U. Volker, A. Teumer, R. Biffar, T. Kocher, S. B. Felix, T. Illig, H. K. Kroemer, C. Gieger, W. Romisch-Margl, and M. Nauck, *Nat. Genet.*, 2011, **43**, 565–569.

Chapter 23

Environmental Metabolomics: NMR Techniques

Myrna J. Simpson[1] and Daniel W. Bearden[2]

[1]*Department of Chemistry and Environmental NMR Centre, University of Toronto, Toronto, Ontario, M1C 1A4, Canada*
[2]*Analytical Chemistry Division, National Institute of Standards and Technology, Hollings Marine Laboratory, Charleston, SC 29412, USA*

23.1 INTRODUCTION

NMR is the primary tool used to measure the metabolome of various environmental organisms in environmental metabolomic studies. It has been an essential discovery platform in environmental metabolomics because most studies are nontargeted or 'discovery' in nature. Nontargeted or comprehensive metabolomics is a common approach when prior knowledge about metabolomic responses is unavailable; something that is often necessary in environmental metabolomics because of the large variety of aquatic and terrestrial organisms that are studied, and in some cases, the organisms' metabolome has not yet been characterized or measured. As such, NMR is ideally suited for environmental metabolomic studies because it is nonselective, nondestructive, rapid, versatile for different sample types (liquid, solid, and gel phases), and quantitative. Its fundamental adaptability translates to numerous opportunities for environmental metabolomics research. Advancements in both hardware and software will continue to facilitate the rapid expansion of NMR-based metabolomic studies with emphasis on environmental organisms. Several recent reviews on NMR-based metabolomics and environmental and ecological metabolomics have been published[1–5] and provide more detail with respect to data analysis methods, experimental design for studies involving different types of organisms, and the other chapters published within this book (see Chapters 22, 24, 25, 26 and 27). Accordingly, this chapter provides an overview

NMR Spectroscopy: A Versatile Tool for Environmental Research
Edited by Myrna J. Simpson and André J. Simpson
© 2014 John Wiley & Sons, Ltd. ISBN: 978-1-118-61647-5

of different NMR methods used in environmental metabolomics.

23.2 ONE-DIMENSIONAL (1D) NMR TECHNIQUES

The vast majority of NMR-based environmental metabolomic studies focus on the ^{1}H NMR analysis of aqueous samples. A buffer solution is often used to ensure that pH variation between samples is minimized so that the resulting chemical shifts are consistent from sample to sample.[6] Brown *et al.*[7] compared different solvents for isolation of metabolites from *Eisenia fetida* earthworm tissues. As shown in Figure 23.1, D_2O-based extractions were observed to isolate the greatest number of metabolites as compared to other solvents tested. Further examination

of water-based extractions revealed a wide variety of endogenous metabolites that are targeted with this extraction method. Water-extractable metabolites include free amino acids, sugars, osmolytes, and energy-related metabolites such as adenosine diphosphate (ADP) and adenosine triphosphate (ATP) and may include as many as 40 identifiable metabolites from the ^{1}H NMR spectrum. Examination of lipids, including various fatty acids, cholesterol, and phosphatidylcholine,[8–10] is performed using chloroform extraction and isolation but water-based extractions are still more prevalent in environmental metabolomics. Accordingly, water suppression NMR techniques are commonly used in NMR-based metabolomics.[4,11,12] Suppression of the water signal is vital for several reasons with the main advantage being the reduction or removal of a large water peak that may mask nearby metabolite peaks.[11,13] Water suppression prevents receiver overload, which can lead to distorted baselines, and permits the use of optimal receiver gain, which is essential for the detection of trace components.[11,13,14]

A number of different water suppression pulse programs are used in NMR-based environmental metabolomics but the most commonly used methods include presaturation, Nuclear Overhauser effect spectroscopy (NOESY) presaturation [also sometimes referred to as water-eliminated Fourier transform (WEFT)], water suppression enhanced through T_1 effects (WET), the Carr-Purcell-Meiboom-Gill (CPMG) spin–echo pulse sequence used with water suppression, and presaturation utilizing relaxation gradients and echoes (PURGE) NMR.[4,8,9,11,12,14–18] McKay[11] compared different water suppression methods using a ubiquitin sample (Figure 23.2) and found that of the methods tested, PURGE resulted in the best quality spectra with the least residual water and baseline distortion. Low baseline distortion is beneficial for data processing and quantification, especially when NMR spectra are binned or bucketed for multivariate analysis such as principal component analysis (PCA).[13,19]

Some NMR-based environmental metabolomic studies have used one-dimensional (1D) projections from two-dimensional (2D) J-resolved (JRES) ^{1}H NMR experiment for improving spectral resolution and metabolite identification.[14,18,20,21] It should be noted that JRES does involve collecting a 2D dataset from which the 1D JRES projection (free of splitting) is created by shearing and a subsequent projection. Viant[18] showed that JRES, in addition to preprocessing with a logarithmic transformation,

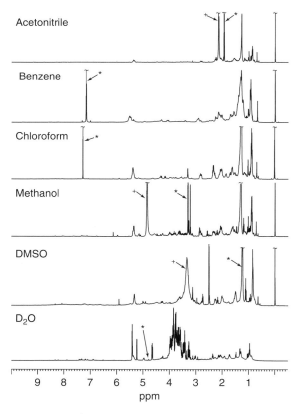

Figure 23.1. ^{1}H NMR spectra for various solvents used to extract *Eisenia fetida* earthworm tissues. *, solvent peaks; +, water; and ∼, scaled peaks. (Reproduced from Ref. 7. © 2008 Wiley)

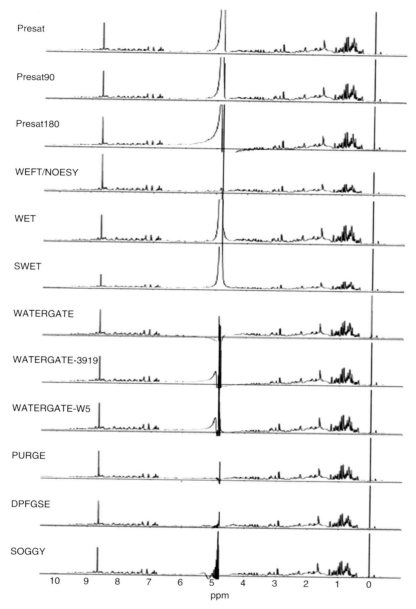

Figure 23.2. Comparison of water suppression methods on a ubiquitin sample. (Reprinted with permission from Ref. 11. © 2009 Elsevier)

can reduce spectral overlap and improve metabolite identification. As shown in Figure 23.3a, the 1D ^1H NMR spectrum includes crowding in the $\delta = 3.0-4.5$ ppm region. The 2D JRES, which separates chemical shift information on one axis and spin–spin (J) coupling information on the other (Figure 23.3b), can be plotted in a 1D skyline projection (Figure 23.3c) as a homonuclear decoupled ^1H spectrum. As a result of suppressing the macromolecular background via T_1 and T_2 discrimination and the strong attenuation of

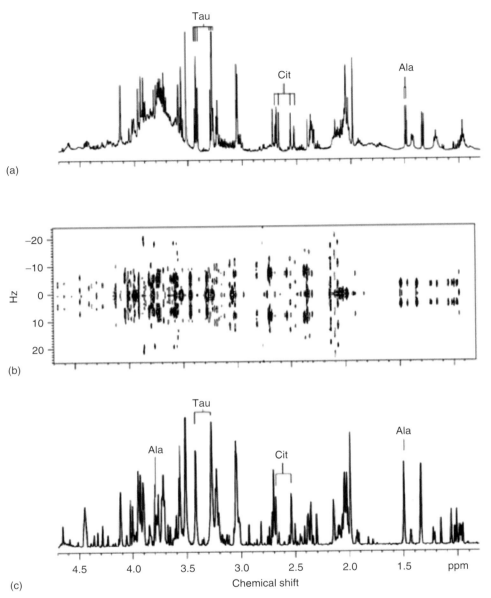

(a)

(b)

(c)

Figure 23.3. Partial NMR (500 MHz) spectra of fish (Medaka) embryo D_2O buffer extracts. (a) ^1H NMR spectrum, (b) two-dimensional (2D) J-resolved (JRES) NMR spectrum, and (c) preprocessed p-JRES NMR spectrum. Tau, Taurine; Cit, Citrate; and Ala, Alanine. (Reprinted with permission from Ref. 18. © 2003 Elsevier)

the homonuclear couplings, JRES projections tend to have excellent baselines and reduced spectral overlap. For example, the $\delta = 3.79$ ppm resonance from the α-CH in alanine is clearly visible in the 1D JRES spectrum (Figure 23.3c) as well as the $\delta = 1.48$ ppm

resonance for β-CH$_3$ in alanine, which is observed in both the 1D JRES projection and the 1D ^1H NMR spectra (Figure 23.3a). This provides more confidence in metabolite assignments because both of the characteristic resonances for alanine ($\delta = 1.48$ and

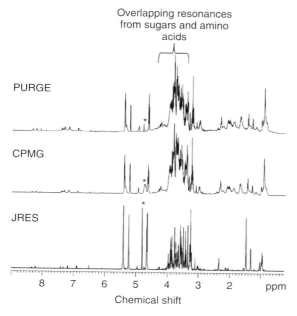

Overlapping resonances from sugars and amino acids

PURGE

CPMG

JRES

Chemical shift

Figure 23.4. Comparison of ^{1}H NMR spectra earthworm tissues extracted with buffer acquired using different NMR techniques: presaturation utilizing relaxation gradients and echos (PURGE), CPMG, and J-resolved spectroscopy (JRES) projections. The asterisk represents residual water signals. (Modified from Yuk *et al.*[14])

3.79 ppm), as an example, can be used for metabolite fingerprinting in comparison to just one ($\delta = 1.48$ ppm) in ^{1}H NMR. Yuk *et al.*[14] investigated different methods for analyzing earthworm tissue-buffer extracts but in addition to ^{1}H NMR (using PURGE for water suppression), compared CPMG spectra to the 1D JRES projections (Figure 23.4). CPMG and PURGE sequences resulted in similar spectra but the PURGE spectrum displayed improved resolution for some metabolites and CPMG resulted in a slightly flatter baseline. The 1D JRES projection provided a considerable improvement in the overlapping resonance region ($\delta = 3.0 - 4.5$ ppm); however, the authors reported lower signal to noise, due to the lower sensitivity of the 2D acquisition method used.[14] An additional benefit of using either ^{1}H NMR or 1D JRES projections is that metabolite databases are readily available to assist with metabolite identification.[20,22]

^{1}H high resolution magic-angle spinning (HR-MAS) NMR (see Chapter 5) may also be used to analyze the metabolite profile of intact tissues[23] but has only been applied sparingly in environmental metabolomics.[24,25] The direct analysis of tissues avoids the need for sample extraction and optimization that may result in less artifacts because of sample extraction; however, a potential trade-off is that the time involved in preparing fresh tissue samples into HR-MAS rotors may represent an uncertainty in stability of the tissue metabolome being measured. This may be particularly acute for bioactive tissues such as liver. Furthermore, the aforementioned NMR experiments can be applied with HR-MAS NMR in addition to a number of 2D experiments. Beckonert *et al.*[23] report on an optimized rotor design for standardizing tissue analysis by HR-MAS NMR, which will likely open the door for future environmental metabolomic analysis on various organism types.

Other recent advances include a recent study by Teng *et al.*[26] who examined the use of direct injection NMR methods for metabolomics applications. With direct injection, a flow probe is connected to an automated sampling unit that can hold large numbers of liquid samples, which removes the need for NMR tubes. This enables the NMR analysis of a large number of samples in full automation, and because the flow cell is fixed within the NMR probe, artifacts related to improper NMR tube positioning from sample to sample are alleviated.[26] However, direct injection may result in sample carryover, loss, contamination, transfer line blockage, or air bubbles in the flow cell. To circumvent these technical challenges, Teng *et al.*[26k] tested a push-through direct injection method on environmental metabolomic samples that includes solvent rinsing of the flow cell as well as compressed air to push solvent through. They reported high reproducibility in the measured NMR spectra of 48 repeated analyses of chicken liver tissue extracts in automation using a 96-well plate configuration.[26] These results are highly promising for future environmental metabolomics analysis of tissue extracts using flow-injection NMR methods. Microcoil probes, which only require low microgram quantities of analytes, may also be valuable for tissue extracts of small organisms or small volumes of biofluids.[27] They have higher magnetic field homogeneity (which is important for consistent shimming across large numbers of samples) and lower detection limits than standard 5 mm probes.[28] In environmental metabolomics, a 1.7 mm microcoil probe was used to analyze metabolites in the hemolymph of the tiny water flea (*Daphnia magna*)[29] and this probe technology enables future applications for small environmental organisms. Cryogenically cooled probes also offer enhanced sensitivity[3,15,30] and are advantageous for examining extracts of tissues from small organisms such as zebra fish.[9]

23.3 TWO-DIMENSIONAL (2D) NMR TECHNIQUES

In general, 2D experiments require more instrument time and may be less sensitive than 1D experiments (Table 23.1) and, as such, are not often used for routine metabolic profiling.[14] However, the information afforded by the second dimension provides additional connectivity information and the resulting dispersion results in a reduction in signal overlap. Because of this, 2D NMR methods are mainly used for structural elucidation and the confirmation of metabolite assignments made using the 1D ^{1}H NMR spectrum with the exception of ^{1}H–^{1}H JRES NMR where the 1D projection has been used for routine metabolomics as discussed previously.[14,18,20,21] The three most commonly used experiments in environmental metabolomics include Correlation Spectroscopy (COSY), Total Correlation Spectroscopy (TOCSY), and ^{1}H–^{13}C heteronuclear single quantum coherence (HSQC) NMR (Table 23.1).[14,17,31–33] A COSY NMR spectrum displays ^{1}H–^{1}H J coupling information of adjacent protons over two dimensions, whereas a TOCSY NMR spectrum shows information for ^{1}H–^{1}H J couplings over the entire spin system. Both of these are helpful when assigning metabolite resonances because the additional coupling information is dispersed into two dimensions and separated from the central diagonal.[14,17] In both COSY and TOCSY NMR spectra, the cross peak represents the chemical shift of the ^{1}H of interest in one dimension and the chemical shift of the ^{1}H to which it is coupled in the second dimension. The ^{1}H–^{13}C HSQC NMR spectrum provides information about ^{1}H connected to ^{13}C nuclei, which is invaluable for metabolite assignment, especially when there are significant overlapping resonances in the 1D ^{1}H NMR spectrum (Figure 23.5).[33] The additional dispersion afforded by the carbon dimension is extremely useful for reducing spectral overlap.

Table 23.1. Summary of different NMR methods used in environmental metabolomics

NMR experiment	Experiment time (min)	Advantages	Disadvantages
^{1}H (with water suppression)	5–10	Rapid and sensitive for routine analysis	Spectral overlap may prohibit metabolite identification and quantification
CPMG	10–15	Removes broad resonances from high molecular weight compounds via T_2 spectral editing	Longer experiment time than routine ^{1}H NMR and spectra may still include overlap from low molecular weight metabolites
^{1}H–^{1}H JRES	10–15	The one-dimensional JRES projection is more resolved than ^{1}H NMR, which facilitates metabolite identification	Longer experiment time and requires higher sample concentration than routine ^{1}H NMR
COSY	~30	Correlation of short-range ^{1}H–^{1}H couplings for additional metabolite information	Longer experiment and data processing times, and may not provide enough resolution for all metabolites
TOCSY	~50	Correlation of long-range ^{1}H–^{1}H couplings for additional metabolite information	Longer experiment and data processing times, and may not provide enough resolution for all metabolites
^{1}H–^{13}C HSQC	~50	Correlation of ^{1}H couplings to adjacent ^{13}C bonds provides additional information for metabolite information	Longer experiment and data processing times

Experiment times are based on extraction of ~100 mg of animal tissue and with the use of 5 mm NMR probes.[6,14]

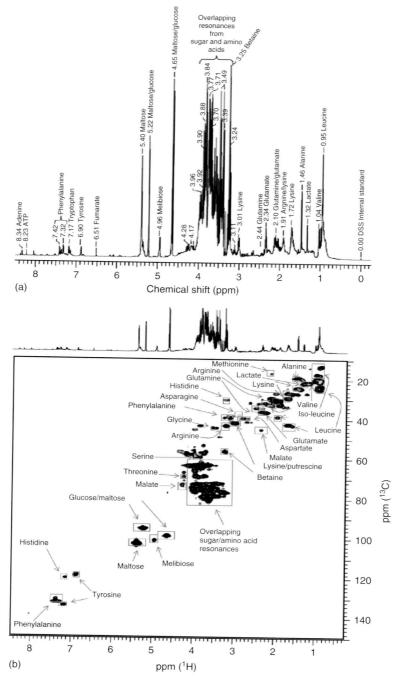

Figure 23.5. (a) ^{1}H NMR (acquired using PURGE at 500 MHz) and (b) ^{1}H–^{13}C HSQC NMR spectra of D$_2$O-based buffer extract of *Eisenia fetida* earthworm tissues. Metabolites identified in the 1D spectrum are confirmed using 2D NMR as illustrated.[34] (Reproduced from Ref 34. © 2012 Springer)

In most cases, 2D NMR techniques are reserved for structural elucidation and metabolite identification rather than for routine monitoring of the metabolome because they require longer experiment times and may have reduced signal to noise (Table 23.1). However, it should be noted that 2D methods (such as the previously discussed 1D projection from $^1H-^1H$ JRES) can also be used and are being developed. For example, Sandusky et al.[35] report the use of a 1D selective TOCSY for targeted analysis of specific metabolites in biofluids. Yuk et al.[14] tested if the additional dispersion afforded by 2D spectra would improve the discrimination quality of metabolomics data from earthworms exposed to organochlorine pesticides. Between $^1H-^1H$ COSY and $^1H-^{13}C$ HSQC NMR, the $^1H-^{13}C$ HSQC NMR spectrum provided an order of magnitude improvement in the statistical resolution between control and exposed groups.[14] In addition to this, $^1H-^{13}C$ HSQC data were used to identify several more statistically significant endogenous metabolite fluctuations that were not observed using $^1H-^1H$ COSY, H^1-H^1 JRES, CPMG, or ^1H NMR (using PURGE water suppression) methods. This study highlights that additional connectivity information and spectral dispersion afforded by $^1H-^{13}C$ HSQC NMR spectra provide improved resolving power for discriminating metabolites in response to environmental stressors.[14] However, as 2D methods require much longer experiment times than 1D experiments (Table 23.1), they are not used routinely in metabolomics studies that focus on environmental stressors. In addition to this, most extraction methods are geared toward

1D experiments and may not result in high enough concentrations for 2D experiments, especially that of small model environmental organisms. This stated probe hardware with improved sensitivity, such as microcoil and cryogenically cooled probes, coupled with the development of 2D experiments with shorter experiment times[36] may facilitate greater use of 2D NMR techniques in routine metabolic profiling studies in the future. Nonetheless, the improved resolving power is highly advantageous for understanding metabolic changes in organisms under environmental stress and is especially beneficial in studies where the metabolome has not yet been fully characterized.

23.4 METABOLITE IDENTIFICATION AND QUANTIFICATION

As 1D ^1H NMR analysis of metabolomic samples is relatively rapid ($\sim 10-15$ min per sample; Table 23.1), it has become a routine tool used in environmental metabolomics, whereas 2D NMR is used mostly for structural elucidation of metabolites. However, the main disadvantage of 1D ^1H NMR, especially for aqueous extracts, is the overlap of resonances in the $\delta = 3.0-4.5$ ppm region of the ^1H spectrum (Figure 23.4).[14,17,18] This may hinder metabolite identification and interpretation of results from multivariate statistics. In Table 23.2, some commonly reported endogenous metabolites are listed along with their chemical shifts and resolvable chemical shifts

Table 23.2. Example metabolites used for monitoring environmental stress responses in environmental metabolomic studies

Metabolite	Chemical shifts (ppm)	Resolvable chemical shift (ppm) window for quantification
Alanine	1.48, 3.79	1.45–1.49 (doublet)
ATP	4.23, 4.28, 4.40, 6.12, 8.21, 8.52	8.21–8.24 (singlet)
Betaine	3.36, 3.91	3.21–3.29 (singlet)
Glutamate	2.10, 2.36, 3.77	2.31–2.37 (multiplet)
Lactate	1.34, 4.15	1.30–1.34 (doublet)
Leucine	0.95, 1.71, 3.73	0.93–0.97 (triplet)
Lysine	1.48, 1.73, 1.91, 3.02, 3.76	2.96–3.04 (triplet)
Tyrosine	3.05, 3.19, 3.95, 6.89, 7.19	7.16–7.21 (doublet)
Phenylalanine	3.13, 3.28, 3.99, 7.33, 7.39, 7.42	7.30–7.34 (doublet)

ATP, adenosine triphosphate.
Owing to overlapping resonances, only selected resonances can be used for metabolite identification and quantification from ^1H NMR spectra (in reference to $\delta = 0.00$ ppm).
Data compiled from Refs 4, 22, 37, 38

within a ^{1}H NMR spectrum of earthworm tissue extracts. For these metabolites, some resonances cannot be used for integration and quantification because of the close proximity to or overlap with resonances of other metabolites in the ^{1}H NMR spectrum.[37] For example, a metabolite such as phenylalanine is best quantified through integration of the $\delta = 7.33$ ppm resonance (Table 23.2). However, quantification, through the use of internal standards such as sodium 3-trimethylsilyltetradeuteropropionate (TSP) or sodium 4,4-dimethyl-4-silapentane-1-sulfonate (DSS) may also be undermined because both TSP and DSS may bind to proteins that may be present in aqueous extracts and result in broadening or intensity reduction of the internal standard.[39,40] This may hinder absolute quantification in some cases, which is why recent studies employ some sort of separation process, often based on organic solvent phases, to remove biomolecules from the sample before addition of internal standard compounds.[30,41,42] Other recently reported variances impact quantification such as sample insertion and shimming errors[43] but these may be overcome with careful experimental design and instrument setup. Furthermore, cross-platform studies have found good agreement between NMR- and mass spectrometry (MS)-based quantifications[44,45] but additional studies are warranted that compare absolute quantification of metabolites with organisms used in environmental metabolomics.

Overall, the analytical error associated with ^{1}H NMR analysis is considerably lower than the biological variation observed in environmental metabolomics.[46] Some studies that have examined the quantitative reliability of ^{1}H NMR for small molecules report high reproducibility over a wide range of concentrations and a relative standard deviation of <1% for the majority of small molecules studied.[47–49] This high level of reproducibility has enabled the detection of small perturbations to the metabolome of environmental organisms as well as studies that quantify the natural variation of endogenous metabolites. For example, Whitfield Åslund et al.[37] reported a standard deviation of 14% based on all ^{1}H NMR signals for the aqueous tissue extract of laboratory-reared *E. fetida* earthworms with individual metabolite relative standard deviation values varied from 7.8% to 48.6%. Szeto et al.[50] reported relative standard deviation values of 29–39% for *Caenorhabditis elegans* and also noted that the extent of variation differed because of different nutrition sources used during culturing,

which highlights the need to maintain consistent food sources during laboratory rearing. The metabolite profile of *Saccharomyces cerevisiae*, a commonly studied yeast model, varied to a lesser extent (relative standard deviation values of 8–12%[50]). It should be noted that the 'quantification' of natural variability of various model environmental organisms is only possible because the very low analytical variation of ^{1}H NMR analysis[46] facilitates the sensitive analysis of physiological variation or reaction to an environmental stressor such as a pollutant or environmental conditions.

Keun et al.[46] examined the analytical reproducibility of ^{1}H NMR analysis of rat urine and reported excellent reproducibility using two different NMR instruments (500 and 600 MHz), suggesting that metabolomics analysis using NMR is directly comparable to different laboratories. Similarly, an international interlaboratory comparison of environmental metabolomic samples found that different laboratories, who measured samples using different hardware and magnetic field strengths, produced highly comparable data.[51] NMR spectra obtained by different laboratories were in good agreement and the resulting multivariate data analysis showed consistent results across all NMR platforms used (Figure 23.6). Consequently, differences in NMR hardware and data processing by various users do not impact the discrimination ability of NMR-based environmental metabolomics. The results of these studies are highly encouraging because they demonstrate that NMR-based metabolomics analyses are reproducible from laboratory to laboratory, which is critical for the development of routine environmental monitoring protocols based on environmental metabolomics data.

23.5 FUTURE DIRECTIONS AND CONCLUSIONS

NMR is a robust and reliable platform for use in environmental metabolomic studies. It is nondestructive, nonselective, sensitive, and reproducible, and requires minimal sample preparation making it the platform of choice in comprehensive metabolomic analysis. The nonselectivity is somewhat essential as the field of environmental metabolomics expands and targets various organisms beyond those already examined to date. Given the immense potential and versatility of NMR, it will undoubtedly continue to be a routine tool used in discovery-type research where no prior

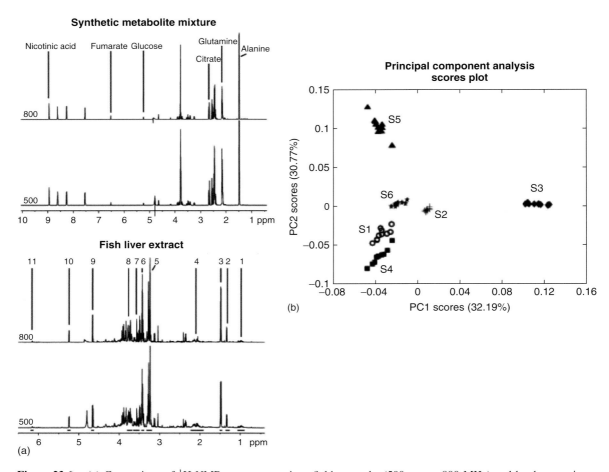

Figure 23.6. (a) Comparison of ¹H NMR spectra at various field strengths (500 versus 800 MHz) and hardware using a synthetic mixture and a fish liver extract (number indicated peaks used for multivariate analysis). (b) Principal component analysis (PCA) of interlaboratory results. The close clustering of samples analyzed by each laboratory (samples S1–S6, $n = 10$) shows the high level of reproducibility using different NMR platforms. (Reprinted from Ref 51. © 2009 American Chemical Society)

knowledge of metabolite perturbations within an organism is known. In addition, the further development of automation methods, novel NMR approaches, and new pulse programs will ensure that environmental metabolomics research is at the forefront. For example, new approaches, such as using supercooled water with NOESY, show promise for enhanced metabolite identification.[52] In addition, the identification of metabolites that fluctuate after exposure to an environmental stressor enables more targeted analyses to be developed for routine environmental monitoring. Future studies should also compare metabolic profiling

results to conventional toxicity endpoints and make comparison using other 'omic' approaches to further understand the ramifications of environmental stressors on organisms. Future environmental metabolomic studies may also include the examination of metabolic pathways and how this may relate to the mode of action of a stressor. NMR-based approaches can also be used for stable isotope (¹³C and ¹⁵N)-resolved metabolomics[53] to study metabolic networks and how environmental stressors may alter these networks. As well, NMR is playing a major role in the development of 'bioindicators' of environmental organism stress, which may be used as an early warning system for potential ecosystem shifts. As such, NMR-based environmental metabolomics will reshape our current practice of environmental monitoring in addition to increasing our fundamental knowledge regarding the impacts of environmental stressors.

ACKNOWLEDGMENTS

M. Simpson thanks the Natural Sciences and Engineering Research Council of Canada, the Canada Foundation for Innovation, the Ontario Research Fund, Bruker BioSpin, the University of Toronto, and the Krembil Foundation for generous support of environmental metabolomics research.

RELATED ARTICLES IN EMAGRES

Environmental NMR: High-resolution Magic-angle Spinning

Environmental Metabolomics

Metabonomics: NMR Techniques

Metabolomics in Environmental Microbiology

Software Tools for NMR Metabolomics

REFERENCES

1. J. G. Bundy, M. P. Davey, and M. R. Viant, *Metabolomics*, 2009, **5**, 3.

2. J. Sardans, J. Penuelas, and A. Rivas-Ubach, *Chemoecology*, 2011, **21**, 191.

3. A. J. Simpson, D. J. McNally, and M. J. Simpson, *Prog. Nucl. Mag. Res. Sp.*, 2011, **58**, 97.

4. M. J. Simpson and J. R. McKelvie, *Anal. Bioanal. Chem.*, 2009, **394**, 137.

5. C. Y. Lin, M. R. Viant, and R. S. Tjeerdema, *J. Pestic. Sci.*, 2006, **31**, 245.

6. M. R. Viant, *Methods Mol. Biol.*, 2007, **358**, 229.

7. S. A. E. Brown, A. J. Simpson, and M. J. Simpson, *Environ. Toxicol. Chem.*, 2008, **27**, 828.

8. B. P. Lankadurai, D. M. Wolfe, M. L. Whitfield Åslund, A. J. Simpson, and M. J. Simpson, *Metabolomics*, 2013, **9**, 44.

9. Q. Teng, D. R. Ekman, W. Huang, and T. W. Collette, *Aquat. Toxicol.*, 2013, **130–131**, 184.

10. D. R. Ekman, Q. N. Teng, D. L. Villeneuve, M. D. Kahl, K. M. Jensen, E. J. Durhan, G. T. Ankley, and T. W. Collette, *Metabolomics*, 2009, **5**, 22.

11. R. T. McKay, *Annu. Rep. NMR*, 2009, **66**, 33.

12. A. J. Simpson and S. A. Brown, *J. Mag. Res.*, 2005, **175**, 340.

13. G. A. Barding, R. Salditos, and C. K. Larive, *Anal. Bioanal. Chem.*, 2012, **404**, 1165.

14. J. Yuk, J. R. McKelvie, M. J. Simpson, M. Spraul, and A. J. Simpson, *Environ. Chem.*, 2010, **7**, 524.

15. J. C. Lindon and J. K. Nicholson, *eMagRes*, 2008. DOI: 10.1002/9780470034590.emrstm1048.

16. S. Stolzenburg, M. B. Lauridsen, H. Toft, P. A. Zalloua, and D. Baunsgaard, *Metabolomics*, 2011, **7**, 270.

17. Q. N. Van, H. J. Issaq, Q. Jiang, Q. Li, G. M. Muschik, T. J. Waybright, H. Lou, M. Dean, J. Uitto, and T. D. Veenstra, *J. Proteome Res.*, 2008, **7**, 630.

18. M. R. Viant, *Biochem. Biophys. Res. Commun.*, 2003, **310**, 943.

19. D. W. Bearden, *eMagRes*, 2012. DOI: 10.1002/9780470034590.emrstm1256.

20. C. Ludwig, J. M. Easton, A. Lodi, S. Tiziani, S. E. Manzoor, A. D. Southam, J. J. Byrne, L. M. Bishop, S. He, T. N. Arvanitis, U. L. Günther, and M. R. Viant, *Metabolomics*, 2012, **8**, 8.

21. C. Ludwig and M. R. Viant, *Phytochem. Anal.*, 2010, **21**, 22.

22. Q. Cui, I. A. Lewis, A. D. Hegeman, M. E. Anderson, J. Li, C. F. Schulte, W. M. Westler, H. R. Eghbalnia, R. Sussman, and J. L. Markley, *Nature Biotechnol.*, 2008, **26**, 162.

23. O. Beckonert, M. Coen, H. C. Keun, Y. Wang, T. M. Ebbels, E. Holmes, J. C. Lindon, and J. K. Nicholson, *Nature Protocols*, 2010, **5**, 1019.

24. S. Fasulo, F. Iacono, T. Cappello, C. Corsaro, M. Maisano, A. D'Agata, A. Giannetto, E. De Domenico, V. Parrino, G. Lo Paro, and A. Mauceri, *Ecotox. Environ. Safe.*, 2012, **84**, 139.

25. B. H. Hansen, D. Altin, A. Booth, S.-V. Vang, M. Frenzel, K. R. Sørheim, O. G. Brakstad, and T. R. Størseth, *Aquat. Toxicol.*, 2010, **99**, 212.

26. Q. Teng, D. R. Ekman, W. L. Huang, and T. W. Collette, *Analyst*, 2012, **137**, 2226.

27. J. L. Griffin, A. W. Nicholls, H. C. Keun, R. J. Mortishire-Smith, J. K. Nicholson, and T. Kuehn, *Analyst*, 2002, **127**, 582.

28. J. H. Grimes and J. M. O'Connell, *J. Biomol. NMR*, 2011, **49**, 297.

29. H. C. Poynton, N. S. Taylor, J. Hicks, K. Colson, S. R. Chan, C. Clark, L. Scanlan, A. V. Loguinov, C. Vulpe, and M. R. Viant, *Environ. Sci. Technol.*, 2011, **45**, 3710.

30. M. R. Viant, *Methods Mol. Biol.*, 2008, **410**, 137.

31. J. Jordan, A. Zare, L. J. Jackson, H. R. Habibi, A. M. Weljie, and J. Proteome, *Res.*, 2012, **11**, 1133.

32. M. R. Viant, E. S. Rosenblum, and R. S. Tjeerdema, *Environ. Sci. Technol.*, 2003, **37**, 4982.

33. J. Yuk, M. J. Simpson, and A. J. Simpson, *Environ. Chem.*, 2011, **8**, 281.

34. J. Yuk, M. J. Simpson, and A. J. Simpson, *Ecotoxicology*, 2012, **21**, 1301.

35. P. Sandusky, E. Appiah-Amponsah, and D. Raftery, *J. Biomol. NMR*, 2011, **49**, 281.

36. T. W.-M. Fan and A. N. Lane, *Prog. Nucl. Mag. Res. Sp.*, 2008, **52**, 69.

37. M. Whitfield Åslund, M. Celejewski, B. P. Lankadurai, A. J. Simpson, and M. J. Simpson, *Chemosphere*, 2011, **83**, 1096.

38. E. M. Lenz, J. M. Weeks, J. C. Lindon, D. Osborn, and J. K. Nicholson, *Metabolomics*, 2005, **1**, 123.

39. M. F. Alum, P. A. Shaw, B. C. Sweatman, B. K. Ubhi, J. N. Haselden, and S. C. Connor, *Metabolomics*, 2008, **4**, 122.

40. J. S. Nowick, O. Khakshoor, M. Hashemzadeh, and J. O. Brower, *Org. Lett.*, 2003, **5**, 3511.

41. M. Liebeke and J. G. Bundy, *Metabolomics*, 2012, **8**, 819.

42. H. Wu, A. D. Southam, A. Hines, and M. R. Viant, *Anal. Biochem.*, 2008, **372**, 204.

43. S. Sokolenko, R. McKay, E. J. M. Blondeel, M. J. Lewis, D. Chang, B. George, and M. G. Aucoin, *Metabolomics*, 2013, **9**, 887.

44. F. M. Geier, E. J. Want, A. M. Leroi, and J. G. Bundy, *Anal. Chem.*, 2011, **83**, 3730.

45. J. R. McKelvie, J. Yuk, Y. Xu, A. J. Simpson, and M. J. Simpson, *Metabolomics*, 2009, **5**, 84.

46. H. C. Keun, T. M. D. Ebbels, H. Antti, M. E. Bollard, O. Beckonert, G. Schlotterbeck, H. Senn, U. Niederhauser, E. Holmes, J. C. Lindon, and J. K. Nicholson, *Chem. Res. Toxicol.*, 2002, **15**, 1380.

47. I. W. Burton, M. A. Quilliam, and J. A. Walter, *Anal. Chem.*, 2005, **77**, 3123.

48. A. Cagliani, D. Acquoitt, G. Palla, and V. Bocci, *Anal. Chim. Acta*, 2007, **585**, 110.

49. A. Moing, M. Maucourt, C. Renaud, M. Gaudillere, R. Brouquisse, B. Lebouteiller, A. Gousset-Dupont, J. Vidal, D. Granot, B. Denoyes-Rothan, E. Lerceteau-Kohler, and D. Rolin, *Funct. Plant Biol.*, 2004, **31**, 889.

50. S. S. W. Szeto, S. N. Reinke, and B. D. Lemire, *J. Biomol. NMR*, 2011, **49**, 245.

51. M. R. Viant, D. W. Bearden, J. G. Bundy, I. W. Burton, T. W. Collette, D. R. Ekman, V. Ezernieks, T. K. Karakach, C. Y. Lin, S. Rochfort, J. S. De Ropp, Q. Teng, R. S. Tieerdema, J. A. Walter, and H. Wu, *Environ. Sci. Technol.*, 2009, **43**, 219.

52. H. Farooq, R. Soong, D. Courtier-Murias, C. Anklin, and A. Simpson, *Anal. Chem.*, 2012, **84**, 6759.

53. T. W.-M. Fan and A. N. Lane, *J. Biomol. NMR*, 2011, **49**, 267.

Chapter 24

Environmental Metabolomics of Soil Organisms

Oliver A.H. Jones[1] and Daniel A. Dias[2]

[1] *School of Applied Sciences, RMIT University, Melbourne, Victoria 3001, Australia*
[2] *Metabolomics Australia, School of Botany, The University of Melbourne, Parkville, Victoria 3010, Australia*

24.1 INTRODUCTION

Metabolomics can be defined as the qualitative and quantitative assessment of all low molecular weight compounds present in cells that are required for maintenance, growth, and cellular normal function. Such molecules include substrates and products of cellular metabolism, such as sugars, organic acids, amino acids, vitamins, lipids, and nucleotides. An organism's metabolome is its full complement of metabolites in the same way that its genome comprises its complete genetic content.[1] Metabolic fluxes are regulated not only by gene and protein expression but also by environmental stresses; hence the metabolome can be used to study the underlying biochemical response of an organism (or population/community in the case of microorganisms) to an environmental stimulus or combination of stimuli.[2] The field is maturing at a rapid pace because of the detailed insights that this approach can bring.

Several, highly synonymous terms are currently in use to describe the process of combining global analytical tools and pattern-recognition analysis to define the metabolic status of a tissue or an organism. Although metabolomics is most common, the terms metabonomics, metabolic fingerprinting, metabolic footprinting, and metabolic profiling are often used interchangeably, despite subtle differences in their exact meaning according to some authors. A good overview is given in Goodacre *et al.*[3] who outlined the following classifications:

- Targeted metabolite analysis, which analyzes a restricted number of metabolites, for example, a certain enzyme system that would be directly affected by a specific nonbiological or biological perturbation.

NMR Spectroscopy: A Versatile Tool for Environmental Research
Edited by Myrna J. Simpson and André J. Simpson
© 2014 John Wiley & Sons, Ltd. ISBN: 978-1-118-61647-5

- Metabolite profiling, which focuses on a group of metabolites such as a class of compounds associated with a specific metabolic pathway.
- Metabolite fingerprinting, which classifies samples on the basis of their biological relevance or origin.
- Metabolic profiling, which is used to trace the fate of a drug or a metabolite.
- Metabonomics, which measures the fingerprint of biochemical perturbations caused by disease, drugs, or toxins.
- General metabolomics, which is a comprehensive analysis of the whole metabolome under a given set of conditions.

In the authors' opinion, the latter term is the one most researchers take to be correct and is the one most commonly used in the literature and at conferences.

There are also several terms in use for the application of metabolomics in ecological/environmental studies. For instance, the term ecotoxicogenomics was proposed by Snape *et al.*[4] to describe the integration of genomic-based science into the field of ecotoxicology. 'EcoGenomics' (or ecological genomics) was used by Chapman[5] to describe the application of genomics-based techniques to ecology. In both cases, the term genomics is taken to encompass all the 'omic sciences'; namely, genomics (genome sequencing and the annotation of function to genes), transcriptomics (gene expression at the transcription level), proteomics (protein expression) and metabolomics. The phrase 'environmental metabolomics' was defined in 2007 by Viant[6] as the 'application of metabolomics to characterize the metabolism of free-living organisms obtained from a natural environment, and of organisms reared under laboratory conditions, where those conditions serve to mimic scenarios encountered in the natural environment.' Perhaps the most useful term is the recently introduced ecological metabolomics, which refers to the application of metabolomics to ecological issues.[7]

NMR spectroscopy has been the primary analytical technique for metabolomics studies for many years. NMR exploits a quantum mechanical property of nuclei called *spin*. Typically, only those nuclei with a spin number of $1/2$ are studied (e.g., ^1H, ^{31}P, ^{13}C, and ^{19}F). When a nucleus is placed in a strong magnetic field aligned along the z-axis, the spin aligns with the external field and precesses with a frequency characteristic of its exact electronic environment. By applying a radio frequency electromagnetic pulse to the sample, the spin can be made to rotate into the x/y plane where it continues to precess. By measuring the oscillation of the signal in the plane and its decay back to alignment with the z-axis, the characteristic frequency of oscillation, termed the chemical shift, can be determined. The intensity of the signal is proportional to the number of identical nuclei, and therefore the concentration of the metabolite, present in the sample. As the chemical shift depends on the exact electronic environment surrounding the nucleus, a chemical shift can be diagnostic for a specific metabolite. Molecules with multiple protons may show multiple peaks in the resulting spectrum but their concentrations (and hence that of the metabolite) will be proportional to the size of the signal and the patterns of intensity and chemical shift greatly aid in identifying a metabolite. By assigning each chemical shift, or group of shifts, to specific metabolites and analyzing the relative changes in signal intensity of those shifts, a wide range of metabolite concentrations can be monitored simultaneously.

NMR has a number of advantages for metabolomics in that it requires minimal sample preparation, and is a highly reproducible, robust and non-destructive technique, which allows the simultaneous measurement of many kinds of small molecule metabolite in solution (see Chapter 23). Coupled with the fact that acquisition of a simple ^1H NMR spectrum takes on the order of 5–10 min using a modern automated system, NMR is a valuable high-throughput tool. Multidimensional NMR spectra can be acquired in order to reduce signal overlap and ease metabolite identification (see Chapter 23). Other NMR-based approaches allow the quantification of metabolite concentrations in intact tissue, either in vivo or ex vivo. For example, magnetic resonance imaging (see Chapter 7) allows the non-invasive assessment of specific metabolite concentrations directly in vivo within a specific localized region,[8] whereas high-resolution magic-angle spinning (HR-MAS) NMR (see Chapter 5) allows spectra to be recorded from intact tissue samples.[9] An overview of the sample processing procedure for environmental metabolomics is given in Figure 24.1.

While robust and reliable, NMR is however, a relatively insensitive technique and lacks the sensitivity offered by, for example, mass spectrometry (MS)-based approaches. This lack of sensitivity can, to some extent, be addressed by recording spectra in a stronger

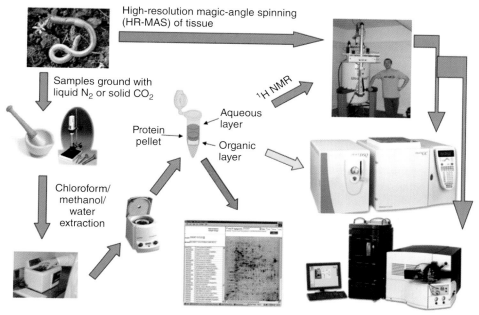

High-resolution magic-angle spinning (HR-MAS) of tissue

Samples ground with liquid N_2 or solid CO_2

Protein pellet

Aqueous layer

Organic layer

^1H NMR

Chloroform/ methanol/ water extraction

Figure 24.1.   Graphical overview of the metabolite extraction process for use in environmental metabolomics (the blue arrows indicate the path for NMR analysis)

applied magnetic field, as NMR signal intensity depends on the strength of the applied field. The use of a cryoprobe greatly improves the signal-to-noise ratio of the spectrum by cooling the receiver coils to cryogenic temperatures, thereby reducing the contribution of thermal noise to the spectrum.[10–12]

Soil species, such as earthworms, potworms, and springtails (Collembola), are widely used to assess the toxicity of soil pollutants and have played a large role in recent forays into the mechanistic side of soil ecotoxicology; ventures in which metabolomics has played a large part. Indeed, metabolomics has great potential for investigating the biochemical effects of a wide range of ecologically important phenomena, such as the impacts of climate change, disease, food restriction, infection, and parasite load as well as other useful applications such as the identification of morphologically similar species and the identification of useful natural products from soil organisms. Such applications are discussed in this chapter.

24.2   NMR-BASED METABOLOMICS IN ENVIRONMENTAL SOIL SCIENCE

24.2.1   Identification of Morphologically Similar Species

Metabolomics offers great potential in comparative biology because low-molecular-weight metabolites are almost universal. Thus, lactate in one species is, for the most, part the same as lactate in another. This means that analytical methods are easily transferred between species and that a fully sequenced genome is not required for the analysis of the metabolome. It is, therefore, possible to perform robust investigations of the biochemistry of genetically uncharacterized species relatively quickly. Indeed, one of the most useful applications of metabolomics to soil organisms is in the identification of cryptic species. By this, we mean two separate species (typically very closely related) that are morphologically indistinguishable to the untrained eye and/or that require a very experienced taxonomist to characterize. With taxonomy rapidly becoming a scarce skill, species identification by metabolomics has great potential. For instance, it has been successfully used to distinguish differing functional/physiological ecotypes of bacterial strains.[13] Interestingly, in that case, screening of the genomic DNA for the presence of genes encoding known toxins gave no candidate genes that were unambiguously able to distinguish the ecotypes. Similar work was carried out by Timmins *et al.*[14] using pyrolysis-MS and Fourier Transform Infrared (FTIR) Spectroscopy in conjunction with cluster analyses to successfully distinguish between three species of *Candida* isolates (*C. albicans, C. dubliniensis*, and *C. stellatoidea*). These studies clearly demonstrated that it was possible to use metabolomic methods as a kind of molecular taxonomy to classify organisms on the basis of their expressed biochemistry even when it is not possible to infer a direct mechanistic or genetic link.

The first application of this approach using NMR and soil organisms was carried out by Bundy *et al.*[15] who used ^{1}H NMR spectroscopy to distinguish closely related earthworm species via their metabolic phenotype (metabotype). Their study utilized three *Eisenia* ('*Oligochaeta, Lumbricidae*') species *E. fetida, E. andrei*, and *E. veneta* using both tissue extracts and coelomic fluid analysis. The authors took care to include environmental histories in the experimental design and were thus able to differentiate species from the contribution of possible confounding environmental factors such as food supply and hydration. It was found that the metabolic profiles of tissue extracts were highly conserved for all three species, with *E. fetida* and *E. andrei* being more similar to each other than either was to *E. veneta*. The coelomic fluid of the different species was found to be highly distinctive. Multivariate analysis [via principal components analysis (PCA) and hierarchical cluster analysis (HCA)] of the NMR spectra allowed unequivocal identification of species. These results show that two morphologically undistinguishable species (*E. fetida* and *E. andrei*) had very different biochemical profiles despite apparently occupying the same ecological niche. Similar outcomes were found by Albani *et al.*[16] using fluorescence spectrometry-based techniques but these did not give as detailed results. Given the decline of taxonomy as a discipline in recent years, this metabolic approach could have numerous useful applications.

24.2.2 Physiological Responses of Soil Organisms to Stress

Because NMR-based metabolomics is relatively quick and simple to carry out and also gives information of the changes of many metabolites at once, it can also be used to look at how the biochemistry of an organism changes in response to various environmental stressors such as drought or changes in temperature/oxygen levels. For instance, Warne *et al.*[17] assessed the biochemical effects of short-term starvation on the earthworms *E. veneta* and *Lumbricus terrestris*. Using ^{1}H-NMR, *E. veneta* showed small metabolic changes for the first 5 days, with clear differences after 6 and 7 days of starvation. These consisted of increases in glutamate, citrate, aspartate, and isoleucine levels, and decreases in lysine, isoleucine, and threonine levels. In contrast, *L. terrestris* showed no consistent metabolite changes over this period suggesting this species to be more suitable for soil toxicity testing and monitoring using this procedure. This is because metabolic changes after toxicant exposure are sometimes misconstrued as related to the toxicant itself rather than a stress event related to reduced feeding rates (calorie restriction) of an organism in response to pollutant exposure.[18]

Similarly, Bundy *et al.*[19] assessed the metabolic changes induced in the freeze-tolerant earthworm, *Dendrobaena octaedra*, compared to four freeze-intolerant earthworm species (*Dendrodrilus rubidus, Aporrectodea icterica, A. caliginosa, and A. longa*) using NMR-based metabolomics. Individuals of each species were frozen at -2 °C and control earthworms were exposed to $+2$ °C. Several, biochemical changes were detected as a result of freezing in all worm species, including an increase in relative free alanine concentrations, and an apparent conversion of adenosine to inosine. A number of biochemical changes that were unique to the freeze-tolerant species, *D. octaedra*, were also observed. Although all species showed an increase in glucose concentrations, the increase was largest in *D. octaedra*, and was coupled with a concomitant decrease in glycogen. These results were confirmed by several other studies in freeze tolerance of this species, which all showed that freeze tolerance was effected by rapid glucose production from glycogen reserves.[20–22] Succinate increased in all species on freezing but the increase was least in *D. octaedra*. Interestingly, there was also no lactate accumulation in *D. octaedra*, indicating that anoxic metabolism was lowest in this freeze-tolerant species.

Biochemical changes in leaf litter as a response to ultraviolet (UV) exposure were studied by Gwynn-Jones *et al.*[23] who showed not only that exposure to UV-B light results in marked changes in the biochemical composition of leaf litter (e.g., in lignin content) but also that *L. rubellus* benefited from the UV-B-treated litter, whereas *L. terrestris* did not. These studies were carried out in the laboratory but the authors suggested that the observed trends could influence competitiveness among these species if these results were found to hold true in populations in the environment. Any resulting changes in the earthworm population composition could have implications for the removal of surface litter and its distribution within soil as well as for cast microbial communities and nutrient cycling. However, further studies would be needed to elucidate the implications of such changes for soil decomposition processes and microbial diversity.

Analyzing the small molecule metabolites, in addition to more abundant macromolecules present in the soil, makes it possible to detect subtle shifts in microbial populations interacting with plant litter. Indeed, the use of ^{13}C NMR techniques has provided valuable data supporting the occurrence, diversity, and extent of carbon cycling in the carbohydrate metabolism of microbes in soil.[24] Targeted studies have also been undertaken in order to measure the flux rates of specific metabolic pathways, for example, using isotope labeling studies.[25]

24.3 ECOTOXICOLOGY

Ecotoxicology (the study of the toxicological effects of pollutants, pesticides and other contaminants in the natural environment) is probably the area of environmental science in which NMR-based metabolomics has had the greatest impact to date.[26] Metabolomics has been widely applied in ecotoxicology, particularly in studies on earthworms.[27] As a crucial part of most soil-based ecosystems, the earthworm is an ideal model for terrestrial ecotoxicology studies. Earthworms are constantly passing soil particles through their gut and are exposed to toxicants both sorbed to soil particles and dissolved in the pore water between them, where the majority of other soil organisms may only be exposed to the pore water toxicants.[28] ^{1}H NMR spectra of earthworm tissue extracts are shown in Figure 24.2.

The use of NMR spectroscopy in ecotoxicology is undeveloped compared with its applications in human toxicology and drug discovery but this is changing rapidly. Particularly, active groups in this area at the time of writing (February 2013) include those of Jake Bundy at Imperial College London (UK), Myrna Simpson at the University of Toronto (Canada) and Jules Griffin at the University of Cambridge (UK).

24.3.1 NMR Studies of Earthworm Metabolomic Responses to Contaminants

^{1}H NMR has been used in earthworms to detect elevated free histidine levels in tissue extracts of *L. rubellus* exposed to heavy metals[29,30] and these were found to be comparable across contaminated sites[31] with changes in sugar and histidine tissue contents were behind the observed statistical separations.[32] NMR has also been used to identify free histidine as a novel biomarker of exposure to copper in earthworms.[30] Increased maltose levels have also been indicated as a potential biomarker for metal toxicity.[32] Recent work also suggests that biomarkers of metal contamination in earthworms are applicable across multiple sites, even those with different physicochemical characteristics.[31] NMR-based metabolomics may thus have a future role in detecting biomarkers related to soil health.

Many studies have utilized NMR to assess the metabolic response of various earthworm species after exposure to an array of organic compounds to determine whether contaminant-specific responses could be identified. For example, the technique was used to define metabolite profiles for *E. veneta* exposed to different substituted anilines[17,29] and to classify the metabolite responses upon exposure to several fluorinated organic compounds.[33,34] The use of metabolomics has also begun to provide a mechanistic component to these studies.[35] For instance, Jones *et al.*[36] looked at the toxicity of pyrene on *L. rubellus* and observed alterations in its metabolic profile that could be observed even when individuals were exposed to concentrations of $40 \, \text{mg kg}^{-1}$, a level that is below the concentration previously found to significantly reduce reproduction. Similarly, earthworm coelomic fluid and tissues have been used to study the earthworm exposure to endosulfan and endosulfan sulfate in soils (these caused significant fluctuations

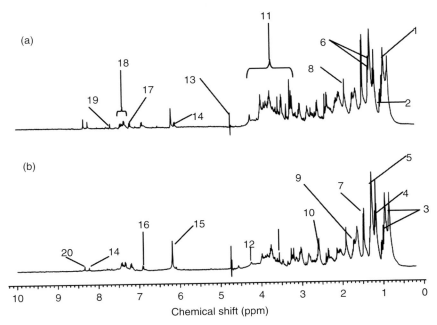

Figure 24.2. 300 MHz ^{1}H NMR spectra of (a) *L. rubellus* control sample and (b) *L. rubellus* exposed to 640 mg kg^{-1} of pyrene with 20 metabolites labeled. Key: 1, leucine; 2, valine; 3, isoleucine; 4, β-hydroxy butyrate; 5, fatty acid; 6, lactate; 7, alanine; 8, acetate; 9, lysine; 10, malate; 11, overlapping resonances from glucose and amino acids; 12, threonine; 13, water (suppressed); 14, adenosine; 15, inosine; 16, 3,4-dihydroxymandelate; 17, tyrosine; 18, Phenylalanine; 19, 4-aminohippurate; and 20, formate

in glutamine/γ-aminobutyric acid (GABA)-glutamate cycle metabolites and spermidine, respectively).[37,38]

Other studies of earthworms exposed to aged polychlorinated biphenyl (PCB) contaminated soils,[39] phenanthrene,[40] and titanium dioxide nanomaterials.[41] One particularly detailed study included two pesticides (carbaryl and chlorpyrifos), three pharmaceuticals and personal care products (carbamazepine, estrone, and caffeine) two persistent organohalogens (Aroclor 1254 and PBDE 209) and two industrial compounds (nonylphenol and dimethyl phthalate)[42] and found that the earthworm response was consistent with the toxic mode of action of the various contaminants. Natural variability in the metabolic profile of earthworms has also been studied using ^{1}H NMR metabolomics.[43] This showed that earthworm extracts exhibited low overall interindividual metabolic variability. These results suggest that NMR based metabolomics offers considerable promise for differentiating between the different toxic modes of action associated with sublethal toxicity to earthworms. It is worth noting, however, that while the metabolites themselves are

similar across species, the metabolic responses are not. For instance, the metabolomic responses of *Caenorhabditis elegans* to cadmium are dominated by changes in cystathionine and phytochelatins and this is deferent from the earthworm response.[44]

24.3.2 Mechanistic Toxicology

While traditional toxicology tests have usually focused on investigating the toxic effects of individual compounds, an important challenge in ecotoxicology is to find a method to reliably study mixtures of sublethal concentrations of different toxicants – a situation that is much more common than the former in the study of natural ecosystems. For example, contaminants may interact to produce synergistic or antagonistic effects, one chemical may interfere with the uptake, metabolism, or excretion of another, or, particularly in samples taken from the natural environment, the presence of another chemical that is not under study may modify the action of the studied chemicals.[2] This issue is particularly relevant to soil environments, as soil organisms are commonly exposed to many different chemicals overtime. For example, many European, agricultural soils have high levels of copper left over from when the metal was used extensively as a pesticide.[45] Even in the present day, the majority of crops (and the underlying soils) grown worldwide are routinely treated with a variety of pesticides throughout the growing season. The extent of pesticide usage varies with each crop and between countries, but many compound classes are applied throughout the year. For example, herbicides are applied on arable crops in the United Kingdom with an average of three applications of four products and five active substances per year.[46]

Two established models currently dominate the field of mixture risk assessment. These are concentration addition (CA) for similarly acting chemicals and independent action (IA) for dissimilarly acting chemicals. Many studies have shown that the CA and IA models can provide good descriptions of the joint changes induced by mixture exposure.[47–49] However, while clearly useful, these models are not universally applicable and a substantial number of studies have highlighted the cases where these models have failed to predict the action of multiple contaminants. Indeed, just considering binary cases, the two models have been suggested to provide predictive power within a factor of 2 in only 50–90% of cases.[49,50] In addition, the limited mechanistic basis of these two models means that alone they cannot provide a comprehensive picture of the physiological changes that relate exposure to phenotype. This is where a metabolomics approach can provide valuable insights.

Baylay *et al.*[51] used NMR-based metabolomics in conjunction with more standard ecotoxicology tests including changes in survival, weight change, cocoon production, and metabolism caused by exposure to two similarly acting (imidacloprid/thiacloprid) and two dissimilarly acting (chlorpyrifos/Nickel) chemicals on the earthworm *L. rubellus*. The study employed CA and IA models, in conjunction with a metabolomics-based approach to elucidate mechanisms of action. For imidacloprid and thiacloprid, the reproductive effects indicated probable additive effect in terms of the metabolic response. Although this suggested joint effects through a similar mechanism, metabolite changes for each pesticide resulted in distinct responses. Further, earthworms exposed to a 0.5 toxic unit equitoxic mixture demonstrated metabolic effects intermediate between those for each pesticide, indicating a non-interactive, independent

joint effect. For higher levels of mixtures (1 and 1.5 toxic units), metabolite changes associated with thiacloprid exposure began to dominate.

Similar work by Jones *et al.*,[52] using NMR (and MS) to study the soil-dwelling nematode *C. elegans* also utilized different concentrations of nickel, chlorpyrifos, and their mixture. Novel metabolic profiles of *C. elegans* were associated with both exposure and exposure level, and the authors concluded that their results could form the basis for a rapid and economically viable toxicity test that defines the molecular effects of pollution/toxicant exposure in a manner that is relevant to higher vertebrates.

In short, the use of NMR-based metabolomics continues to increase our understanding of the molecular responses of terrestrial macroinvertebrate taxa (nematodes, earthworms, and springtails) to both inorganic and organic chemicals and will likely continue to do so for some time to come. Combined with the development of methods for proteomics and transcriptomics, this means that it is now feasible to use soil invertebrates in studies that can advance fundamental knowledge of important aspects of ecotoxicology, such as the biochemical basis of species sensitivity, the prevalence of multiple (and unexpected) modes of action, the basis and consequences of chemical-induced change at the population and community levels, and deriving better understanding of the combined toxic effects of pollutants.[2,35,53,54]

24.4 NMR AND METABOLOMICS IN THE STUDY OF NATURAL PRODUCTS

Synthesized by plants and microorganisms, natural products and their derivatives often have potent physiological activities and therefore play important roles both as frontline treatments for many diseases and as the inspiration for chemically synthesized therapeutics. Natural products may be broadly divided into two main categories; primary metabolites (including fatty acids, sugars, and the common amino acids), which occur in all cells and play a major role in metabolism and secondary metabolites, which are nonessential to 'everyday' metabolism but nevertheless have a specific function and are characteristic of a limited range of species. Examples include polyketides, terpenoids, alkaloids, and range of specialized amino acids, peptides, and carbohydrates. While primary metabolites tend to exert their biological effect within the organism or cell that produced them, secondary metabolites are of great interest because of their potential biological effects on other organisms. This has made them a successful source of potential drug leads for many years.[55–58] Soil organisms are a particularly rich source of useful natural products with antibiotics, antiviral and anticancer agents, immunosuppressors, and antihypertensives being just a few of the applications resulting from the study of this class of compounds.

Despite their use and potential economic value, the detection and synthesis of new therapeutic compounds derived from or inspired by natural compounds have declined in recent years, mostly because of the increased difficulty in identifying and isolating novel active compounds (see Chapter 14). Metabolomics, including both targeted and global metabolite profiling strategies, has the potential to be instrumental in natural products chemistry, as it allows for a systematic study of complex mixtures (such as plant extracts) without the need for prior isolation of natural products (or mixtures thereof).[59] Interestingly, both natural products chemistry and metabolomics have as their goal the identification of active compounds, either as a purified active component (natural products chemistry) or a group of compounds used as an indicator of a particular biological state (metabolomics). Natural products chemistry has a long tradition of sophisticated techniques that allow the identification of complex molecular structures but it often fails when dealing with complex mixtures. In contrast, metabolomics deals well with mixtures and uses the power of multivariate analysis to isolate the driver of a particular effect but it is often limited in the identification of all the compounds involved.[60]

24.4.1 Natural Products of Fungal Origin

In terms of biodiversity and bioprospecting, soil organic matter (SOM) is the home of a plethora of microorganisms, plants, animals, exudates from living organisms and decayed or decomposing material, degraded by microorganisms, and many small organic compounds are often present.[61,62] It is one of the most complex natural mixtures on earth and plays a critical role in ecosystem functions[63] and this complexity has meant that the identification of natural products from soil microorganisms (e.g., bacteria and fungi) remains

a mostly untapped resource. A pertinent example is *Pycnoporus cinnabarinus*, a bright, red-orange, bracket fungus, commonly found growing on deadwood.[64] Gripenberg *et al.*[65–67] have reported that Indigenous Australians have a long history of using this species of fungus to treat sore mouths, oral thrush and sore lips. The constituents responsible for the intense color of the fungus have been previously identified as the phenoxazone alkaloid pigments, cinnabarin, tramesanguin and cinnabarinic acid with the distribution of these pigments being dependent on species and season.[65–67] It was only recently that these notoriously insoluble pigments were isolated by high-performance liquid chromatography (HPLC), and structures elucidated by detailed spectroscopic analyses via a classical natural product isolation approach.[55,68] Figure 24.3 illustrates this general workflow, illustrating (a) HPLC chromatographic isolation and purification of pycnoporin (a phenoxazone pigment), (b) a ^{13}C spectrum of the compound of interest, and (c) the use of a heteronuclear single quantum coherence adiabatic (HSQCAD) NMR to show direct ^{1}H–^{13}C connectivity as the final experiment used to piece structural fragments to elucidate the structure of a molecule.

The earliest report of a natural product derived from soil origin was the isolation of the biologically active natural product koninginin B, from *Trichoderma koningii*. *Trichoderma* is an organism isolated from the roots and soil line of a species of Dieffenbachia (a genus of tropical flowering plants). T. koningii itself is a saprotrophic fungus widely used as a biological control agent against plant pathogens.[69] The compound koninginin B was isolated as a white, finely crystalline compound and its structure was elucidated by detailed spectroscopic analyses using one-dimensional (1D) and two-dimensional (2D) NMR spectroscopy, UV spectroscopy, infrared (IR) spectroscopy, and mass spectrometry (MS). The compound was later found to inhibit the growth of etiolated wheat coleoptiles (the protective sheath covering the emerging shoot).[70]

Natural products can also find use in agriculture. For example, the fungus *Gaeumannomyces graminis* (Sacc.) Arx and Oliver var. *tritici* Walker causes a disease of wheat known as *take-all*. This results in severe yield loss in cereal crops worldwide.[71] Some soils have been shown to suppress the growth of plant pathogens and these are known as *suppressive soils*. The phenomenon of disease suppressive soils has fascinated plant pathologists for decades. The phenomenon is believed to be biological in nature because fumigation or heat-sterilization of the soil

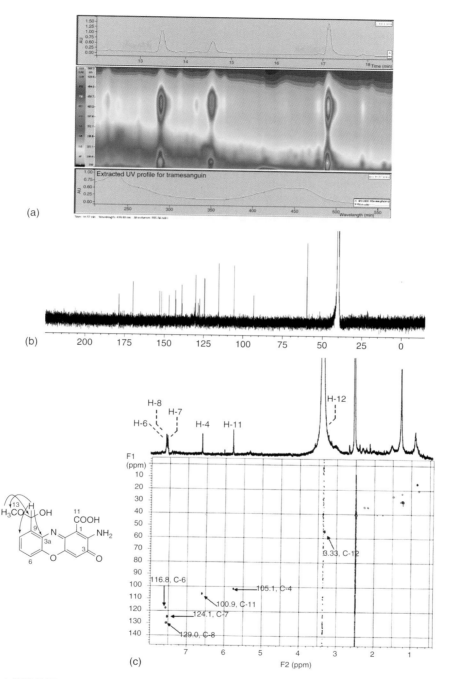

Figure 24.3. (a) HPLC-PDA chromatogram and 2D contour plot of the main pigments of interest identifying unique UV chromophore at 435 nm; (b) ^{13}C spectrum of the novel phenoxazone pigment, pycnoporin; and (c) Far right: HSQCAD 2D-NMR experiment illustrating direct ^{1}H–^{13}C connectivity and far left: arrows on the structure of pycnoporin showing ^{1}H–^{13}C connectivity, that is, 2–3 (J-coupling) bond correlations

usually eliminates the suppressive effect and disease is severe if the pathogen is reintroduced.[71] A classic example is soil suppression of take-all disease. Investigating this phenomenon resulted in the isolation of *Pseudomonas aureofaciens* Q2-87, from the rhizosphere of wheat grown in a *take-all*-suppressive soil. This in turn led to the isolation of an antifungal natural product, 1,3,6-trihydroxy-2,4-diacetophenone, isolated from a chemically defined broth culture of strain Q2-87. The structure was elucidated using ^{1}H and ^{13}C NMR, UV and high-resolution fast atom bombardment mass spectroscopy (HR-FAB-MS) techniques. 1,3,6-trihydroxy-2,4-diacetophenone was also synthesized and demonstrated activity equal to the natural product.[71]

Another plant disease is caused by *Monosporascus cannonballus*, a pyrenomycete fungus that infects the roots of plants. The primary symptoms are necrosis of the small roots and lesions on the larger roots, typically around root junctions, with the most dramatic symptom being the collapse of the plant late in the season.[72] *M. cannonballus* causes severe production losses to muskmelon and watermelon in the United States and other countries.[73] Interestingly, wild-type fungus produces no pigments when grown on potato dextrose agar in the laboratory. However, after long-terms storage on soil/oat hull mix, some of the isolates of the fungus have been observed to produce yellow to brown pigments.[73] Five colored natural products from pigmented cultures of *M. cannonballus* isolate led to the isolation of monosporascone and dehydroxyarthrinone (both previously isolated from other fungi) in addition to demethylcerdarin, monosporascol A and azamonosporascone, which had not previously been reported. The latter three had complex structures that were solved by 1D and 2D NMR analyses.[73]

Fungal natural products may also be useful for humans. For example, Dong *et al.*[74] describe the isolation of four novel 22-membered macrolides, wortmannilactones A–D from the fungus *Talaromyces wortmannii*, isolated from a soil sample collected from Yunnan Province, China. These compounds were elucidated by extensive 1D and 2D NMR and MS spectral analyses. Wortmannilactones A–D were found to exhibit in vitro cytotoxic activity against several human cancer cell lines with IC_{50} values (the concentration of the compound that will inhibit 50% of the activity of a particular biological system) ranging from 28.7 to 130.5 μM.[74]

24.4.2 Natural Products of Bacterial Origin

Streptomyces, the largest genus of Actinobacteria, is found in soil, decaying vegetation and is characterized by complex secondary metabolism.[75] Gerwick *et al.*[76] successfully isolated a novel phytotoxin, pyridazocidin, from the cultures of a *Streptomyces* sp. soil isolate. The structure of pyridazocidin was determined on the basis of NMR and MS experiments. Greenhouse evaluations on weeds with this phytotoxin showed rapid necrosis and chlorosis on multiple species. Isolated chloroplasts were found to consume oxygen in the presence of pyridazocidin over a concentration range also inhibiting plant growth. As such, pyridazocidin represents the first reported natural product that appeared to act via reversible oxidation/reduction linked to photosynthetic electron transport.[76]

The Middle East may also be a productive source of useful natural products. Abdelfattah[77] described the isolation of 30 different *Streptomyces* strains from soil samples collected from several sites in Egypt. The isolated *Streptomyces* sp. Eg23 yielded three novel stereoisomers, (7*R*,9*R*,10*S*)-*e*-rhodomycinone, (7*R*,9*R*,10*S*)-aklavinone, and (9*R*,10*S*)-7-desoxy-*z*-rhodomycinone, and again their structures were solved by 1D and 2D NMR spectroscopic analyses. All three compounds were known synthetically but had not been isolated as natural products and (7*R*,9*R*,10*S*)-*e*-rhodomycinone showed activity against various cancer cell lines.[77]

In a comparable study, Huang *et al.*[78] described the isolation of neaumycin, a novel 30-membered macrolide from the soil actinomycete *Streptomyces* sp. Neaumycin was elucidated on the basis of comprehensive MS and NMR spectroscopic analyses. Similarly, Raju *et al.*[79] described the isolation of a *Streptomyces* sp. Lv3-13, isolated from the rhizosphere soil of the plant *Mespilus germanica*. This *Streptomyces* sp. led to the isolation of three novel pimprinine derivatives, pimprinols A–C, (2-aminophenyl)-(2-ethyloxazol-5-yl)methanone along with the known natural products, 2-ethyl oxazole pimprinine, and 2-propyl oxazole pimprinine. The structures of the compounds were again elucidated based on spectroscopic methods including detailed 1D and 2D NMR analyses.[79] Huang *et al.*[80] also described the isolation of a novel natural product, 7-*O*-methyl-8-chlorogenistein together with eight known isoflavones, 8-chlorogenistein, kakkatin, 7-*O*-methyl genistein, genistein, daidzein, glycitin, genistin, and daidzin, from a *Streptomyces* strain YIM GS3536 isolated from a soil sample collected from Yunnan Province, China. The unambiguous assignments of 7-O-methyl-8-chlorogenistein were elucidated by ^{1}H and ^{13}C NMR analyses. 7-O-Methyl-8-chlorogenistein exhibited appreciable cytotoxicity against human melanoma cell lines and human leukemia cell lines with IC$_{50}$ values of 17.5 and 19.2 μM, respectively. The minimum inhibitory concentration (MIC) values of 7-O-methyl-8-chlorogenistein for *Bacillus subtilis*, *Staphylococcus aureus*, *Escherichia coli*, and *C. albicans* were shown to be in the range 23–35 μM.[80] Metagenomics – the study of metagenomes, genetic material recovered directly from environmental samples – is also of use in natural product research. For example, Wang *et al.*[81] described the generation of combinatorial biosynthetic libraries and the cloning of large fragments of DNA isolated from uncultivable microbes from soil into a *Streptomyces lividans* host, which led to the isolation of terragines A–D from recombinant *436-s4-5b1*. Isolated compounds were elucidated by a combination of spectroscopic techniques including 2D NMR experiments.[81]

Cyanobacteria, a phylum of bacteria, which obtain of their energy from photosynthesis, are common to soil and water and are more commonly known for secreting toxins. However, Prinsep *et al.*[82] isolated several multidrug-resistant reversal agents, tolypodiol (an anti-inflammatory diterpenoid) and tolyporphin A1 along with tolyporphins B–I from *Tolypothrix nodosa*. Further investigations led to the isolation of two new porphinoids, tolyporphins J and K, isolated from the terrestrial cyanobacterium *T. nodosa* (HT-58-2), in which the structures were deduced by NMR and MS analyses.[82] Another important bacterial genus is *Rhodococcus*; it has the ability to catabolize a wide range of aromatic compounds, steroids, and nitriles with application in the bioremediation of contaminated environments. Chatterjee[83] described the microbial degradation of geraniol, a natural monoterpene alcohol using a *Rhodococcus* strain GR3 isolated from soil. The bioconversion product was identified as geranic acid [(2E)-3,7-dimethylocta-2,6-dienoic acid] and its structure was established by ^{1}H NMR, in conjunction with MS.[83]

Han *et al.*[84] described a bacterial strain (*Bacillus* sp. sunhua) antagonistic against *Sarcoptes scabiei* isolated from the soil from a potato-cultivating area. Antibiotics produced by this strain were found to be stable within a broad pH range and at high temperatures. Two compounds were isolated and identified as iturin A and macrolactin A, and structures elucidated by detailed 1D and 2D NMR spectroscopic and MS techniques. The culture broth of *Bacillus* sp. sunhua possessed a suppressive effect on common scab disease, decreasing the infection rate from 75 to 35%, which also suppressed *Fusarium oxysporum*, the pathogen of potato dry rot disease. This was the first report demonstrating macrolactin A and iturin A to inhibit *S. scabiei* and demonstrated the possibility of controlling potato scab disease using *Bacillus* sp. sunhua.[84]

Ratnayake *et al.*[85] also described a novel family of bioactive heterocyclic polyketides known as the *kibdelones* synthesized by a rare soil actinomycete, *Kibdelosporangium* sp. (MST-108465). The relative stereochemistries of kibdelones A–C, kibdelone B rhamnoside, 13-oxokibdelone A, and 25-methoxy-24-oxokibdelone C were assigned on the basis of detailed spectroscopic analysis, chemical interconversions, and by mechanistic and biosynthetic considerations. Kibdelones exhibited potent and selective cytotoxicity against a panel of human tumor cell lines and displayed significant antibacterial and nematocidal activities.[85] In a similar study, Pettit *et al.*[86] described the isolation of kitastatin 1, accompanied by the previously isolated compounds respirantin and its valeryl homolog (cancer cell growth inhibitors) from a *Kitasatospora* sp. collected from a tundra soil sample. These natural products were elucidated by 2D NMR and MS analyses and the stereochemical assignments were confirmed by total synthesis.[86]

Two novel antibiotics, neocitreamicins I and II, were isolated from a fermentation broth of a *Nocardia* strain, a weakly Gram-positive pathogenic bacterium with low virulence.[87] The structures of neocitreamicins I and II were elucidated using UV, MS, and NMR analyses and were found to be related to the polycyclic xanthone antibiotics of the citreamicin class. The neocitreamicins showed *in vitro* activity against Gram-positive bacteria including strains of methicillin-resistant *S. aureus* and the vancomycin-resistant *Enterococcus faecalis*.[87] In another study investigating the isolation of novel bioactive natural products from *Nocardia* sp., a novel natural product nocarsin A, together with seven known compounds lumichrome, cyclo(L-Leu-L-Tyr), cyclo(L-Ala-L-Ile), cyclo(L-Ala-L-Leu), cyclo(L-Val-L-Ala), 5-methyluracil, and uracil were isolated from *Nocardia alba* sp. nov (YIM 30243^T). The gross structures were solved by detailed 1D and 2D NMR spectroscopic analyses.[88]

Fotso *et al.*[89] screened a number of Indonesian microorganisms, which led to the isolation of seven novel bioactive compounds limazepines A, B1, B2 (isomeric mixture), and C–F from the culture broth of a *Micrococcus* strain. These compounds belong to a group of pyrrolo[1,4]benzodiazepine antitumor antibiotics isolated from various soil bacteria. Several known natural products, prothracarcin and 7-*O*-succinylmacrolactin A, as well as two previously reported synthetic compounds, 2-amino-3-hydroxy-4-methoxybenzoic acid methylester and 4-ethylpyrrole-2-carboxaldehyde, were also isolated. The structures of these natural products were solved by spectroscopic methods and by comparison of the NMR data to structurally related compounds. Limazepine D was found to be active against *S. aureus*, and limazepines B1/B2, C, and E were found to be active against the Gram-positive bacterium *S. aureus* and the Gram-negative bacterium *E. coli*.[89] Ivanova *et al.*[90] described the isolation of a *Streptomyces avidinii* strain SB9 from permafrost soil samples. Their work led to the isolation of 2-amino-3-dodecanol, norophthalmic acid, (2*S*)-2-amino-4-[[(1*S*)-1-(carboxymethylcarbamoyl)ethyl-carbamoylbutanoic acid, and a phthalic acid ester in which the structures were elucidated by NMR and MS analyses. These compounds showed antimicrobial activity against important Gram-positive bacteria and fungi.[90]

Eshrat and Aroona[91] described the use of *Penicillium aurantiogriseum* isolated from agricultural soils with the ability to biotransform steroids in particular the side-chain cleavage of progesterone. Incubation of *P. aurantiogriseum* with progesterone for 10 days, followed by extraction, then isolation led to semi-purified metabolites in which NMR, IR, and MS techniques confirmed the identification of androstenedione and androsta-1,4-diene-3,17-dione from progesterone. These products are high-value, steroidal compounds derived from progesterone and were the first report of the bioconversion of progesterone by a natural isolate of *P. aurantiogriseum*.[91]

Wang *et al.*[92] isolated a strain of *Alcaligenes faecalis* from a soil sample collected from Weihe River, which led to the isolation of a bioactive natural product possessing potent immune activity. ^{1}H and ^{13}C NMR spectra (in conjunction with MS) identified for the first time cyclo-(L-Pro-Gly)5. The immune effects of the cyclo-(L-Pro-Gly)5 showed that crucian carp injected with cyclo-(L-Pro-Gly)5 had significantly lower cumulative mortality (13.0%) compared with the control (45.4%) after infection with live *Aeromonas*

hydrophila suggesting that cyclo-(L-Pro-Gly)5 is a possible immune stimulant that may strengthen the immune response and protect the health status of crucian carp against *A. hydrophila*.[92] In a similar study by Fujitake *et al.*,[93] they described the isolation of four bianthraquinones and two monoanthraquinones from a volcanic ash soil sample. 5,5″-Biphyscion (hinakurin) and five known compounds, chrysotalunin, (−)-7,7″-biphyscion, microcarpin, chrysophanol, and physcion, were characterized using MS and 1D and 2D NMR techniques. Although the dimers were rarely found natural products, they were found to be ubiquitous and predominant over other anthraquinones in various soils from Japan and Nepal.[93]

What the studies discussed earlier illustrate is an interest in the functional role played by natural products in the ecological interactions of the strain with other members of the microbial community and how these compounds may be therapeutically and economically useful for humans. Understanding how plants and fungi produce such complex compounds from relatively simple starting points is also of great interest but is a surprisingly complex challenge. NMR plays a crucial role in elucidating the structure of such compounds but the vast number of metabolites typically present in natural products, coupled with their large variation in molecular weight, polarity, and concentration, makes their analysis particularly challenging and there is high potential for many potentially useful compounds to go unnoticed. As such, the skilled NMR spectroscopist will always have a role in this field.

24.5 POTENTIAL LIMITATIONS OF METABOLOMICS

There are several limitations associated with metabolomics of which it is important to be aware. The sensitivity of metabolomics can be a curse as well as a blessing; the metabolic profile of any organism is exceedingly sensitive to external factors such as age, diet, diurnal and reproductive cycles, sex, and parasite load. The response of a particular organism to any form of manipulation can also vary according to the prevailing environmental conditions under which the experiment is carried out including factors such as soil/sediment organic matter content, pH, and temperature. The technique can, therefore, be potentially confounded by inter- and intra-individual differences.

Metabolomics is a static technique providing a snapshot of the prevailing biochemistry at the time of sampling but it cannot directly generate data on dynamic processes. Targeted studies have to be undertaken in order to measure the flux rates of specific metabolic pathways, for example, using isotope labeling studies.[25] Any observed biochemical or metabolic effects will also need to be correlated with established measures of the overall health or development state of individual organisms in order to reliably assess what the metabolic data are actually saying.[94]

24.6 CONCLUSIONS

Metabolomics is a newer technique than other 'omic sciences' but it has already made a major contribution to soil science and will continue to do so for some time yet. The approach offers considerable potential for the rapid assessment of the metabolic status of virtually any organism or biological system and as such is a capable tool for both population- and community-level investigations. As with any investigative tool, metabolomics does have potential limitations of which it is important to be aware, including its sensitivity to external influences. However, providing studies are carefully constructed and monitored, and the results interpreted with respect to the experimental conditions, metabolomics holds great potential. If metabolomics data is integrated within a systems biology framework along with other 'omics'-based technologies such as gene, transcript, and protein expression profiling, it will become possible to obtain a more holistic understanding of the biochemistry underpinning the soil environment and the wide variety of natural products that it contains.

ACKNOWLEDGMENTS

The authors thank Dylan Lukins and Lincoln Jack Dias for useful conversations.

REFERENCES

1. S. G. Oliver, M. K. Winson, D. B. Kell, and F. Baganz, *Trends Biotechnol.*, 1998, **16**, 373.

2. D. J. Spurgeon, O. A. H. Jones, J.-L. C. M. Dorne, C. Svendsen, S. Swain, and S. R. Stürzenbaum, *Sci. Tot. Environ.*, 2010, **408**, 3725.

3. R. Goodacre, S. Vaidyanathan, W. B. Dunn, G. G. Harrigan, and D. B. Kell, *Trends Biotechnol.*, 2004, **22**, 245.

4. J. R. Snape, S. J. Maund, D. B. Pickford, and T. H. Hutchinson, *Aquat. Toxicol.*, 2004, **67**, 143.

5. R. W. Chapman, *Dev. Comp. Immunol.*, 2001, **25**, 549.

6. M. R. Viant, *Mar. Ecol. Prog. Series*, 2007, **332**, 301.

7. M. Macel, N. M. Van Dam, and J. J. B. Keurentjes, *Mol. Ecol. Resour.*, 2010, **10**, 583.

8. U. Flögel, C. Jacoby, A. Gödecke, and J. Schrader, *Magn. Reson. Med.*, 2007, **57**, 50.

9. J. L. Griffin, L. A. Walker, S. Garrod, E. Holmes, R. F. Shore, and J. K. Nicholson, *Comp. Biochem. Physiol. B.*, 2000, **127**, 357.

10. J. Keeler, Understanding NMR spectroscopy, 1st edn, John Wiley and Sons Ltd: Chichester, UK, 2005, 476.

11. R. C. Crouch, W. Llanos, K. G. Mehr, C. E. Hadden, D. J. Russell, and G. E. Martin, *Magn. Reson. Chem.*, 2001, **39**, 555.

12. D. J. Russell, C. E. Hadden, G. E. Martin, A. A. Gibson, A. P. Zens, and J. L. Carolan, *J. Nat. Prod.*, 2000, **63**, 1047.

13. J. G. Bundy, T. L. Willey, R. S. Castell, D. J. Ellar, and K. M. Brindle, *FEMS Lett.*, 2005, **242**, 127.

14. É. M. Timmins, S. A. Howell, B. K. Alsberg, W. C. Noble, and R. Goodacre, *J. Clinical Microbiol.*, 1998, **36**, 367.

15. J. G. Bundy, D. J. Spurgeon, C. Svendsen, P. K. Hankard, D. Osborn, J. C. Lindon, and J. K. Nicholson, *FEBS Lett.*, 2002, **521**, 115.

16. J. R. Albani, S. Demuynck, F. Grumiaux, and A. Leprêtre, *Photochem. Photobiol.*, 2003, **78**, 599.

17. M. A. Warne, E. M. Lenz, D. Osborn, J. M. Weeks, and J. K. Nicholson, *Soil Biol. Biochem.*, 2001, **33**, 1171.

18. S. C. Connor, W. Wu, B. C. Sweatman, J. Manini, J. N. Haselden, D. J. Crowther, and C. J. Waterfield, *Biomarkers*, 2004, **9**, 156.

19. J. G. Bundy, H. Ramlov, and M. Holmstrup, *CryoLetters*, 2003, **24**, 347.

20. S. Calderon, M. Holmstrup, P. Westh, and J. Overgaard, *J. Exp. Biol.*, 2009, **212**, 859.

21. J. Overgaard, M. Tollarova, K. Hedlund, S. O. Petersen, and M. Holmstrup, *J. Comp. Physiol. B*, 2009, **179**, 569.

22. L. M. Rasmussen and M. Holmstrup, *J. Comp. Physiol. B*, 2002, **172**, 691.

23. D. Gwynn-Jones, W. Huang, G. Easton, R. Goodacre, and J. Scullion, *Pedobiologia*, 2003, **47**, 784.

24. J.-C. Portais and A.-M. Delort, *FEMS Microbiol. Rev.*, 2002, **26**, 375.

25. C. Birkemeyer, A. Luedemann, C. Wagner, A. Erban, and J. Kopka, *Trends Biotechnol.*, 2005, **23**, 28.

26. J. Sardans, J. Peñuelas, and A. Rivas-Ubach, *Chemoecology*, 2011, **21**, 191.

27. M. J. Simpson and J. R. McKelvie, *Anal. Bioanal. Chem.*, 2009, **394**, 137.

28. T. C. Van Brummelen, S. A. Verweij, S. A. Wedzinga, and C. A. M. Van Gestel, *Chemosphere*, 1996, **32**, 292.

29. J. G. Bundy, D. Osborn, J. M. Weeks, J. C. Lindon, and J. K. Nicholson, *FEBS Lett.*, 2001, **500**, 31.

30. J. O. T. Gibb, C. Svendsen, J. M. Weeks, and J. K. Nicholson, *Biomarkers*, 1997, **2**, 295.

31. J. G. Bundy, H. C. Keun, J. K. Sidhu, D. J. Spurgeon, C. Svendsen, P. Kille, and A. J. Morgan, *Environ. Sci. Technol.*, 2007, **41**, 4458.

32. J. G. Bundy, D. J. Spurgeon, C. Svendsen, P. K. Hankard, J. M. Weeks, D. Osborn, J. C. Lindon, and J. K. Nicholson, *Ecotoxicology*, 2004, **13**, 797.

33. J. G. Bundy , E. M. Lenz, N. J. Bailey, C. L. Gavaghan, C. Svendsen, D. Spurgeon, P. K. Hankard, D. Osborn, J. M. Weeks, S. A. Trauger, P. Speir, I. Sanders, J. C. Lindon, J. K. Nicholson, and H. Tang, *Environ. Toxicol. Chem.*, 2002, **21**, 1966.

34. J. G. Bundy, E. M. Lenz, D. Osborn, J. M. Weeks, J. C. Lindon, and J. K. Nicholson, *Xenobiotica*, 2002, **32**, 479.

35. D. J. Spurgeon, A. J. Morgan, and P. Kille, in Advances in Experimental Biology, eds H. Christer and K. Peter, Elsevier: Amsterdam, 2008, Vol. 2, Chap. Current research in soil invertebrate ecotoxicogenomics, 133.

36. O. A. H. Jones, D. J. Spurgeon, C. Svendsen, and J. L. Griffin, *Chemosphere*, 2008, **71**, 601.

37. J. Yuk, M. J. Simpson, and A. J. Simpson, *Environ. Pollut.*, 2013, **175**, 35.

38. J. Yuk, M. J. Simpson, and A. J. Simpson, *Ecotoxicology*, 2012, **21**, 1301.

39. M. Whitfield Aslund, M. J. Simpson, A. J. Simpson, B. A. Zeeb, and A. Rutter, *Ecotoxicology*, 2012, **21**, 1947.

40. B. P. Lankadurai, D. M. Wolfe, A. J. Simpson, and M. J. Simpson, *Environ. Pollut.*, 2011, **159**, 2845.

41. M. L. Whitfield Aslund, H. McShane, M. J. Simpson, A. J. Simpson, J. K. Whalen, W. H. Hendershot, and

G. I. Sunahara, *Environ. Sci. Technol.*, 2012, **46**, 1111.

42. J. R. McKelvie, D. M. Wolfe, M. A. Celejewski, M. Alaee, A. J. Simpson, and M. J. Simpson, *Environ. Pollut.*, 2011, **159**, 3620.

43. M. Whitfield Slund, M. Celejewski, B. P. Lankadurai, A. J. Simpson, and M. J. Simpson, *Chemosphere*, 2011, **83**, 1096.

44. S. L. Hughes, J. G. Bundy, E. J. Want, P. Kille, S. R. Stürzenbaum, and J. Proteome, *Res.*, 2009, **8**, 3512.

45. S. Ruyters, P. Salaets, K. Oorts, and E. Smolders, *Sci. Total Environ.*, 2013, **443**, 470.

46. D. G. Garthwaite, M. R. Thomas and S. Dean, Pesticide Usage Survey Outdoor Vegetable Crops in Great Britain 1999, Department for Environment, Food and Rural Affairs & Scottish Executive Environment and Rural Affairs Department.: London, 1999, pp. 65.

47. C. Svendsen, P. Siang, L. J. Lister, A. Rice, and D. J. Spurgeon, *Environ. Toxicol. Chem.*, 2010, **29**, 1182.

48. M. Faust, R. Altenburger, T. Backhaus, H. Blanck, W. Boedeker, P. Gramatica, V. Hamer, M. Scholze, M. Vighi, and L. H. Grimme, *Aquat. Toxicol.*, 2003, **63**, 43.

49. H. L. Martin, C. Svendsen, L. J. Lister, J. L. Gomez-Eyles, and D. J. Spurgeon, *Environ. Toxicol. Chem.*, 2009, **28**, 97.

50. N. Cedergreen, A. M. Christensen, A. Kamper, P. Kudsk, S. K. Mathiassen, J. C. Streibig, and H. Sørensen, *Environ. Toxicol. Chem.*, 2008, 1621.

51. A. Baylay, D. Spurgeon, C. Svendsen, J. L. Griffin, S. Swain, S. R. Sturzenbaum, and O. A. H. Jones, *Ecotoxicology*, 2012, **21**, 1436.

52. O. A. H. Jones, S. C. Swain, C. Svendsen, J. L. Griffin, S. R. Sturzenbaum, and D. J. Spurgeon, *J. Proteome Res.*, 2012, **11**, 1446.

53. T. D. Williams, N. Turan, A. M. Diab, H. Wu, C. Mackenzie, K. L. Bartie, O. Hrydziuszko, B. P. Lyons, G. D. Stentiford, J. M. Herbert, J. K. Abraham, I. Katsiadaki, M. J. Leaver, J. B. Taggart, S. G. George, M. R. Viant, K. J. Chipman, and F. Falciani, *PLoS Comput. Biol.*, 2011, **7**, e1002126.

54. E. J. Perkins, J. K. Chipman, S. Edwards, T. Habib, F. Falciani, R. Taylor, G. Van Aggelen, P. Vulpe, P. Antczak, and A. Loguinov, *Environ. Toxicol. Chem.*, 2011, **30**, 22.

55. D. A. Dias, S. Urban, and U. Roessner, *Metabolites*, 2012, **2**, 303.

56. M. S. Butler, *J. Nat. Prod.*, 2004, **67**, 2141.

57. B. B. Mishra and V. K. Tiwari, *Eur. J. Med. Chem.*, 2011, **46**, 4769.

58. G. M. Cragg and D. J. Newman, *Pure App. Chem.*, 2011, **77**, 7.

59. O. A. H. Jones and H. M. Hügel, in *Methods in Molecular Biology, Metabolomics Tools for Natural Product Discoveries*, eds U. Roessner and D. A. Diaz, Humana Press: New York, USA, 2013, Chap. Bridging the gap: basic metabolic protocols.

60. S. L. Robinette, R. Brüschweiler, F. C. Schroeder, and A. S. Edison, *Acc. Chem. Res.*, 2011, **45**, 288.

61. R. S. Swift, *Soil Sci.*, 2001, **166**, 858.

62. J. I. Hedges and J. M. Oades, *Org. Geochem.*, 1997, **27**, 319.

63. X. Feng and M. J. Simpson, *J. Environ. Monit.*, 2011, **13**, 1246.

64. H. Lepp (2003) Fungi of Australia. In: Gardens, Australian Fungi Website, http://www.anbg.gov.au/fungi/ Australian National Botanic Gardens, Australian National Herbarium (accessed 27 July 2013).

65. J. Gripenberg, E. Honkanen, and O. Patoharju, *Acta Chem. Scand.*, 1957, **11**, 1485.

66. J. Gripenberg, *Acta Chem. Scand.*, 1963, **17**, 703.

67. J. Gripenberg, *Acta Chem. Scand.*, 1951, **5**, 590.

68. D. A. Dias and S. Urban, *Nat. Prod. Commun.*, 2009, **4**, 489.

69. E. L. Ghisalberti and C. Y. Rowland, *J. Nat. Prod.*, 1993, **56**, 1799.

70. H. G. Cutler, D. S. Himmelsbach, B. Yagen, R. F. Arrendale, J. M. Jacyno, P. D. Cole, and R. H. Cox, *J. Agric. Food. Chem.*, 1991, **39**, 977.

71. L. A. Harrison, L. Letendre, P. Kovacevich, E. Pierson, and D. Weller, *Soil Biol. Biochem.*, 1993, **25**, 215.

72. R. D. Martyn and M. E. Miller, *Plant Dis.*, 1996, **80**, 716.

73. R. D. Stipanovic, J. Zhang, B. D. Bruton, and M. H. Wheeler, *J. Agric. Food Chem.*, 2004, **52**, 4109.

74. Y. Dong, J. Yang, H. Zhang, J. Lin, X. Ren, M. Liu, X. Lu, and J. He, *J. Nat Prod.*, 2006, **69**, 128.

75. K. F. Chater, *Philos. Trans. R. Soc. Lond. B Biol. Sci.*, 2006, **361**, 761.

76. B. C. Gerwick, S. S. Fields, P. R. Graupner, J. A. Gray, E. L. Chapin, J. A. Cleveland, and D. R. Heim, *Weed Sci.*, 1997, **45**, 654.

77. M. S. Abdelfattah, *World J. Microbiol. Biotechnol.*, 2008, **24**, 2619.

78. S.-X. Huang, X.-J. Wang, Y. Yan, J.-D. Wang, J. Zhang, C.-X. Liu, W.-S. Xiang, and B. Shen, *Org. Lett.*, 2012, **14**, 1254.

79. R. Raju, O. Gromyko, V. Fedorenko, A. Luzhetskyy, and R. Müller, *Tetrahedron Lett.*, 2012, **53**, 3009.

80. R. Huang, Z.-G. Ding, Y.-F. Long, J.-Y. Zhao, M.-G. Li, X.-L. Cui, and M.-L. Wen, *Chem. Nat. Comp.*, 2013, **48**, 966.

81. G.-Y.-S. Wang, E. Graziani, B. Waters, W. Pan, X. Li, J. McDermott, G. Meurer, G. Saxena, R. J. Anderson, and J. Davies, *Org. Lett.*, 2000, **2**, 2401.

82. M. R. Prinsep, G. M. L. Patterson, L. K. Larsen, and C. D. Smith, *J. Nat. Prod.*, 1998, **61**, 1133.

83. T. Chatterjee, *Biotechnol. Appl. Biochem.*, 2004, **39**, 303.

84. J. S. Han, J. H. Cheng, T. M. Yoon, J. Song, A. Rajkarnikar, W. G. Kim, I. D. Yoo, Y. Y. Yang, and J. W. Suh, *J. Appl. Microbiol.*, 2005, **99**, 213.

85. R. Ratnayake, E. Lacey, S. Tennant, J. H. Gill, and R. J. Capon, *Chem. Eur. J.*, 2007, **13**, 1610.

86. G. R. Pettit, R. Tan, R. K. Pettit, T. H. Smith, S. Feng, D. L. Doubek, L. Richert, J. Hamblin, C. Weber, and J.-C. Chapuis, *J. Nat. Prod.*, 2007, **70**, 1069.

87. A. J. Peoples, Q. Zhang, W. P. Millett, M. T. Rothfeder, B. C. Pescatore, A. A. Madden, L. L. Ling, and C. M. Moore, *J. Antibiot.*, 2008, **61**, 457.

88. Z.-G. Ding, J.-Y. Zhao, P.-W. Yang, M.-G. Li, R. Huang, X.-L. Cui, and M.-L. Wen, *Magn. Reson. Chem.*, 2008, **4**, 366.

89. S. Fotso, T. M. Zabriskie, P. J. Proteau, P. M. Flatt, D. A. Santosa, S. Mahmud, and T. Mahmud, *J. Nat. Prod.*, 2009, **72**, 390.

90. V. Ivanova, D. Lyutskanova, M. Kolarova, K. Aleksieva, V. Raykovska, and M. Stoilova-Disheva, *Biotechnol. Biotechnol.*, 2010, **24**, 2092.

91. G.-F. Eshrat and C. Aroona, *J. Microbiol.*, 2011, **6**, 98.

92. G.-X. Wang, F.-Y. Li, J. Cui, Y. Wang, Y.-T. Liu, J. Han, and Y. Lei, *Scand. J. Immunol.*, 2011, **74**, 14.

93. N. Fujitake, T. Suzuki, M. Fukumoto, and Y. Oji, *J. Nat. Prod.*, 1998, **61**, 189.

94. J. I. Allen and M. N. Moore, *Mar. Environ. Res.*, 2004, **58**, 227.

Chapter 25

Environmental Metabolomics of Aquatic Organisms

Trond R. Størseth and Karen M. Hammer

SINTEF Materials and Chemistry/Environmental Technology, Trondheim NO-7465, Norway

25.1 INTRODUCTION

Aquatic organisms are continuously exposed to natural stressors such as changes in temperature, oxygen level, salinity, and viral and bacterial infections that may all affect their metabolism.[1-4] In addition, release of contaminated wastewater from industry and households, drainage from farmlands, and precipitation of complex compounds in the atmosphere all contribute to making the aquatic environment the ultimate sink of man-made chemicals.[5,6] Aquatic organisms are, therefore, at continuous risk of exposure to both single and complex mixtures of contaminants. As such, there is a constant need for analytical tools to study the effects and monitor the occurrence of both natural and anthropogenic stressors in the aquatic environment.

NMR Spectroscopy: A Versatile Tool for Environmental Research
Edited by Myrna J. Simpson and André J. Simpson
© 2014 John Wiley & Sons, Ltd. ISBN: 978-1-118-61647-5

Metabolomics is a tool that allows simultaneous measurements of multiple metabolites of low-molecular weight. Using methods such as NMR spectroscopy and mass spectrometry (MS), followed by appropriate statistical analysis that typically employs multivariate or other repeated univariate tests, one may detect changes in the metabolic profile, rather than in a single metabolite in response to drugs, toxicants, or diseases.[7,8] In recent years, an increasing number of studies have applied this tool to reveal the metabolic responses of organisms to environmental and anthropogenic stressors, and it holds great promise for environmental risk analysis, gaining increased biological insights, and for developing environmental systems models.[8,9] An increasing number of studies are now including metabolomics to study the aquatic environment[9-34] (see also Chapter 22).

25.2 INITIAL STUDIES AND MAJOR CONTRIBUTORS TO THE FIELD

The first two publications receiving a hit by searching the SCOPUS database with keywords, 'metabolomics', 'environmental', 'NMR', and 'aquatic' (Figure 25.1) were published in 2003, both by Viant and coworkers at the University of

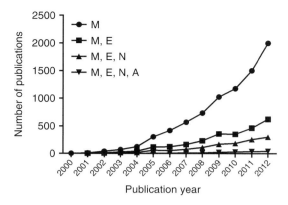

Figure 25.1. The number of publications by year retrieved by searching subsequently for metabolomics (M), environmental (E), NMR (N), and aquatic (A).

California, Davis. One of these papers is a study correlating heat-shock protein induction with metabolic changes in steelhead trout (*Oncorhynchus mykiss*) and the other discusses methods for acquiring and interpreting metabolic data obtained by NMR.[35,36] As this rather simple search is bound to exclude relevant papers with the same topic, a more thorough look into other publications of 2003 from the group at professor Tjeerdema's lab at UC Davis also reveals two other publications of interest with NMR and aquatic organisms.[37,38] One of these deals with withering syndrome in red abalone (*Haliotis rufescens*). In this study, the authors attempted to reveal new biomarker profiles to distinguish healthy from infected animals.[38] Also published was a paper introducing NMR as a tool for environmental research in the journal *Environmental Science and Technology*, with the title 'NMR-Based Metabolomics: A Powerful Approach for Characterizing the Effects of Environmental Stressors on Organism Health'.[39] The latter of these studies have been cited 163 times between 2003 and present. These four studies, two of which represent methodology introduction and guidelines and two of which describe the use of the method, can, in many ways, be considered the start of NMR metabolomics in the environmental sciences, and all had an aquatic focus.

Sorting the publications by author affiliation(s) provides a good overview of what research groups and institutions that have been leading the development of environmental NMR metabolomics of aquatic organisms for the last 10 years. Selecting the top four affiliations coincides with selecting

for those with more than 10 publications within the field (keep in mind that this is just for environmental NMR metabolomics of aquatic organisms). The University of Birmingham produced a substantial part of the published papers (34), with the Chinese Academy of Sciences and Graduate University of Chinese Academy of Sciences next on the list (combined 25), followed by UC Davis (17) and US EPA (17). On the author level, the list of publications is dominated by professor Viant of the University of Birmingham, professor Huifeng Wu and coworkers from the Chinese Academy of Sciences, Key Laboratory of Coastal Zone Environmental Processes, professor Tjeerdema of UC Davis, and Dr Colette and Dr Ekman of the United States Environmental Protection Agency, National Exposure Research Laboratory. All of these have over 10 publications with the restrictions of the search terms.

25.3 ENVIRONMENTAL METABOLOMICS

While metabolomic studies on humans and laboratory-reared model organisms are often clinical studies, those performed on aquatic organisms typically investigate metabolic responses to environmental change.[7,8] Environmental metabolomics is a branch of metabolomics concerning the study of metabolic changes in an organism in response to environmental challenges such as natural stressors and anthropogenic pollution[37] (see also Chapter 22). An even more precise definition is 'the application of metabolomics to the investigation of both free-living organisms obtained directly from the natural environment (whether studied in that environment or transferred to a laboratory for further experimentation) and of organisms reared under laboratory conditions (whether studied in the laboratory or transferred to the environment for further experimentation), where any laboratory experiments specifically serve to mimic scenarios encountered in the natural environment'.[40]

Environmental metabolomics can be divided into studies concerning ecophysiology and ecotoxicology. While ecophysiology comprises the responses of organisms to environmental conditions (i.e., temperature, oxygen level, salinity, and food availability), ecotoxicology only describes the effect of toxic compounds on the structure and the function of ecosystems.[41] Although a large number of published articles have applied metabolomics to study the responses of terrestrial plants and animals to environmental changes (see review by Sardans *et al.*[42]), only a handful of studies have applied metabolomics to study the effects of natural environmental stressors to aquatic organisms.[19,20,30,31,34,35,43,44] Most metabolomics-related studies on aquatic organisms have been ecotoxicological, which focus on chemical exposure in the environment and their impact on the metabolic responses of aquatic organisms.[11–16,18,24–26,28,29,33,45–47]

25.3.1 Natural Stressors of Aquatic Organisms

Abiotic factors such as water temperature, salinity, and concentration of dissolved respiratory gases (O_2 and CO_2) as well as diseases and nutrient availability may all affect the physiology and health status, and thus the metabolic profiles of aquatic organisms. The number of studies applying NMR metabolomics to investigate the effect of natural stressors in aquatic organisms is still relatively low.[20,44,48] However, with the increased attention on factors related to climate change, an increase in these types of studies is expected in the near future.

25.3.1.1 *Temperature and CO₂*

Global warming is an ongoing threat to both terrestrial and aquatic organisms, and may result in a change in geographical distribution of species and ecosystems. While studies related to climate change are steadily increasing in number, only a very few published papers have applied metabolomics as a tool in this field of research.[44,48] Elevated temperatures may induce stress both directly and through reduced oxygen levels for aquatic organisms. Heat stress is one of the most important environmental stressors experienced by anadromous salmonids and may result in lethality if the fish are exposed for a prolonged time. For example, Viant *et al.*[35] subjected juvenile steelhead trout (*O. mykiss*) to elevated temperatures to evaluate metabolism, stress response, and growth rates. ^{1}H NMR metabolomics revealed that in addition to a decrease in the concentration of metabolites related to energy metabolism (energy-rich phosphates, phosphocreatine, and glycogen), it provided a thorough description of the biochemical pathways and metabolites affected during thermal stress. Furthermore, a recent study by Kullgren *et al.*[48] demonstrated that in addition

to a decreased growth rate and condition factor, an index of growth and feeding intensities, a decrease in metabolites involved in energy metabolism was found in Atlantic salmon, *Salmo salar*, subjected to heat stress (18 versus 12 °C).

In addition to global warming, elevated atmospheric CO_2 concentrations are affecting the oceans through a process known as ocean acidification.[49] Another problem may arise because of CO_2 emissions mitigation, leakage from subseabed storage of CO_2, which could potentially create extreme acidification in the vicinity of the leakage sites. Lannig *et al.*[44] studied the synergistic effects of increased temperature and ocean acidification (pH 7.7) on oysters (*Crassostrea gigas*) and observed a depletion of alanine and ATP levels in the mantle, whereas succinate was significantly increased in the gills and hepatopancreas. Their results suggested a shift in the metabolic pathways toward increased gluconeogenesis because of an impairment of glycolysis. Hammer *et al.*[20] found a dose-dependent shift in the metabolic fingerprint of polar extracts of hemolymph, and gill and leg muscle tissues of the shore crab *Carcinus maenas* exposed to high levels of CO_2 (0–30 000 ppm, pH 8.1–6.3). In addition, a time-dependent exposure to pCO₂ of 7600 ppm (pH 6.9) revealed a significant change in the metabolic fingerprints compared to controls after 4 weeks in both hemolymph and gill tissue. Quantification of the NMR data revealed that the metabolic shifts in both tissues were predominantly due to a decrease in amino acids such as glycine and proline, which are important osmolytes in invertebrates. The authors suggested that the changes in metabolic profile could indicate osmotic stress, a compensatory increased catabolism of amino acids to supply the body fluids with proton-buffering ammonia or simply reflect an exhaustive effect of CO_2 exposure.

25.3.1.2 *Hypoxia and Starvation*

During low tide, many intertidal species are isolated from the seawater and, therefore, frequently encounter hypoxia (low O_2) or hypercapnia (high pCO₂) due to restricted gas exchange. In addition, changes in food availability and temperature will also affect the metabolism of aquatic animals.[50] Similar types of environmental stress may also be induced by several chemical compounds that are released into the aquatic environment. Tuffnail *et al.*[31] investigated the potential of ^{1}H NMR metabolomics for identifying the impact of individual pesticides in the bivalve *Mytilus edulis*, and for separating these symptoms from the effects of starvation and hypoxia alone. The results showed a significant separation of the metabolic fingerprint for all treatments, thus confirming that NMR is an appropriate tool to investigate compounds inducing effects similar to natural stressors. Semiquantitative analysis revealed that lindane induced a dose-dependent change in metabolic fingerprint, where exposed animals displayed an increased concentration of alanine, whereas the level of remaining metabolites was reduced compared to controls. Individuals treated with atrazine displayed elevated levels of leucine and isoleucine, consistent with an overshoot in energy output. Hypoxia resulted in an increased level of succinate and valine and a decrease in leucine and isoleucine.

25.3.1.3 Health Status and Diseases

The first NMR-based environmental metabolomic study conducted on aquatic animals was that of Viant et al.[35] concerning withering syndrome in red abalone *H. rufescens*. In this study, the authors attempted to reveal new biomarker profiles to distinguish between healthy and infected individuals. They succeeded in revealing three different metabolic fingerprints for healthy, stunted, and infected animals, for each of the different tissues and body fluids examined. The result was a significantly lower glucose/homarine ratio in infected animals compared to control animals. Animals suffering from withering syndrome also experience starvation as a part of the disease. Marine molluscs rely on amino acids as energy source during starvation. Thus, the intracellular storage of amino acids, which is crucial for maintaining osmotic balance in marine invertebrates, is depleted. The authors suggested that the increased level of homarine may help to balance the loss of osmotically important amino acids.[39]

A recently published study employed ^{1}H NMR metabolomics on serum samples to evaluate the health status of captive whale sharks over several months.[34] The results revealed a significantly different metabolic fingerprint in unhealthy individuals suffering from prolonged inappetence compared to healthy sharks. Sharks with low body mass were found to have lower concentrations of the heteroaromatic metabolite homarine. Homarine is found at relatively high concentrations in many marine invertebrates, particularly crustaceans, which make up a substantial

part of the whale sharks' diet. The reduced homarine concentrations in individuals with lower health status may, therefore, be explained by a reduced food intake. The authors found that although mass spectrometry (MS) gave more information regarding individual compounds, ^{1}H NMR was more useful for distinguishing unhealthy individuals from healthy individuals.

Another environmental stressor that may deteriorate the health of aquatic organisms is bacteria. Crabs infected by *Vibrio* species, Gram-negative bacteria commonly found in saline waters, may experience decreased oxygenation of the hemolymph because of the organism's immune response, where aggregations of hemocytes compromise gas exchange. Schock et al.[30] used ^{1}H NMR metabolomics to investigate bacteria-induced oxygen stress in hemolymph of blue crabs, *Callinectes sapidus*, injected with *Vibrio campbellii*. The results were compared with the metabolic fingerprint of hemolymph from control animals and crabs treated with the uncoupler of oxidative phosphorylation 2,4-dinitrophenol (DNP), which is known to increase aerobic metabolism. The results suggested that crabs injected with *V. campbellii* displayed an increase in glucose, a reliable indicator of biological stress, compared to control animals. DNP treatment also induced an elevated level of lactate, indicating increased anaerobic respiration, in addition to increased glucose levels.

25.3.2 NMR-based Metabolomics in Ecotoxicological Studies of Aquatic Organisms

During the last decade, a number of studies have applied metabolomics in the field of aquatic ecotoxicology.[11,12,30,51–57] These include studies on invertebrates such as marine mussels,[14–18,21,26,28,29,31,46,47,58] crabs,[20,30] and water fleas,[59–62] as well as many species of fish,[12,24,27,63] and have involved studying the effect of toxicants such as pesticides, herbicides, xenoestrogens, heavy metals, and inducers of oxidative stress.

25.3.2.1 Endocrine-disrupting Chemicals

In a study where the effects of two concentrations of 17-α-ethynylestradiol on rainbow trout (*O. mykiss*) were investigated, Samuelsson et al.[64] found that the blood plasma metabolite profile was perturbed by exposure to a high concentration $(10\,\text{ng L}^{-1})$, whereas for a low concentration $(0.87\,\text{ng L}^{-1})$, no effects were found after subjecting the ^{1}H NMR spectra to partial least squares discriminant analysis (PLS-DA). The metabolite response was compared with vitellogenin levels measured by enzyme-linked immunosorbent assay (ELISA), and the major differences were found to be alanine in blood plasma analyzed by the CPMG pulse sequence (see Chapter 23) and phospholipids and cholesterol in an analysis of blood plasma lipid extract. The decrease in alanine in exposed trout was coupled to the fact that it is the major amino acid in the vitellogenin in *O. mykiss*. The increase in phospholipids and cholesterols in the lipid extracts supported this assumption, particularly when seen in context with the increased vitellogenin in the high-dose exposure, as *O. mykiss* vitellogenin contains 18% lipids, of which two-third is phospholipid.[65]

A study conducted by Ekman *et al.*[66] is an excellent example of how ^{1}H NMR metabolomics can be used to construct metabolic response trajectories (first introduced by Viant *et al.*[32]) to study the effects of exposure (Figure 25.2). This study was performed on polar extracts of livers from fathead minnow (*Pimephales promelas*) and using the scores for the first two latent variables of a PLS-DA analysis of the resulting ^{1}H NMR spectra; the authors were able to show a dose-dependent response trajectory over an 8 day recovery period. While the same direction of change (geometry) was observed, the magnitude (distance from the control) varied with the exposure concentration. Furthermore, the authors reported feminization of males based on the metabolic profiles observed. Males exposed to 17-α-ethynylestradiol were more similar to exposed and control females, than to control males. In a follow-up study, the authors used the same sample material but looked at the nonpolar fraction of the liver extracts. One of the important roles of lipids is in reproduction, and, as effects had been observed in the polar fractions, the authors wanted to examine the effects of the estrogen exposure of the lipidome. Although ^{1}H NMR spectra of lipid extracts provide relatively low resolution of individual lipid components[67] compared to other analytical techniques (such as gas chromatography with flame ionization detection (GC-FID) or gas chromatography–mass spectrometry (GC-MS)), several important findings were observed. One response to 17-α-ethynylestradiol exposure in both males and females was elevated vitellogenin levels relative to the control. The ^{1}H NMR lipid profiles revealed that phosphatidylcholine levels responded differently in males and females. One role of phosphatidylcholine is in the synthesis of vitellogenin, and as this lipoprotein is normally only found in females,[68] the presence in males is a sign of feminization. The findings for the high 17-α-ethynylestradiol exposure of $100\,\text{ng L}^{-1}$ identified that on day 1 of the exposure, elevated levels of phosphatidylcholine were not observed in males or females. However, on day 4, the levels of phosphatidylcholine in males dropped, while increasing in females. The authors suggested that this was due to the increased synthesis of vitellogenin disrupting the ability to maintain normal hepatic phosphatidylcholine levels in males, whereas the phosphatidylcholine synthesis in females was compensated. Continued exposure (day 8) revealed that the males' ability to synthesize phosphatidylcholine recovered, and during recovery, the levels were higher in males compared to the control. Observing the trajectory analysis of the complete ^{1}H NMR lipid profiles showed that the highest exposure did not recover by the end of the observed depuration period of 8 days.

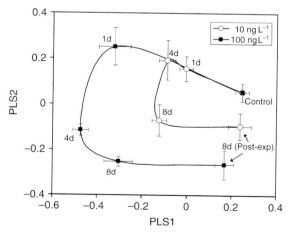

Figure 25.2. Metabolic response trajectories of exposure to male *Pimephales promelas* described by 2D scores for the first two latent variables of a PLS-DA analysis of ^{1}H NMR metabolic profiles of polar liver extracts. (Reprinted with permission from Ref. 66. © 2008 American Chemical Society)

In a study by Katsiadaki *et al.*,[69] the hepatic effects of 17-α-ethynylestradiol on the stickleback (*Gasterosteus aculeatus*) were studied by transcriptomics and metabolomics. Here, a robust approach to ascertaining effects of exposure on the metabolome was followed by applying univariate statistics to test for differences in metabolite levels. The only statistically significant difference in the metabolites was an increase in glucose at low levels of exposure (0.1 and 1.0 ng L^{-1}), which was not seen at higher levels (10, 32, and 100 ng L^{-1}) of 17-α-ethynylestradiol exposure. Furthermore, the authors concluded that their results are suggestive of changes in the amino acid pool, which was supported by differences in expression of protein synthesis genes.

Other examples of endocrine-disrupting compounds include bisphenol A, nonylphenol, and di-(2-ethylhexyl)-phthalate. A study by Jordan *et al.*[63] looked at metabolic stress induced by single compounds and mixtures of these compounds in environmentally relevant concentrations on goldfish (*Carassius auratus*). The importance of metabolomics based on NMR in assessment of toxicants at ambient concentrations to gain insight in mechanisms of exposure is exemplified by this chapter. They quantified 46 metabolites in liver and 48 in gonad extracts (polar fraction) and employed the use of the KEGG (Kyoto encyclopedia of genes and genomes) database to identify metabolic pathways associated with altered metabolites. The responses of the mixtures were not predicted from the responses to the individual contaminants studied. Alterations in tissue metabolites were tissue specific. In liver, they found that perturbations in major energy and nitrogen pathways were largely caused by bisphenol A, whereas in testes, the effects of the mixture were similar to di-(2-ethylhexyl)-phthalate alone causing shifts in the phospholipid metabolism. The authors gave a highly interesting discussion on the contributions from all chemicals to the mixture effects compared to single-compound exposures.

25.3.2.2 Oil

Of the many possible threats to the environment from oil exploration and transport activities, one that has been studied is the potential effects of oil spills on ecologically important species. Using metabolomics, comparisons have been made between the toxicity of weathered crude and dispersed oil to topsmelt,[24] crude versus dispersed oil to Chinook salmon smolts[58] and presmolts.[70] These studies have all followed the same experimental design with comparing the toxicity of the water-accommodated fractions (WAFs) of crude oil and crude oil-added chemical dispersant (CEWAFs, chemically enhanced water-accommodated fractions). Lin *et al.*[58] showed that metabolic effects as observed by NMR-based metabolomics were evident well below the LC$_{50}$ (lethal concentration that causes mortality in 50% of the population) values of the WAF/CEWAFs as measured by total hydrocarbon content.

25.3.2.3 Other Organic Pollutants

In a study exposing embryos of Japanese medaka (*Oryzias latipes*) to trichloroethylene, Viant *et al.*[32] introduced NMR-derived metabolic trajectories as a tool in distinguishing basal and perturbed metabolic changes due to toxicant exposure. Medaka embryos were subjected to constant exposure during embryogenesis by a static nonrenewal system, and metabolic endpoints by NMR were compared to traditional endpoints, i.e., hatching success and mortality. Employing eight (nominal) trichloroethylene concentrations ranging from 11 to 438 mg L^{-1} and a control, the authors established LC$_{50}$ and EC$_{50}$ (effective concentration where 50% of the population elicits a response) values of 379 and 217 mg L^{-1}, respectively. In addition, a no-observable adverse-effect level (NOAEL) for hatching success was determined to be 164 mg L^{-1}. These values were commented on by the authors, as theirs were an order of magnitude higher than what was previously reported in a study that had used a flow through exposure.[71] The metabolomic study consisted of exposing the medaka in the same way with concentrations from 8.76 to 175 mg L^{-1} and in addition, a control. In contrast to the NOAEL of 164 mg L^{-1}, the NMR-observable metabolome subjected to principal component analysis (PCA) was found to be perturbed at all exposure concentrations when comparing score values for the exposure groups. Further, an experiment using low concentrations and sampling throughout the 8 day period showed the increasingly differing developmental metabolic trajectory of the exposed vs control groups. The apparent ability of the methodology to reveal effects of exposure at lower concentrations (20 times lower in this study) than the traditional methods, and the fact that information may be obtained on what metabolites are perturbed, thus having the promise of describing mode of toxicity, makes this an excellent and early example of the possibilities that NMR-based metabolomics would offer to environmental toxicology.

Using NMR-based metabolomics to study the effects of stormwater discharges from farmland, Viant *et al.*[11] chose three pesticides used in the Sacramento Valley to model Chinook salmon exposure during discharge events. This study employed ^{1}H NMR-based metabolomics and high-performance liquid chromatography (HPLC) analysis of phosphorylated nucleotides and phosphocreatine from polar perchloric acid extracts of the samples. The three different pesticides were found to give different effects of exposure by studying the PCA scores and loadings. Diazonin exposure resulted in decreases in ATP, and phosphocreatine (PCr) levels were accompanied with an increase in creatine, which was consistent in affecting oxidative phosphorylation. Further, a disruption of the Krebs cycle through higher levels of succinate was observed, and heightened lactate levels indicated a change to anaerobic metabolism. For diazonin, no changes associated with the known mode of action (MOA) of being a cholinesterase inhibitor were observed, but disruption of amino acid metabolism indicating proteolysis, a known effect, was found. The use of NMR was proved to be a more sensitive method than the targeted HPLC analyses, and as one of the earlier studies of environmental metabolomics, it is pointed out that the unbiased nature of the approach allowed for the detection of changes in metabolism not associated with known MOAs, which the use of a targeted approach would miss.

Diethanolamine is an alkanolamine, which is a class of compounds of interest in CO_2 capture. Addressing the widespread use and scarce information on the ecotoxicological impacts of these compounds, Hansen *et al.*[10] used transcriptomics and ^{1}H high-resolution magic-angle spinning (HR-MAS) NMR (see Chapter 5) to study the effects of diethanolamine on the marine copepod *Calanus finmarchicus*. Using ^{1}H HR-MAS NMR, the NMR metabolic profiles could be obtained directly on a pooled sample of copepods. With this approach, sample preparation consisted of transferring frozen copepods to ^{1}H HR-MAS NMR rotors with the addition of D_2O (with 3-(trimethylsilyl)propionic-2,2,3,3-d_4 acid sodium salt as an internal standard). The authors used the CPMG sequence to filter high levels of lipids found in the copepods and were able to identify 18 metabolites in the ^{1}H HR-MAS NMR spectra. Subsequent analysis showed that nine metabolites were perturbed by exposure to diethanolamine, as well as several unassigned resonances correlating with exposure. The authors concluded that ^{1}H HR-MAS NMR is useful for measuring semiquantitative changes in the metabolome, and this methodology is capable of giving information on both polar and nonpolar metabolites from the analysis of the same sample with the same preparation, albeit at the cost of reduced resolution.

25.3.2.4 Heavy Metals

Heavy metals are introduced to the aquatic environment through effluents from mining activities and are found at concentrations toxic to aquatic organisms in many places in the world.[5,72] While terrestrial organisms mostly accumulate metals through their diets, aquatic organisms also take up metals through their gills via filter feeding. The large volume of water filtered through gills thus makes aquatic organisms particularly susceptible to metal pollution. The toxic effects of different metals vary between organisms and include oxidative stress (i.e., cadmium), organ failure, and neurotoxicity.[17] Although the end effects of most metals such as copper, mercury, cadmium, and zinc are relatively well established, the mechanisms at which metals induce their toxic effects is still poorly known.[73]

In recent years, an increasing number of published studies have used NMR-based metabolomics to investigate the effects of heavy metals in aquatic organisms.[17,18,21,28,29,45,47,74] For example, Santos *et al.*[45] used NMR-based metabolomics in combination with transcriptomics on freshwater-acclimated sticklebacks (*G. aculeatus*) exposed to copper at environmentally relevant concentrations.[45] Transcriptomics revealed the upregulation of genes for enzymes involved in glycolysis, whereas gene expression of enzymes in the electron transport chain was downregulated, suggesting a shift from aerobic to anaerobic metabolisms. This was supported by metabolomics data, which showed that sticklebacks exposed to copper displayed higher levels of the end products of glycolysis: lactate and nicotinamide adenine dinucleotide (NAD^+).

Sessile organisms such as intertidal bivalves are unable to move away from polluted areas and are often used as bioindicators of metal pollution. In the recent years, a large number of studies using ^{1}H NMR-based metabolomics to study toxicological effects of heavy metals on bivalve species found in the Bohai Sea have emerged.[14,15,17,18,21,26,28,29,47] In accordance with previous studies using other established techniques, the metabolic profiles of the mussels imply that metals such as cadmium, mercury, and copper interfere with osmoregulation and energy metabolism, and have neurotoxic effects.

Mercury can induce both neurotoxic effects and disturbances in osmoregulation in aquatic organisms.[77] Liu *et al.*[21] investigated the different metabolic responses of the three main pedigrees of *Ruditapes philippinarum*, Liangdao Red, White, and Zebra, to mercury exposure to discover the biomarkers and the ideal bioindicator pedigree for mercury pollution. In accordance with the previously known toxic effects, the results from ^{1}H NMR analysis indicated neurotoxic effects and disturbance of osmoregulation in exposed clams. However, the metabolic responses varied among the pedigrees. For instance, the digestive glands of mercury-exposed Liangdao Red and White clams displayed elevated alanine and succinate concentrations, indicating anaerobic stress, whereas alanine concentrations in Zebra clams were interpreted as disturbance in osmoregulation. Liangdao Red and White clams also displayed elevated glutamate concentrations, a known symptom of neurotoxic effects of mercury salts, whereas Zebra clams did not. On the basis of the different responses observed, the authors suggested that when using the digestive glands as the target organ, the Liangdao Red pedigree should be the preferred test organism for mercury pollution in marine environments, as it displayed the most sensitive disturbances in osmoregulation, energy metabolism, and neurotoxicity of the *R. philippinarum* pedigrees. However, in a similar study using gills as the target organ, white clams displayed the most prominent metabolic response, suggesting that this pedigree should be used for monitoring if examining the response of gill tissue.

Wu and Wang[14] applied ^{1}H NMR-based metabolomics to the green mussel *Perna viridis* exposed to environmentally relevant concentrations of copper, cadmium, and a mixture of both metals. The results revealed that both metals induced changes in energy metabolism and the concentrations of important osmolytes, such as branched amino acids as well as neurotoxicity. Comparing the different exposure scenarios revealed that copper was the dominant metal in the mixture, and, therefore, seemed to be the most toxic metal to *P. viridis*.

All published studies have found that elevated concentrations of several metals induce changes in the metabolic profile of mussels and that NMR-based metabolomics may, therefore, be applicable for the discovery of biomarkers in the field. The generally clear separation found between intertidal bivalves exposed to heavy metals and control bivalves also suggest that NMR-based metabolomics can be applied for biomonitoring areas with high mining and industrial activities.

25.4 CONCLUSION

By the examples shown here, NMR can be characterized as a highly versatile and sensitive method to study the effects of a multitude of different environmental stressors on aquatic organisms. In particular, two strengths of the technique are important. Compared to classical methods for observing stress in organisms, NMR-based metabolomics frequently detects perturbations at lower levels of exposure, and in addition, insight into the modes of action of stressors may be obtained at the same time. NMR offers highly stable and reproducible analyses with low per-sample costs, which is important for application in high-throughput studies. The inherent quantitative nature of NMR and relative simple sample preparation requirements adds to these strengths.

RELATED ARTICLES IN EMAGRES

Metabonomics: NMR Techniques

Software Tools for NMR Metabolomics

REFERENCES

1. L. E. Burnett, J. D. Holman, D. D. Jorgensen, J. L. Ikerd, and K. G. Burnett, *Biol. Bull.*, 2006, **211**, 50.

2. K. R. Schneider, *Biol. Bull.*, 2008, **215**, 253.

3. A. F. Welker, D. C. Moreira, É. G. Campos, and M. Hermes-Lima, *Comp. Biochem. Phys. A*, 2013, **165**, 384. DOI: http://dx.doi.org/10.1016/j.cbpa.2013.04.003.

4. L. E. Burnett, *Am. Zool.*, 1988, **28**, 125. DOI: 10.1093/icb/28.1.125.

5. J. B. Pritchard, *Environ. Health Persp.*, 1993, **100**, 249.

6. A. J. Mearns, D. J. Reish, P. S. Oshida, T. Ginn, M. A. Rempel-Hester, and C. Arthur, *Water*

Environ. Res., 2012, **84**, 1737. DOI: 10.2175/106143012x13407275695751.

7. M. R. Viant, *Mar. Ecol. Prog. Ser.*, 2007, **332**, 301. DOI: 10.3354/meps332301.

8. J. G. Bundy, M. P. Davey, and M. R. Viant, *Metabolomics*, 2009, **5**, 3. DOI: 10.1007/s11306-008-0152-0.

9. M. R. Viant, *Mol. BioSys.*, 2008, **4**, 980. DOI: 10.1039/b805354e.

10. B. H. Hansen, D. Altin, A. Booth, S. H. Vang, M. Frenzel, K. R. Sorheim, O. G. Brakstad, and T. R. Storseth, *Aquat. Toxicol.*, 2010, **99**, 212. DOI: 10.1016/j.aquatox.2010.04.018.

11. M. R. Viant, C. A. Pincetich, and R. S. T. Eerderna, *Aquat. Toxicol.*, 2006, **77**, 359. DOI: 10.1016/j.aquatox.2006.01.009.

12. M. R. Viant, C. A. Pincetich, D. E. Hinton, and R. S. Tjeerdema, *Aquat. Toxicol.*, 2006, **76**, 329. DOI: http://dx.doi.org/10.1016/j.aquatox.2005.10.007.

13. M. R. Viant, J. H. Walton, P. L. TenBrook, and R. S. Tjeerdema, *Aquat. Toxicol.*, 2002, **57**, 139. DOI: 10.1016/s0166-445x(01)00195-3.

14. H. F. Wu and W. X. Wang, *Aquat. Toxicol.*, 2010, **100**, 339. DOI: 10.1016/j.aquatox.2010.08.005.

15. H. Wu, C. Ji, Q. Wang, X. Liu, J. Zhao, and J. Feng, *Chin. J. Oceanol. Limn.*, 2013, **31**, 65. DOI: 10.1007/s00343-013-2037-y.

16. X. Zhang and H. Chen, *Chin. Sci. Bull.*, 2012, **57**, 3253. DOI: 10.1007/s11434-012-5237-5.

17. X. L. Liu, L. B. Zhang, L. P. You, J. B. Yu, M. Cong, Q. Wang, F. Li, L. Z. Li, J. M. Zhao, C. H. Li, and H. F. Wu, *Clean-Soil Air Water*, 2011, **39**, 759. DOI: 10.1002/clen.201000410.

18. L. B. Zhang, X. L. Liu, L. P. You, D. Zhou, J. B. Yu, J. M. Zhao, J. H. Feng, and H. F. Wu, *Clean-Soil Air Water*, 2011, **39**, 989. DOI: 10.1002/clen.201100208.

19. C. A. Pincetich, M. R. Viant, D. E. Hinton, and R. S. Tjeerdema, *Comp. Biochem. Phys. C*, 2005, **140**, 103. DOI: http://dx.doi.org/10.1016/j.cca.2005.01.010.

20. K. M. Hammer, S. A. Pedersen, and T. R. Størseth, *Comp. Biochem. Phys. D*, 2012, **7**, 292. DOI: http://dx.doi.org/10.1016/j.cbd.2012.06.001.

21. X. L. Liu, L. B. Zhang, L. P. You, J. B. Yu, J. M. Zhao, L. Z. Li, Q. Wang, F. Li, C. H. Li, D. Y. Liu, and H. F. Wu, *Ecotoxicology*, 2011, **20**, 177. DOI: 10.1007/s10646-010-0569-x.

22. S. Fasulo, F. Iacono, T. Cappello, C. Corsaro, M. Maisano, A. D'Agata, A. Giannetto, E. De Domenico, V. Parrino, G. Lo Paro, and A. Mauceri, *Ecotoxicol. Environ. Safe.*, 2012, **84**, 139. DOI: 10.1016/j.ecoenv.2012.07.001.

23. S. L. Shofer and R. S. Tjeerdema, *Ecotoxicol. Environ. Safe.*, 2002, **51**, 155. DOI: 10.1006/eesa.2002.2141.

24. A. R. Van Scoy, B. S. Anderson, B. M. Philips, J. Voorhees, M. McCann, H. De Haro, M. J. Martin, J. McCall, C. R. Todd, D. Crane, M. L. Sowby, and R. S. Tjeerdema, *Ecotoxicol. Environ. Safe.*, 2012, **78**, 99. DOI: 10.1016/j.ecoenv.2011.11.009.

25. A. R. Van Scoy, C. Y. Lin, B. S. Anderson, B. M. Philips, M. J. Martin, J. McCall, C. R. Todd, D. Crane, M. L. Sowby, M. R. Viant, and R. S. Tjeerdema, *Ecotoxicol. Environ. Safe.*, 2010, **73**, 710. DOI: 10.1016/j.ecoenv.2010.03.001.

26. H. F. Wu, X. Y. Zhang, Q. Wang, L. Z. Li, C. L. Ji, X. L. Liu, J. M. Zhao, and X. L. Yin, *Ecotoxicol. Environ. Safe.*, 2013, **90**, 1. DOI: 10.1016/j.ecoenv.2012.02.022.

27. L. M. Samuelsson, B. Björlenius, L. Förlin, and D. G. J. Larsson, *Environ. Sci. Tech.*, 2011, **45**, 1703. DOI: 10.1021/es104111x.

28. H. F. Wu and W. X. Wang, *Environ. Toxicol. Chem.*, 2011, **30**, 806. DOI: 10.1002/etc.446.

29. H. F. Wu, X. L. Liu, J. M. Zhao, and J. B. Yu, *Mar. Drugs*, 2011, **9**, 1566. DOI: 10.3390/md9091566.

30. T. Schock, D. Stancyk, L. Thibodeaux, K. Burnett, L. Burnett, A. B. Boroujerdi, and D. Bearden, *Metabolomics*, 2010, **6**, 250. DOI: 10.1007/s11306-009-0194-y.

31. W. Tuffnail, G. Mills, P. Cary, and R. Greenwood, *Metabolomics*, 2009, **5**, 33. DOI: 10.1007/s11306-008-0143-1.

32. M. R. Viant, J. G. Bundy, C. A. Pincetich, J. S.de Ropp, and R. S. Tjeerdema, *Metabolomics*, 2005, **1**, 149. DOI: 10.1007/s11306-005-4429-2.

33. M. R. Viant, J. H. Walton, and R. S. Tjeerdema, *Pestic. Biochem. Phys.*, 2001, **71**, 40. DOI: 10.1006/pest.2001.2554.

34. A. D. M. Dove, J. Leisen, M. S. Zhou, J. J. Byrne, K. Lim-Hing, H. D. Webb, L. Gelbaum, M. R. Viant, J. Kubanek, and F. M. Fernandez, *PLoS ONE*, 2012, **7**. DOI: e4937910.1371/journal.pone.0049379.

35. M. R. Viant, I. Werner, E. S. Rosenblum, A. S. Gantner, R. S. Tjeerdema, and M. L. Johnson, *Fish Physiol. Biochem.*, 2003, **29**, 159. DOI: 10.1023/b:fish.0000035938.92027.81.

36. M. R. Viant, *Biochem. Biophys. Res. Commun.*, 2003, **310**, 943.

37. M. R. Viant, in Environmental Genomics, ed C. C. Martin, Humana Press: Totowa, 2007, Chap. Environmental metabolomics using ^{1}H-NMR spectroscopy, 137.

38. M. R. Viant, E. S. Rosenblum, and R. S. Tjeerdema, *Toxicol. Sci.*, 2003, **72**, 240.

39. M. R. Viant, E. S. Rosenblum, and R. S. Tjeerdema, *Environ. Sci. Technol.*, 2003, **37**, 4982.

40. N. Morrison, D. Bearden, J. Bundy, T. Collette, F. Currie, M. Davey, N. Haigh, D. Hancock, O. Jones, S. Rochfort, S.-A. Sansone, D. Štys, Q. Teng, D. Field, and M. Viant, *Metabolomics*, 2007, **3**, 203.

41. A. Seitz and H. T. Ratte, *Comparat. Biochem. Physiol. C Comparat. Pharmacol.*, 1991, **100**, 301.

42. J. Sardans, J. Peñuelas, and A. Rivas-Ubach, *Chemoecology*, 2011, **21**, 191. DOI: 10.1007/s00049-011-0083-5.

43. A. Hines, G. S. Oladiran, J. P. Bignell, G. D. Stentiford, and M. R. Viant, *Environ. Sci. Tech.*, 2007, **41**, 3375.

44. G. Lannig, S. Eilers, H. O. Pörtner, I. M. Sokolova, and C. Bock, *Mar. Drugs*, 2010, **8**, 2318.

45. E. M. Santos, J. S. Ball, T. D. Williams, H. Wu, F. Ortega, R.van Aerle, I. Katsiadaki, F. Falciani, M. R. Viant, J. K. Chipman, and C. R. Tyler, *Environ. Sci. Tech.*, 2009, **44**, 820. DOI: 10.1021/es902558k.

46. L. Zhang, X. L. Liu, L. P. You, D. Zhou, Q. Wang, F. Li, M. Cong, L. Z. Li, J. M. Zhao, D. Y. Liu, J. B. Yu, and H. F. Wu, *Environ. Toxicol. Pharm.*, 2011, **32**, 218. DOI: 10.1016/j.etap.2011.05.006.

47. L. B. Zhang, X. L. Liu, L. P. You, D. Zhou, H. F. Wu, L. Z. Li, J. M. Zhao, J. H. Feng, and J. B. Yu, *Mar. Environ. Res.*, 2011, **72**, 33. DOI: 10.1016/j.marenvres.2011.04.002.

48. A. Kullgren, F. Jutfelt, R. Fontanillas, K. Sundell, L. Samuelsson, K. Wiklander, P. Kling, W. Koppe, D. G. J. Larsson, B. T. Bjornsson, and E. Jonsson, *Comp. Biochem. Physiol. A.*, 2013, **164**, 44. DOI: 10.1016/j.cbpa.2012.10.005.

49. S. C. Doney, V. J. Fabry, R. A. Feely, and J. A. Kleypas, *Ann. Rev. Mar. Sci.*, 2009, **1**, 169. DOI: 10.1146/annurev.marine.010908.163834.

50. J. Widdows, *Mar. Biol.*, 1973, **20**, 269. DOI: 10.1007/bf00354270.

51. D. Ekman, Q. Teng, D. Villeneuve, M. Kahl, K. Jensen, E. Durhan, G. Ankley, and T. Collette, *Metabolomics*, 2009, **5**, 22.

52. J. G. Bundy, E. M. Lenz, N. J. Bailey, C. L. Gavaghan, C. Svendsen, D. Spurgeon, P. K. Hankard, D. Osborn, J. M. Weeks, S. A. Trauger, P. Speir, I. Sanders, J. C. Lindon, J. K. Nicholson, and H. Tang, *Environ. Toxicol. Chem.*, 2002, **21**, 1966.

53. J. G. Bundy, D. J. Spurgeon, C. Svendsen, P. K. Hankard, J. M. Weeks, D. Osborn, J. C. Lindon, and J. K. Nicholson, *Ecotoxicology*, 2004, **13**, 797.

54. A. Hines, G. S. Oladiran, J. P. Bignell, G. D. Stentiford, and M. R. Viant, *Environ. Sci. Technol.*, 2007, **41**, 3375.

55. J. McKelvie, J. Yuk, Y. Xu, A. Simpson, and M. Simpson, *Metabolomics*, 2009, **5**, 84.

56. N. Taylor, R. Weber, A. Southam, T. Payne, O. Hrydziuszko, T. Arvanitis, and M. Viant, *Metabolomics*, 2009, **5**, 44.

57. N. S. Taylor, R. J. M. Weber, T. A. White, and M. R. Viant, *Toxicol. Sci.*, 2010, **118**, 307.

58. C. Y. Lin, B. S. Anderson, B. M. Phillips, A. C. Peng, S. Clark, J. Voorhees, H. D. I. Wu, M. J. Martin, J. McCall, C. R. Todd, F. Hsieh, D. Crane, M. R. Viant, M. L. Sowby, and R. S. Tjeerdema, *Aquat. Toxicol.*, 2009, **95**, 230. DOI: 10.1016/j.aquatox.2009.09.006.

59. E. G. Nagato, J. C. D'eon, B. P. Lankadurai, D. G. Poirier, E. J. Reiner, A. J. Simpson, and M. J. Simpson, *Chemosphere*, 2013, **93**. DOI: http://dx.doi.org/10.1016/j.chemosphere.2013.04.085.

60. H. C. Poynton, N. S. Taylor, J. Hicks, K. Colson, S. Chan, C. Clark, L. Scanlan, A. V. Loguinov, C. Vulpe, and M. R. Viant, *Environ. Sci. Tech.*, 2011, **45**, 3710. DOI: 10.1021/es1037222.

61. P. L. TenBrook, S. M. Kendall, M. R. Viant, and R. S. Tjeerdema, *Aquat. Toxicol.*, 2003, **62**, 329.

62. T. Vandenbrouck, O. A. H. Jones, N. Dom, J. L. Griffin, and W. De Coen, *Environ. Inter.*, 2010, **36**, 254. DOI: http://dx.doi.org/10.1016/j.envint.2009.12.006.

63. J. Jordan, A. Zare, L. J. Jackson, H. R. Habibi, and A. M. Weljie, *J. Proteome Res.*, 2012, **11**, 1133. DOI: 10.1021/pr200840b.

64. L. M. Samuelsson, L. Förlin, G. Karlsson, M. Adolfsson-Erici, and D. G. J. Larsson, *Aquat. Toxicol.*, 2006, **78**, 341.

65. C. Silversand and C. Haux, *J. Comparat. Physiol. B Biochem. Syst. Environ. Physiol.*, 1995, **164**, 593.

66. D. R. Ekman, Q. Teng, D. L. Villeneuve, M. D. Kahl, K. M. Jensen, E. J. Durhan, G. T. Ankley, and T. W. Collette, *Environ. Sci. Technol.*, 2008, **42**, 4188.

67. B. W. K. Diehl, *Eur. J. Lipid Sci. Technol.*, 2001, **103**, 830.

68. J. P. Sumpter and S. Jobling, *Environ. Health Perspect.*, 1995, **103**, 173.

69. I. Katsiadaki, T. D. Williams, J. S. Ball, T. P. Bean, M. B. Sanders, H. Wu, E. M. Santos, M. M. Brown, P. Baker, F. Ortega, F. Falciani, J. A. Craft, C. R. Tyler, M. R. Viant, and J. K. Chipman, *Aquat. Toxicol.*, 2010, **97**, 174.

70. A. R. Van Scoy, C. Y. Lin, B. S. Anderson, B. M. Philips, M. J. Martin, J. McCall, C. R. Todd, D. Crane, M. L. Sowby, M. R. Viant, and R. S. Tjeerdema, *Ecotoxicol. Environ. Saf.*, 2010, **73**, 710.

71. C. S. Manning, W. E. Hawkins, D. H. Barnes, W. D. Burke, C. S. Barnes, R. M. Overstreet, and W. W. Walker, *Toxicol. Method.*, 2001, **11**, 147.

72. C.-y. Wang and X.-l. Wang, *J. Environ. Sci.*, 2007, **19**, 1061. DOI: http://dx.doi.org/10.1016/S1001-0742(07)60173-9.

73. S. C. Booth, M. L. Workentine, A. M. Weljie, and J. R. Turner, *Metallomics*, 2011, **3**, 1142. DOI: 10.1039/C1MT00070E.

74. Y.-K. Kwon, Y.-S. Jung, J.-C. Park, J. Seo, M.-S. Choi, and G.-S. Hwang, *Mar. Poll. Bull.*, 2012, **64**, 1874.

75. P. Bjerregaard and T. Vislie, *Comp. Biochem. Phys. C*, 1985, **82**, 227. DOI: http://dx.doi.org/10.1016/0742-8413(85)90235-X.

76. A. Péqueux, A. Bianchini, and R. Gilles, *Comp. Biochem. Phys. C*, 1996, **113**, 149. DOI: http://dx.doi.org/10.1016/0742-8413(95)02081-0.

77. M. Elumalai, C. Antunes, and L. Guilhermino, *Chemosphere*, 2007, **66**, 1249. DOI: http://dx.doi.org/10.1016/j.chemosphere.2006.07.030.

Chapter 26

Metabolomics in Environmental Microbiology

Sean Booth[1], Raymond J. Turner[1] and Aalim Weljie[2]

[1]*Department of Biological Sciences, University of Calgary, Calgary, Alberta, T2N 1N4, Canada*
[2]*Department of Pharmacology, Institute for Translational Medicine and Therapeutics, University of Pennsylvania, Philadelphia, PA 19104, USA*

26.1 INTRODUCTION

The field of environmental microbiology explores how microbes interact with and influence their environment. It encompasses a wide breadth of important endeavors from attempts at cultivating bacteria from lakes below the Antarctic ice[1] to the isolation of novel biotechnology enzymes from geothermal hot springs.[2] Indeed, much of the recent work in the field has focused on studying these so-called extremophiles.[3] How do organisms such as *Deinoccocus radiodurans* withstand huge doses of radiation[4] or how does *Cupriavidus metallidurans* thrive in the midst of toxic heavy metals?[5] Researchers also seek to understand how these microbes interface with their environment.

In the postgenomic era, 'systems biology' tools (the so-called omics approaches), which allow for the high throughput analysis of an organism's functionality, have been widely adopted. Genome sequencing can give an explanation of all the potential gene products an organism *can* make. Next, transcriptomics describes what genes an organism *are* actually expressing. Proteomics takes this a step further by seeking to reveal all the proteins that are present, hence what the cell *could* be doing. Finally, at the pinnacle is metabolomics, which provides a snapshot of exactly what the cell *is* doing. While the former three tools have important uses in environmental microbiology, especially for culture-free techniques, metabolomics is an ideal tool for evaluating the physiological state of a microbe and thus, how it is functioning in response to their environment. (For higher organisms, see Chapters 24 and 25.)

26.1.1 Questions in Environmental Microbiology

The current investigations in environmental microbiology are as diverse as the habitats the organisms come from. Much of the focus of the field in recent years has been directed toward understanding previously uncharacterized and/or unknown metabolic pathways.[6] Another direction is the evaluation of microbes under

NMR Spectroscopy: A Versatile Tool for Environmental Research
Edited by Myrna J. Simpson and André J. Simpson

severe environmental stress.[7] For example, due to rising levels of anthropogenic pollution, microbes capable of degrading organic xenobiotics such as pesticides and plastics as well as those able to survive and potentially immobilize or biotransform the speciation of toxic heavy metals have gained economic value (see Chapters 19 and 20). Thus, researchers are interested in understanding how the hyper-resistors and toxin degraders are capable of such useful metabolic feats.

While much of the field of environmental microbiology involves working with samples from the field, many issues can be addressed with wholly laboratory-based experiments. One of the most important revelations in microbiology has been the acceptance that bacteria naturally grow as biofilms, not as free-swimming planktonic individuals.[8] A biofilm can be defined as an assemblage of cells (single or multispecies) growing attached to a surface in a self-secreted layer of extracellular polymeric substance (EPS).[9] Biofilms have been found to be physiologically different compared to genetically identical planktonic counterparts.[10] Owing to their tolerant properties, and differences from planktonic cultures (which have classically been used to determine inhibitory concentrations of antibacterials), the study of microbial biofilms has evolved into an important subject unto itself. For example, a major reason that biofilms have reached a prominent position in microbiology, and even in society at large, is their much higher tolerance and resistance to antimicrobials of any kind.[11] With respect to environmental microbiology, the ability to withstand much higher concentrations of toxins such as organic pollutants and heavy metals is potentially beneficial, as these abilities can be put to use for bioremediation. This is the process of using living organisms to diminish (completely) the concentration and toxicity of anthropogenic pollutants that have been released into the environment. For organic pollutants, this entails the metabolic conversion of pollutants into innocuous end-products (preferably, carbon dioxide and water). As metals cannot be broken down, their remediation must involve either sequestration or chemical alteration into a less-toxic form. Thus, understanding the changes in physiology of bioremediation-capable organisms under the stress of the pollutant is important for optimization of the process leading to downstream economic benefits.

26.1.2 Metabolomics

Each specific question in microbiology that requires an understanding of the physiological state would be aided by the use of metabolomics (see Chapter 22). The choice of the metabolomics method, however, is key to success. One can evaluate preferential metabolic pathways by simply evaluating the growth of the organism under different reductant, oxidant, and carbon sources. A popular approach is the BioLog Phenotype Microarray technology,[12] which is a high-throughput growth screen method. Although this gives a good overview of metabolic possibilities, it does not provide specific information on pathways used or buildup of intermediates. Evaluation of metabolites present is a good way to obtain such information. In metabolomics, one of the first questions is considering what metabolites and biomolecules one wants to investigate. Many cell-extraction protocols are available;[13] however, it is clear that this step needs to be optimized for each microbial system and question being asked.

Once a metabolite-extraction method is worked out, the next decision is the separation technology combined with the detection method one wishes to use to identify and quantify the metabolites. The most common competing technologies are that of mass spectrometry (MS)- and NMR-based approaches. Chromatography coupled to MS has the advantage of separating hundreds of compounds and detecting with high sensitivity, yet identification of specific metabolites remains a significant challenge.[14] NMR is considered a stronger method for quantitation and identification, yet will yield a much smaller number of metabolites (see Chapter 23).

As with any approach, the specific use of NMR for metabolomics must be refined to particular cases. This is due to the relatively high concentrations required to generate a detectable NMR signal, a nontrivial consideration for the lower cell densities encountered in environmental microbiology. It is worth noting though that NMR can be applied to problems in microbiology in a myriad of useful ways, of which metabolomics is just one.[15] For laboratory-based work, where ease of sample prep and unbiased quantification is of high priority, NMR metabolomics occupies a highly useful laboratory niche. This includes the so-called footprinting or fingerprinting applications where a reproducible and reliable overall pattern of change provides phenotypic information.[16] Specific metabolites of interest can be followed up using additional analytical tools such as MS.

26.2 NMR METHODS FOR ENVIRONMENTAL METABOLOMICS OF MICROBES

To perform a metabolomics experiment using NMR, the bacteria must be cultured and the resulting metabolite sample mixture placed into the NMR magnet. Between these steps, a number of important considerations must be made (see Chapter 23). An excellent set of reviews for developing metabolomics experiments are available.[17,18] Metabolomics inherently works best as a comparative platform, so the relevant control and experimental treatments must be decided. Here, factors such as growth time, method, media, and amount of stress must be carefully controlled to ensure that the metabolic response that is to be observed is relevant to these parameters. For example, to test the effect of a stressor, an undue effect on the doubling time of an organism must be considered, as it may confound the analysis. Additionally, with regard to stress such as exposure to metals, ideally a concentration that is toxic but not lethal should be used.[19] Sampling methods tend to vary widely and are a source of some controversy, as there is no well-accepted standard within the field.[20] The main concern is that a sampling method should stop metabolism and any residual enzymatic activity, thereby providing an accurate snapshot of the metabolic state, without altering the sample in anyway. This is difficult, as metabolite turnover can be below the order of seconds, so most techniques rely on cooling. Techniques used in the studies described in Section 26.3 vary widely.[21-27] Additionally, the method (if any) used for extracting metabolites from a sample has the potential to bias the types of compounds quantified.[13] For the case studies described in Section 26.3,[23-25] a method was developed where cultures are added to cold ($-40\,^\circ$C) 60% methanol to quench metabolism and centrifuged at $-20\,^\circ$C to collect the cells. Metabolites are extracted from sonication-disrupted cells in 2 : 1 methanol chloroform. Compared to other tools used in metabolomics, NMR has the advantage of being able to directly analyze a sample. This can be used to examine the spent media from a culture, which will provide the researcher with information about the so-called exo-metabolome, which is to be contrasted with the intracellular metabolites, deemed the 'endometabolome'. Analysis of each has its benefits. While pooling samples is an option in efforts to increase sensitivity in the NMR, further statistical analysis relies on the variance of each sample, and as such, it is preferable to obtain as many biological replicates as possible.

Once a suitable sample is obtained, NMR spectra are typically acquired for each sample of interest. The choice of NMR acquisition and processing parameters should be considered carefully, as they have a noticeable impact on the resulting data.[28] While a number of NMR nuclei may be and have been considered, we will focus our discussion on the most common ^{1}H NMR experiment, although two-dimensional experiments are useful both for validation and as an experimental approach in their own right (see Chapter 23). Furthermore, quantitative metabolomics work is increasingly important, and as such, specific considerations will be provided in this regard. The readers are also referred to excellent more extensive discussion of these considerations for further information.[29]

26.2.1 Data Preprocessing

Analysis of the data acquired by NMR is a complex process that can be done in several ways (see Chapter 22). An important method used in the case studies discussed in Section 26.3 is the method of 'targeted profiling'.[30] This technique considers information about related resonances from a given molecular structure, depending on a library of metabolites, which can be overlaid to reconstruct the mixture spectrum. This approach, however, requires manual input (Figure 26.1).[31] Newer approaches, which use Bayesian methods, are promising as a means to automate this process, although yet unproven for quantitative work. Moreover, spectral ordering is a promising technique, in which information from one spectrum can be tracked through an entire dataset for quantitative information.[32] Once quantified, the data must undergo further processing. The main processes are normalization, scaling, and centering, which allow for the direct comparison of all samples and metabolites, even if their dynamic ranges are quite different. The proper methods for such procedures are in continuous development and discussion and are described elsewhere.[33]

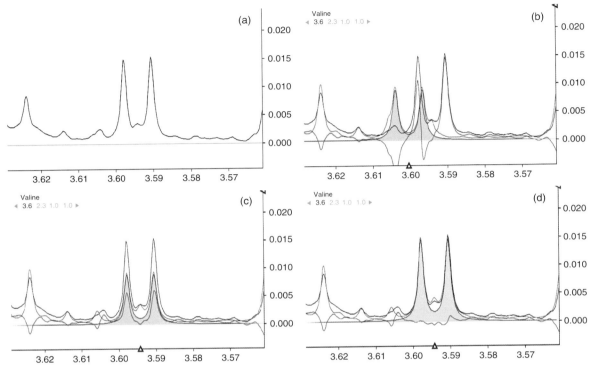

Figure 26.1. Targeted profiling in Chenomx™ of valine in a sample obtained from *P. fluorescens* cells. After phase and baseline corrections, NMR spectra can be viewed in the Chenomx™ Profiler.[30] Initially, just the original spectrum is visible (a), as a black line. Using the provided library of reference compounds, peaks at the appropriate chemical shift can be added (b), in blue. An additive spectrum of all profiled compound's peaks can then be viewed, as a red line. A subtractive spectrum, as a green line, is also provided, which removes the profiled spectrum from the original to ensure that peaks have been properly identified and quantified. Profiled compounds are then matched up to peaks in the original spectrum (c) and manipulated until they match in amplitude to provide quantitative information (d).

26.2.2 Statistical Analysis

Powerful statistical methods have been a necessity for deriving useful biological information from metabolomics datasets.[33] This is the case as cells can contain thousands of different metabolites but only some will be indicative of changes caused by perturbations because of the environmental stress under investigation. As there are likely more variables than samples (especially with spectral binning) as well as numerous confounding factors not the least of which is biological variability and issues of covariance between variables, univariate statistical techniques such as *t*-tests and analysis of variance (ANOVA) are insufficient. Instead, researchers have tended to use tools such as principal component analysis (PCA), partial least squares discriminant analysis (PLS-DA), and clustering, as these techniques take a collection of samples (i.e., replicates and treatments) and observations (metabolites or bins/spectra) together and determine the overall relationships and variances within the entire dataset. PCA and orthogonal partial least squares discriminant analysis (OPLS-DA) are modeling techniques where all the observations for a single sample are condensed and used to calculate its value in a new variation-describing space. Similarly, for each observation, the values in all samples are compressed into a single value, which describes that variable's (metabolite) relationship to the variance between samples. For both techniques, this process is iterated in components, whereby the first component attempts to account for as much variation as possible in the dataset and subsequent components are calculated in the same manner until the remainder of variation cannot be modeled. The difference between the two techniques is that PCA computes the variation inherent within the model, whereas PLS-DA takes class designates (e.g., control, mutant, and treatment) and attempts to maximize variation between these groups. These algorithms are implemented not only in various general mathematical and software packages such as MATLAB, R, Stata, and SPSS but also in specific chemometric packages such as SIMCA-P (Umetrics), AMIX (Bruker), and CAMO.

26.2.3 Biological Interpretation

Once metabolites have been identified, quantified, and those that were significantly altered determined, the biological meaning can begin to be examined. Most interpretations rely on comparisons between control and treated samples, using knowledge of the treatment to make sense of particular metabolite concentration changes. Metabolomics is a very powerful approach to establish the foundation of hypothesis-based research in this regard. Here, NMR is particularly useful, as its accurate quantitation allows for reliable interpretation of metabolic differences between treatments. Deriving biological meaning can be aided by NMR-specific databases such as the HMDB (human metabolome database),[34] BMRB (biological magnetic reference bank),[35] or SDBS (spectral database for organic compounds).[36] Numerous computational approaches have been developed that take a list of metabolites, and automatically, query biochemical databases for associations between these compounds, which are reviewed in Ref. 37. This provides higher level information on which metabolic pathways are likely being affected. Again, the quantitative power of NMR becomes useful, as metabolic flux through these pathways can be observed by mapping concentrations onto the pathways. These computations have increasing power when more metabolites are input, which will not always be the case using an NMR approach.

26.3 CASE STUDIES

26.3.1 Antimicrobial Effects of the Tellurium Oxyanion

Tellurium (Te) is of industrial interest because of its semiconductor properties,[38] and is thus involved in solid-state electronics, batteries, and various nanotechnologies, which result in its release into the environment. Pure Te is insoluble and thus, nontoxic. The oxyanion forms of tellurite (TeO_3^{2-}) and tellurate (TeO_4^{2-}) are highly toxic to microorganisms[39] at concentrations much lower than other metals.[40] Even with considerable effort, a complete picture of the mechanisms of toxicity and resistance is still lacking.[38]

Pseudomonas pseudoalcaligenes KF707 is a model organism capable of degrading polychlorinated biphenyls, which has also been used to study metal resistance.[24] Its natural resistance to tellurite has led to its use in understanding resistance mechanisms to this metalloid.[41] Tremaroli *et al.*[24] isolated a mutant that was hyper-resistant to tellurite, but it was unclear how the mutation was enabling this resistance. As such, it was subjected to metabolomic analysis in order to shed light on the mechanisms of tellurite resistance.

Control and Te-exposed cultures were compared using NMR-based metabolomics. The experiment identified and quantified 28 intracellular metabolites, which accounted for most of the peaks in any single NMR spectrum. To understand the similarities and differences of how tellurite was affecting each strain, several OPLS-DA models were generated. These models respectively compared the wild-type to T5 and independently each strain control and exposed to tellurite.

As was expected, the metabolic effects of tellurite were different between the two strains. A shared and unique structures (SUS) plot of the variable's influence on projection (VIP) highlighted how there were only a few metabolites that were important in both models that compared the control to exposed cultures of the mutant and wild-type (Figure 26.2). Contrarily, there were many metabolites important independently in each model. These differences demonstrated that the T5 mutant had a strong effect on the metabolic response toward tellurite. By examining the coefficients of the important metabolites in each case, it was found that the effect of the T5 mutation was complex. Decreases in glutamate and aspartate suggested their conversion into tricarboxylic acid (TCA) cycle intermediates with tellurite exposure; however, corresponding increases in the α-ketoglutarate and oxaloacetate (also TCA cycle intermediates) were not observed, as these metabolites were not present in the NMR spectra, perhaps because of insufficient sample volumes. This highlights an important consideration in the NMR approach: it is vital to ensure that sufficient quantities of culture are used that lead to concentrations

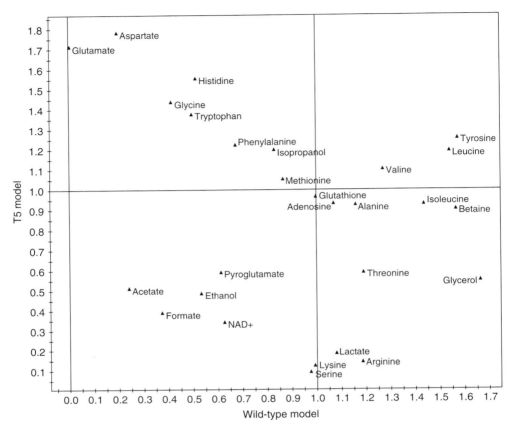

Figure 26.2. Shared and unique structures plot of VIP values obtained from models comparing *P. pseudoalcaligenes* KF707 wild-type with and without tellurite and the T5 hyper-resistant mutant with and without tellurite. The *x*-axis shows the metabolites that are important in the wild-type; the *y*-axis shows those that are important in the T5. Metabolites with a VIP > 1 were considered significant in distinguishing the control from tellurite-exposed samples in each strain. Metabolites in the bottom right were important for the wild-type, in the top left for T5. Those in the top right were important in both and those the bottom left neither. (Adapted from Ref. 24. © 2009 American Society for Microbiology)

of metabolites that are high enough to be detected by a particular NMR set-up. Further comparison of the control samples for each strain found that the T5 hyper-resistant mutant showed significant differences in metabolite levels (glutathione and other antioxidative stress-related compounds) from the wild-type. This was an interesting result, as it indicated that the mutation caused a systematic change, not a reactionary change that was dependent on the exposure to tellurite. Additionally, when both cultures were exposed to tellurite, there were no significant differences between them, as well as between the wild-type exposed to tellurite and the unexposed T5. Together, these results

indicated that the T5 mutant was able to withstand tellurite better through metabolic 'priming'. Specifically, glutathione was observed to be at much higher level in the mutant compared to the wild-type. Tellurite has previously been observed to cause oxidative damage;[42] therefore, this observation along with increased levels of branched-chain amino acids (which have been linked to oxidative stress[43]) led to the conclusion that the T5 hyper-resistance was due to its superior antioxidant abilities. This conclusion was supported by prior observations of antioxidant reactions increasing bacterial tellurite resistance.[42] Another metabolite that was associated with priming resistance to tellurite was betaine. This small molecule is used as an osmoprotectant.[44] Increases to this metabolite correlated with the observation that the membrane potential of T5 cells was less perturbed than that of wild-type. As tellurite interferes with the electron transport chain, stabilization of the membrane was postulated to be mitigating this damage.

This study demonstrated how careful comparison of a few conditions, even when only measuring a relatively low number of metabolites, can lead to a good understanding of a small genetic difference. Examining the metabolites that changed in each strain under tellurite stress did provide some information about why the T5 mutant was hyper-resistant; however, it was by comparing the control samples of both strains as well as the comparison with less differences of exposed wild-type to control T5, which really demonstrated what metabolic factors were responsible for the increased resistance. While it would have been beneficial to use larger amounts of culture to increase the amount of biomass and thus, identify more relevant metabolites, such as the missing TCA cycle intermediates, these observations were not completely necessary to explain the hyper-resistant phenotype. Finally, the certitude in concentrations provided by NMR gave increased confidence on those metabolites that were found to be important, such as betaine and glutathione (Table 26.1).

26.3.2 Evaluation of Phenotypic Variants Arising from a Global Regulator Mutation

Bacterial biofilms have been found to contain phenotypically different subpopulations.[25] While genetically identical to their parents, these phenotypic variants (which are generally described by their colony morphology on solid media) have been found to display a number of altered traits.[45] With *Pseudomonas aeruginosa*, this phenomenon has been implicated in the lung colonization of cystic fibrosis patients, and also in *Pseudomonas fluorescens* rhizospheric interactions.[45,46] These phenotypic variants can be found at higher frequencies from cultures under stress, which along with the observed increases in antimicrobial resistance in some variants indicates that they may be an important component of resistance mechanisms in biofilms.[47] *P. fluorescens* is a soil bacterium, which has been used for many biofilm and metal resistance studies.[48] Strains of this organism harboring an inactive mutation in the global activator of cyanide biosynthesis/regulator of secondary metabolism (gac/rsm) signal transduction pathway have been found to produce phenotypic variants more frequently.[25]

Table 26.1. Statistical terminology

(O)PLS-DA	Partial least squares discriminant analysis. A supervised multivariate modeling technique where the differences among the input classes are maximized and within-class variance is minimized. The orthogonal version computes all variation in the data that is not related to the provided classes in a separate, orthogonal component portion of the model, which makes the model easier to interpret. The discriminant analysis implies that two sample classes are used for discriminating between one another
Coefficient	A value calculated in (O)PLS-DA models for every variable (metabolite), which indicates its relative abundance in each class
Class	A group of one type of samples, for example, control or mutant
Model	A statistical projection of a large dataset with multiple variables. This type of analysis allows for the simultaneous comparison of many variables at once, even in cases where there are more variables measured than samples
VIP	Variable influence on projection. A value calculated in (O)PLS-DA models for every variable (metabolite), which indicates its importance in distinguishing the classes in the model
Supervised analysis	Statistical technique where the differences between defined classes are determined. Compare with unsupervised analysis where the inherent variation within a dataset is determined
SUS plot	Shared and unique structures. A plot that shows the similar and distinct features of two different models. Variables in the top-right and bottom-left quadrants have similar values in both models. Those in the top left are unique in the *y*-axis model and those in the bottom right are unique in the *x*-axis model

To better understand the differences between phenotypic variants, Workentine *et al.*[25] investigated two colony morphology variants from a ΔgacS strain of *P. fluorescens*. The GacS/GacA two-component regulatory system is an important moderator in a number of physiological processes, including biofilm formation.[49] The *P. fluorescens* ΔgacS mutant became a good candidate for studying metal resistance mechanisms in biofilms, as it was found that variant cells in biofilms are a factor of metal tolerance.[50] To this end, Workentine *et al.*[49] isolated two phenotypic variants from cultures of *P. fluorescens* ΔgacS exposed to nonlethal concentrations of metal ions. The small colony variant (SCV) and 'wrinkly spreader' (WS), which can be stably propagated, were separated based on their appearance as well as dye-binding properties.

To further differentiate these phenotypic variants and also to find a biochemical basis for the other observed traits, the two phenotypic variants as well as the parental ΔGacS mutant strain and wild-type *P. fluorescens* were subjected to NMR-based metabolomic analysis.[25] A total of 32 metabolites were identified and quantified. Using PLS-DA (Figure 26.3), it was shown that all four cultures were metabolically distinct from one another. Pair-wise comparisons for all combinations of strains were then performed using OPLS. Metabolite VIPs were used to show which metabolites distinguished the variants from the parental ΔgacS and wild-type strains. It was found that each variant had a distinct set of metabolites that distinguished it from both ancestral strains. Valine, phenylalanine, and glycine were important for the WS, but acetate, pyruvate, aspartate, proline, and glutamate were important for the SCV. The metabolites implicated in both the WS and the SCV phenotypes matched well with prior proteomic observations of a 'large spreader, wrinkly spreader' phenotypic variant.[51] This study demonstrated upregulated catabolic and transport pathways for all the important metabolites from both phenotypes, which helped define the level of similarity between the variants. The identification of these differences in metabolite concentrations made it clear that global metabolic adaptations are part of the physiology of colony morphology variants leading to a better understanding of increased survival in stressed environments.

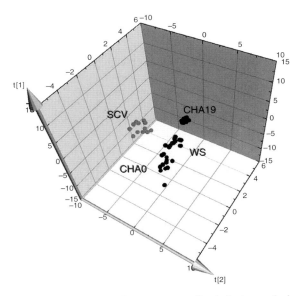

Figure 26.3. Partial least squares discriminate analysis (PLS-DA) 3D scores plot from a model comparing *P. fluorescens* wild-type (CHA0), ΔgacS (CHA19), and the colony morphology phenotypic variants wrinkly spreader (WS) and small colony variant (SCV) obtained from the parental ΔgacS strain. Each data point represents metabolites extracted from a single sample. Each axis represents a component of variation that separates the four classes based on the concentrations of 32 metabolite concentrations. The clustering of all samples in each class indicates the similarity of the samples, and the separation along each axis indicates the difference between classes. (Adapted from Ref. 25. © 2010 Society for Applied Microbiology and Blackwell Publishing Ltd.)

26.3.3 Evaluation of Metal Resistance in Biofilms

Understanding the differences between biofilm and planktonic cultures has important implications medically and also industrially, as biofilms are generally more resistant to toxic antimicrobials.[47] This capability has been attributed to a number of factors inherent to cultures growing as a biofilm, including the aforementioned phenotypic heterogeneity, quorum-sensing-mediated gene regulation, and metal chelation by the EPS that all biofilms produce.[47] The phenotypic variability within biofilms is caused by physical heterogeneity within the biofilm microenvironment.[52] This heterogeneity is considered to be caused by gradients of nutrients, waste products, pH, and oxygen, all which have an influence on cellular metabolism. Thus, it was expected that the metabolism occurring within a biofilm is different than that of a planktonic culture and reasoned that metabolism may be an additional factor responsible for the differences in metal tolerance found in microbial biofilms.[23]

To investigate this possibility, Booth *et al.*[23] used a combined NMR and gas chromatography–mass spectrometry (GC-MS) metabolomics approach to examine the differences between biofilm and planktonic cultures of *P. fluorescens* exposed to copper. Copper is responsible for much environmental contamination because of mining activities, and is also seeing medical uses as an antimicrobial surface[53] and for water sterilization.[54] In both cases, an in-depth understanding of copper toxicity and resistance in microbial biofilms is important, as their metal-tolerating traits could be used in bioremediation.[55]

P. fluorescens was grown as a biofilm and planktonically and then exposed to 1.5 mM $Cu(SO_4)_2$, which is about half the minimum inhibitory concentration,[48] as this amount of metal was expected to cause toxicity effects without being lethal. In this study, 48 metabolites were identified and quantified by NMR, in addition to several hundred features detected by GC-MS, although not quantitatively. Owing to the higher variation between samples in GC-MS data, presumably because of the untargeted methodology employed, significance analysis of microarrays and PLS-DA was used to reduce the number of metabolites to only those that were reliably detected and significantly different. This resulted in 200 named metabolites from the two approaches, which included most of those found by NMR. The relative abundances of these metabolites were determined as an average from the unit-variance scaled and mean-centered values from each method.

Unsupervised statistical analysis using both PCA and hierarchical clustering showed that there were significant metabolic differences between the control and the metal-exposed biofilm cultures but not the planktonic samples. Supervised analysis using OPLS-DA, however, demonstrated that there were significant differences between the control and the metal-exposed planktonic cultures, in addition to major differences between the planktonic and biofilm samples of each type. The goal of this study was to understand how biofilms are resistant to toxic metals; therefore, individual OPLS-DA models were generated for each culture type. Comparison of the models showed that only nicotine adenine dinucleotide, glutathione, and phosphoric acid were important in both culture types.

Further analysis showed that metabolites with a large magnitude of change in one culture type tended to have a much lower magnitude of change in the other.

To understand the actual metabolic differences between toxicity responses in each culture type, hierarchical clustering and enrichment analysis were used. Hierarchical clustering is a technique that organizes a dataset such that the most similar samples are placed adjacent to one another, whereas the most similarly varying metabolites are simultaneously organized in the same way. A heatmap produced using this technique showed a clear division between the biofilm and the planktonic samples, as well as the control and the exposed biofilms. The planktonic control and the exposed samples were also separated, but not as well as the biofilm. As for the metabolites, they were clearly grouped into three categories, those that were altered in the planktonic cultures, in the biofilm, and in both. These lists of metabolites were used as input for the enrichment analysis, a technique that takes a list of metabolites and uses annotation information from a large biochemical database such as the *Kyoto encyclopedia of Genes and Genomes* (KEGG)[56] to compute which metabolic pathways are likely being perturbed.[37] Through this analysis, a number of different metabolic pathways were found to be altered in each culture type (Table 26.2). The metabolic pathways altered in the biofilms in response to copper stress appeared mostly to be involved in EPS production. The EPS is composed of both proteins and polysaccharides. Alterations to starch, sucrose, and galactose metabolism indicated that changes were being made to polysaccharide production. Changes were also observed in a large number of amino acid metabolic pathways, which suggested wholesale alterations in protein production. As the EPS of biofilms has been found to chelate metals, protecting the cells inside, it was concluded that the observed changes were to thicken the EPS to chelate the copper and prevent it from poisoning the biofilm cells.[57] In the planktonic cultures exposed to copper, changes to many of the same amino acid metabolic pathways were also observed. As planktonic cells do not produce EPS, it was suspected that these changes were due to greater protein turnover, as those damaged by copper were recycled and stress proteins synthesized. Additionally, the TCA cycle, pyruvate metabolism, and nicotinate/nicotinamide metabolism were perturbed in planktonic cultures.

Table 26.2. Metabolic pathways enriched in response to copper exposure determined by enrichment analysis

Planktonic Metabolic pathway	*p*-Value	Both Metabolic pathway	*p*-Value	Biofilm Metabolic pathway	*p*-Value
Nicotinate and nicotinamide metabolism	0.01	Glycine, serine, and threonine metabolism	0.01	Arginine and proline metabolism	0.01
Pyruvate metabolism	0.01	Cyanoamino acid metabolism	0.01	Phosphotransferase system	0.01
Glycolysis/ gluconeogenesis	0.01	Nitrogen metabolism	0.01	Lysine degradation	0.01
Purine metabolism	0.01	Phenylalanine metabolism	0.01	Galactose metabolism	0.01
TCA cycle	0.01	Glutathione metabolism	0.01	Glutathione metabolism	0.01
Alanine, aspartate, and glutamate metabolism	0.01	Alanine, aspartate, and glutamate metabolism	0.01	Starch and sucrose metabolism	0.015
Valine, leucine, and isoleucine biosynthesis	0.01	Valine, leucine, and isoleucine biosynthesis	0.01	—	—
—	—	Phenylalanine, tyrosine, and tryptophan biosynthesis	0.035	—	—

Only pathways with a *p*-value < 0.05 were included, and life-general metabolic pathways (such as aminoacyl-tRNA synthesis) and plant-specific pathways were excluded.
Reprinted with permission from Ref. 23. © 2011 American Chemical Society.

These observations corresponded to similar physiological changes produced in *P. fluorescens* cultures exposed to aluminum.[58] These studies demonstrated that central metabolism was rearranged to deal with the oxidative stress caused by aluminum. While the metabolites altered by copper exposure did not match exactly with those affected by aluminum, it was concluded that copper induces a similar stress response. The metabolomics investigation from the Booth *et al.*[23] study demonstrated for the first time a clear physiological difference between the planktonic and the biofilm modes of growth.

26.3.4 Evaluation of Metabolic Effects of Osmotic Stress

P. aeruginosa is a well-studied opportunistic pathogen. Isolates obtained from long-term infections have been found to overproduce alginate, a polysaccharide with an unknown environmental function. Strains with a mutated *MucA* gene overproduce this compound, which is not normally an important component of wild-type biofilms. To gain a better understanding of the mutant *MucA* phenotype, a temporal analysis of spent media from wild-type and *mucA*22 mutant *P. aeruginosa* as well as an end-point analysis of

both cells and spent media under osmotic stress were performed using NMR-based metabolomics.[22]

The initial screening experiment comparing exogenous metabolites from the two strains indicated differences in the concentrations over time of valine, methionine, glycine-betaine, and trehalose.[22] These last two metabolites are known bacterial osmolytes suggesting that the *mucA*22 mutant may have an effect on osmotic stress, which was determined over the course of the study. This metabolomics-derived hypothesis was confirmed by growing the two strains under a range of salt concentrations, with the wild-type consistently outgrowing the mutant. At this point, the researchers returned to metabolomics in order to identify which metabolites might be responsible for the differences in osmotolerance. In this secondary experiment, both strains were either (i) exposed to various salt concentrations and an end-point measurement taken or (ii) exposed to a single high-salt concentration and samples were taken over the entire course of growth. Four metabolites were found to increase inside the cells of both strains when salt concentration was increased. *N*-Acetlyglutaminylglutamine amide, glutamate, glyine-betaine, and trehalose were found at comparable levels in both strains at most concentrations of salt. At the highest concentrations tested

(0.7 and 0.8 M NaCl), the wild-type showed significantly higher levels of all these osmolytes. When analyzing the differences between the two strains at a single concentration of salt (0.4 M) over time, PCA analysis showed that the main difference between samples was growth time; however, the second component separated the control from salt-exposed samples. This was surprising, as these data did not support the previously observed differences between the wild-type and *mucA*22 mutant. As such, a more sophisticated analysis was undertaken, which integrated the observations over time for each treatment. The hierarchical PCA that was then obtained showed clear differences among the strains, as well as between the control and the exposed samples. This emphasizes the importance of understanding when metabolic differences will be maximized in a metablomics experiment, as merely observing an endpoint in this case would not have demonstrated the effect of the *mucA*22 mutation on osmotic stress response. Additionally, from this analysis, differences in the uptake of several metabolites were observed. Relevant to the osmotic stress hypothesis, when under salt stress, both glycine-betaine and glutamate were taken up later by the mutant compared to the wild-type. These results, along with the data from the salt concentration experiments, clearly showed that *mucA* has a role in the osmotic stress response. This study demonstrated how NMR-based metabolomics, even when only quantifying a small number of metabolites, but under a wide variety of conditions, can generate and subsequently support testable hypotheses.

Strains of *Burkholderia cenocepacia* have been found to promote plant growth in the rhizosphere, an environment that may be prone to osmotic stress.[59] To understand the differences in osmotic stress tolerance of several strains, including two highly pathogenic strains, both intracellular and extracellular metabolites were separately quantified under control and hypersaline conditions using NMR metabolomics.[21] Of the metabolites that were detected, five were deemed osmoresponsive in at least some strains. Glycine-betaine and trehalose are well-known bacterial osmolytes; the other three were the amino acids alanine, glutamate, and phenylalanine. Contrary to expectations, the concentrations of these metabolites were not altered under increased salt in the same way in the highly pathogenic strains. In fact, three different responses were observed. Three strains, including one of the highly pathogenic strains, showed minimally altered levels of osmolytes, but increased levels of the three amino acids. These strains were the most

impaired by hypersalinity, and so it was postulated that these changes were due to a general stress response. The remaining two strains, one which was highly pathogenic, displayed different metabolite alterations, which granted some protection. Resistance in one strain was attributed to its constitutively high levels of glycine-betaine. The other, highly pathogenic strain, demonstrated marked increases in all five metabolites, including trehalose and glycine-betaine. Intriguingly, the changes in these metabolites in spent media did not perfectly mirror the intracellular changes. NMR quantification showed decreased usage of the amino acids, likely because of the decreased growth, whereas the osmolytes showed only slight increases in use in the corresponding strains. Still this study highlighted the diversity of osmotic stress responses among five closely related strains. Additionally, the lack of congruity between the osmoprotective strategies of the highly pathogenic strains as well as the differences between the other strains support the importance of environmental stress responses in clinically relevant bacteria.

26.3.5 Other Environmental Stressors

Coral reef ecosystems have become endangered because of human activities.[60] Normally, corals live symbiotically with bacteria such as *Vibrio corallilyticus*, but this interaction can turn pathogenic, as temperatures above 24 °C induce virulent coral bleaching and lysis.[61] Boroujerdi *et al.*[27] set out to understand this temperature-dependent pathogenicity by comparing the metabolic profiles of *V. corallilyticus* cultures grown at 24 and 27 °C using NMR-based metabolomics. While much of this work focused on understanding inter- and intrabatch effects caused by natural biological variation, PCA identified a number of osmoprotectants that distinguished the high-temperature grown pathogenic cultures from the controls. Additionally, the observed dissimilarity between metabolic profiles obtained at each temperature has led to continued research into this phenomenon.[43]

Staphylococcus epidermis biofilms have been associated with hospital-acquired infections.[62] As the transition to growth as a biofilm requires a large modification of physiology and hence, gene regulation, the methods by which bacteria sense, interpret, and respond to signals in their environment must be understood.[9] The TCA cycle is central to a correctly metabolically functional cell, which, because of the propensity of its enzymes to be affected by a wide variety of environmental and nutritional factors, led to its implication in transducing signals interpreted from beyond the cell.[63] To further investigate this function, Sadykov *et al.*[64] used NMR-based metabolomics to first compare the independent effects of iron limitation and ethanol stress on cultures of *S. epidermis*[64] then followed up on this work by examining and contrasting these results with the metabolic effects of an aconitase mutant as well as the effects of salt, glucose, and subinhibitory levels of antibiotics.[65] Using PCA, the initial results showed that cultures stressed with either iron limitation or ethanol were metabolically similar to the unstressed aconitase mutant strain. While the ethanol stress caused some additional unrelated effects, this similarity clearly linked the two stressors with TCA cycle inhibition. To control for a generalized stress response, comparably treated cultures of a *sigB* mutant (a general stress response transcription factor) were also compared. PCA showed that these cultures were different from the control and stressed wild-type cultures as well as the aconitase mutant, indicating that the TCA inhibition was independent from the *sigB* controlled stress response. In the stressed cultures and the aconitase mutant, glucose-6-phosphate as well as lactate, acetate, acetaldehyde, and acetyl phosphate were increased in concentration relative to the wild-type. These metabolites are indicative of fermentative metabolism, glucose-6-phosphate as an upstream inducer, and the rest as downstream end-products. The inhibited TCA cycle associated with these changes was hypothesized to be inducing the production of the biofilm molecule polysaccharide intracellular adhesion (PIA). Using the metabolomic data as a starting point, this was confirmed using appropriate regulatory mutants. Thus, it was confirmed that the TCA cycle was being used to regulate responses to environmental stress.

This was followed up by comparing the metabolic effects of additional glucose, ethanol, and iron limitation, all of which were expected to be sensed by the TCA cycle.[65] Cultures of both the wild-type and the aconitase mutant were exposed separately to each of the stressors and harvested during and after exponential phase making for 12 different experimental conditions. Using spectral binning, concentrations of 55 metabolites were altered under these various conditions. PCA, OPLS-DA, and clustering demonstrated that after 6 h of growth (postexponential phase), the metabolomes of all these cultures, with the exception

of the wild-type control, were highly similar. As the cultures of the stress-exposed TCA-inactivated aconitase mutants did not differ from their unexposed control, this further supported the notion that the TCA cycle interprets multiple stressors and modulates them into a common response. The concentrations of aspartate, asparagine, glutamate, and glutamine, which are biosynthetically derived from the TCA cycle, were found to be lower in stressed cultures as would be expected from an inhibited TCA cycle. Phosphoenolpyruvate, acetaldehyde, and fructose-6-phosphate, however, were found to be increased, which was indicative of active glycolytic metabolism, which would be necessary to produce ATP in the absence of the TCA cycle. Increased levels of several amino sugars were observed including UDP-*N*-acetylglucosamine, which is a precursor to PIA, again linking TCA cycle modulation with biofilm formation. Overall, this study clearly demonstrated how different environmental stressors can affect the TCA cycle.

26.4 CONCLUSIONS AND PROSPECTS

NMR-based metabolomics has been successfully applied to a number of problems in environmental microbiology of diverse nature, but overall, it is an underutilized technique. This may be due to the perception of NMR as a structural tool or due to the comparatively low sensitivity. Improvements in NMR technology, such as increasingly accessible cryo/cold probes, microcoil probes (see Chapter 23), flow probes, improved detector technology, dynamic nuclear polarization, and hyphenation with tools such as solid phase extraction (see Chapter 3), will serve to overcome some limitations inherent to the technique. Another possible reason for the lack of NMR-based metabolomics studies in this field is the lack of consensus in quenching metabolism and extracting metabolites for microbial metabolomics.[20] This issue only serves to further confound environmental microbiologists already working with field samples of unknown concentration. Additionally, many researchers are already fully engaged in keeping up with the rapid advances in genetic sequencing of whole communities, that is, metagenomics, and optimizing their analysis pipeline from DNA extraction to functional annotation.[66] Compared to other metabolomics technologies, analysis of NMR data requires considerably more expert chemical knowledge; however, various efforts toward automating metabolite identification are underway.[32] In situations where identification of compounds is not necessary, such as in fingerprinting or footprinting analysis, NMR can be quite powerful. Spent media and filtered aquatic samples can essentially be directly analyzed, allowing for rapid screening for metabolic differences between treatments or environments. The use of such metabolic profiling (as in nonidentification centered) for ecological metabolomics of higher organisms has been touted as a rapid method for classifying samples.[67] While such an approach would be useful for preliminary research, the use of NMR would also be beneficial for elucidating structures from samples containing unknown compounds.[68] It is well established by now that NMR and MS are complementary tools for high-throughput applications. For example, compounds from GC-MS are confounded by derivatizing agents and potentially different fragmentation between samples, but detected with greater sensitivity. Tandem mass spectrometry and tandem separation technologies [two-dimensional liquid chromatography (LC-LC), LC-GC, or GC-GC] would also be an excellent tool for this type of application, although are not unbiased to the same extent as NMR. For compounds with known structures, NMR is superior for distinguishing ^{13}C isotopomers.[69] NMR's simple sample preparation and quantitative strength makes it an ideal tool for monitoring carbon flux,[70] for example, in a chemostat[71] or even a bioreactor. Furthermore, the ability of NMR to be applied to other NMR-active nuclei can lead to convenient biodegradation studies of fluorinated compounds.[72] While MS has its advantages and will remain a useful tool for metabolomics, NMR has great potential to be further exploited in the field of environmental microbiology. With continuing developments to NMR hardware and software for data analysis, the potential for this technology to help solve the myriad of questions in environmental microbiology will only rise.

ACKNOWLEDGMENTS

SB is funded by a graduate student scholarship from Alberta Innovates. RJT and AW both hold funding from Natural Sciences Engineering Research Council of Canada. RJT is also supported by the Canadian Institutes of Health Research.

RELATED ARTICLES IN EMAGRES

Metals in the Environment

Organic Pollutants in the Environment

Environmental Metabolomics

Environmental Metabolomics: NMR Techniques

Environmental Metabolomics of Soil Organisms

Environmental Metabolomics of Aquatic Organisms

REFERENCES

1. A. E. Murray, F. Kenig, C. H. Fritsen, C. P. McKay, K. M. Cawley, R. Edwards, E. Kuhn, D. M. McKnight, N. E. Ostrom, V. Peng, A. Ponce, J. C. Priscu, V. Samarkin, A. T. Townsend, P. Wagh, S. A. Young, P. T. Yung, and P. T. Doran, *Proc. Natl. Acad. Sci. U. S. A.*, 2012, **109**, 20626.

2. P. Fuciños, R. González, E. Atanes, A. B. F. Sestelo, N. Pérez-Guerra, L. Pastrana, and M. L. Rúa, *Methods Mol. Biol.*, 2012, **861**, 239.

3. F. Canganella and J. Wiegel, *Naturwissenschaften*, 2011, **98**, 253.

4. M. M. Cox and J. R. Battista, *Nat. Rev. Microbiol.*, 2005, **3**, 882.

5. D. H. Nies, *Extremophiles*, 2000, **4**, 77.

6. D. R. Lovley, *Nat. Rev. Microbiol.*, 2003, **1**, 35.

7. L. J. Rothschild and R. L. Mancinelli, *Nature*, 2001, **409**, 1092.

8. A. Jain, Y. Gupta, R. Agrawal, P. Khare, and S. K. Jain, *Crit. Rev. Ther. Drug Carrier Syst.*, 2007, **24**, 393.

9. D. López, H. Vlamakis, and R. Kolter, *Cold Spring Harb. Perspect. Biol.*, 2010, **2**, a000398.

10. A. M. Spormann, *Curr. Top. Microbiol. Immunol.*, 2008, **322**, 17.

11. H. Ceri, M. E. Olson, and R. J. Turner, *Expert Opin. Pharmacother.*, 2010, **11**, 1233.

12. B. R. Bochner, P. Gadzinski, and E. Panomitros, *Genome Res.*, 2001, **11**, 1246.

13. M. J.v. d. Werf, K. M. Overkamp, B. Muilwijk, L. Coulier, and T. Hankemeier, *Anal. Biochem.*, 2007, **370**, 17.

14. E. Want and P. Masson, *Methods Mol. Biol. (Clifton, NJ)*, 2011, **708**, 277.

15. J. P. Grivet and A. M. Delort, *Prog. Nucl. Magn. Reson. Spectrosc.*, 2009, **54**, 1.

16. K. L. Resmer and R. L. White, *Mol. Biosyst.*, 2011, **7**, 2220.

17. B. Álvarez-Sánchez, F. Priego-Capote, and M. D. L. d. Castro, *Trends Anal. Chem.*, 2010, **29**, 120.

18. B. Álvarez-Sánchez, F. Priego-Capote, and M. D. Luque de Castro, *Trends Anal. Chem.*, 2010, **29**, 111.

19. S. C. Booth, M. L. Workentine, A. M. Weljie, and R. J. Turner, *Metallomics*, 2011, **3**, 1142.

20. W. M. van Gulik, *Curr. Opin. Biotechnol.*, 2010, **21**, 27.

21. V. Behrends, J. G. Bundy, and H. D. Williams, *Lett. Appl. Microbiol.*, 2011, **52**, 619.

22. V. Behrends, B. Ryall, X. Wang, J. G. Bundy, and H. D. Williams, *Mol. Biosyst.*, 2010, **6**, 562.

23. S. C. Booth, M. L. Workentine, J. Wen, R. Shaykhutdinov, H. J. Vogel, H. Ceri, R. J. Turner, and A. M. Weljie, *J. Proteome Res.*, 2011, **10**, 3190.

24. V. Tremaroli, M. L. Workentine, A. M. Weljie, H. J. Vogel, H. Ceri, C. Viti, E. Tatti, P. Zhang, A. P. Hynes, R. J. Turner, and D. Zannoni, *Appl. Environ. Microbiol.*, 2009, **75**, 719.

25. M. L. Workentine, J. J. Harrison, A. M. Weljie, V. A. Tran, P. U. Stenroos, V. Tremaroli, H. J. Vogel, H. Ceri, and R. J. Turner, *Environ. Microbiol.*, 2010, **12**, 1565.

26. A. F. B. Boroujerdi, S. S. Jones, and D. W. Bearden, *Lett. Appl. Microbiol.*, 2012, **54**, 209.

27. A. F. B. Boroujerdi, M. I. Vizcaino, A. Meyers, E. C. Pollock, S. L. Huynh, T. B. Schock, P. J. Morris, and D. W. Bearden, *Environ. Sci. Tech.*, 2009, **43**, 7658.

28. K. A. Kaiser, C. E. Merrywell, F. Fang, and C. K. Larive, in NMR Spectroscopy in Pharmaceutical Analysis, eds H. Ulrike, W. Iwona and D. Bernd, Elsevier: Amsterdam, 2008, Chap. Metabolic profiling, 233 Chapter 5.

29. R. Powers, *Magn. Reson. Chem.*, 2009, **47**, S2.

30. A. M. Weljie, J. Newton, P. Mercier, E. Carlson, and C. M. Slupsky, *Anal. Chem.*, 2006, **78**, 4430.

31. Chang, D., Weljie, A., Newton, J. Leveraging Latent Information in NMR Spectra for Robust Predictive Models, In *Pacific Symposium on Biocomputing*, 2007, pp 115.

32. E. Alm, T. Slagbrand, K. M. Åberg, E. Wahlström, I. Gustafsson, and J. Lindberg, *Anal. Bioanal. Chem.*, 2012, **403**, 443.

33. H. J. Issaq, Q. N. Van, T. J. Waybright, G. M. Muschik, and T. D. Veenstra, *J. Sep. Sci.*, 2009, **32**, 2183.

34. D. S. Wishart, D. Tzur, C. Knox, R. Eisner, A. C. Guo, N. Young, D. Cheng, K. Jewell, D. Arndt, S. Sawhney, C. Fung, L. Nikolai, M. Lewis, M. A. Coutouly, I. Forsythe, P. Tang, S. Shrivastava, K. Jeroncic, P. Stothard, G. Amegbey, D. Block, D. D. Hau, J. Wagner, J. Miniaci, M. Clements, M. Gebremedhin, N. Guo, Y. Zhang, G. E. Duggan, G. D. MacInnis, A. M. Weljie, R. Dowlatabadi, F. Bamforth, D. Clive, R. Greiner, L. Li, T. Marrie, B. D. Sykes, H. J. Vogel, and L. Querengesser, *Nucleic Acids Res.*, 2007, **35**, D521.

35. Q. Cui, I. A. Lewis, A. D. Hegeman, M. E. Anderson, J. Li, C. F. Schulte, W. M. Westler, H. R. Eghbalnia, M. R. Sussman, and J. L. Markley, *Nat. Biotechnol.*, 2008, **26**, 162.

36. T. Saito and S. Kinugasa, Development and release of a spectral database for organic compounds - key to the continual services and success of a large-scale database, *Synthesiology*, 2011, **4**, 26.

37. S. C. Booth, A. M. Weljie, and R. J. Turner, *Comput. Struct. Biotechnol. J.*, 2013, **4**, e201301003.

38. R. J. Turner, R. Borghese, and D. Zannoni, *Biotechnol. Adv.*, 2012, **30**, 954.

39. T. G. Chasteen, D. E. Fuentes, J. C. Tantaleán, and C. C. Vásquez, *FEMS Microbiol. Rev.*, 2009, **33**, 820.

40. D. Zannoni, F. Borsetti, J. J. Harrison, and R. J. Turner, *Adv. Microb. Physiol.*, 2007, **53**, 1.

41. G. Di Tomaso, S. Fedi, M. Carnevali, M. Manegatti, C. Taddei, and D. Zannoni, *Microbiology*, 2002, **148**, 1699.

42. V. Tremaroli, S. Fedi, and D. Zannoni, *Arch. Microbiol.*, 2006, **187**, 127.

43. H. Tweeddale, L. Notley-McRobb, and T. Ferenci, *Redox Rep.*, 1999, **4**, 237.

44. M. Roeßler and V. Müller, *Environ. Microbiol.*, 2001, **3**, 743.

45. M. Starkey, J. H. Hickman, L. Ma, N. Zhang, S. De Long, A. Hinz, S. Palacios, C. Manoil, M. J. Kirisits, T. D. Starner, D. J. Wozniak, C. S. Harwood, and M. R. Parsek, *J. Bacteriol.*, 2009, **191**, 3492.

46. M. Sánchez-Contreras, M. Martín, M. Villacieros, F. O'Gara, I. Bonilla, and R. Rivilla, *J. Bacteriol.*, 2002, **184**, 1587.

47. J. J. Harrison, H. Ceri, and R. J. Turner, *Nat. Rev. Microbiol.*, 2007, **5**, 928.

48. M. L. Workentine, J. J. Harrison, P. U. Stenroos, H. Ceri, and R. J. Turner, *Environ. Microbiol.*, 2008, **10**, 238.

49. M. L. Workentine, L. Chang, H. Ceri, and R. J. Turner, *FEMS Microbiol. Lett.*, 2009, **292**, 50.

50. J. J. Harrison, H. Ceri, C. A. Stremick, and R. J. Turner, *Environ. Microbiol.*, 2004, **6**, 1220.

51. C. G. Knight, N. Zitzmann, S. Prabhakar, R. Antrobus, R. Dwek, H. Hebestreit, and P. B. Rainey, *Nat. Genet.*, 2006, **38**, 1015.

52. P. S. Stewart and M. J. Franklin, *Nat. Rev. Microbiol.*, 2008, **6**, 199.

53. J. Elguindi, X. Hao, Y. Lin, H. A. Alwathnani, G. Wei, and C. Rensing, *Appl. Microbiol. Biotechnol.*, 2011, **91**, 237.

54. S. S. Martínez, A. A. Gallegos, and E. Martínez, *Int. J. Hydrogen Energy*, 2004, **29**, 921.

55. R. Singh, D. Paul, and R. Jain, *Trends Microbiol.*, 2006, **14**, 389.

56. M. Kanehisa and S. Goto, *Nucleic Acids Res.*, 2000, **28**, 27.

57. H. C. Flemming, T. R. Neu, and D. J. Wozniak, *J. Bacteriol.*, 2007, **189**, 7945.

58. J. Lemire, R. Mailloux, C. Auger, D. Whalen, and V. D. Appanna, *Environ. Microbiol.*, 2010, **12**, 1384.

59. A. Bevivino, V. Peggion, L. Chiarini, S. Tabacchioni, C. Cantale, and C. Dalmastri, *Res. Microbiol.*, 2005, **156**, 974.

60. O. Hoegh-Guldberg, *Regional Environ. Change*, 2011, **11**, 215.

61. Y. Ben-Haim, M. Zicherman-Keren, and E. Rosenberg, *Appl. Environ. Microbiol.*, 2003, **69**, 4236.

62. C. Von Eiff, G. Peters, and C. Heilmann, *Lancet Infect. Diseases*, 2002, **2**, 677.

63. G. A. Somerville and R. A. Proctor, *Microbiol. Mol. Biol. Rev.*, 2009, **73**, 233.

64. M. R. Sadykov, B. Zhang, S. Halouska, J. L. Nelson, L. W. Kreimer, Y. Zhu, R. Powers, and G. A. Somerville, *J. Biol. Chem.*, 2010, **285**, 36616.

65. B. Zhang, S. Halouska, C. E. Schiaffo, M. R. Sadykov, G. A. Somerville, and R. Powers, *J. Proteome Res.*, 2011, **10**, 3743.

66. D. Chmolowska, *Central Eur. J. Biol.*, 2013, **8**, 399.

67. J. Sardans, J. Peñuelas, and A. Rivas-Ubach, *Chemoecology*, 2011, 1.

68. K. A. Leiss, Y. H. Choi, R. Verpoorte, and P. G. L. Klinkhamer, *Phytochem. Rev.*, 2011, **10**, 205.

69. A. N. Lane, T. W. M. Fan, and R. M. Higashi, *Methods Cell Biol.*, 2008, **84**, 541.

70. C. L. Winder, W. B. Dunn, and R. Goodacre, *Trends Microbiol.*, 2011, **19**, 315.

71. D. J. V. Beste, B. Bonde, N. Hawkins, J. L. Ward, M. H. Beale, S. Noack, K. Nöh, N. J. Kruger, R. G. Ratcliffe, and J. McFadden, *PLoS Pathog.*, 2011, **7**, e1002091.

72. M. G. Boersma, I. P. Solyanikova, W. J. H. Van Berkel, J. Vervoort, L. Golovleva, and I. M. C. M. Rietjens, *J. Ind. Microbiol. Biotechnol.*, 2001, **26**, 22.

Chapter 27
Plant Metabolomics

Gregory A. Barding Jr, Daniel J. Orr and Cynthia K. Larive

Department of Chemistry, University of California, Riverside, Riverside, CA, 92521, USA

27.1 INTRODUCTION

Metabolomics focuses on the description of small molecule metabolites in complex biological samples. Recent advances in analytical platforms, sample preparation protocols, hyphenated techniques, throughput, and statistical analyses have allowed for the application of metabolic measurements to increasingly complex problems in plant biology. These advances are making metabolomics approaches

NMR Spectroscopy: A Versatile Tool for Environmental Research
Edited by Myrna J. Simpson and André J. Simpson
© 2014 John Wiley & Sons, Ltd. ISBN: 978-1-118-61647-5

invaluable to furthering our understanding of the metabolic responses of plants and other organisms to genetic and environmental perturbations.

The study of metabolism can be pursued through a variety of strategies.[1] One strategy is to adopt a "targeted analysis" approach that studies a specific genetic alteration and the resulting metabolic product. "Metabolite profiling" experiments seek to understand the function of an entire pathway or interactions between pathways by targeting a specific class of molecules for analysis often using enzymatic assays or labeled compounds. In contrast to targeted approaches, "metabolite fingerprinting" does not necessarily focus on the identification of specific metabolites but rather examines global metabolic changes as a function of a known perturbation such as genetic alteration and biotic or abiotic stress. Finally, the comprehensive analysis of the metabolites in a biological system is termed *metabolomics*, although this term is sometimes mistakenly applied to the above strategies. Regardless of the approach chosen, delineating metabolic responses to stress conditions, genetic modifications, and environment perturbations are pivotal to understanding plant biology.

A variety of analytical platforms have been adopted for plant metabolomics studies. These platforms include hyphenated mass spectrometry (MS) techniques such as gas chromatography,[2] capillary electrophoresis,[3] and liquid chromatography (LC)[4] as well as nuclear magnetic resonance (NMR).[5,6] MS is thus far the favored platform for plant metabolomics because it offers greater sensitivity and simple coupling to chromatographic separations. Although less sensitive than MS-based analyses, NMR is nonde-structive, inherently quantitative, and requires no sample derivatization (see Chapter 23).[7]

With these advantages, NMR was used for metabolite profiling as early as the 1970s when Brown and coworkers applied NMR to study human erythrocyte metabolism.[8] Additional metabolite profiling experiments were reported in the early 1980s by Moore *et al.*[9] and Nicholson and coworkers.[10] Recent work using NMR in mammalian systems has led to clinical application of NMR metabolite profiling of intact tissues,[11] pharmacogenomics,[12] and the geographical dependence of cardiovascular risks.[13]

Plant metabolomics, however, poses a unique challenge compared with mammalian systems because of the complex and sessile nature of plants. Plants are estimated to have as many as 200000 metabolites, many of which are not described compounds.[1,14] Plant metabolomics utilizing NMR has been applied to understand the response of plants to hypoxia stress,[15] improve crop stress-resistance,[5,16] identify the botanical origins of a food supply,[17] and evaluate wine quality.[18] Recent developments in NMR metabolomics and examples of its application for understanding plant metabolism are summarized in this chapter.

27.2   SAMPLE PREPARATION FOR NMR ANALYSIS

Although NMR methods allow for minimal sample handling prior to detection, typically some sample preparation prior to NMR analysis is required. Solution-state NMR experiments require extraction of soluble metabolites from the plant tissue. Deuterated extraction solvents are commonly used for NMR experiments and spectra may be acquired directly on the extract supernatant,[19] or after additional sample preparation steps, for example, lipid removal.[6] The extraction solvent composition is another important consideration. Some solvent systems, such as acetonitrile–water, can contribute to unwanted side effects such as resonances from lipids, which can interfere with the analysis of the aliphatic spectral regions.[6] Additionally, extraction solvents can contribute to line broadening of the NMR resonances, decreasing spectral resolution and increasing the difficulty of metabolite assignment.[20] As illustrated in the ^1H–^{13}C heteronuclear single quantum coherence (HSQC) spectrum (Figure 27.1a), the hexafluoroacetone/4-(2-hydroxyethyl)-1-piperazineethanesulfonic acid-d_{18}

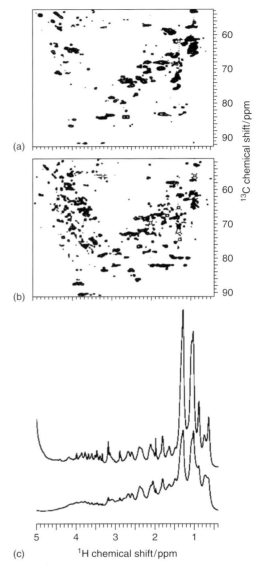

Figure 27.1.   Example of the line broadening effects of solvents on NMR spectra. 700 MHz ^1H–^{13}C HSQC spectra of (a) HFA/HEPES-d_{18} extract and (b) the same extract after drying to remove HFA and reconstitution in MeOD-d_4. Spectra after reconstitution in MeOD-d_4 are sharper and more peaks can be detected. (c) Comparison of the 700 MHz CPMG spectrum (above) with the single pulse 700 MHz ^1H NMR spectrum (below) of the HFA extract, indicating that line broadening results from short T_2 values. (Reprinted with permission from Ref. 20. © 2011 American Chemical Society)

(HFA/HEPES) solvent system contributes significantly to line broadening compared with the spectrum measured in methanol-d_4 (MeOD) (Figure 27.1b). As an NMR solvent, HFA is useful for small molecule analysis[21] and can increase the resolution of ^{13}C-bondomers in HSQC spectra of plant extracts incorporating an isotopic label.[22] Because line broadening in the ^{1}H NMR spectrum of the extract in HFA/HEPES (Figure 27.1c) was reduced in the Carr-Purcell-Meiboom-Gill (CPMG) spectrum, the source of the broadening was attributed to the extraction of paramagnetic ions and macromolecules by HFA, shortening T_2 values and broadening resonances.

Another important consideration in plant metabolomics experiments is the selection of the chemical shift reference standard. A commonly used ^{1}H chemical shift reference for aqueous solutions is sodium-3-trimethylsilylpropionate-d_4 (TMSP); however, its ^{1}H chemical shift is pH sensitive around the pK_a of the carboxylate group. Problems have also been reported due to the adsorption of TMSP to the glass of the NMR tube over time.[23] Sodium-2,2-dimethyl-2-silapentane-5-sulfonate-d_6 (DSS) can also be used as a ^{1}H chemical shift reference. Although DSS is more expensive than TMSP, it is not affected by pH changes and is a more suitable internal standard for concentration determinations. Another method for quantitation is the electronic reference to access in vivo concentration (ERETIC).[24] This method applies an electronic reference signal in experiments where internal standards cannot be used for quantitation such as in vivo metabolite measurements.

27.3 SPECTRAL NORMALIZATION AND ALIGNMENT

Many factors can introduce variance into plant metabolomics data sets such as differences in growth conditions, biological variance, and the imprecision of the sample extraction and analysis steps. Biological variance can be minimized through careful control of the plant's environment, tissue pooling, and measurements for replicate biological samples. Analytical precision can be optimized through careful evaluation of the analytical protocols for all steps prior to data collection (e.g., homogenization and extraction), and the use of technical replicates.

A common normalization method is referred to as sum normalization or constant-sum normalization.[25,26] Sum normalization is achieved by dividing the spectrum into small bins (usually 0.02–0.04 ppm) and integrating each bin. After removal of the bins in regions containing impurities or solvents, the bins are then summed. The integral measured for each bin is divided by this sum, providing a simple mathematical method to normalize against dilution, pipette errors, and instrument gain differences. Although widely used, sum normalization may not be the best normalization method because it can introduce bias toward more abundant metabolites and does not account for variation in chemical shifts. Several different normalization techniques have been created to address problems of dilution in NMR data sets. One of the more widely used methods is partial-quotient normalization (PQN), which scales spectra according to the most probable dilution factor determined by comparison to reference spectra.[27]

Another problem encountered in the analysis of biological samples is that pH differences can occur between the data sets, resulting in chemical shift changes for pH-sensitive resonances.[28] Although pH-dependent chemical shift changes can be minimized by use of a buffer, even small differences between data sets can become problematic with automated data analysis or unsupervised statistical treatments. To correct for chemical shift differences between spectra, several methods have been developed for spectral alignment. One such method is an adaptive binning technique using a wavelet transform to detect the peaks in a spectrum.[28] This technique is advantageous over traditional binning in that data is classified based on the detected peaks instead of a bin. Other spectral alignment and binning techniques rely on different mathematical approaches for the intelligent alignment of data sets.[29] Recently, the combination of PQN to account for dilution along with the adaptive intelligent binning technique for spectral alignment was shown to be a robust and reproducible approach for normalization of NMR data.[30] Other normalization and alignment techniques have also been evaluated to improve the reliability of statistical analysis.[31] An appropriate alignment and normalization approach must be determined based on the variability of the data set being explored. Although normalization and alignment are important aspects of NMR metabolomics studies, many of the current approaches were developed for mammalian systems and have yet to be applied to plant metabolomics data sets.

27.4 HYPHENATED NMR APPROACHES FOR PLANT METABOLOMICS

Although NMR metabolomics studies are often performed using whole extracts, the complexity of the plant metabolome produces ^{1}H spectra with many overlapped resonances, hindering the detection and quantification of low-level metabolites. Increasing the resolution of metabolite signals can greatly increase the sensitivity, selectivity, and identification capabilities of NMR. Recently, several different hyphenated approaches have been applied for the study of plant metabolism.

27.4.1 Solid-phase Extraction

Solid-phase extraction (SPE) is a useful technique for sample cleanup or isolation of desired classes of molecules. The availability of several different SPE stationary phases including weak/strong anion exchange, reverse-phase C18, and cation exchange provide the possibility for the isolation and separation of diverse classes of compounds. For example, Beretta and coworkers applied reverse-phase SPE-NMR for the identification of the botanical origin of honey.[17] Application of SPE allowed the isolation of hydrophobic markers that could be used for the identification of the type of honey and any contaminants that might be present. Another recent example of SPE-NMR includes the enrichment of phenolic compounds from grape extracts during various grape berry development stages.[32] These examples demonstrate that SPE-NMR has a wide variety of applications in characterizing the plant metabolome.

27.4.2 Multi-platform Analysis for Compound Identification and Characterization

Multi-platform analysis can provide more comprehensive coverage of the metabolome than can be achieved by a single technique. Hyphenation of NMR and MS can be valuable in evaluating a plant's metabolome (see Chapter 3). Online liquid chromatography (LC)-SPE-NMR-MS (or -MS-NMR) allows for the identification and detection of metabolites using multiple detection platforms in a single experiment. For example, Exarchou and coworkers

demonstrated that online LC-SPE-NMR-MS is capable of identifying flavinoids in Greek oregano.[33] This was possible by first performing a separation using high performance liquid chromatography (HPLC) with protonated solvents and then splitting 5% of the peak for MS analysis with 95% directed to SPE-NMR. Prior to NMR analysis, peaks were trapped on SPE cartridges, dried, and eluted with deuterated solvents. SPE provided a method for online concentration of the sample for NMR detection. Peaks were trapped multiple times in the SPE cartridge prior to elution for NMR analysis. Agnolet and coworkers compared online HPLC-MS-SPE-NMR with traditional ^{1}H-NMR metabolite profiling for the evaluation of commercial *Gingko biloba* extracts.[34] The purpose of these experiments was to evaluate ^{1}H NMR as a replacement for MS techniques for the analysis of glycosides and terpene trilactones (TTLs) in *G. biloba*. Using HPLC-MS-SPE-NMR, the authors were able to identify and characterize the composition of several unique glycosides and TTLs in commercial extracts and validate ^{1}H NMR for global profiling of extracts for quality control purposes. Additionally, these techniques allowed the identification of potentially harmful molecules in the commercial extracts. Hyphenated NMR analysis can also be completed offline. Rezzi and coworkers demonstrated an automated SPE-HPLC fractionation method for biofluids followed by NMR analysis.[35] Although MS detection was not employed, it can be readily adapted for further analysis of the biological system, taking advantage of the inherently nondestructive nature of NMR in combination with the sensitivity of MS. Multi-platform approaches should be considered in the design of plant metabolomics experiments as they have shown to be advantageous for sample enrichment, component identification, and method verification. The combination of LC, MS, SPE, and NMR allowed researchers to simultaneously reduce sample complexity, identify contaminants, target specific classes of compounds, and evaluate biological diversity quantitatively using a single analysis. Because of the complexity of plant systems and the large number of predicted metabolites, further development and application of hyphenated NMR techniques will be necessary for plant metabolomics studies.

27.5 2D NMR APPROACHES FOR QUANTITATIVE PLANT METABOLOMICS

2D NMR methods are often used for the identification of resonances in ^{1}H survey spectra acquired for metabolomics studies. Recent developments in the use of 2D NMR for metabolomics measurements have extended their application to the quantitative evaluation of data sets. An advantage of 2D NMR experiments is that they spread information into a second dimension, increasing spectral resolution, and improving quantitation, especially in crowded regions of the NMR spectrum.[36] For example, the 2D J-resolved spectroscopy experiment reduces convolution from spectral crowding by providing a proton-decoupled spectrum without a significant increase in the overall experiment time.[37] Mustafa and coworkers demonstrated the utility of the 2D J-resolved spectroscopy experiment in plant metabolomics by exploring the effect of salicylic acid in plant cell cultures.[38] The ^{1}H–^{13}C HSQC experiment has also been applied to metabolomics studies, using automated peak fitting routines to quantify metabolite resonances.[39] Other 2D NMR techniques have also been adapted to metabolomics experiments, including total correlation spectroscopy (TOCSY),[40] correlation spectroscopy (COSY),[41] and projection NMR.[42]

An alternative approach is to use statistical analysis to extend metabolomics data sets based on 1D ^{1}H-NMR spectra to produce a pseudo-2D NMR spectrum. Statistical correlation spectroscopy (STOCSY), developed by Cloarec and coworkers, uses the statistical comparison of resonances in a series of related NMR spectra to calculate a pseudo-TOCSY spectrum, facilitating the assignment of resonances to specific molecules.[43] This technique builds on the concept that resonances of all the protons within a molecule will be altered in a similar way by the metabolomics perturbation. STOCSY also allows the correlation of metabolites that exhibit similar changes within the biological system by statistically monitoring the response of all of the resonances in the spectrum. STOCSY has been further developed for metabolomics to assist in interpretation of diffusion-edited NMR experiments and has been implemented with other statistical techniques to increase the information obtained from the pseudo-2D experiment.[44] STOCSY has been shown to be useful for exploring metabolic profiles and correlating resonances in metabolomics experiments to our knowledge it has not yet been incorporated into a plant metabolomics study.

27.6 STABLE ISOTOPE MONITORING AND FLUX ANALYSIS

An important feature of NMR is its sensitivity to specific isotopes particularly spin 1/2 nuclei, providing researchers a means to study environmental effects and specific biological pathways using stable isotope tracers. With the exception of ^{1}H and ^{31}P, most spin 1/2 nuclei of biological relevance are present at low natural abundance. Therefore, the use of compounds incorporating a stable isotope label such as ^{13}C or ^{15}N can be used to probe metabolic pathways by following transfer of the label in metabolic flux studies.[45] Approaches and strategies for stable isotope labeling have been discussed extensively,[46,47] therefore, the primary focus of this section is to discuss recent applications of stable isotope labeling in plant metabolomics.

Labboun and coworkers elucidated the fate of ^{15}NH$_4^+$ and [^{15}N] glutamate in transgenic tobacco plants using in vivo real-time NMR spectroscopy.[48] Using a 10 mm NMR probe, these authors demonstrated that glutamate dehydrogenase contributes to the control of glutamate homeostasis in tobacco leaves through a cycling pathway in which the ammonium released by glutamate was reassimilated by glutamine synthase. These experiments reinforce the unique ability of NMR for real-time imaging of metabolic flux.

Because of concerns about global climate change, the effect of high CO_2 levels on plant metabolism is of significant interest to the plant biology community. Yu and coworkers used the rotational-echo double resonance (REDOR) solid-state NMR experiment to evaluate the effects of high and low $^{13}CO_2$ on a crop (soybean) and an invasive plant species (cheatgrass).[49] The REDOR experiment allows the restoration of dipolar couplings between pairs of heteronuclear spins, giving spatial and connectivity information for ^{15}N- and ^{13}C-labeled molecules. In this study, the authors applied $^{13}CO_2$ gas at high (600 ppm) and low (200 ppm) concentrations to elucidate the differential carbon assimilation of soybean and cheatgrass grown on ^{15}N-labeled ammonium nitrate as a function of CO_2 concentration. Figure 27.2 compares the soybean and cheatgrass REDOR spectra after exposure to 600 ppm $^{13}CO_2$ gas, revealing that cheatgrass has a better capacity for assimilation and distribution of

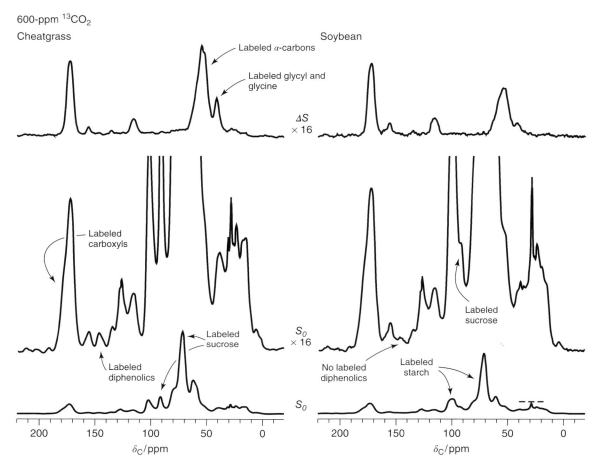

Figure 27.2. Example of an experiment to measure the impact of high CO_2 levels on two different plants using solid-state NMR to follow a ^{13}C label. The experiments shown are 125 MHz $^{13}C[^{15}N]$ REDOR NMR spectra of soybean (right) and cheatgrass (left). Plants were exposed to 600 ppm $^{13}CO_2$ and grown on ^{15}N-labeled fertilizer. S_0 spectra (full-echo) are shown at the bottom and the ΔS (REDOR difference spectra) are shown above. (Reprinted with permission from Ref. 49. © 2010 American Chemical Society)

carbon. This is indicated by the presence of a wider variety of labeled compounds, such as proteins and protein precursors, in cheatgrass compared with soybean. Under low CO_2 conditions, however, soybean is capable of assimilating carbon into a variety of molecules, suggesting that carbon assimilation in soybean is concentration dependent. Although these REDOR experiments were performed on frozen tissue, it is possible that in the future they can be extended to measurements of live tissue or cell cultures during a treatment regime, facilitating metabolic flux studies.

Of growing interest to researchers is the metabolic fate of molecules and the metabolic composition of organs, compartments, or cell types. It has been estimated that there are approximately 40 different cell types in a single plant[50] and metabolic flux into and out of these cell types can differ based on cellular or organ function. Current metabolomics experiments typically monitor the metabolism of the whole plant or specific tissues because of the difficulty of unbiased isolation of specific cells or compartments without changing their metabolic profile. One approach for overcoming this problem is the use of laser microdissection (LMD) to isolate specific organs, organelles, or cells.[51] Schneider and Holscher demonstrated the potential of LMD for metabolomics by removing secretory cavities from the leaves and flowers of several species of *Dilatris*.[52] Using LMD, ^{1}H NMR spectroscopy, and HPLC, these authors confirmed the presence of two novel natural products in the isolated secretory cavities and elucidated their structures. These experiments show the potential of LMD for cell- or organelle-specific isolation to enhance metabolic flux experiments by monitoring the fate of stable isotope tracers in specific compartments or cells. Although not yet widely adopted in plant metabolomics, LMD can play an important role in understanding how plants react to stresses or genetic alterations at the cellular and subcellular level.

27.7 NMR IMAGING AND MRI FOR PLANT METABOLISM STUDIES

As described in the previous section, data acquired with better spatial resolution can lead to a greater understanding of metabolic regulation. NMR and related magnetic resonance imaging (MRI) techniques are uniquely suited for the noninvasive and nondestructive acquisition of physiological information. Imaging techniques can provide much greater spatial resolution than traditional extraction methods and nondestructive measurements offer the potential for temporal resolution. Spatial mapping of the concentrations of major metabolites has been demonstrated in *Ricinus communis* seedlings[53] and a limited number of other studies.[54–56] Observation of carbohydrate flow in *R. communis* has also been demonstrated.[57] In comparison to animal studies, direct in vivo measurements of metabolites in plants using magnetic resonance techniques are uncommon. The poor resolution caused by the air space within many plant tissues and the lack of dedicated imaging instrumentation contribute to the lag in application of MRI techniques to plants.[53,58]

Although monitoring specific metabolites is possible, imaging of water in plants is more common.[59] Because metabolite concentrations are dependent on water content and the transport of photosynthetic products and other metabolites between tissues is modulated by the phloem and xylem, understanding water status is essential for gaining a deeper analysis of metabolic processes. With careful consideration of relaxation times, standard MRI experiments can provide information about water density or maps of water mobility and relaxation times. The relaxation time of water varies between plant tissues and subcellular compartments and has been used to characterize tissue structures.[60,61] Water density of various tissues can be assessed on a pixel-to-pixel basis; however, pixel size and sensitivity typically limit the image resolution to groups of cells.[62,63] Similar experiments have been used to observe changes in water distribution during drought.[64] Diffusion in the xylem and phloem can be measured using pulsed-field gradient experiments[65] or alternatively, using isotope tracers as demonstrated by Da Ines and coworkers.[66] The combination of MRI techniques with conventional NMR metabolomics studies has potential to improve our fundamental understanding of water flux in plants, especially in conjunction with water stress.

ACKNOWLEDGMENTS

National Science Foundation Integrative Graduate Education Research and Training Program fellowships DGE-0504249 to GAB and DJO and National Science Foundation grant CHE 0848976 to CKL are gratefully acknowledged.

RELATED ARTICLES IN EMAGRES

Metabonomics: NMR Techniques

Metabolite Quantification in MRS and Pattern Recognition

Plant Physiology

Plants, Seeds, Roots, and Soils as Applications of Magnetic Resonance Microscopy

REFERENCES

1. O. Fiehn, *Plant Mol. Biol.*, 2002, **48**, 155–171.

2. O. Fiehn, J. Kopka, R. Trethewey, and L. Willmitzer, *Anal. Chem.*, 2000, **72**, 3573–3580.

3. P. Britz-McKibbin and S. Terabe, *J. Chromatogr., A*, 2003, **1000**, 917–934.

4. R. t'Kindt, L. De, Veylder, M. Storme, D. Deforce, and J. Van, Bocxlaer, *J. Chromatogr., B*, 2008, **871**, 37–43.

5. A. J. Charlton, J. A. Donarski, M. Harrison, S. A. Jones, J. Godward, S. Oehlschlager, J. L. Arques, M. Ambrose, C. Chinoy, P. M. Mullineaux, and C. Domoney, *Metabolomics*, 2008, **4**, 312–327.

6. K. A. Kaiser, G. A. Barding, and C. K. Larive, *Magn. Reson. Chem.*, 2009, **47**, S147–S156.

7. J. Gullberg, P. Jonsson, A. Nordstrom, M. Sjostrom, and T. Moritz, *Anal. Biochem.*, 2004, **331**, 283–295.

8. F. F. Brown, I. D. Campbell, P. W. Kuchel, and D. L. Rabenstein, *FEBS Lett.*, 1977, **82**, 12–16.

9. G. R. Moore, R. G. Ratcliffe, and R. J. P. Williams, *Essays Biochem.*, 1983, **19**, 142–195.

10. J. K. Nicholson, M. J. Buckingham, and P. J. Sadler, *Biochem. J.*, 1983, **211**, 605–615.

11. O. Beckonert, M. Coen, H. C. Keun, Y. L. Wang, T. M. D. Ebbels, E. Holmes, J. C. Lindon, and J. K. Nicholson, *Nat. Protoc.*, 2010, **5**, 1019–1032.

12. J. K. Nicholson, I. D. Wilson, and J. C. Lindon, *Pharmacogenomics*, 2011, **12**, 103–111.

13. I. K. S. Yap, I. J. Brown, Q. Chan, A. Wijeyesekera, I. Garcia-Perez, M. Bictash, R. L. Loo, M. Chadeau-Hyam, T. Ebbels, M. De, Iorio, E. Maibaum, L. C. Zhao, H. Kesteloot, M. L. Daviglus, J. Stamler, J. K. Nicholson, P. Elliott, and E. Holmes, *J. Proteome Res.*, 2010, **9**, 6647–6654.

14. R. A. Dixon, *Nature*, 2001, **411**, 843–847.

15. C. Branco-Price, K. A. Kaiser, C. J. H. Jang, C. K. Larive, and J. Bailey-Serres, *Plant J.*, 2008, **56**, 743–755.

16. R. Narsai, K. A. Howell, A. Carroll, A. Ivanova, A. H. Millar, and J. Whelan, *Plant Physiol.*, 2009, **151**, 306–322.

17. G. Beretta, E. Caneva, L. Regazzoni, N. G. Bakhtyari, and R. M. Facino, *Anal. Chim. Acta*, 2008, **620**, 176–182.

18. A. Cuadros-Inostroza, P. Giavalisco, J. Hummel, A. Eckardt, L. Willmitzer, and H. Peña-Cortés, *Anal. Chem.*, 2010, **82**, 3573–3580.

19. Y. Sekiyama, E. Chikayama, and J. Kikuchi, *Anal. Chem.*, 2010, **82**, 1643–1652.

20. Y. Sekiyama, E. Chikayama, and J. Kikuchi, *Anal. Chem.*, 2011, **83**, 719–726.

21. G. R. Leader, *Appl. Spectrosc. Rev.*, 1976, **11**, 287–317.

22. Y. Sekiyama and J. Kikuchi, *Phytochemistry*, 2007, **68**, 2320–2329.

23. C. K. Larive, D. Jayawickrama, and L. Orfi, *Appl. Spectrosc.*, 1997, **51**, 1531–1536.

24. S. Akoka, L. Barantin, and M. Trierweiler, *Anal. Chem.*, 1999, **71**, 2554–2557.

25. M. E. Bollard, E. G. Stanley, J. C. Lindon, J. K. Nicholson, and E. Holmes, *NMR Biomed.*, 2005, **18**, 143–162.

26. A. Craig, O. Cloareo, E. Holmes, J. K. Nicholson, and J. C. Lindon, *Anal. Chem.*, 2006, **78**, 2262–2267.

27. F. Dieterle, A. Ross, G. Schlotterbeck, and H. Senn, *Anal. Chem.*, 2006, **78**, 4281–4290.

28. R. A. Davis, A. J. Charlton, J. Godward, S. A. Jones, M. Harrison, and J. C. Wilson, *Chemometrics Intell. Lab. Syst.*, 2007, **85**, 144–154.

29. T. De, Meyer, D. Sinnaeve, B. Van, Gasse, E. Tsiporkova, E. R. Rietzschel, M. L. De, Buyzere, T. C. Gillebert, S. Bekaert, J. C. Martins, and W. Van, Criekinge, *Anal. Chem.*, 2008, **80**, 3783–3790.

30. T. De, Meyer, D. Sinnaeve, B. Van, Gasse, E. R. Rietzschel, M. L. De, Buyzere, M. R. Langlois, S. Bekaert, J. C. Martins, and W. Van, Criekinge, *Anal. Bioanal. Chem.*, 2010, **398**, 1781–1790.

31. S. C. Zhang, C. Zheng, I. R. Lanza, K. S. Nair, D. Raftery, and O. Vitek, *Anal. Chem.*, 2009, **81**, 6080–6088.

32. K. Ali, F. Maltese, A. M. Fortes, M. S. Pais, Y. H. Choi, and R. Verpoorte, *Food Chem.*, 2011, **124**, 1760–1769.

33. V. Exarchou, M. Godejohann, T. A. van, Beek, I. P. Gerothanassis, and J. Vervoort, *Anal. Chem.*, 2003, **75**, 6288–6294.

34. S. Agnolet, J. W. Jaroszewski, R. Verpoorte, and D. Staerk, *Metabolomics*, 2010, **6**, 292–302.

35. S. Rezzi, F. A. Vera, F. P. J. Martin, S. Wang, D. Lawler, and S. Kochhar, *J. Chromatogr., B*, 2008, **871**, 271–278.

36. C. Ludwig and M. R. Viant, *Phytochem. Anal.*, 2010, **21**, 22–32.

37. M. R. Viant, *Biochem. Biophys. Res. Commun.*, 2003, **310**, 943–948.

38. N. R. Mustafa, H. K. Kim, Y. H. Choi, and R. Verpoorte, *Biotechnol. Lett.*, 2009, **31**, 1967–1974.

39. J. S. McKenzie, A. J. Charlton, J. A. Donarski, A. D. MacNicoll, and J. C. Wilson, *Metabolomics*, 2010, **6**, 574–582.

40. A. N. Lane and T. W. M. Fan, *Metabolomics*, 2007, **3**, 79–86.

41. Y. X. Xi, J. S. deRopp, M. R. Viant, D. L. Woodruff, and P. Yu, *Metabolomics*, 2006, **2**, 221–233.

42. C. Pontoizeau, T. Herrmann, P. Toulhoat, de B. Elena-Herrmann, and L. Emsley, *Magn. Reson. Chem.*, 2010, **48**, 727–733.

43. O. Cloarec, M. E. Dumas, A. Craig, R. H. Barton, J. Trygg, J. Hudson, C. Blancher, D. Gauguier, J. C. Lindon, E. Holmes, and J. K. Nicholson, *Anal. Chem.*, 2005, **77**, 1282–1289.

44. B. J. Blaise, V. Navratil, C. Domange, L. Shintu, M. E. Dumas, B. Elena-Herrmann, L. Emsley, and P. Toulhoat, *J. Proteome Res.*, 2010, **9**, 4513–4520.

45. F. Mesnard and R. G. Ratcliffe, *Photosynth. Res.*, 2005, **83**, 163–180.

46. A. Roscher, N. J. Kruger, and R. G. Ratcliffe, *J. Biotechnol.*, 2000, **77**, 81–102.

47. N. J. Kruger and R. G. Ratcliffe, *Biochimie*, 2009, **91**, 697–702.

48. S. Labboun, T. Terce-Laforgue, A. Roscher, M. Bedu, F. M. Restivo, C. N. Velanis, D. S. Skopelitis, P. N. Moshou, K. A. Roubelakis-Angelakis, A. Suzuki, and B. Hirel, *Plant Cell Physiol.*, 2009, **50**, 1761–1773.

49. T. Y. Yu, M. Singh, S. Matsuoka, G. J. Patti, G. S. Potter, and J. Schaefer, *J. Am. Chem. Soc.*, 2010, **132**, 6335–6341.

50. C. Martin, K. Bhatt, and K. Baumann, *Curr. Opin. Plant Biol.*, 2001, **4**, 540–549.

51. M. R. Emmert-Buck, R. F. Bonner, P. D. Smith, R. F. Chuaqui, Z. P. Zhuang, S. R. Goldstein, R. A. Weiss, and L. A. Liotta, *Science*, 1996, **274**, 998–1001.

52. B. Schneider and D. Holscher, *Planta*, 2007, **225**, 763–770.

53. W. Kockenberger, C. De, Panfilis, D. Santoro, P. Dahiya, and S. Rawsthorne, *J. Microsc.-Oxford*, 2004, **214**, 182–189.

54. T. Y. Tse, R. M. Spanswick, and L. W. Jelinski, *Protoplasma*, 1996, **194**, 54–62.

55. B. A. Goodman, B. Williamson, and J. A. Chudek, *Magn. Reson. Imaging*, 1993, **11**, 1039–1041.

56. J. M. Pope, D. Jonas, and R. R. Walker, *Protoplasma*, 1993, **173**, 177–186.

57. M. Szimtenings, S. Olt, and A. Haase, *J. Magn. Reson.*, 2003, **161**, 70–76.

58. H. Van, As, T. Scheenen, and F. J. Vergeldt, *Photosynth. Res.*, 2009, **102**, 213–222.

59. H. Van, As, *J. Exp. Bot.*, 2007, **58**, 743–756.

60. J. E. M. Snaar and H. Van, As, *Biophys. J.*, 1992, **63**, 1654–1658.

61. D. Vandusschoten, P. A. Dejager, and H. Van, As, *J. Magn. Reson., Ser. A*, 1995, **116**, 22–28.

62. H. C. W. Donker, H. Van, As, H. J. Snijder, and H. T. Edzes, *Magn. Reson. Imaging*, 1997, **15**, 113–121.

63. L. Van der, Weerd, M. M. A. E. Claessens, T. Ruttink, F. J. Vergeldt, T. J. Schaafsma, and H. Van, As, *J. Exp. Bot.*, 2001, **52**, 2333–2343.

64. J. Sardans, J. Penuelas, and S. Lope-Piedrafita, *BMC Plant Biol.*, 2010, **10**, 1–12 (article 188), DOI: 10.1186/1471-2229-10-188.

65. A. D. Peuke, C. Windt, and H. Van, As, *Plant Cell Environ.*, 2006, **29**, 15–25.

66. O. Da Ines, W. Graf, K. I. Franck, A. Albert, J. B. Winkler, H. Scherb, W. Stichler, and A. R. Schaffner, *Plant Biol.*, 2010, **12**, 129–139.

Index